Essentials of
Biotechnology

Essentials of
Biotechnology

Dr R. C. Sobti

M.Sc. (Hons. Sch.), Ph.D., F.N.A.Sc.,
F.A.M.S., F.Z.S., F.P.A.S., F.S.C.G.
Dept. of Biotechnology
Panjab University, Chandigarh

&

Dr Suparna S. Pachouri

M.Sc., Ph.D.
Consultant, WHO Project
National Institute of Health & Family Welfare
New Delhi

Taylor & Francis

Taylor & Francis Group
Boca Raton London New York

CRC is an imprint of the Taylor & Francis Group,
an informa business

Ane Books Pvt. Ltd.

Essentials of Biotechnology

© Ane Books Pvt. Ltd.

First Published in 2009 by

Ane Books Pvt. Ltd.

4821 Parwana Bhawan, 1st Floor
24 Ansari Road, Darya Ganj, New Delhi -110 002, India
Tel: +91 (011) 2327 6843-44, 2324 6385
Fax: +91 (011) 2327 6863
e-mail: kapoor@anebooks.com
Website: www.anebooks.com

For

CRC Press
Taylor & Francis Group
6000 Broken Sound Parkway, NW, Suite 300
Boca Raton, FL 33487 U.S.A.
Tel : 561 998 2541
Fax : 561 997 7249 or 561 998 2559
Web : www.taylorandfrancis.com

For distribution in rest of the world other than the Indian sub-continent

ISBN-10 : 1 42008 284 1
ISBN-13 : 978 1 42008 284 5

British Library Cataloguing in Publication Data
A catalogue record for this book is available from the British Library

Printed at Ajanta Offsets & Packagings Ltd., Delhi

Preface

The Science of Biotechnology concerns the use of cells and biological molecules and their function in meeting the multifarious challenges faced by the Society. It has provided us with an incredible set of research tools that are responsible for the breakneck speed of discoveries in biology today. It is a silent revolution and has the potential to touch every aspect of living organisms. Biotechnology does not imply a single technology, but a collection of technologies concerning the disciplines of genetics, biochemistry, chemical engineering, bioprocessing, animal and plant cell cultures. What in the past was unthinkable is now possible especially in the manufacture of essentials of life on a large scale in industrial and research sectors, such as agricultural production, pharmaceutical manufacturing and waste treatment. Consequently, biotechnology, like all technologies developed over the last two centuries has been responsible for stimulating with the increase of industrial output.

Biological and biomedical research is in the midst of a significant transition that is being driven by two primary factors: the massive increase in the number of DNA sequence technologies to exploit its use and consequently, the multiplication of new types of experiments on an unprecedented scale. Over the past few years, significant research has enabled the complete sequencing of genomes of more than 30 organisms, with another 100 or so in progress. At least partial sequence has been obtained for tens of thousands of mice, rats and human genes, as well as the sequences of entire human chromosomes has been determined by both public and private efforts. The complete sequence of other animal and plant genomes will undoubtedly be the next logical step. Unfortunately, billions of bases of DNA sequence do not tell us about the function of genes, how cells work, how cells form organisms, what goes wrong in disease, how we age or how to develop a drug. This is where functional genomics comes into play. The aim is not simply to provide a catalogue of all the genes and information about their functions, but to understand how the components coherently comprise the functioning of cells and organisms.

To take full advantage of the bulky and rapidly increasing body of sequence information, new and innovative technologies are required. The most powerful and versatile tools for genomics include high-density arrays of complementary DNAs. Nucleic acid arrays work by hybridization with probe attached at specific locations on the surface. The hybridization of a sample to an array is, in effect, a highly parallel search by each molecule for a matching partner on an 'affinity matrix', with the eventual pairing of molecules on the surface. This is determined by the principles of molecular recognition.

Interestingly we now have arrays with more than 250,000 different oligonucleotide probes. It is even possible to produce 10,000 different cDNAs per square centimeter. Although it is possible to synthesize or deposit DNA fragments of unknown sequence, the most common implementation is to design arrays based on specific sequence information. This process is sometimes referred to as 'downloading the genome onto a chip'. There are several variations on this basic technical theme: the hybridization reaction may be driven, for example, by an electric field and the surface may be made of materials other than glass such as plastic, silicon, gold, a gel or membrane, or may even be comprised of beads at the ends of fiber-optic bundle. The key elements of parallel hybridization are the localized, surface-bound nucleic acid probes and subsequent counting of bound molecules.

Indeed protein arrays are poised to become a central proteomics technology important both in basic research and commercially for enterprises pertaining to biotechnology. It has been recognized that the complexity of the human proteome far exceeds that of the genomic variables such as alternative gene splicing events and post-translational modifications that result into different molecular protein species. In a human body, it is likely to be of the order of magnitude greater than the number of genes *i.e.*, about 500,000 proteins. These investigations are at the leading edge of functional genomics today and the development of protein arrays reflects the realization that functional genomic discoveries will depend on progress in defining the expression of, and interactions among proteins. Conventional proteome analyses by 2 D gel electrophoresis and mass spectrometry, while highly effective, have limitations and in particular, may overlook many proteins of interest. Protein arrays are rapidly becoming established as a powerful means for the detection of proteins and monitoring their expression levels, and investigating protein interactions and functions. The objective behind protein array is to achieve efficient and sensitive high throughput protein analysis, carrying out large numbers of determinations in parallel by automated means.

Myriads of new regulatory and catalytic functions for RNA have now been discovered. It has been demonstrated that certain RNA sequences can directly sense ambient temperature or any of a variety of small molecular metabolites. These sensors allow without the need for regulatory proteins associated mRNAs to regulate their own transcription or translation accordingly. A type of natural genetic control elements, the iboswitches, uses untranslated sequence in an mRNA to form a binding pocket for a metabolite that regulates expression of that gene. Innumerable riboswitches have been discovered. This mechanism of genetic regulation is widespread in bacteria. These natural RNA switches provide us illustration of how functional RNAs are engaged in fundamental cellular processes.

It has now been discovered that mRNAs have certain *cis*-acting elements that sense external signals for the regulation of gene expression. Both sensing and regulatory functions are performed by these RNAs without the need for any proteins. These elements act as sensors of external signals in both prokaryotic and eukaryotic genes, through mechanism of molecular recognition that goes beyond simple base pairing. Nutrients, temperature and stress within the cell are sensed by these RNA sensors.

Double-stranded RNA-mediated interference (RNAi) has recently emerged as a powerful reverse genetic tool to silence gene expression in multi-cellular organisms including plants, *Caenorhabditis elegans*, and *Drosophila*. The discovery that synthetic double-stranded, 21-nt small interfering RNA triggers gene-specific silencing in mammalian cells has further expanded the utility of RNAi into mammalian systems. There is a technology that allows synthesis of small interfering RNAs from DNA templates *in vivo* to efficiently inhibit endogenous gene expression.

Development is astoundingly an orderly process, which gets ensued with totipotent zygote and ends with an assortment of differentiated cell types in adults. Going by the outcome of the Human Genome Project, it appears that approximately 35,000 genes that are needed to build a human being harbour 200 histologically distinct cell types, which on further subdivision form numerous specialized cell types. These different cell types perform specialized functions that are as diverse as digestion, absorption, filtration, visual perception, and developing body's defence system, *etc*. Fully differentiated state of a particular cell type represents a strikingly stable state of that cell. Starting from a single large totipotent embryo, the cells divide and differentiate to form the pleuripotent stem cells. These stem cells have the remarkable property of both self-renewal and cell type differentiation.

Stem cells hold the key to replace cells, lost in many degenerating diseases like Parkinson's disease, Alzheimer's disease, stroke, diabetes, chronic heart disease, spinal cord injury, muscular dystrophy and liver failure. The challenging question is, whether and how stem cells can be tamed to develop a cell-based medicine to repair malformed, damaged or ageing tissues? Because of these characteristics as well as their potential of becoming futuristic medicine, we all are drawn in one direction or another to understand the intricacies of stem cell at various levels.

All living things are nanofoundries. Nature has perfected the artform of biological nanotechnology for billions of years. We are at a juncture where emerging technology is poised to present a toolkit from which new life forms can be developed/designed, the inner molecular workings of living cells can be directly manipulated, even aging may be treated not as a disease, but as a reversible pathology. The very definition of life itself is virtually at the threshold of the next great revolution in medicine: nanobiology. Currently, the existing areas are technologies and applications in the arenas of biomolecular components and biocompatible surfaces integrated into micro-scale systems, implantable biochip devices, synthetically engineered quasi-viral components, modified DNA, structured proteomics, pseudoproteins, biomolecular "devices". NASA has a computational molecular nanotechnology research group examining the ways in which this technology can be used to advance the exploration and human habitation of space. IBM is engaged in pathbreaking research to revolutionize computing. Storing one bit in a few atoms no longer seems outlandish, and molecular switches will shortly replace the bulky device made today with optical lithography technique.

Biolithography, the process of photonically instigating the binding of biomolecules to a variety of target surfaces, such as metals, glass, plastic, even teflon, was originally patented by BSI (now

Surmodics) over a decade ago. At present, the procedure has become very robust, enabling the precise "attachment" of proteins, lipids, enzymes, even entire living cells, with molecular precision to almost any type of surface. There are many novel uses of existing biopolymers that could provide us with new tools. DNA, for example, is known primarily for its ability to encode information. But it can also produce structures as complex as a truncated octahedron and even provide power when its chemical conformation changes in response to changes in its environment. The great diversity of proposals, ideas, and experimental capabilities makes it very difficult to predict the tranjectory of future procedures essential for reaching general goals of nanotechnology.

Tissue engineering is a biomedical engineering discipline integrating biology with engineering to create tissues or cellular products outside the body (*ex vivo*) or to make use of gained knowledge to better manage the repair of tissues within the body (*in vivo*). This discipline requires understanding of diverse biological fields, including cell and molecular biology, physiology and system integration, stem cell proliferation and differentiation with lineage attributes, extracellular matrix chemistry and compounds and endocrinology. It also requires knowledge of many engineering fields, including biochemical and mechanical engineering, polymer sciences, bioreactor design and application, mass transfer analysis of gas and liquid metabolites, and biomaterials. Translation of tissue engineering constructs to clinical applications will involve other scientific disciplines so that novel engineered tissues will be easily accepted and used by clinicians. The combination of these sciences has spawned the field of regenerative medicine.

Indeed, it is very clear that biotechnology is a powerful technology with the potential to create remarkable new products which may raise ethical issues. The society has to prepare itself to face the next generation biotechnology products.

In order to make the student conversant with the basics of biotechnology, the present volume has been compiled. It has 14 Chapters starting from the overall review of biotechnology to various techniques and their role in health, agriculture, environment, aquaculture and food. It has chapters concerning information on the status of biotechnology in developing countries as well as the debate on ethical consideration and intellectual property rights.

It is hoped that the volume will help to sensitize students to various aspects of biotechnology.

The compilation of such a volume having diverse topics has not been an easy task. Many colleagues and friends have been of immense help. We would like to mention particularly Dr. (Mrs.) V.L. Sharma whose untiring help has made the compilation of the volume possible. We are thankful to her and of course, our special thanks to our spouses and children for their forebearance.

R.C. Sobti
Suparna S. Pachouri

Contents

CHAPTER-2 : Basic Methodologies and Tools in Biotechnology   55–114

CHAPTER-4 : Specialized Techniques in Biotechnology-II :   191–134
Nucleic Acid Based Technologies

CHAPTER-7 : Applications of Biotechnology in Health   447–536

CHAPTER-8 : Biotechnology in Agriculture   537-592

CHAPTER-9 : Biotechnology and Environment   593-564

CHAPTER-10 : Biotechnology and Aquaculture   655-672

CHAPTER-11 : Biotechnology and Food   673-694

CHAPTER-13 : The Ethical and Social Implications of Genetic Engineering — 729–742

Chapter-14 : Intellectual Property Rights in Biotechnology — 743–756

Biotechnology

The term biotechnology was coined in 1919 by Karl Ereky, a Hungarian Engineer. At that time, biotechnology meant the science by which products are manufactured from raw materials with the aid of living organisms. In fact, Ereky envisioned a biochemical age similar to the stone and iron ages. The use of the term biotechnology may imply a single subject, but its essence is multidisciplinary, requiring a wide range of science and engineering inputs.

Biotechnology today, therefore, is the integrated use of biochemistry, microbiology and engineering sciences in order to achieve technological (industrial) applications of capabilities of microorganisms, cultured tissue cells and parts thereof (European Federation of Biotechnology). In short, biotechnology is the application of biological agents either in manufacturing industry or service operations.

1.1 BACKGROUND

Biotechnology is not new. It began without being known by users. Man has always been manipulating living organisms to solve problems and improve his way of life. The origins of the biotechnology revolution may be traced back to 10,000 years ago when farmers domesticated plants and animals. Domestication is inseparable from genetic modifications. About 8,000 years ago, man extended the use of living organisms to microorganisms to produce beer, vinegar, yoghurt and cheese. Sumerians and Egyptians were among pioneers in using biotechnology around 6000-8000 B.C. The ancient Egyptians used yeast to brew beer and to bake bread. Some 7,000 years ago in Mesopotamia, people used bacteria to convert wine into vinegar. The Roman civilization probably had ale-houses. The Romans also liked wine. In fact they tried to introduce grape-vines into southern Britain for the express purpose of wine making. There is an evidence of the use of wine and vinegar in the Bible also. All these facts indicate that their production dates back to the ancient times. The ancient civilizations exploited tiny organisms that live in the earth by the rotation of crops to increase their yields, out thay did not know why it worked. Theophrastus an ancient Greek who lived 2,300 years ago indicated that broad beans leave something in the soil. It took another more than two thousand years before a French Chemist suggested in 1885 that some soil organisms might be able to 'fix' atmospheric nitrogen into a form that plants could use as a fertilizer.

It has long been recognized worldwide that genetic resources such as plants are linked to economic growth. Since ancient times, rulers have sent plant-collectors to gather prized exotic species. In 1495 B.C., Queen Hatshepsut of Egypt sent a team to the Land of Punt (modern Somalia and Ethiopia) to gather specimens of plants that produced valuable frankincense. Similarly, in modern times, colonial powers made huge plant-collecting expeditions across Latin America, Asia and Africa. They installed their findings in botanical gardens and were able to establish agricultural monocultures around the globe. It was from these early gene banks, a single coffee tree from Arabia was planted in the Amsterdam Botanic a Garden in 1706. This was the origin of most of the coffee grown in South America. In their first big attempt to establish an economic base using an exotic plant in the colonies, the British colonizers started growing Chinese tea plants in the foothills of the Himalayas. By the late 19th century, India had replaced China as the main exporter of tea to Britain.

Earlier agriculture concentrated on producing food. Plants and animals were selectively bred and microorganisms were used to make food items such as beverages, cheese and bread. In the late eighteenth century and the beginning of the nineteenth century, there was an advent of vaccination, crop rotations involving leguminous crops and animal drawn machinery.

The end of the nineteenth century saw a milestone in biology. Microorganisms were discovered, Mendel's work on genetics was accomplished and institutes for investigating fermentation and other microbial processes were established by Koch, Pasteur and Lister.

Then in the late twentieth century genetic engineering was added to biotechnology. Genetic modification has opened up many new areas and allowed the modification of plants, animals and even humans on a molecular level.

Timeline of Notable Biotechnology Events	
8000 BC :	Collection of seeds for replanting. Evidence that Mesopotamian people used selective breeding (artificial selection) practices to improve livestock.
6000 BC :	Brewing of beer, fermention of wine, baking of bread with the help of yeast.
4000 BC :	Making of yoghurt and cheese with lactic acid producing bacteria by Chinese.
1500 AD :	Collection of plants around the world.
1590 AD :	Invention of microscope by Zacharias Janssen.
1675 AD :	Discovery of microorganisms using first microscope.
1856 AD :	Discovery of the laws of inheritance by Gregor Mendel.
1862 AD :	Discovery of the bacterial origin of fermentation by Pasteur.
1919 AD :	Usage of the word biotechnology by Karl Ereky, a Hungarian agricultural engineer.
1928 AD :	Discovery that a certain mould could stop the development of bacteria by Alexander Fleming.
1953 AD :	Description of the structure of Deoxyribonucleic acid, DNA by James D. Watson and Francis Crick.
1972 AD :	Discovery of the DNA composition of chimpanzees and gorillas to be 99% similar to that of humans.

1975 AD : Development of method of producing monoclonal antibody by Köhler and Milstein.

1980 AD :

o   Characterization of modern biotechnology by recombinant DNA technology. *E.coli* was used as prokaryotic model to produce insulin and other medicine.

o   Discovery of a viable brewing yeast strain *Saccharomyces cerevisiae* 1026 as a modifier of the microflora in the rumen of cows and digestive tract of horses.

o   Ruling of the United States Supreme Court 447 U.S. 303 (1980) in favour of microbiologist Ananda Chakrabarty in the case of a USPTO request for the first patent granted to a genetically modified living organism (GMO) in history.

1984 AD : Consideration of nutrigenomics as applied science in animal nutrition.

1994 AD : Approval of the first GM food from Calgene: "Flavr Savr" tomato by FDA.

1997 AD : A sheep called Dolly was cloned using DNA from two adult sheep cells by British scientists from the Roslin Institute.

2000 AD : Completion of a rough draft of the genome in the Human Genome Project.

2002 AD : Sequencing of the DNA of rice, the main food source for two-thirds of the world's population. Rice is the first crop to have its genome decoded.

2003 AD : Introduction of the first biotechnological pet, GloFish to the North American market. It was specially bred to detect water pollutants, the fish glows red under black light. It was achieved by the addition of a natural bioluminescence gene.

1.1.1 Discovery of Microorganisms

In many instances, microbial contamination of food results in spoilage, although what is not liked by one person may be a delicacy to another. But in certain instances, microbial growth has resulted in beneficial changes such as improved flavour and texture and perhaps improvement in quality of storage. Once these desirable changes had occurred, they were observed to be self-perpetuating. When there was no knowledge of microbiology, storage vessels that were not properly cleaned, carried the residual food, had probably acted as an inoculum. Similarly little amount of curd had been in use that acts as seed to make more curd. There is not much difference in the modern production of fermented foods. Open vessels are still used and the residue from one batch is used to inoculate the next one.

Although baking bread, brewing beer and making cheese have been going on for centuries, the scientific study of the biochemical processes is not older than 200 years. Clues to understanding fermentation emerged in the 17th century when the Dutch experimentalist Anton van Leeuwenhoek examined scrapings (discovered microbes) from his teeth under the microscope.

The first observation of microorganisms by Leeuwenhoek was published by the Royal Society, London in 1684. The experiments of Spallanzani in 1799 and Schwann in 1837 not only discarded the idea of spontaneous generation of microorganisms, but also provided a means of sterilization of liquid and air. Schwann's findings suggested that alcoholic fermentation was due to a finger mold, *i.e.,* yeast and inoculation resulted in quicker fermentation

1.1.2 Early Evidence of Involvement of Microorganisms in Fermentation

Until just over a century ago, the realization had not yet dawned, that microorganisms were involved in the production of alcohol and vinegar. When a group of French merchants were looking for a method that would prevent wine and beer from souring, while being shipped over long distances, help was sought from Louis Pasteur, the great Microbiologist. It was believed at that time that air acted on the sugars to convert these fluids into alcohol. Pasteur showed that growth and physiology of yeast (and hence the accumulation of fermentation product, alcohol) differs depending on the presence or absence of oxygen. This phenomenon was named as **Pasteur effect**. It is applicable to other microorganisms as well. Such an anaerobic process is known as **fermentation.** Souring and spoilage occur later and are due to activities of the acetic acid bacteria that converts alcohol into vinegar (acetic acid). Pasteur provided a solution to this. According to him, most of the microorganisms present were killed on heating the alcohol. This process did not greatly affected the flavour of wine or beer and was called a **pasteurization**. A similar technique was used for the manufacture of **sake** in the Orient over 300 years earlier.

1.1.3 Evolution of Induced Biotechnology from Natural Biotechnology

Wars: An Impetus to Natural Biotechnology

Though alcohol was produced with the aid of natural biotechnology, the technology of distillation has been in use from the pre-Christian era until the early twentieth century.

In fact ethanol was the first chemical to be produced with the aid of biological materials. The origin of distillation is not clear, but by the fourteenth century, it was widely used to increase the alcoholic content of wines and beers. During this time the French brandy manufacturers taught the Scottish brewers the method to distill their beer to produce whisky. It was a small step towards the production of neat alcohol. This biological route is still used to produce approximately 24% of world ethanol. The impetus to natural biotechnology, however, was provided by war.

Production of glycerol: The import of the vegetable oils required for the production of glycerol for manufacture of explosives by Germans was prevented by the British naval blockade at the start of the First World War. As a result, the Germans tried the microbial production of glycerol by yeast and they succeeded in increasing the yield to over 1000 tons per month.

Production of acetone and butanol: Germany provided the British a source of acetone and butanol, the former was required for ammunitions and the latter for artificial rubber. The British using *Clostridium acetobutylicum*, performed acetone-butanol fermentation, a process that survived until the early 1950s. Present-day citric acid manufacturing also goes back to the first World War. Until then, citric acid had been extracted from citrus fruits and Italy was a major producer. As men were made to enlist in the war, citrus groves were not taken care of and hence were spoiled. When hostilities were over, the industry was in ruins and the price of citric acid had escalated. Therefore, the microbial process for the production of citric acid was introduced in 1923. *Aspergillus niger,* an obligate aerobe was used for this purpose. Initially, large-scale culture of this microbe was achieved by placing liquid medium in shallow metal pans and allowing the organism to grow on the surface. Later this method of surface culture was improved by absorbing the nutrient medium on to an inert granular support.

Production of Penicillin and Fine Chemicals

Penicillin was the name given by Fleming to the antibacterial substance produced by the mould *Penicillium notatum.* By 1940 when the first purified preparations became available, the phenomenal curative properties of penicillin were known. *P. notatum* an obligate aerobe grown in surface culture was used for this purpose. The culture of the microbe was labour-intensive. Besides, the cultures were also prone to contamination. This reduced the yield of penicillin. The need for aseptic operation led to the development of the stirred tank reactor, which even today is the preferred method for the large-scale cultivation of microbes. Aseptic operation is achieved by sterilizing all the equipment with steam prior to inoculation and by keeping the pressure inside the vessel higher than atmospheric pressure. To meet the oxygen demand of culture, sterile air is blown into the vessel and distributed throughout the medium by agitation.

Another contribution of the Penicillin programme was the development of strain selection procedures. The original *P. notatum* culture yielded only 2 mg of penicillin per litre of culture fluid. By screening many different *Penicillium* isolates, a higher-yielding variant, *P. chrysogenum,* was identified. In an attempt to improve the yield still further, the *P. chrysogenum* was systematically exposed to a variety of mutagens such as nitrogen mustard, ultraviolet radiation and X-irradiation. After each round of exposure, the survivors were screened and the highest-yielding variant carried forward for the next round of exposure. By combining fermentation improvements with the use of mutants, the titre has been increased to over 20g/l.

Once the clinical utility of penicillin was established, pharmaceutical companies began to consider antibiotics seriously. Following the discovery that *Streptomyces griseus* also produces a clinically useful antibiotic, streptomycin, it became standard practice to screen large number of environmental isolates for their ability to produce antibiotics. It has since then been shown that the filamentous bacteria known as actinomycetes, a group to which *S. griseus* belongs, produces many hundreds of different antibiotics, including at least 90% of those known today. Gradually the screening became more and more sophisticated and the need to test microorganisms extensively led to their isolation from increasingly exotic sources. In recent times it has become increasingly difficult to find antibiotics which are both novel and useful; the number of new antibiotics discovered per year has remained constant, but over the last 20 years, the clinical success rate has dropped from 5 to less than 1%. Consequently pharmaceutical companies have redirected their screening efforts towards the identification of pharmacologically active fermentation products. As with the penicillin fermentation, once a microbe is identified which produces a useful metabolite, strain improvement and fermentation development are undertaken.

Development of semicontinuous fermentation: During the Second World War some interesting innovations were made. These included introduction of semi-continuous fermentation. Initially *Clostridium acetobutylicum* was cultured in a large volume of medium until growth ceased. With such a large inoculum of cells, complete conversion of sugar to solvents occurred only in 12 hours. It was later on realized that even the non-sterile medium could be used; although the yield of solvent was lower due to the growth of contaminating bacteria. However, substantial amount of fuel was saved.

In the immediate post-war period, many organic chemicals, including acetone and butanol, became readily available from the by-products of the petroleum industry and hence the fermentation process was discontinued. The acetone-butanol fermentation, however, is an exceedingly simple process which is of benefit to many third world countries which cannot afford to spend large amount of money on either petroleum itself or petrochemical-based products. Since these countries often have an abundance of the necessary cheap raw materials such as sugar or starch, this fermentation process continues.

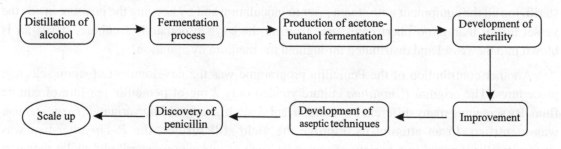

Land Marks in Fermentation Technology

(i)   A dilution technique for obtaining the first pure microbial culture of lactic acid bacterium was described by Lister in 1878.

(ii) Robert Koch in 1881 described a simpler and more effective technique of obtaining pure cultures from isolated separate colonies developed on solidified medium. Even today this technique is widely followed.

(iii) In 1876, Cohn showed that bacterial spores have a high level of heat resistance and developed the technique of 'intermittent sterilization' for their inactivation.

(iv) Petri in 1887 devised the Petri plate.

(v)   In 1897, Büchner demonstrated alcohol fermentation by cell-free yeast juice. He suggested that a proteinacious enzyme was responsible for fermentation.

(vi) Wildiers in 1901 demonstrated that yeast requires growth factors (Vitamins) for growth, especially at low inoculum levels. Vitamins are also used in fermentation even today.

(vii) In 1917, d' Herelle described the biological properties of the parasite of dysentry, and called it **bacteriophage**; phages are important since they pose a constant threat to bacteria and actinomycetes used for fermentation.

Earlier Milestones in Biotechnology

From the above, the following milestones in biotechnology are evident (Table 1.1).

Table 1.1: An Overview of Historical Developments of Biotechnology

- Biotechnological production of foods and beverages in Ancient times.
- Drinking of beer by Sumerians and Babylonians in 6000 B.C. 4000 B.C., baking of leavened bread by Egyptians is well known.
- In the 17th century, microorganisms were first seen by Antony von Leeuwenhoek, who developed the simple microscope.
- Between 1857 and 1876, Pasteur showed fermentation ability of microorganisms. Cheese production and mushroom cultivation have ancient origins.

(a) **Non-Sterile conditions for the development of Biotechnological Processes:** By the end of 19th century, ethanol, acetic acid, butanol and acetone were produced by open microbial fermentation processes. This capacity was used for waste-water treatment and municipal composting of solid wastes.

(b) **Development of sterile conditions:** This involved the introduction of complicated engineering techniques for the mass cultivation of useful microbes. For this, techniques were developed to kill contaminating microorganisms. Attempts were made to yield antibiotics, amino acids, organic acids, enzymes, steroids, polysaccharides, vaccines and monoclonal antibodies.

(c) **Introduction of applied genetics and recombinant DNA technology:** Improvement of strains of important industrial organisms was done. Recombinant DNA techniques together with protoplast fusion were introduced to allow new programming of the biological properties of organisms.

In the early 20th century, the modern biotechnology movement gained momentum again. New sciences continued to emerge, particularly immunology and genetics. There was a forward leap for biotechnology when Gregor Mendel announced his observations about the 'units of heredity' – later known as **genes.** These did not change their identity from generation to generation, but only recombined. In 1906, the science of genetics was born. Genetics (from the term Genesis) relates to the origin of a thing. It tries to explain how organisms resemble and differ from their parents. It was believed that every gene directly corresponded to a specific trait. It was used to promote the theory of genetic determinism, whereby life-forms are viewed as machines controlled by genes in linear chains of cause and effect. By the 1920s, genetics was helping plant breeders improve their crops. But when applied to people, the correspondence between genes and traits led to simplistic interpretations of human behaviour as resulting largely from genetic make-up. Thomas Hunt Morgan and his group of *Drosophila* (fruit fly) researchers made significant contributions to genetics. They showed that the basic units of Mendel's heredity, genes, were physically located on chromosomes. The first cancer-causing virus was discovered by Peyton Rous and it was in England that the first bioremediation project that utilized bacteria to treat raw sewage was started.

By the 1940s, the US was transformed into a leader of monoculture agriculture by genetics. It relied on a narrow range of genetic material and considerable high-technology. In the 1960s, the Green Revolution took place. Artificial fertilizers were extensively used. Seedless varieties became popular, especially fruit and vegetables, and so did hybrids which reduced the viability of crops over generations. This forced farmers to buy fresh seed every year. Ever since, the world's gene pool has been rapidly shrinking and many varieties have vanished for ever. Power and resources became concentrated among fewer farms. The farmers had to depend on the suppliers of seeds and inputs to provide them with crop varieties and fertilizers. The farmers being cut-off from knowledge and thus control over the raw materials of farming, became less autonomous, and there was a reduction in their capacity to conduct adaptive experiments. On the failure of new varieties, they had to wait for solutions from research stations. Alongside the dissolution of expertise among farmers, especially in the Third World, global seed industry flourished.

The discovery of DNA (or deoxyribonucleic acid) then transformed the science of genetics. DNA carries the hereditary information in the cells. Though the chemical nature of DNA had already been discovered in 1869, the chemical basis of genes was unraveled only in 1953. Francis Crick and James Watson, discovered that the DNA structure was a double helix: two strands twisted around each other. The structure, function and composition of DNA is virtually identical in all living organisms. Each individual is made unique by the precise ordering of the chemical bases in the DNA molecule. A further understanding of cellular processes, delving into the biochemistry of metabolic disorders and diseases, was made in 1950s and 1960s. The genetic code was understood for the first time in 1961. In the late 1960s and early 1970s, DNA restriction enzymes, that can cut segments of DNA, were discovered. This ushered in to an era of genetic engineering and cloning.

The next step was the realization that man could change the ordering of bases in DNA, and thereby, modify life-forms and research started moving in that direction. This was the beginning of the alteration of the genetic make-up of living things by transferring specific genes from one organism to another. The first successful recombinant DNA experiment was perfomed in 1973 by Paul Berg, Herbert Boyer and Stanley Cohen. They stitched together different bacterial genes from the common human gut bacterium, *Escherichia coli.* The US company Genentech became the first to develop technology to rearrange DNA in the mid-1970s, and then the others followed. With the success of this experiment, much progress in genetic engineering was made in the subsequent years. The birth of the biotechnology industry took place in 1970s. Revolutions in forensics and biomedicine were made by new lab methods such as DNA sequencing and protein analysis, and later the polymerase chain reaction (PCR), which makes unlimited copies of genes.

The maturation and growth of the biotech industry continued unabated in the 1980s as the first genetically engineered products were approved by the FDA. Genentech's Humulin, became the first new treatment for diabetes. It was produced from genetically engineered bacteria. Soon after thus, methods to genetically engineer plants were discovered and the first field tests of genetically engineered tobacco plants were performed. Later on, the Flavr Savr, a genetically engineered tomato resistant to rotting, was approved for sale. The late 1980s, saw the launching of the Human Genome Project. This international effort had a fifteen year goal of mapping and sequencing the 3 billion letters of the human genome. In 1990s the applications of DNA sequence toward health and medicine were analyzed. These were made possible by the identification of genes responsible for cystic fibrosis, breast cancer and Huntington's disease.

Towards the end of the twentieth century, the world was introduced to the first sheep to be cloned from DNA derived from adult cells, named Dolly. One year later, John Gearhart and James Thomson, published independent results showing their ability to isolate human stem cells. As the millennium ended, vigorous debates over the ethics related to biotechnology, genetic testing, stem cell research and genetically modified organisms ensued.

The dawn of the new millennium began with the presentation of a rough draft of the human genome. It was completed by Celera Genomics and the Human Genome Project. In 2001, the sequence of the human genome was published. This initiated a future era of genomics, proteomics, bioinformatics and personalized medicines.

1.1.4 Current Biotechnology

Biotechnology is currently being used in many areas including agriculture, bioremediation, food processing, and energy production. DNA fingerprinting is becoming a common practice in forensics. Similar techniques were used to identify the mummies of the last Czar of Russia and several members of his family. Production of insulin and other medicines is accomplished through cloning of vectors that now carry the chosen gene. Immunoassays are used not only in medicine for drug level and pregnancy testing, but also by farmers to aid in the detection of unsafe levels of pesticides, herbicides and toxins in crops and in animal products. These assays also provide rapid field tests for industrial chemicals in ground water, sediment and soil. In agriculture, genetic engineering is being used to produce plants that are resistant to insects, weeds and plant diseases.

Biotechnology today, in fact, is a combination of all the newer methodologies such as RDT, cell culture, protein engineering, *etc*. They can be used individually and in combination. It is leading to the development of a large number of new products and improved methods for well-established processes. With all these developments, the definition of biotechnology has undergone rapid evolution.

Evolution of Biotechnology Timeline		
Prior to 1750		Plants were used for food
		Animals were used for food and for work
		Plants were domesticated, selectively bred for desired characteristics
		Microorganisms were used to make cheese, beverages and bread by fermentation
1797	Edward Jenner	Living microorganisms were used to protect people from disease
1750-1850		There was an increased cultivation of leguminous crops and crop rotation to increase yield and land use
1820		Animal-drawn machines were introduced
1850s		Use of horse drawn harrows, seed drills, corn planters, horse hoes, 2-row cultivators, hay mowers. Rakes and industrially processed animal feed were used as organic fertilizer
1859	Charles Darwin	It was hypothesized that animal and plant populations adapt over time to best fit the environment
1864	Louis Pasteur	The existence of microorganisms was proved. It was shown that all living things are produced by other living things
1865	Gregor Mendel	Description of how traits are passed from generation to generation was given. They were called factors

1869	Johann Friedrich Meischer	DNA from the nuclei of white blood cells were isolated
1880		Steam engine was used to drive combine harvesters
1890		Ammonia synthesis took place
1893	Koch, Pasteur	Fermentation process was patented
	Lister	Diphtheria antitoxin was isolated
1902	Walter Sutton	The term "gene", was coined and it was proposed that chromosomes carry genes (factors which Mendel said that could be passed from generation to generation)
1904		Development of artificial "silks"
1910	Thomas H. Morgan	It was proved that genes carried on chromosomes
1912	Sir Henry Bragg and L Bragg	It was discovered that the atomic structure of crystal can be deduced by their X-ray diffraction.
1918		Acetone produced by plants to make bombs was used by Germans Yeast was grown in large quantities Activated sludge for sewerage treatment process was made
1919	Karl Ereky	The term "Biotechnology" was coined
1919		Phoebus Aaron Levene The "tetranucleotide" structure of DNA was proposed
1924		It was shown by microscopic studies that DNA and proteins are present in chromosomes.
1927	Herman Muller	The mutation rate in fruit flies was increased by exposing them to X-rays
1928	Frederick Griffths	It was noticed that a rough kind of bacterium changed to a smooth type when unknown "transforming principle" from smooth type was passed to it
1930	Alexander Fleming	Antibiotic properties of certain molds were discovered
1938	Fleming	Penicillin was discovered Proteins and DNA were studied by X-ray crystallography The term 'molecular biology' was coined
1938	R Signer Torbjorns Caspersson and E Hammarsten	Molecular weight for DNA between 500,000 and 1,000,000 daltons was determined
1941	George Beadle Edward Tatum	'One gene-one enzyme' hypothesis was proposed
1944		Large Scale production of Penicillin took place
1943-1953	Linus Pauling	Sickle cell aneamia was described. It was called molecular disease Cortisone made in large amounts
1944	Oswald Avery	Transformation experiment with Griffith's bacterium was performed
1945	Max Delbruck	A course to study a type of bacterial virus that consists of a protein coat containing DNA was organized Transition from animal power to mechanical power on farms was made
1949	Roger and Colette Vendrely with Andre Boivin	It was observed that sex cells contain half as much DNA as body cells

1950	Erwin Chargaff	It was determined that there is always a ratio of 1:1 adenine to thymine in DNA of many different organisms
1951	Rosalin Franklin	Two forms of DNA, the paracrystalline B and the crystalline A forms were discovered.
1952	Alfred Hershey Margaret Chase	Radioactive labeling was used to determine that it is the DNA, not protein, which carries the instructions for assembling new Phages
1952	Rosalin Franklin and Raymond Gosling	A magnificient X-ray diffraction pattern of the B-form of DNA was produced
1953	James Watson Francis Crick	The double helix structure of DNA was determined
1954	George Garmow	DNA code for the synthesis of proteins was suggested
1956	Dangr	Insulin (protein) from pork was sequenced
1957	Francis Crick George Gamov	The sequence hypothesis and the central dogma was proposed. During a dysentery epidemic in Japan, it was discovered that some strains of bacterium are resistant to antibiotics.
1958-59		DNA polymerase was discovered
1958	Fritz Miescher	Nucleic acid named "nuclein" was seen in PUS Cell
1960		mRNA was isolated
1959	Arthur Korenberg	The enzyme DNA polymerase was isolated
1961	Sidney Brenner Francis Crick	The group of three nucleotide bases or codons was used to specify individual amino acids.
1961	Marshall Nirenberg Johann Heinrich	A sequence of nucleotide was reported to encode a particular amino acid
1965		Classification of the plasmids took place
1966	Marshall Nirenberg Severo Ochoa	A sequence of three nucleotides code for each of the 20 amino acids was determined
1970		Reverse transcriptase was isolated
	Hamilton Smith	The first restriction enzyme, an enzyme that cuts DNA at a very specific nucleotide sequence was isolated.
1972	Paul Berg	Sections of viral DNA and bacterial DNA with same restriction enzyme were cut
	Stanley Cohn and Herbert Boyer	Efforts were made to create recombinant DNA. The viral DNA was spliced to the bacterial DNA
1973	Stanley Cohn Herbert Boyer	First recombinant DNA organism was produced Beginning of genetic engineering was made
1975		Moratorium on recombinant DNA techniques was made. Monoclonal antibodies were produced
1976	Herbert Boyer	The first firm was founded in the United States to apply recombinant DNA technology National Institute of Health guidelines was developed for the study of recombinant DNA
1977		First practical application of genetic engineering: First human gene was cloned Human growth hormone was produced by bacterial cells

1978	Genentech Inc.	Genetic engineering techniques were used to produce human insulin in *E. coli*
		First Biotech company entered in NY stock exchange
	Stanford University	Frist successful transplantation of mammalian gene was made
		Nobel Prize in Medicine was received by discoverers of restriction enzymes
1979	Genentech Inc.	Human growth hormone and two kinds of interferon DNA from malignant cells transformed a strain of cultured mouse cells
1980		It was decided by the US Supreme Court that man-made microbes could be patented
1981		The monoclonal antibodies were used for diagnosis in USA.
1983	Genetech, Inc.	Eli Lily got license to make insulin
		First transfer of foreign gene in plants was made
1985		Plants were allowed to be patented
1986		First field trial of DNA recombinant plants resistant to insects, viruses, bacteria was made
1988		First living mammal was patented
1993		Flavr savr tomatoes were sold to public
1993-2003		Very rapid developments took place which included the mapping of genome of various organisms including man.

1.2 WHAT IS BIOTECHNOLOGY?

From the above preliminary discussions, now the question arises what is biotechnology? In fact there is no clear cut definition. It has been defined in many forms. It involves the use of biological material (microbial, animal or plant cells or enzymes) to synthesize, break down or transform materials of marketable value.

Scientists, as per their own perception, have tried to define biotechnology. Some of these definitions have been reproduced in Table 1.2.

Table 1.2: Various Definitions of Biotechnology

Biotechnology is defined as

1. the application of biological organisms, systems or processes to manufacturing and service industries.
2. the integrated use of biochemistry, microbiology and engineering sciences in order to achieve technological (industrial) application capabilities of microorganisms, cultured tissue cells and parts thereof.
3. a technology using biological phenomena of copying and manufacturing various kinds of useful substances.
4. the application of scientific and engineering principles to the processing of materials by biological agents to provide goods and services.
5. the science of production processes based on the action of microorganisms and their active components and of production processes involving the use of cells and tissues from higher organisms. Medical technology, agriculture and traditional crop breeding are not generally regarded as biotechnology.
6. the name given to a set of techniques and processes.
7. the use of living organisms and their components in agriculture, food and other industrial processes.
8. the deciphering and use of biological knowledge is biotechnology.

The European Federation of Biotechnology (EFB) considers it 'the integration of natural sciences and organisms, cells, parts thereof and molecular analogues for products and services'. Both 'traditional or old' and 'new or modern' biotechnology comes within the fold of this definition.

(a) **Traditional Biotechnology:** It includes the conventional techniques that have been used for many centuries to produce beer, wine, cheese and many other foods.

(b) **New Biotechnology:** It includes all methods of genetic modifications by recombinant DNA and cell culture and fusion techniques, together with modern developments of 'traditional' biotechnological processes.

New biotechnology is distinguished from the traditional by scientific understanding. Earlier the organisms were used for brewing, baking, wine making production, the production of oriental foods such as soya sauce and tempeh. The analysis of these processes and modern scientifically-defined practices are now replacing traditional methods. Biotechnology therefore, is a culmination of more than 8000 years of human experience using cells, tissues and even whole organisms and the process of fermentation, cell and tissue culture, genetic engineering to make products of use by mankind directly or indirectly.

In other words, we can say that biotechnology is applied to manufacturing processes used in health care, food and agriculture, industrial and environmental clean-up, among other applications.

According to Maria *et al.* (2004), Biotechnology's broadest definition can be given as "the application of all natural sciences and engineering in the direct or indirect use of living organisms or parts of organisms, in their natural or modified forms, in an innovative manner in the production of goods and services and/or to improve existing industrial processes. Modern biotechnology involves understanding, mapping, manipulation or change of the genetic patrimony of a living organism."

According to Jayant (2005), a widely accepted definition of Biotechnology, therefore, is "Application of scientific and engineering principles to processing of materials by biological agents and to provide goods and service". Some other definitions replace the rather ambiguous word 'biological agents' with more specific words such as microorganisms, cells (plant and animal) and enzymes.

It is now being felt that the term biotechnology may be abandoned, because it is quite difficult to exactly define it and remove misunderstandings caused by these definitions. It needs to be replaced by the precise term of whatever specific technology or application is being used.

A major part of all biotechnology financial profits (in terms of money) represented by the huge and expanding world market has been established by traditional biotechnology. But 'new' aspects of biotechnology based on advances in molecular biology, genetic engineering and fermentation process technology are now increasingly finding acceptance by the industry.

The necessary biological and engineering knowledge and expertise is now being put to productive use. But this is limited by scientific or technical considerations and more by the following factors:

(i) Adequate investment by the relevant industries

(ii) Improved systems of biological patenting

(iii)   Marketing skills

(iv)   The economics of the new methods in relation to current technologies

(v)   Importance of public perception and acceptance

The new techniques are bound to revolutionize many aspects of medicine, agriculture and food.

1.2.1 Core Components of Biotechnological Processes

Biotechnological processes have the following core components :

(i)   Development or selection of the best biological catalyst or accelerator for a specific function or process

(ii)   Creation of (by construction and technical operation), the best possible environment for the functioning of catalyst

(iii)   Downstream processing, *i.e.,* separation and purification of an essential product or products from a fermentation process

1.2.1.1 Catalyst/Accelerator

The most effective, stable and convenient form of catalyst for a biotechnological process has been the whole organism. Because of this, much of biotechnology in the beginning revolved around microbial processes.

Microorganisms act as primary fixers of photosynthetic energy. They also bring about chemical changes in almost all types of natural and synthetic organic molecules. This is because of the following reasons :

(i)   Immense gene pool that offers almost unlimited synthetic and derivative potential

(ii)   Extremely rapid growth rates, much more than any of the higher organisms (plants and animals)

These allow the production of large quantities of products under the right environmental conditions in short time periods.

The following methodologies are used:

(i)   The selection of improved microorganisms from the natural environmental pool

(ii)   Modification by mutation or transfer and mobilization of a large number of new techniques, developed with the increase in our knowledge at molecular level. These lead to the construction of microorganisms, plants and animals with totally novel biochemical potentials

(iii)   Maintenance of these manipulated and improved organisms in a substantially unchanged form. A spectrum of techniques are required for the following :

(a) The preservation of organisms,

(b) Retaining essential features during industrial processes,

(c) Retaining long-term vigour and viability (*e.g.,* cryopreservation).

In fact the catalyst is used in a separated and purified form, named as an enzyme. Attempts are being made on the large-scale production, isolation and purification of individual enzymes.

Today, biotechnological processes also involve cells in culture alone or in combination with whole organisms *i.e., in vitro* and *in vivo* settings have been merged. Plant and animal cell cultures are also playing important roles in biotechnology.

1.2.1.2 Containment System (Bioreactor) for the Proper Function

In recent years, newer technologies and instruments have been designed for the maintenance and control of the physicochemical environment such as temperature, aeration, pH, *etc*. This is done by innovative advances made by bioscientists and bioprocess engineers. This allows the optimum expression of biological properties of the catalyst.

1.2.1.3 Downstream Processing

In most of the cases, once the product is obtained in the bioreactor (*e.g.,* biomass or biochemical product), there is a need to separate organic products from the predominantly aqueous environment. This is done by **downstream processing.** It is a technically difficult, expensive and cumbersome procedure. It involves the following steps:

(i)   Initial separation of the bioreactor broth or medium into a liquid and solid phases that contain cells. It is usually done by filtration or centrifugation

(ii) Subsequent concentration and purification of the product.

The processing requires more than one stage. The cost of downstream processing of fermentation products considerably vary.

The biotechnology has now become part of our living system. There is hardly any area where it does not have applications.

1.3 BIOTECHNOLOGY: AN INTER-DISCIPLINARY PRODUCT- ORIENTED APPROACH

It is now clear that biotechnology is not new, but represents a developing and expanding series of technologies dating back (in many cases) to thousands of years, when man first began to use microbes to produce foods and beverages such as bread and beer and to modify plants and animals through progressive selection for desired traits.

Today's biotechnology, however, is a combination of bewildering subdisciplines, *viz.,* microbiology, genetics, biochemistry, immunology, cell biology, molecular biology, enzymology, plant and animal anatomy and physiology, morphogenesis, systematics, ecology, bioprocess technology and many others (Fig.1.1). The increasing diversity of modern biology, now involves the application of principles of physics, chemistry and mathematics. This has mainly occurred in the post war scenario due to emergence of newer and increased needs. All these have together made the description of life processes at the cellular and molecular levels possible.

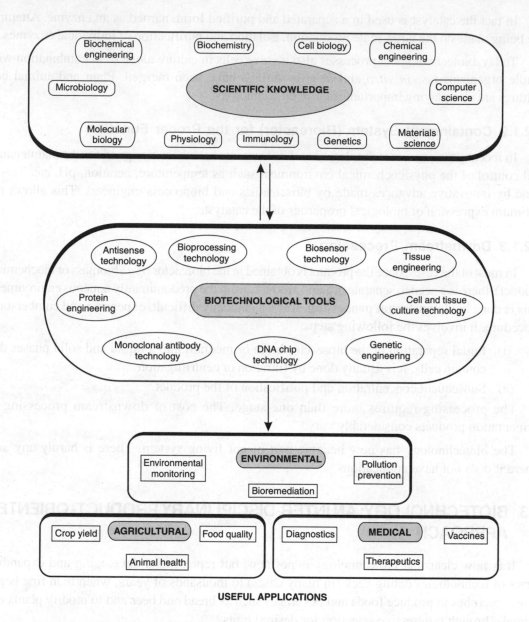

Fig. 1.1. Biotechnology : An interactive science emerged as a combination of technologies. (Adapted and modified from Kreuzer, 2004).

It is beyond doubt that the developments in biotechnology are proceeding at a speed similar to that of microelectronics in the mid-1970s. There is, however, a difference because biotechnology requires more caution. It has now been fully realized that biotechnology has a considerable impact across all industrial uses of life sciences. In each case, the relative merits of competing means of production influence the economics of a biotechnological route.

1.4 TECHNOLOGIES THAT RESULTED IN THE DEVELOPMENT OF CONTEMPORARY BIOTECHNOLOGY

It has already been made clear that biotechnology is not a science, but a combination of technologies. The current biotechnology which was started as fermentation technology received impetus with the development of technologies such as SCP, fermentation, cell culture and recombinant DNA technology.

This was followed by the development of newer technologies not only using knowledge of biology, but combining it with information technology, electronics, *etc.*

Biotechnology, therefore, is not in itself a product or range of products, rather it involves a range of enabling technologies that have significant applications in many industrial sectors. This technology looks for new applications and the main benefits for the future. New biotechnological processes in many instances will have following advantages:

(i) They will function at low temperature

(ii) They will consume little energy

(iii) They will rely mainly on inexpensive substrates for biosynthesis

1.4.1 Basic Technologies: Advent of Modern Biotechnology

1.4.1.1 Single Cell Proteins (SCPs)

The development of aseptic technique led to improvements in the large-scale culturing of cells. In the late 1960s, it was thought that microbial cells (or biomass) could be used as a source of proteins. It generated much excitement. This was, however, short lived. In fact since ancient times, people near Lake Chad in Africa and the Aztecs near the Lake Texaco in Mexico harvested the filamentous blue-green alga, S*pirulina* from the lake, dried it in the sun and used the same as food. But the first industrial production of SCP dates back to World War I when 'Torula Yeast' *(Candida utilis)* was produced in Germany and used in soups and sausages. Interest in SCP peaked again during World War II and then during mid-1950s. By 1967, British Petroleum was producing SCP at an industrial scale.

The term "Single Cell Protein" (SCP) refers to cells or protein extracts of microorganisms grown in large quantities for use as human (food) or animal (feed) protein supplements. The outline of the process for the production of SCPs in bacteria is given in Fig 1.2. SCP has a high protein content. Besides, it also contains fats, carbohydrates, nucleic acids, vitamins and minerals. So the SCPs were developed to meet the problem of shortage of proteins caused due to an increase in population. The interest in SCP was expected to relieve the deficiency in the following two ways:

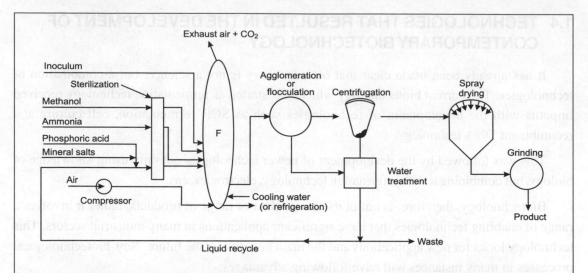

Fig. 1.2. Schematic diagram of the process for the continuous production of bacterial single cell protein. The inoculum containing the bacteria is introduced into the fermentation tank (F). To prevent the introduction of bacteria that might contaminate the product. The necessary nutrients (methanol, ammonia, phosphoric acid and mineral salts) must be sterilized before they are added. The fermentor contents must be supplied with air and cooled to dissipate the heat generated by the growing bacteria. In continuous processes, the nutrients are replenished as they are consumed to maintain the concentrations needed by the bacteria. The solution containing the bacteria is drawn off, treated to cause the bacteria to agglomerate or flocculate, and centrifuged. While the bacteria are spray-dried and ground to yield the final product the liquid may then be recycled to the fermentor. (Modified from Feral Marx, 1989).

(i) Utilization of substrates of poor value that are easy to produce. SCP may be used directly as human food supplement.

(ii) It may be used in animal feed to replace at least partially the currently used protein rich soybean meal and fish protein and even cereals, which can be diverted for human consumption.

At the same time, there was a rapid increase in oil prices and the introduction of improved high-yielding varieties of crops. The developed countries did not prefer to have SCPs, as they could obtain a large amount of proteins from conventional sources. On the other hand, the underdeveloped countries could not afford to buy SCPs or even to build and run SCP plants.

There are certain disadvantages of using microbial biomass as supplement to diet. Many microorganisms produce toxic substances *e.g.,* aflatoxin produced by fungi, and it has to be made sure that biomass does not contain any such substance. Sometimes, the microorganisms, when taken in, lead to indigestion and allergenic reactions due to the slow digestion of microbial cells in the digestive tract. The increased amount of nucleic acid contained in many microbial biomass products is also undesirable.

Despite all these, concurrently the interest in SCPs has been rejuvenated.

Regardless of the type of substrate or organism employed, the production of SCP involves following basic steps :

    (i)   Preparation of suitable medium with suitable carbon source

   (ii)   Prevention of contamination of medium

  (iii)   Production of the desired microorganism

  (iv)   Separation of microbial biomass and its processing

1.4.1.2 Bioprocess Technology

Bioprocess technology involves the use of living cells or components of their biochemical machinery doing what they normally do: synthesizing products, breaking down substances and releasing energy. The one-celled microorganisms are most frequently used. These include bacteria and yeast or mammalian cells; the cellular components most often used are proteins called **enzymes.**

1.4.1.3 Protein Engineering Technology

This technology aims to improve existing proteins and create proteins not present in nature. In fact, protein engineering technology in combination with genetic engineering can give us any protein of our choice. It involves the following :

1.4.1.3.1 Enzyme Engineering

This technology is being used for the following purposes :

    (i)   For catalysis of extremely specific chemical reactions

   (ii)   For immobilization of enzymes

  (iii)   For creating specific molecular converters (bioreactors)

The processes may involve modification of enzymes both directly through chemical manipulations and indirectly by specifically manipulating genes that code for enzymes

    (i) to increase their stability under extreme manufacturing conditions,

   (ii) to broaden their substrate specificity, and

  (iii) to improve their catalytic power.

Attempts have been made to produce L-amino acids, high fructose syrup, semi-synthetic penicillins, starch and cellulose hydrolysis enzyme probes, *etc.* for bioassays. Enzyme technology is going to play a crucial role in biotechnology. The native and genetically engineered enzymes are likely to function in environments previously thought to be hostile. A new role is expected to be played by engineering new metabolic pathways in organisms. Enzymes have been developed for use in detergents and production of biofuels, vitamins, amino acids and fine chemicals.

Novozyme has developed enzymes for use in animal feed, food, textiles, leather, oil/fat and meat processing among others. It has over 700 products and 100 different types of enzymes and microbes. These are replacing chemical products that pollute the environment.

Another industry Genencor has produced enzymes with improved performance in detergents and vitamin C, biofuel, sugar and biopolymer production (Reverchon, 2002), while Prodigene is manufacturing TrypZen. It is a recombinant trypsin used in wound care and food processing. Multifarious benefits of these innovations expected in the near future are :

    (i)   cut in the cost of production,

   (ii)   reduction in the number of processing steps,

  (iii)   less expense on energy,

  (iv)   possible reduction in the cost of investment,

   (v)   reduction in the environmental pollution and demand for high-grade feedstock.

The chemical synthesis of vitamin B_2 is a complex eight-step process. It has, however, been reduced to single step by BASF AGs new biotechnology process. Besides, this process reduces overall costs by about 40%, carbon dioxide emissions by 30%, resource consumption by 60% and waste by 95%.

Similarly synthesis of the antibiotic Cephalexin is a multistep chemical process. It has been currently reduced to a mild biotransformation. It is being isolated from biotechnological process which involves less energy and less input of chemicals. It is water-based and generates less waste (OECD, 2001).

Some of the enzyme producing organisms live in hostile environments, *i.e.,* as extremophiles, which are found in the following environments :

    (i)   Hot springs

   (ii)   Salty waters

  (iii)   Polluted surroundings

The organisms survive in these environments because they possess unique enzymes that support life-saving pathways, whereas in such environments most other organisms would be killed. The enzymes could be harnessed for industrial use, such as in detergents, textile industry, pharmaceutical and bioremediation processes. Genencor and Maxygen are interested in extremophiles for their peculiar metabolism and evolution.

Abzyme Engineering: Although most of the protein engineering work has been directed at changing the catalytic properties of existing enzymes, a way has been found to synthesize novel catalysts : antibodies with catalytic ability or abzymes. Antibodies resemble enzymes because both are proteins that bind to specific molecules. But the similarity ends there. Antibodies bind for the sake of binding; enzymes bind to make reactions happen.

Attempts are being made to custom design abzymes to catalyze reactions for which there are no known enzymes. This is opening up a new world of exciting possibility. This may allow us to create antibodies that can function as protease, the enzymes that break down proteins. If we design the proteolytic abzymes so that they break peptide bonds with great specificity, they will be the protease equivalent of a restriction endonuclease, a type of enzyme that is an invaluable tool in genetic engineering. These "restriction endoproteases" might allow us to cut protein molecules with great specificity.

1.4.1.4 Cell Culture Technologies

These involve growing of the cells in appropriate nutrients in laboratory conditions or in bioreactors.

1.4.1.4.1 Microbial Cell Culture

As mentioned earlier, the microbes are exploited to increase their efficiency and manufacture products of our choice. Various media have been developed for this purpose and many modifications in culture conditions have been made. These have been described in the next chapter on specialized techniques in biotechnology. These cultures, as mentioned above, are exploited for various purposes including production of enzymes and other useful products either as such or by genetically engineering the microbes.

1.4.1.4.2 Animal Cell Culture

Some of the emphasis shifted to animal cell culture, but it had to face unique problems which did not exist in the culture of microbial cells. All the cells required a surface for growth and could not be grown in suspension. This requirement was met by the use of the following:

   (i) Novel microcarrier beads which were maintained in suspension

   (ii) Novel multiple well plates which were fitted inside the fermenter

The application of mass animal cell culture, had been limited largely to the production of vaccines. But today, its horizon has widened. Attempts are now being made to produce a large number of molecules useful to mankind, *e.g.,* interferons and antiviral and anticancer proteins. Most of the studies have been carried out on cell lines derived from tumours. Animal cell culture has played a great role in the development of technologies such as *in vitro* fertilization, cloning, stem cell, tissue engineering, *etc.*

1.4.1.4.3 Cell Fusion and Hybridoma Technology

Kohler and Milstein in the mid eighties of the last century revolutionized the biological sciences by developing hybridoma technology. Certain tumour cell lines are the basis of hybridoma production. **Hybridomas** are hybrid cells. They are created by fusing myeloma (a type of tumour) cells with antibody-producing spleen lymphocytes. The lymphocyte on fusion with the myeloma cell, acquires immortality and can indefinitely be grown in cell culture, but it continues to secrete antibody. Since any given lymphocyte synthesizes a single antibody species hence all the antibody molecules made by culture of any particular hybridoma are identical. It is because of this, they are called **monoclonal,** *i.e.,* they are derived from a single clone of lymphocytes (Fig. 1.3). Monoclonal antibodies can be easily purified. They have applications in many different areas from diagnostic kits to cancer therapy and protein purification.

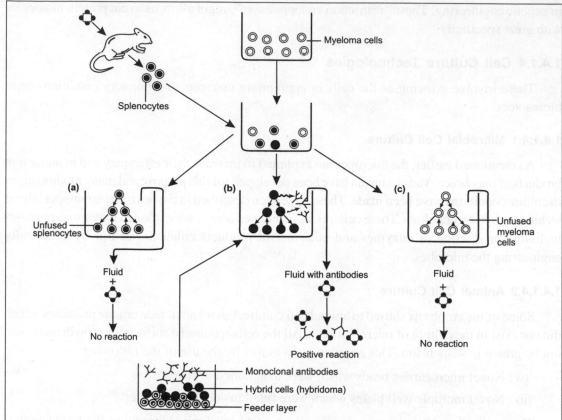

Fig. 1.3. Hybridoma technology and monoclonal antibody production : Short-lived antibody-producing cells from the mouse spleen are fused with immortal, continuously dividing tumour cells to form hybridomas. These genetically uniform cell lines can be kept and propagated in cell cultures, each cell line producing a single type of (monoclonal) antibody. The uniform structure and composition of monoclonal antibodies (MAbs) makes them suitable for research, diagnostic, and therapeutic purposes. Humanised MAbs can be used as highly effective drugs in the treatment of cancer.

1.4.1.4.4 Plant Cell Culture

In vitro techniques were initially developed to demonstrate the totipotency of plant cells predicted by Haberlandt in 1902. Totipotency is the ability of a plant cell to perform all the functions of development which are characteristic of a zygote, *i.e.,* its ability to develop into a complete plant. In 1902, Haberlandt reported the culture of isolated single palisade cells from leaves in Knop's salt solution enriched with sucrose. The cells remained alive for upto one month, increased in size, accumulated starch, but failed to divide. Efforts to demonstrate totipotency led to the development of techniques for cultivation of plant cells under defined conditions. This was made possible by the brilliant contribution of R.J. Gautheret in France and P.R. White in USA during the third and the fourth decades of the 20th century. Most of the modern tissue culture media have been derived from the work of Skoog and coworkers during 1950s and 1960s.

Many attempts have been made to cultivate plant cells. Whole plants serve as valuable source of agricultural chemicals, drugs, flavourings and colourings. Plant cell culture is a direct result of development of methods for the regeneration of plants from individual cells. This is of much importance today and has two-fold benefits :

(i) In a single experiment, regeneration enables hundreds of plants to be produced. During regeneration, genetic variants are formed at a high frequency (somaclonal variation) and these can be very useful to the plant breeder.

(ii) It is possible to isolate virus-free cells that can be used to produce virus-free crops, which will give an increased yield per acre.

1.4.1.4.5 Tissue Engineering Technology

With the knowledge of cell biology and material science, it is now possible to make semisynthetic tissues in the laboratory. These tissues consist of biodegradable scaffolding material. For this purpose, natural biological materials such as collagens are used for scaffolding. It is expected that tissue engineering/regenerative medicine may revolutionize the area of health care. This may improve the health and quality of life of a large number of people worldwide as it will restore, maintain or enhance, tissue and organ function. It has therapeutic applications, where the tissue is either grown in a patient or outside the patient and transplanted. In addition, it has diagnostic applications where the tissue is made *in vitro* and used for testing drug metabolism and uptake, toxicity and pathogenicity.

1.4.1.4.6 Embryonic Stem Cell Culture

With the successful culture of a unique type of human cells (embryonic stem cells), the use of animal cell culture has entered a new era.

Embryonic stem cells can give rise to virtually any type of cell. This complete developmental plasticity sets them apart from other stem cells. These can be used for therapeutic purposes. For example, if we develop the capability to control the differentiation of human embryonic stem cells, we may be able to produce replacement cells to treat disease like diabetes, Parkinson's and heart diseases.

Besides the embryonic stem cells, there are adult stem cells too. A number of tissue-specific cells in our body are not specialized to perform a specific function. These permanently immature cells, *i.e.,* stem cells, but under certain conditions they become specialized for a particular function. Unspecialized cells in the liver can differentiate into one of a number of specialized liver cells, such as cells that produce bile or epithelial cells that line the bile duct. Stem cells found in the bone-marrow can produce all blood cell types as well as muscle, cartilage and bone cells. But liver cells will not differentiate into white blood cells, nor will bone-marrow cells become specialized for bile production. The attempts are now being made to achieve this also *i.e.,* bone-marrow adult stems are being made capable of forming any type of cells (Fig. 1.4).

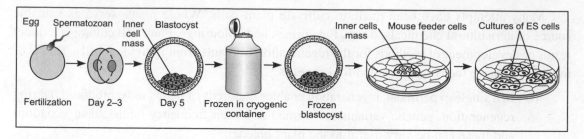

Fig. 1.4. Creation of human embryogenic stem cell cultures. Cells from the inner cell mass are cultured following excision of oviducts and isolation of blastocysts from a suitable mouse strain. Such embryonic stem (ES) cells retain the capacity to differentiate into different types of tissues in the adult mouse. By insertion of foreign DNA or by introducing a subtle mutation, ES cells while in culture can be genetically modified (Modified from Kreuzer and Massey 2005).

1.4.1.4.7 Recombinant DNA Technology (RDT)

Though the development of penicillin in 1960s and use of microbial cells as a source of

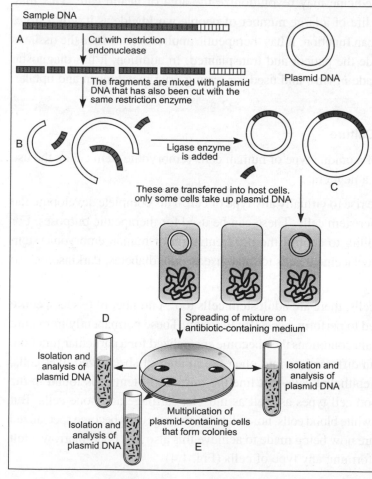

Fig. 1.5. Use of restriction enzymes in building a human DNA library (A) Some restriction enzymes is used to cut human chromosomes and bacterial plasmids. Therefore both types of DNA, have the same sticky ends. **(B)** The cleaved human and plasmid DNA molecules are mixed. Plasmid-human combinations are temporarily held together by complementary base pairing of sticky ends. **(C)** To bind together, the backbones of the plasmid-human combination DNA ligase is used **(D)** The new plasmid-human rings are mixed with bacteria the treatment of bacteria with calcium salts make them permeable to DNA. Some bacteria are thus able to take up a plasmid human ring. **(E)** Bacteria are placed on culture dish filled with medium containing the antibiotic ampicillin. Bacteria without plasmids die and those with plasmids survive. Within the bacteria, a human DNA library is constituted by plasmids.

protein were the hallmarks in biotechnology, the real breakthrough came in the 1980s with the discovery of the ability to *in vitro* splice the DNA molecules derived from different sources. For many, the term "biotechnology" is often equated with the manipulation of genes. For the more specific technique of gene manipulation, the term "genetic engineering" or "recombinant DNA technology" is more appropriate. It dates back to the 1970s. At that time, molecular biologists devised methods to isolate, identify and clone genes as well as to mutate, manipulate and insert them into other species. One of the key elements in such research was the discovery of restriction enzymes (Fig. 1.5). This gene splicing ability is referred to as gene manipulation. This involves recombining pre-existing genetic sequences to create a novel combination and that is why the term recombinant DNA technology (RDT) is also used. Using these new techniques, it is possible to manipulate directly the DNA of cells between different types of organisms. This creates new combinations of characters and abilities, not present earlier. These techniques now have a great industrial potential, but at the same time, have raised many ethical questions and posed inherent dangers of tampering with nature.

Theoretically, it is possible to transfer a particular gene from one organism into any other microorganism, plant or animal. But in reality, there had been numerous constraints. These included:

(i) which genes are to be cloned and how to select them ?

(ii) lack of basic scientific knowledge of gene structure and function.

But now with the sequencing of genomes of various organisms (both pro- and eu-karyotes including humans), both these constraints have been taken care of. New applications are emerging in the areas of healthcare, agriculture and food technology. Exciting new medical treatments and drugs based on biotechnology are appearing with ever-increasing regularity.

Prior to 1982, insulin for human diabetics was derived from cow and pig pancreas. The gene for human insulin was isolated and cloned into a microorganism, which was then mass-produced by fermentation. This genetically engineered human insulin, identical to the natural human hormone, was the first commercial pharmaceutical product of recombinant DNA technology. It is now commercially available. The principle of production of recombinant proteins has been highlighted in Fig. 1.6.

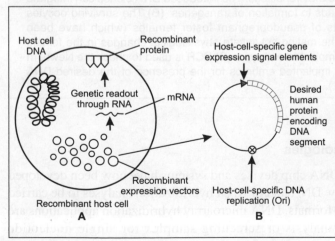

Fig. 1.6. Production of recombinant proteins. The gene encoding the desired protein is inserted into an expression vector (enlarged in fig. B). It is usually either a plasmid or virus. The expression vector allows the gene product to be made in the desired host cells, whether these be bacterial, yeast, mammalian or plant in origin. Necessary signals for its replication in the host cells are product in the vector. It also has the gene control signals needed for transcription of the vector DNA into messenger RNA and the subsequent translation of the messenger RNA into protein (Modified from, Feral Marx, 1989)

This technology did not remain confined to microorganisms, plant and animal cells, but even intact plants and animals have been made targets for modification.

In the beginning, commercial applications of recombinant DNA technology were limited to the production of proteins such as those of therapeutic interest. Attempts are now being made to achieve the following :

(i)   To clone all the genes associated with biosynthetic pathways. All the genes encoding the 29 steps in erythromycin biosynthesis have been cloned on a single DNA fragment;

(ii) To reduce the range of different organisms used in commercial biotechnology. When low-molecular-weight compounds are required, the pathways may be introduced to a limited selection of bacteria and fungi, *e.g., Escherichia coli, Streptomyces* sp., *Saccharomyces cerevisiae* and *Aspergillus* sp. An even more restricted range of organisms and animal cells can be used for the production of large-molecular-weight compounds. The cells in culture are used because they can affect post-translational modifications of proteins in an identical fashion to the intact animal (Fig. 1.7).

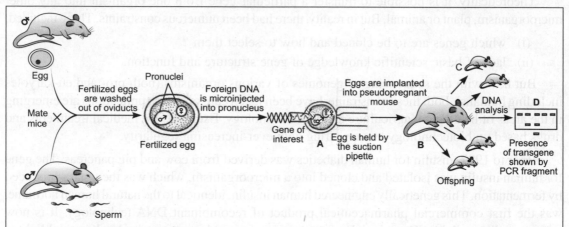

Fig. 1.7. Construction of transgenic mice by pronuclear microinjection (A). An aqueous solution of the desired DNA is pipetted directly into the pronucleus. The introduced DNA clones can integrate into chromosomal DNA at nicks. This leads to formation of transgenes. **(B)** The surviving oocytes are then re-implanted into the oviducts of pseudopregnant foster females (which have been mated with a vasectomized male). **(C)** The mating act initiate physiological changes in the female which stimulate the development of the implanted embryos. **(D)** PCR is used to check the Newborn mice resulting from development of the implanted embryos for the presence of the desired DNA sequence (Modified from Gordon, 1992).

1.4.2 Newer Technologies

1.4.2.1 Microarray or Chip Technologies

A variety of DNA microarray and DNA chip devices and systems have now been developed and commercialized. These devices allow DNA and/or RNA hybridization analyses to be carried out in microminiaturized highly parallel formats. DNA microarray hybridization applications are usually directed at gene expression analysis or screening samples for single nucleotide polymorphisms (SNPs). In addition to the molecular biology related analyses and genomic research applications, these microarray systems are also being used for pharmacogenomic research, infectious, genetic disease and cancer diagnostics, forensic and genetic identification purposes. Microarray technology is being continuously improved in sensitivity and selectivity and also attempts are being made to make it a more economical research tool. The use of DNA microarrays will continue to revolutionize genetic analysis and many important diagnostic areas. Additionally, microarray technology that has been developed for DNA analysis is now also being applied to new areas of proteomics and cellular analysis (Fig. 1.8).

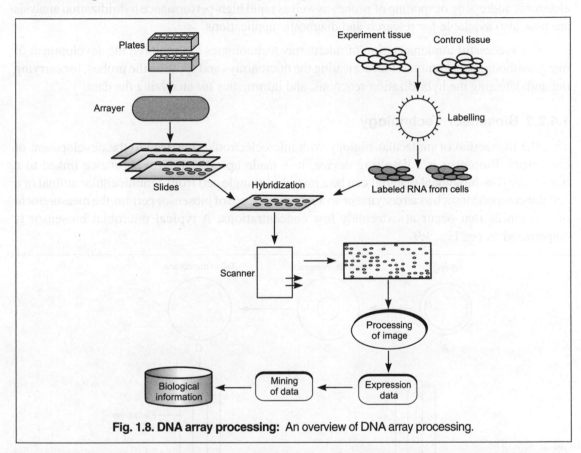

Fig. 1.8. DNA array processing: An overview of DNA array processing.

DNA hybridization microarrays are generally fabricated on glass, silicon, or plastic substrates. The microarrays may have from a hundred to many thousands of test sites that can range in size from 10 to 500 microns. High-density microarrays may have up to 10^6 test sites in a 1–2-cm^2 area. A variety of techniques are used to selectively spot DNA probes. Probes can include synthetic oligonucleotides, amplicons, or larger DNA/RNA fragments. DNA hybridization analysis on microarrays usually involves detecting the signal generated by the binding of a reporter probe (fluorescent, chemiluminescent, colorimetric, radioisotope, *etc.*) to the target DNA sequence. The microarray is scanned or imaged to obtain the complete hybridization profile.

The actual construction of microarrays involves the immobilization or *in situ* synthesis of DNA probes onto the specific test sites of the solid support or substrate material.

High-density DNA arrays can be fabricatcd using physical delivery techniques (*e.g.*, inkjet or microjet deposition technology) that allow the dispensing and spotting of nano/picoliter volumes onto the specific test site locations on the microarray. In some cases, the probes or oligonucleotides on the microarrays are synthesized *in situ* using a photolithographic process.

Microarray devices with lower densities of test sites have been developed that provide direct electronic detection of the hybridization reactions. Active electronic microarray devices that provide electronic addressing or spotting of probes as well as rapid high-performance hybridization analysis are now also available for research and diagnostic applications.

The successful implementation of microarray technologies has required the development of many methods and techniques for fabricating the microarrays and spotting the probes, for carrying out and detecting the hybridization reactions, and informatics for analyzing the data.

1.4.2.2 Biosensor Technology

The interaction of molecular biology with microelectronics has lead to the development of biosensors. Biosensor is a detecting device. It is made up of biological substance linked to a transducer. The biological material can be a microbe, a single cell from a multicellular animal or a cellular component such as an enzyme or an antibody. The use of biosensor permits the measurement of substances that occur at extremely low concentrations. A typical microbial biosensor is constructed as per Fig. 1.9.

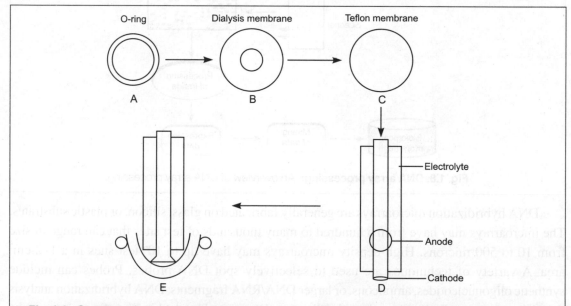

Fig 1.9. Construction of a biosensor. (A) O-ring, **(B)** Dialysis membrane impregnated with immobilised microorganisms, **(C)** Teflon membrane, **(D)** Electrode and **(E)** Biosensor (Schmander, 2006).

Construction of a microbial biosensor and its uses

The microbial strain is properly grown for using as biosensor. The cells are harvested by centrifugation and suspension is made from cell pellet in phosphate buffer and polyvinyl alcohol. This cell suspension is spotted (transferred) on a dialysis membrane. This membrane is then dried keeping at 4°C for 24 hours. The membrane is then placed on a Teflon membrane of an oxygen electrode in such a way that the microbial spot gets aligned above the cathode. The O-ring helps to keep membrane in place.

The electrode is filled with KCI solution which serves as an electrolyte. The electrode is connected to the measuring and recording devices.

The culture is placed in glass beaker in a buffer, thus, constituting the measuring cell. This is maintained at 37°C or as per requirements of culture and is stirred continuously to maintain O_2 supply. The newly constructed microbial sensors require 2-12 hours for activation which can be checked by measuring base current adding substance to be checked and testing the effect.

The substrate whose effect is to be seen is added rapidly to get the sharp signal. The electrode is rinsed as soon as reading is taken. The effects can be calculated from recorded measurements.

The microorganisms used for the construction of biosensor are produced and immobilized as per Fig. 1.10.

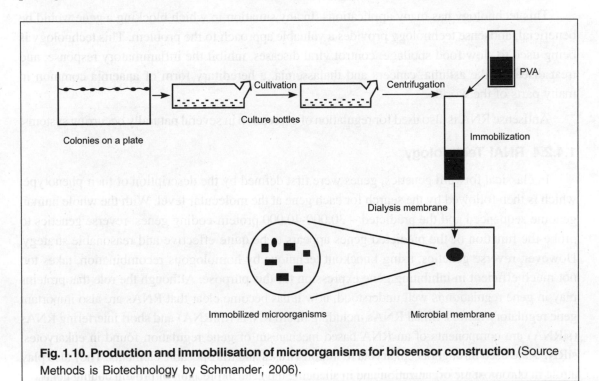

Fig. 1.10. Production and immobilisation of microorganisms for biosensor construction (Source Methods is Biotechnology by Schmander, 2006).

1.4.2.3 Antisense Technology

Antisense technology is being used to block or decrease the production of certain proteins. This is accomplished by using small complementary nucleic acids (oligonucleotides) that prevent translation of the information encoded in mRNA for a protein by binding with the Shine-Dalgarno and AUG initiation region in a mRNA and block ribosome access (Fig.1.11).

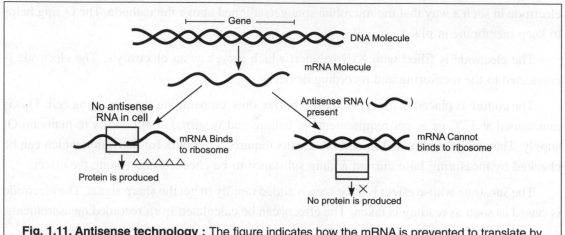

Fig. 1.11. Antisense technology : The figure indicates how the mRNA is prevented to translate by making it double stranded.

This technology has many applications. In any situation in which blocking a gene would be beneficial, antisense technology provides a valuable approach to the problem. This technology is being used to slow food spoilage, control viral diseases, inhibit the inflammatory response, and treat diseases like asthma, cancers and thalassemia, a hereditary form of anaemia common in many parts of the world.

Antisense RNA is also used for regulation of expression in several naturally occurring systems.

1.4.2.4 RNAi Technology

In classical forward genetics, genes were first defined by the description of their phenotype, which is then followed by the search for each gene at the molecular level. With the whole human genome sequenced and the predicted ~ 30,000-40,000 protein-coding genes, reverse genetics to probe the function of the predicted genes appears to be quite effective and reasonable strategy. However, reverse genetics, using knockout technique by homologous recombination, takes too not much efficient in inhibiting gene expression for this purpose. Although the role that proteins play in gene regulation is well understood, now it has become clear that RNAs are also important gene regulatory factors. Small RNAs including micro RNAs (miRNA) and short interfering RNAs (siRNA) are components of an RNA based mechanism of gene regulation found in eukaryotes. siRNAs are used throughout eukaryota to inhibit viruses and transposable elements. They also play a role in chromosome organization and in silencing the gene expression of protein coding genes.

A breakthrough was made with the discovery of RNAi in *C. elegans* in 1998 when Fire *et al.* reported that dsRNA can induce strong and specific silencing of homologous genes. RNAi can be induced in this nematode worm by injection of dsRNA into gonad, by soaking the worm in dsRNA or by simply feeding the worm on bacteria engineered to express dsRNA. RNAi is now being used for studies of individual genes as well as for genome-wise genetic screening. A bacterial library for inactivation of 16,757 of the worm's predicted 19,757 genes was developed and the corresponding phenotypes were listed. The bacterial clones are reusable and have been used in screening for genes with more specific functions such as body fat regulation, longevity and genome stability (Fig. 1.12).

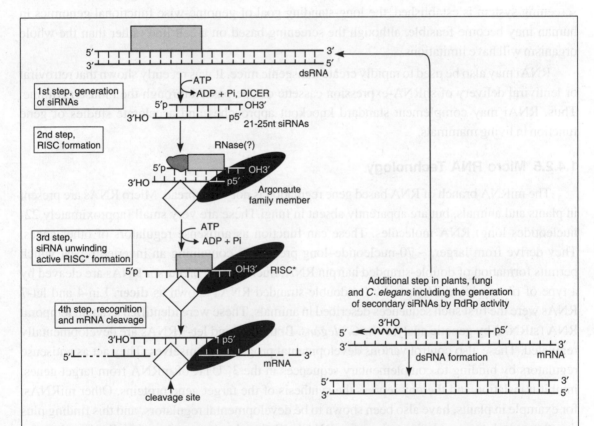

Fig. 1.12. Schematic representation of four-step gene silencing pathway: RNAi mechanism can be divided into four stages (1) Cleavage of double-stranded RNA, (2) Formation of silencing complex, (3) Activation of silencing complex and (4) Degradation of mRNA . The first one includes ATP-dependent, processive dsRNA cleavage into double-stranded fragments 21 to 25 nucleotides long. They contain 5' phosphate and 3' hydroxyl termini, and two additional overhanging nucleotides on their 3' ends (Abbreviation: dsRNA, double-stranded RNA; siRNA, small interfering RNA; RISC, RNAi-induced silencing complex. RdRp, RNA-dependent RNA polymerase. (Hamilton and Baulcombe, 1999; Elbashir *et al.,* 2001).

Drosophila is another popular model organism where RNAi has been successfully used to study functions of individual genes. Unlike in *C. elegans,* and the plants, RNAi in *Drosophila* is not systematic, that is RNAi does not spread into other cells or tissues. This property allows cell-specific RNAi in *Drosophila.*

RNAi in cultured mammalian cells is quickly becoming a standard laboratory technique to study functions of individual genes. Transfection of synthetic siRNA has been most frequently used. Screening of human genome in a wider scale is also being considered. This would be much more painstaking than screening in *C. elegans* because the number of human genes is about two times to that of *C. elegans* and RNAi technique in human is more complicated. However, efforts are being made by developing libraries of siRNAs and automatic screening systems. Once the screening system is established, the long-standing goal of genome-wise functional genomics in human may become feasible, although the screening based on a cell line rather than the whole organism will have limitations.

RNAi may also be used to rapidly create transgenic mice. It was recently shown that retroviral or lentiviral delivery of shRNA-expression cassette can be passed through the mouse germline. Thus, RNAi may complement standard knockout approaches and accelerate studies of gene function in living mammals.

1.4.2.5 Micro RNA Technology

The miRNA branch of RNA based gene regulation is not widespread. Micro RNAs are present in plants and animals, but are apparently absent in fungi.These are very small (approximately 22-nucleotides long) RNA molecules. These can function as antisense regulators of other genes. They derive from larger, ~ 70-nucleotide–long precursors containing an inverted repeat which permits formation of double-stranded hairpin RNA. Such hairpin precursor RNAs are cleaved by a type of ribonuclease III (specific for double-stranded RNA) known as **dicer**. Lin-4 and let-7 RNAs were the first such sequences described in animals. These were identified as small temporal RNA (stRNA) by genetic analyses in *C. elegans.* Both lin-4 and let-7 RNAs are developmentally regulated. These also control various developmental programs themselves. They act as antisense regulators by binding to complementary sequences in the 3´ UTR of mRNA from target genes, inhibit translation and thereby repress the synthesis of the target gene proteins. Other miRNAs, for example in plants, have also been shown to be developmental regulators, and this finding plus the strong evolutionary conservation of miRNA has led to the expectation of similar functions for mammalian miRNAs (Stratchen, 2003).

Novel miRNAs have been discovered in both humans and mice. Humans have atleast 326 miRNA genes. Although dispersed on many chromosomes there is evidence for some clustering, notably a cluster of atleast seven miRNA genes within a 0.8 kb region on chromosomes 13 micro RNAs are transcribed by RNA polymerase II and the primary transcripts contain hairpin-loop domains that fold back into duplex-like structures. The fully mature miRNA accumulates as a single-stranded species and incorporates into a ribonucleoprotein complex that carries out its function of silencing gene expression. It has been thought that micro RNA inhibits a target mRNA by base pairing to complementary sequences.

Since involvement of micro RNA in gene regulation provided an indispensable tool to manipulate the gene function and, thereby, animal development and physiology and other cellular activities like proliferation, morphogenesis, apoptosis and differentiation, the potential of rapidly evolving miRNA regulation could be important for evolving new regulatory circuits and ultimately, new patterns within body plans.

1.4.2.6 Bioinformatics Technology

It involves the use and organization of information about biology. Bioinformatics technology in fact exists at the interface of computer science, mathematics and molecular biology.

There have been enormous developments in molecular biology. A huge amount of sequence data for genes and proteins, information on the three dimensional structure of proteins, carbohydrates, and glycoproteins, and genetic maps for many species have been accumulated. There is, thus, a need for organizing all this information. To analyze and interpret the information, we need tools and methods for organizing, entering, processing, storing, accessing and integrating data from different sources. These tools and methods must, however, be consistently used by laboratories involved in molecular biology research all over the world. This is expected from bioinformatics technology, which uses computational tools such as algorithms, graphics, artificial intelligence, statistical software, stimulation and database management. All these will help in the following areas :

   (i) Help us map and compare genomes

  (ii) Determine protein structure

 (iii) Stimulate molecular binding

 (iv) Pursue structure-based drug design

  (v) Identify genes, assess the effects of virtual mutations

 (vi) Determine phylogenetic relationship

1.4.3 Applications of Biotechnology

Biotechnology, therefore, is a set of revolutionary techniques. It has been the subject of public policy aspirations for the last more than two decades. According to agenda 21, the work programme adopted by the 1992 United Nations Conference on Environment and Development, biotechnology promises to make a significant contribution in the following areas :

   (i) Better health care

  (ii) Enhanced food security through sustainable agricultural practices

 (iii) Improved supplies of potable water

(iv) More efficient industrial development processes for transforming raw materials

(v) Sustainable methods of afforestation and reforestation

(vi) Detoxification of hazardous wastes

A number of organisms have had their genomes (genetic composition) sequenced or decoded. The human, mosquito, *Drosophila, Saccharomyces* and malarial parasite genomes are among hundreds of those which have already been sequenced. These activities are expected to increase the number and pace of drug and vaccine discoveries.

It becomes apparent from above that biotechnology has applications in almost all areas such as agriculture (including food), medicine and environment (Table 1.3).

Table 1.3. End-use Based Categorization of Biotechnology

Agricultural Biotechnology (Green Biotechnology)	Biopesticides Biofertilizers Plant extraction
Medical Biotechnology (Red Biotechnology)	Medicine Vaccines Diagnostics Gene therapy
Industrial Biotechnology	Hybrid seeds Industrial enzymes Polymers
Environmental Biotechnology	Fermentation products Biofuels Effluent and waste water management Bioremediation Biosensors Creation of germplasms

(*Source* : Ernst and Young "Biotechnology a Primer")

1.4.3.1 Agricultural Biotechnology (Green Biotechnology)

Human history indicates that early man lived in small bands of nomadic hunters and gatherers but with the advent of agriculture, approximately 10,000 years ago, they settled into permanent, self-supporting communities. Genetically engineered plants and animals are being produced to improve nutrition, disease resistance, maintenance of quality, improved yields and stress tolerance.

Plant biotechnology includes :

(i) microbiology (*e.g.,* biopesticides and biofertilizers)

(ii) tissue culture (*e.g.,* the clonal multiplication and production of planting material)

(iii) marker-assisted breeding and disease identification

(iv) genetic engineering (*e.g.,* the transfer of genes from one organism to other and deactivating or activating gene expression)

Plant biotechnology, therefore, may be defined as generation of useful products or services from plant cells, tissues and often organs. Such cells, tissues and organs are either continuously maintained *in vitro* or pass through a variable phase to enable them to regenerate complete plants that are ultimately transferred to the field. Therefore, plant tissue culture forms an integral part of any biotechnological activity. Plants are genetically modified for having products of our choice. For this purpose, gene vector mediated direct gene transfer can be carried out (Fig. 1.13).

Animal cell cultures have been and are being used to generate valuable products based on their genetic information or due to genes transferred into them (transgenes) using recombinant DNA technology (Fig. 1.13). On the other hand, biotechnology approaches are used either to rapidly multiply animals of desired genotypes or to introduce specific alterations in their genotypes to achieve certain useful goals.

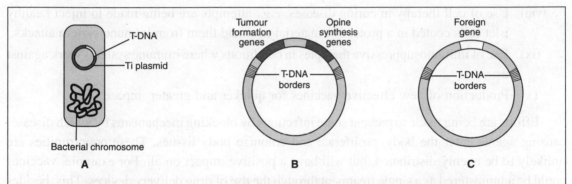

Fig. 1.13. Structure of *Agrobacterium* and its *T-DNA* before and after the integration of foreign gene. (A) *Agrobacterium tumifaciens* (B) Ti plasmid (C) Altered Ti plasmid containing foreign gene.

1.4.3.2 Medical Biotechnology (Red Biotechnology)

Many new developments have taken place in the area of medicine. They may help to increase the span and the quality of life. For example, ailing organs that shorten life could be replaced. Diagnostic systems could be improved and made easily accessible.

It is now possible to detect many diseases and medical conditions using new sensitive diagnostic tools and techniques such as monoclonal antibodies, DNA probes, PCR, DNA chips, restriction fragment length polymorphisms (RFLPs) *etc*. The **human genome project** has facilitated the identification and sequencing of genes so as to arrive at precise diagnosis. The development of biosensors is allowing us to carry out tests and interpret results in the minimum possible time.

Biotechnology provides us new therapeutics. These include the following :

(i) Natural products as pharmaceuticals (*e.g.,* use of chemical extracts of a few trees in the treatment of breast and ovarian cancers; ticks and leeches, for potential anticoagulant compound, exploration of marine fauna and flora for their therapeutic values),

(ii) Endogenous therapeutic products such as interleukin-2, erythropoietin and tissue plasminogen activator.

(iii) Biopolymers such as hyaluronate, adhesive protein polymers, chitin for various purposes in the treatment and healing, *etc.*

(iv) Designer drugs, *i.e.,* drugs designed by computer molecular modelling.

(v) Replacement therapies wherein many substances are being produced by genetic engineering to replace/supplement the ones produced by defective genes.

(vi) Use of transgenic plants/animals for the production of pharmaceuticals. These are being used to produce complex therapeutically active human proteins.

(vii) Potential of gene therapy in treating some diseases such as adenosine deaminase deficiency.

(viii) Use of cell therapy in curing diseases, *e.g.,* attempts are being made to inject healthy islet cells coated in a protective material to shield them from immune system attacks.

(ix) Use of immunosuppressive therapies in conditions where immune systems work against us.

(x) Production of new effective vaccines for quicker and greater impact.

Efforts are being made to prevent some infections by blocking mechanisms by which disease-causing agents enter the body, proliferate and colonize body tissues. These opportunities are unlikely to be evenly distributed, but will have a positive impact on all. For example, vaccines could be administered as a single treatment through the use of drug delivery devices. This, besides other inbuilt benefits, will reduce the cost complexity and energy needs of diagnostic systems and the technology may reach the lower strata and rural society.

Besides these, new research tools are being developed to understand the various aspects of disease management in detail. These include the development of knockout organisms.

1.4.3.3 Environmental Biotechnology

A special area of growing concern amongst the scientific community is environmental protection and conservation, and the need for a policy of sustainable development in harmony with the environment. The Stockholm Conference in 1972, and the UNCED Conference in Rio de Janeiro in 1992, both focused world attention on areas of pollution, biodiversity conservation and sustainable development. Plants and microbes are becoming important factors in pollution control. World Bank estimates show that pollution in India costs almost US$80 billion, as well as the human cost in terms of sickness and death. The priority research areas include new developments such as bioindicators, phytoremediation methods, bioleaching, development of biosensors and identification and isolation of microbial consortia.

We are in the midst of an environmental crisis. The air, soil and water of the earth are contaminated as a result of human activities. The earth's non-renewable resources are being depleted and we are generating a large amount of non-degradable waste. Today, however, biotechnology is being used to reclaim wasteland through the use of microorganisms and plants that remove and/or degrade toxic compounds.

Bioremediation: It uses living organisms, such as microbes and plants, to extract, eliminate and/or bind toxins in forms that are not harmful to the environment. The processes used include biostimulation, biotransformation, biostabilization and biofiltration. These processes have been used for many years to eliminate pollutants. The organisms used include :

(i) *Microalgae*: These are used in ponds to eliminate nitrogen and phosphorous

(ii) *Aquatic plants* (*e.g.*, water lentils): These are used to extract heavy metals in industrial effluents.

Modern biotechnology techniques have enhanced the performance of these natural processes in pollution control. For example, mercury is a highly toxic metal that accumulates in the food chain when released in water. In the Minamata accident, inhabitants of the Japanese island of Kyushu suffered the toxic effects of fish poisoned by mercury-rich industrial effluents. Since naturally thriving mercury-tolerant bacteria are rare and cannot be grown easily in culture, the metallothionein gene has been inserted into *Escherichia coli* at Cornell University, USA. Such genetically modified bacteria grow well in culture. The genetically engineered bacteria are placed inside a bioreactor that efficiently removes mercury from water. The bacteria are incinerated. The accumulated mercury is then recovered (European Commission, 2002). Existing techniques of mercury removal are expensive and inefficient.

Phytoremediation and smart plants: Phytoremediation is the use of plants to remove pollutants from water and soils. There are about 1.4 million polluted sites in Western Europe alone. Current techniques are costly and destroy soil structure. The use of plants that can store 10 to 500 times more pollutants in their leaves and stems is cheaper and stabilizes the soil structure.

Now "smart plants" (plants that can detect and/or remove toxins and pollutants) have been developed. They are expected to have a role in :

(i) improving environment,

(ii) enhancing health,

(iii) promoting national security.

"Smart plants" in combination with bioremediation can also be used to detect heavy metals, microorganisms and other environmental changes.

The metals can be recovered from ashes and reused. There are many hyperaccumulating plants. These plants accumulate lead, zinc, nickel, copper and cobalt, among others, at levels toxic to other plants. For example, *Sebertia acuminata* can contain up to 20% of nickel in its sap (nickel is generally toxic to plants at a concentration of 0.005%). Similarly, the fern *Pteris vittata* accumulates arsenium, while conserving a very rapid growth and a high biomass. The commercialization rights of the fern (now called Edenfern™) for use in phytoremediation have been obtained by the firm Edenspace. The potential market for phytoremediation in the United States alone was estimated at $100 million in 2002 (Tastemain, 2002).

Genetic engineering is being used to enhance the ability of plants to accumulate heavy metals. Two *Escherichia coli* genes have been inserted into the genome of *Arabidopsis thaliana* which encodes an enzyme involved in transformation of arsenate into arsenite. This enables the plant to accumulate three to four times more arsenium than the normal plants (Tastemain, 2002). Similarly, Aresao Biodetection (www.aresa.dk/), a biotechnology company in Denmark, has engineered *Arabidopsis thaliana* for the detection of landmines and/or explosives. The plants on coming into contact with explosive materials, change colour from green to red.

Besides their use in phytoremediation, they can help in monitoring the changes in quantities or presence of toxins or pollutants. Tobacco plants with bacterial genes that control the synthesis of abzyme-detoxifying trinitrotoluene, (TNT) have been developed. Through the introduction of genes controlling specific degradation pathways, these initiatives could improve bioremediation processes.

Biotechnology is developing clean and efficient technologies for the sustainability of industries. It is developing changes in production processes, practices or products that make production cleaner and efficient per unit of production or consumption, a move towards sustainability. According to Kreuzer and Massay (2003), in practical terms, industrial sustainability means utilizing technologies and know how to :

(a) reduce material and energy inputs, while maximizing renewable resources as inputs,

(b) minimize the generation of harmful pollutants or wastes during product manufacture and use,

(c) produce recyclable or biodegradable products.

Leather and textiles are among the major environment polluting industries. The industrial discharge can be reduced by the use of enzymes through recycling of water. It will cut down electrical and water bills and improve the quality of final products. For this purpose, only simple adjustments are required in the plants besides replacements of 140 harsh chemicals with biological systems. Enzymes and microbes could easily be locally produced with minor additions. The industry will be set to become competitive with a reduced clean-up bill, increased earnings and turnover.

Biomining and Bioleaching: The process of biomining has also benefitted from biotechnology. Bioleaching is a common technology in mines of developing countries. Bioleaching technology can be used by the small mining sector which often targets small mineral deposits to improve the quality of final products and reduce wastes associated with mechanical cracking. In other cases, in the mining of amethyst, agate, diamond and gold, harmful chemicals are still used. The availability of biotechnological solutions will enhance the value and earnings from this sector. It will also help to reduce environmental degradation.

Bioplastics: Currently biodegradable plastics are being developed to reduce the pollution caused by synthetic plastics. Renewable resources are being used to produce bioplastics. It is because they are biodegradable and, thus, environment friendly. Of the 40 billion tons of global production of plastics, bioplastics account for only 500 million tons (roughly 1.25%). Attempts are being made to reduce their production costs to half so that the amount of bioplastics produced in 2010 is trebled (Reverchon, 2002). Microbes, plants and animals are used to produce desired plastic polymers, as the microbes and/or enzymes convert carbohydrates and/or proteins into desired plastics (*The Economist*, 2003). For example, Cargill-Dow Chemical Company uses enzymes to produce Ingeo, a polylactic acid (PLA) product made from glucose. It has commissioned a $300-million plant that can manufacture 140,000 tons of Ingeo. It is mainly used in packaging.

A transgenic bacterium containing biochemical pathways from three different microorganisms is used to convert (maize) glucose syrup to 1,3-propan-diol, for manufacturing a polyester called Sorona, a copolymer. It is made from 1,3-propan-diol and terephthalate (oil product) by DuPont. Some bacteria synthesize and accumulate polyhydroxyalkanoate (PHA) to make bioplastic up to 80% of their weight.

Plastics are being developed from PHA by the firm Metabolix. They have genetically engineered plants to produce PHA. Metabolix has started commercial production of PHA.

Biodiversity Conservation : The global biosphere can survive only if resource utilization is about 1 and not 10 per cent. The global environment is regulated by climate changes and biosphere dynamics. Knowledge about biodiversity accumulated in the last 250 years is being used by scientists throughout the world. There are many gene banks, botanical gardens, and herbaria for conservation purposes. There are also molecular approaches including DNA fingerprinting for plant conservation. The totality of gene species and ecosystems has become exceedingly important not only to understand the global environment but also from the viewpoint of the enormous commercial significance of the biodiversity.

12% of vascular plants are threatened with extinction. Over 5,000 animal species are threatened worldwide, including 563 Indian species. Biotechnology is therefore becoming a major tool in conservation biology.

1.4.3.4 Food Biotechnology

Much progress has been made in the improvement of foods, both quantitatively and qualitatively. The development of neutraceuticals has revolutionized this area. There are great expectations concerning the possibilities of finding new food products, enzymes and other biological proteins by bioprospecting.

Biotechnology is expected to provide consumers with plant products that are specifically designed to be healthier and more nutritive. A variety of healthier cooking oils is being produced by the application of biotechnology. These have a low concentration of saturated fatty acids. Nutritionally beneficial products are being engineered.These include the following:

(i)   Grains to have a complete protein profile

(ii) Fruits, vegetable and grains to have vitamin and mineral content

(iii) Potatoes to have higher solid content to reduce the amount of oil absorbed when fried

(iv) Plants that will modify fat contents (Fig. 1.14).

Functional foods containing significant levels of biologically active components are being prepared.

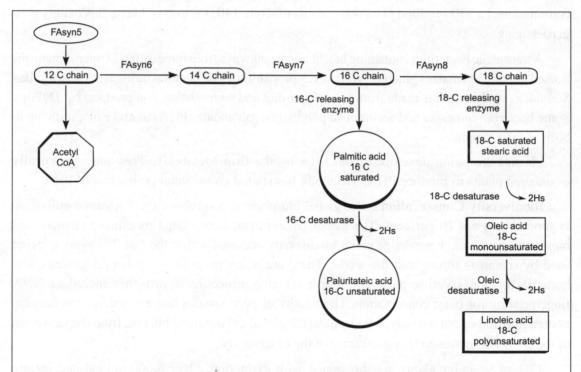

Fig. 1.14. Synthesis of saturated and unsaturated fatty acids. 16-carbon saturated fatty acid (palmitic acid) is synthesized by a number of FAsyn enzymes that are organized into a complex. In case there is a release of palmitic acid from FAsyn7 by the 16-carbon releasing enzyme, it can be used as such or is converted to a monounsaturated fatty acid by the 16-carbon desaturase enzyme. The latter removes two hydrogen atoms by FAsyn8. In case palmitic acid is not released, the 16-carbon fatty acid is converted to stearic acid, an 18-carbon fatty acid. FAsyn9 converts stearic acid to a 20-carbon fatty acid. Stearic acid may be released from FAsyn 8 by the 18-carbon releasing enzyme. Stearic acid is converted to oleic, linoleic and ultimately linolenic (not shown by desaturase enzymes) acids. The latter has three double bonds (Modified from Kreuzer and Massay, 2003).

1.4.3.5 Other Applications

Designer wood : It can be produced by genetic engineering. This will grow faster and when processed, requires few steps. It will result in extra savings and improved quality of paper. Many of the paper-manufacturing plants that are currently uncompetitive could soon become exporters of paper. Application of biotechnology to this sector will provide more value to their raw materials.

Others : Attempts are being made to convert cassava into export products (*e.g.,* plastics, sweeteners or fibres). This may help many poor farmers who currently do not have an international market for their products. These fibres or polymers can be used to generate bags, plates and other utensils that have a higher value than the raw materials.

1.4.3.6 Industrial Biotechnology

Sustainable development to industry means continuous innovation, improvement and use of "clean" technologies to make fundamental changes in pollution levels and resource consumption. Biotechnology could present a means to indirectly increase marketing of products that are currently difficult to sell. In many developing countries with a market for tubers, their production could soon exceed that of cereals. To achieve sustainability "clean" and "efficient" are the key words. Any change in production processes, practices or products that makes production cleaner and more efficient per unit of production or consumption is a move towards sustainability. Advances in biotechnology can be translated into products, processes, and technologies by creating an interdisciplinary team. The pharmaceutical sector has had a major impact in this field, as rare therapeutic molecules in pure form become available. Now over 600 biotechnology-based diagnostics (valued at about US$20 billion worldwide) are available in clinical practice. Amongst these, the polymerase chain reaction (PCR)-based diagnostics are the most common.

According to the Organization for Economic Cooperation and Development (OECD), industrial sustainability is the continuous innovation, improvement and use of clean technology to reduce pollution levels and consumption of resources. Modern biotechnology provides avenues for achieving these goals.

In recent years, policy makers, corporate executives, private citizens and environmentalists have become more concerned about sustainable development. In response to that concern, many leading industrial companies are doing more than meeting their legal minimums. Many are developing policies and implementation plans for sustainability that include guidelines for environmental health and safety as well as product stewardship.

Living systems manage their chemistry more efficiently than man-made chemical plants and the wastes that are generated are recyclable or biodegradable. Biological products that provide us the greatest potential for decreasing the environmental impact of industrial manufacturing processes are the biocatalysts which are living organisms or simply their enzymes. Biocatalysts and particularly enzyme-based processes, operate at lower temperatures and produce less toxic waste, fewer byproducts and less emissions than conventional chemical processes. They may also use less purified raw materials (selectivity). Unlike many chemical reactions that require very high temperatures and pressures, reactions using biological molecules work best under conditions that are compatible with life that is, temperatures under 100°F, atmospheric pressure and water-based solutions. Therefore, manufacturing processes that use biological molecules can lower the amount of energy needed to carry on reactions.

Finally, just as biotechnology provides us with new tools for diagnosing health problems and detecting harmful contaminants in food, it also yields new methods of monitoring environmental conditions and detecting pollutants.

Biotechnology in industry employs the techniques of modern molecular biology to reduce the environmental impact of manufacturing. Industrial biotechnology also works to make manufacturing processes more efficient for industries such as textiles, paper and pulp, and specially chemicals. It is predicted that biotechnology will transform the industrial manufacturing sector in much the same way that it has changed the pharmaceutical, agricultural and food sectors. Industrial biotechnology will be a key to achieve industrial and environmental sustainability.

Material and Energy Inputs

Biotechnology also has an impact on two sources of energy:

(i)   Fossil fuels

(ii) New biomass-based fuels.

Manufacturing processes have long relied on petroleum, a non-renewable resource that generates pollution and solid waste, as a source of material and energy. Biotechnological technique leads to cleaner products and processes by reducing the use of petroleum inputs. Industrial biotechnology on the other hand uses natural sugars as feedstocks.

The use of renewable, biomass based feedstocks will increase through biotechnology. Biofeed stocks offer two environmental advantages over petroleum-based production. There will be clean production in most cases and the generation of waste will be less. When the biomass source is agricultural refuse, we have double gains. We will enjoy all the advantages of bio-feedstocks, while reducing wastes generated from another human endeavour, that is agriculture. The use of plant biomass as feedstock helps in the growth of feedstock crop accompanied by consumption of CO_2-one of the greenhouse gases.

Using plant biomass as the primary feedstock, atleast 5 billion kilograms of commodity chemicals are produced annually in the United States.

Biotechnology is also used to remove sulfur from fossil fuels, thereby, significantly decreasing their polluting power. Using biomass for energy has the same environmental advantages as using biomass feedstocks. Attempts are being made to improve the economic feasibility of biomass-derived energy by applying genetic engineering techniques.

In addition to working towards sustainability by using biomass-based material and energy inputs, biotechnology also helps by providing measures to minimize the environmental impact of manufacturing processes by decreasing energy use and replacing harsh chemicals with biodegradable molecules produced by living things.

Manufacturing processes that use biodegradable molecules as biocatalysts, solvents or surfactants are also less polluting. For decades, some very important industrial solvents, such as ethanol and acetic acid have been provided by microbial fermentation systems. Many surfactants used in chemical manufacturing processes are biological molecules that microorganisms produce naturally. These include emulsan and sophorolipids. A surfactant produced by marine microorganisms that may replace chemical solvents has been discovered.

New enzymes, biocatalysts, to be used in manufacturing processes of other industries are being developed by industrial biotechnology companies.

Attempts are on to discover out biocatalysts with industrial value in the natural environment, improve the biocatalysts to meet very specific needs, and manufacture them in commercial quantities using fermentation systems similar to those that produce human therapeutic proteins or bulk yeast for brewing and baking industries. In some cases, genetically altered microbes (bacteria, yeast, *etc.*) are used. In other cases, fermentation is carried out by either naturally occurring microbes or those genetically modified with other techniques.

Chemical processes, including paper manufacturing, textile processing and specially chemical synthesis, sometimes require very high or very low temperatures or very acidic or alkaline conditions.

Incorporation of biocatalysts into manufacturing processes carried out under extreme conditions requires organisms that can survive in those conditions. Attempts are being made to find out such organisms in natural environments that mimic extreme manufacturing conditions. This is because since the dawn of life, microbes have adapted to every imaginable environment. Despite the harsh environment, some microbes have found a way to make a living in hostile conditions. Life in unusual habitats makes unique biocatalysts, and the great majority of that biochemical potential has not been tapped. A small fraction, less than one per cent of the microorganisms in the world has been cultured and characterized. Scientists, through bioprospecting, are discovering novel biocatalysts that will optimally function at relatively extreme levels of acidity, salinity, temperature or pressure found in some industrial manufacturing processes.

Genomic studies of microbes are providing information that helps us capitalize on the wealth of genetic diversity in microbial populations. DNA probes are used to find out genes that express enzymes with specific biocatalytic capabilities. These enzymes can be identified and characterized for their ability to function in industrial processes. Biotechnology techniques can be used to improve them.

To improve the productivity-to-cost ratio, attempts are being made to modify genes to increase enzyme productivity in microorganisms currently used in enzyme production.

Biotechnology techniques of protein engineering and directed protein evolution help to maximize the effectiveness and efficiency of enzymes. They are being used to modify the specificity of enzymes, improve catalytic properties or broaden the conditions under which enzymes can function. It may make them more compatible with existing industrial processes.

Attempts are also being made to move towards a biobased economy in which agricultural operations will be the energy and natural resource fields of tomorrow.

The first commercial shipment of bioethanol-ethanol made from wheat straw using biotech enzymes brought a major breakthrough industrial microbiology in April 2004. According to Burrill and Co, some 10 to 15 billion gallons of ethanol could be produced each year from corn stalks and husks and wheat straw. Another 50 billion could be made using such raw materials as wood-product manufacturing residues, municipal solid waste and garden waste.

Efforts are being made to establish biobased products and biobased fuels initiatives to encourage industrial use of plant matter "with specific attention to rural economic interests, energy security and environmental sustainability." The motive is to develop enzyme biomass conversion technologies and the construction of green biorefineries.

Attempts are also being made to replace petroleum derived polymers with biological polymers derived from grain or agricultural biomass.

In 2001, Cargill Dow opened a biorefinery in Blair, Nebraska, to convert sugars from field corn into polylactic acid (PLA)—a compostable biopolymer that can be used to produce packaging materials, clothing and bedding products. Price and performance are competitive with petroleum-based plastics and polyesters.

DuPont and its development partners, Genencor and Tate and Lyle, in 2001 created the high-performance polymer Sorona from the bioprocessing of corn sugar at a biorefinery in Decatur, Illinois. The Sorona fibre has been used to create clothing.

Industrial scientists have also genetically modified both plants and microbes to produce polyhydroxybutyrate, a feedstock for producing biodegradable plastics.

Abundant amounts of natural protein polymers, such as spider silk and adhesives from barnacles are being produced through microbial fermentation.

In place of petroleum-based chemicals to create plastics and polyesters, biotechnology uses sugar from plant material. Almost all the giant chemical companies are building partnerships with biotech companies to develop enzymes that can breakdown plant sugars.

It is, therefore, apparent that biotechnology provides tools, techniques and know-how to move beyond regulatory compliance to proactive pollution prevention and resource conservation strategies that are the hallmarks of industrial sustainability.

According to Kreuzer, industrial biotechnology is poised to merge its applications with carbon and silicon, a merger that could catapult industrial biotech companies from nanospace into financial hyperspace.

1.4.4 Economics and Biotechnology: Advent of Econo-Biotechnology

As is apparent from above, biotechnological processes have already made important contributions to the health and welfare of mankind. The life sciences in 1990s effected over 30%

of global economic turnover by way of healthcare, food and energy, agriculture and forestry; its economic impact has now grown multifold as biotechnology provides new ways of influencing raw material processing. There is an increased impact of biotechnology on the efficiency of all fields involving the life sciences leading to the contribution of many trillions of dollars to world markets.

According to Ernst and Young (2002), the biotechnology industry was estimated to have generated at least $34.8 billion in revenue and employed about 190,000 in traded firms worldwide. An estimated 4,200 public and private biotechnology firms were in operation in 2002. It is quite impressive because in 1992, the biotechnology industry was estimated to have contributed about $8.1 billion. The number of modern biotechnology-based drugs and vaccines has also considerably increased - from about 23 in 1990 to over 130 by 2001. There are more than 350 biotechnology-derived drugs and vaccines in clinical trials targeting over 200 diseases.

The use of biological catalysts or enzymes has entered almost every industry. Atleast 600 different products and more than 75 types of enzymes are being used in industries. The global market for industrial enzymes is about $1.6 billion. The demand for other biotechnology-related products, such as feed additives continues to grow. Vitamins and amino acids account for about $3 billion and digestive enhancers $1.3 billion (UNCTA, 2002). These steps are expected to lead to an improved environment, sustainable use of resources and increased productivity.

In fact, all sectors of economy have been affected by biotechnology-related applications and products. The technology has started to overcome bottlenecks that, in last century, favoured chemical substitutes against biological ones.

As the knowledge base of biotechnology is consolidating, the number of platforms that will depend on it are multiplying to generate new fields.

Despite these developments, biotechnology does not seem to have taken root in many developing countries and the goals have not been attained. A huge section of the human population, mainly in developing countries faces food insecurity, disease and poverty. In some regions of the world, public attitudes and political viewpoints towards biotechnology have changed. Similarly, wider applications of biotechnology are buried in the debate surrounding genetically engineered crops.

Rich dividends are being paid by the coming together of biotechnology and informatics. Genome projects, drug design and molecular taxonomy are all becoming increasingly dependent on information technology. There is a rapid accumulation of information on nucleotides and protein sequences. The number of genes characterized from a variety of organisms and the number of evolved protein structures is rapidly increasing.

According to the film "Fantastic Voyage" there is a possibility of having technology to shrink a full-size submarine and its human passengers to microscopic size. Industrial biotech companies are now trying to move towards submicroscopic worlds of biotechnology and nanotechnology. Physio-chemical activities of cells are being exploited to accomplish tasks at nanoscale.

In fact the future trends in biotechnology will be influenced by advances in the other technological fields. Technologies such as information and communication, material, cognitive and nano are bound to make great impressions on research in biotechnology and development. The coming together of any of these or all of these and other fields with biotechnology will probably give rise to new technologies having such impact on industrial competitiveness and quality of life. Many exciting and diverse materials and products are expected to come out from the combination of these new technologies.

According to Kruzer (2005) the knowledge from genomics and proteomics is being applied to the inorganic world of carbon and silicon. For example, Genencor International and Dow-Corning have joined together to combine their respective expertise in protein-engineered systems and silicon. Their strategic alliance seeks to apply the biotech business model to a third outlet of creativity where products can be developed for other companies based on specific needs.

Pre-engineered nanostructures for specific tasks are being prepared with the knowledge of protein-engineering by the nano-biotech applications. It is now known that certain genes in aquatic microorganisms code for proteins that govern the construction of inorganic exoskeletons. It should in theory, be possible to elucidate these gene functions and re-engineer them to code for nanostructures that could be commercially important, such as specific silicon chips or micro-transistors. A carbon-silicon compound has been discovered in freshwater diatoms. It may be possible to understand the molecular process of biosilicification, or the ways plants and animals build natural structures, by this discovery. This understanding may have usefulness in the low cost synthesis of advanced biomaterials for the treatment for osteoporosis. NASA and some companies are also looking at bioactive ceramics found to have exceptional bio-adhesive properties. These properties may also help in the provision of new ways to purify water since bacteria and viruses adhere to these ceramic fibres.

The development of new micro-optical switches and optical micro-processing platforms lies in the area of photonics. In the field of catalysis, the use of inorganic carbon or silicon substrates embedded with biocatalysts has high commercial potential.

Attempts are being made to develop techniques to express spider silk in goats. It may be possible in the near future to build polymers in the lab that are even stronger and that will not need living expression systems for large-scale production. We may have commercially attractive polymers developed by using recombinant DNA technology.

Carbon nanotube being recently developed has a great tensile strength. These nanotubes are perfect for use in new high-tech composites, for switching in computers, and for the storage of hydrogen energy for transportation or power-generation applications. Carbon nanotubes can be coated with reaction-specific biocatalysts and other proteins for specialized applications. Biotechnology may help to make carbon nanotubes even more economically attractive. DNA fragments along with materials may be used for electronic switching. The number of possible new nano-bio combinations is large. These allow the development of smarter, smaller, productive, efficient, durable, multi-purpose and user-friendly materials. Such materials will help the construction of tiny devices that could be embedded in animal or plant tissues for controlled drug delivery, monitoring tissue activity, performance and location of the organism.

Genomics may allow us to predict not only the chances of having/developing a given disease, but also physical appearance and perhaps, levels of intelligence. With the ability to produce replacement organs and with better delivery and diagnostic systems, health will improve. These technologies, therefore, will allow production of materials that will interact (identify, communicate, analyse and protect) with the user. Hence they will all have industrial application.

1.4.5 Safety of Product

The product thus produced comes under certain regulations. Critical determinants are the time and the total costs of bringing a product to market by the governmental regulations. Regulatory agencies act as 'gate-keepers' for the development and availability of new biotechnology products.

There are certain barriers in the industrial development. These include :

   (i)   the costs of testing products,

  (ii)   meeting regulatory standards,

 (iii)   possible delays,

 (iv)   uncertainties in regulatory approval and

  (v)   even outright disapproval of new products on grounds of safety.

According to Anton (2001), the following technologies are going to have an impact on us (Table 1.4).

Table 1.4. Some of the Future Technologies and their Benefits

Technology	Purpose	Firms involved
Biofuels	To provide alternative fuels and energy from biomass	Cargill-Dow, Iogen (Canada)
Biosensors	To locate and monitor the molecular and human/animal activities/position	Biacore, Digital Angel, Oxford Biosensors, Sensatex
Bionics	Produce devices for neural transmission and stimulation, and functional artificial body parts	Advanced Bionics, Bionic Technologies, Oplobionics Corp
Cognitronics	Develop communication platforms between intelligent beings and machines (*e.g.*, robots and computers)	Bionic Technologies, Iguana Robotics, Neural Signals
Combinatory	Develop platforms for quick evaluation of drug/ target and possible allergens	Aurora Biosciences, Bioanalytical, Genetech, Symyx
Molecular farming	Provide smart and precise tools for molecular design, production and manipulation	Molecular Nanosystem, Nanowave, Zyvex
Stem cell (cloning)	To clone animals and organs/tissues from the person/ patient's own cells	Advanced Cell Technology, Biotransplant, Geron,
Genomic	Profiling individuals based on their genetic materials; its analysis for the development of vaccines, drugs and diagnostic systems; and prediction of developing known conditions; silent or active genetic information	Celera Genomics Incyte Genomics DeCode Genetics

Source . Anton, P.S. et. al. (2001).

In fact, the insurance of product safety is costly and sometimes it may discourage new research or curtail product development if a future new product is not likely to have a high financial market return. The use of recombinant DNA technology has created the greatest areas of possible safety concern.

1.5 PUBLIC PERCEPTION AND SOCIAL ACCEPTANCE OF BIOTECHNOLOGY

As we have seen by now, biotechnology has a great potential in almost all the areas, *i.e.,* medicine, agriculture, environment, food, *etc.* These new biotechnological processes have more implication than their technical benefits. The implementation of new techniques depends upon their acceptance by consumers. According to the Advisory Committee on Science and Technology report, Developments in Biotechnology of US, public perception of biotechnology will have a major influence on the rate and direction of developments and there is growing concern about genetically modified products. Questions like that of safety, ethics and welfare are associated with genetic manipulation.

There is a need to address policy issues related to these technologies as they develop. For example, the rights, ownership and integration of clones in human society are still controversial and have not yet been resolved. Similarly, the rights, choice, incentives and selection of surrogate mothers that would carry these clones may also be controversial. As tools get smaller, smarter, more productive and efficient, new productions will be expected by the mass education of entire populations. It will be easy to access systems, easy to configure and produce. There is, thus, going to be an increase in the responsibility of individuals. It is, therefore, quite possible that these technology platforms may be misused. At the same time, the security of individuals will increase as detection tools and protective materials become cheaper. The management of benefit and risk of technology are going to be part of our future governance.

Though there have been initial hiccups, during the last couple of years, changes have occurred in the public perception. The understanding of these new technologies has well accelerated public acceptance. Today it is being accepted as part of our future developments.

There is positive social acceptance of biotechnology even at the industrial level. This is because it is environment friendly and contributes to sustainable development. The following are the reasons of this acceptance:

   (i) It provides new materials and fuels that are not derived from petrochemical processes
   (ii) It improves and enhances the bioremediation of water, soils and ecosystems and is trying to use less fossil-fuel energy
   (iii) It is not dependent on extensive use of natural resources leading to their exhaustion
   (iv) It aims at removing problems of human beings

Industrial biotechnology is acceptable because of the following reasons :

(i) It simplifies processing

(ii) It improves efficiency

(iii) It reduces waste production and increases productivity

It meets both the demands of shareholders by being cost-effective and profitable, and those of environmental advocates by being environment friendly (OECD, 2001). Therefore, firms may be compelled to adopt industrial biotechnology applications because of :

(i) their simplicity

(ii) reduced initial investment capital and flexibility or

(iii) legal requirements

1.6 SCOPE AND IMPORTANCE

Recent developments in biotechnology and its other offshoots have been par excellence. Development in biotechnology is a continuous process. Research and Development at the very frontiers of current knowledge and techniques is going on. In the late 1970s, vague promises about the wonders of this scientific discipline were put forward by molecular biologists. Today, however, with extensive inputs in research, biotechnologists are making predictions with more confidence since many of the hitherto complex problems have been more easily overcome than had been predicted and many transitions from laboratory experiments to large-scale industrial processes have been achieved. New biotechnology has come of age.

It is now clear to the world over that biotechnology has a marked effect as well as a potential impact on virtually all domains of human welfare. For instance:

(i) it has now become an integral part of the education system both at primary and secondary levels,

(ii) it now plays a very important role in employment, production and productivity, trade, economics and economy, human health and quality of human life throughout the world,

(iii) numerous biotechnology companies have emerged throughout the world, including India, and the noted scientists, even some of the Nobel Laureates are moving to some of these companies,

(iv) the total volume of trade in biotechnology is sharply increasing every year, and it is expected to soon become the major contributor to world trade.

Biotechnology methods are now offering a 'green' alternative promoting a positive public image and also avoiding new environmental penalties. Knowledge of biotechnology innovations are being translated to all sectors of industry.

The new biotechnology, however, needs very high expertise and continued heavy funding coupled with dedicated effort. Therefore, highly industrialized countries have a dramatic edge over the less industrialized ones and in fact the 21ˢᵗ century is being considered as the century of biotechnology, just as the 20ᵗʰ century was that of electronics.

Biotechnology-based industries are not labour-intensive, although they are creating valuable new employments. Much of the modern biotechnology has been developed and utilized by large companies and corporations.

New biotechnology companies have certain special features (Table 1.5). A unique need for abstracting information from a wide range of sources has been created as the new biotechnology is at the interface between academia and industry. For this reason, companies are spending large sums on information management.

Table 1.5. Some Unique Features of Biotechnology Companies

1.	Technology driven and multidisciplinary approach	Product development involves molecular biologists, clinical researchers, product sales force.
2.	Management	It requires regulatory authorities, consideration of public perception, issues of health and safety, risk assessment.
3.	Rapid adaptation to new innovations	As the field is fast developing, one biotechnology innovation may quickly supercede another, so the industry should always be prepared for a change.
4.	Dependence on venture capital	The companies need exceptionally high level of funding before the gains start pouring in after having a zero gain period in between.

(*Source* : Promises of Biotechnology-Internet)

The commercial success and exploitation however depend upon a specialist workforce for understanding technology and its application in a wide range of areas. These areas include laws, patents, medicine, agriculture, engineering, *etc.* This necessitates a new look on higher education for the supply of a range of specialists in disciplines encompassing biotechnology. Some courses will endeavour to produce 'biotechnology' graduates who have covered many of the specialist areas. Thereafter, they need to be exposed to the practical and competitive world. Their success lies in the thinking power and flexibility in adopting new changes and then interactive power, not only with others in the field, but also those who matter in promoting, manufacture, regulation and its scale.

1.7 GLOBAL SCENARIO

Biotechnology has been and is still dominated by countries like USA, Japan and those in Europe. In USA alone, there are more than 400 such companies including Genentech, Cetus, Hybritech, Biogen, *etc.* These companies have undertaken various commercial products such as automated bioscreening, improved production of vitamin B_{12}, manufacturing fructose from inexpensive forms of glucose, production of human insulin, microbiological production of human interferon, transgenic animals, production of biopesticides and biofertilizers, and human gene therapy, *etc.*

An International Center for Genetic Engineering and Biotechnology (ICGEB) established by the United Nations Organization (UNO) has two centres: one is located at Trieste, Italy, while the other is in New Delhi.

At its fourth regular session held in Geneva from 17 to 22 May 1999, the United Nations Commission on Science and Technology for Development (CSTD) selected "National capacity building in biotechnology" as the substantive theme for the inter-sessional period 1999-2001.

This theme included the following :

(i) Human resource development through basic science education, research and development as well as their interdisciplinary aspects

(ii) The transfer, commercialization and diffusion of technology

(iii) Increasing public awareness and participation in science policy-making, and bioethics, biosafety and biodiversity

(iv) Awareness in society about the legal and regulatory matters affecting the above issues to ensure equitable treatment.

It was recognized that developing countries were deriving only limited benefits from biotechnology due to declining investments in public agricultural research and development. Furthermore, the dominant role of the private sector of developed countries in biotechnology makes it difficult for public sector research to benefit from the new innovations.

Agricultural biotechnology has offered the potential for increasing and improving food production capacity and promoting sustainability. However, most of the agricultural biotechnology innovations are owned by few countries and private firms. The investment in public agriculture research systems in developing countries has declined. The objective of the meeting was to identify areas of concern and recommend possible strategies that could promote equitable use of resources. A planning meeting was held in Cambridge, Massachusetts, from 2nd to 3rd September, 1999, in conjunction with the international conference on "Biotechnology in the Global Economy". It was co-organized with Centre for International Development at Harvard University. Thereafter, the CSTD Bureau decided that three panels would be organized to address the main aspects of biotechnology. These panels were:

(i) Capacity building,

(ii) Legal and regulatory issues,

(iii) Public awareness and participation.

(A) The First CSTD Panel "Capacity-Building in Biotechnology"

It was held in Tehran, Islamic Republic of Iran, from 11th to 13th April 2000. It had the following objectives :

(i)   To identify key priorities and steps for developing countries and countries with economies in transition to build their capacity to monitor,

(ii) To assess, regulate and manage the impact of biotechnology applications,

(iii) To ensure their safety as well as generation of knowledge for the development of biotechnology by developing human resources through education, training and research.

(B) Second Panel : Legal and Regulatory Issues in Biotechnology

It was convened in Geneva, Switzerland from 3rd to 5th July 2000. This panel examined issues related to :

(i)   intellectual property rights (IPR)

(ii) biosafety, bioethics

(iii) other regulatory policy areas relating to the transfer and diffusion of biotechnology in the key sectors of agriculture, health and environment

The objective of the meeting was to identify the key issues and capacity-building needs necessary for building legal and regulatory frameworks for equitable access and protection of innovations as well as safe use of biotechnology products and services.

(C) Third Panel : Public Awareness and Participation in Science Policy

It was held in Tunis, Tunisia (14th-16th November, 2000). The aim was to analyze and devise a process for building public awareness about the opportunities and challenges presented by biotechnology through the development and promotion of dialogue amongst the following :

(i)   Scientists

(ii) The biotechnology industry

(iii) Policy makers

(iv) The public.

It was recognized that the public does not sufficiently trust many national regulatory regimes as providers of balanced and accurate information on complex issues in science and technology. It was also noted that as the public understanding of biotechnology issues is very low, it is must to find an alternative communication mechanisms for public participation in policy development to achieve its goal.

The issues discussed above indicate that biotechnology has become a part of our system. It is there in individual, social and political arenas, and needs to be exploited, but with caution, for the betterment of mankind.

Questions

1. What is Biotechnology ? Explain in detail.
2. Highlight the historical aspects of biotechnology.
3. What led to the emergence of modern biotechnology ? Give its global scenario.
4. Describe in detail the application of biotechnology.
5. Describe various major technologies involved in biotechnology.
6. Give the historical account of biotechnology. Illustrate your answer by giving examples that how the traditional biotechnology gave way to modern/new biotechnology ?
7. What are the core components of biotechnology ? Describe in detail.
8. In what way Government of India has tried to advance biotechnology in the country ? Explain in detail
9. Write short notes on :
   (a) Emergence of modern biotechnology.
   (b) Product safety.
   (c) Global scenario of biotechnology.
10. What is the scope and importance of biotechnology ? Explain
11. Describe the following :
    (a) Impact of wars on the development of Biotechnology.
    (b) Emerging areas in biotechnology.
    (c) Some unique features of biotech industries.

Basic Methodologies and Tools in Biotechnology

Biotechnology is the integrated use of biochemistry, microbiology, genetic and engineering sciences in order to achieve technological (industrial) applications of capabilities of cells, tissues and organisms.

2.1 INTRODUCTION

As the name indicates biotechnology is a combination of two words: Bio (Biology) and technology. It is an experiment based subject. A student of biotechnology has to be conversant with various theoretical and technology based subjects and methodologies right from the preparation of solution to synthesis and purification of products. Herein are given some of the basic concepts, equipments and technologies/methodologies involved in biotechnology.

2.1.1 Basic Concepts

2.1.1.1 Acids and Bases

According to Bronsted concept of acids and bases, an acid is defined as a substance which has the tendency to give a proton (H^+) and a base is a substance which has the tendency to accept a proton. In other words, an acid is a proton donor and base is a proton acceptor, *e.g.*,

$$HCl + H_2O \rightleftharpoons H_3O^+ + Cl^- \text{ (Here HCl acts as Bronsted acid)}$$

$$CH_3COOH + H_2O \rightleftharpoons H_3O^+ + CH_3COO^- \text{ (Water indicates a Bronsted base here)}$$

$$HCl + NH_3 \rightleftharpoons NH_4^+ + Cl^-$$

Water acts both as an acid as well as a base and is called **amphoteric** or amphiprotic.

$$H_2O = [H^+] [OH^-]$$

$$k_D = \frac{[H^+][OH^-][C_2H_5COO^-][H_3O^+]}{[H_2O] \quad [C_2H_5COOH][H_2O]}$$

k_D = Dissociation constant, which represents the tendency of a compound to dissociate into a solution.

By Bronsted concept

$$H_2O + H_2O \rightleftharpoons H_3O^+ + OH^-$$

$$k_i = \frac{[H_3O^+][OH^-]}{[H_2O]^2}$$

k_i = Ionization Constant

2.1.1.2 pH and pOH

pH may be defined as negative logarithm of H^+ concentration.

$$pH = \log [H^+]^{-1}$$

$$= \log \frac{1}{[H^+]}$$

Similarly, pOH is the negative logarithm of the $[OH]^-$ activity. Ionic product of water (k_w) is

$$k_w = [H^+] [OH^-]$$

Thus, if $[H^+]$ is known, $[OH^-]$ can easily be calculated

$$pH + pOH = 14$$

If any one of the values $[H^+]$, $[OH^-]$, pH or pOH is known, the other three can easily be calculated.

2.1.1.3 Buffer

A buffer solution is defined as a solution whose pH remains practically constant even when small amounts of acids or bases are added to it.

Ionization of Strong acids and bases

A strong acid is a substance that ionizes hundred per cent in aqueous solution, *e.g.*,

$$HCl + H_2O \longrightarrow H_3O^+ + Cl^-$$

Ionization of weak acid or weak base

Weak acid ionizes in aqueous solution as follows:-

$$HA + H_2O \rightleftharpoons H_3O^+ + A^-$$

The ionization constant k_i is

$$k_i = \frac{[H_3O]^+[A]}{[HA][H_2O]}$$

$$k_i [H_2O] = \frac{[H^+][OH^-]}{[H_2O]}$$

e.g., when weak acid C_2H_5COOH is dissolved in water then

$$C_2H_5COOH + H_2O \rightleftharpoons C_2H_5COO^- + H_3O^+$$

$$k = \frac{[C_2H_5COO^-][H_3O^+]}{[C_2H_5COOH][H_2O]}$$

$$k\,[H_2O] = \frac{[C_2H_5COO^-][H_3O^+]}{[C_2H_5COOH]}$$

$$k_a = \frac{[C_2H_5COO^-][H_3O^+]}{[C_2H_5COOH]} \qquad pH = pk_a + \log \frac{[Conjugate\ base]}{[Conjugate\ acid]}$$

$$k_a = \text{Dissociation constant}$$

pH of a solution of weak acid

$$HA \rightleftharpoons H^+ + A^-$$

$$k_i = \frac{[H^+]+[A]}{[HA]}$$

$$\text{Rearranging} = [H^+] = k_a \frac{[HA]}{[A^-]}$$

On taking logarithm :

$$\log [H^+] = \log k_a + \log \frac{[HA]}{[A^-]}$$

Taking negative on both sides

$$-\log [H^+] = \log k_a - \log \frac{[HA]}{[A^-]}$$

if $-\log k_a$ is defined pk_a and $\log [A^-]/[HA]$ is substituted for $-\log, \frac{[HA]}{[A^-]}$ we get

$$pH = pk_a + \log \frac{[A^-]}{[HA]}$$

This forms Henderson-Hasselbalch equation can be written in general expression:

$$pH = pk_a + \log \frac{[Conjugate\ base]}{[Conjugate\ acid]}$$

The important equation that reveals the titration curve of all the weak acids and at the mid point of the titration curve of weak acids, the concentration of proton donor or acceptor (conjugated base or acid) is equal, therefore:

$[A^-] = [HA]$ and the equation becomes

$pH = pk_a + \log 1$

$pH = pk_a = 0$ [$\log 1 = 0$]

$pH = pk_a$

Buffering capacity

Buffer capacity (β) or Buffer index may be defined as the change in the concentration of the base or the acid required in bringing about the change in pH of the buffer by one unit.

$$\beta = \frac{db}{d(pH)}$$

Where d (pH) is the increase in pH resulting from addition of db of base. The buffering capacity of particular acid and its conjugate base will be maximum when their concentrations are equal i.e. when pH = pk_a. Buffering capacity too depends upon total concentration as well as ratio of acid and salt *e.g.* greater the total concentration, the greater buffer capacity.

pk_a values of some common *in vitro* buffers

Name	pk_a	Name	pk_a
Acetic acid	4.75	Phosphoric acid	7.2 (pk_a 2)
Citric acid	5.40 (pk_a 3)	HEPES	7.5
Succinic acid	5.57 (pk_a 2)	Tris	8.3
Imidazole	7.0	Glycylglycine	8.4
		Boric acid	9.2

2.1.2 Basic Equipments Needed in a Biotechnology Laboratory

The following equipment is required in the biotechnology laboratory :

(i) Laminar Flow

Sterile operation in the cell culture or laboratory are normally carried out in the cabinet (called Hood), which serves the purpose of minimizing the risk of contamination and ensures the safety of the operator. For media preparation or when handling non-primate cell lines, culture manipulation can be conducted in a small front-opening cabinet which is equipped with an ultraviolet light source to prevent contamination of working area or surface. It is essential to have a sterile work table. Laminar air flow cabinets are commercially available in various sizes and shapes. Various types of laminar hoods are available. The use of a particular type of laminar hood depends upon the needs of the laboratory (Fig. 2.1). Under climatic conditions where atmospheric dust is very high, it is advisable to use the airflow cabinet in a culture room fitted with double doors. An important precaution is that a laminar airflow cabinet should never face a window or door that is frequently used. It may also be convenient to maintain an aseptic area by installing a High Efficiency Particulate Air (HEPA) filter ventilation unit. Laminar airflow hoods are usually sterilized by switching on the hood and wiping the working surface with 70% ethyl alcohol for 15 min before initiating any operation under the hood. Sterilization can also be done by exposure to UV light. Since UV radiation is harmful to the eyes, sterilization should be done when there are no experiments in progress. Inside the hood, only sterile instruments, appliances and culture tubes or vessels should be used. During flame sterilization of small instruments, spirit lamp should be kept at a sufficient distance from the alcohol bottle to avoid fire. After using the hood, its surface should be cleaned with ethanol, all appliances should be removed and the light and blower motor should be switched off. The hood should be closed with an air tight screen.

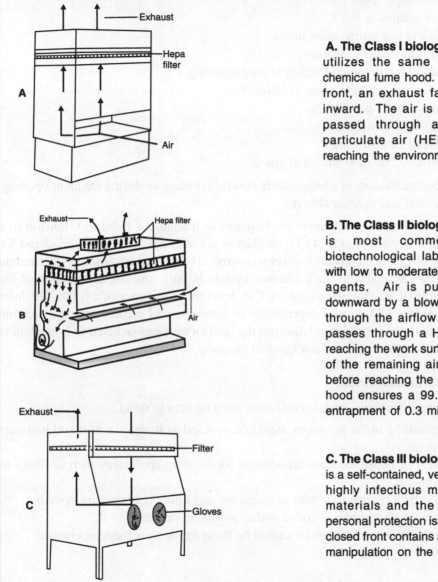

A. The Class I biologic safety hood. It utilizes the same principle as the chemical fume hood. Through the open front, an exhaust fan moves the air inward. The air is circulated and is passed through a high-efficiency particulate air (HEPA) filter before reaching the environment.

B. The Class II biologic safety hood. It is most commonly used in biotechnological laboratories to work with low to moderate toxic or infectious agents. Air is pulled inward and downward by a blower and passed up through the airflow plenum where it passes through a HEPA filter before reaching the work surface. A percentage of the remaining air is HEPA filtered before reaching the environment. This hood ensures a 99.99% efficiency of entrapment of 0.3 micrometer particle.

C. The Class III biologic safety hood. It is a self-contained, ventilated system for highly infectious microorganisms or materials and the highest level of personal protection is provided by it. The closed front contains attached gloves for manipulation on the work surface.

Fig. 2.1. Various types of laminar Hoods : Three types of hoods have been shown with their functional roles.

(ii) Incubators

Incubators and walk-in environmental rooms have generally the range of control and flexibility desirable for growth and development of plant tissues and microbial cultures, *etc*. Besides this, they occupy less space and are readily available in the market. The general characteristics of commercially available incubators or growth chambers are:

  (i)   Temperature range, 2-40°C.
  (ii) Temperature control, ± 0.5°C.
  (iii) Highly safe and low temperature limits.
  (iv) Continuous temperature recorder.
  (v)   Twenty four hour temperature and light programming.
  (vi) Adjustable fluorescent lighting up to 10,000 lux.
  (vii) Relative humidity range, 20-98%.
  (viii) Relative humidity control, ± 3%.
  (ix) Uniform forced air distribution.
  (x)   Capacity up to $0.7m^3$ of $0.5m^2$ shelf space.

It is useful to keep incubators in a temperature controlled because during frequent opening, incubators lose more heat and recover slowly.

For animal cell culture, CO_2 incubators are required to maintain a fixed CO_2 tension in a humidified atmosphere. Several brands of CO_2 incubators are also available. In the modern CO_2 incubators, percentage of CO_2 in incubator chamber is controlled by a valve governed by intermittent reading of an infra red gas analyzer. Such a sensor system is built into the control box of the incubator and is capable of maintaining a constant CO_2 level to an accuracy of $± 0.1\%$. The infra red controller measures the CO_2 level independently of humidity and incorporates a correction for temperature. It is advisable to clean and disinfect the interior of water jacketed CO_2 incubators as there are chances of contamination if not handled properly.

(iii) Temperature Controlled Room

While constructing a hot room, certain conditions must be kept in mind :

  (a) Wooden furnishing in the hot room should be avoided as it absorbs heat and harbours microorganisms.
  (b) Stainless steel or plastic benches are suitable for keeping appliances such as flasks or minor equipments.
  (c) There should be a twenty-four hour temperature and light programming system.
  (d) The temperature should be controlled within a definite range.
  (e) A continuous temperature recorder should be there to monitor the temperature.

(iv) Cold Room

A cold room with low temperature will reduce cost on other items such as refrigerators, deep freezers, cold centrifuge and other instruments. The shelves should be made of plastic and stainless steel to keep chemicals, media and samples. The cold room should be provided with proper gas kit to avoid heating.

In the absence of a cold room, domestic refrigerators and freezers are sufficient for a medium biotechnology laboratory. For keeping blood and tissue samples, –20°C freezer is sufficient. It is important to defrost freezers atleast at a 3 month interval. Refrigerators and freezers require less maintenance than the cold rooms.

(v) Sterilization and Drying Equipment

Glass culture vials, metal instruments and aluminum foil can be sterilized by exposure to hot dry air for 2-4 hours in a hot air oven. Generally sterilization by autoclaving is recommended. Glassware and other accessories such as cotton plugs, gauze, plastic caps, filters or pipettes and proteins less media can be sterilized in a commercially available autoclave at a temperature 121°C and preasure of 15 psi for 15-20 minutes. Normally, instruments such as foreceps, scalpels, needles and spatulae are sterilized by dipping in 95% ethanol, followed by flaming and cooling. This technique is called **flame sterilization.** Now-a-days, in place of flame sterilization, dry sterilization of instruments is practiced.

(vi) Weighing Facility

An electronic balance is a must for the biotechnology laboratory. The balance should be kept at a place where there is minimum (disturbance). It should be calibrated and placed on an even surface. It should be cleaned daily.

(vii) Media Preparation Facilities

The area marked for media preparation should have ample storage and bench space for chemical, labware, culture vessels and other equipments. There should be provision for placing stirrers, pH meters, balances, water baths and Bunsen burners. Microwave oven, autoclave or domestic pressure cooker for sterilizing media and vessels are needed for media preparation. Other requirements include vacuum sources, refrigerators and freezers for storage of chemicals and media stocks.

(viii) Washing and Storage Facilities

For washing purposes, large sinks with provision for hot and cold running water, racks, distilled and double-distilled water apparatus are required. Sufficient space should be available to set up drying ovens, washing machines, plastic or steel buckets for soaking labware, acid or detergent baths, pipette washers, driers and cleaning brushes. The washing area must be provided with dust-proof cupboards.

(ix) Containment Facility

For handling infectious organisms, the laboratory should have a separate containment facility. The workers should be protected with masks, gloves and disposable aprons.

(x) Thermal Cycler (PCR machine)

These are required to amplify DNA sequences for research/diagnostic purposes. There are three types of thermal cyclers (PCR machines) :

   (i)   Regular thermal cyclers
   (ii) Gradient thermal cyclers
   (iii) Real time thermal cyclers

Most of the experiments are carried out on regular thermal cyclers. All the thermal cyclers have sufficient memory to store programmes. There are either 24 or 96 well blocks in a standard thermal cycler. It is better to use thermal cyclers under laminar flow hood to avoid contamination. Gradient thermal cyclers are especially useful in standardization procedures, whereas real time thermal cyclers give us the virtual progression of a reaction.

(xi) Electrophoretic Apparatus

In order to analyze nucleic acids (DNA, RNA) and proteins, electrophoretic apparatus is a must in a biotechnology laboratory. It separates charged molecules on the basis of their movement in an electric field. Dissimilar molecules move at different rates and the components of a mixture will separate when an electric field is applied. It is a widely used technique, particularly for the analysis of complex mixtures or for the verification of purity (homogeneity) of isolated biomolecules. This is most commonly used to separate DNA, protein and amino acids.

(xii) UV Transilluminator

After electrophoresis, DNA/RNA bands can be observed on UV transilluminator. It is a cheap device with UV bulb or tube. When viewing the gel, it is better to protect hands and eyes from UV light. But these days, more sophisticated computerized gel documentation system is available in which gels can be viewed and images stored in the computer for future use. But this instrument is very expensive.

(xiii) Centrifuge Machine

A centrifuge is an instrument designed to produce a centrifugal force far greater than earth's gravity, by spinning the sample about a central axis. Particles of different size, shape or density will, thereby, sediment at different rates, depending on the speed of rotation and their distance from the central axis. There are several kinds of commonly used centrifuges:

    (i)   Low speed centrifuges

   (ii)   Microcentrifuges

  (iii)   Continuous flow centrifuges

  (iv)   High speed centrifuges

   (v)   Ultracentrifuges

2.1.3 Safety

There are some rules and regulations for safety:

(i) Risk assessment

The following points should be considered during risk assessment :

   (a)   Type and the method of using hazardous material

   (b)   General environment, extent and frequency of using that material.

(c) The person who is using them should be trained so that he/she knows how to handle that hazardous material.

(ii) Standard operating procedures

Basically, there are four different stages of standard operating procedures:

(a) Procurement

(b) Storage

(c) Handling

(d) Disposal

All these operating procedures should be considered for genetically manipulated products.

(iii) Safety regulations

The local and central authorities or Safety Commission have framed general safety regulations. These safety regulations should be considered before starting any biotechnology laboratory. Many books and manuals are also available on this topic. Certain guidelines can also be obtained from internet.

(iv) Containment and disposal

Toxic chemicals, carcinogens and mutagens must be used after wearing masks and gloves. Fume hood should be used in case of volatile or toxic chemicals, *etc.* Before disposal, hazardous material should be sterilized and autoclaved. Disinfectants such as hypochlorite should be used to disinfect the reusable equipment.

(v) Incineration facilities

In order to dispose of biological hazardous materials, incineration facility is required.

2.2 TECHNOLOGIES COMMONLY USED IN BIOTECHNOLOGY

Biotechnology is mainly dependent on the development of new instruments and technology. It is, thus, not possible understand it without knowing the technology. A number of instruments have enabled biologists to discover the very existence of cells. These provide the starting point of all information.

2.2.1 Microscopy

2.2.1.1 The Light Microscope

The purpose of a microscope is to resolve objects of the size of single cells or smaller and to create a suitably magnified image that can be comfortably viewed by the human eye. It, therefore, adapts the dimensions of the objects to the limitations of the observing eye.

The basic set up of a compound light microscope consists of (Fig. 2.2a and b).

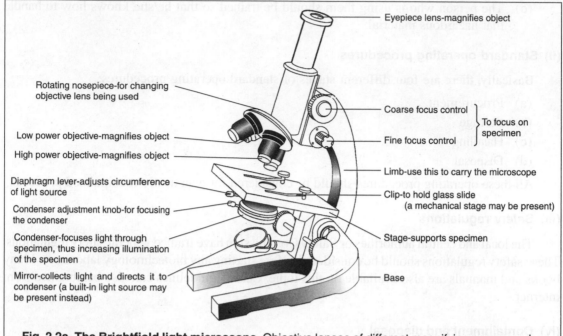

Eyepiece lens-magnifies object

Rotating nosepiece-for changing objective lens being used

Coarse focus control

To focus on specimen

Fine focus control

Low power objective-magnifies object

High power objective-magnifies object

Limb-use this to carry the microscope

Diaphragm lever-adjusts circumference of light source

Condenser adjustment knob-for focusing the condenser

Clip-to hold glass slide (a mechanical stage may be present)

Condenser-focuses light through specimen, thus increasing illumination of the specimen

Stage-supports specimen

Mirror-collects light and directs it to condenser (a built-in light source may be present instead)

Base

Fig. 2.2a. The Brightfield light microscope. Objective lenses of different magnifying power can be selected by rotating the nosepiece.

(a) a **light source,**

(b) a **condenser,**

(c) an **objective** and

(d) an **eyepiece,** also called an **ocular lens.**

The microscope can be considered in terms of illumination and detection components with the specimen lying in between.

(a) Light source : Owing to the extreme magnification involved, the first requirement for microscopic imaging is sufficient illumination for the sample. For this purpose, light rays from an illumination source such as the sun or an incandescent bulb are focused on the specimen by the condenser lens.

On the detection side, structures within the specimen are resolved by the objective and an intermediate magnified image is formed. This image is further magnified by the eyepiece.

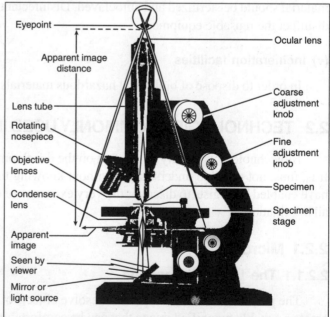

Eyepoint

Ocular lens

Apparent image distance

Lens tube

Coarse adjustment knob

Rotating nosepiece

Fine adjustment knob

Objective lenses

Specimen

Condenser lens

Specimen stage

Apparent image

Seen by viewer

Mirror or light source

Fig. 2.2b. A Microscope's light path. The light path in an advanced bright-field microscope and the location of the virtual image.

Research microscopes can be

(a) upright

(b) inverted

Upright microscopes: Upright microscopes have the transmission light source at the bottom, the light passes through the sample from below to where it is collected by the objective above the specimen stage.

Inverted microscopes: An inverted microscope has the light source on top and the objective below the specimen.

Both the designs are equally well-suited for the observation of slides, but also have distinct properties. Inverted microscopes are used in cell culture because Petri dishes and multiwell plates can be directly observed through the base. On the other hand, upright microscopes work well with 'dip' lenses that can be lowered into the medium and image cells without a coverslip in between.

(b) Objective lens: The light rays focused on the specimen by the condenser are then collected by the microscope's objective lens (Fig. 2.3).

Two sets of light rays enter the objective lens:

(i) those that the specimen has altered and

(ii) those that it has not

The latter group constitutes a cone of light from the condenser lens that passes directly into the objective lens and forms the background light of the visual field. The former group of light rays are ones that emanate, in a sense, from the many points of which the specimen is composed. Light rays from the specimen are brought to focus by the objective lens to form a real, enlarged image of the object within the column of the microscope.

(c) Ocular lens: The image formed by the objective lens is then used as an object by a second lens system, the ocular lens, to form an enlarged and virtual image.

(d) Eye pieces: A third lens system is located in the front part of the eye. It uses the virtual image produced by the ocular lens as an object to generate a real image on the retina. When the focusing knob

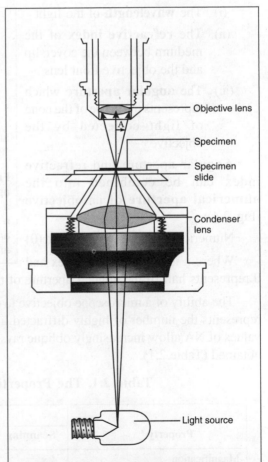

Fig. 2.3. Pathway of light-ray. Light path from the condenser to the objective lens. For maximum resolution the half-angle (α) of the cone of light entering the objective lens should be as large as possible. The maximum half-angle obtainable is about 70°.

of the light microscope is turned on, the relative distance between the specimen and the objective lens changes and allows the final image to become focused on the plane of the retina. The overall magnification attained by the microscope is the product of the magnification produced by the objective lens and that produced by the ocular lens.

Resolution

Although magnification is the most obvious characteristic of microscopes, the ability to resolve fine details of the specimen is the most crucial. Resolution is defined as the ability to see two neighbouring points in the visual field as distinct entities. If two distinct parts of an object are not separated by sufficient distance, they will be seen as one structure, that is, they will not be resolved. The resolution of an objective is determined by three properties :

(i)   The **wavelength** of the light

(ii) The **refractive index** of the medium between the coverslip and the objective front lens

(iii) The **angular aperture** which represents the angle of the cone of light collected by the objective.

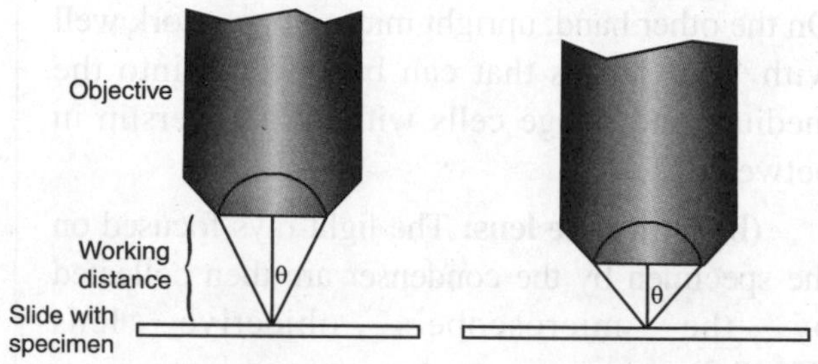

Fig. 2.4. Numerical aperture in microscopy. The angular aperture θ is ½ the angle of the cone of light that enters a lens from a specimen, and the numerical aperture is n sin θ. In the right-hand illustration the lens has larger angular and numerical apertures; its resolution is greater and working distance is smaller.

Angular aperture and refractive index can be combined into the **numerical aperture** of an objective (Fig. 2.4):

Numerical aperture $(NA) = n, \sin(\theta)$

Where n is the refractive index and θ represents half of the angular aperture of the objective.

The ability of a microscope objective to collect light for imaging is determined by its NA. It represents the number of highly diffracted light rays that are captured by the objective. Higher values of NA allow increasingly oblique rays to enter the objective and a more resolved image is obtained (Table 2.1).

Table 2.1. The Properties of Microscope Objectives

Property	Objective			
	Scanning	Low Power	High Power	Oil Immersion
Magnification	4 ×	10 ×	40–45 ×	90–100 ×
Numerical aperture	0.10	0.25	0.55–0.65	1.25–1.4
Approximate focal length (f)	40 mm	16 mm	4 mm	1.8–2.0 mm
Working distance	17–20 mm	4–8 mm	0.5–0.7 mm	0.1 mm
Approximate resolving power 4.50 nm (blue light)	2.3 µm	0.9 µm	0.35 µm	0.18 µm

As a thumb rule, magnification should be at least 500 X the NA and not exceed 1000 X the NA.

The minimum distance (d) at which two points can be distinguished as separate is given by the Rayleigh criterion as:

$$d = \frac{0.61\lambda}{n \sin\theta} = \frac{0.61\lambda}{N}$$

$$N = n \sin\theta = NA$$

$$d = \frac{0.61\lambda}{NA}$$

where λ is the wavelength of light.

The numerical aperture is a constant for each lens, a measure of its light-gathering qualities. For an objective that is designed for use in air, the maximum possible NA is 1.0, since the sine of the maximum angle of theta possible, 90°, is 1, and the refractive index of air is 1.0. For an objective designed to be immersed in oil, the maximum NA is approximately 1.5. Oil immersion objectives with very short working distances have the highest NA and accordingly have the highest resolving power (Fig. 2.5). As a direct consequence of the imaging properties, the depth of field of an objective decreases with increasing NA. A common rule of thumb is that a useful magnification for a microscope is approximately 1000 times the numerical aperture of the objective lens being used. Attempts to enlarge the image beyond this point result in empty magnification and the quality of the image deteriorates. High numerical aperture is achieved by using lenses with a short focal length, which allows the lens to be placed very close to the specimen.

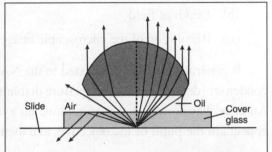

Fig. 2.5. The Oil immersion objective. An oil immersion objective operating in air and with immersion oil. (Adopted from Molecular and Cellular Biology by Stephen L. Wolfe, 1993 Publ. Wadsworth Inc.)

Limit of Resolution

If the minimum possible wavelength of illumination and the greatest possible numerical aperture in the preceding equation are substituted, the limit of resolution of the light microscope can be determined. When these substitutions are made, a value of slightly less than 0.2 mm (or 200 nm) is obtained. It is sufficient to resolve large cellular organelles, such as nuclei and mitochondria. On the other hand, the limit of resolution of the naked eye, which has a numerical aperture of about 0.004, is approximately 0.1 mm (0.0036 inches).

A number of aberrations of the lens also affect its resolving power. Lens makers must overcome them to produce objective lenses whose actual resolving approaches are of theoretical limit.

Magnification

The total magnification of a microscopic image consists of the product of magnification of the objective and the eyepiece. The purpose is to adjust the details resolved by the objective to a size where they can be comfortably viewed by the eye. The total magnification should be adjusted so that all details resolvable by the objective are clearly detectable by eye or camera. Increasing magnification beyond this point adds no detail and leads to empty magnification.

Illumination

A uniformly bright and glare-free illumination of a specimen is setting up Kohler illumination on a microscope. To illuminate the field of view for Kohler illumination, the field diaphragm is opened just wide enough. The field diaphragm protects the sample from unnecessary heat and prevents light not needed for imaging from entering the specimen. It is necessary to adjust the field diaphragm accordingly when changing the objective.

The aperture diaphragm of the condenser determines the following:

(a) The contrast

(b) Depth of field

(c) Resolution of the microscopic image

It generally should be adjusted to the NA of the objective. A lower value for the NA of the condenser (determined by the aperture diaphragm) than that of the objective degrades the image. An easy way to adjust it is to remove an eyepiece adjust the aperture diaphragm until it just appears in the pupil of the objective, and then replace the eyepiece.

Contrast Methods for Transmission Imaging

A microscope no doubt resolves and magnifies small structures, these are often imaged with little contrast. To detect differences in brightness (amplitude) that are mainly created by absorption and the interference of direct and diffracted light in the sample, very little contrast is created by thin samples needed for transmission microscopy. Staining reagents are, therefore, used to enhance the visibility of specific structures by changing their intensity and/or colouring them. Colour filters in the light path often enhance the contrast thus gained in stained samples.

Dark Field Illumination

Light passing through a sample is often diffracted. These changes cannot be perceived because of the total amount of light in bright field illumination. In dark field microscopy, the central light passing through the condenser is blocked out and only oblique illumination gets into the condenser and goes to the sample. Rays diffracted by the sample get into the objective and the sample appears as a bright structure against a dark background (Fig. 2.6).

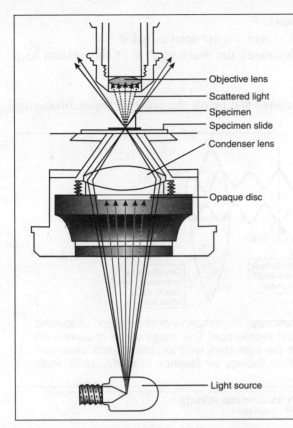

Objective lens
Scattered light
Specimen
Specimen slide
Condenser lens
Opaque disc
Light source

Fig. 2.6. Light path from the condenser to the objective lens in dark field illumination. The central region of the condenser is blocked by an opaque disc so that rays transmitted through the specimen without scattering cannot enter the objective lens. As a result, only scattered rays contribute to image formation. Specimen points that scatter light into the objective lens, therefore, appear bright against a dark background.

2.2.1.2 Phase Contrast Microscope

No amplitude difference easily detectable by the human eye is shown by unstained objects that do not absorb light. Light diffracted by such objects does, however, have alterations in its phase by approximately ¼ of its wavelength (phase object) (Fig. 2.7). These are transformed by phase contrast into amplitude differences that are detectable by the eye. These amplitude differences are detected either by:

(i) increasing this phase difference to ½ of its wavelength and, thus, creating destructive interference of diffracted and non-diffracted light (positive, dark phase contrast) or by reducing it for constructive interference (negative, bright phase contrast).

(ii) inserting a ring annulus in front focal plane of the condenser, the non-diffracted direct light is separated from the diffracted light (Fig. 2.8). This leads to a representation of the non-deviated light in the objective near focal plane as a distinct ring, while the fainter diffracted light is spread more widely. This distinct spatial separation of direct and diffracted light at the rear focal plane is used by a phase plate located there. The thickness of the phase plate is different for the direct (ring-like) and the diffracted light. This further increases (for dark phase contrast) the phase difference to approximately ½ of the wavelength. When the intermediate image is formed, the increased phase difference creates interference between the diffracted and non-diffracted light and the phase object is visualized in high (amplitude) contrast.

This is associated with the following drawbacks:
(a) Around the outlines of structures, optical artifacts are seen as halos.
(b) The ring annulus in the condenser decreases the working NA of the system to a certain extent.
(c) This also reduces the resolution.
(d) In thick specimens, phase shifts from above and below the plane of focus, obscuring the image.

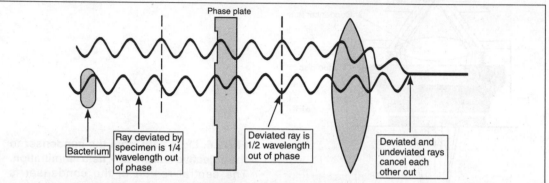

Fig. 2.7. The production of contrast in phase microscopy. The behaviour of deviated and undeviated or undiffracted light rays in the dark-phase-contrast microscope. The image of the specimen will be dark against a brighter background because the light rays tend to cancel each other out. (Adopted and modified from Molecular and Cellular Biology by Stephen L. Wolfe, 1993, Publ. Wadsworth Inc.)

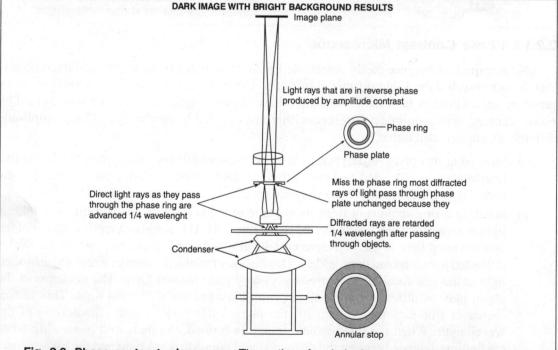

Fig. 2.8. Phase-contrast microscopy. The optics of a dark-phase-contrast microscope have been shown.

2.2.1.3 Differential Interference Contrast (DIC)

To gain contrast in unstained material, concurrently the polarization property of light has been used. Light is polarized by passing through a polarizer before reaching the condenser. Here, a specialized Wollaston prism splits a polarized beam into two beams with slightly different paths and with 90 degrees of difference in polarization to each other. The path difference (shear) is very small, below the resolving power of the objective. The passage through the specimen on separate paths alters the characteristics of the beams and leads to differences between them. Above the rare focal plane of the objective, the shear of the beams is removed by a second prism, which is horizontally movable for adjusting optical path difference. The beams now travel along the same path, but still with perpendicular polarization.

The polarization of the beams is brought into the same plane subsequently by an analyzer in the light path. This causes interference between them that can be perceived as differences in brightness and colour. These variations reflect changes in refractive index, specimen thickness or both.

In DIC, one side of a structure appears brighter, while the opposing side appears darker. This gives the microscopic image a relief-like, pseudo three-dimensional effect.

Compared to phase contrast, full use of the NA of the microscope (condenser + objective) can be made because there is no ring annulus in the light path. Resolution, therefore, is better. The use of full objective aperture allows the visualization of a thinner section of the sample and 'optical sectioning' by focusing through the sample.

This technique is not suited for birefringent and some very thin specimens. The image is degraded by the plastic in tissue culture vessels. For this application, therefore, other imaging methods should be used.

2.2.1.4 Fluorescence Microscopy

This is the most commonly used method of imaging in modern cell biology. In all modes of transmission microscopy, the light emitted by the light source carries information along to the detector (eye or camera). In fluorescence imaging, the light going into the sample that being detected, are spectrally different (Fig. 2.9).

Upon excitation of fluorescent molecules (fluorophores) with light of a shorter wavelength, fluorescence emission occurs at a longer wavelength that is determined by the **Stokes shift** of the molecule. A fluorophore is excited with green light at wavelengths around 550 nm and emitted in the red range around 610 nm. The spectra of the excitation and emission light are very specific for each fluorophore. For fluorescence imaging, light at the emission wavelengths is only detected, while that at the excitation wavelengths is rejected. The image acquired with the microscope is not obscured by the high amount of diffraction and refraction of the excitation light.

A high intensity mercury lamp is used as the light source and emits white light. The exciter filter transmits only blue light to the specimen and blocks out all other colours. The blue light is reflected downward to the specimen by a dichroic-mirror (which reflect light of certain colour, but transmits light of other colour). The specimen is stained with a fluorescent dye-certain portions of specimen retain dye but others do not. The stained portions absorb blue light and emit green colour light, which passes upward, penetrates the dichroic mirror and reaches the barrier-filter.

This filter allows the green light to pass to the eye, however, it blocks out any residual blue light from the specimen which may not be completely deflected by the dichroic-mirror. Thus the eye perceives the stained portions of the specimen as glowing green against a jet black ground whereas the unstained portions of the specimen are invisible.

Confocal microscopy: Fluorescence microscopy is usually performed using whole cells because of embedding media used for sectioning often is fluorescent and obscures the fluorescence generated from the sample. As the eukaryotic cells have thickness, the fluorescence we observe is coming not only from molecules in the plane of focus but also from molecules above and below. To circumvent this problem, confocal microscopy uses a laser light source to produce the excitation light beam that can be focused into a narrow focal plane permitting only a thin optical section of a sample to be illuminated. The position and intensity of emitted light is recorded and information sent for computer analysis. The result of these scanned are combine to generate a composite digital image of fluorescence from a sample.

Immunofluorescence microscopy: It involves fluorophore conjugated antibodies to specifically highlight intracellular structures.

Fluorescence imaging is a very sensitive method. It can detect objects below the resolution limit of the objective. Even single fluorophore can be detected. Such objects are not optically resolved, but the light emitted by them is registered. To be able to detect even faint signals, as much of the emitted light as possible has to be collected, so that in fluorescence microscopy, objectives with the highest possible numerical aperture are used.

Fluorescence microscopy differs from transmission imaging. In fluorescence imaging, the excitation light is directed onto the sample through the objective instead of passing through a condenser on the opposite side of the sample.

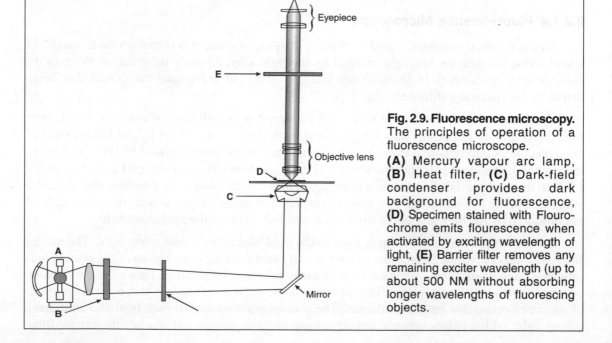

Fig. 2.9. Fluorescence microscopy. The principles of operation of a fluorescence microscope. **(A)** Mercury vapour arc lamp, **(B)** Heat filter, **(C)** Dark-field condenser provides dark background for fluorescence, **(D)** Specimen stained with Flourochrome emits flourescence when activated by exciting wavelength of light, **(E)** Barrier filter removes any remaining exciter wavelength (up to about 500 NM without absorbing longer wavelengths of fluorescing objects.

It is also possible to simultaneously image two or four distinct fluorophores by using filters and dichroics with several spectral 'windows'.

2.2.1.5 Three Dimensional Microscopy

In imaging thick samples, the problem is out-of-focus light blurring the samples. The original biological samples for light microscopy consisted of thin sections, single cells or films of material only a few microns in height. Current research, however, also requires the visualization of deeper and more complex samples. Sectioning microscopes allows insight into tissues or even whole embryos. It is now possible to three-dimensionally resolve biological details smaller than a micrometer.

2.2.1.6 Electron Microscopy

Electrons are the illumination source of electron microscopes. These take advantage of the very short wavelengths attainable in electron beams. This allows a considerable improvement in resolution over the light microscope. To produce an image, the following points are taken care of:

(i) The electron beams are focused by magnetic or electrostatic fields.

(ii) Precisely controlled electric currents are passed through massive coils of wire to generate magnetic focusing fields (Fig. 2.10).

(iii) The magnetic field is shaped into the three-dimensional configuration required for focusing electrons by an iron pole piece inserted into the axis of a wire coil. Because electron lenses can be varied in focus by altering the current applied to their wire coils, electron microscopes are focused by changing lens current rather than by moving the lenses as in light microscopy.

All the space traversed by electrons inside the electron microscope must be kept under a high vacuum

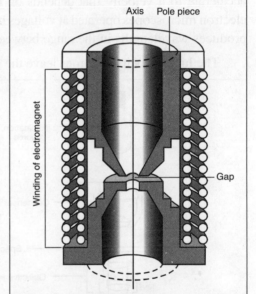

Fig. 2.10. A magnetic lens. The metallic pole piece that focuses electrons gives the magnetic field, generated by a current passed through the winding, a three dimensional form. In shaping the magnetic field, the gap in the pole piece is important. (Redrawn from R.B. Setlow and E.C. Pollard, Molecular Biophysics)

because the illumination source is a beam of electrons. Otherwise, the electrons of the beam, which have relatively poor power to pass through matter, would be completely scattered and absorbed by gas molecules in the microscope. The specimen, therefore, must be dry and non-volatile.

The common types of electron microscopes are:

(i) Transmission electron microscope (TEM)

(ii) Scanning electron microscope (SEM)

2.2.1.6.1 Transmission Electron Microscope (TEM)

It is so called because the electron beams forming the image pass through the specimen. A TEM resembles an inverted light microscope. It has the following components (Fig. 2.11):

An electronic gun : It is present at the top of the central column and is the illumination source. It consists of a filament and an anode. An electric current heats the filament, a thin tungsten wire, to a high temperature which causes electrons to be driven from its surface. The filament and its holder, which are well insulated from the rest of the column are maintained at a high negative voltage, –50,000 to –100,000 volts (V) in most microscopes. The anode is grounded and is, thus, positive with respect to the filament. As a result, electrons leaving the filament are strongly attracted to the anode. As they travel from the filament to the anode, the electrons accelerate to a velocity that depends on the voltage difference between the two locations. In electron microscopes operated at voltages between – 50,000 and – 100,000 V, the velocity attained produces wavelengths in the range between 0.005 and 0.003 nm.

The high velocity electrons leave the gun through a hole in the centre of the anode.

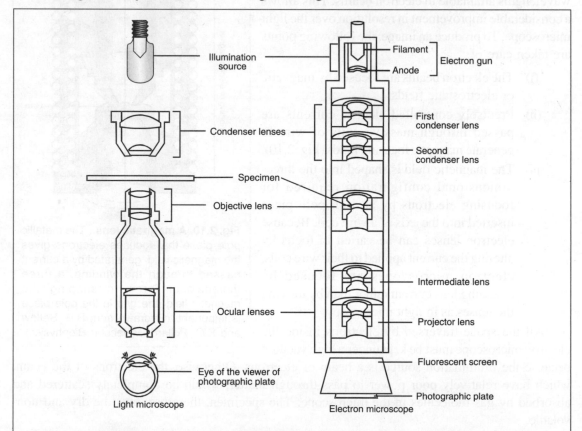

Fig. 2.11. Comparison of inverted and transmission electron microscopes. The arrangement of the illumination source, lenses, specimen and viewing screen in a transmission electron microscope (right), resembles an inverted light microscope (left).

Condenser Lenses : Just below the gun, a series of two condenser lenses focus a very small, intense spot of electrons on the specimen. The specimen can be moved, allowing different regions to be illuminated by the spot. It can also be held in a stable position, with movements no greater than an Angstrom or so, to allow photographs to be made.

Other Lenses : The electrons passing through the specimen are focused by another set of lenses usually including an objective, intermediate and projector lenses.

Each lens in the train forms successively magnified image. Total magnification varies from a few thousand to 300,000 times or more depending on the current applied to the lenses (Fig. 2.12).

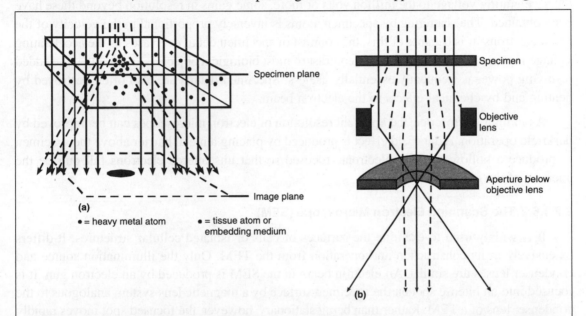

Fig. 2.12. Image formation in the TEM (a) Scattering of electrons by an object point, represented by a cluster of heavy metal atoms in the specimen. As a result of scattering, fewer electrons fall on the corresponding region of the focused image. It produces a "shadow" of the object point. **(b)** Elimination of scattered electrons (dashed lines) by an aperture below the objective lens. The aperture is so small that a large proportion of the scattered electrons, which otherwise would fall on the image plane as a general background fog, are screened out (Redrawn from F.S. Sjostrand, Electron Microscopy of Cells and Tissues, 1967. Courtesy of F.S. Sjostrand and Academic Press, Inc.).

The Projector : It focuses the magnified image onto a fluorescent screen at the bottom of the column. This screen (similar to the screen of a television tube) is coated with crystals that respond to electron bombardment by emitting visible light.

By this process, the electron image is converted to a visual image (Fig. 2.11). The image is permanently recorded by exposure of a photographic plate to the electron beam at the level of the screen. For this purpose, ordinary films and plates can be used because photographic emulsions respond to electrons and light in essentially the same way. There is a provision of airlocks so that specimens and photographic plates can be exchanged without disturbing the microscope vacuum.

TEM helps in the resolution of structures such as ribosomes, microtubules, microfilaments and large molecules such as proteins. However, it differs from the bright-field light microscope because the interaction between specimen points and the illuminating beam is significantly different. In the brightfield light microscope, development of contrast in the image depends primarily on differential absorption of light by structures within cells, whereas in the TEM, electron scattering rather than differences in absorbance produces contrast in the image. An interaction between specimen atoms and electrons of the illuminating beam results in scattering. Even images of individual heavy metal atoms have been produced under special operating conditions. By increasing the accelerating voltage to the million volts or more, some gains in resolution beyond these have been obtained. The scattering by specimen atoms is inversely proportional to the velocity of the beam electrons. It is because of this, the contrast of specimen detail decreases as the accelerating voltage increases. At 1,000,000 V the contrast of most biological objects is so poor that the added resolving power is frequently essentially useless. Specimen details are also often destroyed by heating and by chemical effects to the electron beam.

As in light microscope, the apparent resolution of electron microscopes can be improved by darkfield operation. A darkfield effect is produced by placing a metal barrier above the specimen to produce a hollow beam of electrons, focused so that unscattered electrons fall outside the aperture of the objective lens.

2.2.1.6.2 The Scanning Electron Microscope (SEM)

It is widely used to examine the surfaces of cells or isolated cellular structures. It differs extensively in its construction and operation from the TEM. Only the illumination source and condenser lenses are similar. An electron beam in the SEM is produced by an electron gun. It is focused into an intense spot on the specimen surface by a magnetic lens system analogous to the condenser lens of a TEM. Rather than being stationary, however, the focused spot moves rapidly or scans back and forth over the specimen. The scanning movement is accomplished by beam deflectors, charged plates between the condenser lenses and the specimen.

Molecules on the surface of the specimen being examined are excited by the intense spot of electrons to high energy levels (Fig. 2.13). The excited molecules release this energy in several forms, including high-energy electrons called **secondary electrons.** The image in SEM is formed from the secondary electrons rather than the illuminating beam.

No other lenses are present in the SEM. The secondary electrons leaving a particular spot on the specimen surface are picked up by a **detector** at one side of the specimen.

The detector It is constituted by three elements. These are:

(i) A small fluorescent screen,

(ii) A photomultiplier and

(iii) A photoelectric cell.

Photomultiplier : The emitted light is amplified by photomultiplier. It is projected into the photoelectric cell.

The photoelectric cell : It works on the same principle as the light meter on a camera, emits a current proportional to the amount of light striking it. An image is created by the current leaving the photoelectric cell television tube.

An electron gun in the tube produces a narrow beam of electrons that is focused into a spot at the front of the tube. Deflectors inside the tube, connected to the same electronic circuits scanning the beam in the microscope, move the spot back and forth across the screen at the same rate and in the same direction as the spot scanning the specimen. As a result, the spots on the screen and specimen are at corresponding points at any instant.

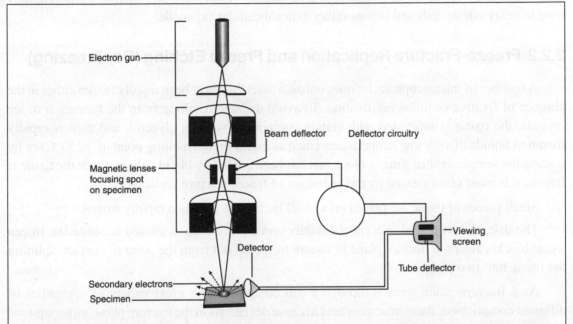

Fig. 2.13. The scanning electron microscope : The sample is scanned by the electron beam. The surface of a CRT however is scanned by a spot. No paths of rays connect the image with points on the sample as in a light microscope or TEM. The image is thus not a true image, but instead a map of the surface.

The resolution in SEM primarily depends on the size of the spot scanning the specimen. As the spot becomes smaller, the number of secondary electrons emitted will be determined by smaller details on the specimen surface, allowing these to be recognized as separate points with different degrees of brightness on the viewing screen. A scanning spot 5 to 10 nm in diameter, typical of contemporary SEMs, resolves surface details of about the same dimensions.

Relatively thick specimens can be observed in this microscope, because the SEM uses electrons emitted from surfaces rather than transmitted electrons for image formation.

In order to avoid scattering, the absorption of the secondary electrons emitted by the specimen surface, the lens system, specimen and detector, thus, must be kept at high vacuum. This operating restriction requires that the specimen must be dry so that it releases gas or water molecules too slowly to disturb the vacuum. Due to this it is not possible to examine most living organisms in the SEM. A few organisms that are highly resistant to desiccation (such as Tardigrades), however, have survived the vacuum and electron bombardment long enough to produce a usable image in the SEM.

Excellent images of object surfaces ranging in size from whole cells to small insects have been produced by SEM. Since its resolution is about 20 times better than the light microscope, the surfaces of objects in this size range are imaged with significantly greater fidelity. The limited resolution of SEM as compared to TEM, makes it less useful for observing structures smaller than whole cells. However, SEM provides a 3D picture. As a result, this microscope is generally used to study whole cells and tissues rather than subcellular organelle.

2.2.2 Freeze-Fracture Replication and Freeze Etching (Dry Freezing)

A number of microscopic techniques utilize tissues that have been rapidly frozen either in the absence of fixative or following fixation. To avoid damage resulting from the formation of ice crystals, the tissue is infiltrated with certain compounds such as glycerol, and then is rapidly frozen in liquids of very low temperatures (such as liquid Freon, melting point of −150°C) or by placing the sample against a metal block that has been cooled by liquid helium. Once the tissue is frozen, it is most often viewed by the technique of freeze-fracture replication.

Small pieces of tissue are placed on a small metal disc and then rapidly frozen.

The disc is then placed in a special holder and a knife-edge is caused to strike the frozen tissue block, causing a fracture plane or fissure to spread out from the point of contact, splitting the tissue into two pieces.

As a fracture plane spreads through a cell composed of a great variety of organelles of different composition, these structures tend to cause deviations in the fracture plane, either upward or downward. This causes the surfaces produced to have elevations, depressions and ridges that reflect the contours of the protoplasm traversed. In other words, the surfaces exposed by the fracture contain information about the contents of the cell.

The goal is to make this information visible. The replication process accomplishes this by using the fractured surface as a template upon which a heavy-metal layer is deposited. The heavy metal is deposited on the newly exposed surface of the frozen tissue in the same chamber as the fracturing was carried out. The metal is deposited at an angle to provide shadows that accentuate local topography.

(i) On top of the metal layer, a carbon layer is then deposited from directly overhead, rather than at an angle, so as to form a uniform layer of carbon that cements the patches of metal into a solid layer.

(ii) Now that a cast of the surface is formed, the tissue that provided the template could be thawed, removed and discarded; the metal-carbon **replica** is placed on the specimen grid and viewed in the electron beam.

(iii) Variations in the thickness of the metal in different parts of the replica cause variations in the number of penetrating electrons to reach the viewing screen, producing the necessary contrast in the image.

(iv) Fracture planes take the path of least resistance through the frozen block, which often carries them through the centre of cellular membranes.

This technique is particularly well suited for examining the interior of membranes.

Freeze-fracture replication by itself is an extremely valuable technique, but by including a step called **freeze-etching,** it can be made even more informative. In this step, the frozen fractured specimen, while still in place within the cold chamber, is exposed to a vacuum at an elevated temperature for one to a few minutes, during which a layer of ice evaporate (sublime) from the surface. As the surrounding water has been removed, the surface of the structure can be coated with heavy metal.

2.2.3 Centrifugation

The stability of a solution (or suspension) depends on the nature of its components. A large number of factors determine whether or not a given component will settle through a liquid medium. This includes:

(a) the size, shape and density of the substance

(b) the density and viscosity of the medium

For a substance to sediment towards the bottom of a tube, it must have a greater density than the surrounding medium. Even if the substance meets the density requirement, the sedimentation process, which tends to concentrate the molecules, is counteracted by their tendency to be redistributed uniformly as a result of diffusion. Other factors remaining constant, the sedimentation of a particular population of molecules depends on its rate of diffusion compared to the opposing centrifugal force being applied. Larger proteins (or nucleic acids) diffuse more slowly than smaller species. A commonly used format where this occurs, is sedimentation in a centrifugal field, a process which is also called **centrifugation.** Centrifugation has following uses :

(i) To separate complex mixtures present in biological samples

(ii) To determine mass, shape or density of particles

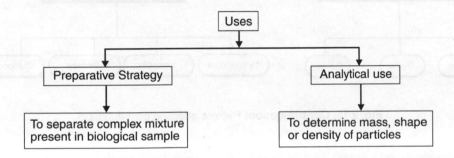

A centrifuge consists of a rotor driven at high speed by an electric motor (Fig. 2.14). The rotor holds tubes containing solutions or suspensions of the materials to be centrifuged. For centrifugation at higher speeds the centrifuge is enclosed in an armoured chamber that can be cooled and held at a vacuum to reduce heat and friction caused by air resistance. The centrifugal forces generated by spinning rotor in the most powerful instruments can reach 500,000 times the force of gravity.

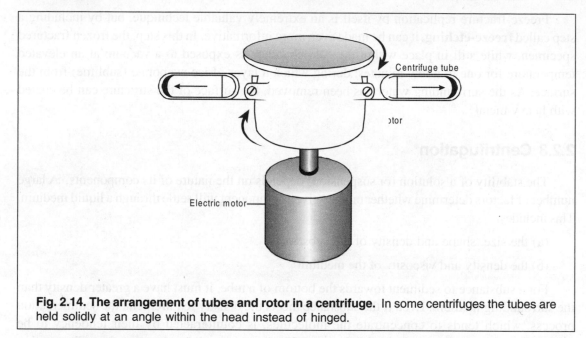

Fig. 2.14. The arrangement of tubes and rotor in a centrifuge. In some centrifuges the tubes are held solidly at an angle within the head instead of hinged.

As mentioned above, movement in response to centrifugal force is related to the mass, density and shape of the structures or molecules being centrifuged and to the mass, density and viscosity of the surrounding solution (Fig. 2.15).

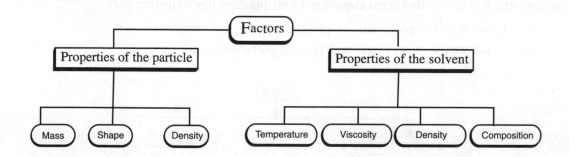

Fig. 2.15. Centrifugation: Factors affecting centrifugation.

(i) If a structure or molecule has greater mass or is denser than the surrounding solution, it moves downwards to the centrifuge tube.

(ii) If it has lesser mass or is less dense than the surrounding solution, it remains at the top of the tube or if present at lower levels, it is displaced by molecules of the solution and moves upward in the tube.

(iii) The velocity of movement is modified by the shape of the particle with rodlike or elliptical forms moving more slowly than spherical particles.

(iv) Increasing the viscosity of the suspending solution also slows the rate of movement of structures or molecules in the centrifuge tubes, particularly for non-spherical forms.

With the development of ultracentrifuges, it is now possible to generate centrifugal forces (as high as 500,000 times the force of gravity), great enough to cause the sedimentation of macromolecules.

The velocity of a given particle during centrifugation in a liquid medium is directly proportional to the square of the angular velocity (ω) and the distance of the particle from the centre of the rotor (r) :

$$V = s(\omega^2 r)$$

In this equation, $\omega^2 r$ is termed the radial acceleration and is given in centimeters per second square. The term 's', which is given in seconds, is the sedimentation coefficient, which is equivalent to the average velocity per unit acceleration. Centrifugal accelerations are generally expressed relative to the earth's gravitational acceleration, which has a value of 980 cm/sec^2. For example, a value for radial acceleration of 4.9×10^7 cm/sec^2 is equivalent to 50,000 times the force of gravity, that is, 50,000 g.

The unit S (or Svedberg, after the inventor of the ultracentrifuge) is equivalent to a sedimentation coefficient of 10^{-13} sec. Since the velocity at which a particle moves through a liquid column depends on a number of factors, including shape, determination of the sedimentation coefficient does not, by itself, provide the molecular weight. However, as long as one is dealing with the same type of molecule, the S value provides a good measure of relative size. For example, the three ribosomal RNAs of E. coli, the 5S, 16S, and 23S molecules, have nucleotide lengths of 120, 1600, and 3200 bases, respectively.

Based on their use, two types of ultracentrifuges are available :

(i) Preparative models

(ii) Analytic models

(i) **Preparative type centrifuges.** These are designed simply to generate large centrifugal forces for a set period of time. The centrifugation proceeds in a near vacuum to minimize frictional resistance. In some cases, the rotors (centrifuge heads) are constructed to allow the tubes to swing out, causing the particles to move in a direction parallel to the walls of the tube. This type of rotor is said to contain swinging buckets.

In contrast, in the fixed-angle variety, the tubes are maintained at a specific oblique angle. In preparative ultracentrifugation, all determinations are made after the tube is removed from the centrifuge. The term preparative refers to the use of this type of centrifuge to purify components for further study.

(ii) Analytic ultracentrifuges. It contains instrumentation that allows one to follow the progress of the substances in the tubes (called cells) during the centrifugation. Various estimates of sedimentation velocity can be made during centrifugation, and measurements of molecular weight, purity, and so forth can be determined.

Relationship between speed (rpm) and acceleration creative centrifugal field (CF) for a typical bench centrifuge with an average radius of rotation r_{av} = 15mm

rpm	RCF*
500	30
1000	130
1500	290
2000	510
2500	800
3000	1160
3500	1570
4000	2060
4500	2600
5000	3210
5500	3890
6000	4630

(*RCF value rounded to near 10).

2.2.3.1 Differential Centrifugation

Most cells contain a wide variety of different types of organelles. When we are to study a particular function of an organelle or to isolate a particular enzyme from the Golgi complex, it is useful to isolate the relevant organelle in a purified state. The isolation of a particular organelle in bulk quantity is generally accomplished by the technique of differential centrifugation.

Principle : Based on the principle that as long as the organelles are more dense than the surrounding medium, particles of different size and shape travel towards the bottom of a centrifuge tube at different rates when placed in a centrifugal field (Fig. 2.16). The homogenate is then subjected to a series of sequential centrifugations.

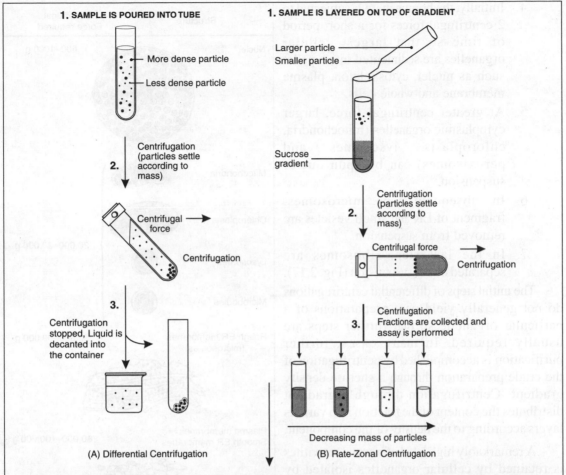

1. SAMPLE IS POURED INTO TUBE

More dense particle

Less dense particle

2. Centrifugation (particles settle according to mass)

Centrifugal force

Centrifugation

3. Centrifugation stopped, Liquid is decanted into the container

(A) Differential Centrifugation

1. SAMPLE IS LAYERED ON TOP OF GRADIENT

Larger particle

Smaller particle

Sucrose gradient

2. Centrifugation (particles settle according to mass)

Centrifugal force

Centrifugation

3. Centrifugation Fractions are collected and assay is performed

Decreasing mass of particles

(B) Rate-Zonal Centrifugation

Fig. 2.16. Separation of particles that differ in mass or density by centrifugation techniques. (A) In differential centrifugation, a cell homogenate or other mixture is spun long enough to sediment the denser particles (*e.g.,* cell organelles, cells). These collect as a pellet at the bottom of the tube (step 2). The less dense particles (*e.g.,* soluble proteins, nucleic acids) remain in the liquid supernatant. These can be transferred to another tube (step 3). **(B)** In rate-zonal centrifugation, a mixture is spun just long enough to separate molecules that differ in mass but may be similar in shape and density (*e.g.,* globular proteins, RNA molecules) into discrete zones within a density gradient commonly formed by a concentrated sucrose solution (step 2). Fractions are removed from the bottom of the tube and assayed (step 3).

To carry out this technique, the following steps are followed :

1. Cells are first broken open by mechanical disruption using a mechanical homogenizer or by exposing cells to high frequency sound (sonication).

2. Cells are homogenized in an isotonic buffered solution, *i.e.,* 0.25 M sucrose, which prevents rupture of membrane vesicles due to osmosis.

3. The homogenate is then subjected to a series of sequential centrifugations at increasing centrifugal forces.

4. Initially, the homogenate is subjected to 2 centrifugal forces for a short period of time so that largest cellular organelles are sedimented into a pellet such as nuclei, cytoskeleton, plasma membrane and whole cells.

5. At greater centrifugal force, larger cytoplasmic organelles (mitochondria, chloroplasts, lysosomes and peroxisomes) can be spun out of suspension.

6. In subsequent steps, microsomes, fragment of ER and small vesicles are removed from suspension.

7. In the last step, ribosomes are separated out of suspension (Fig. 2.17).

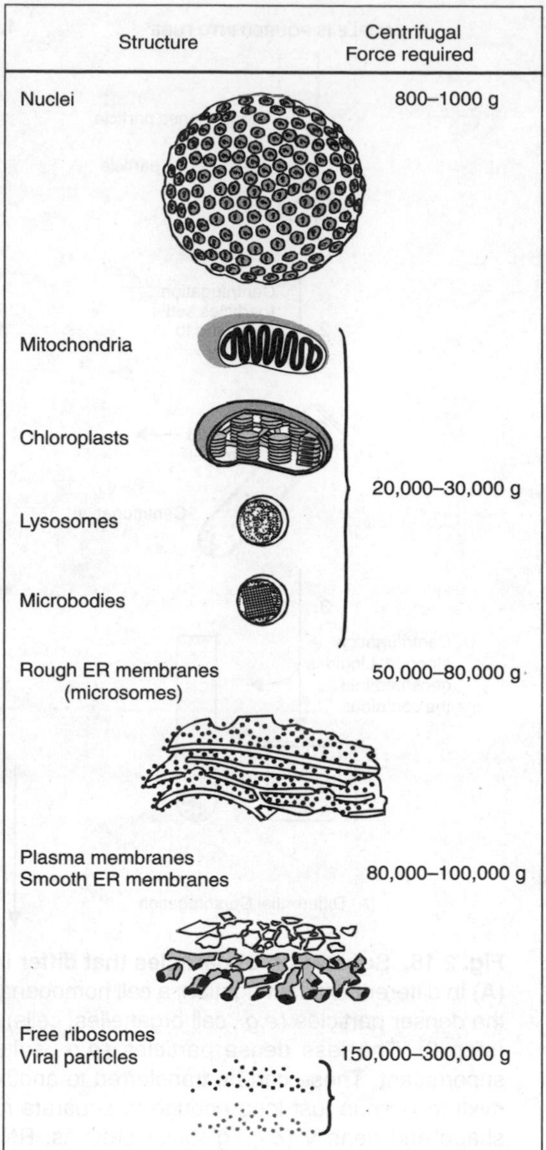

The initial steps of differential centrifugations do not generally yield pure preparations of a particular organelle so that further steps are usually required. In many cases, further purification is accomplished by centrifugation of the crude preparation through a sucrose density gradient. Centrifugation through a gradient distributes the content of the fraction into various layers according to the density of the component.

A remarkably high level of normal activitites is retained by cellular organelles isolated by differential centrifugation as long as they are not exposed to denaturing conditions during their isolation.

Organelles isolated by this procedure can be used in cell-free-system to study a wide variety of activities that occur within the living cells. These include the following:

(i) The synthesis of membrane bound proteins

(ii) The transport of solutes

(iii) The development of ionic gradients and oxidative phosphorylation

Structure	Centrifugal Force required
Nuclei	800–1000 g
Mitochondria	
Chloroplasts	
Lysosomes	20,000–30,000 g
Microbodies	
Rough ER membranes (microsomes)	50,000–80,000 g
Plasma membranes Smooth ER membranes	80,000–100,000 g
Free ribosomes Viral particles	150,000–300,000 g

Fig. 2.17. Ultracentrifugation. Centrifugal fractionation of cell organelle

2.2.3.2 Density Gradient Centrifugation

Particles of different densities can be separated by using density gradient centrifugation, also called **isopycnic centrifugation.**

(i) This is achieved by establishing a density gradient between a high density at the bottom of the centrifugation tube and a region of low density at the top.

(ii) The density gradient can be established by :

(a) zonal centrifugation

(b) isopycnic centrifugation

(a) Zonal centrifugation: Density gradient is established before particle centrifugation.

(b) Isopycnic centrifugation: Density gradient is established during centrifugation. The separation in this is made on the basis of buoyant density.

The zonal centrifugation experiments involve creation of a density gradient before centrifugation by mixing high and low density solutions of materials such as sucrose and glycerol.

This gradient is created in a centrifuge tube prior to centrifugation and sample is then layered on top. Particles sediment through gradient with gradually decreasing velocity.

In isopycnic centrifugation, the metal salt $CsCl_2$ is used because of the high density possible with this material (1.8 gm/ml).

(i) A solution of $CsCl_2$ is mixed with the sample and then centrifuged which forms a $CsCl_2$ density gradient from low density at the top of the centrifuge tube to high density at the bottom.

(ii) Sample particles in the region of highest density at the bottom of the tube are less dense than the surrounding medium.

(iii) They, therefore, tend to float-towards the top of the tube to a region of low density.

(iv) The driving force for this is the centrifugal force that drives salt molecules towards the bottom of the tube, thus, exerting upward pressure on particles of lower density.

(v) Eventually each particle collects and concentrates in a narrow band at its isopycnic density.

(vi) With the exception of lipoproteins, the densities of protein molecules are generally similar and so this technique is not very useful in separating proteins from each other.

(vii) This technique is widely used for :

(a) separating different forms of DNA e.g., plasmid, chromosomal, supercoiled and uncoiled

(b) preparation of macromolecular assemblies e.g., viruses and individual cell types (e.g., viruses).

2.2.3.3 Density Barrier Centrifugation

A single step density barrier can be used to separate cells from their surrounding fluid. Blood cell type can be separated using a density barrier of, e.g., Ficoll.

2.2.4 Chromatography

Chromatography is a term used for a wide variety of techniques in which a mixture of dissolved components is fractionated as it moves through some type of porous matrix.

Chromatographic techniques offer two alternative phases to components in the mixture for association of :

(*i*)   a mobile phase consisting of the moving solvent, and;

(*ii*) an immobile phase consisting of the matrix through which the solvent is moving

In the chromatographic procedures, the immobile phase consists of materials that are packed into mobile phase (consisting of materials that are packed into a column). The proteins to be fractionated are dissolved in the solvent and then passed through the column. The immobile phase has the materials that contain sites to which the proteins in solution can bind. As individual molecules interact with the materials of the matrix, their progress through the column is retarded. Thus, the greater the affinity of a particular molecule for the matrix material, the slower its passage through the column. Since different components of the mixture have a differential affinity for the matrix, they are retarded by different degrees. As the solvent passes through the column and drips out at the bottom, it is collected as fractions in a series of tubes. The components in the mixture with the least affinity for the column appear in the first fractions to emerge from the column.

With the development of high-performance liquid chromatography (HPLC), the resolution of many chromatographic procedures has been improved. In this process long, narrow columns are used and the mobile phase is forced through a tightly packed matrix under high pressure.

2.2.4.1 Ion-Exchange Chromatography

Proteins are large, polyvalent electrolytes, and many proteins in a partially purified preparation do not have the same overall charge. In a variety of techniques, ionic charge is used as a basis for purification (or analytic fractionation). These include ion-exchange chromatography (Fig. 2.18). The overall charge of a protein is a summation of all the individual charges of its component amino acids. Since the charge of each amino acid is dependent on the pH of the medium, the charge of each protein is also dependent on the pH. As the pH is lowered, negatively charged groups become neutralized and positively charged groups increase in number. By increasing the pH, the opposite result is obtained. A pH exists for each protein at which the negative charges equal the positive charges. This pH is the **isoelectric point,** at which the protein is neutral. The isoelectric point of most proteins is below pH 7.

Ion-exchange chromatography depends on the ionic association of proteins with charged groups bound to an inert supporting material such as **cellulose.** The two most commonly employed ion-exchange resins are diethylaminoethyl (DEAE)-cellulose and carboxymethyl (CM)-cellulose. DEAE-cellulose is positively charged and, thus, it acts by binding negatively charged molecules; it is an **anion exchanger.** CM-cellulose is negatively charged and acts as a **cation exchanger.** The resin is packed into a column, and the protein solution allowed to percolate through the column in a buffer whose composition promotes the binding of some or all of the proteins to the resin. By increasing the ionic strength of the buffer (which adds small ions to compete with the charged groups of the macromolecules for sites on the resin) or changing its pH. Both proteins bound to the resin can then be displaced. Displacement of the bound proteins can be carried out in a stepwise manner by a sequential addition of a series of different buffers or in a continuous manner by adding a solution of continually changing ionic strength or pH (a gradient).

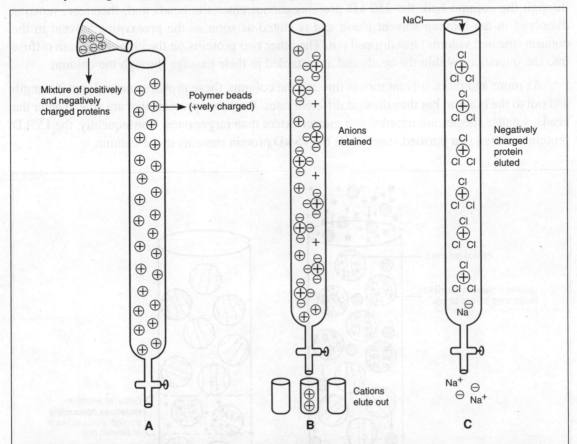

Fig. 2.18. Ion exchange chromatography. The column is packed with beads opposite in charge to the proteins of interest. A mixture of positively and negatively charged proteins is added to the column. **(A)** The proteins of opposite charge stick to the beads; those with the same charge pass rapidly through the column. **(B)** Displacement of the proteins attached to the beads by NaCl solution. The salt ions compete for charged groups on the surface of the beads, causing release of the proteins of interest. **(C)** By gradually increasing the salt concentration, proteins of increasingly greater charge can be released in succession. (Adopted and Modified from Molecular Cell Biology by Lodish *et.al.*, 2003)

2.2.4.2 Gel Filtration Chromatography

Proteins (or nucleic acids) are separated by gel filtration primarily on the basis of molecular weight (Fig. 2.19). The separation material consists of tiny beads. These beads are packed into a column through which the solution of proteins slowly passes. The beads used in gel filtration are composed of cross-linked polysaccharides (dextrans or agarose) of different porosity.

If a protein to be purified has a molecular weight of 155,000 daltons and is present in a solution with two contaminating proteins of 250,000 daltons and 75,000 daltons, the mixture is passed through a column of Sephadex G150, which consists of beads that allow only molecules of less than 200 kD to penetrate their interiors. When a solution containing these proteins passes through the column bed, the 250 kD proteins cannot enter the beads and, therefore, remains dissolved in the moving solvent phase and is eluted as soon as the preexisting solvent in the column (the bed volume) has dripped out. The other two proteins on the other hand, can diffuse into the interstices within the beads and are retarded in their passage through the column.

As more and more solvent moves through the column, these proteins move down its length and out to the bottom, but they do so at different rates. For those proteins that are able to enter the beads, smaller species are retarded to a greater extent than larger ones. Consequently, the 155kD protein is eluted in a purified state, while the 75 kD protein remains in the column.

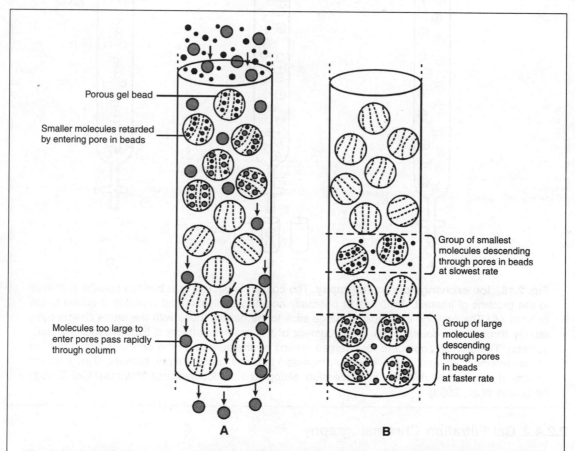

Fig. 2.19. Gel filtration chromatography. (A) A mixture of proteins of different sizes is poured into a gel filtration column. Proteins too large to enter pores in the gel are excluded and rapidly pass through the column. Smaller molecules enter the pores in the gel and are retarded. **(B)** The molecules after traveling some distance through the column fit into the pores are retarded differentially according to size, with the smallest molecules traveling most slowly. The differential retardation separates the proteins into distinct groups, or bands, descending through the column at different rates. (Adopted and modified from Moleculer Cell Biology by Lodish *et.al.*, 2003)

2.2.4.3 Affinity Chromatography

A protein has unique structural properties. It allows the molecule to be specifically withdrawn from solution, while all other molecules remain in solution. This property has been exploited by affinity chromatography. Proteins interact with specific compounds: (a) enzymes with substrates, (b) receptors with ligands, (c) antibodies with antigens, and so forth. These types of proteins can be removed from solution by passing a mixture of proteins through a column in which the specific interacting molecule (substrate, ligand, antigen, *etc.*) is immobilized by linkage to an inert material (the matrix) (Fig. 2.20).

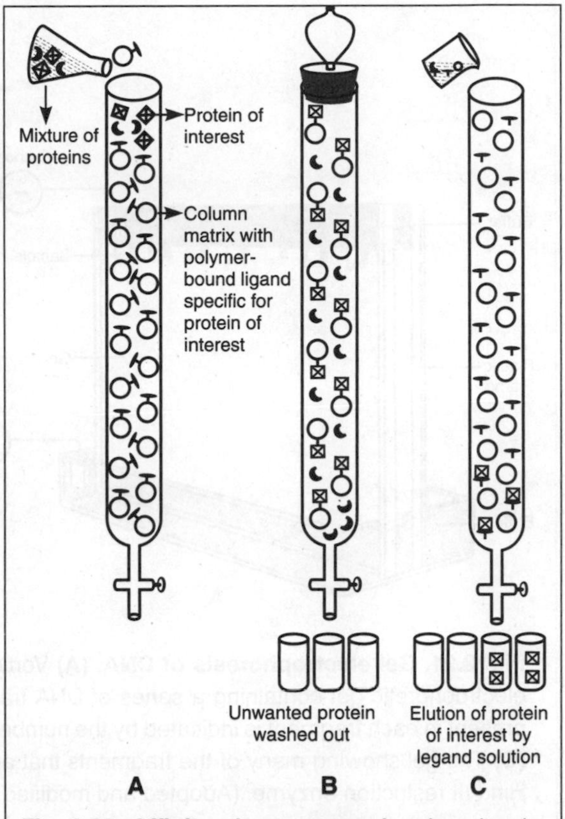

When an impure preparation of an insulin receptor is passed through a column containing agarose beads to which the protein insulin was attached, the receptor will specifically bind to the beads as long as the conditions in the column are suitable to promote the interaction. Once all the contaminating proteins have passed through the column and out from the bottom end, the insulin receptor molecules can be displaced from the matrix by changing the ionic composition of the solvent in the column.

Affinity chromatography is helpful to remove one specific protein from a complex mixture of proteins. It is possible to achieve a near-total purification of the desired molecule in a single step.

Fig. 2.20. Affinity chromatography. A molecule recognized and bound by the protein of interest is covalently linked to the beads in the column. A mixture of proteins is added to the column **(A)** Only the protein binding the molecule attached to the beads is trapped by the beads; other proteins pass through the column without hindrance. **(B)** The attached proteins are released by adding a solution of the molecules recognized and bound by the protein of interest. The added molecules compete for the binding site on the protein. The protein released from the beads **(C)** (Adopted from Moleculer Cell Biology by Lodish *et.al.*, 2003)

2.2.5 Gel Electrophoresis

Electrophoresis is a powerful technique that is widely used to fractionate proteins and nucleic acid. There are many different variations among electrophoretic techniques, all of which depend on the ability of charged molecules to migrate when placed in an electric field. Charge of a molecule is influenced by the following factors:

    (i)   The type, concentration and pH of buffer

   (ii)   Temperature

  (iii)   Strength of the electric field

  (iv)   Nature of the support material (matrix) used for electrophoresis

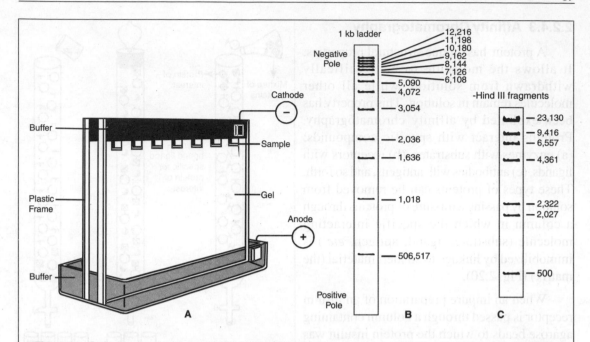

Fig. 2.21. Gel electrophoresis of DNA. (A) Vertical gel apparatus. **(B)** The 1 kilobase ladder in an electrophoretic gel containing a series of DNA fragments of known size. The number of base pairs present in each fragment is indicated by the number. The smallest fragments have moved the farthest. **(C)** The gel showing many of the fragments that arise when lambda phage DNA is digested with the Hind III restriction enzyme. (Adopted and modified from Molecular Cell Biology by Lodish *et.al.* 2003)

Gel electrophoresis is also widely used in the separation of nucleic acids of different molecular weights (nucleotide length) (Fig. 2.21). Generally small RNA or DNA molecules of a few hundred nucleotides or less are separated by polyacrylamide gel electrophoresis (PAGE). As it is difficult for the larger molecules to move through the highly cross-linked polyacrylamide, they are generally fractionated on agarose gels, which have greater porosity.

Agarose is a polysaccharide extracted from seaweed. It is dissolved in hot buffer and is poured into a mold. It is then made to gelate simply by lowering the temperature. Gels of lower agarose concentration (as low as 0.3 per cent) are used. It is made to separate larger DNA fragments.

All nucleic acid molecules unlike proteins, regardless of their length, have a similar charge density (the number of negative charges per unit of mass). All of them, thus, have an equivalent potential for migration in an electric field. The gelated polyacrylamide or agarose provides the necessary resistance to migration, so that the greater the molecular weight of RNA or DNA molecule, the more slowly it makes its way through the gel. The gel electrophoresis is very sensitive and it is because of this, DNA or RNA molecules that differ by only a single nucleotide can be separated from one another by this technique.

2.2.5.1 Polyacrylamide Gel Electrophoresis (PAGE)

The electrophoretic separation of proteins is usually accomplished using **polyacrylamide gel electrophoresis (PAGE).** In this the proteins are driven by an applied current through a gel composed of a small organic molecule (acrylamide) that is cross-linked to form a molecular sieve. The following steps are taken for this purpose:

(i) The gel may be formed as a thin slab between two glass plates or as a cylinder within a glass tube.

(ii) Once the gel has polymerized, the slab (or tube) is suspended between two compartments containing sucrose or glycerol, whose density prevents the sample that is added to the gel from mixing with the buffer in the upper compartment.

(iii) A voltage is then applied between the buffer compartments and current flows across the slab.

(iv) This causes the proteins to move towards the oppositely charged electrode.

(v) Alkaline buffers are used for separations.

These make the proteins negatively charged and cause them to migrate towards the positively charged anode at the opposite end of the gel.

The factors that influence the relative movement of proteins through a polyacrylamide gel are the size, shape and charge density (charge per unit of mass) of the molecules.

(i) **Size:** Since the polyacrylamide forms a cross-linked molecular sieve, the larger the protein, the more it becomes entangled as it migrates and, thus, the slower its rate of migration.

(ii) **Charge Density:** The greater the charge density, the more forcefully the protein is driven through the gel, and thus, the more rapid its rate of migration.

The ability of a given-sized protein to migrate through the gel depends on the concentration of polyacrylamide used to make it. The less the concentration (down to a minimum of about 2 per cent) of acrylamide, the less the gel will be cross-linked, and the more rapidly a given protein molecule will migrate.

(iii) **Shape:** The compact globular proteins move more rapidly than elongated fibrous proteins of comparable molecular weight or linear DNA compared with circular DNA.

(iv) **Electric Field Strength:** Mobility increases with the increasing field strength (voltage), but there are practical limitations in using high voltages, especially due to heating effects.

The combine influence of net charge : mass ratio is given according to the formula:

$$\mu = \rho E/r; \qquad \text{where} \qquad \rho = \text{net charge on the molecule,}$$
$$r = \text{is the molecular radius}$$
$$E = \text{is the field strength}$$

The progress of the electrophoresis is indicated by watching the migration of a charged tracking dye that moves just ahead of the fastest proteins.

(i)   After a period of time, the current is turned off, and the gel is removed from its container

(ii)   The gel that emerges can be stained to reveal the location of the proteins

(iii) If the proteins are radioactively labelled, their location can be detected by pressing the gel against a piece of X-ray film to produce an autoradiograph

(iv)   The gel can be sliced into fractions, and individual proteins isolated, or the gel can be pressed against a piece of nitrocellulose filter paper and subjected to a further electrophoretic treatment that causes the proteins to move out of the gel. They become absorbed in the surface of the nitrocellulose.

This procedure is known as **Western blot**. It can be used to identify the proteins in individual bands by their interaction with specific antibodies and transferred to nylon filter.

2.2.5.2 Sodium Dodecyl-sulphate Polyacrylamide Gel Electrophoresis (SDS-PAGE)

SDS has an anionic head group and a lipophilic tail. It non-covalently binds to proteins with ratio around one SDS molecule per two amino acids.

PAGE is usually carried out in the presence of negatively charged detergent sodium dodecyl-sulphate (Fig. 2.22). The latter binds in large number to all types of protein molecules. The electrostatic repulsion between the bound SDS molecules causes the proteins to unfold into a similar rodlike shape. This eliminates differences in shape as a factor in separation. The number of SDS molecules that bind to a protein is roughly proportional to the protein's molecular weight. Each protein species, regardless of its size, has an equivalent charge density and is driven through the gel with the same force. However, because the polyacrylamide is highly cross-linked, larger proteins are held up to a greater degree than smaller proteins. As a result proteins become separated on the basis of a single property – their molecular weights.

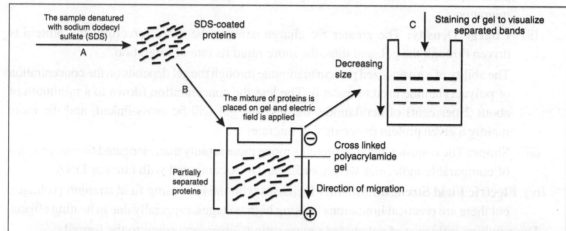

Fig. 2.22. SDS polyacrylamide gel electrophoresis (PAGE). (A) Multimeric proteins are dissociated and all the polypeptide chains are denatured by treatment with SDS (a negatively charged detergent). The SDS-protein complexes migrate through the polyacrylamide gel **(B)** Small proteins are able to move through the pores more easily, and faster, than larger proteins. **(C)** The proteins, therefore, separate into bands according to their sizes as they migrate through the gel. The separated protein bands are visualized by staining with a dye. (Adopted and modified from Molecular Cell Biology, Lodish *et.al.*,2003)

Besides separating the proteins in a mixture, SDS-PAGE can also be used to determine the molecular weights of the various proteins by comparing the positions of the bands to those produced by proteins of known size.

2.2.5.3 Isoelectric Focusing

In the supporting medium, isoelectric focusing consists of a gel containing a mixture of ampholytes, low-molecular-weight polymers having varying ratios of positively charged amino groups and negatively charged carboxyl groups.

When a voltage is applied across the gel, the ampholyte molecules become redistributed. Most negative ones move close to the anode and the most positive ones to the cathode. This movement sets up a stable pH gradient within the gel from one end to the other. As proteins migrate through the gel in response to the electric field, they are exposed to a continually changing pH. It produces a continuous change in their ionic charge. Each protein at some point along the gel encounters a pH that is equal to its isoelectric point. It converts the protein into a neutral molecule and causes its migration to stop. As a result, each species of protein becomes focused in a very sharp band at a predictable position along the length of the gel.

2.2.5.4 Two-Dimensional Gel Electrophoresis

This technique was developed in 1975. It separates proteins in two dimensions based on different properties of the molecules. Proteins are first separated according to their isoelectric focusing within a tubular gel. The gel after separation is removed and placed onto a slab of SDS-saturated polyacrylamide and subjected to SDS-PAGE. The proteins move into the slab gel and become separated according to their molecular weight (Fig. 2.23). The technique has very high resolution. This allows virtually all the proteins present in a cell in detectable amounts to be distinguished. This procedure has been used to resolve a large number of different proteins.

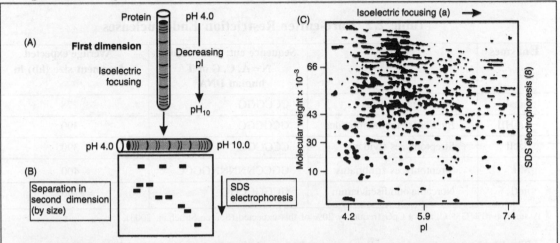

Fig. 2.23: Two dimensional gel electrophoresis (A) The proteins are first separated on the basis of their charges by isoelectric focusing **(B)**. The resulting gel strip thus produced is applied to an SDS-polyacrylamide gel **(C)**. The proteins are separated into bands by mass (Adopted from Molecular Cell Biology by Lodish *et al.*, 2003).

2.2.5.5 Pulse Field Gel Electrophoresis

The standard agarose gel electrophoresis as a result of a sieving effect can help to separate the molecules of a limited size ranging upto 30 kb. The DNA molecules pass through pores in the agarose gel and small molecules are able to migrate more quickly through the tham. Above a certain size of DNA fragment, the sieving effect, however, is no longer effective and thus resolution of DNA fragments above 40 kb is extremely limited. An alternative electrophoresis method, therefore is required for separating very large fragments of DNA, mammalian genes and other functional sequence units that are very large.

Pulsed field gel electrophoresis (PFGE) can resolve DNA fragments in a size range from about 20 kb to several Mb in length. The rare-cutter restriction endonucleases, specialized restriction nucleases, can cleave vertebrate DNA and produce large restriction fragments which can be size separated by PFGE. These enzymes often recognize GC-rich recognition sequences which contain one or more CpG dinucleotides. Because the CpG dinucleotide occurs at low frequencies in vertebrate DNA, human DNA and other vertebrate DNAs have comparatively few recognition sequences for restriction enzymes which cut at sequences containing CpGs.

The procedures involves lysing the cells and purifying the DNA. It result in shear forces that cause considerable fragmentation of the DNA conventionally prepared genomic DNA is not suitable for PFGE. In PFGE, the DNA is isolated in such a way as to minimize artificial breakage of the large molecules. It is then digested with appropriate rare cutter restriction endonucleases. To prepare high molecular weight DNA, samples of cells, for example, white blood cells, are mixed with molten agarose and then transferred into wells in a block-former and allowed to cool. The cells then become entrapped in solid agarose blocks. The agarose blocks are removed and incubated with hydrolytic agent and cellular components are digested. It leaves the high molecular containing purified high molecular weight DNA can then be incubated in a buffer containing a rare-cutter restriction endonuclease Table 2.2.

Table 2.2. Rare-cutter Restriction Endonucleases

Enzymes	Source	Sequence cut: CG = CpG: N = A, C, G or T human DNA*	Average expected fragment size (kb) in
SmaI	Serratia marcescens	CCCGGG	78
bssHII	*Bacillus* stearothermo hilus	GCGCGC	390
SacII	Streptomyces lividans	CCGCGG	390
SfiI	Streptomyces fimbriatus	GGCCNNNNNGGCC	400
notI	Norcadia otitidiscaviarum	GCGGCCGC	9766

* Assuming 40% G + C, and a CpG frequency 20% of that expected (Source Strachan, 2003).

The blocks are placed in wells at one end of an agarose gel contained within a PFGE apparatus to size-separate the large restriction fragments contained within the agarose blocks. The negatively charged DNA is repelled *from* the negative electrode and migrates in the electric field. The relative orientation of the gel and the electric field during a PFGE run is periodically altered. It is done by setting a switch to deliver brief pulses of power. It alternatively activates two differently oriented fields. Variants of the technique use a single electric field but with periodic reversals of the polarity (field inversion gel electrophoresis) or periodic rotation of the gel or electrodes (Stratchen, 2003).

There are so many variants of methods the principle of a discontinuous electric field is common to each of the variant methods. By this the DNA molecules are intermittently forced to change their conformation and direction of migration during their passage through the gel. The time taken for a DNA molecule to alter its confirmation and re-orient itself in the direction of the new electric field is strictly size dependent. Thereby, the DNA fragments up to several Mb in size can be efficiently fractionated. These also include intact DNA from whole yeast chromosomes (Schwartz and Cantor, 1984).

2.2.6 Colorimetry/Spectrophotometry

2.2.6.1 Basic Principle

This method involves measuring the **intensity of a colour** in a solution and relating it to the concentration of the analyte. While some materials of interest are already coloured, most of these analyses require the analyst to add some chemical reagents (reacting chemicals) to a sample to produce a characteristic colour.

Visual comparison : It is the simplest type of measurement of the intensity of colour to a set of colour standards which represent various concentrations of the analyte. This method does not require any expensive equipment, but perception of colour is quite subjective. A more precise measurement can be made using a colorimeter.

A colorimeter works on the principle of light absorption.

Basic Laws of Light Absorption

For a uniform absorbing medium (solution: solvent and solute molecules that absorb light), the proportion of light radiation passing through it is called the **transmittance (T)** and the proportion of light absorbed by molecules in the medium is **absorbance (Abs)**.

Transmittance is defined as :

$$T = I/I_o$$

where

I_o = intensity of the incident radiation entering the medium.

I = intensity of the transmitted radiation leaving the medium.

T is usually expressed as per cent transmittance, %**T**:

$$\%T = I/I_o \times 100$$

The relationship between per cent transmittance (%**T**) and absorbance (**A**) is given by the following equation :

A = 2-log (%T)

A = log (1/%T)

Two scales %T and Abs are present on most spectrophotometers. Absorbance has no units and varies from 0 to 2 (linear region for most substances is from 0.05 to 0.7).

The **Beer-Lambert Law** states that **Abs** is proportional to the concentration (**c**) of the absorbing molecules, the length of light-path through the medium and the molar extinction coefficient:

A = åλcl

where

   åλ = molar extinction coefficient for the absorbing material at wavelength (in units of mol/ cm)

   c = concentration of the absorbing solution (molar)

   l = light path in the absorbing material (l = 1 cm for our purposes)

The Beer-Lambert Law may not be applicable to all solutions since they can ionize/polymerize at higher concentrations, or precipitate to give a turbid suspension that may increase or decrease the apparent absorbance. The Beer-Lambert Law is most accurate between Abs of 0.05 to 0.70. Above 0.70, the measured Abs tends to underestimate the real Abs. Below 0.02 Abs, many instruments are not accurate (as Abs is the log of a ratio). A protein solution for example had an Abs of 1.42, but when it was diluted 1:4, it had an Abs of 0.45 meaning that the original, undiluted solution had a real Abs of 4 × 0.45 = 1.80 Abs.

Accordingly the Beer's Law tells us that there is a linear relationship between absorbance and concentration. That is, if the absorbances of a number of standards are measured and plotted against the solution concentrations, a straight line results. A possible calibration curve is shown below:

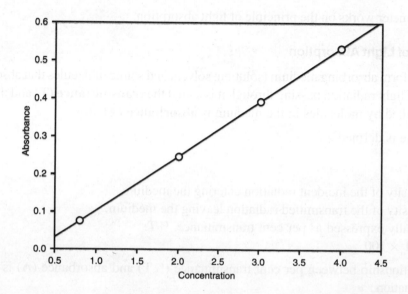

The data points may not fall exactly on the line due to experimental errors.

2.2.6.2 Colorimeter (Instrumentation)

It is an instrument that is capable of measuring how much light of a given wavelength is absorbed by a solution.

A colorimeter is a device that consists of the following:

  (i) A light source, which can be as simple as a tungsten-filament light bulb

  (ii) Some optics for focusing the light

 (iii) A coloured filter, which passes light of specific colour that is absorbed by the treated sample

 (iv) A sample compartment to hold a transparent tube or cell containing the sample

  (v) A light-sensitive detector, like the light metre on a camera, which converts the light intensity into an electric current

 (vi) Electronics for measuring and displaying the output of the detector (Fig. 2.25).

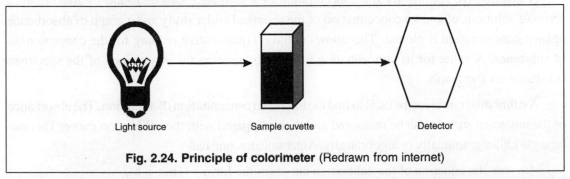

Light source   Sample cuvette   Detector

Fig. 2.24. Principle of colorimeter (Redrawn from internet)

In discussing colorimetry, two terms, *i.e.,* transmittance and absorbance are useful. The transmittance of a sample measures how much light is able to pass through it. A very intensely coloured sample has a low transmittance, while plain water has high transmittance. While carrying out the experiments, to measure the transmittance of a test sample, the absorbance is set to zero. This indicates 100% transmittance, which is taken as the reference point.

Absorbance (Abs) is inversely proportional to transmittance and measures how much light is absorbed by a sample. An intensely coloured solution has a high absorbance value, whereas a faintly coloured solution has low absorbance. Sometimes we can visually distinguish between highly concentrated and unconcentrated solutions by the depth of their colour. However, a colorimeter is able to attach numbers to these observations and allows us to quantitatively measure concentrations.

Some colorimeters may be designed to read out directly in concentration units, while others may show the results in units of light absorbance which need to be compared to a calibration curve. The filter used is not of the same colour as the solution being tested, but rather of a complementary colour. A filter transmitting the light of the colour which the solution absorbs, is to be used. A yellow solution looks yellow because it absorbs blue light (so a blue filter would be used).

A colorimeter is generally any tool that characterizes coloured samples to provide an objective measure of colour characteristics. In chemistry, the colorimeter is an apparatus that allows the absorbance of a solution at a particular frequency (colour) of visual light. Colorimeters hence make it possible to ascertain the concentration of a known solute, since it is proportional to the absorbance.

Different chemical substances absorb varying frequencies of the visible spectrum. Colorimeters rely on the principle that the absorbance of a substance is proportional to its concentration, *i.e.,* a more concentrated solution gives a higher absorbance reading. A filter in the colorimeter is used to select the colour of light which the solute absorbs the most, in order to maximize the accuracy of the experiment. The amount of light which has passed through the solution, compared to the amount entering, is measured by a sensor. The amount absorbed is displayed on a sensor.

A quantitative reading for the concentration of a substance can be found by making up a series of solutions of known concentration of the chemical under study and a graph of absorbance against concentration is plotted. This allows to have a quantitative reading for the concentration of substance. A value for its concentration is found by reading the absorbance of the specimen substance on the graph.

A calibration curve can be used to find the unknown concentration of a solution. The absorbance of the unknown sample can be measured and then compared with the calibration curve. This can be done either graphically or algebraically. Algebraically, one can :

   (i)   use the equation of the calibration curve (in the form y = mx + b),

   (ii) substitute the measured absorbance for y, and

   (iii) solve for the concentration x.

This method, therefore, can be used to find out the concentration of a solution of unknown concentration.

It should, however, be noted that the linear behaviour of the absorbance-concentration relationship may be relied on only for absorbance values of approximately 1 or below. Therefore, the absorbance of an unknown solution needs to be within this range for the method to be reliable. One of the challenges of analyzing **"real-world"** samples is to perform appropriate dilutions so that absorbances fall in the 0-1 range. It has been observed that liquid vitamin supplements are needed to be diluted by a factor of 1000 in order to have an absorbance in this range.

Colorimeters can also be used for the following functions :

   (i)   To characterize and correct colour response in video monitors

   (ii) To calibrate colours in a photographic print (by photographers)

(iii) To characterize colours by disabled people who suffer from blindness or colour-blindness, where subjective colour names are announced based on objective measurements of colour parameters (*e.g.,* hue, saturation and luminance).

While carrying out colorimetry, the following points must be kept in mind :

(i) Colorimetry is a destructive technique *i.e.,* once reacted, the sample cannot be recovered.

(ii) A chromophore reflects the complementary colour (s) that it absorbs, *i.e.,* a yellow compound appears yellow because it absorbs blue light and, therefore, it must be estimated in the blue region of the spectrum.

(iii) Colorimetric assays are usually most sensitive at the λ_{max} of the chromophore produced. If this is not known, before commencing the assay, the spectrum needs to be determined.

(iv) Reference cuvettes (*i.e.,* blanks) should contain everything except the substance being assayed, *i.e.,* all the reagents in the same concentrations as in the test cuvettes.

(v) The reference cuvette and its contents require more care because any error in these will be reflected in all of the values obtained.

(vi) Assays should normally be performed in duplicate or triplicate, and individual values, not means, should be plotted. This procedure allows the experimentor to justifiably omit erroneous values from the calibration curve.

(vii) The "line of best fit" should be drawn through the data points, not necessarily the line that passes through the origin and other points.

(viii) As batches of reagents and standards vary, calibration curves may also vary. Therefore, each time an assay is run, new calibration curves should be prepared.

(ix) Calibration curves should never be extrapolated beyond the highest absorbance value measured. It is always more accurate to repeat an assay at a concentration which falls within the most accurate region of the calibration curve (*i.e.,* to accomplish this, there may be a need to dilute the solution). This is usually from 0.05 to 0.70 nm.

2.2.6.3 Spectrophotometry

2.2.6.3.1 Introduction

If more precise and more interference-free measurements are required a **spectrophotometer** can be used. This is very similar to a colorimeter, except that instead of using a filter to select the colour of light to pass through the sample, a prism or a diffraction grating the white light up into a spectrum (rainbow) of colours, is used.

(i) The light is passed through a narrow opening (slit) before reaching the sample

(ii) By rotating the prism or grating, the colour (wavelength) of light can be more precisely selected

(iii) Colour is better matched with that absorbed by the sample.

Absorption spectra in the ultraviolet (200 to 400 nm) and visible regions are due to energy transitions of both bonding and non-bonding outer electrons of the molecule. Usually delocalized electrons such as the B bonding electrons of $C = C$ and the lone pairs of nitrogen and oxygen are involved. Since most of the electrons in a molecule are in the ground state at room temperature, spectra in this region give information about this state and the next higher one. As the wavelengths of light absorbed are determined by the actual transitions occurring, specific absorption peaks may be recorded and related to known molecular substructures. The term chromophore is given to that part of a molecule that independently gives rise to distinct parts of an absorption spectrum, for example the carbonyl group. The energy required for electronic transition is lowered by conjugated double bond. It results in an increase in the wavelength at which a chromophore absorbs. This is referred to as a bathochromic shift, whereas a decrease in conjugation, caused for example by protonating a ring nitrogen atom, causes a hypochromic shift which leads to a decrease in wavelength.

Hyperchromic and hypochromic effects refer to an increase and a decrease in absorbance respectively.

2.2.6.3.2 Instrumentation

To obtain an absorption spectrum, the absorbance of a substance must be measured at a series of wavelengths. Absorption in the visible and ultraviolet regions can be measured by a UV/visible spectrophotometer.

There are three basic components of UV / visible spectrophotometers:

(i) **A light source** and a mechanism to select a specific wavelength of light in the UV/visible region of the spectrum

(ii) **A chamber** where a cuvette containing a test solution can be introduced into the light path

(iii) **A photocell** that can determine the amount of light absorbed by the sample (or the intensity of light transmitted through the sample).

The light source: It is usually a tungsten lamp for the visible region of the spectrum, and either a hydrogen or deuterium lamp for ultraviolet wavelengths.

Cuvettes: These are optically transparent cells that hold the material(s) under study and are used to introduce samples into the light path. A reference cuvette optically identical to, and containing the same solvent (and impurities) as the test cuvette is always required for setting the spectrophotometer to read zero absorbance at each wavelength used. For accurate work, the optical matching of the two cuvettes should always be checked. Glass and plastic strongly absorb below 310 nm and are not useful for measuring absorbance below that wavelength. When measuring absorption of ultraviolet wavelengths by a solution, quartz or silica cells are used, since they are transparent to wavelengths greater than 180 nm.

When a cuvette is positioned in the light path, it becomes an integral part of the instrument's optical system. It should, therefore, be treated with the same care given to other optical components. The use of scratched or contaminated cuvettes should be avoided since they reflect and/or absorb radiation that will give inaccurate measurements. Also, bubbles, turbidity, fingerprints, or condensation on, or inside cuvettes should be avoided since they will diminish the accuracy of readings.

The commonly used cuvettes for accurate work have an optical path length of 1 cm and require 2.5 to 3.0 ml of sample for all accurate reading.

2.2.6.4 Atomic Spectroscopy

While many tests are carried out using visible light, in some analyses, invisible ultraviolet or infrared portions of the spectrum are also used. This type of testing, usually referred to as **atomic spectroscopy** is mostly used for trace metal analysis.

Atomic absorption spectrometry : The sample is converted to a gas by one of the several methods - usually involving heating. Then the light from a lamp containing the same metal is passed through the gas and the absorbance measured as with a liquid sample.

Atomic emission spectrophotometry : The concentration can also be measured from the intensity of the light emitted from the heated atoms of the metal in the gas.

Inductively coupled plasma spectrometry (ICP) : The sample is carried in a stream of argon gas surrounded by coils which emit radio frequency energy that converts some of the gas into a very hot, ionized (electrically charged) form. This matter is advantageous because many elements can be measured simultaneously, or in rapid succession.

2.2.7 Radioisotopes and Radioisotopy

2.2.7.1 Radioisotopes

Three main forms of radiation can be released by the atoms during their disintegration:

(a) **Alpha particle :** It consists of two protons and two neutrons and is equivalent to the nucleus of a helium atom

(b) **Beta particle :** It is equivalent to an electron

(c) **Gamma radiation :** It consists of electromagnetic radiation or photons

Amongst these, alpha particles are the least, and gamma "rays" the most energetic. Most of the biologically important isotopes emit beta particles, which are the easiest types of radiation to detect and quantitate.

Beta particles can range in energy from very weak ones, such as those emitted by a tritium atom (0.018 million eV), to highly energetic ones, such as those from a ^{32}P atom (1.71 million eV) (Table 2.3). This difference in energy content is manifested in the distance that the beta particle travels before coming to a stop, a value known as its **path length**, and in its **penetrating power**. Whereas the beta particles emitted by an ^3H atom is only capable of traveling a few micrometers and is not able to penetrate the walls of a glass container, the beta particle from a ^{32}P atom can travel across a room, so that highly radioactive solutions are kept in the laboratory behind a lead shield to protect workers.

Table 2.3 : Properties of a Variety of Radioisotopes Commonly Used in Biological Research

Atomic number	Symbol and atomic weight	Half-life	Type of particle (s) emitted	Energy of particle (million eV)
1	^{3}H	12.3 yr	Beta	0.018
6	^{11}C	20 min	Beta	0.981
	^{14}C	5700 yr	Beta	0.155
11	^{24}Na	15.1 hr	Beta	1.39
			Gamma	2.57, 1.37
15	^{32}P	14.3 d	Beta	1.71
16	^{35}S	87.1 d	Beta	0.167
19	^{42}K	12.4 hr	Beta	3.58, 2.04
			Gamma	1.395
20	^{45}Ca	152 d	Beta	0.260
26	^{59}Fe	45 d	Beta	0.460, 0.26
			Gamma	1.30, 1.10
27	^{60}Co	5.3 yr	Beta	0.308
			Gamma	1.317, 1.115
29	^{64}Cu	12.8 hr	Beta	0.657, 0.571
			Gamma	1.35
30	^{65}Zn	250 d	Beta	0.32
			Gamma	1.11
53	^{131}I	8.0 d	Beta	0.605, 0.250
			Gamma	0.164, *etc.*

The half-life ($t_{1/2}$) of a radioisotope : It is a measure of its instability. The more unstable a particular isotope, the greater the likelihood that a given atom will disintegrate in a given amount of time. If one starts with one Curie of tritium, half that amount of radioactive material will be left after approximately 12 years (which is the $t_{1/2}$ of this radioisotope). The half-life of a substance is not usually very important in biological investigations as long as it is sufficiently long to be used in individual studies. In the early years of research in photosynthesis and other metabolic pathways, the only available radioisotope of carbon was ^{11}C, which has a half-life of approximately 20 minutes. Experiments with ^{11}C were literally carried out on the run so that the amount of incorporated isotope could be measured before the substance had virtually disappeared.

2.2.7.2 Radioisotopy

2.2.7.2.1 Process

Radioactive isotopes have found widespread use because of the ease with which they can be developed and accurately quantified. The most commonly used isotopes are beta emitters which are monitored by either of the two different methodologies:

(a) **Liquid scintillation spectrometry**

(b) **Autoradiography**

When one wants to determine the amount of radioactivity in a given sample, such as the contents of a fraction taken from the sucrose gradient, a liquid scintillation counter is used. In contrast, autoradiography is used when one wants to know where a particular isotope is located, whether in a cell, in a polyacrylamide gel, or on a nitrocellulose filter.

(a) Liquid Scintillation Spectrometry

It is based on the property of certain molecules, termed phosphors or scintillants, to absorb some of the energy of an emitted particle and release that energy in the form of light. In preparing a sample for liquid scintillation counting, the sample is mixed with a solution of phosphors in a glass or plastic scintillation vial. This brings the phosphor and radioactive isotopes in very close contact with one another so that radiation from even the weakest beta emitters can be efficiently measured. Once mixed, the vial is placed into the counting instrument where it is lowered into a well whose walls contain an extremely sensitive photodetection device. The disintegration of radioactive atoms within the vial leads to the emission of particles that activate the scintillants, causing them to emit flashes of light.

The light is detected and the signal is amplified by a photomultiplier tube within the counter. After being subjected to electronic screening to eliminate background noise, the amount of radioactivity present in the vial is revealed in the form of counts per minute on the display apparatus of the machine.

(b) Autoradiography

It is a broad-based technique to visualize sites in which radioactivity is located in various types of specimens, including a fixed cell, an electrophoretic gel, or a slide containing a DNA-RNA hybrid.

The principle of autoradiography induces the ability of a particle emitted from a radioactive atom to activate a photographic emulsion, much like light or X-rays activate the emulsion that coats a piece of film. If the photographic emulsion is brought into close contact with a radioactive source, the particles emitted by the source leave tiny, black silver grains in the emulsion after photographic development.

Autoradiography is used to localize radioisotopes within sections of cells and tissues that have been immobilized on a slide or EM grid. The emulsion is applied to the sections on the slide or grid as a very thin overlying layer and the specimen is put into a lightproof container to allow the emulsion to be exposed by the emissions. The longer the specimen is left before development, the greater the number of silver grains that are formed. When the developed slide or grid is examined in the microscope, the location of silver grains in the layer of emulsion just above the tissue indicates the location of radioactivity in the cells.

2.2.7.2.2 Limitations

The most serious limitation of electron microscopic autoradiography in the localization of radioisotopes within cells is the resolution. The location of radioisotopes in the cells being examined is determined by the positions of the overlying silver grains. If the silver grains are too large, relative to the organelle from which the radioactivity is originating, it can become impossible to identify the site from which the particles are being emitted.

Similarly, the particle emitted by the radioactive source can travel a certain distance from its point of origin before it strikes the emulsion. As a result, the location of silver grain, that is, where the particle interacted with the emulsion, may be some distance from the source of emission within the tissue. Because of this, only those radioisotopes that emit particles with short path lengths, such as ^3H and ^{14}C, are suitable for widespread use in microscopic autoradiography.

2.2.7.2.3 X-Ray Diffraction Analysis

In order to analyze a particular protein by any technique, it is necessary to obtain the most purified preparation of that protein possible. It can be done as follows:

X-ray diffraction (or X-ray) crystallography: It utilizes protein crystals, the preparation of which is one of the best guaranteed molecular purity. A crystal is composed of a regularly repeating arrangement of a unit cell, *e.g.,* an individual protein molecule.

For X-ray diffraction analysis, a crystal is bombarded with a thin beam of X-rays of a single (monochromatic) wavelength. The radiation that is scattered (diffracted) by the electrons of the protein atoms, strikes a photographic plate placed behind the crystal. Just as each protein molecule is repeated in a periodic manner, the diffraction pattern produced by a crystal is determined by the structure within the protein; the large number of molecules in the crystal reinforces the reflections, causing it to behave as if it were one giant molecule.

The positions and intensities of the spots on the plates can be mathematically related to the electron densities within the proteins. Because of the way the crystal is bombarded, each plate represents a slice of the molecule and many plates must be analyzed. A single photographic plate is used in the determination. The spots closer to the centre of the pattern result from X-rays scattered at smaller angles from the crystal and provide information about the grosser aspects of the protein, that is, the long spacing within the molecule. The periphery contains information about the closely spaced aspects of the molecules within the crystal.

The resolution obtained by X-ray diffraction depends on the number of spots that are analyzed.

The first protein whose structure was determined by X-ray diffraction was **myoglobin.** The protein was analyzed successively at 6, 2 and 1.4 Å, with years elapsing between each completed determination. Considering that covalent bonds are between 1 and 1.5 Å in length and non-covalent bonds between 2.8 and 4 Å, the picture of the protein obtained was greatly dependent on the resolution. This was illustrated by a comparison of the electron density of a small organic molecule at four levels of resolution. In myoglobin, a resolution of 6 Å is sufficient to show the manner in which the polypeptide chain is folded and the location of the haeme moiety, but it is not sufficient to show structure within the chain. At a resolution of 2 Å, groups of atoms can be separated from one another, whereas at 1.4 Å, individual atoms can be seen.

X-ray diffraction analysis can also be used to study DNA. It served as one of the most important sources of data that led Watson and Crick to give the structure of the double helix in 1953. The analysis of DNA is made using fibres containing oriented DNA molecules rather than crystals and does not provide as high a level of resolution as that obtained with proteins.

2.2.8 Fractionation and Purification of Cell Structures and Molecules

Once obtained in quantity from either cultures or tissues removed from an experimental organism, cells can be fractionated to yield cell organelles, structures such as ribosomes or microtubules or individual molecular types. The methods available for this work are capable of producing highly purified cellular and molecular fractions.

Cells are disrupted by breaking the plasma membrane in a buffered solution at physiological pH. Among the several techniques used for breaking plasma membranes are :

(i)   sonication (exposure to high-frequency sound waves),

(ii) grinding in fine glass beads or other abrasive materials,

(iii) forcing the cells through a narrow orifice,

(iv) osmotic pressure, and

(v)   exposure to detergents.

The suspension of cell organelles, structures and molecules produced by the breakage is usually separated into fractions by spinning it in a centrifuge. Centrifugation can separate organelles as large as nuclei or mitochondria or molecules as small as proteins into separate groups (This has already been explained under centrifugation). For molecular studies, the fractions obtained by centrifugation are often further separated into pure samples by either of the two methods – gel electrophoresis or chromatography.

Mechanical methods : All mechanical procedures for cell disruption generate heat, and this may denature proteins. It is, therefore, very important to cool the starting material, the homogenizing medium and if possible, the homogenizer itself ($\approx 4°C$). The homogenization should be carried out in short bursts, and the homogenate should be cooled in an ice bath between each burst. Cooling will also reduce the activity of any degradative enzymes in the homogenate. Ideally, homogenization is carried out in a walk-in cooler ('cold room').

Equipment commonly used include :

- *Mixer and blenders.* These are similar to domestic liquidizers, with a static vessel and rotating blades. The Waring blender is widely used: it has a stainless steel vessel that will stay cool, if pre-chilled. The vessel and blades are designed to maximize turbulence, both disrupting and homogenizing tissues and cells.

- *Ball mills (e.g.,* Retch mixer mill, Mickle mill). These devices contain glass beads that vibrate and collide with each other and with tissues/cells, leading to disruption.

- *Liquid extrusion devices* (*e.g.,* French pressure cell). Cells are forced from a vessel to the outside, through a very narrow orifice at high pressures ($\approx$ 100 M Pa). The resulting pressure changes are a powerful means of disrupting cells.

- *Solid extrusion* (*e.g.,* Hughes press). Here, a frozen cell paste is forced through a narrow orifice, where the shear forces and the abrasive properties of the ice crystals cause cell disruption.

- *Rotor-stators* (*e.g.,* Ultra-turrax^R homogenizer). These have a rotor (a set of stainless steel blades) and a stator (a slotted stainless steel cylinder) at the tip of a stainless steel shaft, immersed in the homogenizing medium.The high speed of the rotor blades causes material in the homogenizing fluid to be sucked into the dispersing head, where it is pressed radially through the slots in the stator. Along with the cutting action of the rotor blades, the material is subjected to very high shear and effective mixing is given by thrust and the resulting turbulence in the gap between rotor and stator. The vigour of the homogenization process can be altered by varying the rotor speed setting. Various sizes of rotor-stators are available, with typical diameters in the range 8-65 mm: the smaller sizes are particularly useful for small-scale preparation.

- *Sonicators*. Ultrasonic waves are transmitted to an aqueous suspension of cells via a metal probe. The ultrasound creates bubbles within the liquid and these produce shock waves when they collapse. Successful disruption depends on the correct choice of power and incubation time, together with pH, temperature and ionic strength of the suspension medium, often obtained by trial and error. The effects of heating during ultrasonication can be reduced by using short 'bursts' of power (10-30 s), with rests of 30-60s in between, and by keeping the cell suspension on ice during disruption. An ultrasonic water bath provides a more gentle means of disrupting certain types of cells, *e.g.*, some bacterial and animal cells.

- *Homogenizers*. These involve the reciprocating movement of a ground glass or Teflon® pestle within a glass tube. Cells are forced against the walls of the tube, releasing their content. For glass pestles, the tubes also have ground glass homogenizing surfaces and may have an overflow chamber. The homogenizer can either be hand operated (*e.g.,*Dounce) or motorized (*e.g.,* Potter-Elvejham). The clearance between the pestle and the tube (range 0.05 mm to 0.5 mm) must be chosen to suit the particular application.

Cell fractionation and the isolation of organelles

The fractionation and separation of organelles from a cell homogenate is achieved by differential centrifugation. Particular organelles can be obtained by appropriate choice of source tissue and homogenization method, as illustrated in Table 2.4 for the major types of organelle.

Table : 2.4: Isolation and Fractionation Procedures for Various Organelles

Stage	Nuclei	Mitochondria	Microsomes	Chloroplasts*
Source	Thymus tissue, which has little cytoplasm, giving high yields.	Beef heart, with fat and connective tissue removed, then cubed & minced. Keep at pH 7.5 using TRIS buffer.	Rat liver, stored over night to reduce glyco-gen content.	Spinach leaves, de-ribbed and cut into 1 cm strips
Pre-treatment	Rinse with buffered physiological saline. Suspend in homogenizing medium.	Suspend in 2 X volume of ice-cold homogenizing medium. Squeeze through muslin.	Chop finely with scissors and thaw in 2X volume of homo-genizing medium.	Rinse, then suspend in 3X volume of pre-chill homo-genizing medium.
Homo-genizing medium	250 mmol 1^{-1} sucrose; 10 mmol 1^{-1} TRIS/HCl buffer, pH 7.6; 5 mmol 1^{-1} MgCl$_2$ 0.2-0.5% v/v Triton X-100	250 mmol 1^{-1} sucrose; 10 mmol 1^{-1} TRIS/HCl buffer pH 7.7, containing 1 mmol 1^{-1} succinic acid and 0.2 mmol 1^{-1} EDTA Bring to pH 7.8 using 2 mmol 1^{-1} TRIS base:	250 mmol 1^{-1} sucrose; 25 mmol 1^{-1}TRIS/HCl buffer, pH 7.5; 25 mmol 1^{-1} KCl; 5 mmol 1^{-1} MgCl$_2$	400 mmol 1^{-1} sucrose; 25 mmol 1^{-1} HEPES/NaOH buffer at pH 7.6; 2 mmol 1^{-1} EDTA
Homogeni-zation	Waring blender, low speed, 3 min.	Waring blender, high speed, 15s: check and adjust pH to 7.8 min at 800 rpm and repeat blending step, 5s.	Potter-Elvejham glass homogenizer with a Teflon pestle-3 x 5	Pre-chilled waring blender or rotor-stator.
Before use	Resuspend in homogenization medium without Triton X-100	Resuspend pellet in buffer pH 7.8 and either use immediately, or store at freeging temp.	Resuspend in buffer solution at pH 8.0.	Resuspend pellet appropriate incubation –20°C overnight medium, containing surcrose, e.g., for CO$_2$/O$_2$ studie

* Alternatively plant protoplasts can be used as the starting material, releasing the chloroplasts by gentle lysis–diluting the medium with water.

2.2.8.1 Sedimentation Behaviour of Nucleic Acids

DNA (and RNA) molecules are extensively analyzed by utilizing the ultracentrifuge.

Velocity sedimentation: Nucleic acid molecules in this technique are separated according to nucleotide length. The sample containing the mixture of nucleic acid molecules is carefully layered over a solution containing an increasing concentration of sucrose (or other suitable substance). This preformed gradient increases in density (and viscosity) from the top to the bottom. The molecules when subjected to high centrifugal forces move through the gradient at a rate determined by their sedimentation coefficient. The greater the sedimentation coefficient, the farther a molecule moves in a given period of centrifugation. Since the density of the medium is less than that of the nucleic acid molecules, even at the bottom of the tube (approximately 1.2 g/

ml for the sucrose solution and 1.7 g/ml for the nucleic acid), these molecules continue to sediment as long as the tube is being centrifuged. In other words, centrifugation never reaches equilibrium.

After a prescribed period, the tube is removed from the centrifuge, its content are fractionated, and the relative positions of the various molecules are determined. By the presence of viscous sucrose, the contents of the tube are prevented from mixing as a result of convection or during handling. This allows molecules of identical S value to remain in place in the form of a band. The S value of unknown components can be determined by using marker molecules of known sedimentation coefficient.

Equilibrium (or isopycnic) sedimentation: In this, procedure nucleic acid molecules are separated on the basis of their buoyant density. A highly concentrated solution of the salt of the heavy metal cesium is employed. The analysis is begun by mixing the DNA with the solution of cesium chloride or cesium sulfate in the centrifuge tube and then the tube is subjected to extended centrifugation (*e.g.*, 2 to 3 days at high forces). During centrifugation, the heavy cesium ions are very slowly driven towards the bottom of the tube. They form a continuous density gradient through the liquid column. The tendency for cesium ions to be concentrated toward the bottom of the tube is counterbalanced after a time by the opposing tendency for them to become redistributed by diffusion, and the gradient becomes stabilized. As the cesium is driven downward or moves buoyantly upward in the density equivalent to their own, at which point they are no longer subject to further movement, the molecules of equivalent density within the tube form narrow bands in the tube. This technique is quite sensitive and able to separate DNA molecules having different base composition or ones containing different isotopes of nitrogen (^{15}N versus ^{14}N).

2.2.8.2 Isolation, Purification and Fractionation of Proteins

Since most cells contain thousands of different proteins, the purification of a single species, particularly if it is present as a relatively minor component, can be a challenging goal.

Protein purification methodology can be divided into two broad categories. These are :

(i) Preparative

(ii) Analytic

Preparative techniques : These are used to isolate and purify a particular protein by removal of contaminating molecules. The term preparative implies use of the material beyond simply its isolation. It allows one to estimate the molecular diversity present in a given preparation and to determine the properties of individual species of molecules.

Analytic techniques : The term analytic implies the disassembly of the whole so that its parts can be examined. If, for example, one wanted to determine the number of different types of macromolecules present in bacterial ribosomes and their various molecular weights, the first step would be preparative so that a purified sample of ribosomes could be obtained from which the components could be solubilized. Subsequent steps would require that the components be spread out in some manner so that their diversity in the molecular weights could be assessed.

Analytic Procedures

These require the mixture to be fractionated into its component ingredients. In many cases, the same types of techniques accomplish both the aims, since the fractionation of a mixture simultaneously accomplishes a purification of each substance.

The purification of a protein is generally performed by the stepwise removal of contaminants. Two proteins may be very similar in one property, such as overall charge, whereas they may be very different in another property, such as molecular weight. Consequently, the complete purification of a given protein usually requires the use of successive techniques that take advantage of different properties of the proteins being separated.

Purification is measured as an increase in specific activity, which is the ratio of the amount of that protein to the total amount of protein present in the sample. Some identifiable features of the specific protein must be utilized to provide an assay to determine the relative amount of protein in the sample. If the protein is an enzyme, its catalytic activity may be used as an assay to monitor purification. Alternatively, assays can be based on immunologic, electrophoretic, electron microscopic or other criteria. Measurements of total proteins in a sample can be made in various ways, including total nitrogen, which can be very accurately measured and is quite constant at about 16 per cent of the dry weight for all proteins.

Selective Precipitation

The first step in purification should be one that can be carried out on a highly impure preparation and can be accomplished with a large increase in specific activity. This step takes advantage of solubility differences among proteins by selectively precipitating the desired protein. The solubility of a protein depends on the relative balance between protein-solvent interactions, which tend to cause it to aggregate and precipitate. The ionic strength of the solution is particularly important in determining which of these types of interactions will predominate.

Low ionic strength : At this the interactions with the solvent are enhanced, and proteins tend to remain in solution; they are said to be salted in.

Higher ionic strengths : At high ionic strength, the solubility of proteins can rapidly decrease. The most commonly employed salt for selective protein precipitation is ammonium sulfate, which is highly soluble in water and has high ionic strength.

Purification is achieved by gradually adding a solution of saturated ammonium sulfate to the crude protein extract. As the addition of salt continues precipitation of contaminating proteins increases and the precipitate can be discarded. A point is ultimately reached at which the protein being studied comes out of solution. This point is recognized by the loss of activity in the soluble fraction when tested by the particular assay being used. Once the desired protein is precipitated, contaminating proteins are left behind in solution and can be discarded while the protein being sought can be redissolved and subjected to further purification and analysis.

A pure preparation of protein is essential before its properties, amino acid composition and sequence can be determined. The source of a protein is generally a tissue or microbial cells. The following steps are applied in the purification of proteins:

(i) The first step in any protein purification procedure is to break open the cells, releasing their proteins into a solution called crude extract.

(ii) If necessary, differential centrifugation can be used to prepare sub-cellular fractions or to isolate specific organelle.

(iii) Once the extract/organelle preparation is ready, various methods are available for purifying one or more proteins present in it.

(iv) The extract is subjected to treatments that separate the proteins into different fractions based on some property such as size or charge, a process called **fractionation.**

(v) Early fractionation steps utilize differences in protein solubility.

(vi) The solubility of protein is generally lowered at high salt concentration, an effect called **salting out**.

(vii) The addition of salt in the right amount can selectively precipitate some proteins, while others remain in solution.

(viii) Ammonium sulfate because of its high solubility in H_2O is often used for this purpose.

(ix) A solution containing protein of interest often must be further altered before subsequent steps are possible.

(x) Dialysis is a procedure that separates proteins from solvents by taking advantage of the large size of a protein.

The partially purified extract is placed in a bag/tube made up of semi-permeable membrane. When this is suspended in a larger volume of buffered solution of an appropriate ionic strength, the membrane allows exchange of salt but buffer and not of proteins.

Thus, dialysis retains large protein within the membranous bags or tube, while lowering the concentration of other solutes in the protein preparation to change until they come into equilibrium with the solution outside the membrane. Dialysis can be used to remove ammonium sulfate from protein preparation.

2.2.8.3 Nucleic Acid Purification and Fractionation

The steps followed in the purification of nucleic acids are very different from those applied in the purification of protein as there are basic differences in the structure of these two types of macromolecules. For purification of nucleic acids, the cells are homogenized and the DNA from nuclei is extracted. The extraction medium usually contains a detergent, such as SDS, which serves to lyse the nuclei and release DNA. This causes the solution's viscosity to noticeably increase. The detergent also inhibits any nuclease activity present in the preparation.

The DNA is separated from contaminating materials, such as RNA and proteins.

Deproteinization

It is usually accomplished by shaking the mixture with a volume of phenol. Phenol (or, alternatively, chloroform) makes an active protein in the preparation to lose its solubility and precipitate out of solution. Since phenol and buffered saline solutions are immiscible, the suspension is simply centrifuged to separate the phases, which leave the DNA (and RNA) in solution in the upper aqueous phase and the protein present as a flocculent precipitate concentrated at the boundary between the two phases.

The aqueous phase is removed from the tube and subjected to repeated cycles of shaking with phenol and centrifugation, until no further protein is removed from solution.

The nucleic acids are then precipitated from solution by the addition of cold ethanol.

The cold ethanol is often layered on top of the aqueous DNA solution, and the DNA is wound on a glass rod as it comes out of solution at the interface between the ethanol and saline. The RNA on the other hand comes out of solution as a flocculent precipitate that settles at the bottom of the vessel.

Removal of RNA

The DNA is then redissolved and treated with ribonuclease to remove contaminating RNA. The ribonuclease is then destroyed by treatment with a protease; the protease is then removed by deproteinization with phenol and heating at 56°C and the DNA is reprecipitated with ethanol.

Isolation of RNA

Measurement of concentration of proteins, Nucleic acids.

RNA is generally purified in a similar manner using DNase in the final purification steps rather than ribonuclease.

The solution is placed in a special flat-sided quartz container (quartz is used because, unlike glass, it does not absorb ultraviolet light), termed a cuvette, which is then placed in the light beam of the spectrophotometer. The amount of light that passes through the solution unabsorbed, that is, the transmitted light, is measured by photocells, on the other side of the cuvette.

Of the 20 amino acids incorporated into proteins, two of them, tyrosine and phenylalanine, absorb light in the ultraviolet range, both having an absorbance maximum at about 280nm. Thus, if the proteins being studied have a typical percentage of these amino acids, the absorbance of the solution at this wavelength provides a sensitive measure of protein concentration. A variety of chemical assays, such as the Lowry or Biuret technique, in which the protein in solution is engaged in a reaction that produces a coloured product whose concentration is proportional to the concentration of protein can also be used. Nucleic acids have their maximal absorption at 260 nm which is thus, used as the wavelength of choice for measurement of DNA or RNA concentrations.

Nucleic Acid Hybridization

A variety of related techniques are based on the observation that two single-stranded nucleic acid molecules of complementary base sequence will form a double-stranded hybrid. Supposing there is a mixture of hundreds of fragments of DNA of identical length and overall base composition from one another they solely in their base sequences. For example, one of the DNA fragments constitutes a portion of a β-globin gene and the other fragments contain unrelated genes. The two can be distinguished by carrying out a molecular hybridization experiment using complementary molecules as probes (Fig. 2.25).

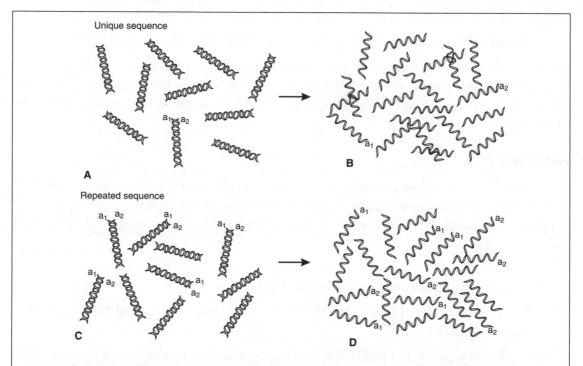

Fig. 2.25. Hybridization of DNA fragments from unique or repeated sequences. The a_1 and a_2 nucleotide chains are complementary. **(A)** DNA containing unique, nonrepeated sequences. **(B)** After unwinding by heating, each nucleotide chain has only one pairing partner in the solution. On cooling, the random collisions of complementary sequences required for pairing are relatively rare and rewinding proceeds slowly. **(C)** DNA fragments containing repeated sequences. **(D)** After unwinding by heating, each single chain has many possible pairing partners in the solution, numbering in proportion to the degree of sequence repetition. The probability of a collision involving complementary chains is greatly increased and rewinding proceeds relatively rapidly. Adopted from Molecular and Cellular Biology by Stephen L. Wolfe, 1993 Publ. Wadsworth Inc.).

In the present example, the mixture of denatured DNA fragments is incubated with an excess number of β-globin mRNAs. The latter would drive the globin fragments into formation of double-stranded DNA-RNA hybrids. The other DNA fragments will remain single-stranded. The hybrid can be separated from the single-stranded fragments. For this the mixture could be passed through a column of hydroxylapatite under the ionic conditions in which the hybrids would bind to the calcium phosphate salts in the column, while the non-hybridized DNA molecules would pass through unbound. The hybrids could then be released from the column by changing the concentration of the eluting buffer.

Similarily, nucleotide sequence similarity between two samples of DNA, from two different organisms can be known by nucleic acid hybridization.

If the two species are far apart from each other, they will show greater differences. If purified DNAs from species A and B are mixed together, denatured and allowed to reanneal, a percentage of the DNA duplexes will be formed by DNA strands from the two species. Because they contain mismatched bases, such duplexes are less stable than those formed by DNA strands of the same species, and this instability is reflected by the lower temperature at which they melt. When DNAs from different species are allowed to reanneal in different combinations, the melting temperature (T_m), of the hybrid duplexes provides a measure of the evolutionary distance between the organisms.

Questions

1. What is microscopy? Write a note on light microscope.
2. Write notes on:
    (i) Fluorescence Microscope
    (ii) Darkfield Microscope
    (iii) Phase contrast Microscopy
3. Write detailed note on EM.
4. What is electrophoresis? Discuss its applications.
5. Write a note on chromatography.
6. Discuss spectrophotometry in detail.
7. Explain the various techniques to separate and purify cellular molecules.

Chapter $\boxed{3}$

Specialized Tools in Biotechnology-I:
Cell Culture Technologies

According to Krüzer and Massay (2002) biotechnology is a collection of technologies and thus some of the confusion surrounding the word 'biotechnology' could be eliminated by simply changing the singular noun to its plural form 'biotechnologies' because biotechnology is not a singular entity, instead it is a collection of technologies, all of which utilize cells and biological molecules. The most important technologies/tools are involved in it :

   (a) Cell and tissue culture including hybridoma technology, stem cell culture and tissue engineering
   (b) Genetic engineering and its outcome technologies including antisense technology and DNA chip technology
   (c) Protein engineering
   (d) Bioprocess technology

Some of the technologies/tools have been outlined in the following pages.

3.1 CELL AND TISSUE CULTURE TECHNOLOGY

The development of techniques for isolating cells and maintaining them in culture vessels under conditions in which they can grow and divide have tremendously contributed to the molecular revolution. Many unicellular microorganisms such as yeast, protozoa, algae and bacteria can be cultured in large quantities relatively easily. For some, membrane filtration technique can be used (Fig. 3.1). The cells of higher eukaryotes, including plants and animals have more stringent requirements. But many animal and plant cell types have been successfully grown in culture. These are indispensable as experimental subjects as well as sources of cellular organelles and molecules.

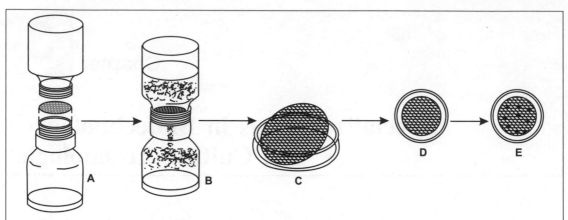

Fig. 3.1. The membrane filtration procedure: Membrane with different pore sizes are used to trap different microorganisms. Incubation times for membranes also vary with the medium and microorganism **(A)** Membrane filter on a filter support **(B)** Water sample filtered through membrane filter (0.45 μm) **(C)** Membrane filter removed and placed in plate containing the appropriate medium **(D)** Incubation for 24 hours **(E)** Typical colonies (Modified from Cultures of Animal Cell by Ian Freshney, 2000).

3.1.1 Culture of Microorganisms

Microbes have been used in different ways by mankind since time immemorial. The culture of microorganisms involves the use of sterile technique. So that foreign microorganisms do not contaminate the culture. Production of alcohol, vinegar, solvents, antibiotics, vitamins and many more therapeutically important recombinant proteins is possible by exploiting the metabolism of different microorganisms. A microbial culture works as a factory in converting the raw material into product. Each microbial cell has a maximal capacity to do so in a given period of time. For this purpose, it is necessary to calculate the number of cells. The calculation must include the doubling time for that particular bacterium. *E. coli* under optimum conditions completes a cycle of cell growth and division every 20 minutes. Thus a very large number of organisms can be obtained in a relatively short time the following basic steps are used to grow, examine and characterize microorganisms, in the laboratory :

(a) Inoculation   (b) Incubation   (c) Isolation   (d) Inspection

For culturing microorganisms (such as yeast and bacteria), relatively simple growth media and conditions are required. All bacteria have three major nutritional needs :

(i)   Carbon source such as glucose,
(ii) Source of nitrogen,
(iii) Source of energy (ATP) for carrying out cellular functions.

In fact following nutrients are the components of most of the media used for microbial cultures.

Carbohydrates. Carbohydrates are capable of being used by all microorganisms, although in no case, there is an absolute requirement for this group of organic compounds. Glucose is the most readily metabolizable sugar. Most fungi can use disaccharides.

Lipids. Long-chain fatty acids like linoleic and oleic acids are required by bacteria and fungi.

Generally steroids, other than cholesterol, are not required or utilized by microorganisms. In all fungi including yeast, ergosterol is a nutritional requirement.

N-Bases (Purines and Pyrimidines). Purine and pyrimidine metabolisms have been reported in bacteria. Algae do not utilize these compounds at all.

Vitamins and growth factors. There is a considerable species variation in the requirements of vitamins and related factors by the microorganisms. Generally, vitamins A, C, D and K are not necessary for growth.

Amino acids. Amino acids are generally not required by algae, although several algal species are capable of utilizing them. Species of other microorganisms are capable of utilizing all amino acids, except for yeast, where there is no evidence of citrulline being used. Usually L-form of the amino acids are biologically active are required by bacteria, but unlike higher animals, some bacteria can also utilize the D-amino acids.

Nitrogen sources. Not all species require or utilize these compounds. Some species that are able to utilize these compounds have been identified. Fungi require ammonia, nitrate and nitrite.

Sulfur sources. Some species of yeasts can utilize elemental sulfur and sulfate. Generally yeasts do not require or utilize sulfur containing organic compounds. Bacteria require glutathione and thio-acetic acid, while yeasts require sulfonic acid, amides, thioacetate, thiocarbonate, thioglycolate and glutathione.

Chemical elements and inorganic ions. Mineral nutrients required by microorganisms are species dependent, but generally they need Fe, K, Mg, Mn. Sometimes elements like S, N, Ca, Co, Cu, P, Zn are also required.

The cells may be grown as follow:

  (i)   In liquid suspension
  (ii) On the surface of a solid growth medium such as an agar gel (agar is a polysaccharide extracted from an alga).

For many experimental purposes, genetically uniform culture of bacterial or yeast cells is required.

Smaller amounts of molecules such as phosphate (for nucleic acids) and a variety of metals and ions (for enzymatic activity) must also be present. The basic building blocks required for growth are the same for all cells, but bacteria vary widely in their ability to use different sources of these molecules.

3.1.1.1 Classification of Microbes on the Basis of Nutritional Requirements for Growth

The main determinants of a microbe's nutritional type are its sources of carbon and energy. According to the mode they meet their nutritional needs, bacteria are classified into two basic groups:

  (i)   **The autotrophs (lithotrophs) :** They are able to grow simply. They use CO_2 as the sole source of carbon. Besides, water and inorganic salts are required. These obtain energy either photosynthetically (phototrophs) or by oxidation of inorganic compounds (chomolithotrophs). These occur in environmental millieus.

(ii) **The heterotrophs :** These require more complex substances for growth. They need an organic source of carbon, such as glucose. They obtain energy by oxidizing or fermenting organic substances. Often the same substance (for example, glucose) is used as both carbon and energy source.

All bacteria that inhabit the human body belong to heterotrophic group. Within this group, however, nutritional needs greatly vary. A wide variety of organic compounds is used as carbon sources by bacteria such as *E. coli* and *Pseudomonas aeruginosa*. They, therefore, grow on most simple laboratory media. Other pathogenic bacteria, such as *Haemophilus influenzae* and the anaerobes are fastidious. Additional metabolites such as vitamins, purines, pyrimidines and haemoglobin are required by them. They are supplied in the growth medium.

Some pathogenic bacteria, such as *Chlamydia,* cannot be cultured on laboratory media at all and must be grown in tissue culture or detected by other means. All bacteria can be subdivided according to their energy source as phototrophs or chemotrophs. Microbes that photosynthesize are phototrophs and those which gain energy from chemical compounds are **chemotrophs**.

Accordingly we have photoautotrophs and chemoautotrophs.

Photoautotrophs. These are photosynthetic that is they capture the energy of light rays and transform it into chemical energy that can be used in cell metabolism.

Chemoautotrophs. These are of two types :

(a) *Chemoorganic autotrophs.* These use organic compounds for energy and inorganic compounds as a carbon source.

(b) *Lithoautotrophs.* These require neither sunlight nor organic nutrients. They totally rely on inorganic materials. These bacteria derive energy in diverse and amazing ways. They remove electron from inorganic substrates such as hydrogen gas, hydrogen sulfide, sulfur or iron and combine them with CO_2 and hydrogen. This reaction provides simple organic molecules and a modest amount of energy to drive the synthetic processes of the cell. Lithoautotrophic bacteria play an important role in recycling inorganic nutrients.

3.1.1.2 Types of Growth Media

(a) **Minimal medium.** This is a laboratory growth medium whose contents are simple and complete. This type of medium is usually not used in the diagnostic microbiology laboratory.

(b) **Standardized media.** These include the following :

(i) *Nutrient media.* These are more complex and made up of extracts of meat or soybeans *e.g.,* nutrient broth, trypticase, soya broth.

(ii) *Enriched media.* These growth media contain added growth factors, such as blood, vitamins and yeast extract *e.g.,* blood, agar, chocolate agar.

(iii) *Selective media.* These contain additives that inhibit the growth of some bacteria, but allow others to grow *e.g.,* McConkey agar.

(iv) *Differential media.* These allow virtualization of metabolic differences between groups or species of bacteria.

(v) *Transport media.* When a delay between collection of the specimen and to culture is necessary, a transport medium is used. It is a holding medium designed to preserve the viability of microorganisms in the specimen, but does not allow multiplication. Common examples of such media are Stuart Broth, Ames and Cary Blain.

3.1.1.3 Factors Influencing Growth

The growth rate of bacteria is influenced by three environmental factors. While culturing bacteria in the laboratory, they must be taken into account. These are :

- pH
- Temperature
- Composition of the gaseous atmosphere

3.1.1.3.1 pH

Most pathogenic bacteria grow best at a neutral pH. Diagnostic laboratory media for bacteria are usually adjusted to a final pH between 7.0 and 7.5.

3.1.1.3.2 Temperature

It influences the rate of growth of a bacterial culture. According to their optimal temperature for growth, microorganisms have been categorized as follows :

    (a) *Psychrophiles.* These grow best at cold temperatures (optimal growth 10° to 20°C) and are found in arctics.

    (b) *Mesophiles.* These optimally grow at moderate temperatures (optimal growth at 20° to 40°C). Most bacteria that have adapted to humans, are mesophiles and grow best near human body temperature (37°C).

    (c) *Thermophiles.* These grow best at high temperatures (optimal growth at 50° to 60°C). These are found in hot springs.

Cultures for bacterial growth in diagnostic laboratories are routinely incubated at 35°C. Some pathogenic species, however, prefer a lower temperature for growth. When there is a suspicion of the presence of organisms, the specimen plate is incubated at a lower temperature. Fungal cultures are incubated at 30°C. For some bacteria, the ability to grow at room temperature (25°C) or at an elevated temperature (42°C) is used as a diagnostic characteristic.

3.1.1.3.3 Composition of Gaseous Atmosphere

The bacteria that grow on humans vary in their atmospheric requirements for growth and hence they have been classified as follows :

    (a) **Obligate aerobes.** These require oxygen. Approximately 21% oxygen and 1% carbon dioxide is present in air. When the carbon dioxide content of an aerobic incubator is increased to 10%, the oxygen content of the incubator is lowered to approximately 18%. Obligate aerobes must have oxygen to grow. Their oxygen recruitment is satisfied by incubation in air or in an aerobic incubator with 10% CO_2.

    (b) **Obligate anaerobes.** Obligate anaerobes must be grown in an atmosphere either devoid of oxygen or with a very reduced oxygen content. These cannot grow in the presence of oxygen.

    (c) **Facultative anaerobes.** Facultative anaerobes are routinely cultured in an aerobic atmosphere because aerobic culture is easier and less expensive than anaerobic culture *e.g., E. coli.* These can grow either with or without oxygen.

    (d) **Capnophilic.** These bacteria require extra carbon dioxide (5 to 10%) for growth *e.g., H. influenzae.* Because many bacteria grow better in the presence of increased carbon dioxide, the aerobic incubators of diagnostic microbiology laboratories are often maintained at a 5% to 10% carbon dioxide level. These can grow better when the atmosphere is enriched with extra carbon dioxide.

(e) **Microaerophilic bacteria.** These require a reduced level of oxygen to grow *e.g.*, *Campylobacter*. It requires 5 to 6% oxygen. Using a commercially available microaerophilic atmosphere generating system, this type of atmosphere can be generated in culture jars or pouches.

3.1.1.4 Growth of Bacteria

3.1.1.4.1 Determination of Cell Number

The number of bacterial cells present is determined by any of the following methods:

(i) *Direct counting under the microscope.* The number of bacteria present in a sample can be estimated by this method. It does not distinguish between live and dead cells. Most commonly the microbal number is determined through direct counting. The use of a counting chamber is easy, inexpensive, and relatively quick. It also gives information about the size and morphology of microorganism. For counting prokaryotes, Petroff-Hausser counting chambers can be used. Haemocytometers can be used for both pro- and eukaryotes. Prokaryotic cells are more easily counted in these chambers if they are stained, or when a phase-contrast or a fluorescence microscope is employed. These specially designed slides have chambers of known depth with an etched grid on the chamber bottom (Fig. 3.2).

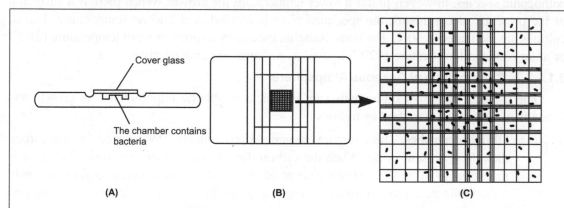

Fig. 3.2. The Petroff-Hausser counting chamber. (A) Side view of the chamber showing the cover glass and the space beneath it that holds a bacterial suspension. **(B)** Top view of the chamber. The grid is located in the centre of the slide. **(C)** An enlarged view of the grid. The bacteria in several of the central squares are counted, usually at X 400 to X 500 magnification. The average number of bacteria in these squares is used to calculate the concentration of cells in the original sample. Since there are 25 squares covering an area of 1mm^2, the total number of bacteria in 1mm^2 of the chamber is calculated as (number/square) (25 squares). The chamber is 0.023 mm deep and therefore, bacteria/mm^3 = (bacteria/square) (25 squares) (50). The number of bacteria per cm^3 is 10^3 times this value. For example, suppose the average count per square is 28 bacteria: bacteria/mm^3 = (28 bacteria) (25 squares) (50) (10^3) = $3.5 \times 10^{7.}$

(ii) ***Direct plate count.*** By growing dilutions of broth cultures on agar plates, one can determine the number of colony-forming units per ml (CFU/mL). This provides a count of viable cells only. This method is used in determining the bacterial cell count in urine cultures.

(iii) ***Density measurement.*** The density of a bacterial broth culture in log phase can be correlated to CFU/ml of the culture (Fig. 3.3). This method is used to prepare a standard inoculum for antimicrobial susceptibility testing.

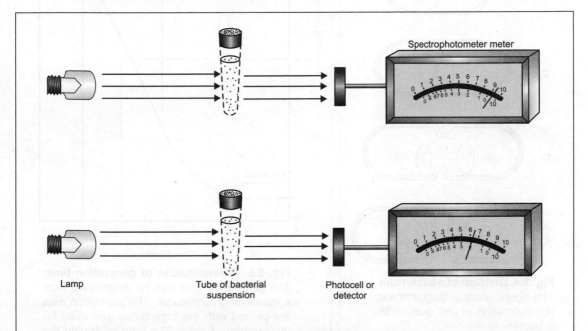

Fig. 3.3. Turbidity and microbial mass measurement: Determination of microbial mass by measurement of light absorption. As the population and turbidity increase, more light is scattered and there is an increase in absorbance reading given by the spectrophotometer. There are two scales on the meter of the spectrophotometer. The lower scale displays absorbance, and the upper scale, percent transmittance. As percent transmittance decreases, absorbance increases.

3.1.1.4.2 Generation Time

Replication of bacteria takes place by binary fission, with one cell dividing into two (Fig. 3.4). The time required for one cell to divide into two cells is called the **generation** or **doubling time** (Fig. 3.5). The generation time of a bacterium in culture can be as little as 20 minutes for a fast-growing bacterium such as *E. coli* or as long as 24 hours for a slow-growing bacterium such as *Mycobacterium tuberculosis*. The generation time of some of the bacteria has been summarized in Table 3.1.

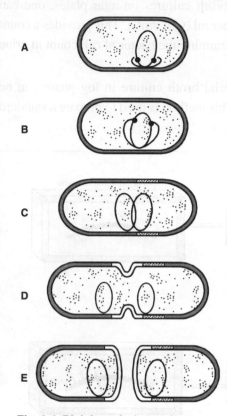

Fig. 3.4. Division of a bacterium
The figure gives a diagrammatic representation of cell division in a bacterium

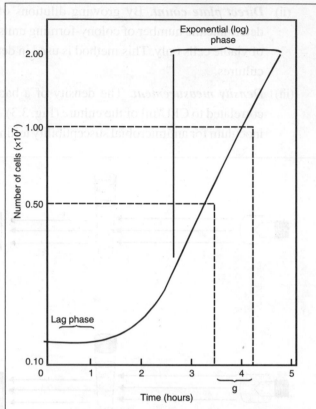

Fig. 3.5. Determination of generation time. The generation time can be determined from a microbial growth curve. The population data are plotted with the logarithmic axis used for the number of cells. The time to double the population number is then directly read from the plot. The log of the population number can also be plotted against time on regular axes.

Table 3.1: Generation Times for Selected Microorganisms

Microorganism	Temperature (°C)	Generation Time (Hours)
Bacteria		
Escherichia coli	40	0.35
Bacillus subtilis	40	0.43
Staphylococcus aureus	37	0.47
Pseudomonas aeruginosa	37	0.58
Clostridium botulinum	37	0.58
Anabaena cylindrica	25	10.6
Mycobacterium tuberculosis	37	≈ 12

Algae		
Scenedesmus quadricauda	25	5.9
Chlorella pyrenoidosa	25	7.75
Asterionella formosa	20	9.6
Ceratium tripos	20	82.8
Protozoa		
Leishmania donovani	26	10-12
Paramecium caudatum	26	10.4
Acanthamoeba castellanii	30	11-12
Giardia lamblia	37	18
Fungi		
Saccharomyces cerevisiae	30	2
Monilinia fructicola	25	30

3.1.1.4.3 Growth Curve

If bacteria are in a balanced growth state with enough nutrients and no toxic products present, the increase in the number of bacteria is proportional to the increase in their other properties. These include mass, protein and nucleic acid content.

The measurement of any of these properties indicates the bacterial growth. In fact after the inoculation of a sterile nutrient solution with microorganism and cultivation under physiological conditions, four typical phases of growth are observed as indicated in Fig. 3.6.

(i) **A lag phase.** During this, bacteria are prepare themselves to divide. It represents the physico-chemical equilibrium between microorganism and environment following inoculation with very little growth.

(ii) **A log phase.** During this, number of bacteria logarithmically increases. By the end of the lag phase, cells have adapted to the new conditions of growth. Growth of the cell mass can now be described quantitatively as a doubling of cell number per unit time for bacteria and yeasts or a doubling of biomass per unit time for filamentous organisms such as fungi. By plotting the number of cells or biomass against time on a semilogarithmic graph, a straight line results, hence the term log phase. Although the cells alter the medium through uptake of substrates and excretion of metabolic products, the growth rate remains constant during the log phase. Growth rate is independent of substrate concentration as long as excess substrate is present.

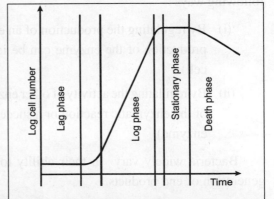

Fig. 3.6. The growth curve of a bacterial culture. It indicates various phases of growth cycles mentioned in the text (Modified from Schaechter M. Medoff G. Eisenstein Bl. Mechanisms of microbial diseases, ed., Baltimore, 1993, Williams & Wilkins.)

(iii) **A stationary phase.** In this, nutrients become limited and the number of bacteria remains constant (although viability may decrease). In fact as soon as the substrate is metabolized or toxic substances have been formed, growth slows down or is completely stopped. During this stationary phase, the biomass increases only gradually or remains constant, although the composition of cells may change. Due to lysis, new substrates are released which may then serve as energy source for this slow growth of survivors. The various metabolites formed in the stationary phase are often of great biotechnological interest.

(iv) **A death phase.** In this, the number of non-viable bacterial cells exceeds that of viable cells. The energy reserves of the cells are exhausted in this phase. A straight line may be obtained when a semilogarithmic plot is made of survivors versus time, indicating that the cells are dying at an exponential rate. The length of time between the stationary phase and the death phase is dependent on the microorganism and the process used. The fermentation is usually interrupted at the end of the log phase or before the death phase begins.

3.1.1.5 Metabolism of Bacteria

It consists of the biochemical reactions that are used by bacteria for :

(i) breaking down organic compounds,

(ii) synthesizing new bacterial parts from the resulting carbon skeletons.

Energy for the new constructions is generated during the metabolic breakdown of substrate.

The presence and activity of specific enzymes determine the occurrence of all biochemical reactions in the cell. The metabolism, therefore, can be regulated in the cell by either of the following ways :

(i) By regulating the production of an enzyme itself (a genetic type of regulation in which production of the enzyme can be induced or suppressed by molecules present in the cell).

(ii) By regulating the activity of other enzyme (via feed-back inhibition, in which the products of the enzymatic reaction or a succeeding enzymatic reaction inhibit the activity of the enzyme).

Bacteria widely vary in their ability to use various compounds as substrates and in the generation of end products.

(i) In the microbial world, a variety of biochemical pathways exist for substrate breakdown. The end product and final pH of the medium depends upon the particular pathway used.

(ii) These metabolic differences are used as phenotypic markers in the identification of bacteria.

Diagnostic schemes analyze each unknown microorganism for the following activities:

(i) Utilization of a variety of substrates as carbon sources.

(ii) Production of specific end products from various substrates (Fig. 3.6). Some of the products yielded by microbes have been given in Table 3.2.

(iii) Production of an acidic or alkaline pH in the test medium.

Table 3.2 : Chemicals Commonly Produced by Microorganisms

Organic chemical	Microbial source (s)	Industrial uses
Ethanol	*Saccharomyces*	Industrial solvent, fuel, beverages
Acetic acid	*Acetobacter*	Industrial solvent, rubber, plastics, food acidulant (vinegar)
Citric acid	*Aspergillus*	Food, pharmaceuticals, cosmetics, detergents
Gluconic acid	*Aspergillus*	Pharmaceuticals, food, detergent
Glycerol	*Saccharomyces*	Solvent, sweetener, printing, cosmetics, soaps, antifreezes
Acetone	*Clostridium*	Industrial solvent, intermediate for many organic chemicals
Lactic acid	*Lactobacillus,* *Streptococcus*	Food acidulant, fruit juice, soft drinks, dyeing, leather treatment, pharmaceuticals, plastic
Butanol	*Clostridium*	Industrial solvent, intermediate for many organic chemicals
Fumaric acid	*Rhizopus*	Intermediate for synthetic resins, dyeing, acidulant, antioxidant
Succinic acid	*Rhizopus*	Manufacture of lacquers, dyes and esters for perfumes
Malic acid	*Aspergillus*	Acidulant
Tartaric acid	*Acetobacter*	Acidulant, tanning, commercial esters for lacquers, printing
Itaconic acid	*Aspergillus*	Textiles, paper manufacture, paint.

3.1.1.5.1 Fermentation and Respiration

To catabolize (break down) carbohydrates and produce energy, bacteria use two mechanisms involving biochemical pathways. These are **fermentation** and **respiration** (commonly referred to as oxidation).

3.1.1.5.1.1 Fermentation

It is an anaerobic process. It is carried out both by obligate and facultative anaerobes. In fermentation, an organic compound is the electron acceptor. Fermentation is less efficient in energy generation than respiration (oxidation) because the beginning substrate is not completely reduced, and therefore all the energy in the substrate is not released.

The oldest and most familiar bioprocessing technology is microbial fermentation. Originally, the microbial fermentation products we use were derived from a series of enzyme-catalyzed reactions that microbes use to break down glucose. In the process of metabolizing glucose to acquire energy, microbes synthesize by-products we can use : carbon dioxide for leavening bread, ethanol for brewing wine and beer, lactic acid for making yoghurt and acetic acid (vinegar for pickling foods) (Fig. 3.7).

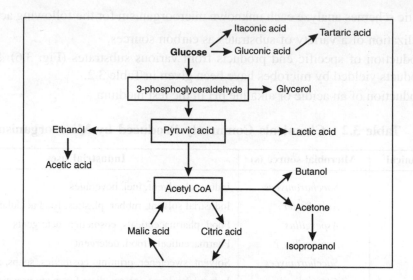

Fig. 3.7. Glucose metabolism: Different products yielded by various microorganisms during metabolism of glucose.

When fermentation occurs, a mixture of end products (such as lactate, butyrate, ethanol and acetoin) accumulates in the medium. Analysis of these end products is particularly useful for the identification of anaerobic bacteria. End-product determination is also used in two important diagnostic tests : the Voges-Proskauer (VP) and methyl red test. These are used in the identification of the Enterobacteriaceae. The term fermentation is often loosely used in the diagnostic microbiology laboratory to indicate any type of utilization – fermentative or oxidative – of a carbohydrate – sugar with the resulting production of an acidic pH (Fig. 3.8).

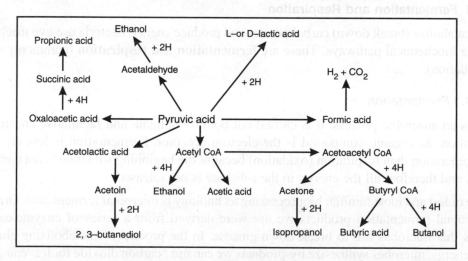

Fig. 3.8. The fate of pyruvate in major fermentation pathways by microorganisms.
(From Joklik *et al* : Zinsser Microbiology, ed. 20. Norwalk, Conn, 1992, Appleton & Lange).

During fermentation, NADH from glycolysis is reoxidized by being used to reduce pyruvate or a pyruvate derivative. It results in lactate or reduced product (Fig. 3.9).

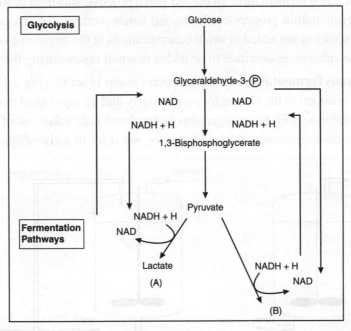

Fig. 3.9. Reoxidation of NADH during fermentation. (A) NADH from glycolysis is reoxidized by being used to reduce pyruvate or a pyruvate derivative (B). It results in the formation of either lactate or reduced product.

The fermentation can be of following types :

   (i) Batch fermentation

   (ii) Fed batch fermentation

   (iii) Continuous fermentation

   (i) Batch fermentation. It can be considered to be a closed system (Fig. 3.10). At time t = 0, the sterilized nutrient solution in the fermenter is inoculated with microorganisms and incubation is allowed to proceed. In the course of entire fermentation, nothing is added, except oxygen (in case of aerobic microorganisms), an antifoam agent, and acid or base to control the pH. Generally as a result of metabolism of the cells, the composition of the culture medium, the biomass concentration, and the metabolite concentration constantly change.

Fig. 3.10. Batch fermentation. In the batch mode the microorganisms and the necessary nutrients are simply mixed in the fermentor until the reaction has reached the desired stage of completion. At that time the contents of the vessel are withdrawn and the product recovered.

(ii) **Fed batch fermentation.** In the conventional batch process, all of the substrate is added at the beginning of the fermentation. An enhancement of the closed batch process is the fed batch fermentation. In the fed-batch process, substrate is added in increments as the fermentation progresses. In the fed batch method, the critical elements of the nutrient solution are added in small concentrations at the beginning of the fermentation and these substances continue to be added in small doses during the production phase.

(iii) **Continuous fermentation.** In this an open system is set up (Fig 3.11). Sterile nutrient solution is added to the bioreactor continuously and an equivalent amount of converted nutrient solution with microorganisms is simultaneously taken out of the system. In the case of a homogeneously mixed bioreactor, we refer to a chemostat or a turbidostat.

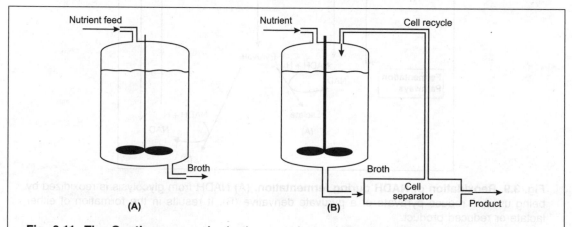

Fig. 3.11. The Continuous mode, In the a nutrient solution is fed slowly into the fermentor while the broth with the product is slowly withdrawn. **(A)** Some continuous operations allow for the cells that are removed with the broth to be recycled back into the fermentor **(B)**. Continuous with cell recycling.

In the chemostat in the steady state, cell growth is controlled by adjusting the concentration of one substrate (Fig. 3.12).

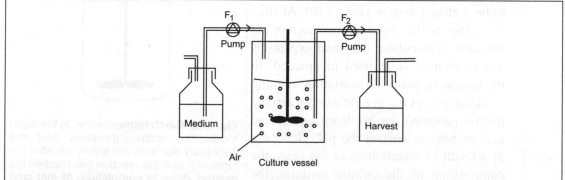

Fig. 3.12. A chemostat culture system (Adopted from Culture of Animal Cells by Ian Freshney, 2000).

In the turbidostat, cell growth is kept constant by using turbidity to monitor the biomass concentration and the rate of feed of nutrient solution is appropriately adjusted (Fig. 3.13).

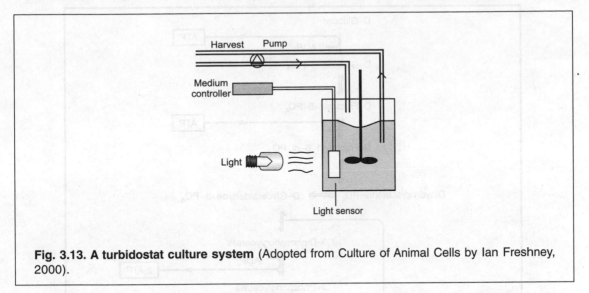

Fig. 3.13. A turbidostat culture system (Adopted from Culture of Animal Cells by Ian Freshney, 2000).

3.1.1.5.1.2 Respiration

It is an efficient energy-generating process. In this molecular oxygen is the final electron acceptor. Aerobic respiration is carried out by obligate aerobes and facultative anaerobes. Anaerobic respiration can be carried out by certain anaerobes. In the final reaction, electron acceptors are inorganic forms of oxygen, such as nitrate and sulfate.

3.1.1.5.2 Some Biochemical Pathways

3.1.1.5.2.1 Breakdown of Glucose to Pyruvic Acid

For bacterial fermentations or oxidations, the starting carbohydrate is glucose. When bacteria use other sugars as a carbon source, the following steps take place :

(i)   They first convert the sugar to glucose.

(ii) It is then processed by one of the three pathways. These pathways are designed to generate pyruvic acid, which is a key three carbon intermediate.

(iii) Pyruvate can then be further processed either fermentatively or oxidatively.

These three major biochemical pathways used by bacteria for breaking down glucose to pyruvic acid are :

(i) the Embden-Meyerhof-Parnas (EMP) or glycolytic pathway (Fig. 3.14).

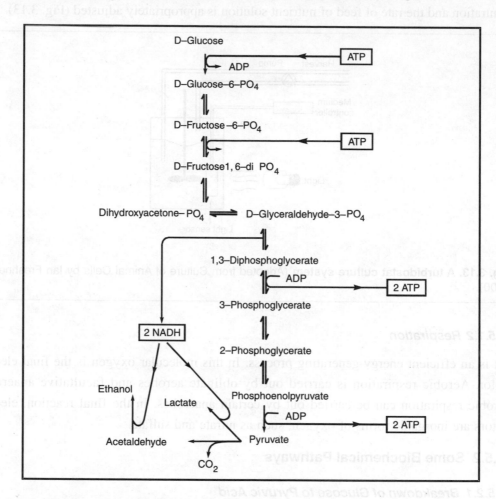

Fig. 3.14. The Embden-Meyerhof-Parnas (EMP) glycolytic pathway (From Joklik WK *et al* : Zinsser Microbiology, ed 20, East Norwalk, Conn. 1992, Appleton & Lange.)

(ii) the pentose phosphate pathway (Fig. 3.14)

(iii) the Entner-Doudoroff pathway (Fig. 3.15)

3.1.1.5.2.2 Anaerobic utilization of pyruvic acid (Fermentation)

Pyruvic acid is a key metabolic intermediate. Bacteria process pyruvic acid further using a variety of fermentation pathways (Fig. 3.15). Each pathway results into different end products. These can be analysed and used as phenotypic markers. Microbes that inhabit the human body use the following fermentation pathways :

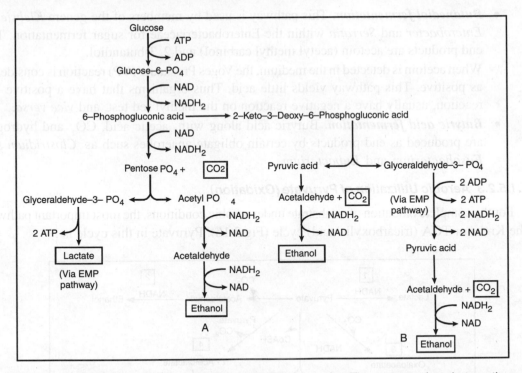

Fig .3.15. Alternate biochemical pathways used by bacteria. The pentose phosphate pathway (A) and the Entner-Doudoroff pathway (B) for Glucose Fermentation through EMP pathway (Adopted from Joklik W.K. *et al.* Zinser Microbiology, ed 20, Norwalk, Conn 1992, Appleton and Large and also published in textbook of Diagnostic Microbiology by Mahon *et.al.* 2000).

- *Alcoholic fermentation.* The major end product is ethanol. Yeasts use this pathway to ferment glucose to produce ethanol.

- *Homolactic fermentation.* The end product is almost exclusively lactic acid. This pathway is used by *Streptococcus* sp. and many members of the *Lactobacillus* sp. to ferment pyruvate.

- *Heterolactic fermentation.* This mixed fermentation pathway is used by some lactobacilli. In this, in addition to lactic acid, the end products include CO_2, alcohols, formic acid and acetic acid.

- *Propionic acid fermentation.* The major end product of fermentations carried out by *Propionibacterium acnes* and some anaerobic non-spore-forming gram-positive bacilli is propionic acid.

- *Mixed acid fermentation.* This pathway is used by members of the genera *Escherichia*, *Salmonella* and *Shigella* within the Enterobacteriaceae. This is used for sugar fermentation and produce a number of acids as end products. These include lactic, acetic, succinic and formic acids. The strong acid produced is the basis for the positive reaction on the methyl red test exhibited by these organisms

- **Butanediol fermentation.** This pathway is used by members of the genera *Klebsiella, Enterobacter* and *Serratia* within the Enterobacteriaceae for sugar fermentation. The end products are acetoin (acetyl methyl carbinol) and 2, 3-butandiol.

  When acetoin is detected in the medium, the Voges Proskauer (VP) reaction is considered as positive. This pathway yields little acid. Thus organisms that have a positive VP reaction, usually have a negative reaction on the methyl red test, and *vice versa.*

- **Butyric acid fermentation.** Butyric acid along with acetic acid, CO_2 and hydrogen are produced as end products by certain obligate anaerobes such as *Clostridium sp., Fusobacterium* and *Eubacterium.*

3.1.1.5.2.3 Aerobic Utilization of Pyruvate (Oxidation)

For the complete oxidation of a substrate under aerobic conditions, the most important pathway is the Krebs or TCA (tricarboxylic acid) cycle (Fig 3.16). Pyruvate in this cycle is :

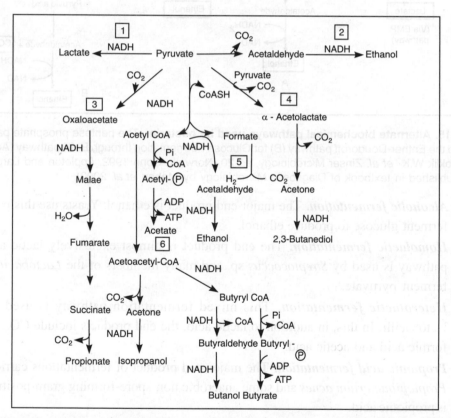

Fig. 3.16. Some common microbial fermentations. Only the pyruvate fermentations are depicted in the figure. In order to simplify the things in most of these pathways, one or more steps and intermediates have been omitted. (1) Lactic acid bacteria (*Streptococcus, Lactobacillus*), *Bacillius,* (2) Yeast, *Zygomonas,* (3) Propionic acid bacteria (*Propionibacterium*) (4) *Enterobacter, Serratia, Bacillus,* (5) Enteric bacteria (*Escherichia, Enterobacter, Salmonella, Proteus*), (6) *Clostridium* (Modified from Diagnostic Microbiology by Mahon *et.al.,* 2000).

(i) oxidized

(ii) carbon skeletons for biosynthetic reactions are created

(iii) the electrons donated by pyruvate are passed through an electron transport chain and used to generate energy in the form of ATP

3.1.1.5.2.4 *Utilization of Carbohydrate and Fermentation of Lactose*

Most diagnostic identification schemes exploit the ability of microorganisms to use various "sugars" (carbohydrates) for growth. The fermentation of sugar is usually detected by the following parameters :

(i) Production of an acid

(ii) A concomitant change of colour indicated by a pH indicator present in the culture medium.

In general, bacteria preferentially ferment glucose over other sugars, and therefore, if the ability to ferment another sugar is being tested, glucose must not be present.

Determination of ability of the microorganisms to ferment lactose is one of the important steps in classifying members of the family Enterobacteriaceae. These bacteria are classified as :

(i) Lactose fermenters

(ii) Lactose non-fermenters

Lactose is a disaccharide. It consists of one molecule each of glucose and galactose linked together by a galactosidic bond. The following two steps are involved in the utilization of lactose by a bacterium.

(i) Requirement of an enzyme, α-galactoside permease, for the transport of lactose across the cell wall into the bacterial cytoplasm.

(ii) Breakdown of the galactoside bond by β-galactosidase inside the cell.

3.1.2 Culturing Cells from Higher Eukaryotes

In contrast to the culture of microbial cells, it is quite difficult to culture higher eukaryotic cells. The approach of cell biology is to understand particular processes by their analysis in a simplified controlled *in vitro* system. The same approach can be applied to the study of cells themselves, since they too can be removed from the influences to which they are normally subject to within a complex multicellular organism. The ability to grow cells outside the organism that is **cell culture** has proved to be one of the most valuable technical achievements in the entire study of biology. The cells can be easily obtained in large quantity. The culture of cells is preferred because of the following reasons :

(a) Most cultures contain only a single type of cell.

(b) A wide variety of different types of cells can be grown in culture.

(c) Many different cellular activities, including endocytosis, cell movement, cell division, membrane trafficking and macromolecular synthesis can be studied.

(d) There is an opportunity to study cell differentiation. In this process undifferentiated, embryonic cells are converted into highly specialized cell types.

(e) Cultured cells will respond to treatment with drugs, hormones, growth factors and other active substances.

3.1.2.1 Animal Cell Culture

The first successful attempt to culture living vertebrate cells outside the body was made in 1907 by Harrison. Over the next few decades, a number of researchers worked out optimal conditions for growing animal cells outside the body and keeping cell cultures free from contamination by microorganisms.

Now the cultivation of animal cells is widely used technique in many different disciplines from basic science of cell and molecular biology to biotechnology. This is being used to answer biochemical, physiological and morphological questions.

Individual cells released from complex tissues or organs can be maintained in artificial conditions and treated as discrete organisms *in vitro*. With the development of animal cell culture techniques, the annual demand of more than 280 millions of experimental animals worldwide has reduced. The range of different cell types which can now be grown in culture is quite extensive and includes connective tissue elements, such as fibroblasts, skeletal tissue (bone and cartilage, skeletal, cardiac and smooth muscles), epithelial tissues (*e.g.*, liver, lung, breast, skin, bladder and kidney *etc.*) neural cells (glial and neurons, though neurons do not proliferate), endocrine cells (adrenal, pituitary, pancreatic cells), melanocytes and many different types of tumours. Mammalian cells are now being cultured to produce a variety of pharmaceutically important macromolecules. Many are being made to transfer animal cells from laboratory to the production level.

3.1.2.1.1 Types of Cell Culture

Different types of culture have been summarized in Fig 3.17 (Freshney, 2005).

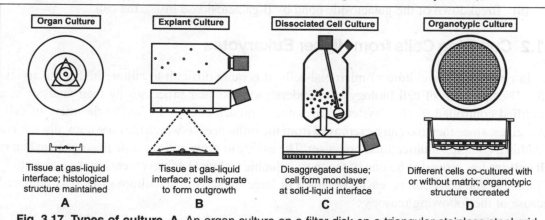

| Organ Culture | Explant Culture | Dissociated Cell Culture | Organotypic Culture |

Tissue at gas-liquid interface; histological structure maintained

A

Tissue at gas-liquid interface; cells migrate to form outgrowth

B

Disaggregated tissue; cell form monolayer at solid-liquid interface

C

Different cells co-cultured with or without matrix; organotypic structure recreated

D

Fig. 3.17. Types of culture. A. An organ culture on a filter disk on a triangular stainless steel grid over a well of medium, seen in section in the lower diagram. **B.** Explant cultures in a flask, with section below and with an enlarged detail in section in the lowest diagram, showing the explant and radial outgrowth under the arrows. **C.** A stirred vessel with an enzymatic disaggregation generating a cell suspension seeded as a monolayer in the lower diagram. **D.** A filter well showing an array of cells, seen in section in the lower diagram, combined with matrix and stromal cells. (From Freshney, 2005).

3.1.2.1.1.1 Primary Culture

It can be established either an explant culture or by dissociating organized tissues into single cell suspension. In the former a simples price of tissue or biopsy is adhered to the surface either spontaneanly or aided by mechanical means, a plasma clot or an extra cellula constituent e.g., collagen, will gise rise to an outgrowth of cells, (Freshney, 2005). The cells proliferating is the explant are selected by subculturing, and form a primary culture.

Enzymatic, chemical and **mechanical treatments** can be used to dissociate cells. Tissues are dissociated by the use of hydrolytic enzymes that breakdown the extracellular matrix. This method though efficient can damage the cell membrane by removing its components. The exposure time to these chemicals is very critical. The enzymes used for this purpose are trypsin (0.1-0.25%), pronase (0.05%), collagenase (0.01-0.15%), elastase (0.05%), hyaluronidase (0.1%) *etc.*

Sometimes the tissues are disaggregated with the use of chelating agents such as EDTA or EGTA. Some tissues are disaggregated by **mechanical disruption.**

The cells, thus, separated are put in primary culture and they remain so until are passaged or subcultured. They are usually heterogenous and have a low growth fraction, but have more representatives of different cell types in the tissue from which they were derived and in the expression of tissue specific properties. According to Freshney (2005) tissue disaggregation larger cultures more rapidly than explant culture, but explant culture may still be preferable where only small tragment of tissue are available.

Once the cells have been dissociated into a single-cell suspension, they can either be cultured directly or separated according to cell type and then placed in culture. A number of techniques including **differential centrifugation** or **the fluorescence-activated cell sorter** can be used to separate cells. In this later approach, the cell suspension is treated with a fluorescent antibody that specifically binds to the surface of the cell type to be cultured, and the suspension is passed through an electronic instrument that is able to deflect the cells lacking the fluorescent label.

3.1.2.1.1.2 Cell Line

Normal (non-malignant) cells are capable of a limited number of cell divisions (typically 50 to 100) before they undergo a process of senescence and death. Because of this, many of the cells that are commonly used in tissue culture studies have undergone genetic modifications that allow them to grow indefinitely. Cells of this type are referred to as a **cell line.** The origin of cell lines has been beautifully explained by Freshney, 2005 (Fig. 3.18).

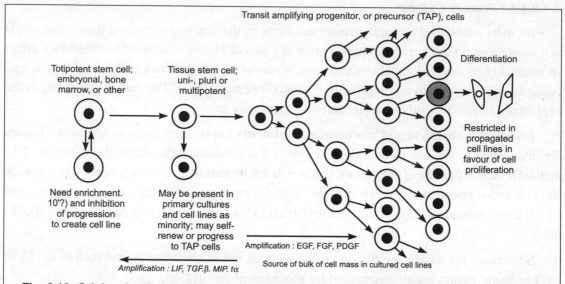

Fig. 3.18. Origin of cell lines. Diagrammatic representation of progression from totipotent stem cell, through tissue stem cell (single or multiple lineage committed) to transit amplifying progenitor cell compartment. A pressure on the progenitor compartment to proliferate limit the exit from this compartment to the differentiated cell pool. (From Ian Freshney, 2006).

It is impossible to determine the moment when the transition to continuous cell line has taken place. The result of culture alteration is generally called "**transformation**". This term should, however, be used only in those cases in which the alteration can be unambiguously ascribed to the introduction of foreign genetic material. The frequency with which a normal cell growing in culture becomes spontaneously transformed into a cell line, is related to the species from which it was derived. For example, mouse cells will frequently become transformed in this way; human cells, only rarely do so. Normal cells can transform into continuous cell lines without becoming malignant and malignant tumours can give rise to cultures which transform and become more tumorigenic. It allows cloning, characterization and preservation of cells with greater uniformity. It may, however, cause a loss of specialized cells and differentiated properties unless the correct lineage is carefully selected.

The appearance of continuous cell line is usually marked by following characteristics:

   (i)   An alteration in cytomorphology

   (ii)   An increase in growth rate

   (iii)   A reduction in serum dependence

   (iv)   An increase in cloning efficiency

   (v)   A reduction in charge dependence

   (vi)   An increase in heteroploidy.

The continuous cell lines have several advantages. These include the following:

(a)   Increased growth rates to higher cell densities and resultant higher yield

(b)   Lower serum requirement

(c)   General ease of maintenance in simple media

(d)   Ability to grow in suspension

Organ and Organotypic Cultures

There is a loss of histologic characteristics after disaggregation. The cells within a primary explant may retain some of the histology of the tissue. This will however soom be lost because of flattening of the explant with cell migration and some dgree of central necrosis due to poor oxygenation. The air/medium interface retention of histologic structure, and its associated differentiated properties, may be enhanced. It is because here the gas exchange is optimized and cell migration minimized, as distinct from the substrate/medium interface, where dispersed cell cultures and primary outgrowths are maintained. This so-called *organ culture*, will survive for up to 3 weeks, normally. It however cannot be propagated organotypic culture is the amplification of the cell stock by generation of cell lines from specific cell types and the subsequent recombination in *organotypic culture*. This allows the synthesis of a tissue equivalent or construct on demand for basic studies on cell-cell and cell-matrix interaction and for *in vivo* implantation. The fidelity of the construct in terms of its real tissue equivalence naturally depends on identification of all the participating cell types in the tissue *in vivo* and the ability to culture and recombine them in the correct proportions with the correct matrix and juxtaposition. According to Freshney (2005) this has so far worked best for skin (Michel *et al.*, 1999, Schaller *et at.*, 2002), but even then, melanocytes have only recently been added to the construct, and islet of Langerhans cells are still absent, as are sweat glands and hair follicles, although some progress has been made in this area (Regnier *et al.*, 1997; Laning *et al.*, 1999).

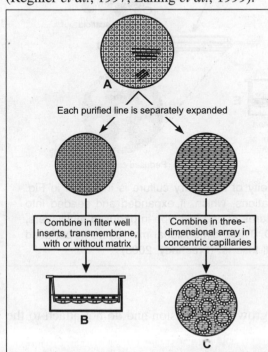

Each purified line is separately expanded

Combine in filter well inserts, transmembrane, with or without matrix

Combine in three-dimensional array in concentric capillaries

3.19. Organotypic Culture. Organotypic cultures can be generated by the expansion of purified populations and recombination from hetero genous primary culture **(A)** in filter well inserts **(B)** or on concentric microcapillaries **(C)** this seems to be suggested by the architecture of the device (CellGro Triac), but the author has no knowledge of its use in this capacity. (Freshney, 2006),

3.1.2.1.1.4 *Histotypic Cultures*

Tissues have been simulated by combining cells is various way so far. It ranges from simply allowing the cells to multilayer by perfusing a monolayer (Kruse *et al.*, 1970) to highly complex perfused membrane (Membroferm (Klement *et al.*, 1987) or capillary beds (Knazek *et al.*, 1972). These are termed *histotypic cultures*. The purpose is to attain the density of cells found in the tissue from which the cells were derived. For this purpose selective media, cloning or physical separation methods are used to isolate purified cell strains from disaggregated tissue or primary culture or at first subculture. these purified cell populations can then be combined in organotypic culture to recreate both the tissue cell density and, also the cell interactions and generations of matrix as is found in the tissue. Fig. 3.19, explaines the development of histotype cultures from heterogeneity primary culture (Freshney 2005).

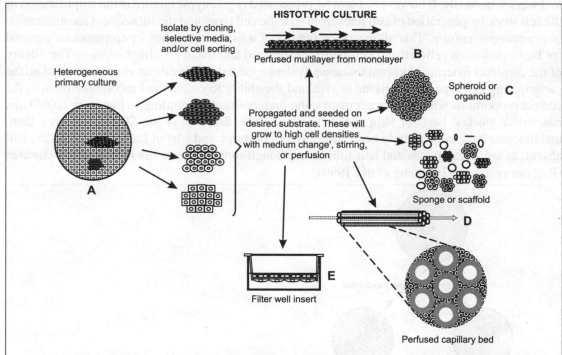

Fig. 3.20. Histotypic and culture. The heterogeneity of a primary culture is indicated in Fig. **A.** This can be purified to give defined cell populations, which, if expanded and seeded into appropriate conditions can give **B.** High-density cultures of one cell type in perfused multilayers, **C.** Spheroids or organoids in stirred suspension. **D.** Three-dimensional multilayers in perfused capillaries **E.** Monolayers or multilayers in filter well inserts. (Freshney, 2005)

Types of Cultures

(a) **Suspension cultures.** In these, the cells grow in suspension and do not attach to the substratum.

(b) **Monolayer cultures.** In these, a large number of cells are added to a culture dish; they settle and attach to the bottom and form a relatively uniform layer of cells. Those cells that survive will grow and divide and, after a number of generations, form a monolayer of cells that covers the bottom of dish.

(c) **Clonal cultures.** In this, a relatively small number of cells is added to the dish, each of which, after settling and attaching to the surface, is at some distance from its neighbours. In this case, the proliferation of cells generates individual colonies or clones of cells.

3.1.2.1.2 Requirement of Cell Culture Laboratory

The diagrammatic plan of tissue culture laboratory is given in Fig. 3.21.

1. *General requirements.* The major requirements for setting and maintenance of culture laboratory include clean aseptic laboratory area furnished with laminar flow, incubators (including CO_2 incubator), sterilizing equipment, inverted microscopes *etc.* All these depend upon the nature, amount, duration *etc.* of the cells to be cultured.

2. *Substrate.* Some cells require a definite substrate for their growth. The nature of substrate depends on the type of cells and the use to which they will be put.

   Diploid cells require for their growth a solid substrate with a defined surface charge. This demand is satisfied by glass, but in its cleaning, detergents must be avoided and be replaced, *e.g.,* by polyphosphates. Polystyrene equipment which has been subjected to treatments that adjust the desired negative charged density is increasingly being used. This has already been taken into account in commercial "tissue-culture grade" flasks and dishes. In contrast, polystyrene "bacteriological grade" vessels are not suitable.

   In special cases (*e.g.,* culture of neurons, muscle cells and epithelial cells), the plastic is precoated with gelatin, polylysine or collagen.

   To give a net positive charge, variable sized culture vessels are used. They range from multiwell plates (1 mm^2 square area) and microtitration plates (30 mm^2) to dishes and flasks to 180 cm^2 and multi surface propagators.

3. *Incubators.* The cell lines that are processed now-a-days are mostly derived from warm-blooded animals and require a cultivation temperature of 37°C. Climatized rooms are quite complex and have the added disadvantage that practically only closed system can be operated since otherwise, the partial pressure of carbon dioxide necessary for a controlled metabolism would not be ensured. In order to meet these problems, incubators with an adjustable supply of carbon dioxide and oxygen are used.

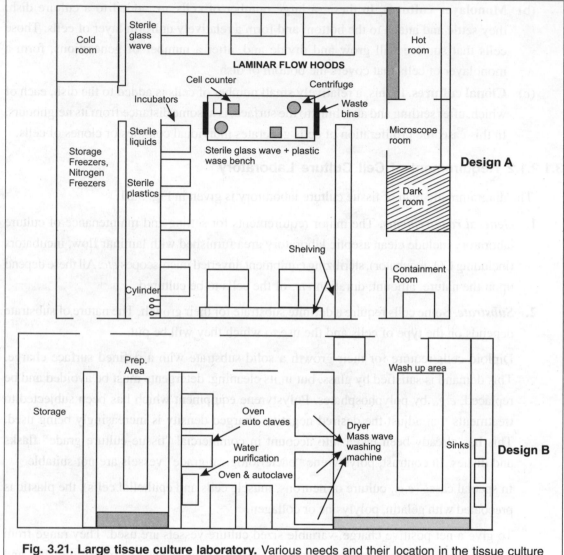

Fig. 3.21. Large tissue culture laboratory. Various needs and their location in the tissue culture lab have been highlighted (Design A, Design B).

3.1.2.1.3 Media

The media are the source of nutrients, pH adjustors, energy *etc*. The choice of medium depends on the type of cell culture. In the beginning, blood plasma was used as medium. It was followed by the advent of all purpose medium and then came minimal essential media (MEM).

The early tissue culture studies employed media containing a great variety of unknown substances. Cell growth was accomplished by adding fluids obtained from living systems, such as **lymph, blood serum** or **embryo homogenates.** It was found that cells require a considerable variety of nutrients, hormones, growth factors and cofactors to remain healthy and grow. Even today, most culture media contain large amounts of serum. During the last couple of years, the prerequisites for the maintenance and propagation of animal cells in culture have been systematically worked out.

Almost all the nutrient media used for animal cultures contain thermolabile components, so it is impossible to sterilize them by autoclaving. The media are, therefore, sterilized by high pressure filtration through membrane filters with pore widths of about 0.2 μm. Such procedures can be carried out on clean benches with a horizontal laminar air flow which offers the best protection against contamination from outside.

The media contain amino acids, energy source-glucose or its alternative, trace elements, vitamins, lipids, buffer, major inorganics, nucleic acid precursors, gases (O_2, CO_2) *etc.* and sometimes antibiotics too.

The use of serum in the medium has several advantages. It provides growth factors and other ill defined, but essential things required for growth. Now-a-days, however, many serum free media have been developed.

3.1.2.1.3.1 Serum Free Media

One of the primary goal of cell culturists has been to develop defined, serum-free media that can support the growth of cells. Using a pragmatic approach in which combinations of various ingredients are tested for their ability to support cell growth and proliferation, a growing number of cell types have been successfully cultured in "artificial" media completely lacking serum or other natural fluids. The compositions of these chemically defined media are relatively complex and consist of a mixture of nutrients and vitamins, together with a variety of purified proteins, including insulin, epidermal growth factor and transferrin (which provides the cells with most commonly usable iron).

3.1.2.1.3.2 Medium Design and Type of Culture

There is a direct effect of culture system on medium design as all the three types of cultures (suspension, monolayer, clonal) have their own requirements and limitations.

A. Growth Media. These are used for the growth of cell.

(i) **Suspension culture.** The medium should be able to support the relatively high yield, prevent cell aggregation and chemical damage to the cells and also the precipitation of serum proteins.

(ii) **Monolayer cultures.** The medium should be enriched with calcium and magnesium ions.

(iii) **The clonal cultures.** These have special needs. In these, feeder layers are frequently used. Use of conditioned medium supports the cell growth in a clone. Conditioned medium is a medium that has been removed from active growing or confluent cultures and clarified by centrifugation. It may be used after typically 3:1 or 1:1 dilution with fresh medium for cells that are very difficult to grow in culture or where very low cell densities are involved.

B. The maintenance medium. It is used for maintaining the cells in non-growing state. Such media have low levels of serum.

3.1.2.1.4 Cryopreservation

When not in use or to be used after a long time, cells are preserved at ultra low temperature to avoid the wastage of media *etc.* used in their maintenance. It is called **cryopreservation.** It is the storing of cells at very low temperatures where they remain in a state of suspended animation for as long as they are not required. The technique is very important for a cell worker as it saves time, efforts and money. Cells are also cryopreserved to avoid genetic drift in continuous cell lines and senescence in non-continuous cell lines and guard against contamination. The ability of cryopreservation of cells is very critical. For the critical temperature interval between +20 and –70°C, programmable apparatuses have been developed, but the graduated use of refrigerators in combination with insulating materials fulfills the same purpose. For long-term preservation, temperatures below –70°C are required, and for this purpose, Dewar vessels with liquid nitrogen or more sophisticated "Liquid nitrogen refrigerators" are used. The procedure requires a cryoprotective reagent. The preferred additive is 10% glycerol except for those cells that are permeated too slowly and may require a cytotoxic agent like 7.5% dimethyl sulfoxide. The critical lymphoid cells are preserved using 10% polyethyleneglycol (PEG 1000- 20000).

3.1.2.1.4.1 Importance of Cryopreservation

Under culturing conditions, cells have an ability to change their properties from time-to-time. Differences are found to occur between cells from the same source handled in different laboratories. To overcome this problem, cryopreservation is required. Preservation at low temperature is essential for the following reasons:

(a) Research laboratories where consistent data are required over a period of time
(b) It is applicable for genome conservation in genetic engineering
(c) It is used for preserving the seed stock
(d) It makes the availability of low-passage stock of finite-life cells *e.g.,* human diploid lines for prolonged period
(e) It is used to preserve parental stock for genetic studies
(f) It is also helpful in acting as a reserve against any loss due to contamination
(g) It avoids the repeated subcultivation of cells that are not required immediately
(h) It reduces labour, risk of contamination and media costs *etc.* Thus clones of fusion experiments can be investigated at ease

3.1.2.1.4.2 Cooling and Warming Processes

The ability of cryopreservation of cell is very critical. They are first grown in a proliferating medium and kept dormant. They are then put in a preserving medium. For this purpose, serum is used, however, cells in the preserving medium, do not grow. The procedure is very critical because there is always a fear of cell death. The cooling process must, therefore, start from inside of the cell.

Depending on the cooling rate. The following three events are supposed to occur during the process of cryopreservation.

(i) Formation of ice-crystals

(ii) The removal of water

(iii) Increase in solute concentration

The intracellular ice formation occurs at fast cooling rates *e.g.,* 100°C/min and high salt concentration which occurs at slower cooling rates *e.g.,* <1°C min. This causes cell damage and loss of viability. The optimum cooling rate therefore, is a bridge between salt damage on one side and intracellular ice damage on the other.

During the process of cooling a cell, the temperature falls to freezing point of water, but due to the presence of salts, there is depression of this freezing point and the system is supercooled. During the formation of ice crystals, due to reduction in energy of water molecules, there is a release of latent heat of fusion. The movement of water molecules is inhibited. This results in rise in temperature to above the melting point which is the ice crystallization temperature. This confirms that no more nuclei are available for initiating crystallization and those that have already formed, will continue to grow until all free water is frozen. The temperature then starts declining continuously. This phenomenon gives rise to a long phase transition period during which the cell is subjected to fluctuations of temperatures above and below the eutectic point. In remaining water, the salt concentration continuously increases. There is thus, osmotically induced loss of water and shrinkage of cell.

During this process, the cell membrane remains intact. It also acts as a barrier for ice-crystals. This process is efficient, if proceeded at correct rate with high recovery and minimum damage. If the rate of cooling is too slow, excessive exposure to hypertonic conditions will cause damage to the cells. If it is too fast, then the latent heat of fusion is absorbed and the temperature will continue to fall, thus exposing many more nuclei for crystallization. Thus a large number of ice-crystals are formed and convert all the free-water to ice and there will not be enough time for controlled dehydration of the cell to occur.

At the critical temperature, the water still within the cell will freeze and intracellular ice crystals will be formed. As intracellular ice is far more damaging than solute concentration, it is desirable to achieve fastest cooling rate that avoids any risk of this happening. The rate varies between cell types, but as a generalisation, the following are generally recommended when cryoprotectants are present, 0.5–2°C/min for lymphocyte cells, 1°C–3°C/min for fibroblasts and 2°C to 10°C/min for epithelial cells. The damage to cell during cooling is manifested only until the cell is thawed. During rewarming, the cells are subjected to hypertonic stress, followed by rehydration. The rehydration of cells occurs when they take in water to return to their normal size. Whatever damage occurs during cooling, is compensated during rewarming process. It determines whether the cell will survive or the injury is sufficient to lead to dilution/shock damage caused by excessive entry of water.

However, the above mentioned effect can be overcome by using warming rate as fast as possible in order to minimize the time spent in hypertonic condition and to prevent growth of small ice-crystals into larger ones.

3.1.2.1.4.3 Experimental Aspects of Cryopreservation

Volume and container used. The volume used should be as small as possible to avoid a significant temperature gradient through the freezing mixture which would cause variable freezing and thawing rates. The glass ampules are most suitable. The volume used should not be so large that the cooling rate is slower in the centre than that in the periphery. Screw cap plastic vials are not used, because leakage can occur, which may lead to a loss of material, a risk of contamination and could be potential biological hazard.

Cryoprotectants. For general use, the cells are suspended in a mixture of growth medium and cryoprotectant at a concentration between 3×10^6 and 1×10^7/ml. Cryoprotective agents (of which there are many, but all of them have a high solubility in water) remain in solution at temperature well below the freezing point of water.

Cryoprotectants are of two types :

(i) *Penetrating.* Dimethyl sulphoxide (DMSO) and glycerol

(ii) *Non penetrating.* Polyvinyl pyrrolidine and hydroxyethyl starch

The penetrating type of cryopreservatives have the ability to reduce the temperature at which ice is formed. This postpones active leakage to higher osmolarities so that ice when formed is glass rather than crystalline and prevent overshrinkage of the cells. These effects are due to the impact of cryoprotectant that reduces the proportion of the system that converts to ice. They thus protect against solution, rather than damage by ice and allow cooling rates to be used which are too slow to lead to the formation of ice.

The action of the non-penetrating cryoprotectants is not so clear. DMSO being 10%, a small lipid soluble molecule, enters the cell quickly by diffusion across the lipid bilayer of the plasma membrane where it is believed to alter the permeability characteristics of membranes. In the presence of DMSO, ice crystals which would otherwise rupture cell membranes causing cells to lyse, do not form. Glycerol has a similar effect. Polyvinyl pyrrolidene is used for lymphocytic cell lines. If DMSO is used, the freezing mixture must be diluted at least 40 fold immediately on thawing to reduce its toxicity and a medium is changed as soon as the cells are attached to the substrate.

Freezing Mixture. A double concentrated freezing mixture comprising 40% v/v growth medium (containing 10% serum, 40% v/v FCS and 20% DMSO or glycerol) is used. It should be mixed well and made sterile by passing through a 0.2 mm sterile filter.

The high serum concentration probably contributes to cell integrity by maintaining the intercellular protein concentration of cells, made permeable by DMSO. Glycerol also has the same effect.

The freezing mixture is mixed with an equal volume of cell suspension in complete medium which can then be used for cryopreservation.

Freezing Down. Cells which are healthy and growing in exponential phase should be cryopreserved. Contaminated, confluent or overgrown cells should not be used.

1. After counting, the cells should be centrifuged at 150–200 g at 4°C for 5 min, the supernatant discarded and the pellet resuspended in the residual medium.
2. The concentration is adjusted to double than that required finally with fresh ice-cold growth medium containing 10% FCS.
3. Cell suspension should be placed on ice and an equal volume of freezing mixture added.
4. 1 ml aliquots of this should be put into cold prepared cryotubes.
5. The final concentration is 2×10^6 for adherent cells and 5×10^6 for suspension per vial. Care should be taken as not to overtighten the screw caps of the cryotubes. No antibiotic is added.

Cooling Rates. To obtain good results, a commercial biological freezer should be used. The cooling rates in these are well set which are maintained throughout phase transition without permitting prolonged eutectic point fluctuations to occur. The cooling rate should be controlled to at least 5°C, at which it is permissible to gradually increase the cooling rate. At −120° to −140°C, the cells can be taken from the cooling chamber and put into liquid nitrogen.

The biological freezers work on the principle that the temperature difference between freezing chamber and inside of the ampule can be programmed. The larger the difference, the greater the input of N_2 vapours and faster the fall in temperature. The cells can be driven straight through to the phase-transition point by increasing the temperature difference, thus, absorbing the latent heat of fusion.

The alternative to biological freezers is to lower the vials/ampules slowly through nitrogen vapours into the liquid nitrogen.

A freezing plug allows cells to be cooled at a diffused rate in the vapour above the liquid nitrogen in the storage tank. The rate of cooling depends upon the number of vials and the type of plug used. A polystyrene box with a capacity of 10–20 vials and 5–10 min is used. The dead space in the box can be packed with tissue paper. This sealed box should be placed in a −70°C freezer. Cells are cooled at 1°C/min. The cells in 3 hours reach the temperature of liquid nitrogen and can then be placed in the liquid nitrogen tank.

Resuscitation. As mentioned earlier, cultures should be rewarmed as fast as possible. Transferring a one ml ampoule from liquid nitrogen into water at 37°C gives a good warming rate. Proper safety clothing over the ampoule in a container should be used.

The vial is then wiped with a tissue soaked in 70% alcohol and taken to a hood. The contents of the vials are emptied into a 10 ml sterile centrifuge tube containing 9 ml complete medium which has been prewarmed to 37°C and centrifuged at 150–200g for 5 min. The freezing mixture is then immediately removed and the cells resuspended in 10 ml complete medium. Even the most optimally freeze-thawed cells are extremely vulnerable at this time due to membrane damage which has to be repaired. Thus the use of supplemented (rather than minimal) medium and even conditioned medium, results in higher recovery and faster establishment in culture.

A sample should be taken for counting and also to check its viability and then incubated at a subconfluent concentration for 24 hrs. The culture should be observed and the cells counted again for confirmation.

3.1.2.1.4.4 Special Methods of Cryopreservation

For cells of special importance, however, alternative cryopreservation method can be used to increase the chances of high recovery rate.

A two-step freezing method of Farrant can be used.

In this procedure, the cell suspension is frozen as rapidly as possible to a temperature upto –40°C (the exact temperature has to be found out emperically for a given cell and system), held at this temperature for 10 minutes and then frozen rapidly to –196°C. The basis of this technique is that the cell is rapidly pushed through the hypertonic phase, stopped before intracellular ice is formed, dehydrated, then frozen rapidly again with the knowledge that no intracellular ice can be formed in a dehydrated cell.

3.1.2.1.4.5 Cryopreservation of Embryos

Many advantages have been offered by the freezing of embroys. It is possible to preserve all viable embryos till recipients at a matched physiological state are available. These embryos can be transported any time or to any place where recipients are available. To freeze embryos, the following steps are applied:

1. The embryo is placed in a culture medium containing a cryopreservative such as glycerol
2. The embryo is cooled to –7°C, below the freezing point of the solution
3. The vial containing the embryo is seeded with ice crystals using a very cold rod held away from the embryo
4. It is cooled to –30°C at a rate of 0.3°C/min.
5. It is then transferred to liquid nitrogen

3.1.2.1.5 Contamination

The contamination may be biological or chemical.

3.1.2.1.5.1 Biological Contamination

A wide variety of organisms such as bacteria, fungi, yeast and mycoplasma can contaminate media and cell culture. Sources of contamination include the cells that are cultured, serum, reagents, the environment and the person handling the cultures.

Bacteria and Fungi. Bacteria can be seen as round bodies which exhibit Brownian motion or they may be mobile and rod like under a microscope. If they are not detected prior to incubation, then the culture medium will become extremely acidic (yellow) and cloudy (Morgan and Darling, 1993). The nutrients in the medium will be used up by bacteria, the latter will excrete waste products.

It is quite easy to see fungal growth because many of them reveal long hyphal growths. It gives at the macroscopic level, a fluffy, fuzzy appearance. It may be possible to prevent large scale contamination before it becomes macroscopic. At this level, spores are produced at a great rate which spread rapidly and aggravate the contamination.

To test for contamination by bacteria and fungi, the following procedure is used :

(i) Samples of media from freshly prepared batches without antibiotics are incubated at room temperature at 35°C for 7 days.

(ii) One millimeter aliquots from these samples can be inoculated into thioglycollate broth and Sabourand's liquid medium (SLM) and a few drops are streaked onto a blood agar plate, a Sabourand's agar plate and deoxycholate plate. An anaerobic culture procedure is also in use.

(iii) If contamination is suspected in a culture, some cells can be dislodged into the medium and the suspension is used as the inoculum.

(iv) One half of the SLM tubes can be kept at room temperature and the other half incubated at 35°C. The thioglycollate broth tubes and the blood agar plates are incubated at 35°C. The cultures should be examined daily for 14 days.

The cultures with bacterial contamination must be discarded and precautions be taken to prevent the spread of bacteria to other cultures.

Mycoplasma. The *Mycoplasma* are characterized by their small size (0.25–1.0 μm) and lack a cell wall. These agents can infect cell cultures because they are present in the serum used as a component of the culture medium or introduced during the handling of culture in the laboratory. In some instances, the mycoplasma that have been contracted do not induce any change in cultured cells. In other situations, the cytopathogenic reaction can vary from minimal to total effect. The mycoplasma can often be isolated from contaminated cell cultures in cellular medium, however, culture methods are not always reliable since many of these organisms become well adapted to a cellular environment. It is because of this that morphological and biochemical methods have been introduced to supplement the culture procedure. Usually a biochemical technique is preferred, since it does not depend upon a subjective interpretation.

The liquid and solid media used for the isolation of mycoplasma have been developed by Hayflick. The mycoplasma infection can be prevented as follows :

- Periodic cleaning and disinfecting the incubators
- Water baths should be emptied and cleaned on a regular basis.
- Waste containers should be emptied daily and wastes disposed off safely.
- The cooling coils on refrigerators and freezers should be vacuumed at least yearly.
- Pest management program to reduce presence of mice, ants *etc.*

Sterility maintenance. The sterility can be maintained as follows.

- All autoclaves and dry heat ovens used to sterilize glasswares, solutions *etc.* must be regularly maintained.

- Thermometers and chart recorders should be tested and periodically calibrated to ensure their accuracy.

- Samples of all-in-house filter sterilized solutions should be tested for sterility each time.

- Cell culture media, especially those that are outdated or no longer used in the laboratory can provide a very rich, readily available and useful substitute for standard microbiological media.

Control of cell culture contamination. Cell cultures can be managed to reduce both the frequency and seriousness of culture related problems, especially contamination. Lack of basic culture management procedures frequently leads to long-term problems. There is a need to actively manage cultures to reduce problems.

Detection of biological contaminants in cultures. The direct culture tests and indirect DNA fluorochrome test for mycoplasma will also detect most bacteria, yeast and fungi, including intracellular forms. Several other important quality control tests should be used to both identify and characterize the cell cultures used in research. Besides the serious problems of cross contamination by other cell lines, cells are also continually evolving in culture *i.e.,*

(a) Important characteristics can be lost

(b) Mutations can occur

(c) Chromosomes can undergo rearrangements and change in number.

For monitoring cell cultures, the following characterization methods may be used:

- **Karyotyping.** It is done to determine the modal chromosome number, presence of any unique marker chromosome.

- **Electrophoresis and isoenzyme analysis.** These help to generate a protein "fingerprint" that can be used to determine species.

- **Immunological or biochemical techniques.** These are used to detect markers that are unique to tissues, cell lines or species from which it is derived.

- **DNA fingerprinting.** It is used to detect both intra- and inter-species contamination.

3.1.2.1.5.2 Chemical Contaminants

To determine that a chemical contaminant is the cause of a cell culture problem, is usually much more difficult than with biological contaminants. Generally, the first signs are alterations in the growth, behaviour or morphology of these cultures in laboratory, however, it may take weeks before these changes are noticed. To eliminate each source of problem, simple comparison experiments can be performed. Attempts may be made to avoid chemical contamination by testing all new lots of reagents, media and especially sera and testing the water purity.

3.1.2.1.5.3 *Steps to be Applied to Reduce Contamination Problems*

To prevent contamination, the following points must be kept in mind.

- Good aseptic techniques to be used
- Accidents to be avoided
- The laboratory to be kept clean
- Monitoring for contamination to be routinely carried out
- Frozen cell repository to be strategically used
- Antibiotics to be sparingly used
- Characteristics of cell lines to be frequently checked

Strategic use of a frozen cell repository. A cryogenic cell repository is commonly used in laboratory to reduce the need to carry large number of cultures and to provide replacements for cultures lost to contamination or accidents. Freezing cultures also stop culture loss due to contamination or biological time for them, preventing them from acquiring the altered characteristics that can normally occur in actively growing cells as a result of environmental or age related changes.

A cell repository can also be strategically used to convert continuously carried cultures into a series of short-term cultures, thereby, greatly reducing both the amount of quality control testing required and potential problems from cryptic contaminants. Test stocks should be set up in the cell repository for each culture that is routinely used in the laboratory. Cultures should be grown for at least two weeks in antibiotic-free media and then thoroughly tested to check their viability.

Use of antibiotics. A convenient means of growing cells from tissues that are likely to be contaminated by microorganisms is provided by antibiotics. Penicillin, streptomycin, gentamycin (inhibition of mycoplasma, active against gram +ve and gram −ve bacteria) and others such as chlorotetracycline, hydrochloride, kanamycin, neomycin and polymycin B are used. Antimycotic compounds like fungisone, the sodium deoxycholate complex of amphotericin B form a dispersion in aqueous media and is effective in suppressing growth of many yeasts and fungi. Antibiotics in some laboratories are used as prophylactic agents. If employed routinely, they should be withdrawn at frequent intervals to determine, if the cultures are contaminated. In fact with the exception of gentamycin, it is preferable not to grow cell lines with these antimicrobial agents on a permanent basis. Anti PPLO agent (GIBCO) and mycoplasma reduction agent or MRA (Flow Laboratories) are both claimed to cure contaminated culture. They are, however, best used as short-term prophylactic treatment. As the inhibitory agents against fungi and yeast are toxic to cells, they should not be used for long-term.

Use of fumigation. Sometimes, but rarely, fumigation is needed. Incubators are to be fumigated when the build up of contamination seriously hampers work.

3.1.2.1.6 Use of Animal Cell Cultures

Animal cells are the source of macromolecules of medical interest that are not available by chemical methods. The only available source of such substances previously was urine, animal or even human tissues when human specific molecules such as growth hormones were required.

Now the possibilities of continuous production based on cell lines are being developed for them. In many cases, these processes have to compete with technologies derived from the possibilities of genetic engineering in bacteria and yeasts.

The most important commercial utilization of animal cell cultures now is the multiplication of viruses for manufacture of vaccines. The animal cells are also being genetically engineered to produce molecules of interest. These also form the basis of cloning, *in vitro* fertilization and stem cell research *etc.*

3.1.2.1.7 Stem Cells

3.1.2.1.7.1 *General & Historical*

Development is astoundingly an orderly process, which gets ensued with totipotent zygote and ends with an assortment of differentiated cell types in adults. Going by the outcome of the Human Genome Project, it appears that approximately 35,000 genes are needed to build a human being, that harbour 200 histologically distinct cell types, which on further subdivision form numerous specialized cell types. These different cell types perform specialized functions that are as diverse as digestion, absorption, filteration, visual perception, and developing body's defense system *etc.* Fully differentiated state of a particular cell type represents a strikingly stable state of that cell. Starting from a single large totipotent embryo, the cells divide and differentiate to form the pluripotent stem cells. These stem cells have remarkable property of both self-renewal and cell type differentiation. Regardless of how different a cell type is, most cells retain an intact genome with the full complement of genes that are present at the beginning in the zygote. The distinguishing features of cells arise from an orderly selection of genes that are expressed, while the rest are switched off. The extensive proliferative potential of early mammalian embryos, along with an ability to give rise to one or more differentiated cell types, declines upon initiation of the differentiation and repression of the growth promoting signals. By adulthood, the few remaining stem cells are dispersed and virtually invisible, while the surviving adult stem cells equip themselves with a remarkable ability to maneuover at 'steady state' This allows stem cells to spawn on average one replacement stem cell and one tissue cell, at each division with no apparent limit. The machinery, that stem cells adopt to achieve self-renewal, assures fundamental insight into the origin and design of multicellular organisms. Stem cells have a potential to replace cells, lost in many degenerating diseases like Parkinson's disease, Alzheimer's disease, stroke, diabetes, chronic heart disease, spinal cord injury, muscular dystrophy and liver failure. The attempts are being made to tame the stem cells to develop a cell-based medicine to repair malformed, damaged or ageing tissues. It is becoming increasingly clear, however, that the specialized functions required to ensure proper stem cell function are vested in neighbouring differentiated cells. Thus, location as well as special patterns of gene expression might provide signatures to the stem cells. Because of these characteristics as well as their potential of becoming futuristic medicine, we all are drawn in one or the other way to understand the intricacies of stem cells at different levels. Methodologies are being developed to isolate, characterize and culture stem cells for use as therapeutics.

Key stem cell research events (Source : Internet)

- **1960s** - Joseph Altman and presented evidence of adult neurogenesis, ongoing stem cell activity in the brain; their reports contradicted "no new neurons" dogma and were largely ignored.
- **1963** - McCulloch and Till illustrated the presence of self-renewing cells in mouse bone marrow.
- **1968** - Bone marrow transplant between two siblings successfully treated SCID.
- **1978** - Haematopoietic stem cells were discovered in human cord blood.
- **1981** - Mouse embryonic stem cells were derived from the inner cell mass by scientists Martin Evans, and Gail R. Martin. Gail Martin is attributed for coining the term "Embryonic Stem Cell".
- **1992** - Neural stem cells are cultured *in vitro* as neurospheres.
- **1995** - U.S. President Bill Clinton signs into law which prohibited federally appropriated funds to be used for research where human embryos would be either created or destroyed.
- **1997** - Leukemia was shown to originate from a haematopoietic stem cell, the first direct evidence for cancer stem cells.
- **1998** - James Thomson and coworkers derived the first human embryonic stem cell line at the University of Wisconsin-Madison.
- **2000s** - Several reports of adult stem cell plasticity were published.
- **2003** - Dr. Songtao Shi of NIH discovered new source of adult stem cells in children's primary teeth.
- **2002 November, 2004** - California voters approved Proposition 71, which provided $3 billion in state funds over ten years to human embryonic stem cell research.
- **2004-2005** - Korean researcher Hwang Woo-Suk claimed to have created several human embryonic stem cell lines from unfertilised human oocytes. The lines were later shown to be fabricated.
- **2005** - Researchers at Kingston University in claim to have discovered a third category of stem cell, dubbed cord-blood-derived embryonic-like stem cells (CBEs), derived from umbilical cord blood. The group claimed these cells were able to differentiate into more types of tissue than adult stem cells.
- **2001-2006** - U.S. President George W. Bush endorsed the Congress in providing federal funding for embryonic stem cell research of approximately $100 million as well as $250 million for research on adult and animal stem cells. He also enacted laws that restricted federally-funded stem cell research on embryonic stem cells to the already derived cell lines.
- **5 May, 2006** - Senator Rick Santorum introduced bill number S. 2754, or the Alternative Pluripotent Stem Cell Therapies Enhancement Act, into the U.S. Senate.
- **18 July, 2006** - The U.S. Senate passed the Stem Cell Research Enhancement Act and voted down Senator Santorum's S. 2754.
- **19 July, 2006** - President votoed H.R. 810 (Stem Cell Research Enhancement Act), a bill that would have reversed the Clinton-era law which made it illegal for federal money to be used for research where stem cells are derived from the destruction of an embryo.
- **August 2006** - *Cell Journal* published, "Induction of Pluripotent Stem Cells from Mouse Embryonic and Adult Fibroblast Cultures by Defined Factors" by Kazutoshi Takahashi and Shinya Yamanaka.
- **January 2007** - Scientists at Wake Forest University led by Dr. Anthony Atala reported discovery of a new type of stem cell in amniotic fluid. This may potentially provide an alternative to embryonic stem cells for use in research and therapy.
- **16 February, 2007** - The California Institute for Regenerative Medicine became the biggest financial backer of human embryonic stem cell research in the United States when they awarded nearly $45 million in research grants.

3.1.2.1.7.2 *What are Stem Cells?*

Stem cells are primarel cells found in all multi-cellular organisms. They retain the ability to renew themselves through mitotic division and can differentiate into a diverse range of specialized cell types. Research in the human stem cell field grew out of findings by Canadian scientists Ernest A. McCulloch and James Till in the 1960s.

The two broad categories of mammalian stem cells are:

(a) Embryonic stem cells, derived from blastocysts.

(b) Adult stem cells, which are found in adult tissues.

In a developing embryo, stem cells can differentiate into all of the specialized embryonic tissues. In adult organisms, stem cells and progenitor cells act as a repair system for the body, replenishing specialized cells (Fig. 3.22).

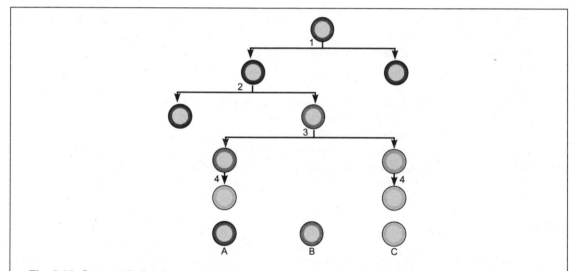

Fig. 3.22. Stem cell division and differentiation. A - stem cell; B - progenitor cell; C - differentiated cell; (1) symmetric stem cell division; (2) asymmetric stem cell division; (3) progenitor division; (4) terminal differentiation.

As stem cells can be grown and transformed into specialized cells with characteristics consistent with cells of various tissues such as muscles or nervous through cell culture, they have potential for use in medical therapies. In particular, embryonic cell lines, autologous embryonic stem cells generated through therapeutic cloning, and highly plastic adult stem cells from the umbilical cord blood or bone marrow, are touted as promising candidates.

3.1.2.1.7.3 *Properties of Stem Cells*

The stem cell possesses the following two properties:

- *Self-renewal* - It is an ability to go through numerous cycles division of while maintaining the undifferentiated state

- *Unlimited potency* - It is the capacity to differentiate into specialized cell types. In the strictest sense, this requires stem cells to be either **totipotent** or **pluripotent** - to be able to give rise to any mature cell type, although some **multipotent** and/or **unipotent** progenitor cells are sometimes referred to as stem cells.

  Using methods such as clonogenic assays, these properties can be illustrated *in vitro*. In these assays the progeny of single cells is characterized. However, *in vitro* culture conditions can alter the behaviour of cells, making it unclear whether the cells will behave in a similar manner *in vivo*.

3.1.2.1.7.4 Definitions of Potency

Potency specifies the differentiation potential (the potential to differentiate into different cell types) of the stem cell.

- **Totipotent.** These stem cells are produced from the fusion of an egg and sperm cell. Cells produced by the first few divisions of the fertilized egg are also totipotent. These cells can differentiate into embryonic and extraembryonic cell types.
- **Pluripotent.** These stem cells are the descendants of totipotent cells and can differentiate into cells derived from any of the three germ layers.
- **Multipotent.** These stem cells can produce only cells of a closely related family of cells (*e.g.,* haematopoietic stem cells differentiate into red blood cells, white blood cells, platelets, *etc.*).
- **Unipotent.** These cells can produce only one cell type, but have the property of self-renewal which distinguishes them from non-stem cells (*e.g.,* muscle stem cells).

Pluripotent, embryonic stem cells originate as inner mass cells within a blastocyst. The stem cells can become any tissue in the body, excluding a placenta. Only the cells of morulla are totipotent, able to become all tissues and a placenta.

3.1.2.1.7.5 Types of Stem Cells

A. Embryonic Stem Cells

Since their discovery, 18 years ago, the embryonic stem cells from the mouse have been intensely investigated .

ES cells are valuable scientifically as they have three main properties not found together in other cell lines.

(i) They appear to replicate indefinitely without undergoing senescence or mutation of the genetic material.

(ii) They appear genetically normal, both by a series of genetic tests and functionally as shown by the creation of mice with genomes derived entirely from ES cells.

(iii) These are developmentally totipotent in mice. ES cells can also differentiate into many cell types in tissue culture like neurons, blood cells, cardiac and skeletal muscle etc.

The isolation, culture and partial characterization of stem cells obtained from human embryos was reported in November 1998. It was demonstrated that they retain the ability to maintain their pluripotent character even after 4-5 months of culturing. It is felt that this feature of these cells could lead to cancerous growth and there is an evidence of occurrence of benign hyperproliferation.

Embryonic stem cell lines (ES cell lines). These are cultures of cells derived from the epiblast tissue of the inner cell mass (ICM) of a blastocyst or earlier morula stage embryos (Fig 3.23). A blastocyst is an early stage embryo. It is approximately four to five days old in humans and consists of 50–150 cells. ES cells are pluripotent and give rise during development to all derivatives of the three primary germ layers: ectoderm, endoderm and mesoderm. They can develop into each of the more than 200 cell types of the adult when given sufficient and necessary stimulation for a specific cell type. They do not contribute to the extra-embryonic membranes or the placenta.

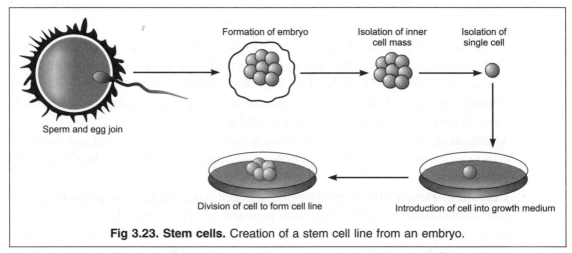

Fig 3.23. Stem cells. Creation of a stem cell line from an embryo.

Almost all research to date has taken place using mouse embryonic stem cells (mES) or human embryonic stem cells (hES). Both have the essential stem cell characteristics, yet they require very different environments in order to maintain an undifferentiated state. Mouse ES cells are grown on a layer of gelatin and require the presence of Leukemia Inhibitory Factor (LIF). Human ES cells are grown on a feeder layer of mouse embryonic fibroblasts (MEFs) and require the presence of basic Fibroblast Growth Factor (bFGF or FGF-2). Without optimal culture conditions or genetic manipulation, embryonic stem cells will rapidly differentiate.

A human embryonic stem cell is also defined by the presence of several transcription factors and cell surface proteins. The core regulatory network is formed by the transcription factors, Nanog, and Sox2. It ensures the suppression of genes that lead to differentiation and the maintenance of pluripotency. The glycolipids SSEA3 and SSEA4 and the keratan sulfate antigens Tra-1-60 and Tra-1-81 are the cell surface antigens most commonly used to identify hES. The molecular definition of a stem cell includes many more proteins and continues to be a topic of research.

After twenty years of research, there are no approved treatments or human trials using embryonic stem cells. ES cells, being totipotent require specific signals for correct differentiation —if injected directly into the body, ES cells will differentiate into many different types of cells.

B. Human Embryonic Germ Cells (HGCs)

Unlike EGCs, which are derived from the inner cell mass of a blastocyst, these are derived from the foetal tissue at somewhat later stage of development, from a region called the **gonodal ridge.** Cell types into which they can develop, may be slightly limited. There have been limited animal experiments on embryonic germ cells. These were isolated, cultured and partially characterized in November 1998. These were found to be capable of forming the three germ layers that make all the specific body organs.

The EGCs pose the following problems:

(i) They arrive further along in development, *i.e.*, 5-9 weeks. Therefore, it is difficult to tackle ethical questions and isolation is quite difficult.

(ii) They may provide committed neural progenitor, but feasibility of large scale sourcing and manufacture of products of these cells is questionable.

(iii) Behaviour of these cells, *in vivo* is not fully understood.

(iv) To avoid unwanted outcomes like ectopic tissue formation, tumour induction or abnormal development during their *in vitro* growth, significant research is required.

C. Adult Stem Cells

They are undifferentiated cells found throughout the body that divide to replenish dying cells and regenerate damaged tissues.

A great deal of adult stem cell research has focused on clarifying their capacity to divide or self-renew indefinitely and their differentiation potential. Many adult stem cells may be better classified as progenitor cells, due to their limited capacity for cellular differentiation (Fig. 3.24).

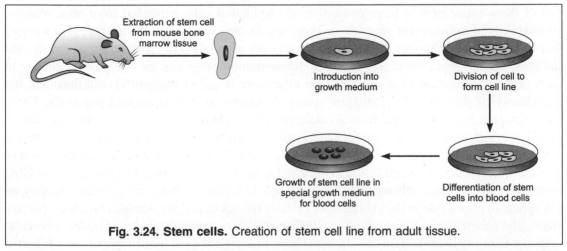

Fig. 3.24. Stem cells. Creation of stem cell line from adult tissue.

Stem cells found in developed tissue, regardless of the age of the organism, are known as ASCs, *e.g.,* the human haematopoietic stem cells of blood. Mesenchymal stem cells required for the maintenance of bone, muscle *etc.* have also been discovered. These are multipotent. They, however, have lower regeneracy than embryonic/germ stem cells, which are pluripotent.

At present, the use of mesenchymal stem cells in bone and cartilage replacement is undergoing FDA-approved clinical trials.

Nevertheless, specific multipotent or even unipotent adult progenitors may have potential utility in regenerative medicine. The use of adult stem cells in research and therapy is not as controversial as embryonic stem cells, because the production of adult stem cells does not require the destruction of an embryo. In contrast with the embryonic stem cell research, more US government funding has been provided for adult stem cell research. Adult stem cells can be isolated from a tissue sample obtained from an adult. They have mainly been studied in humans and model organisms such as mice and rats.

HSCs. There has been great interest in studying bone marrow stem cells because of its easy accessibility and the need to replace hematopoietic cells in many clinical situations. The bone marrow contains haematopoietic stem cells (HSCs) as well as stromal cells capable of differentiation into various lineages. HSCs generate all of the blood cells and can reconstitute the bone marrow after depletion caused by disease or irradiation. It is possible to collect HSCs directly from the bone marrow, from umbilical cord blood, and from circulating blood of individuals receiving cytokines, such as granulocyte-macrophage colony stimulating factor, which mobilizes HSCs. Depending on the tissue environment, bone marrow stromal cells, can generate chondrocytes, osteoblasts, adipocytes, myoblasts, and endothelial cell precursors.

The HSCs may be capable of giving rise to neurons, hepatocytes, and other cell types. Adult bone marrow cells injected into mice can contribute, in variable proportions, to hepatocyte repopulation of injured livers and to muscle cell production in injured muscle. When injected into the heart, a small proportion of these cells acquire a cardiac myoblast phenotype. It has been evidenced that a small number of hepatocytes in transplanted livers, and cardiac myocytes in transplanted hearts, may be derived from cells from the recipient's bone marrow. In the vascular bed of these transplants, a large proportion of endothelial cells generated from bone marrow stromal cells of the recipient are present. These results are contradictory to the accepted wisdom that cells of adult organisms, including stem cells, are committed to the generation of restricted lineages, and suggest instead that stem cell differentiation programs are not fixed. A change in stem cell differentiation from one cell type to another is called **transdifferentiation**, and the multiplicity of stem cell differentiation options is known as **developmental plasticity.** There have, however, been certain questions about the plasticity of HSCs. In some situations, transplanted HSCs fuse with host cells and transfer genetic material to them, thus, giving the false appearance of having transdifferentiated with generation of new cells in the host. The relative contribution of true transdifferentiation or cell fusion to the development of various mature cell types from HSC is not clear as yet. Also, although HSCs may be able to replace cells in damaged tissues, they do not appear to play a role in the maintenance of these tissues under physiologic conditions (steady state). The generation of tissue cells from HSCs probably occurs only at sites of injury, where the response to injury recruits stem cells from the bone marrow for local tissue repopulation. It is also possible that the main contribution of bone marrow-derived cells to the repair of non-haematopoietic tissues is not the generation of cells for these tissues. Instead, stem cells may produce growth factors and cytokines that act on the cells of the tissue to which they migrate, promoting injury repair and cell replication.

MAPCs. The adult bone marrow also harbors a heterogeneous population of stem cells, which appears to have very broad developmental capabilities. These cells, called **multipotent adult progenitor cells or MAPCs.** These have been isolated from postnatal human and rodent bone marrow. They proliferate in culture without senescence. They can differentiate into mesodermal, endodermal, and neuroectodermal cell types. MAPCs in fact are not confined to the bone marrow. They have been isolated from muscle, brain and skin, and can be made to differentiate into endothelium, neurons, hepatocytes and other cell types. MAPCs isolated from bone marrow, muscle, and brain have very similar gene expression profiles. This suggests that they may have a common origin. It has been proposed that MAPCs constitute a population of stem cells derived from, or closely related to, ES cell (*i.e.,* they may be the adult counterparts of ES cells). If this view is true the "trans differentiation" and "plasticity" of stem cells in adult tissues may actually represent the process of differentiation of multipotent ES-like cells into specific lineages. It has been observed that mouse bone marrow MAPCs, injected into blastocysts contribute to all somatic cell types. This demonstrates their pluripotency. It is not known whether a single type of adult bone marrow stem cell is capable of generating all tissue lineages or there are multiple types of bone marrow stem cells, each committed to differentiate into a specific tissue or a group of related tissues.

3.1.2.1.7.6 Role of Stem Cells in Tissue Homeostasis

Adult stem cells in addition to bone marrow cells that may migrate to various tissues after injury, reside permanently in most organs. These cells (known as tissue stem cells) can generate the mature cells of the organs in which they reside. However, their differentiation commitment can change when they are transplanted into a different tissue. The role of stem cells in various tissues such as liver, brain, muscle and renewing epithelium is given below:

- **Liver.** The liver contains stem cells in the canals of Hering. These are the junction between the biliary ductular system and parenchymal hepatocytes. These cells can give rise to a population of precursor cells known as **oval cells.** These are bipotential progenitors, capable of differentiating into hepatocytes and biliary cells. Liver stem cells in contrast to stem cells in proliferating tissues function as a secondary or reserve compartment. The cells get activated only when hepatocyte proliferation is blocked. In hepatic growth, processes such as liver regeneration after partial hepatectomy and in liver growth after most types of acute necrotizing injury, hepatocytes themselves readily replicate and the stem cell compartment is not activated. On the other hand, in the livers of patients recovering from fulminant hepatic failure, in liver carcinogenesis, and in some cases of chronic hepatitis and advanced liver cirrhosis, situations in which hepatocyte proliferation may be slow blocked oval, cell proliferation and differentiation are prominent.

- **Brain.** The brain contains nonproliferative tissue in mammals. However, the earlier dogma that no new neurons are generated in the brain of normal adult mammals is now known to be incorrect, because neurogenesis does occur in some areas of the adult brain. Neural stem cells (also known as neural precursor cells) have been identified in two areas of adult rodent brains, **the olfactory bulb,** and **the dentate gyrus** of the hippocampus. To identify these cells by histochemical methods, the intermediate filament protein *In situ* can be used as a marker. In some species of birds, particularly in canaries, in which neurogenesis in adult brains was first described, vocal center neurogenesis is required for the bird's ability to sing. The question arises whether, as in birds, newly generated neurons in the adult mammalian brain are functional and, more broadly, what the purpose of adult neurogenesis may be. It has now been evidenced that newly minted neurons in the mammalian hippocampus are functionally integrated into neural circuits. However, it is yet to be seen that neurogenesis in the adult brain increases "brain power" or improves the ability to sing.

- **Skeletal and cardiac muscle**. Even after injury the myocytes of skeletal muscle do not divide. Growth and regeneration of injured skeletal muscle occur instead by replication of **satellite cells.** These cells are located beneath the myocyte basal lamina. They constitute a reserve pool of stem cells that can generate differentiated myocytes after injury. Satellite cells placed in different tissue environments, can be osteogenic and adipogenic. Although it has been proposed that the heart may contain progenitor-like cells/stem cells have not been found in cardiac muscle.

- **Epithelial tissue.** Stem cells are also present in self-renewing epithelia. These are highly proliferative intermediate cells that constitute an amplifying compartment, and cells at various stages of differentiation. At the external surface of the epithelium terminally differentiated cells do not divide and are continuously lost. After injury, self-renewing epithelia reconstitute themselves by following three nonmutually exclusive strategies:

  (i)   Increasing the number of actively dividing stem cells

  (ii) Increasing the number of replications of cells in the amplifying compartment

  (iii) Decreasing the cell-cycle time for cell replication

3.1.2.1.7.7 Advantages and Disadvantages of ASCs

Advantages. These are useful in autologous and allogenic transplants. It is very easy to direct them to a desired fate. No ethical questions are raised in their use.

Disadvantages. It is not possible for them to produce virtually every cell type like ESC. They can complement but not replace ESC therapies. It is difficult to multiply them in large numbers.

3.1.2.1.7.8 Stem Cell Divisions

Stem cells are able to self-renew and produce cells that can differentiate. Stem cells to ensure self-renewal, undergo two types of cell divisions. **Symmetric division** gives rise to two identical daughter cells both endowed with stem cell properties. **Asymmetric division,** on the other hand, produces only one stem cell and a progenitor cell with limited self-renewal potential. Progenitors can go through several rounds of cell division before terminally differentiating into a mature cell. It is possible that the molecular distinction between symmetric and asymmetric divisions lies in differential segregation of cell membrane proteins between the daughter cells, however, there is no evidence for this mechanism. They simultaneously perpetuate themselves (self-renew) and generate differentiated progeny by asymmetric cell division. Many stem cells, however, can divide symmetrically, particularly during development and injury. The asymmetric division is, thus, not necessary for the identity of stem cells. It, however, is a tool that stem cells can use to maintain appropriate number of progeny. A key adaptation that is crucial for adult regenerative capacity may be the facultative use of symmetric or asymmetric divisions by stem cells.

Though asymmetric division allows stem cells to self-renew and produce differentiated progeny, it leaves the stem cell unable to expand in number. This can only be achieved by symmetric division.

Developmental and environmental signals to produce appropriate number of stem cells and differentiated daughter cells is controlled by balance between these two modes of division. It has been suggested that stem cells have the ability to switch between asymmetric and symmetric modes of divisions and in some diseased states, this becomes defective.

An alternative theory is that stem cells remain undifferentiated due to environmental cues in their particular niche. Stem cells differentiate when they leave that niche or no longer receive those signals. Studies in *Drosophila* germarium have identified the signals 'dpp' and 'adherin' junctions that prevent germarium stem cells from differentiating.

The signals that lead to reprogramming of cells to an embryonic-like state are also being investigated. These signal pathways involve several genes including the oncogene *c-Myc*. It has been indicated that transformation of mice cells with a combination of these anti-differentiation signals can reverse differentiation and may allow adult cells to become pluripotent. However, the need to transform these cells with an oncogene may prevent the use of this approach in therapy.

Asymmetric cell division. Two mechanisms control the asymmetric cell division:

(a) *Intrinsic.* It depends on the asymmetric partitioning of cell components that determine cell fate. To this category belong the regulated assembly of cell polarity factors and regulated segregation of cell fate determinants.

When the only difference between the daughter cells is their position relative to the stem cell niche, the format may initially have equivalent developmental potential. They may, however, acquire different fates because of exposure to varying external signals. There is no doubt that division is intrinsically symmetric, but it is asymmetric with respect to the fate of daughter cells. Such a model is provided by the asymmetric divisions of the *C. elegans* zygote, *e.g.,* asymmetric divisions of *Drosophila* neuroblast. Numb (evolutionary conserved cell fate determinant) is asymmetrically localized to daughter cells that are destined to differentiate.

(b) *Extrinsic. Drosophila* germline stem cells provide a classical example of an asymmetric division that is controlled by an extrinsic mechanism. This cell divides with a reproducible orientation to generate one daughter cell that remains in the stem cell niche retaining stem cell identity (these niches are created by cap cells in *Drosophila* ovary and hub cells in *Drosophila* testis) and another which is placed away from the niche and undergoes differentiation. Orientation of these asymmetric stem cell divisions controls the location of daughter cells and, thus, their access to extrinsic signals that regulate cell identity.

The division of mammalian neural progenitors also shows asymmetric division. Undifferentiated neural progenitors in the developing rodent cortex distribute Numb asymmetrically to precursors destined for neurogenesis. The inhibition of Notch signalling by Numb is crucial for neurogenesis in flies and in mammalian asymmetric division. Asymmetric segregation of Numb is found to be a common mode of control.

Symmetric cell divisions. These are common during wound healing and regeneration. These lead to an increase in cell number, *e.g.,* symmetric cell divisions during development occur in *C. elegans* germ line. The larval nematode has only two germline stem cells, but subsequently they proliferate to produce roughly 2,000 descendants in the adult gonad, including a pool of undifferentiated gametes. The germline stem cells during larval development and in adults are maintained by signalling from a niche formed by the "distil tip cell".

3.1.2.1.7.9 Stem Cell Niche

The populations of stem cells are established in 'niches'-specific anatomic locations that regulate how they participate in tissue generation, maintenance and repair.The niche differs among various tissues. For instance, in the gastrointestinal tract, they are located at the isthmus of stomach glands and at the base of the crypts of the colon (each colonic crypt is the clonal product of a single stem cell). Niches have been identified in other tissues, such as the bulge area of hair follicles and the limbus of the cornea. We first consider bone marrow stem cells and then discuss stem cells located in other tissues (tissue stem cells).

The niche saves stem cells from depletion, at the same time protect the host from over-exuberant stem cell proliferation.

Niche constitutes basic unit of tissue physiology and integrates signals that mediate the balanced response of stem cells to the needs of organisms. However, the niche may also induce pathologies by imposing aberrant function on stem cells or targets. The dynamic system necessary for sustaining tissues and for the ultimate design of stem cell therapeutics is created by the interplay between stem cells and their niche.

3.1.2.1.7.10 Role of Stem Cells in Medicine

The progress in stem cell research has led to better understanding of development potential of various types of stem cells. Adult stem cells have limited potential for differentiation. Embryonic stem cells are thought to be able to differentiate into almost any tissue. Some of the potential stem cell applications have been discussed below:

(i) Type I diabetes in children

Type I diabetes is an auto-immune disease characterized by the destruction of insulin producing cells in the pancreas. Experiments are being proformed to transplant cells from the pancreases of deceased organ donors into people with Type I diabetes. In some cases, the transplants relieve the recipients of the need to give themselves daily injections of insulin. But the effect wears off for most patients by two years.

Donated pancreases are scarce, so scientists hope to use stem cells to create insulin-producing cells. Moreover there are problems with respect to the toxicity of immunosuppressive drugs. People with Type I diabetes and their families were among the biggest backers of the effort to create a $3 billion program of stem cell research in California. The program's Chairman, the real estate developer Robert N. Klein, has a son with diabetes.

This problem can be solved by pleuripotent stem cells, instructed to differentiate into b-pancreatic cells. Scientists at a small California biotechnology company have developed a process to turn human embryonic into pancreatic cells that can produce insulin. The work by the company, Novocell, based in San Diego, is a step toward using embryonic stem cells to replace the insulin-producing cells that are destroyed by the body's immune system in people with Type I, or juvenile, diabetes.

According to Dr. Mark A. Magnuson, a professor at Vanderbilt University, it is now being thought that it will be possible to make insulin-producing pancreatic beta cells from human E.S. cells in a culture dish.

Dr. Magnuson, however, also said that in laboratory experiments the cells had not varied their insulin production much in response to the level of glucose, a key requirement for a beta cell. So more work is needed.

According to Emmanuel Baetge, insulin-producing cells had been derived by taking the embryonic stem cells and adding and subtracting various growth factors in a series of stages that mimicked the process that cells in an embryo go through to become a pancreatic cell. The process takes 16 to 20 days, he said.

According to Dr. Baetge the company hoped to begin testing its cells in animals in 2008 and that if all goes well clinical trials in human patients will begin in 2009.

(ii) Liver

Scientists have, for the first time, used adult bone marrow stem cells to regenerate healthy human liver tissue. In Dusseldorf, Germany, stem cells have been used to help quickly regenerate liver tissue in cancer patients unable to undergo surgery because removing the cancerous tissue would leave little liver to support the body.

According to Dr. Gunther Furst, a Professor of Radiology and co-author of the investigations, liver stem cells harvested from the patient's own bone marrow can further augment and accelerate the liver's natural capacity to regenerate itself.

The study included 13 patients with large central liver malignancies who were unable to undergo surgery because resection would leave less than 25 per cent of their total liver volume.

(iii) Nervous system diseases

The loss of nerve cells is the main cause of many diseases of nervous system. Mature nerve cells cannot divide to replace the lost ones. In Parkinson's disease, nerve cells that synthesize dopamine, die. In Alzheimers's disease, cells that are responsible for the production of certain neurotransmitters die. In amyotrophic lateral sclerosis, the motor nerve cells that activate muscles die. In spinal cord injury, trauma/stroke, many nerve cells die or are lost. In multiple sclerosis, glial cells are lost.

The only hope for treating individuals suffering from one or more of the above mentioned diseases is through creating new nerve tissue from pluripotent stem cells and thus, restore the function.

A project to treat human brain and other diseases by plundering the secrets of regeneration from creatures with remarkable powers of self-renewal, such as salamanders, newts, starfish and flatworms is being carried out by researchers at the McKnight Brain Institute of the University of Florida.

A salamander can be injured so that it loses its limbs or part of its spinal column. A few weeks later small growth can be seen at the place. The Regeneration project is to focus on unlocking the mysteries in living, simple organisms that sustain successful tissue and organ regeneration following injury and disease, and then applying this knowledge towards encouraging repair in the more complex human, where regeneration is not so simple. In terms of brain diseases, scientists may look at ways to mobilize and reinforce the body's own supply of adult stem cells to protect against or fight Alzheimer's and Parkinson's diseases, cancer, multiple sclerosis and traumatic injury.

Our body can heal itself, and that's why many of us live to be 80. Our arms or fingers cannot grow as had happended in the early stages of our development. There is a need to know what makes this to happen. It has been shown that humans possess some of the same genes and communication pathways used by some of nature's most remarkably regenerative animals. The scientists of UF McKnight Brain Institute have discovered more than 100 genes associated with all major human neurological diseases in a simple marine snail, as well as more than 600 genes that control development. It has been shown by the Brain Institute researchers that ordinary human brain cells can generate new brain tissue in mice.

(iv) Primary immunodeficiency diseases

More than 70 different forms of congenital and inherited deficiencies of the immune system have now been recognized. These include the diseases like SCID, Wiskott-Adrich syndrome and the autoimmune diseases. The immune deficiencies suffered as a result of infection with AIDS virus are also relevant here. These diseases are characterized by increased susceptibility to infections. These are often associated with anaemia, arthritis, diarrhoea and selected malignancies. The normal immune function can be restored by transplantation of stem cells reconstituted with the normal gene.

(v) Diseases of bone and cartilage

The bone and cartilage diseases due to cells that are either defective in structure or function can be corrected by appropriately differentiated stem cells. The genetic disorders like osteogenesis imperfecta, chondrodysplasias, damaged areas of joint cartilage, osteoarthritis or filling of large gaps in bone from fractures/surgery can be treated in this way.

(vi) *In vitro* model system

The lack of *in vitro* models has hampered the investigation of a number of human diseases. There are a number of pathogenic viruses *e.g.,* AIDS and hepatitis C, which grow on human or chimpanzee cells only. ES cells are expected to provide cell and tissue types that will help in this direction.

The bone marrow stem cells, representing a more committed stem cell, are used to secure patients following high dose chemotherapy. These recovered cells, however, are limited in their capacity to restore immune function completely.

(vii) Transplantation

To create an unlimited supply of cells, tissues even organs that could be used to restore function, pluripotent stem cells can be used. This will not require toxic immunosuppression or tissue matching compatibility. Such cells can be used in "transplantation therapies". These would be suitable for "universal" donation. Thus, transplantation would become safe, cost-effective and for a wide range of clinical disorders and all major organs.

It is presumed that injections of properly differentiated stem cells would restore the complete repertoire of immune response to patients undergoing bone marrow transplantation. This would in the long run allow the use of toxic chemotherapeutic regimens. Due to lack of an ability to restore marrow and immune functions, these have not been used as yet

The therapeutic effect can be gained by introducing genetic material that provides a missing or necessary protein, or causes a clinically relevant biochemical process, into an organ for therapeutic effect. It is important for gene-based therapies, that the desired gene be introduced into organ stem cells to achieve long-term expression and therapeutic effect. The stem cell should in turn differentiate and target correctly.

The world's first human-sheep chimera which has the body of a sheep and half human organ created by Professor Esmail Zanjani, of the University of Nevada was produced by injecting adult human cells into sheep's foetus. The sheep have 15% human cells and 85% animal cells and their evolution brings the prospect of animal organs being transplanted into humans. The development did revive criticisms about scientists playing God, with the possibility of silent viruses, which are harmless in animals, being introduced into the human race.

Dr. Zanjani also created a sheep liver which had a large proportion of human cells and eventually hopes to precisely match a sheep to a transplant patient, using their own stem cells to create their own flock of sheep.

The process involves extracting stem cells from the donor's bone marrow and injecting them into the peritoneum of a sheep's foetus. When the lamb is born, two months later, it would have a liver, heart, lungs and brain that are partly human and available for transplant.

Scientists at King's College, London, and the North East Stem Cell Institute in Newcastle have now applied to the HFEA, the Government's fertility watchdog, for permission to start work on the chimaeras. At present more than 8000 patients are waiting for an organ transplant in Britain alone, and two thirds of them are expected to die before an organ becomes available (Daily Mail, London).

(viii) Creation of Heart Valves

One percent of all newborns, or more than one million babies born worldwide each year, have heart problems. Such defects kill more babies in the United States in the first year of life than any other birth defect.

Defects in heart valves can be detected during pregnancy with ultrasound tests at about 20 weeks. According to Dr. Hoerstrup at least one-third of afflicted infants have problems that could be treated with replacement valves.

Conventional procedures to fix faulty heart valves have drawbacks. Artificial valves are prone to blood clots, and patients must take anti-clotting drugs for life.

Valves from human cadavers or animals can deteriorate, their replacement require repeated open-heart surgeries. That is especially true in children because such valves do not grow along with the body.

Human heart valves using fluid that cushions babies in the womb have been created. This provides an approach that may be used to repair defective hearts. About 250,000 patients worldwide have surgery to replace heart parts each year.

It is being tried to create new valves in the laboratory, while a pregnancy progresses and have them ready to implant in a baby with heart defects after it is born.

It has been predicted that people may one day be able to grow their own replacement heart parts, in some cases, even before they are born. These could lead to home grown heart valves that are more durable and effective than artificial or cadaver valves.

According to Dr. Simon Hoerstrup of the University of Zurich, this may open a whole new therapy concept to the treatment of congenital heart defects.

Japanese have also grown new heart valves in rabbits using cells from the animal's own tissue. According to Dr. Hayashida, who is with the National Cardiovascular Research Institute in Osaka, valves made from the patient's own cells might be able to grow with the patient.

The Swiss scientists use the foetus cells shed in amniotic fluid. This avoids controversy because it does not involve destroying embryos to get stem cells. For this purpose amniotic fluid was obtained through a needle inserted into the womb during amniocentesis. Foetal stem cells were isolated from the fluid, cultured in a laboratory dish, then placed on a mold shaped like a small ink pen and made of biodegradable plastic. It took four to six weeks to grow each of the 12 valves created in the experiment.

The next step was to see if they work in sheep, a two-year experiment.

According to Dr. John E. Mayer Jr., a Children's Hospital Boston heart surgeon the research would revolutionize care for people with valve diseases.

In one of Dr. Mayer's experiments, heart valves fashioned from stem cells harvested from sheep bone marrow appeared to function normally when implanted in sheep. A similar experiment used cells harvested from sheep arteries.

Dr. Hoerstrup said amniotic fluid was potentially a richer source of stem cells than other sources.

(ix) Improvement of vision

Attempts are being made to improve vision in rats suffering from a disease similar to age-related macular degeneration.

The scientists, at Oregon Health and Science University, used human embryonic stem cells that had spontaneously converted into the special cells that line the base of the retina. The cells, which support the light-sensing rod and cone cells above them, are damaged in some forms of macular degeneration.

Raymond Lund and his team injected the human cells into the retina of a special breed of rat in which the retina degenerates shortly after birth. The cells rescued the vision of rats, for three months after birth.

The injected human cells seemed to behave as retinal cells should, and the treated rats retained some six layers of rods and cones in their retinas, as much as half the normal value.

The human retinal lining cells, derived from different cultures of embryonic stem cells, were supplied by Advanced Cell Technology, a company with laboratories in Worcester, Mass.

Macular degeneration is a good candidate for embryonic stem cell therapy, at least in principle, because the eye is not closely monitored by the immune system. The cells therefore from another individual be could grafted with lesser risk of rejection than at other sites.

According to Robert Lanza of Advanced Cell Technology a bank of retinal cells derived from 100 embryonic stem cells of different immunological backgrounds should suffice to provide a good match to most of the American population .

Although the experiment suggested a possible treatment for age-related macular degeneration, a disease that affects one-third of the population older than 75, many uncertainties remain to be addressed. The disease treated in the rats is caused by a genetic defect and is not exactly the same as the human disease.

(x) Stems cells engineered from fats used to fight cancer

Mesenchymal stem cell have been derived from human adipose (fat) tissue and engineered into suicide genes that seek out and kill tumours similar to tiny homing missiles. This gene therapy approach is a novel way to attack small tumour metastases that evade current detection techniques and treatments (Cancer Research, July 1, 2007).

These fat derived stem cells could be exploited for personalized cell based therapeutics. Accordingly Cestmir Altaner. An Associate Professor in the Cancer Research Institute of the Slovak Academy of Sciences. Nearly everyone has some fat tissue they can spare and this tissue could be a source of cells for cancer treatment that can be adapted into specific vehicles for drug transport.

Mesenchymal stem cells help repair damaged tissue and organs by renewing injured cells. They are also found in the mass of normal cells that mix with cancer cells to make up a solid tumour. It is believed that mesenchymal stem cells "see" a tumour as a damaged organ and migrate to it, and so might be utilized as a "vehicle" for treatment that can find both primary tumours and small metastases. These stem cells also have some plasticity, which means they can be converted by the microenvironment of a given tissue into specialized cells.

After extracting the stem cells from human fat tissue, the researchers worked to find a less toxic way to treat colon cancer than the standard-of-care chemotherapy agent, 5-fluorouracil (5-FU), which can produce toxic side effects in normal cells. The number of mesenchymal stem cells was expanded in the laboratory and then a retrovirus vector was used to insert the gene cytosine deaminase into the cell. This gene can convert a less toxic drug, 5-fluorocytosine (5-FC), to 5-FU inside the stem cells, and the chemotherapy can then seep out into the tumour, producing a lethal by-stander effect.

In nude mice-animals with an inhibited immune system—engrafted with human colon cancer were given the engineered mesenchymal stem cells and then 5- FC was injected. It was found tumour growth was inhibited by up to 68.5 per cent in the animals, and none of the mice exhibited any signs of toxic side effects.

However, none of the animals remained tumour-free. "The procedure was quite effective even though they applied the stem cells just once. Obviously, repeated treatment will increase the efficacy, as would using this strategy in combination with other treatments," Altaner said.

Normal mesenchymal cells can be isolated from various sources, including bone marrow, but the yield is not nearly as great as what the researchers derived from fat tissue. Removal of fat tissue during surgery to remove a tumour would be simple, says Altaner. Liposuction could also be used to isolate mesenchymal stem cells can also be gathered and isolated through liposuction, and the cells frozen in liquid nitrogen for future therapeutic use.

3.1.2.1.7.11 Ethical Concerns of Stem Cell Research

Many moral and ethical questions have been raised by certain sections of society especially with regard to the use of human embryonic stem cells for research purposes. The question is that is it morally appropriate to destroy an embryo and do the benefits of reserach provide a justification for doing so?

The question arises that whether the human embryo has a significant moral status and thus needs to be protected from harm ? Another question is whether the cellular materials produced in these ways from embryos can be ethically utilize ?

According to some people, foetus is a human being from the time of conception, as it is genetically human and has the potential for development into a human individual. But other people, however, believe that embryos only represent human life and their life may be taken for the sake of saving and preserving the other lives in future.

There is also fear of the risks of stem cell implants, potential harms and misuses of stem cell research in the society. Many more questions, that are being raised have very confusing answers. These provide hicups for the progress of stem cell research. They, however, need to be answered.

3.1.2.1.7.12 Guidelines for the use of stem cells

Stem cells for research, whether from children, adults, aborted foetuses or embryos must be obtained under straight conditions to protect the interests of donors and also to assure the society that important boundaries are not being over-stepped to keep the research parameters ethically right and to assure highest quality of research and outcomes.

No ethical problems are posed by obtaining adult stem cells. Adequate protection regulation exists and the research with human subjects is regulated by various guidelines.

In case of the stem cells derived from aborted foetuses, regulations clearly separate the woman's decision to have an abortion from her decision to donate the tissue.

In case of isolation of ESC from the about to be discarded embryos of the *in vitro* fertilization (IVF) clinics, the formal, written consent is needed from the couples. As eggs cannot be frozen, the embryos are available which would at times be discarded off. There is an option for the persons with excess embryos of donating them to other infertile couples, destroying them or donating them for research purposes.

Informed consent requires that the woman or couple, with substantial understanding, without inappropriate influences, authorize the use of their spare embryo (s) for research.

To secure embryos, the policies should include the following:

(i) Women should not undergo extra cycles of ovulation and retrieval in order to produce spare embryos for research.

(ii) Analogous to the practice of organ donation, there must be a solid "wall" between personnel working with a couple who hope to get a baby and the personnel requesting embryos for stem cell purpose.

(iii) Women and men should not be paid to produce embryos.

(iv) To obtain the consent of both gamete donors, all reasonable efforts should be made.

If these norms are adhered to, the procurement of embryos for derivation of stem cells does not raise ethical problems.

3.1.2.1.7.13 Future of stem cell research

Stem cell research promises a vistas of opportunities in the field of medicine to cure the most fatal and wide-spread diseases.

The stem cells have the following roles :

(a) In curing diseases

(b) Studying developmental biology and gene therapy

(c) Tremendous potential to better our understanding of life as such.

The following are the emerging areas:

(i) **Nuclear reprogramming:** To achieve stem cells by de-differentiation of differentiated cells.

(ii) **Stem cells and cancer:** A better understanding of the disease and its connection with stem cells.

(iii) **Stem cells and the fight against aging:** The human quest for immortality has given rise to this field which promises to better our understanding of aging related diseases and their possible cures.

Now to work with human embryonic stem cells (hES) low passage cell lines are commercially available Millipore Corp. (Billerica, MA, USA; www.milligore.com) has begun marketing two hES cell lines, MEL-1 and MEL-2. Both lines have been tested extensively with Millipore's HEScGRO media for human embryonic stem cell culture. The cell lines are provided at early passage (p10-p12), ideal for maximizing the stable lifespan of the cell line and ensuring extended research time in a stable, pluripotent state. The cells grow as well-defined colonies, with compact cells displaying high nuclear to cytoplasmic ratios and prominent nucleoli.

Millipore has developed the MEL lines in collaboration with the Australian Stem Cell Centre (Clayton, Victoria; www.stemcelicentre.edu.au) under license from the Australian National Health and Medical Research Council. A June 2006 agreement between Millipore and the Australian Stem Cell Centre allows Millipore to distribute and market these stem cell lines in all countries outside of Australia.

3.1.2.1.8 Tissue Engineering

The human body is a superbly engineered biological specimen. It is a community of trillions of cells. From inception, through embryogenesis and development from the neonate to toddler, teenager, adult and the geriatric in sickness and in health, the human body is a phenomenon and mystery. Tissue engineering focuses on ways to make our lives better by developing products to help people. Tissue engineering has been variously defined as follows:

(i) Use of a combination of cells, engineering materials and suitable biochemical factors to improve or replace biological functions in an effort to effect the advancement of medicine.

(ii) Langer and Vacanti stated it to be "an interdisciplinary field that applies the principles of engineering and life sciences towards the development of biological substitutes that restore, maintain or improve tissue function".

(iii) The employment of natural biology of the system allowing for greater success in developing therapeutic strategies aimed at the replacement, repair, maintenance, and/or enhancement of tissue function.

(iv) Tissue engineering is the development and manipulation of laboratory-grown molecules, cells, tissues and organs to replace or support the function of defective or injured body parts.

Tissue engineering is also remarkable for its role in the fundamental and applied biomedical research, in areas such as cell and developmental biology, basic medical and veterinary sciences, transplantation science, biomaterials, biophysics and biomechanics and biomedical engineering.

3.1.2.1.8.1 Historical

The pioneer of the term and concept of tissue engineering was Y.C. Fung of the University of California at San Diego (UCDS), who led the USCD team that submitted an unsuccessful proposal to National Science Foundation (NSF) in 1985 for an Engineering Research Centre Program Award under the title "Centre for the Engineering of Living Tissue". Fung proposed the term again at a 1987 panel meeting considering future direction on Engineering, Bioengineering and Research to Aid the Handicapped Program. Strong interest in the concept led to a special panel meeting on tissue engineering at NSF in the fall of 1987 and then to the Lake Granlibakken, CA workshop of 1988, the first formal scientific meeting of this emerging field. This workshop, and succeeding symposia in 1990 and 1992, helped "seed" the scientific literature with this new concept. More widespread awareness of the term tissue engineering appears to have followed with the 1993 publication of a review article in Science by Robert Langer and Joseph Vacanti.

Tissue engineering is closely related to the field of cell transplantation and represent applications that repair or replace structural tissues (*i.e.,* bone, cartilage, blood vessels, bladder *etc.*). These are tissues that function by virtue of their mechanical properties. The term 'Reparative Medicine' is often used synonymously with Tissue Engineering, Reparative medicine sometimes referred to as regenerative medicine or tissue engineering, is the regeneration and remodeling of tissue *in vivo* for the purpose of repairing, replacing, maintaining or enhancing organ function and the engineering and growing of functional tissue substitutes *in vitro* for implantation *in vivo* as a biological substitute for damaged or diseased tissues and organs.

3.1.2.1.8.2 Strategies

The following three general strategies are employed in tissue engineering :

(i)   Use of isolated cells or cell substitutes.

(ii) Use of tissue-inducing substances.

(iii) Use of cells placed on or within matrices.

Tissue engineering encompasses cells and the engineering materials called **scaffolds.**

A. Cells

Tissue engineering solves problems by using living cells as engineering materials. These could be artificial skin that includes living fibroblasts, cartilage repaired with living chondrocytes, or other types of cells. From fluid tissues such as blood, cells are extracted by bulk methods, usually centrifugation or aspheresis. From solid tissues, extraction is more difficult. Usually the tissue is minced, and then digested with the enzymes such as trypsin or collagenase to remove the cellular matrix that holds the cells. After that the cells are free floating and extracted using centrifugation or aspheresis.

Digestion with trypsin is dependent on temperature. Higher temperatures digest the matrix faster, but create more damage. Collagenase is less temperature dependent, and damages fewer cells, but takes longer and is a more expensive reagent.

Categorization of cells: These are often categorized by their source.

- **Autologous** cells are obtained from the same individual to which they will be reimplanted.

Autologous cells do not pose much problems with regard to rejection and pathogen transmission, however in some cases, these might not be available. For example in genetic diseases suitable autologous cells are not available. Seriously ill or elderly persons, and the patients suffering from severe burns, may also not have sufficient quantities of autologous cells to establish useful cell lines. Moreover in such cases surgical operations might lead to infection or chronic pain at the donor site. Autologous solutions are not very quick as culturing of samples takes time before they can be used. Alternativly mesenchymal stem cells from bone marrow and fat can be used. These cells can differentiate into a variety of tissue types, including bone, cartilage, fat, and nerve. A large number of cells can be easily and quickly isolated from fat, thus, opening the potential for a large numbers of cells to be quickly and easily obtained. Several companies have been founded to capitalize on this technology, the most successful at this time being Cytori Therapeutics.

- **Allogenic** cells come from the body of a donor of the same species. While there are some ethical constraints in the use of human cells for *in vitro* studies, the employment of dermal fibroblasts from human foreskin has been demonstrated to be immunologically safe and, thus, a viable choice for tissue engineering of skin.
- **Xenogenic** cells are those isolated from individuals of another species. In particular animal cells have been used quite extensively in experiments aimed at the construction of cardiovascular implants.
- **Syngenic or isogenic** cells are isolated from genetically identical organisms, such as twins, clones, or highly inbred research animal models.
- **Primary** cells are obtained from an organism.
- **Secondary** cells are obtained from a cell bank.
- **Stem cells** are undifferentiated cells with the ability to divide in culture and give rise to different forms of specialized cells.

B. Engineering materials

Cells as found above are generally implanted or 'seeded' into an artificial structure capable of supporting three-dimensional tissue formation. Such devices, usually refered to as scaffolds, serve at least one of the following purposes:

- Enhance structural properties
- Deliver biochemical factors
- Deliver or allow delivery of vital cell nutrients
- Exert certain mechanical and biological influences to modify the behaviour of the cell.

In order to achieve the goal of tissue reconstruction, scaffolds must meet some specific requirements.

(i) A high porosity and an adequate pore size are necessary to facilitate cell seeding and diffusion throughout the whole structure of both cells and nutrients.

(ii) Scaffolds must be biodegradable as it needs to be absorbed by the surrounding tissues without the necessity of a surgical removal.

(iii) The rate at which degradation occurs has to coincide as much as possible with the rate of tissue formation: this means that while cells are fabricating their own natural matrix structure around themselves, the scaffold is able to provide structural integrity within the body and eventually it will break down leaving the neo-tissue, (newly formed tissue), which will take over the mechanical load.

Many different materials (natural and synthetic, biodegradable and permanent) have been investigated. Examples of these materials are collagen or some linear aliphatic polyesters like PLA (polylactic acid), PGA (polyglycolic acid) and PCL (polycapro lactone).

Scaffolds may also be constructed from natural materials: in particular different derivatives of the extracellular matrix have been studied to evaluate their ability to support cell growth. Proteic materials, such as collagen or fibrin, and polysaccharidic materials, like chitosan or glycosaminoglycans (GAGs), have all proved suitable in terms of cell compatibility, but some issues with potential immunogenicity still remains. Among GAGs, hyaluronic acid, possibly in combination with cross linking agents (*e.g.*, glutaraldehyde, water soluble carbodiimide, *etc.*), is one of the possible choices as scaffold material. Functionalized groups of scaffolds may be useful in the delivery of small molecules (drugs) to specific tissues.

A number of different methods has been described in literature for preparing porous structures to be employed as tissue engineering scaffolds. Each of these techniques presents its own advantages, but none is devoid of drawbacks.

Textile technology. These techniques include all the approaches that have been succesfully employed for the preparation of non-woven meshes of different polymers. The principal drawbacks are related to the difficulties of obtaining high porosity and regular pore size.

Solvent casing and particulate leaching (SCPL). This approach allows the preparation of porous structures with regular porosity, but with a limited thickness. Other than the small thickness range that can be obtained, another drawback of SCPL lies in its use of organic solvents which must be fully removed to avoid any possible damage to the cells seeded on the scaffold.

Gas foaming. To overcome the neccessity to use organic solvents and solid porogens, a technique using gas as a porogen has been developed. The main problems related to such techniques are caused by the excessive heat used during compression molding (which prohibits the incorporation of any temperature labile material into the polymer matrix) and by the fact that the pores do not form an interconnected structure.

Emulsification/Freeze-drying. This technique does not require the use of a solid porogen like SCPL. Emulsification and freeze-drying allow a faster preparation, if compared to SCPL because it does not require a time consuming leaching step. It, however, still requires the use of solvents, moreover pore size is relatively small and porosity is often irregular. Freeze-drying by itself is also a commonly employed technique for the fabrication of scaffolds. It is particularly used to prepare collagen sponges: collagen is dissolved into acidic solutions of acetic acid or hydrochloric acid that are cast into a mold, frozen with liquid nitrogen and then lyophilized.

Liquid-liquid phase separation. Similar to the previous technique, this procedure requires the use of a solvent with a low melting point that is easy to sublime. Liquid-liquid phase separation presents the same drawbacks of emulsification/freeze-drying.

CAD/CAM technology. Since most of the above described approaches are limited, when it comes to the control of porosity and pore size, computer assisted design and manufacturing techniques have been introduced to tissue engineering.

3.1.2.1.8.3 Developments in tissue engineering

It takes many years to develop protocols to solve many problems in clinical medicine. In some clinical domains, physicians and non-clinician researchers reached the stage of incorporating living cells into prototype tissue engineered clinical solutions years before the emergence of a generalized concept of tissue engineering. The following illustrate the depth and breadth of activity prior to the year 1987 :

Vascular grafts. It was more than a century ago that the surgeons explored the possibility of transplanting blood vessels. The renowned surgical researcher Alexis Carrel was awarded the 1912 Nobel prize in physiology or medicine for his demonstration of successful techniques for the anastomosis of blood vessels. The technique was then extended from the transplantation of vessels to the transplantation of entire solid organs. In the early 1950s, Voorhees demonstrated the first use of tubes of synthetic fabric as arterial prostheses. The concept of a resorbable vascular graft was introduced in the 1960s and the first fully-resorbable graft was reported in 1979. Improvement in the healing process of Dacron vascular grafts via pre-seeding with endothelial cells was reported in 1978. In 1982, the first attempt to create entirely biologic vascular structures *in vitro,* using collagen and cultured vascular cells was reported.

Progress todate in the development of tissue engineered vascular grafts has focused on mimicking the three layers of the normal muscular artery, using combinations of live cells, bioresorbable and non-bioresorbable scaffolding constructs. At present, there are no FDA approved live vascular replacement therapies. Several techniques are in pre-clinical trial, but face challenges that may prevent their widespread use/application in the near future.

Skin grafts. For centuries, physicians have attempted to cover severe wounds with grafts from a variety of sources, including cadavers and living humans. The early 20th century marked the beginning of immunologic basis for rejection of skin allografts and world war II for the research on skin replacement. It was during this period distinguished immunologist Peter Medawar made exemplanary contributions towards our understanding of the immunology of graft rejection, and to the *in vitro* culture of epithelial cells drawn from a patient. Despite these early successes, more efficient means of cultivation were needed to provide enough cells to sustain transplantation. Overgrowth of certain cell types, such as fibroblasts, suggested that existing culture techniques would not be effective in producing large quantities of cells. The growth of cultured cells in sheets in a Petri dish was demonstrated by Green and his Colleagues in 1979. They transferred intact rather than disaggregated cells to a graft wound bed. Bell and colleagues in the same year described the use of fibroblasts to condense a hydrated collagen lattice to a tissue-like structure potentially suitable for wound healing. These findings led to the first functional living skin equivalent (LSE) in 1981, consisting of fibroblasts suspended in a collagen-glycosaminoglycan matrix. All these lines of research, led to the development of commercial products by the end of 1980.

Skin has been the most successful of the tissue engineered therapies. Several products have completed clinical trials and have got the approval of FDA. Some have made transition to market. In 1997, the FDA approved TransCyst. It is a skin replacement tissue made up of dermal keratinocytes grown on a biodegradable polymer. TransCyste serves as a temporary wound cover for burns as new tissue forms. Apligraf, utilizes live human skin cells to form a dual layer skin to treat diabetic leg and foot ulcers.

Recent advances in skin tissue engineering have resulted in the following examples of products in the last couples of years:

- *EpiDex*. For treatment of chronic skin ulcers; EpiDex grafts are grown hair follicle stem cells.

- *Dermagraft*. Dermagraft is a cryopreserved human fibroblast-derived dermal substitute, composed of fibroblasts, extracellular matrix, and a bioabsorbable scaffold.

- *Integra*. It is a two-layered dressing and is completely acellular. The top layer serves as a temporary synthetic epidermis; the layer below serves as a foundation for re-growth of dermal tissue. The underlying layer is made of collagen fibres that act as a lattice through which the body can begin to align cells to recreate its own dermal tissue.

- *Epicel*. It is manufactured by Genzyme Biosurgery. It is the only autologous skin graft that can permanently close a burn wound. Its developement was based on original research done by Howard Green.

- *Alloderm*. It is a cell-seeded allogenic skin replacement. It consists of human dermal collagen seeded with allogenic fibroblasts. The material has recently been launched in the US - initially for patients with third degree burns and limited donor-site tissue.

- *Xenoderm*. It consists of porcine dermis. It is used as a replacement for burn wounds. LifeCell claims that there is a consistent incorporation of the matrix into the wound bed, low immunogenicity, and re-population with host cells.

Kidney. A number of investigators experimented sporadically with kidney transplantation during the early part of the 20th century. The real planned renal transplantation efforts, however, began only in the late 1940s. The first dialysis machine was developed by Kolff during the war years in the Netherlands. Its design was refined at the Peter Bent Brigham Hospital in Boston. It was used in patients for the first time in 1948. The availability of effective short-term dialysis facilitated progress on transplantation, that resulted in the successful transplant of a donated kidney from a twin by Murray and colleagues at Brigham and Women's Hospital in 1954. The advances in immunosuppression for transplantation and further developments in Kolff's dialysis machine, made these techniques appreciable for widespread, routine use. All those transformed the management of end-stage renal disease. There was, however, a limited supply of organs suitable for transplantation. This was coupled with therapeutic limitations of the dialysis machine. Both of these limitations motivated the formulation of the concept of the bioartificial kidney, which would mimic more faithfully the physiologic functions of the kidney and, thus, avoid the debilitating side effects of chronic dialysis.

During the late 1960s and early 1970s, Wolf experimented with combinations of kidney cells with hollow synthetic fibers as conduits for nutrients and waste. Subsequent work on growing liver cells on the outside of hollow fibers by Wolf and independently by Knazek led to the demonstration of hollow-fibre bioreactors. Galletti *et al.* (1980) furthered the development of bioartificial kidney concept through their research on hollow-fibre bioreactors employing renal epithelial cells.

Now the progress has been made to develop temporary replacement devices, such as extracorporeal kidney assist devices. Dr. David Humes, Chairman of the Department of Internal Medicine at the University of Michigan, Ann Arbor, has successfully completed *in vivo* testing of a Renal Tubule Assist Device (RAD) for treating acute renal failure. The only other treatments currently available for acute renal failure are haemofiltration and dialysis. Extracorporeal devices may improve the outcomes of these patients, while making treatment much less costly.

Pancreas/Islet Cells. The introduction of insulin more than 70 years ago had a miraculous impact on the lives of diabetes patients. It, however, became apparent that the highly imperfect glycemic control typical of routine insulin therapy was associated with severe long-term complications for a variety of organ systems. Though the first insulin pumps appeared in the 1960s, the sensor and control technology required for a complete "closed loop" system to mimic the adaptive character of physiologic glucose control was not available. The first pancreas transplant, in conjunction with a simultaneous kidney transplant, was performed by Lillehei in 1966.

Lacy reported a method for isolation of intact islets in 1967, and isolated islet cells were first transplanted in 1970. It was, however, without a solution to the problem of immune rejection. The use of microencapsulated islets as artificial beta cells was proposed by Chang in as early as the mid-1960s. During the 1970s, building on the work of Knazek, Chick and colleagues, developed a "hybrid artificial pancreas". It consisted of beta cells cultured on synthetic semipermeable hollow fibres. They demonstrated the ability of this device to restore glucose homeostasis in rats when connected to the circulatory system via shunt. Sun and colleagues reported similar work in the 1970s. It was followed by studies of implanted microencapsulated islets beginning in 1980. Since then investigation of different ways of "packaging" islet cells to provide effective and durable glycemic control continued.

The subsequent transplantation techniques of cells consisted of two major approaches:

(i) **Perfusion devices:** It was developed in 1970, but failed to make it to the clinical trial stage due to long-term biocompatibility issues, membrane breakage, and size limitations (a problem which plagues bioartificial implantable livers as well).

(ii) **Microencapsulation:** It has also been used for several decades. Refinements to this technique over time involved improving the biocompatibility of the encapsulating materials. It is being thought that widespread clinical application of microencapsulation techniques is just around the corner.

Several commercial tissue engineering approaches to repair/replace pancreatic function describe the current state of the field:

Metabolex is developing proprietary technologies for the microencapsulation of insulin-producing tissues using thin, conforming, biocompatible coatings.

BetaGene is developing innovative strategies for the detection and treatment of diabetes. This company was formed for the purpose of developing proprietary technology originating at the University of Texas Southwestern Medical Centre. BetaGene retains exclusive license to aspects of this technology including the use of engineered cell lines for the treatment of type I and type II diabetes and the use of these cells for bulk insulin production.

Circe Biomedical has developed the *PancreAssist System*. It consists of a single tubular membrane surrounded by insulin-producing islets, which are, in turn, enclosed within a disc-shaped housing. The tubular membrane is porous and permeable to glucose and insulin.

Liver. The first successful liver transplant was carried out by Starzl in 1967, but in the absence of an adequate supply of transplantable organs, persuites for alternative approaches to the replacement of hepatic function, continued. Over more than four decades, attempts were made to provide extracorporeal support to patients suffering from liver failure. Non-biological approaches that have been explored include haemodialysis, haemoperfusion over charcoal or resins or immobilized enzymes, plasmapheresis, and plasma exchange. However, these approaches resulted into limited success. It was possibly because of the complex synthetic and metabolic functions these systems could not adequately replace the liver.

Concurrently, several bioartificial liver (BAL) bioreactor designs have been developed in the laboratory to replace liver function. The basic design of a BAL device consists of circulating patient plasma extracoporeally through a bioreactor that houses/maintains liver cells (hepatocytes) sandwiched between artificial plates or capillaries. Bioreactor materials have either a spherical shape, large surface area, large pores or high porosity, or are hydrophilic and biocompatible. These features can help to achieve the required high density cultures of hepatocytes. However, there is no one material that possesses all of these desired properties. Attempts are being made to develop a support matrix that could provide all these properties in order to have a BAL with improved efficiency and effectiveness. Clinical trials of some BAL devices are already underway in the United States and the UK.

Bone and cartilage. A variety of materials generally perceived as chemically inert, such as various metals and alloys, have been used for many years to replace damaged bone or to provide support for healing bones. It, however, soon became clear that non-biologic materials did not remain biologically inert in the environment of the human body, rather elicit reactions. The intensity of these reactions was related to a variety of factors such as implantation site, the type of trauma at the time of surgery, and the precise material in use. These led to the development of bioactive materials, such as porous glass and hydroxyapatite ceramic in 1970s. They were examined as alternatives, as they elicited the formation of normal tissue on their surfaces.

In the recent years, tissue engineering of bone and cartilage has experienced relative success as compared to other tissue engineered products. Current strategies consist of two major approaches: transplantation of osteochondral grafts and transplantation of chondrocytes. Cell populations from cultured periosteum have the ability to form new bone and cartilage under the appropriate conditions and with the addition of the appropriate growth factors. Transplantation of osteochondral grafts, however, runs a possible risk of rejection in the recipient.

Current products and strategies include the following developments:

- *Carticel* by Genzyme Biosurgery of Cambridge, MA, which has received FDA approval to replace damaged knee cartilage. The product uses autologous chondrocytes and grows them in a biodegradable matrix, which is then transplanted in place of the damaged tissue.

- Stryker Biotech of Hopkinton, MA has an FDA approved OP-1 Implant under the Humanitarian Device Exemption (HDE). The OP-1 Implant is now available across the country and is indicated for use as an alternative to patient's own bone in recalcitrant long bone nonunions where an autograft is not feasible and alternative treatments have failed.

- Arnold Caplan of Case Western Reserve has performed mesenchymal stem cell (MSC), transplants in animals, and is working on similar transplants in humans. MSCs have been found to induce bone and connective tissue growth.

- Antonios Mikos at Rice University has developed an injectable copolymer that hardens quickly in the body and provides a surface to guide severed long bone regeneration.

Besides these, there had been attempts to exploit the growth and regenerative capacities inherent in bone. In a 1945 publication in Nature, Lacroix hypothesized that osteogenin, a substance in bone, was responsible for its growth. In 1965, Marshall Urist proved that some substance or combination of substances present in demineralized bone, which when transplanted, could induce growth of new bone. This led to the investigation of the precise factors responsible for the triggering of bone induction. During the 1970s and 80s, it was demonstrated that growth factors, termed bone morphogenetic proteins or BMPs, mediate the process. These act in a multistep cascade, highly reminiscent of embryonic bone morphogenesis. Reddi and colleagues developed techniques to isolate these proteins from the extracellular matrix of bone. The findings were promising for induction of cartilage as BMPs initially induce a cascade of chondrogenesis and could easily be called cartilage morphogenetic proteins (CCMI). There were continuous efforts to isolate, purify and proliferate BMPs in 1980s.

3.1.3 Plant Cell Culture

The origin of plant cell culture methods dates back to early 1900s when Gotleib Haberlandt showed that it is possible to maintain certain types of plant cells in a healthy condition in culture. Although the cells did not divide, Haberlandt's work set the direction for future research. The tissue culture methodology did not enter the modern era until the 1950s. In the middle of that decade, Folk Skoog Wisconsin in Madison discovered the cytokines. In addition, Robert Gauthert of Paris, found that auxins stimulate the division of callus cells. Since then, various mitogens have been discovered.

Today, a large number of commercially valuable chemicals are directly derived from plant material. These include about 25% of all prescription pharmaceuticals. These chemicals fall into five broad categories of applications *viz.,* drugs, flavours, perfumes, pigments and agriculturally useful chemicals. Biochemically, the great majority of compounds are secondary metabolites. Specific secondary products accumulate only in a restricted range of species and sometimes only in a single species or genus (*e.g.,* capsaicin, the pungent principle in chilli pepper, *Capsicum frutescence*), and often within a specific organ or tissue or at a specific stage of development in that species.

Tiny replicas of single parent are produced by growing small part of plant explant aseptically on a nutritional medium in a test tube. It happens in the absence of microorganisms and in the presence of a balanced diet of chemicals. Plants multiplied by tissue culture generally have identical genetic make ups, they will, thus, prove to be true clones. Tissue culture techniques can be used for most plants, provided that right chemical solution and processes are used. Generally, through tissue culture procedures, diseases transmitted from parent to offspring can be eliminated. This also helps in saving of time and space and can be cheaper than other means of vegetative propagation.

Plant tissue culture techniques are being used for the following purposes :

(i) To improve stress, salt, cold and chemical tolerance and to increase the nutrient contents in various crops

(ii) For the production of virus, pesticide and herbicide resistant plants

(iii) To bypass growing whole plants to get desired products.

The concept of totipotency is important in tissue culture. Many plant species can be regenerated from single cells or explants via tissue culture. This process can itself produce genetic change.

Individual wall-less cells (protoplasts) can be fused together to produce novel hybrids and it is also possible to produce genetically modified transgenic plants by the introduction of foreign DNA into protoplasts of cells from which the modified plants can be regenerated.

3.1.3.1 Methods of Plant Cell Culture

The area of plant biotechnology is now burgeoning. Various plant cell and tissue culture methods are available. The following techniques are used to develop new plant varieties :

(1) Clonal propagation

(2) Somaclonal variation

(3) Gametoclonal variation

(4) Protoplast fusion.

3.1.3.1.1 Clonal Propagation

It helps in the large-scale production of genetically uniform plant population. Many plant species can be clonally propagated from tissues using suitable culture media. The clones are generated at both the cellular and the organ levels.

Donor plant cells can be induced to multiply in culture and then the plants can be regenerated. Many cells or plants can be produced in a shorter span of time. Plants are produced from these cells by the following methods:

(i) *Organogenesis.* It is the induction of shoots or roots from roots and shoots respectively

(ii) *Somatic embryogenesis.* It is the induction of embryos that directly develop into plants. Somatic embryos develop from somatic cells without the involvement of gametic fusion and have shoot and root axes.

Meristematic cells are located at the tips of stems and roots, in leaf axils, in stems as cambium, on leaf margins, and in callus tissue. These usually initiate new growth by differentiating into leaves, stems, roots and other organs and tissues. However, this will only occur under the influence of factors like light, temperature, nutrients, hormones *etc*.

For culture purposes, the meristems are cut from an elite stock plant and sterilized to remove contaminating organisms. With the appropriate balance of plant growth regulators, temperature and light, the meristem will develop into a rooted plantlet. A major advantage of micropropagation is the elimination of viruses and other pathogens in commercial stock.

Tissue explants including pieces of leaves, shoots, roots or immature fruits are transferred to a solid medium. It contains nutrients and various growth regulators. The cells proliferate to form a non-differentiated mass called **callus**. By transferring the callus tissue to a suitable growth medium, embryo-formation can be induced. To select mutant varieties that demonstrate resistance to agent used, selective agents can be incorporated into the growth medium. Callus tissue can be transferred into a liquid culture medium and grown as a suspension culture in bioreactors for the production of secondary metabolites.

3.1.3.1.2 Somaclonal Variations

Somaclonal variants are created by regeneration of plants from callus, leaf tissue explants or plant protoplasts (wall-less cells). This permits modification of genes in cultured plant cells. This is used to develop new breeding lines with improved characteristics. This variability may provide a way of producing desirable new characteristics in established varieties of crop species.

Somaclonal variations have been widely described both in cultured cells and plants regenerated from culture. The somaclonal variation was recognized as ubiquitous only in 1981. Although the causes of somaclonal variations have not yet been fully elucidated, they arise as a consequence of DNA transposition. Certain examples involve changes in number and structure of chromosomes of regenerated plants resulting in morphologically abnormal plants and these are generally of no useful value.

It is, however, possible to obtain variants with apparently normal chromosome complements which show useful differences in agronomic characters. The common examples are crops such as potato, tomato, cereals, carrot and celery. The usefulness of somaclonal variants includes the fact that the changes occur at high frequency, some of the changes cannot be achieved by conventional technology. The frequency of variations that is induced by cell culture is greater than obtainable by subjecting plants to radiation or chemical treatments.

In corn, a new fully functional electrophoretic variant at the alcohol dehydrogenase locus has been found among somaclones and subsequently characterized as resulting from a single nucleotide substitution. In wheat, variants have been analyzed which affect traits such as height and alcohol dehydrogenase synthesis, as well as complex gene loci involved in the synthesis of grain amylases. Somaclonal analysis in corn and alfalfa has shown that cell culture greatly enhances the activation of transposable elements.

3.1.3.1.3 Gametoclonal Variations

These are created by regenerating plants from cultured microspore or pollen cells or pollen still contained within the anther. This genetic variation results from both recombination during cell meiosis and mutation induced by tissue culture.

3.1.3.1.4 Protoplast Fusion

On removal of the wall of the plant cells, protoplasts are produced. It is possible by fusion technologies to obtain hybrid cells of widely different origins and then to culture the heterokaryones to produce somatic hybrid plants. To produce hybrid plants, protoplasts from similar or different plant cells can be used. Potential applications are in tomato, potato, lettuce and alfalfa.

Protoplasts from cells of two-parent plants can be fused together with a multi-step chemical treatment using polyethylene glycol. It is now possible to use electrofusion technique to fuse protoplasts at relatively high frequencies. The procedures of electrofusion involve the following steps :

(a) The protoplasts are placed between electrodes in a non-conductive medium.

(b) First a high frequency alternating field is applied which causes protoplasts to align in chains by dielectrophoresis.

(c) Then direct current pulse is applied.

Fused protoplasts can be grown in a culture medium suitable for callus formation and plant generation. This is a suitable breeding technique for the species that can be regenerated from protoplasts. This procedure is useful for the following purposes:

(i) The combination of complete genomes *e.g.,* of sexually incompatible species,

(ii) Partial genome transfer from a donor to a recipient,

(iii) Manipulation of organelles of interest

Genes can now also be introduced in protoplasts. After introducing the required genes, the protoplasts are regrown in culture medium. To modify gene expression, sequences are first identified which regulate gene expression, and then it is exactly established which changes in nucleotide sequences are required to increase or decrease the level of expression. These sequences are generally present in the upstream (5′) to the coding region in DNA.

3.1.3.2 Materials and Equipment

Glassware, instruments, equipment *etc.* are required for tissue culture. Cultures are incubated under controlled temperature, relative humidity and light conditions. For this purpose, controlled environment chambers are used.

3.1.3.2.1 Sterilization

Before starting culture of cells, all instruments, glassware, equipment and nutrient media must be sterilized. Dry heat can be used for the sterilization of glassware and many instruments. A minimum exposure of 3 hours at 160 to 180°C will be sufficient. In plastic labware, only polypropylene can be autoclaved at 121°C (122 kPa gauge). Materials like polystyrene, polyvinyl chloride, polyethylene and styrene acrylonitrile cannot be autoclaved.

The hands of investigators should be relatively aseptically cleaned. It can be accomplished by washing with an antibacterial detergent followed by spraying with 40 to 70% ethanol. A variety of agents can be used for the sterilization of plant material.

3.1.3.2.2 Media Requirements

For tissue culture, various nutrients are required. They are inorganic salts, vitamins, carbon and energy sources and phytohormones. Organic acids, organic nitrogen compounds and complex substances are some of the important optional components.

 (i) *Inorganic salts.* Salts of Na, K, P, Ca, S and Mg are required in millimole quantities. The optimum concentration of each nutrient tissue and growth rate considerably differs for different types of cells. For most purposes, the medium should contain at least 25mM nitrate and potassium.

 (ii) *Carbon and energy source.* The standard source of carbon is glucose. However, in certain studies, other carbohydrates such as maltose have also been found to be suitable.

 (iii) *Vitamins.* Vitamins are required for growth and development. Thiamine is an essential requirement. Other vitamins required include ascorbic acid, choline chloride, foliate, p-amino-benzoic acid and riboflavin.

 (iv) *Phytohormones.* Plant tissue culture also requires cytokinins and auxins for growth.

 (v) *Organic nitrogen.* It is available from amino acids, such as asparagine and glutamine.

 (vi) *Organic acids.* Plant cells cannot utilize organic acids as carbon source. Supply of acids such as citrate, fumarate, malate or succinate allows plant cell growth on ammonium as the sole nitrogen source.

 (vii) *Complex substances.* These include malt extracts, protein hydrolysates, yeast extracts and many other plant preparations such as coconut milk and potato starch.

The substances mentioned above are mixed in various combinations. The required chemicals are mixed with demineralized water, then desired stock solutions are added and finally needed pH is adjusted. The medium is autoclaved at 121°C for 15 minutes cooled at room temperature and stored at 10°C.

Optimum nutrients and growth regulators for specific cultivars are provided by standard media. The commonly used medium is known as Murashige and Skoog (MS) medium. Specific media for each variety and three stages of culture have been developed by Murashige and his associates.

Several companies sell premixed powdered media. These premixes are available with or without sugar, agar or hormones. They are added to water, heated to dissolve agar, supplemented, pH-adjusted, where necessary and dispensed.

3.1.3.3 Setting up of Culture

Five stages have been proposed by Debergh and Maene (1981):

Stage 0 : Preparation and conditioning of mother plant

Stage 1 : Explant preparation for incubation.

Stage 2 : Multiplication and induction of shooting.

Stage 3 : Induction of roots

Stage 4 : Hardening of plant and transfer to soil

3.1.3.3.1 Explant Material

Cells from any plant species can be aseptically cultured on or in a nutrient medium. The materials in order of priority are sections of seedlings, swelling buds, stems and leaves.

By germinating sterilized seeds under sterile conditions, sterilized seedlings are prepared. For this purpose, following procedure is used :

(i)   Seeds are first soaked in 70% ethanol for 2 min.

(ii) After removing the alcohol, 20% commercial bleach containing 5% hypochlorite or 2% sodium hypochlorite is added and stirred for 15 to 20 min.

(iii) The floating seeds are removed and second bleaching treatment, if required, is repeated.

(iv) The bleached seeds are rinsed in sterilized distilled water and placed on double layers of sterilized filter paper in Petri dishes.

(v)   Sterilized distilled water is added and the dishes are sealed with parafilm.

(vi) The dishes are placed in dark at 25 to 28°C for seed germination. Any part of the seedling can be used as explant for callus formation.

Similarly, dipping in 70% ethanol for 1-3 min followed by 2-3 washings in sterilized distilled water will sterilize bud, leaf and stem section of plants. Bulbs, roots and tuber sections can also be sterilized in 70% ethanol followed by washing in 20% commercial bleach (5% sodium hypochlorite) solution.

3.1.3.3.2 Cell Culture

The cell culture is started as follows :

(i) 3-5 pieces (of about 0.5 cm each) from explant materials are placed into nutrient medium containing 2,4-D and 0.6-0.8% Difco Bacto agar in glass jars or flasks.

(ii) Generally plant cells in culture require an acidic pH (5.5–5.8).

(iii) A temperature between 26–28°C is considered optimum and proper aeration by shaking (100–150 rpm) is needed for optimum growth.

(iv) Generally light is not required for growing cell cultures. Thus, the sections should be incubated in the dark at 26–28°C for 3–4 weeks.

(v) The callus from 2–3 containers is then transferred into 20 ml of liquid medium in a flask and incubated at 26°C in the presence of light with continuous shaking at 150 rpm.

(vi) This cell suspension is used to regenerate plants.

3.1.3.3.3 Plant Regeneration from Cell Suspension Cultures

(i) 5–10 ml of cell suspensions is washed 2–4 times in hormone-free medium, placed on agar plates and incubated at 25–27°C in light.

(ii) Plantlets are then placed in agar medium in tubes or jars, and incubated at 25°C with 16.8 h photoperiod.

(iii) Rooted plants are transferred to soil or peat-vermiculite and provided with nutrient solution.

(iv) Plants are then grown in greenhouse and polyhouses.

3.1.3.3.4 Embryogenesis and Organogenesis

Somatic embryogenesis is the formation of embryo like structures capable of growing into new plants from cells. It can be artificially induced in tissue culture.

Embryogenesis can occur in either callus or suspension cultures, but the cells should be closer to the zygotic condition and show negligible differentiation.

In organogenesis, shoots can be directly induced on differentiated explants by culturing them on a medium having low concentrations of auxin and cytokinin in the suitable ratio.

Artificial Seeds

These can be produced by somatic embryogenesis. This involves the following steps :

(i) Somatic tissue is cultured on a growth medium, usually containing the synthetic plant growth regulator, 2, 4- D.

(ii) The cells pass through a period of rapid division and growth leads to the formation of a callus.

(iii) In this, some cells develop like fertilized egg into embryos, which are termed **"somatic embryos"** because they originate from vegetative, non-reproductive tissue.

(iv) These embryos under appropriate physical environments, with proper nutrients and growth regulators, can be separated from the callus mass and will develop like zygotic (fertilized) embryos.

For long-term storage, somatic embryos can be converted into artificial seeds by encapsulating in a protective covering. Embroys are mixed with Na-alginate and dropped in solution of Ca $(NO_3)_2$ or $CaCl_2$. This leads to encapsulation of embryo by coating of Ca alginate.

To modify the development of somatic embryos to prevent early germination, the technique has been refined. In this, the embryo is inhibited at the early cotyledonary stage of development with medium having 6–9% sucrose. This blocks early germination.

Somatic embryogenesis has been reported for a range of tropical plants including papaya, lemon, coffee, sweet potato, mango, cassava and sugarcane. Artificial seed would be a viable alternative for the propagation, exchange and germplasm preservation of these species.

3.1.3.3.5 Protoplast Fusion

The protoplast isolation and culturing can be performed by following the method in which aseptic conditions in a laminar airflow cabinet are required for all the operations (Fig. 3.25):

(i) Leaves are sterilized in 70% ethanol for 1 minute and placed in 15% CloroxTM bleach and Tween-80TM (1 drop/25 mL) for 15 minutes in a Petri dish. Washing is carried out with sterile distilled water (3x).

(ii) The lower epidermis is peeled off with a suitable forceps, the remaining portion is sectioned (1cm^2 surface area) and incubated in a 1:1 mixture of enzyme solution (Cellulase and Pectinase) and protoplast culture medium in a Petri dish. After sealing the dish with parafilm, incubation is carried out for 5–15h at 22°–24°C with shaking (50 rpm).

(iii) Protoplast release can be observed with an inverted microscope. Using a Pasteur pipette, protoplast suspension is removed and sieved through 60–80mm mesh in a centrifuge tube. Centrifugation is carried out at 50g for 6 min. and the supernatant is removed with a Pasteur pipette.

(iv) The protoplasts are immersed in 5–10ml of S1 solution, centrifugation is carried out at 50xg for 5 min. Washing is repeated once for fusion and twice for culturing.

(v) The protoplast concentration can be measured with a Coulter counter. About 10 protoplasts are rinsed in 1 ml of S1 solution for fusion or in protoplast culture medium for culturing.

(vi) For culturing, 150ml (6–8 droplets) protoplast suspension is placed in each Petri dish, and sealed with parafilm. Incubation is carried out at 25°C in diffused light in a box humidified by a blotting paper in a 1% $CuSO_4$ solution.

If suspension cell culture is available for protoplast fusion, then equal amounts of cell culture and enzyme solution are mixed in a Petri dish after sealing the dish with parafilm

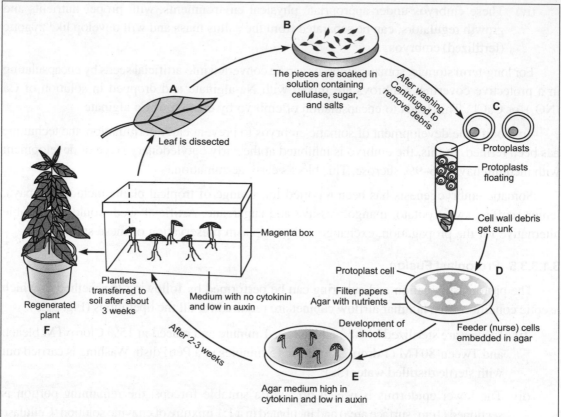

Fig. 3.25. Protoplast culture. Leaf cells have cytoplasmic compartment containing numerous chloroplasts, a large vacuole, and a nucleus. Plant tissue is incubated in a solution containing cellulose to remove a tough cellulase wall surrounding the plasma membrane. Sugars and salts are added to the solution to maintain osmotic balance and prevent the protoplasts from lysing. The protoplasts then obtained are placed on filter paper covering a layer of nurse factors and other molecules produced by the nurse cells can diffuse into the protoplasts. Microcolonies are formed by the division of protoplasts. The microcolonies are carefully transferred to a medium high in cytokinin and low in auxin. In about two to four weeks shoots appear. Then the cultured cells are transferred to a Magenta box, which contains root-inducing medium lacking cytokinin and low in auxin. Once the roots appear, the plantlets can be placed in soil, where they develop into regenerated plants (Modified from Watson, 2002).

Alternative Method

(i)   A 3 ml drop of sterile silicone-200 fluid is placed in a Petri dish and a cover slip is put on the drop.

(ii) Cell culture protoplasts are mixed with an equal volume of mesophyll protoplasts, and a drop of mixture is placed on the cover slip. This may settle onto the slip in about 5 min.

(iii) 450 ml of polyethylene glycol solution is added to the protoplast suspension drop by drop (about 6 drops). Protoplast agglutination or adhesion can be observed on an inverted microscope. This is incubated for 5-25 min at the room temperature.

(iv)   0.5 ml of eluting solution of the mixture is slowly added. After about 10 min, 1 ml of more solution is added.

(v)   500 ml of culture medium is left on the cover slip and 0.5 to 1 ml of medium is placed on the dish to maintain humidity. The dish is sealed and incubated.

3.1.3.4 Transformation

Plant cells exhibit a variety of characteristics that distinguish them from animal cells. These characteristics include the presence of a large central vacuole and cell wall *etc.* Along with these physical differences, another factor that distinguishes plant cells from animal cells, is that many varieties of full-grown adult plants can regenerate from single, modified protoplasts. More specifically when some species of plant cells are subjected to the removal of the cell wall by enzymatic treatment, they respond by synthesizing a new cell wall and eventually undergoing a series of cell divisions and developmental processes that result in the formation of a new adult plant.

Plants that can be cloned with relative ease include carrots, tomatoes, potatoes, petunias, and cabbage. The capability to grow a whole plant from a single cell means that it is possible to do the following :

(i)   The genetic manipulation of the cell

(ii)   Let the cell develops into a completely mature plant

(iii)   Examine the whole spectrum of physical and growth effects of the genetic manipulation within a relatively short period of time.

Such a process is far more straight forward than the parallel process in animal cells, which cannot be cloned into full-grown adults. Therefore, the results of any genetic manipulation are usually easier to examine in plants than in animals.

Various methods are used for the introduction of foreign genes into plant cells or protoplasts. These can be grouped as follows:

(a)   Vector mediated

(b)   Direct gene transfer (DGT)

3.1.3.4.1 Vector Mediated Gene Transfer

Most commonly *Agrobacterium* is used as gene vector. The natural ability of the soil bacteria *Agrobacterium tumefaciens* and *A. rhizogenes* is used to transfer genes into plant cells. Wild type *A. tumefaciens* induces tumours at sites of wounds of dicotyledonous plants. This process depends on the presence of a tumour inducing (T_i) plasmid in the bacteria (Fig. 3.26). The presence of phenolic compounds from the wounded cell induces transcription of genes in the virulence region of the plasmid.

(i)   Normally, wounded plant tissues release chemicals that stimulate cells to divide to form a **callus** which quickly covers the wound.

(ii)   Chemicals released by the wounded cells also stimulate *Agrobacterium* to infect the wound.

(iii) Once it has infected the plant, the plasmid in *Agrobacterium* causes development of the gall.

(iv) The plasmid (known as the T_i plasmid *i.e,* tumour-inducing), contains a short piece of DNA called **T-DNA.**

(v) This leaves the bacterium and enters the plant cells where it is inserted into the plant's own DNA.

(vi) As a result, it brings about the unregulated growth in plant cell.

(vii) The bacterium itself does not enter the cells, but can live between them, feeding on new products which the T-DNA directs the plant cells to make.

(viii) The plant cells are said to be transformed.

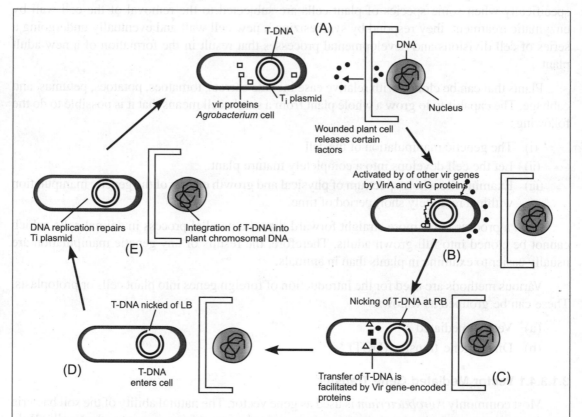

Fig. 3.26. Transfer of DNA from *Agrobacterium* into a plant cell. When a plant cell is wounded certain factors that stimulate transcription of the Vir genes on the T_i plasmid are released. These play a role in the transfer of T-DNA into the plant cell. Only the T-DNA region of the T_i plasmid is transferred to the plant cell. T-DNA is bordered by 25-bp imperfect repeats termed the *left border* (LB) and the *right border* (RB). Transfer begins with a nick in the DNA strand in the RB, then a nick occurs at the LB. It leads to the production of a single-stranded T-DNA molecule. By a mechanism that is still not completely worked out, the T-DNA molecule enters the plant cells. Here it integrates randomly into the chromosomal DNA. The single-stranded T-DNA region of the T_i plasmid is repaired by DNA replication, so the *Agrobacterium* has not lost any information by transferring DNA to the plant cell (After Watson, 1994).

Whole plants can be grown from single transformed cells using the **cloning techniques** (Fig. 3.27).

(i) The first stage involves growing the cells in a culture medium to produce an undifferentiated mass of cells called a **callus**.

(ii) The callus can be plated out onto nutrient agar and with the correct balance of hormones will produce shoots and roots and grow into a new plant. *Agrobacterium* can be made to infect the cut edges of discs punched out from leaves and to culture the discs on nutrient agar.

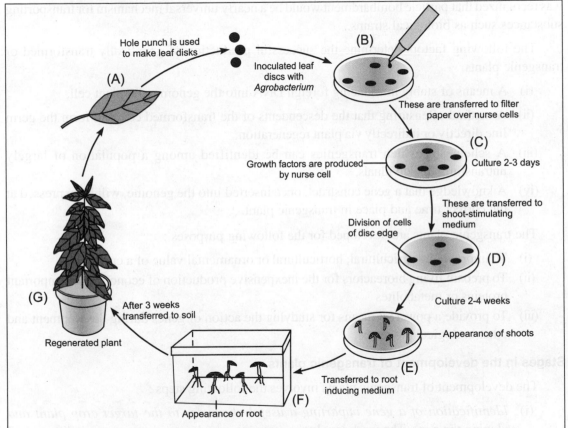

Fig. 3.27. Regeneration of *Agrobacterium* infected leaf discs. (A-B) Leaf disks are cut out and placed in a shallow dish. These are exposed to the solution of agrobacteria. **(C)** After some time, the leaf disks are transferred onto nurse cell medium. Wounded cells at the edge of the disk release, some compounds that allow the agrobacteria to infect the cells. **(D)** The plant discs are cultured in a medium containing an antibiotic such as cefotaxime that kills *Agrobacterium*. **(F)** Growth of plantlets is a chamber. **(G)** Plantlets transferred to a pot.

3.1.3.4.2 Direct Gene Transfer

Efforts have also been made to develop direct gene transfer (DGT) techniques to introduce DNA into plant cells without the need of an intermediate host. One DGT method involves firing small particles coated with DNA through the cell wall and the plasma membrane.

Most of the DGT methods require the removal of plant cell wall to form protoplasts.

Direct gene transfer has also proved to be a powerful tool for tailoring foreign genes for expression in plants, using a method that allows evaluation within days without the need to use cells capable of further divisions. Researchers at Cornell University developed a range of devices to accelerate tungsten microprojectiles to velocities (approximately 150 m/s) sufficient to penetrate plant cell walls and membranes. Onion epidermal cells penetrated by small number of projectiles remained viable and up to 40% of cells penetrated by tungsten microprojectiles coated with TMV RNA developed distinctive crystalline inclusions, indicating expression of viral nucleic acids. It was recognized that particle bombardment would be a nearly universal mechanism for transporting substances such as biological strains.

The following factors determine the successful production of genetically transformed or transgenic plants.

(i) A means of stably introducing foreign DNA into the genome of a plant cell,

(ii) A method of ensuring that the descendents of the transformed cell will form the germ line directly or indirectly via plant regeneration,

(iii) A scheme by which transgenics can be identified among a population of largely untransformed individuals,

(iv) A knowledge that a gene construct, once inserted into the genome, will be expressed at the correct time and place in transgenic plant.

The transgenic plants are developed for the following purposes :

(i) To improve the agricultural, horticultural or ornamental value of a crop plant

(ii) To prepare living bioreactors for the inexpensive production of economically important proteins or metabolites

(iii) To provide a powerful means for studying the action of genes during development and other biological processes.

Stages in the development of transgenic plants

The development of transgenic plants involves the following steps :

(i) *Identification of a gene imparting a useful character to the target crop plant and cloning the gene.* The work involves constructing and screening a genomic library and identifying the gene with a probe/marker DNA using standard protocols.

(ii) *Modification of the target gene for expression in crop plants.* This involves selecting and cloning a promoter and selecting the appropriate gene terminator sequences. The steps in the modification of target gene involve incorporation of all these sequences into the target plasmid and marker genes along with the right positioning of the promoter and the terminator sequences.

(iii) *Incorporation of the modified gene construct into the target plant genome.* This involves transferring the modified gene into the plant through a suitable method: *Agrobacterium*, particle gun or direct gene uptake through protoplasts.

(iv) *Regeneration of whole plants capable of transmitting the incorporated target gene to the next generation.* For this process, a selection agent, usually an antibiotic-resistant marker gene is required. This is initially incorporated into the modified gene construct to facilitate the selection process. Cells not containing the resistant marker gene would be destroyed under the selection pressure deliberately built into the experimental design for the selection of target plants.

(v) *Confirmation of integration of insert.* Molecular analysis techniques such as PCR, Southern, Western blotting *etc.* are used for confirming the integration of the gene of interest.

(vi) Inheritance pattern of gene of interest in the transgenic progeny.

(vii) Performance of transgenic plants under containment.

(viii) Field trial of transgenic plants.

Questions

1. Describe in detail the requirements of microbial cultures.

2. Describe the following :
   (a) Generation time
   (b) Growth pattern
   (c) Metabolism

3. Describe the biochemical pathways involved in glucose metabolism.

4. What do you mean by fermentation and respiration ? Describe in detail.

5. How will you culture animal cells in laboratory ? Describe

6. Explain the following in detail
   (a) Cryopreservation
   (b) Contamination
   (c) Medium components used for animal cell culture

7. Why do we culture plant cells ? Give the requirements for culturing plant cells.

8. Describe the following :
   (a) Protoplast culture
   (b) Plant cell transformation

4

Specialized Techniques in Biotechnology-II: Nucleic Acid Based Technologies

4.1 RECOMBINANT DNA TECHNOLOGY

The basic technique of genetic engineering was started in the early 1970s. It involves introducing a new gene into the DNA of another cell. The gene may be newly synthesized or transferred from another organism. The genetic engineering turns the recipient cell (bacterium, eukaryotic) into a living factory for the production of protein, the inserted gene encodes.

Today it is a routine to obtain copies of any gene. In some cases, only a single original molecule is required. It in turn is used to make many identical copies of a molecule. This process is called **cloning**.

This process is based on the use of some vector/vehicle such as plasmids or bacteriophages. The plasmid or phage is known as the **'vector'** or **'cloning vector'**, because it acts as a carrier for the DNA to be cloned in the host. A number of vectors have been developed for this purpose.

Recombinant DNA

The piece of DNA to be cloned is combined with either a plasmid or the DNA of a phage. This modified plasmid or phage DNA is called **recombinant DNA.**

If it is inserted into a bacterium, it will replicate (clone) itself. The multiplication of bacterium also replicates the recombinant DNA. It is possible to separate the cloned DNA from the plasmid or phage. The new gene while inside the bacterium, may be active and used to make a useful protein *e.g.,* human insulin.

The logical sequence of activities for this would be to :

   (i) produce a "DNA library" of an organism

   (ii) identify individual genes of interest

(iii)   produce a copy or (preferably) many copies of the gene

(iv)   insert the gene into the desired organism and regulate the expression of gene in a useful way

---

MILESTONES IN RECOMBINANT DNA

Many researchers have contributed to advances in the science of recombinant DNA. Listed below are only a few of the key developments (Modified fromWitkowski, 1988).

1944	:	Oswald Avery discovered that DNA carries the genetic code
Late 40s	:	Electron microscope became available to scientists at reasonable cost.
1952	:	Alfred Hershey and Martha Chase applied radioactive labeling to confirm that DNA carries the genetic code.
1953	:	James Watson and Francis Crick discovered the double helix structure of DNA.
1955	:	Severo Ochoa synthesized RNA
1956	:	Arthur Korenberg synthesized DNA
1960	:	mRNA discovered by researchers at Pasteur Institute in Paris
1970	:	Temin and Mizutani and Baltimore independently discovered reverse transcriptase.
1971	:	Danna and Nathans made the 1st practical use of restriction enzyme for mapping SV40 DNA.
1972	:	The Recombinant DNA Era began.
	:	Paul Berg, Symons and Jackson constructed the Ist man-made recombinant DNA-circular SV40 DNA molecules containing lambda phage genes and the galactose operon of *E. coli*.
	:	Mertz and Davis used T4 ligase to join DNA molecules cut with a restriction endonuclease.
	:	Anand Chakrabarty created oil – digesting *Pseudomonas* through cell fusion of plasmids. Although not rDNA, a major genetic engineering achievement which ultimately led to landmark (1980) Supreme court USA decision.
1973	:	Cohen, Chang, Boyer and Helling cloned a gene using the plasmid method of rDNA.
	:	Paul Berg demonstrated viral method of rDNA
	:	Sharp, Sugden and Sambrook deviced a simple technique to lighten the burden of rDNA experiment – agarose – ethidium bromide electrophoresis.
	:	O. Wesley and McBride and Ozer demonstrated that chromosomes from one species can be made to function in another.
	:	Graham and Erb demonstrated that the mammalian cells can be transfected with naked DNA.
	:	Lobban and Kaiser showed how to produce 3 extensions of homopolymeric tails and how to ligate the resulting molecules.
1974	:	Fears of possible dangerous consequences of rDNA technology leads to the suggestion of a self-imposed moratorium on certain experiments by a group of eminent molecular biologists ("the Berg letter")
1975	:	A year of techniques

	:	Southern gave the Southern Blotting technique for DNA detection.
	:	Kohler and Milstein gave the technique for monoclonal antibody production.
	:	Asilomar conference set stage for self-regulation of gene-splicing experiments by scientific community.
1976	:	Maniatis *et al.*, produced rabbit (eukaryotic) B-globin gene by cDNA cloning
	:	Struhl *et al.*, studied expression of eukaryotic genes (yeast) in *E. coli*
	:	Kan, Dozy and Golbus applied molecular hybridization technique for diagnosis of a human inherited disorder – β-thalassemia.
	:	The Ist guideline for rDNA research were published by the department of Health, Education and Welfare.
1977	:	Maxam Gilbert and Sanger *et al.*, gave methods of DNA sequencing – marked the start of a new epoch in the study of genetics.
	:	Genetech produced human brain hormone somatostatin through application of rDNA.
1978	:	Collins and Hohn developed COSMIDS.
	:	Hutchison *et al.*, demonstrated introduction of site-specific mutation.
	:	Harvard researchers produced rat insulin through RDT.
	:	Genetech produced human insulin via plasmid method of rDNA.
	:	NIH proposed loosening guidelines.
1979	:	Solomon and Bodmer suggested that as few as 200 RFLPs could be used as the basis for a linkage map of the human genome.
	:	Genetech produced HGH and thymosin –1 through RDT.
1980	:	Wigler *et al.*, demonstrated introduction of non-selectable genes into mammalian cells by co-transfection with a selectable marker.
	:	A US patent for cloning was awarded to Cohn, Boyer and the Board of Trustees of Stanford University.
1981	:	3 firms announce plans to market automated gene-synthesis machines.
1983	:	Herrera *et al.*, and Bevan *et al.*, used Ti plasmids as vectors for transformation of plant cells.
1984	:	Schwartz and Cantor demonstrated separation of yeast chromosome sized DNAs (~ 500Kbp) by pulsed field gradient gel electrophoresis.
1985	:	Saiki *et al.*, introduced PCR in print.
	:	Smithies *et al.*, demonstrated insertion of DNA sequences into specific sites in the mammalian genome.
1986	:	Burke, Carle and Oslon used YACs for cloning very large DNA fragments.
	:	Hoffman, Brown and Kunkel used RDT to study human inherited disease – Duchenne muscular dystrophy.
1993	:	Kary Mullis, who developed PCR, was awarded Nobel Prize.

4.1.1 A DNA Library

DNA library is the first task in recombinant DNA technology. It is a readily accessible, easily duplicable assemblage of all the DNA of a particular organism. All the genes of the organism are there, but it is quite difficult to find and study them. A DNA library organizes the DNA in a way that researchers can use. A DNA library also allows "molecular photocopying" of the DNA, so that the thousands to billions of gene copies are obtained for the experiments. Restriction enzymes, plasmids and bacteria have provided us with the filing systems, catalogues and shelf space needed for the DNA libraries.

Many bacteria produce restriction enzymes that cleave DNA at particular nucleotide sequences. Restriction enzymes in nature defend bacteria against invasion bacteriophage by that cuts apart the phage DNA. The host bacteria protect their own DNA against being cut by the enzymes, probably by methylation of some of the DNA bases. Restriction enzymes, thus, "restrict" phage infections to the types of phage whose DNA is not cut apart by the enzymes.

A number of restriction enzymes have been isolated from various species of bacteria. Each enzyme cuts DNA apart at a different nucleotide sequence. The specificity is utilized by geneticists to identify and isolate segments of DNA from many organisms, including humans (Table 4.1).

More than 200 different enzymes are known today. These attack 230 different sequences.

Table 4.1. Some Examples of the Frequently Used Restriction Endonucleases. (The sequences shown are that of the positive strand of DNA from the 5′ to 3′ direction).

Enzyme	Recognition site	Blunt or cohesive end
Hind III	A/AGCTT	cohesive
Bam H I	G/GATCC	cohesive
Bgl II	A/GATCT	cohesive
Pst I	CTGCA/G	cohesive
Sma I	CCC/GGG	blunt
Pvu II	CAG/CTG	blunt
Alu I	AG/CT	blunt
Not I	GC/GGCCGC	cohesive

The name of each enzyme is after the bacterium from which it comes. Generally, the base sequence is six bases long and palindromic *i.e.,* it reads the same in both the directions. A palindrome in English is a word that reads the same forward and backward such as MADAM. The two complementary strands of DNA run in opposite directions. Some restriction enzymes make a staggered cut with single-stranded ends (*e.g., EcoRI*). These are 'sticky ends' as they can be used to rejoin fragments of DNA. They stick together by forming hydrogen bonds to complementary sticky ends from other DNA molecules cut by the same restriction enzyme. The sticky ends TTAA are produced by *EcoRI* (Fig. 4.1)   Some restriction enzymes produce **blunt ends** *e.g.,* Hind II.

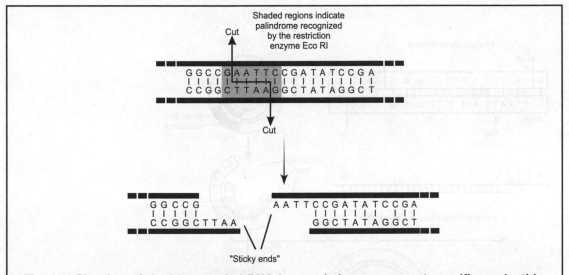

Fig. 4.1. Cleaving of double-stranded DNA by restriction enzymes at specific nucleotide sequence. DNA is cut by most useful restriction enzymes, such as *EcoRI* at palindromes. The nucleotide sequences are read the same on one DNA strand (GAATTC) in the diagram as it does in the opposite direction on the other strand. When a double-stranded DNA molecule is cut by *EcoRI*, two short, single-stranded "sticky ends" remain : AATT on one strand and TTAA on the other. Because these sticky ends are complementary, they can base pair, temporarily holding these strands, or any strand with the same sticky end sequences, together again.

In this way the DNA of any organism can be cut into pieces of different sizes. The pieces are known as **restriction fragments**. The different lengths of these fragments depend on the restriction enzyme used and on where the particular base sequences recognized by the enzymes are located.

Each nucleotide in a piece of DNA carries a phosphate group which is negatively charged. Thus, different lengths of DNA carry different total charges. On the basis of these differences, pieces of DNA of different lengths can be separated by placing them in an electrical field and allowing them to migrate to the positive electrode. This is done in a gel. This technique is known as **gel electrophoresis.** The gel is made up of agarose (for very large fragments) or polyacrylamide (for smaller fragments). As the DNA is colourless, its final position is revealed by staining or by using radioactive DNA and carrying out **autoradiography.** It involves exposure of the gel to photographic film. The film is blackened by radiation showing the location of DNA.

4.1.1.1 Human Gene Library

Suppose the DNA is isolated from a human source, say white blood cells and is cut apart with the restriction enzyme *EcoRI*, the region where GAATTC / CTTAAG, pairings are there, the human DNA will be cleaved leaving single-stranded AATT and TTAA "sticky ends" protruding.

The next step is to isolate many copies of a bacterial plasmid that has an easily identifiable "marker gene," such as a gene that confers resistance to an antibiotic called **ampicillin**. The plasmids are then exposed to the same restriction enzyme. The plasmids, too, will be cut open with AATT and TTAA sticky ends protruding (Fig. 4.2).

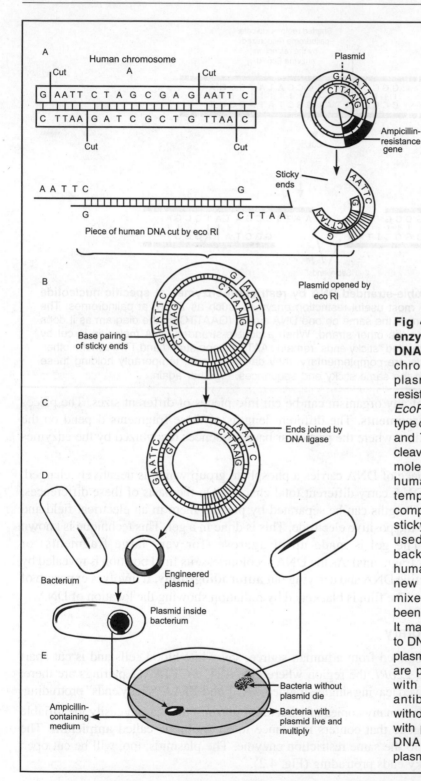

Fig 4.2. Role of restriction enzymes in building a human DNA library. (A) Both human chromosomes and bacterial plasmids (with the ampicillin-resistance gene) are cut with the *EcoRI* restriction enzyme. Both type of DNA, therefore, have AATT and TTAA sticky ends. **(B)** The cleaved human and plasmid DNA molecules are mixed. Plasmid-human combinations are temporarily held together by complementary base pairing of sticky ends. **(C)** DNA ligase is used to bind together the backbones of the plasmid-human combinations. **(D)** The new plasmid-human rings are mixed with bacteria that have been treated with calcium salts. It makes the bacteria permeable to DNA. Some bacteria take up a plasmid human ring. **(E)** Bacteria are placed on culture dish filled with medium containing the antibiotic ampicillin. Bacteria without plasmids die and those with plasmids survive. A human DNA library is constituted by plasmids within the bacteria.

The human DNA fragments and opened plasmids are now mixed together. The sticky ends will form hydrogen bonds, forming following combinations

(a) human-human,

(b) plasmid-plasmid and

(c) plasmid-human DNA combinations.

Then DNA ligase enzymes are added. DNA ligase binds the sugar-phosphate backbones together, inserting human DNA into plasmids (Fig. 4.3). Ideally, only a relatively small piece of human DNA is received by each plasmid. Millions or billions of plasmids collectively would incorporate DNA from the entire human genome.

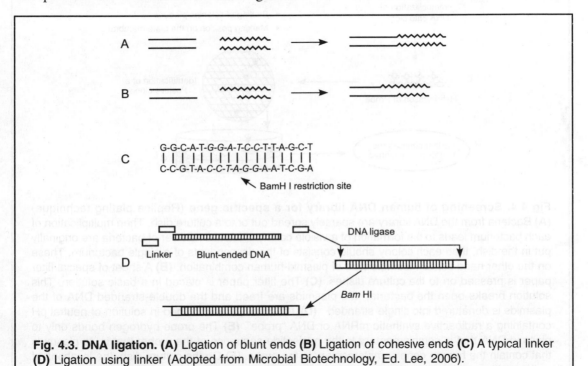

Fig. 4.3. DNA ligation. (A) Ligation of blunt ends **(B)** Ligation of cohesive ends **(C)** A typical linker **(D)** Ligation using linker (Adopted from Microbial Biotechnology, Ed. Lee, 2006).

(i) The new rings of plasmid-plus-human DNA are mixed with bacteria treated with calcium salts to make them permeable to DNA.

(ii) The bacteria take up the plasmid-human DNA. Usually 100 to 1000 times more bacteria than plasmids are used, so that no individual bacterium ends up with more than one plasmid-human DNA molecule.

(iii) This procedure also ensures that most of the bacteria don't have any plasmid-human DNA combination at all.

(iv) The ampicillin-resistance marker gene becomes useful at this moment. The bacteria are placed on a culture dish-containing medium with ampicillin (Fig. 4.4).

(v) The antibiotic kills off all the bacteria that have not taken up a plasmid.

(vi) These result in a population of bacteria, all with a plasmid and some with recombined plasmid-human DNA.

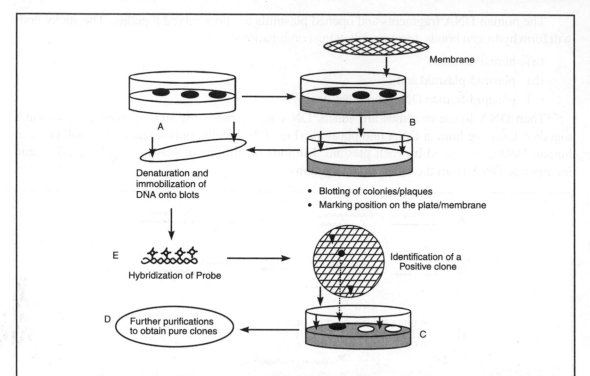

Fig 4.4. Screening of human DNA library for a specific gene (Replica plating technique)
(A) Bacteria from the DNA library are sparsely spread out onto a culture dish. Then multiplication of each bacterium leads to the formation of a visible colony. If a small number of bacteria are originally put in the dish, then each colony should consists of the descendants of a single bacterium. These on the other hand contain a single type of plasmid-human combination. **(B)** A sheet of special filter paper is pressed on to the culture dishes. **(C)** The filter paper is placed in a basic solution. This solution breaks open the bacteria, the plasmids are freed and the double-stranded DNA of the plasmids is denatured into single stranded. **(D)** The paper is now bathed in solution of neutral pH containing a radioactive synthetic mRNA or DNA "probe". **(E)** The probe hydrogen bonds only to plasmid DNA that is complementary to the nucleotide sequence of the probe; that is, to plasmids that contain the human gene complementary to the probe. **(F)** On the culture dish, the locations of radioactivities on the paper are matched with bacterial colonies (Not shown). Colonies in the same position consist of bacteria containing plasmids with the desired gene. Samples of these bacteria are cultured in their own dishes (Modified from Microbial Biotechnology, Ed Lee, 2006).

This constitutes our human DNA library.

4.1.2 Identification of Genes of Interest

The human genome, consisting of all the genes in an individual, contains about 6.6 billion nucleotides (3.3 billion pairs). The average protein of 300 to 400 amino acids (so at three nucleotides per amino acid) needs only about a thousand nucleotides to encode its information. Identifying or finding a particular gene within the human genome, then, would appear to be a search for a very small needle within a very large haystack. To identify or locate genes, several "tricks" can be used.

If we first know the amino acid sequences of its encoded protein, it is easier to find a gene. For example, the amino acid sequences of most of the major human hormones are known. From the amino acid sequence, we can work backward through the genetic code to determine likely DNA sequences. The problem, however, is that a single amino acid may be encoded by several nucleotide sequences and, thus, complicates the things a little.

Machines that can synthesize DNA chains from nucleotide sequences typed on a keyboard are now commercially available. The synthesizers cannot make really long chains, but they can accurately synthesize a chain of a few dozen nucleotides in length.

Another way to identify a gene is to find a cell that we know synthesizes vast amounts of a particular protein. Immature red blood cells, for example, synthesize lot of haemoglobin. This means that they have lots of messenger RNA (mRNA) for haemoglobin. This mRNA is complementary to the DNA of the haemoglobin gene and can also serves to locate the gene.

Use of Gene Probes to Search the DNA Library of Bacteria Containing Plasmids Bearing the Desired Gene : DNA library obtained in this form is useless, because it lacks a card file or computer index that would tell us which bacterium contains the plasmid with the gene of interest. For this purpose, synthetic DNA or isolated mRNA, labeled with radioactive isotopes is used.

4.1.3 Amplification of Selected DNA Sequences in the Library

Once the gene is located in the library, it can be amplified by making millions or billions of copies for further use. This process is called **cloning**. We can pick up bacteria from the appropriate colonies located during the search and grow them in appropriate culture conditions. In fact now huge vats (bioreactors) are used to produce many kilograms of bacteria, all containing a specific human gene.

The polymerase chain reaction (PCR) is another method for making copies of specific stretches of DNA, not only from DNA libraries, but also even from single cells.

4.1.4 Vectors

Vector is a DNA molecule that is used as a vehicle to carry/transport/insert a foreign piece of DNA molecule into a host cell

An ideal cloning vehicle should have :

- an origin of replication
- low molecular weight
- readily selectable markers
- multiple cloning sites

4.1.4.1 Plasmids

Plasmids are autonomously replicating, extrachromosomal, double-stranded, closed circular and dispensable DNA molecules, ranging in size from 1-200 Kb.

They confer some selective advantages to the host *e.g.,*

- resistance to antibiotics, heavy metals
- production of antibiotics
- degradation of complex organic compounds
- production of colicins, enterotoxins, restriction and modification enzymes

Plasmids to which phenotypic traits have not yet been ascribed are called **cryptic.**

4.1.4.1.1 Essential Features of Plasmids

(i) **Replication.** To propagate in bacterial host, plasmid must have specific DNA sequences that allow it to replicate within the host cell. DNA polymerase and other proteins required to initiate DNA synthesis bind to this region, called the origin of replication *(ori).*

(ii) **Most plasmids exist as double–stranded circular DNA molecules.** If both strands are intact circles, the molecules are described as covalently closed circles or CCDNA. If only one strand is intact, then they are described as open circles or OC DNA.

CCDNA often exists in supercoiled form. Addition of excess intercalating agent causes CCDNA plasmid to rewind in opposite direction and this property is made use of in plasmid separation.

(a) **Copy number:** Plasmids under "relaxed" control have copy number 10-200 and plasmids under "stringent" control have 1-10.

(b) **Conjugation:** Plasmids containing *"tra"* genes, that promote conjugation are called **conjugative plasmids.** Plasmids lacking *"tra"* genes are called non-conjugative plasmids.

(c) **Plasmid incompatibility :** It is the inability of two different plasmids to coexist in the same host cell in the absence of selection pressure.

DNA fragments of up to 5-10 kb are efficiently cloned in plasmid vectors.

4.1.4.1.2 Features of Plasmid Cloning Vector

(i) *An ori sequence :* It allows its replication in host cells.

(ii) *A dominant selectable marker :* It enables cells that carry the plasmid to be easily distinguished from cells that lack the plasmid *e.g.,* genes conferring antibiotic resistance *amp, tet.*

(iii) *Unique restriction enzyme cleavage sites*: These sites are present just once in a vector – for the insertion of the DNA molecules that are to be cloned. One of the first cloning vectors designed was pBR 322 (Fig. 4.5)

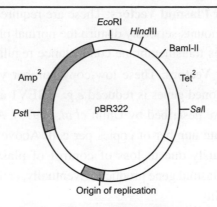

Fig. 4.5. Genetic map of the plasmid cloning vector pBR322. Unique *Hind III, Sa II, Bam HI* and *Pst I* recognition sites and the genes for tetracycline resistance (Tet^r) and ampicillin resistant (Amp^r) are present. The unique *EcoRI* site is just outside the tetracycline resistance gene. The origin of replication functions in the bacterium *E. coli.* The complete DNA sequence of pBR322 consists of 4,361 bp.

4.1.4.1.3 Natural Plasmids as Cloning Vehicles

Plasmids that occur naturally in organisms *e.g., Col*EI occur in *E. coli* and produces colicin. EI RSF 2124 occurs in *E. coli* and produces colicin EI and encodes amp resistance, pSC101 occurs in *Salmonella panama* and encodes tetracycline resistance. To clone DNA in these, plasmid and the foreign DNA are digested with *EcoR1,* mixed, and treated with DNA ligase and used to transform a host pSC101 has been used for cloning of *S. aureus* plasmid genes and for cloning of *Xenopus* DNA in *E. coli.*

4.1.4.1.4 Artificial Plasmids

Considerable efforts have been expended on constructing *in vitro* superior cloning vehicles. Most versatile and widely used of these artificial vectors is pBR322. It contains the *Ap* and *Tc* genes of RSF 2124 and pSC 101 respectively combined with replication elements of pMB1, a ColE I - like plasmid. Its size is 4363 bp and has over 20 unique restriction sites. Some improvements of pBR322 are pBR 325, pAT 153 pUC vectors and *pBluescript* II vectors. They are also being widely used.

    (a) **Direct Selection Plasmids**: These are recombinant plasmids that can be directly selected after transformation *e.g.,* plasmid pLX100, in which a gene for xylose isomerase, containing contiguous unique sites for *Hind* III, *Pst* I, *Bam* HI and *XhoI*, is placed under the control of *lac* promoter. *E. coli* transformants containing pLX100 cannot grow in minimal medium with xylose unless a DNA fragment is inserted into one of the unique restriction sites.

(b) **Low-copy-Number Plasmid Vectors**: These are required for cloning of genes whose products in large amounts seriously disturb the normal physiology of the cell *e.g. omp* A, the cystic fibrosis transmembrane conductance regulator *e.g.,* pSC101.

(c) **Runaway Plasmid Vectors**: These low-copy-number vectors avoid cell killing, but expression of the cloned genes is reduced *e.g.,* pBEV1 and pBEV2, the first runaway plasmid vectors were described by Uhlin *et al.* (1979). At 30°C the plasmid vector is present in a moderate number of copies per cell. Above 35°C, plasmid copy number increases continuously due to loss of control of plasmid replication resulting in overproduction of plasmid gene products. Eventually, cell growth inhibition occurs and cells lose their viability.

4.1.4.2 Phage Cloning Vectors

Commonly used phage cloning vectors are derivatives of bacteriophage l which have been genetically engineered so that the lytic cycle is possible, but lysogeny is not possible (Russel, 1998). Wild type l DNA contains several sites for most of the commonly used REs and, therefore, is not itself suitable for use as a vector. Derivatives of the wild type phage have, therefore, been produced that either have a single target site at which foreign DNA can be inserted (insertional vectors), or have a pair of sites defining a fragment that can be removed (therefore, it does not contain any gene needed for phage propagation), and replaced by foreign DNA (replacement vectors).

The only splicing events that produce viable, functional chromosomes are those in which the foreign DNA is inserted within/ between the target site(s) and in which the total length of the recombined DNA fragment is approximately 37 to 52 kb (75-105% the size of lambda phage DNA), since, only DNA fragments of about 37-52 kb can be packaged into lambda particles.

DNA molecules that are either too long or too short are unable to be packaged. Therefore, only about 15 kb of DNA can be inserted into an alpha replacement vector. Vectors are commonly used for initial cloning of genes of interest, since it is easier to screen for plaques with the desired sequence than it is to screen bacterial colonies.

Sternberg (1990) has developed a bacteriophage PI cloning system that is able to insert upto 100 kb in size (Monaco and Larin, 1994) and has cloning efficiency of 10^5 clones per 1-2 mg vector and insert DNA (Figs. 4.6 and 4.7). This system packages both cloned DNA and vector into phage particles that injects them into *E. coli* and circularizes the DNA, using the PI *cox*P recombination sites and a host expressing the PI Cre recombinase. The vector carries a gene for kanamycin resistance for selection and PI plasmid replicon for maintaining the vector at one copy per cell, but enabling the DNA to be amplified before re-isolation (Sternberg,1992). The details of bacteriophage λ cloning are given in Fig. 4.8.

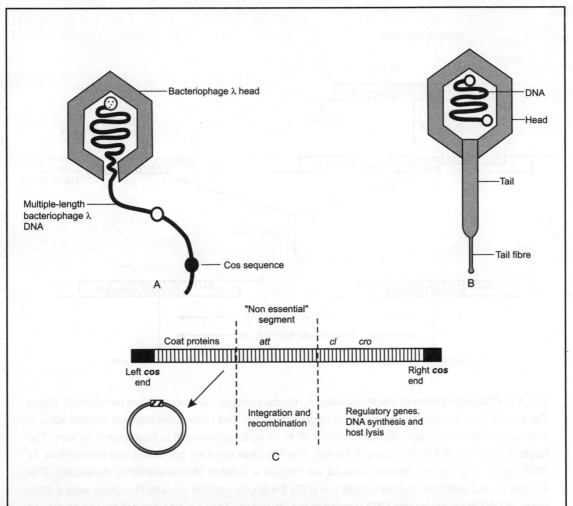

Fig. 4.6. The bacteriophage λ as a cloning vector. (A) It shows the multiple lengths bacteriophage λ DNA. (B) The structure of intact λ phage particle (Adopted from Glick, 2004). (C) The detailed structure of bacteriophage λ genome has been shown (Adopted and modified from Microbial Biotechnology, Ed. Lee, 2006).

Uses : It is useful in :

• chromosome mapping

• generating contigs

• It can be modified so that the cloned DNA is transferred back to its host of origin

• The increased cloning capacity may improve the efficiency of homologous gene targeting in mammalian cells.

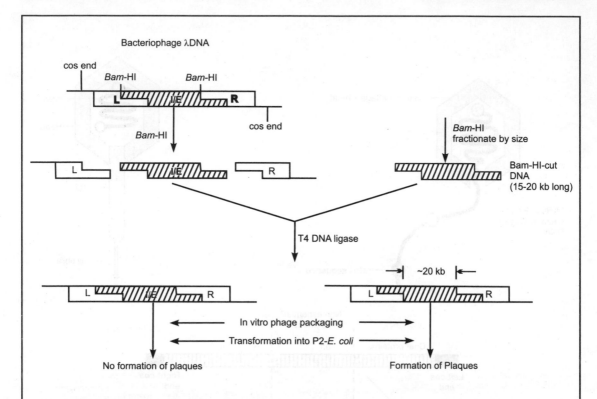

Fig. 4.7. Cloning system of Bacteriophage λ. Bacteriophage λ is made to have two *Bam*HI. These Parts flank the I/E region of the bacteriophage λ genome. The extensions indicate the cos sites λ. The source DNA is cut with *Bam*HI to isolate 15 to 20 kb long pieces. It is fractionated by size. The bacteriophage λ DNA is also cut with *Bam*HI. The two DNA samples are mixed and treated with T4 DNA ligase. The ligation reaction mixture will contain a number of different DNA molecules. This include (1) reconstituted bacteriophage λ and (2) the bacteriophage λ L and R regions with a 20kb piece of DNA from the source DNA instead of the I/E region. These molecules are *in vitro* packaged into bacteriophage λ heads and after the addition of tail assemblies infective particles are formed. After infection of *E. coli* cells having P2 bacteriophage DNA integrated in their chromosomes, only the molecules with the R and L regions and a cloned ~20-kb piece of DNA can replicate and form infectious bacteriophage λ. In this way, only the bacteriophage λ containing a DNA insert are perpetuated (Adopted from Molecular Biotechnology Glick and Pasternak, 2004).

Many vector derivatives of both insertional and replacement type were produced by several groups of reasearchers early in the development of RDT (Thomas *et al.,*1974; Murray and Murray 1975; Blattner *et al.*, 1977). Blather *et al.* (1977) constructed Charon phages. Charon 16 A is an insertional vector with a single EcoRI sites located in the gene for β-galactosidase (lac Z) which provides a convenient colour test for the production of β-galactosidase, when a chromogenic substrate is included in the plating medium. Improved phage vector derivatives have also been developed.

4.1.4.3 Cosmids

Plasmids have been constructed which contain a fragment of DNA including the *cos* site of λ phage (Fig. 4.8). These plasmids have been termed cosmids and can be used as gene cloning vectors in conjunction with the *in vitro* packaging system (Hohn and Collins, 1988).The cosmids are analysed as per procedure given Fig. 4.9.

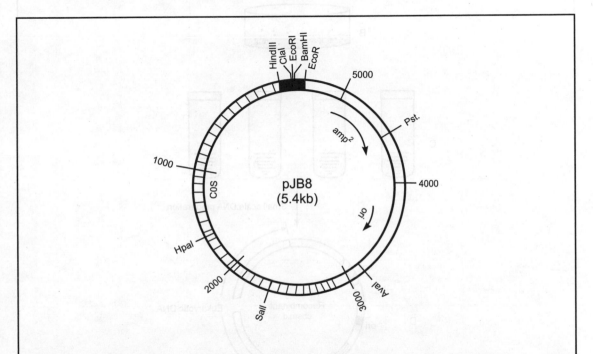

Fig. 4.8. Cosmid pJB8. The detailed structure and restriction enzyme cutting sites have been shown.

Principles of cosmid cloning

The initial requirements for a cosmid vector were, that it contained:

- a replication system that functions in *E. coli*
- a selective marker
- a *cos* site of phage to allow packaging
- a restriction enzyme cleavage site

The details of cloning by cosmid is given in Fig. 4.10.

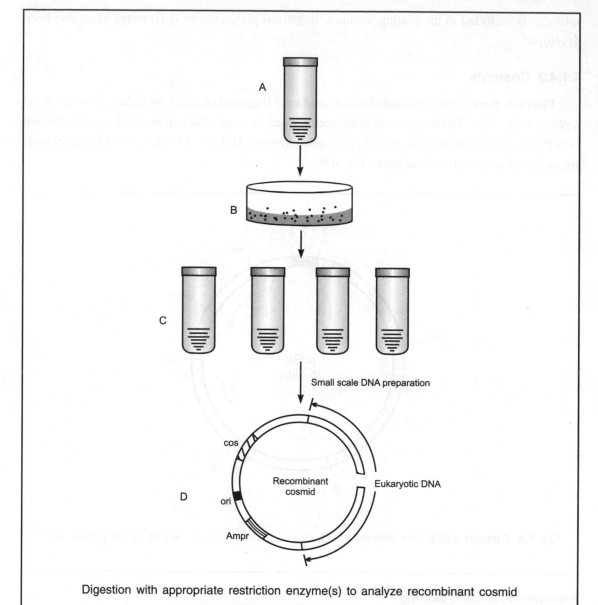

Digestion with appropriate restriction enzyme(s) to analyze recombinant cosmid

Fig. 4.9. Cloning in cosmid vector. (A) Infected bacterial culture is spread on agar plates containing ampicillin **(B)** The ampicillin resistant bacterial colonies are obtained by replica plating method **(C)** The bacterial colonies are made to grow **(D)** Small preparations of DNA are made and cosmid isolated to analyze. (Modified from Food Biotechnology).

Advantages of cosmid cloning method

- A direct physical selection for hybrid plasmids, so is imposed by the packaging process the vectors without insert are not obtained.

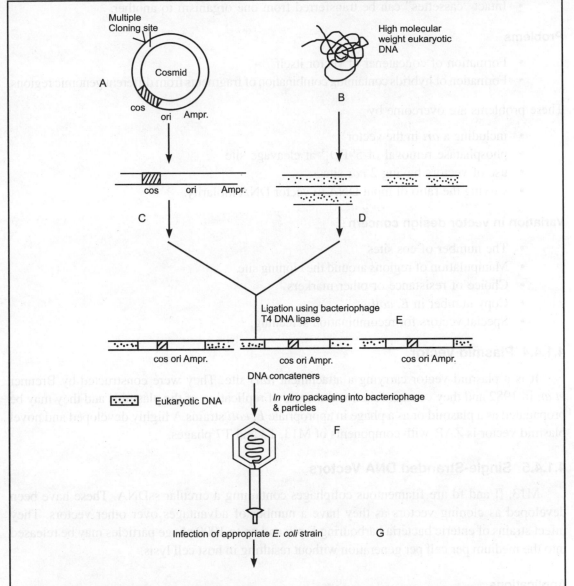

Fig. 4.10. Transfection by cosmid. (A) Cleavage of cosmid DNA with restriction enzyme (s) **(B)** Partial digestion of high molecular DNA with restriction enzyme (s) to produce ends with linearized cosmid terminal **(C)** Treated with alkaline phosphatase to remove 5' terminal phosphate groups **(D)** DNA fragments that are within 35-45 mb range **(E)** Ligation using bacteriophage T4 DNA ligase **(F)** *In vitro* packaging into bacteriophage and particle **(G)** Infection of appropriate *E. coli* strain. (Modified from Food Biotechnology).

- The use of small vectors allows the cloning of genomic fragments of up to 46kb in size. Thus, most eukaryotic genes or prokaryotic operons can be recovered intact. These also minimize the number of analyses and manipulations required to establish a restriction map of entire genomes by identifying contiguous clones.

- Intact "cassettes" can be transferred from one organism to another.

Problems

- Formation of concatener of vector itself
- Formation of hybrids containing combination of fragments from different genomic regions.

These problems are overcome by

- including a *ori* in the vector
- phosphatase removal of 5'-PO_4^{3-} at cleavage site
- use of vectors having 2 *cos* sites
- varying the ratio of input DNA to vector DNA molarity.

Variation in vector design concern

- The number of cos sites
- Manipulation of regions around the cloning site
- Choice of resistance or other markers
- Copy number in *E. coli*
- Special vectors for recombination screening.

4.1.4.4 Plasmid Vector

It is a plasmid vector carrying a attachment (*att*) site. They were constructed by Brenner *et al.* in 1982 and they contain functional *ori*gins of replication of the plasmids and they may be propagated as a plasmid or as a phage in appropriate *E. coli* strains. A highly developed and novel plasmid vector is ZAP, with components of M13, T3 and T7 phages.

4.1.4.5 Single-Stranded DNA Vectors

M13, fI and fd are filamentous coliphages containing a circular ssDNA. These have been developed as cloning vectors as they have a number of advantages over other vectors. They infect strains of enteric bacteria harbouring F-pili and up to 1000 phage particles may be released into the medium per cell per generation without resulting in host cell lysis.

Applications

ssDNA is required for several applications of cloned DNA. These include

: sequencing by di deoxy method ,
: oligonucleotide directed mutagenesis like a plasmid,
: methods of probe preparation.

Advantages

: Phage DNA replicates via a ds-circular DNA replicative form (RF). This can be purified and manipulated like a plasmid,

: both RF and ssDNA transfect competent *E. coli* cells to yield either plaques or infected colonies,

: no packaging constraints are these upto 6 times the length of M13 DNA has been packaged,

: orientation of an insert can be easily determined.

4.1.4.6 ss/ds DNA vectors : pEMBL (Cesaseni and Murray,1987)

In this, the ability to prepare ssDNA added to plasmids by inserting into a plasmid vector, a small vector carrying the replication and packaging signals from a filamentous ss phage (fI, M13). This second origin is normally functionally silent, but whenever necessary can be activated by superinfection with helper phage. It provides the function required in *trans* to replicate the viral strand and package it into phage rods. This results in the extrusion of viral particles carrying on plasmid strand into the culture medium from which they can be readily purified. The orientation of the phage origin determines which plasmid strand is packaged. Such plasmids maintain virtually all the advantages of ss phage vectors. Although there is no upper limit for DNA packaging, increase in phage size is associated with a higher rate of insert instability caused by spontaneous deletion. This makes phage vectors unsuitable for cloning large inserts, or for the more complex construction required for shuttle vectors that can replicate in different hosts.

4.1.4.7 Shuttle Vectors

These vectors can replicate in two or more host organisms. They are used for experiments in which recombinant DNA is to be introduced into organisms other than *E. coli* (Russel, 1998).

Yeast –*E. coli* shuttle vector, *Yep 24* has an *ori* sequence. With this it can replicate in *E. coli*. It also has dominant selectable markers that confer ampicillin resistance and tetracycline resistance. A selectable marker URA3, enzyme required for uracil biosynthesis is also present. It allows selection of *ura3* mutant yeast cells containing Yep24. It also carries a yeast specific sequence, that allows it to autonomously replicate in a yeast cell. The first Yep24 was constructed by Beggs in 1978 using naturally occurring yeast plasmid.

Often *E. coli* genes are satisfactorily expressed in other Gram-negative organisms. The reverse however is usually not true. Genes from *Pseudomonads,* for example, are only poorly expressed in *E. coli*. They thus have to be analyzed in their original hosts. In many cases, the techniques available for these organisms are limited. For example, transformation (or electroporation) frequencies may be relatively low, and the methods for screening gene libraries may be difficult. It is because of this the initial cloning of a gene can be done by using *E. coli* as the host organism. After the desired DNA construct has been obtained, it can be recovered and then inserted into the target host. This procedure requires a vector, which can replicate in either organism. Certain plasmids are capable of existing in a variety of hosts and can function as shuttle vectors. In a shuttle vector replication origins for two different plasmids are present on the same vector. This will facilitate preliminary manipulations in a well-characterized system, such as *E. coli* and then transfer the clone to the organism of interest. In case the clone is to be expressed in both hosts, then it may also be needed to provide two promoters as well.

4.1.4.8 Artificial Chromosomes

The artificial chromosomes have the following characteristics :

(i) The chromosome should have a segregation efficiency approaching 100% in order to be useful in a cell population undergoing multiple rounds of cell divisions.

(ii) The chromosome should have a defined structure for regulatory and practical reasons. A defined structure is needed to maximize the control of expression of the genes that it contains.

(iii) The chromosome should not be so large that delivery becomes a problem.

(iv) Chromosomal effects such as centromeric or telomeric silencing should not dominate the expression of genes contained in an artificial chromosome.

The cloning of high molecular weight genomic DNA promises to provide the means of mapping chromosomes, isolating genes and understanding long-range effects on gene expression.

4.1.4.8.1 Yeast Artificial Chromosomes

As the upper limit of cosmid cloning is about 35.0455 kb, to cover the whole human genome in a theoretical contiguous chain would require about 70,000 clones. It is because of this YACS have been constructed. Construction of the first artificial chromosome about 55 kb long was done in Szostaki laboratories by Berk *et al.* in yeast. This behaved fairly stably in mitosis and meiosis in much the same way as natural chromosomes (Daves,1983). By a combination of *in vitro* construction and *in vivo* recombination, two linear plasmids of about 55 kb were made containing telomeres from *Tetrahymena*, phage DNA and yeast UR3A. TRPI, ARSI and CEN3 sequences. These large linear structures were stable during mitosis and meiosis and were present in 1 or 2 copies per cell and the introduction of large inserts at least ten-fold larger than those that could be accommodated by phages and cosmids in bacterial host was made possible (Fig. 4.11).

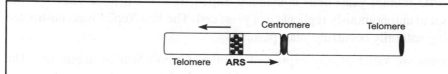

Fig. 4.11. Yeast artificial chromosome as a cloning vector. The detailed structure of YAC is shown.

YACs are being extensively used in physical mapping and positional cloning projects (Monaco and Larin, 1994).

Problem with YAC clones

* 40-60% of these clones are chimaeric
* These clones are unstable
* It is difficult to separate host chromosome from YACs.

YAC vector incorporates essential features of linear chromosomes and includes 4 types of genetic elements.

(i)   A cloning site

(ii) A yeast centromere, replication origins and genetic markers that are selectable in yeast

(iii) An *E. coli* origin of replication and genetic markers that are selectable in *E. coli*

(iv) A pair of telomeric sequences from the ciliated protozoan *Tetrahymena*. YAC is, therefore, a shuttle vector that can replicate and be selected both in yeast and in *E. coli*. YAC chromosome inserts as large as 1Mb can be recovered intact in yeast cells.

4.1.4.8.2 Bacterial Artificial Chromosomes (BAC)

BAC vector : It is based on the f factor of *E. coli*. The essential functions included in the 6.8 kb vector are genes,

a) for replication (*rep*E and *ori*S),

b) for regulating copy number (*par* A and *par* B),

c) for chloramphenicol resistance.

BAC vectors with inserts of 300kb can be maintained. They have low copy number due to strict control on replication (Fig. 4.12).

BACs can be transformed into *E. coli* very efficiently by electroporation. However, they lack a positive selection system for clones containing inserts and give a very low yield of DNA (Monaco and Larin,1994). As analyzed by Simon (1997), the ability to easily retrofit BAC clones and reintroduce them into cells and animals has come to the rescue of the post human genome project field of inquiry – referred to as functional genomics (Simon,1997).

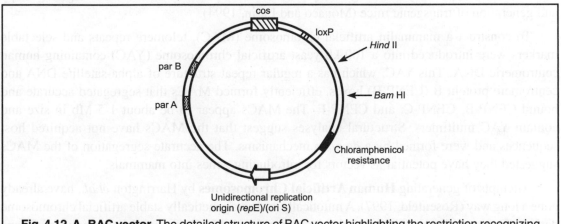

Fig. 4.12. A BAC vector. The detailed structure of BAC vector highlighting the restriction recognizing sites has been shown.

4.1.4.8.3 Derived Artificial Chromosomes (PAC)

λ1 is a bacteriophage that infects cells of *E. coli*. The genome of the wild type PI phage is ~100kb, but 17kb DNA is sufficient for replication of the vector as a plasmid, amplification of the vector using the lytic replicon and selection of the plasmid using kanamycin resistance. Selection for DNA inserts occurs through interruption of the *cacB* gene. When *cacB* gene is expressed, the cells die if grown in medium containing sucrose. Insert of DNA at the cloning site separates the *cacB* gene from its promoter, the gene can not be expressed and so vectors with inserts in cloning site allow growth in insert size of 145 kb and largest inserts are '300kb'. PAC contains a plasmid replicon that results in 1/2 copies/cell, but there is also an inducible lytic replicon by which the number of copies can be amplified.

PACs may be important in complementing YAC contigs for gene isolation and sequencing projects in the future.

P1 vector (pCYPAC-1) is designed for the introduction of recombinant DNA into *E. coli* using electroporation procedures. The new cloning system, P1-derived artificial chromosomes (PACs), was used to establish an initial 15,000 clone library with an average insert size of 130-150 kilobase pairs (kb). No chimaerism was observed in 34 clones by fluorescence *in situ* hybridization. Similarly, no insert instability was observed after extended culturing, for 20 clones. It was concluded that the PAC cloning system will be useful in the mapping and detailed analysis of complex genomes.

4.1.4.8.4 Mammalian Artificial Chromosome (MAC)

If all the elements involved in the structure of mammalian chromosomes were identified, then MAC could be constructed. These would provide an experimental system for defining the size requirements for accurate mitotic and meiotic segregation, and for studying chromosome function in mammalian cells. In addition, they would provide an autonomously replication transfer vector for functional analysis of large complex genes and a potential tool for somatic gene therapy and generation of transgenic mice (Monaco and Larin, 1994).

To construct a mammalin artificial chromosome (MAC), telomere repeats and selectable markers were introduced into a 100 kb yeast artificial chromosome (YAC) containing human centromeric DNA. This YAC, which has a regular repeat structure of alpha-satellite DNA and centromere protein B (CENP-B) boxes, efficiently formed MACs that segregated accurate and bound CENP-B, CENP-C, and CENP-E. The MACs appear to be about 1-5 Mb in size and contain YAC multimers. Structural analyses suggest that the MACs have not acquired host sequences and were formed by a *de novo* mechanisms. The accurate segregation of the MACs suggested they have potential as vectors for introducing genes into mammals.

Attempts of generating **Human Artificial Chromosomes** by Harrington *et al.,* have already gone a long way (Rosenfeld, 1997). A mitotically and cytogenetically stable artificial chromosome derived from transfected DNA has also been generated. MACs are useful for analysis of gene expression and gene therapy.

Human artificial chromosome (HAC) technology has developed rapidly over the past four years. HACs are useful gene transfer vectors in expression studies and important tools for determining human chromosome function. HACs have been used to complement gene deficiencies in human cultured cells by transfer of large genomic loci also containing the regulatory elements for appropriate expression. and, they now offer the possibility to express large human transgenes in animals, especially in mouse models of human genetic diseases.

4.1.4.8.5 Plant Artificial Chromosome

Plant artificial chromosomes using centromeres in *Arabidopsis* have been constructed by utilizing the fact that its telomeres are similar to that in yeast (Somerville and Somerville, 1999).

4.1.5 Cloning

4.1.5.1 Cloning in *Bacillus subtilis*
Cloning in B.subtilis **has the following advantages**

- These are gram positive and generally obligate aerobes.
- They are able to sporulate and, therefore, used to study prokaryotic differentiation.
- They are widely used in fermentation industry particularly for production of exoenzymes.
- They can be tailored to secrete the products of cloned eukaryotic genes.
- It is an extremely safe organism (non pathogenic)
- It is naturally transformable

Plasmid pCI94 from *S. aureus* and pUB110, pEI194 and pT127 have been used in studies for the development of cloning vectors in *B. subtilis*.

Hybrid plasmids that can replicate in both *E. coli* and *B. subtilis*, are constructed by fusion between pBR322 and pUB110 or pCI194. With such plasmids, *E. coli* can be used as an efficient intermediate host for cloning *e.g.*, pHV14, pHP3, pLB5.

Improved vectors : To overcome the structural and segregational instability of plasmids in *B. subtilis,* improvements have been made. One approach is to use endogenous plasmids as vectors to increase copy number and, thus stability *e.g.*, pTA1060 derivatives. Use of plasmids with mode of replication provide structural stability *e.g.*, pAMb1 and pTB19 derivatives.

4.1.5.2 Cloning in Fungi

A target for gene cloning in fungi has been a vector capable of isolating, propagating and transferring cloned genes via *E. coli*. Cloning vector must be able to replicate in the fungus and in *E. coli,* the replication should be stable, and the vector should confer a selectable or detectable phenotype on both hosts.

4.1.5.3 Cloning in *Saccharomyces cerevisiae*

The advances in vector development are primarily due to 2 discoveries:

(i)   Identification of high copy-number *S. cerevisiae* plasmid, the 2 µm circles.

(ii) The isolation of DNA sequence capable of conferring the property of independent replication on a chimaeric plasmid in yeast, the autonomously replicating sequence (ARS).

The 2 µm circle 6318 bp in length, has 50 copies per cell. It resides in nucleus and is very stable.

The basic types of vectors that have been developed for genetic engineering of yeast are:

(i) yIp–These vectors integrate into the chromosome. They have low transformation frequency, but are most stable and behave as an ordinary genetic marker.

(ii) yRp –These vectors replicate autonomously and carry a chromosomal ARS, transformants are very unstable, but transformation frequency is high. They have high copy number and are very useful for complementation studies and are easy to recover from yeast.

(iii) yEp–These vectors replicate autonomously and carry origin of replication from the 2μm circle.They have high copy number and high transformation frequency but are slightly unstable. They are useful for complementation studies. A similar plasmid from *Zygosacchromyces cerevisiae* as a vector has been described by Awane (1992).

(iv) yCp–These vectors carry stabilizing centromere (CEN) sequences which allow them to function as comparatively stable mini-chromosomes.They have low copy number with high transformation frequency and stability.

(v) yLp–These are linear vectors containing homologous or heterologous telomeric sequences at their ends.They have high copy number with high transformation frequency and low stability.

(vi) Ty–Transposable elements of 5.9 kb found in *S.cerevisae* (30-40 copies) have promoters in the terminal regions which can be replaced by promoters of galactose-utilization gene *etc*. When these are introduced into a yeast on a high copy number vector, Ty elements are overexpressed.

Cloning in Yeast has the following advantages

• Potential use of yeast as a cloning host for over production of proteins of commercial value. Yeast has the ability to glycosylate proteins during secretion and the absence of pyrogenic toxins.

• Ability to clone large pieces of DNA

• Used in surrogate genetics of yeast *i.e.,* to understand function *in vivo*

• Natural yeast secretion signals have been used in expression vectors which also direct the export of the recombinant product outside the cell, thereby aiding in the purification of product.

4.1.5.4 In Yeast other than *S. cerevisiae*

High levels of expression have been obtained in yeasts like *Kluyveromyces lactis, Pichia pastoris and Hansenula polymorpha.*

4.1.5.5 Cloning in Filamentous Fungi

These are important in understanding the biochemical mechanisms involved in conidiation in fungi, to understand the basis of plant pathogenicity. In *Aspergillus,* there is no evidence for replicating extra-chromosomal plasmids, but they have been identified in *Neurospora*. Genes of interest can be cloned by complementation of mutations or by making use of positive selection. Heterologous protein production and secretion at levels as high as 3 g/l have been obtained in *A. oryzae.*

4.1.5.6 Mammalian Cloning Vectors

Most eukaryotic vectors to date, contain DNA sequences derived from viruses which have following characteristics. They may or may not be infectious.

a) They may or may not replicate autonomously;

b) They may permit the expression of an inserted gene transiently or indefinitely.

c) They have been derived from papovavirus *e.g.* SV40 and polyoma adeno virus (Graham,1990) parvo virus, herpes virus *e.g.,* HSV-I, V2V, EBV pox virus, *e.g.* vaccinia retro virus (Temin,1986).

Insect cells are also important hosts for baculoviruses.

It has now become possible to enhance retroviral titers further by including the retroviral vector, alongwith the helper genome(s) within the same adenovirus vector, an advance that should become feasible. These adenoviral vectors liberate more space for even larger, multiple inserts (Vile, 1997).

4.1.5.7 Plant Cloning Vectors

It is now possible to transfer any gene into a plant as a routine procedure through vector systems based on *Agrobacterium tumefaciens* or based on transfection of plant protoplasts or plant viruses as vector systems. Zaenen *et al.* (1974) first noted that tumour forming strains of *A. tumefaciens* harbour large plasmids (140-235 kb) Ti-plasmid and, therefore, could be used as vector. However, wild type Ti-plasmids are not suitable as general gene vectors. They, therefore, cause disorganized growth of the recipient plant cells owing to the effects of the oncogenes in the T-DNA. Ti-plasmids, in which T-DNA has been disarmed by making it non-oncogenic are used.

4.1.5.8 Algal Cloning Vectors

The understanding of the molecular genetics of Cyanobacteria has provided the basis for genetic modification of industrially important strains. The development of gene cloning systems involves the selection of suitable host strains and construction of vectors capable of maintaining and expressing foreign genes in host cells. Host-vector systems have been developed for both unicellular and filamentous Cyanobacteria. A general method for construction of Cyanobacterial cloning vector, according to Craig and Reichelt (1986) is to form small indigenous plasmids with *E. coli* plasmid vectors. Such hybrid plasmids have *ori* and genetic markers for both hosts and allocate restriction sites for insertion of foreign DNA (Craig and Rechelt,1986).

4.2 DETAILED TECHNIQUE FOR GENETIC ENGINEERING IN BACTERIA

The process can be divided into five steps :

- *Stage* 1: To obtain a copy of the required gene from whole genome of the donor organism.
- *Stage* 2: Insertion of the gene in a vector.
- *Stage* 3: Transfer of the inserted gene through vector into the host cell.
- *Stage* 4: Selection of the transformed cells (having the foreign DNA).
- *Stage* 5: Cloning of the gene.

Sometimes this sequence of events may be altered. These steps have been explained in Fig. 4.13.

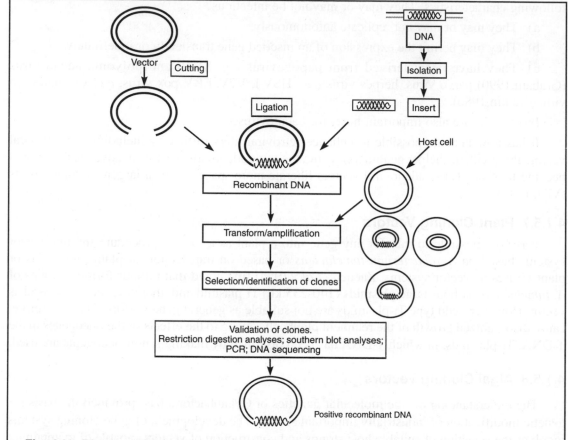

Fig. 4.13. Diagrammatic representation of basic steps in gene cloning. The detailed steps involved in cloning have been highlighted (Modified from Microbial Biotechnology by Lee Yuan Kun, 2006)

4.2.1 Obtaining a Copy of the Required Gene

The human genome contains about 3.3 billion base pairs and 30-35 thousand genes. A typical gene is several thousand base pairs long, so even to find a particular gene, is very difficult. The following methods are used to get a copy of a gene :

(a) **cDNA Strategy :** A copy of the DNA (cDNA) is made from its mRNA, using reverse transcriptase.

(b) **Artificial gene synthesis**: A gene can be synthesized using nucleotides and joining them together in the right order.

(c) **'Shotgun' approach** *i.e.,* chopping up of the DNA with 'restriction enzymes' and searching for the piece with the required gene.

4.2.1.1 *cDNA Preparation*

In a diploid cell, there are only two copies (one each on a chromosome from the female and male parent) of each gene. If the gene is active, usually thousands of mRNA molecules are produced by it. These are complementary to the gene. From experimental evidences it is often known in which cells the gene is active. The gene for insulin is active in the β cells of the pancreas. An enzyme which can make a complementary DNA copy of mRNA molecule is present in retroviruses. This is called **reverse transcriptase.** For some genes, it is relatively easy to isolate mRNA from the particular gene in a particular type of cell. Once this has been done, this is made suitable for the gene coding for the required protein. The DNA formed in this way is called **complementary DNA** or **cDNA** (Fig. 4.14).

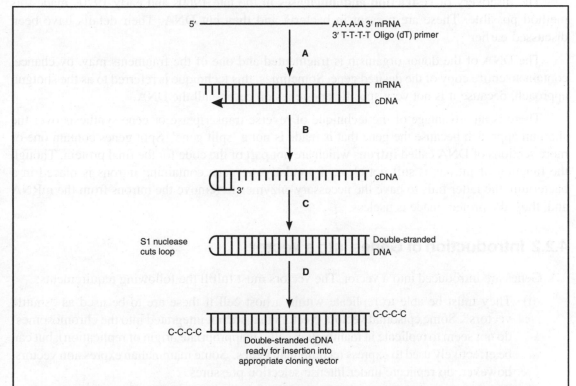

Fig. 4.14. cDNA : Synthesis of DNA from mRNA. (A) Messenger-RNA is incubated with reverse transcriptase. It uses the mRNA as a template for the synthesis of a complementary DNA (cDNA) strand. Oligo (dT), a short chain of thymine deoxyribonucleotides, can be used as a primer, because eukaryotic mRNA always has a stretch of adenine nucleotides at its 3′ end. **(B)** The resulting mRNA-cDNA hybrid is treated with alkali or an enzyme to hydrolyze the RNA. It leaves the single-stranded cDNA. **(C)** The complementary DNA strand can also be synthesized by DNA polymerase. Generally, the looped-around 3′ end of the first DNA strand can be used as a primer. To cleave the loop, an enzyme S1 nuclease is then used. **(D)** The double-stranded DNA must have single-stranded tails that are complementary to those of the vector for efficient insertion in a cloning vector. For this purpose an enzyme terminal transferase can be added. This enzyme adds nucleotides one at a time to the ends of the molecule.

4.2.1.2 Synthesizing a Gene

It is possible to directly know the base sequence of a gene. Alternatively suitable base sequence can be worked out from the amino acid sequence of the protein made by it. A gene can then be constructed using nucleotides (each base is a part of one nucleotide) and joining them together in the right order. This at present is possible only for short genes. However, this could become a routine possibility for any gene. It has been used for the synthesis of **proinsulin** and **somatostatin genes.** Somatostatin (otherwise known as growth hormone inhibitory hormone) is a protein hormone which contains only 14 amino acids.

4.2.1.3 The Shotgun Approach

The discovery of **restriction endonucleases** in the late 1960s and early 1970s made this method possible. These are present in bacteria and they cut DNA. Their details have been discussed earlier.

The DNA of the donor organism is fragmented and one of the fragments may, by chance, contain an entire copy of the desired gene. Sometimes, this technique is referred to as the shotgun approach, because it is not very specific and relies on cutting all the DNA.

There is an advantage of the technique of reverse transcriptase or gene synthesis over the shotgun approach because the gene that is made is not a 'split gene'. Split genes contain one or more sections of DNA called **introns** which are not part of the code for the final protein. Though the function of introns is still not clear, if a eukaryotic gene containing introns is placed in a bacterium, the latter fails to have the necessary enzymes to remove the introns from the mRNA and, thus, the protein made is useless.

4.2.2 Introduction of Genes into a Vector

Genes are introduced into a vector. The vectors must fulfill the following requirements:

(i)   They must be able to replicate within a host cell if these are to be used as "shuttle vectors". Some episomal vectors (vectors that are not integrated into the chromosomes) do not seem to replicate in mammalian cells (inappropriate origin of replication), but can be effectively used to express recombinant genes. Some mammalian expression vectors, however, do replicate under intense selection pressures.

(ii) There must be efficient ways of introducing vector DNA into a host cell.

(iii) There must be ways to isolate the vector DNA from the host DNA.

(iv) The presence of specific restriction sites in the vector that can be used for cloning an insert DNA.

(v)   The vector should have atleast a selectable marker to indicate its presence in host cell. This can be antibiotic or other xenobiotic marker, or simply reporter groups *e.g.,* luciferase, green fluorescent protein (GFP) and β-galactosidase.

(vi) It is desirable to have unique restriction enzyme sites within a selectable or screenable marker so that the presence of a DNA insert in the cloning vector can be selected or screened.

A number of these vectors are commercially available. They can be used for cloning in prokaryotes, yeast, baculovirus, mammalian and other cells (Table 4.2).

Table 4.2. Some Examples of Vectors and their Maximum Insert Sizes

Vectors	Size of inserts (kb)
Plasmids (pBR 322)/Phagemids (pGEM, pBSK)	0–10
Bacteriophage (insertion vectors; *e.g.,* gt 10, gt 11, Lambda Zap, Unizap XR, Zap-express)	0–10
Bacteriophage (replacement vectors; Lambda Fix II, Lambda DASH II, EMBL, Charon)	9–23
Cosmid vectors ??(SuperCos I, pWE15, COS)	25–45
Bacteriophage P1	60–100
PAC (PI artificial chromosomes)	100–150
BAC (bacterial artificial chromosomes)	100–300
YAC (yeast artificial chromosomes)	200–2000
HAECs	
NACS	

Plasmids and phage DNAs are commonly used as vectors.

Plasmid DNA: Besides the larger chromosomal DNA, bacteria have much smaller circular plasmid DNA molecules. They can easily be separated on the basis of size. The bacterial cells are lysed open and by centrifugation, chromosomal DNA is centrifuged down. The plasmid DNA remains in the supernatant above the pellet. The plasmids are then purified before cutting with a restriction enzyme. One of the first plasmids used for RDT was pBR322 (Fig. 4.15).

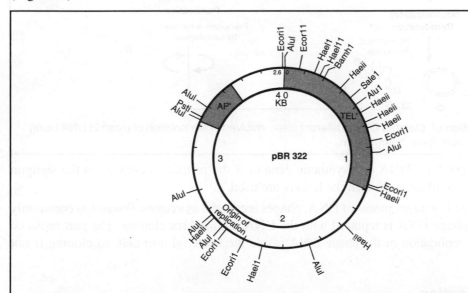

Fig. 4.15. The pBR322 plasmid. A map of the *E. coli* plasmid pBR322. The map is marked off in units of 1×10^5 daltons (outer circle) and 0.1 kilobases (inner circle). The locations of some restriction enzyme sites are indicated. The plasmid has resistance genes for ampicillin (AP^r) and tetracycline (Tet^r).

In case restriction enzyme has been used to isolate the donor DNA (*i.e.,* the shotgun approach), the same restriction enzyme must be used for the plasmid DNA. The restriction fragments from the donor DNA, including those containing the wanted gene, are then mixed with the plasmid DNA and joined by their sticky ends. For example, the sticky end –GGTT will bind to the complementary sticky end –CCAA. The initial attraction is due to hydrogen bonding, but the sugar-phosphate backbones are then joined using an enzyme DNA ligase. The plasmid DNA is Isolated by using the protocol summarized in Fig. 4.16.

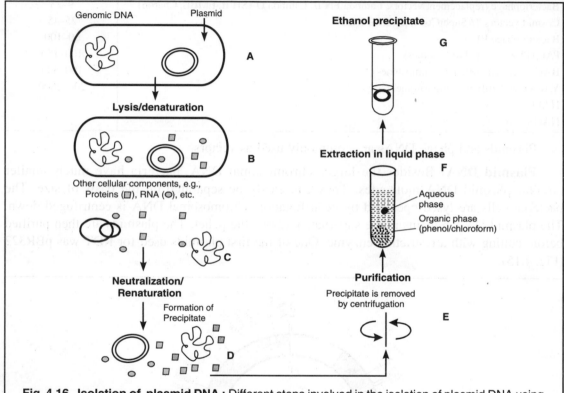

Fig. 4.16. Isolation of plasmid DNA : Different steps involved in the isolation of plasmid DNA using alkalilysis technique have been shown.

If the donor DNA is cDNA or a synthetic gene or if the restriction enzyme in the shotgun method has produced blunt ends, then the linkers are used.

Phage vector : For larger pieces of DNA, phages are useful as vectors. Phage λ is commonly used. Part of the phage DNA is replaced with the DNA required for cloning. The part replaced is not needed for replication of the phage DNA inside the bacterial host cell, so cloning is not affected.

4.2.3 DNA Ligation

After cutting DNA molecules, they have to be joined to create artificially recombinant molecules by 3 methods.

   (i)   Join covalently the annealed cohesive ends using *E. coil* ligase.

(ii)   Join covalently the blunt-ended fragments using T4 ligase.

(iii)   Homopolymer tailing using deoxynucleotidyl transferase

The discovery of DNA ligases was one of the greatest "happenings" in 1967 (Singer, 1979).

It was another event important to the growing congruence of the DNA revolution and the enzyme revolution; and the atmosphere was enlivened because the discovery was made essentially simultaneously by several independent groups working in the laboratories of Gellert, Richardson, Lehman, Hurwitz and Korenberg.

The importance of the enzyme was emphasized by its presence in *E. coli* itself, and by the synthesis of new ligase upon infection of cells with bacteriophage T4. Although the role of enzyme in replication itself was not yet suspected, it was seen as an attractive tool to join the DNA chains during genetic recombination.

DNA ligase seals single stranded nicks between adjacent nucleotides in a duplex DNA chain. The *E. coli* ligase and T4 enzymes differ significantly in two ways.

(i)   The *E. coli* ligase requires diphosphopyrimidine nucleotide as a co-factor (NAD^+), while the T4 ligase requires ATP even though each uses it's co-factor to form analogous, ligase-adenylated and DNA-adenylate intermediates.

(ii)   The T4-ligase does not require overlapping complementarity on the single-stranded ends of the two chains to be joined as does the *E. coli* ligase.

When termini created by a RE that creates a cohesive ends associate, the joint has nicks a few base-pairs apart in opposite strands. *E. coli* DNA ligase can repair these nicks to produce intact duplex.

The optimum temperature for ligation is 37°, but at this temperature the H bonded joint between the sticky ends is unstable. The optimum temperature for ligation of cohesive termini has been found to be in the range of 4-15°, which is a compromize between the rate of enzyme action and association of the termini.

The ligation reaction can be performed to favour the formation of recombinants. The population of recombinants can be increased by performing the reaction at a high DNA concentration, thereby preventing circularization of linear fragments. By treating linearized plasmid vector DNA with *alkaline* **phosphatase** to remove 5' terminal phosphate groups, both recircularization and plasmid dimer formation are prevented.

Alkaline phosphatases catalyze the hydrolysis of phosphomonoesters. They were discovered by Horiuchi *et al.*, and Torriani in 1950's. They are used for removing phosphate from monoester ends that are either at nicks in duplex DNA, or otherwise covered by an overhanging complementary strand.

T4 ligase is used to join blunt-ended DNA molecules, "blunt-end" ligation, the only requirement being the presence of a phosphomonoester end group at the 5' terminal and a hydroxy group at the 3' terminal.

Blunt-end ligation is most usefully applied to join blunt-ended fragments via linker molecules *e.g.,* self-complementary decameric oligonucleotides are synthesized containing one or more restriction sites. The molecule is then ligated to both ends of the foreign DNA to be cloned, then treated with restriction endonucleases to produce a sticky – ended fragment which can be incorporated into a vector molecule that has been cut with the same RE. Insertion by means of the linker creates RE target sites at each end of the foreign DNA and so enables the foreign DNA to be excised and recovered after cloning and amplification in the host bacterium.

4.2.3.1 Double Linkers

Plasmid vectors possessing a set of closely clustered cloning sites (*e.g.,* pUC8) have been used to clone duplex cDNA molecules by the double linker approach, in which different linker molecules are added to the opposite ends of the cDNA. By this method problem of vector borne promoter is sought. Double linkers are short, blunt-ended pieces of double-stranded DNA of known sequences, with one or more restriction sequences inserted within it. The linkers, which can be synthesized in large quantities are attached to blunt-ended DNA molecules by DNA ligase. Although this is a blunt end ligation, this reaction can be performed with high efficiency as the linkers can be added in high concentrations. It is important to realize that the use of high linker concentrations would inevitably cause multiple linkers to be attached to either side of the DNA molecule. However, restriction digestion with the appropriate enzymes will create cohesive ends within the linkers. This modfied DNA-linker molecules will now be ready to be ligated to a cloning vector restricted with the appropriate restriction enzymes.

4.2.3.2 Adapters

Sometimes the RE used to generate the cohesive ends in the linker may also cut the foreign DNA at internal sites, thus DNA will be cloned as sub-fragments. One solution is to choose another RE, but there may not be a suitable choice if the foreign DNA is large and has multiple restriction sites. Another solution is to methylate internal restriction sites with methylase. Alternatively, chemically synthesized adapter molecules which have preformed cohesive end, can be used (Fig. 4.17).

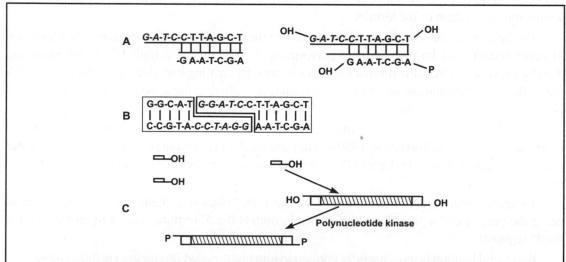

Fig. 4.17. Use of adaptors in the cloning of DNA. (A) An example of a typical adaptor and a 5'-modified adaptor, **(B)** Typical adaptors could ligate to one another, **(C)** Ligation with the modified adapter. (Adopted from Microbial Biotechnology Ed. Lee, 2006).

4.2.3.3 Homopolymer Tailing

A general method for joining DNA molecules makes use of the annealing of complementary homopolymer sequences. Thus, by adding oligo (dA) sequence to the 3'-ends of one population of DNA molecules, and oligo (dT) to the 3'-ends of another population, the 2-types of molecules can be annealed to form mixed dimeric circles. Typically 10-40 homopolymeric residues are added to each end.

Homopolymeric extensions can be synthesized by enzyme **terminal deoxynucleotidyl transferase,** it was first purified from calf thymus (Singer,1979). It repeatedly adds nucleotides to the 3'OH termini of DNA molecules exposed due to treatment with phage exonuclease or by restriction with enzyme *Pst I.*

In 1972, Jackson *et al.* were among the first to apply the homopolymer method when they constructed a recombinant in which a fragment of phage DNA was inserted into SV40 DNA. The annealed circles are directly used for transformation with repair of gaps occurring *in vivo*. They provide important application in cDNA cloning.

4.2.4 Introduction of Vector DNA into the Host Cell

The plasmid or phage vector is then introduced into a bacterial cell wherein the vector multiplies. Commonly the bacterium *Escherichia coli* is used as its genetics is known and it rapidly grows with a doubling time of 30 minutes. For genetic engineering, a mutant form of *E. coli* has especially been developed. This form can only survive in special laboratory conditions. Therefore if it escapes with foreign genes inserted, it cannot infect humans.

If a plasmid vector is being used, it is added to a flask containing a culture of *E. coli*. Calcium ions, usually in the form of calcium chloride, are added to the flask. Then brief heat shock is given. This creates holes in the cell surface membranes of the *E. coli*. These make it permeable to DNA and allow the plasmids to enter. This process of adding new DNA to a bacterial cell is called **transformation** (Fig. 4.18). Phage vectors are introduced by infection of a bacterial lawn growing on an agar plate.

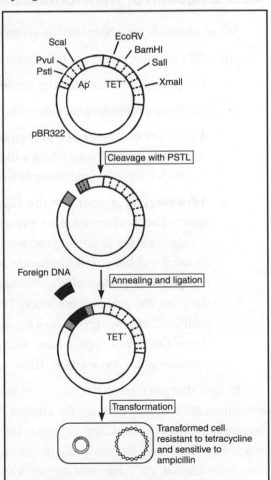

Fig. 4.18. Transformation of bacterial cells using plasmid DNA : Various steps involved in the transformation have been shown.

4.2.5 Cloning the DNA

More than 10^{12} identical copies are formed by a single phage containing one recombinant DNA molecule in less than one day. *E. coli* cells containing plasmids are usually plated out onto nutrient agar in Petri dishes. They can grow and divide once every 30 minutes. Ultimately visible colonies are found. In this way, at least as many copies of the required DNA are obtained as on this phage vector. Bacteria can also contain hundreds of copies of a plasmid and these will be copied each time the bacterium divides. In a very short time, billions of clones can, thus, be produced with both the techniques. Before further cloning, the transformed bacteria must now be selected.

4.2.6 Selection of Transformed Bacteria

When plasmids have been used as vectors, two problems arise on their mixing with bacteria.

(i)   All the bacteria will not be transformed (take up plasmids).

(ii) The plasmids will not take up the foreign donor DNA.

To avoid these problems, plasmids with the following characteristics are chosen :

- *A gene for resistance to a particular antibiotic:* The bacteria are grown on a medium containing that antibiotic which will allow only the transformed cells (the ones containing plasmids) to survive and multiply to form colonies.

- *Advantage of a gene for the β-galactosidase which has a group of restriction sites:* The performance of the gene is not affected by these restriction sites. β-galactosidase is an enzyme which breaks down lactose to galactose and glucose (it breaks down any disaccharide and glucose). It can also break down a colourless compound X-gal to a blue compound. If foreign DNA is inserted at a restriction site in the gene, the gene will not work. Therefore, if the bacteria that survive growing on the antibiotic, are then grown on a medium containing X-gal, those colonies which lack the donor DNA will appear blue. Bacteria which form colourless colonies are the ones containing the donor DNA. These can be isolated for further cloning.

In case shotgun method has been used the bacteria which are successful in cloning the DNA containing the required gene are the clones. This is because the donor DNA was a mixture of a very large number of restriction fragments (up to a million in the case of human DNA). Only one or a few of these may be containing the piece of DNA or gene required for cloning, but all will have been cloned. As mentioned earlier, a mixture of clones like this is called **a library.** If a mixture of mRNAs is used in the step I. A library will also be produced by the reverse transcriptase method. Sometimes this is necessary if the desired mRNA cannot be isolated in pure form.

Using a gene probe : For selecting the required bacteria, **gene probe** is used. This, however, is possible if some or all of the base sequence of the DNA being looked for is known. The base sequence (or one very similar) can also be predicted from a knowledge of the amino acid sequence of the protein it codes. The DNA or RNA probe is a short sequence of nucleotides which is complementary to part of the required DNA and will, therefore, bind to it. For example, the probe GATGGA would find and hydrogen bond to CTACCT. Probes can be as short as 15 to 20 nucleotides or much longer. Usually the probes are made from radioactively labeled nucleotides using the radioactive element ^{32}P. When this binds to the DNA, the radioactivity acts as a marker which can be detected by autoradiography. But now probes can be labeled non radioactively using biotin.

4.3 Interference and miRNA

One of the most exciting findings in the recent years has been the discovery of RNA interference (RNAi) methodologies which hold the promise to selectively inhibit gene expression in mammals. RNAi is an innate cellular process activated when a double-stranded RNA (dsRNA) molecule of greater than 19 duplex nucleotides enters the cell, causing the degradation of not only the invading dsRNA molecule, but also single-stranded RNAs (ssRNAs) of identical sequences, including endogenous mRNAs. The use of RNAi for genetic-based therapies has been widely studied, especially in viral infections, cancers, and inherited genetic disorders. As such, RNAi technology is a potentially useful method to develop highly specific dsRNA-based gene-silencing therapeutics.

RNA interference (RNAi) is a mechanism in the cell biology of many eukaryotes in which fragments of double-stranded ribonucleic acid (ds RNA) interfere with the expression of a particular gene whose sequence is complementary to the dsRNA. RNAi is mediated by the same cellular machinery that processes microRNA, small RNA molecules involved in large-scale gene regulation in the cell. In 2006, American scientists Andrew Fire and Craig C. Mello shared the Nobel Prize in Physiology/Medicine for their work on RNA interference in the nematode worm *Caenorhabditis elegans*, which they initially described in a seminar in 1998 (paper published in the journal Nature).

Before RNA interference was well characterized, the phenomenon was known by other names, including post transcriptional gene silencing, transgene silencing, and quelling. Single stranded antisense RNA was introduced into plant cells and hybridized to the homologous single-stranded "sense" messenger RNA. It is now clear that the resulting dsRNA was responsible for reducing gene expression.

The ability of RNAi to dramatically and selectively reduce the expression of an individual protein in a cell makes RNAi a valuable laboratory research tool, both in cell culture and *in vivo* in livnig organisms. It is particularly useful in certain organisms such as *C. elegans*. Large-scale screens that systematically shut down each protein in the cell can aid in identifying the necessary components for a particular cellular process or event and can identify which proteins are required for cell survival and replication. RNAi also holds promise as a therapeutic technique in human disease (Fig. 4.19).

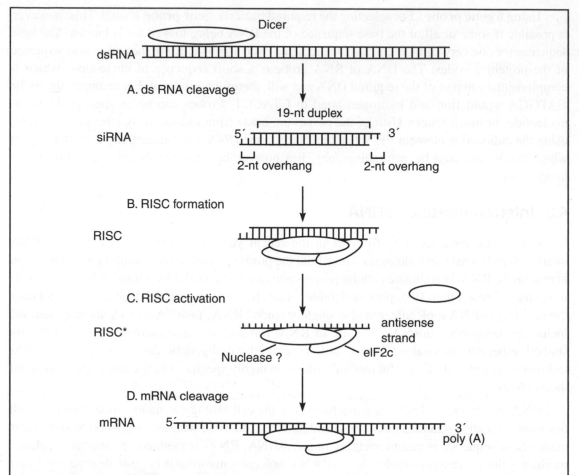

Fig. 4.19. Current model for RNA interference. RNAi process can be divided into four stages **(A)** dsRNA cleavage by Dicer and generation of siRNA duplex, **(B)** recruitment of RNAi factors and formation of RISC (RNA-induced silencing complex. **(C)** siRNA unwinding and RISC activation and **(D)** mRNA targeting and degradation (Kim, 2003).

RNA interference (RNAi) in eukaryotes is a recently identified phenomenon in which small double stranded RNA molecules called short interfering RNA (siRNA) interact with messenger RNA (mRNA) containing homologous sequences in a sequence-specific manner. Ultimately, this interaction results in degradation of the target mRNA. Because of the high sequence specificity of the RNAi process, and the apparently ubiquitous expression of the endogenous protein components necessary for RNAi, there appears to be little limitation to the genes that can be targeted for silencing by RNAi. Thus, RNAi has enormous potential, both as a research tool and as a mode of therapy. Several recent patents have described advances in RNAi technology that are likely to lead to new treatments for cardiovascular disease. These patents have described methods for increased delivery of siRNA to cardiovascular target tissues, chemical modifications of siRNA that improve their pharmacokinetic characteristics, and expression vectors capable of expressing RNAi effectors *in situ*. Though RNAi has only recently been demonstrated to occur in mammalian tissues, work has advanced rapidly in the development of RNAi-based therapeutics. Recently, therapeutic silencing of apolipoprotein B, the ligand for the low density lipoprotein receptor, has been demonstrated in adult mice by systemic administration of chemically modified siRNA. This demonstrates the potential for RNAi-based therapeutics and suggests that the future for RNAi in the treatment of cardiovascular disease is bright.

There has been lot of discussion on miRNA being playing role in the expression of pattern of genes (Fig. 4.20).

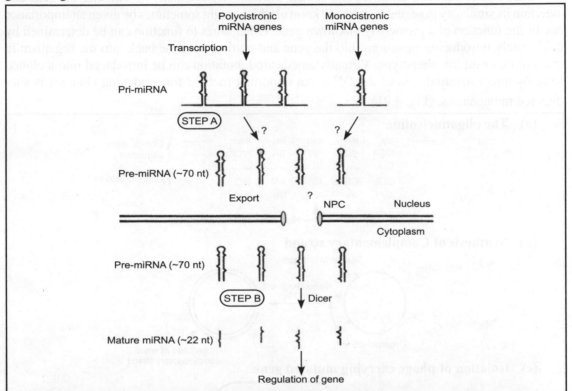

Fig. 4.20. A model for miRNA biogenesis and function. miRNA genes are transcribed by an unidentified polymerase to generate the primary transcripts, referred to as pri-miRNAs. Illustrated in the upper left is the clustered miRNA such as miR-23~27~24-2 of which the pri-miRNA is polycistronic. Illustrated in the upper right is the miRNA such as miRN-30-a of which the pri-miRNA is monocistronic. The first step processing (Step A) releases pri-miRNAs of ~ 70-nt that is recognized and exported to the cytoplasm. The processing enzyme for the Step 1 and the export factor are unidentified. Upon export, Dicer and possibly other factors participate in the second-step processing (Step B) to produce mature miRNAs. The final product may function in a variety of regulatory pathways, such as translational control of certain mRNAs. The question marks indicate unidentified factors (Kim, 2003).

Single base substitutions or short segment substitutions a DNA sequence. DNA cloned into a plasmid containing an F! origin of replication is able to replicate single-strands of plasmid DNA. These single-stranded DNAs are primed with an oligonucleotide that has the mismatched base (S) and extended with the Klenow fragment of DNA polymerase I. The plasmid is re-transformed into a bacterial host and allowed to replicate. Half of the progeny will contain the mutation. In modification of this techniques two mutating primers, one to mutate the base of interest and a separate primer that mutates a selectable marker, such as a drug resistance gene, to be either functional or non-functional are used with this approach. Mutants can be identified by selection instead of being screened by sequencing or restriction site analysis.

4.4 SITE SPECIFIC MUTAGENESIS

Once a gene has been identified and cloned, the next step is usually to sequence the gene and ascertain its similarity to sequences already known, which might sometimes be given an importance due to the function of a previously unknown gene. Other clues to function can be determined by deliberately introducing mutations into the gene and putting the gene back into the organism to observe the resulting phenotype. Virtually any desired mutation can be introduced into a cloned gene by direct manipulation of the DNA. An important method for producing changes is site-directed mutagenesis. (Fig. 4.21)

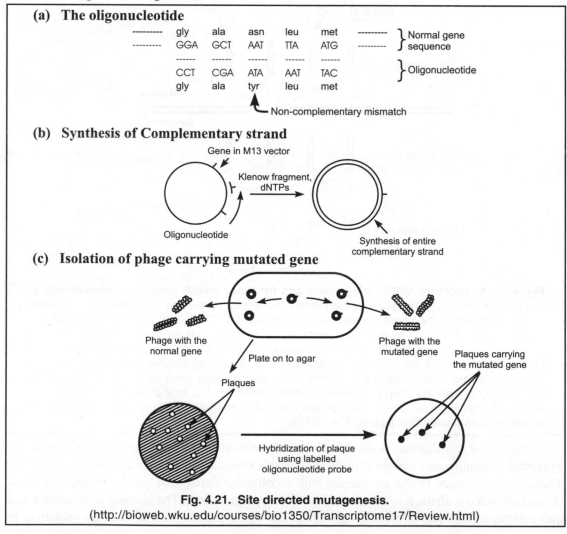

Fig. 4.21. Site directed mutagenesis.
(http://bioweb.wku.edu/courses/bio1350/Transcriptome17/Review.html)

Three different methods have been devised:

(1) Cassette mutagenesis: A synthetic DNA fragment containing the desired mutation is used to replace the corresponding sequence in the wild-type gene. This method was originally used to generate improved variants of the enzyme subtilisin by Wells *et al.* in 1985. It is a simple method and 100% efficient. The disadvantages are the requirement for unique restriction sites flanking the region of interest and the limitation on the realistic number of different oligonucleotide replacements which can be synthesized (Botstein and Shortle, 1985; Old and Primrose, 1994).

(II) Primer extension: There are no constraints on the types of mutations that can be induced by this method.

Single-primer extension – strategy developed for incorporating a short mutant primer (7-20 nucleotides long), into a longer segment of DNA. DNA polymerase from T4 and T7 phages is used to copy wild-type sequences onto both ends of the primer.

Double primer extension – A disadvantage of all the primer extension methods is that they require a ss-template. Oslen and Eckstein in 1990 have adapted the phosphorothionate method for use with a ds-template. Another plasmid based method was developed by Deng and Nickeloff (1992), which relies on denaturation of ds-template to allow annealing of primer followed by extension. Amplification of the mutant is performed by two rounds of transformation.

(III) PCR-Methods of Site-Directed Mutagenesis

The advantage of a PCR-based mutagenic protocol is that the desired mutation is obtained with 100% efficiency.

Disadvantages are that the

(I) PCR product usually needs to be ligated into a vector, although Sarkar and Sommer (1990) have generated the mutant protein directly using coupled *in vitro* transcription and translation.

(II) Taq polymerase copies DNA with low fidelity. Therefore, sequence of the entire amplified segment generated by PCR mutagenesis must be determined to ensure that there are no extraneous mutations. Two thermostable polymerases with improved fidelity have been described. *Thermococcus litoralis* polymerase (Vent) and polymerase isolated from P*yrococcus furiosus* by Cariello *et al.,* (1991) and Lundberg *et al.,* (1991).

Finally, all the methods of *in vitro* mutagenesis, including those that require synthetic oligonucleotides can generate nucleotide sequence changes at sites other than those targeted for mutations. Therefore, even though a specific mutation of defined nucleotide sequence has been isolated and sequenced, an observed change in phenotype cannot immediately be attributed to the constructed mutation. Except for determining the entire nucleotide sequence of the gene or genetic element that has undergone mutagenesis. Genetic mapping of the mutation is the only conclusive means of establishing the connection between a change in nucleotide sequence and its phenotypic consequence (Botstein and Shortle, 1985).

4.5 THE POLYMERASE CHAIN REACTION

The generation of large number of identical copies of DNA by the construction and cloning of rDNA molecules was made possible in the 1970s. However, the cloning of DNA is time consuming, involving insertion of DNA into vectors and typically the screening of libraries to detect specific DNA sequences. In the mid-1980s the POLYMERASE CHAIN REACTION (PCR) was developed and this has resulted in yet a new revolution in the way genes may be analyzed. PCR is an *in vitro* method for the primer-directed enzymatic synthesis of millions of copies of specific DNA segment (amplification). Kary Mullis, who developed PCR, was awarded the Nobel Prize in 1993.

4.5.1 Principle

The PCR involves the enzymatic amplification of DNA *in vitro*. The method is capable of increasing the amount of a target DNA sequence in a sample by synthesizing many copies of the DNA segment. The reaction is based on annealing and extension of 2 oligonucleotide primers that flank the target region in duplex DNA. After the denaturation of DNA, achieved by heating, each primer hybridizes to one of the two separated strands such that extension from each 3- hydroxy end is directed towards the other. The annealed primers are then extended on the template strand by using DNA polymerase. Three steps (denaturation, primer binding and primer extension) constitute a PCR cycle (Fig. 4.22). If the newly synthesized strands extend to or beyond the region complementary to the other primer, it can serve as a primer-binding site and template for subsequent primer annealing and primer extension results in the exponential accumulation of a discrete fragment whose termini are defined by the 5 ends of the primers. The exponential amplification occurs because, under appropriate conditions, the primer extension products synthesized during a given cycle, function as templates for the other primer in subsequent cycles. Thus, each cycle of PCR essentially doubles the amount of DNA in the region of interest. The length of the products that accumulate during PCR is equal to the sum of the lengths of the two primers plus the distance of PCR amplification. PCR can amplify double-stranded or single-stranded DNA and with the reverse transcription of RNA into a cDNA copy, RNA can also serve as a target. Because the primers become incorporated into the PCR product and some base pair mismatches (away from the 3' -end) between the primer and the *ori*ginal genomic template can be tolerated, new sequence information (*e.g.,* specific mutations, restriction sites regulatory elements) and labels can be introduced via the primers into the amplified DNA fragments.

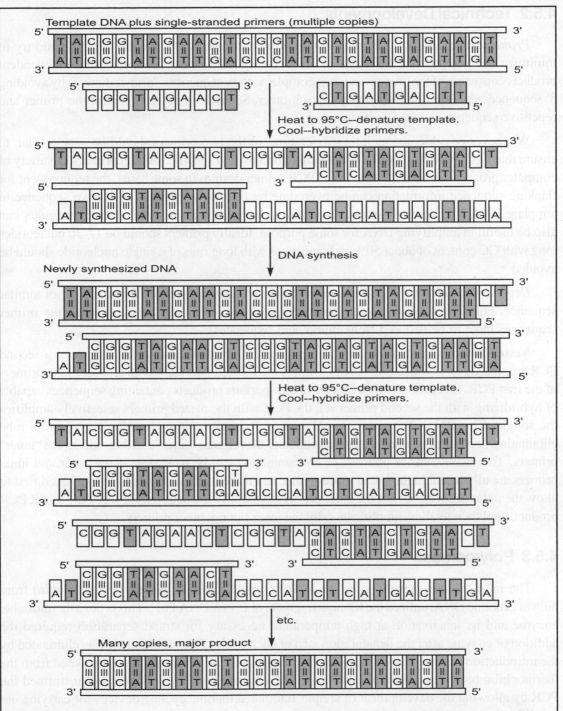

Fig. 4.22. PCR produces many copies of a DNA segment. The DNA segment lies between and includes the sequences at which two single-stranded primers hybridize to the template DNA molecule.

4.5.2 Technical Developments

Primers: The design of oligonucleotide primers to amplify a specific target should try to minimize the potential for PCR primer artifacts (*e.g.,* the synthesis of template independent products consisting of the primers and their complementary sequences-"primer dimer") by avoiding 3' sequences that are complementary to each other. Secondary structure within the primer and repetitive sequence should also be avoided.

Whenever possible, the melting temperatures of the two primers should be very similar, to ensure that a given thermal profile is optimally efficient and specific for both primers. A variety of computer programs are available to aid in PCR primer design. In some cases, the requirement for flanking sequence information can be overcome by ligating a fragment of known sequence to template DNA fragments to serve as a priming site. PCR with random sequence primers can also be useful in amplifying DNA for some purpose. Ideally primers should be 17-30 nucleotides long with GC content of about 50% and sequences with long runs of a single nucleotide should be avoided.

Degenerate Primers: A degenerate primer is actually a mixture of primers all of similar sequence, but with variations at one or more positions (especially required when the primer sequences have to be deduced from amino acid sequences).

Nested Primers: The products of an initial PCR amplification are used to seed a second PCR amplification in which one or both primers are located internally with respect to the primers of the first PCR. Since there is little chance of the spurious products containing sequences capable of hybridizing with the second primer set, the PCR with the nested primers selectively amplifies the sought after DNA. The conventional nesting strategy requires opening the reaction tube eliminating/decreasing/diluting the concentration of the original "outer" and then adding the "inner" primers. The inconvenience and risk of contamination can be overcome if the outer and inner primers are all present in the initial reaction mix and if the thermal profile is programmed first to allow the outer primers, but not the inner primers to amplify the targeted subset of the initial PCR products and then to allow amplification by the inner but not outer primers.

4.5.3 Polymerases

The initial studies that relied on the PCR to amplify specific targets (*e.g.,* β-globin) from human genomic DNA, utilized the Klenow fragment of *E. coli* DNA Pol I. This is not a thermostable enzyme and its inactivation at high temperature necessary for strand separation required the addition of enzyme after the denaturation step of each cycle. This requirement was eliminated by the introduction of a thermostable DNA polymerase, the Taq DNA polymerase isolated from the thermophilic bacterium *Thermus aquaticus*. The use of Taq DNA polymerase transformed the PCR by allowing the development of simple automated thermal cycling devices for carrying out amplification reaction in a single tube containing all the necessary reagents. It has also increased the specificity and yield of the amplification reaction by allowing the primers to be annealed and extented at a temperature much higher (*e.g.* 60° C) than was possible with the Klenow fragment.

Moreover, longer PCR products could be amplified from genomic DNA probably due to reduction in 2°C structures of template strands at the elevated temperature.

The Taq DNA polymerase lacks a 3'–5' proof reading exonuclease activity. This lack appears to contribute to errors during PCR amplification due to misincorporation of nucleotides. To overcome this problem, other thermostable DNA polymerases with improved fidelity have been sought e.g, by genetic engineering.

Many of new thermostable polymerases have additional useful activities. The thermostable DNA polymerase from *Thermus thermophilus* (T th) can reverse transcribes RNA efficiently in the presence of $MnCl_2$ at high temperatures. Under appropriate conditions, the DNA polymerase activity can also occur in the presence of $MnCl_2$ allowing both cDNA synthesis and PCR amplification to be carried out in a single-enzyme, single tube reaction. Two thermostable polymerases with improved fidelity have been described: *Thermococcus litoralis* polymerase (Vent) by Cariello *et al.* (1991) and *Pyrococcus furicans* polymerase by Lundberg *et al.* (1991).

4.5.4 Protocols

PCR amplifications with Klenow fragments were not highly specific, although a unique DNA fragment could be amplified approximately 200,000 – fold from genomic DNA, only about 1% of the product was, in fact, the targeted sequence. Amplification with the Taq DNA polymerase greatly increased the specificity. Conditions that increase the stringency of primer hybridization, such as higher annealing temperatures and lower $MgCl_2$ enhance specific amplification. In addition the specificity of the PCR can be affected by the enzyme concentration, primer concentration, annealing time, extension time and the number of cycles. The relative homogeneity of the PCR products is influenced by the concentration of specific sequence in a sample.

The PCR specificity is being improved on recognizing that the Taq DNA polymerase retains considerable enzymatic activity at temperatures well below the optimum for DNA synthesis (72°C). Therefore, if the DNA polymerase is activated only after the reaction has reached high (70°C) temperatures, non-target amplification can be minimized. This can be accomplished by "hot start": the manual addition to the reaction tube of an essential reagent at elevated temperatures. This also increases specific amplification. "Hot Start" also minimizes the formation of primer-dimer. PCR specificity can also be increased using nested-primer approach.

4.5.5 Problems

Contamination : Amplification by PCR is extremely rapid. Twenty five cycles can be carried out in just over one hour. Beginning with 1µg of human DNA, which contains about 300,000 copies of each unique sequence, 25 cycles of PCR can generate upto several mg of a specific product several hundred bp in length. Contamination of reaction with products of an earlier PCR reaction or with exogenous DNA or cellular material can create problems. Precautions must be taken to minimize the risk of contamination. These include :

(a) attention to careful laboratory procedures,

(b) the prealiquoting of reagents,

(c) the physical separation of the reaction preparation from the area of reaction product analysis and

(d) the use of dedicated pippets, positive–displacement pippets, or tips with aerosol barriers. To monitor the process and to reveal any contamination multiple negative controls are necessary.

The product carryover can be minimized by utilizing the principles of the restriction-modification and excision repair systems of bacteria to pretreat amplification reactions and selectively destroy DNA synthesized in an earlier PCR.

Misincorporation and "Recombinant PCR" : During PCR amplification, in the newly synthesized strand a mismatched nucleotide can become incorporated. The rate of misincorporation depends on the reaction conditions (*e.g.,* nucleotide concentration, pH). The rate for Taq polymerase is 10 nucleotide/cycle. In addition by *in vitro* recombination or template strand switching the hybrid sequences may be generated .

It is now evident that PCR has had major impact on molecular biology and many other fields and several accounts of the concept and its development from an idea into a routine technique have appeared (Arnheim, 1990; White,1996).

It is now possible to amplify longer fragments by changes in DNA extraction protocols and use of modified buffers. The latter prevents single-strand nicks and provide longer and intact templates. In amplificantion of longer fragments longer extension times (15minutes) relatively high denaturation temperatures, addition of cosolvents as glycerol and DMSO, hot start technique as well as choice of thermostable polymerase are critical.

Automated thermal cyclers have now been developed. These thermal cyclers have increased rates of heating, cooling and heat transfer to modified reaction vessels and can accommodate more samples (*e.g.,* 96-well array). For on-line monitoring of amplification reactions ("Kinetic PCR"), instruments that detect the fluorescence accumulating in reaction vessels have also been designed using either fibre-optics or video cameras.

The *in situ* PCR, has applications in diagnostic as well as basic research potential.

4.5.6 Diversification of PCR Technology (Arnheim, 1990)

(1) *DNA Cloning:* To simplify cloning, the PCR primers can be constructed so that in addition to the DNA sequence homologous to the region flanking the target, additional sequences containing a restriction endonuclease cleavage site can be appended to the primers 5' -end. Cutting the PCR product with appropriate RE, allows it to be ligated to an appropriate cloning vector.

(2) *Inverse PCR:* Even when DNA sequence information is known on only one side of the target it is possible to amplify a target sequence. Consider two primers which lie within a region of known sequence. The unknown sequences on either side of this

region comprise the target region. The primers are designed so that their 3'-ends face away from each other. The linear DNA molecule containing the target and the primer sequences is first circularized by restriction enzyme digestion followed by ligation and amplification by PCR. This technique allows amplification of unknown DNA sequences to one or the other or both sides of a known DNA segment.

(3) *Whole genome PCR:* Starting from specific segments of human chromosomes dissected from whole chromosomes a DNA library is constructed. This is used for the analysis of protein-DNA interactions. In Sept.1989, Olson *et al.* proposed to use PCR as the basis for translating all types of physical mapping land marks into a common language of short fragments of single-copy DNA sequence called 'sequence-tagged sites' (White, 1996).

(4) *Analysis of RNA populations by cDNA amplification:* Even if the available sequence data for constructing primers are limited, PCR of reverse transcribed copies of specific RNAs can be carried out. This approach is termed "RACE" or "single-sided PCR". This method makes use of the fact that most mRNAs contain a polyA tail at the 3'-end.

(5) *PCR cloning of cDNA with degenerate primers:* PCR amplification has also been used to clone specific cDNA in the complete absence of any available nucleic acid sequence information. In these experiments, protein sequence data were used to design the primers for PCR in a manner analogous to the design of mixed oligonucleotides for screening cDNA libraries by hybridization.

(6) *The purposeful modification of DNA by PCR:* With the use of a primer with a base-pair substitution, a deletion or an insertion relative to the target sequence, large amounts of mutant PCR product can be generated.

(7) *Analysis of DNA-protein interaction:* PCR has been used to select, among all of the sequences in the genome, those that can bind to a specific protein.

(8) *Genome footprinting:* Genome footprinting with PCR has also been tried.

(9) *Digital PCR:* Single molecules are isolated by dilution and individually amplified by PCR; each product is separately analyzed for an expected mutation by using fluorescent probes. This approach, transforms the exponential, analog nature of PCR into linear, digital signal (Vogelstein and Kinzler, 1999).

4.5.7 Applications

According to White (1996) PCR has the following applications.

(1) *Humun genome project:* PCR has been used as basis for translating all types of physical mapping landmarks into STSs, for amplification of short tandem repeat (STR) and has helped in the analysis of cDNA clones to generate expressed sequence tags (EST).

(2) *Single sperm analysis:* Sensitivity of PCR allows amplification of DNA target from a single molecule of template derived from a single haploid cell or gamete. PCR is used in sperm typing, to co-amplify polymorphic regions at two loci for which the individual is

heterozygous. The alleles at each locus are then identified and the genotype of each meiotic product is directly obtained. Recombination frequencies can then be estimated by comparing the number of recombinant sperm to the total number examined.

(3) *Molecular archaeology and ancient DNA:* PCR has made it possible to extend the analysis of molecules back in time to directly examine species and populations that may now be extinct and to determine their relationship to living people/ organisms, because of its ability to amplify regions of DNA from very small amounts of degraded, archaic samples.

(4) *Molecular ecology and behaviour:* PCR has an important role in the integration of genetic data with historical or field observations of various species to address topics such as sex determination hybrid zones and gene flow *etc.* It is also used to obtain multilocus genotypes from samples such as hair, feathers *etc.*

(5) *Diagnosis of infectious disease and molecular epidemiology:* PCR has played a seminal role in the field of study of slow, fastidious microorganisms as well as in the field of emerging disease via its use in the discovery and rapid identification of the pathogens. PCR can be further used in epidemiological research and in the characterization of likely modes of disease transmission. It is used to establish a rapid and highly sensitive prenatal diagnostic test for sickle cell anaemia (Saiki *et al.,* 1985).

(6) *Drug discovery:* PCR is used in positional cloning to identify genes that cause disease and also to identify potential new targets for drug screening. PCR based telomerase assay can potentially be used both for screening for telomerase inhibitors as well as for cancer diagnosis. PCR also has potential application in gene therapy.

(7) *Forensic science:* PCR has been used to analyze the genetic patterns of polymorphic DNA sequences amplified from biological evidence found at crime scenes. It can also be used in paternity testing as well as in the monitoring of bone marrow engraftment by distinguishing donor and recipient cells. Potential mix-ups of cell lines and clinical specimens have also been-detected quickly by simple PCR-based arrays.

4.5.8 Q-Beta Amplification

An alternative approach to PCR, Q-beta amplification, has been deviced which exploits the nucleic acid replication, whose exponential nature produces a gain of 10 fold or more in a short period of time (Lizardi and Kramer, 1991). Exponential amplification methods are based on either DNA or RNA replication using DNA polymerase I or RNA replicase respectively. This technique enables the designing of ultra-sensitive diagnostic assays for infections assays.

4.6 ANALYSING DNA SEQUENCES

DNA sequencing is a fundamental capability for modern gene manipulation. Techniques for large-scale DNA sequencing became available in the late 1970s (Fig. 4.23). (1) The chemical degradation method of Maxam and Gilbert (1977), and (2) the enzymatic dideoxynucleotide chain termination method of Sanger *et.al.* (1977).

The basic principle of these two techniques is different. Both of these methods generate different species of radiolabeled oligonucleotides that start from a fixed point but terminate randomly at a fixed residue. These randomly generated oligonucleotides are resolved by electrophoresis for different nucleotides loaded into adjacent lanes of a sequencing gel. It is possible to read the nucleotide sequence directly from the image of the gel on an X-ray film from the bottom of the gel towards the top.

The Maxam and Gilbert technique, relies on base-specific chemistry, was popular for a time. But chain-terminator techniques soon gained popularity. Since there have been many modifications and new technological developments. It could soon be possible to read one's DNA in a day instead of a decade (Voss, 1999). The international race to sequence the human genome has turned gene sequencing into a high-speed and high profile endeavour (Voss, 1999).

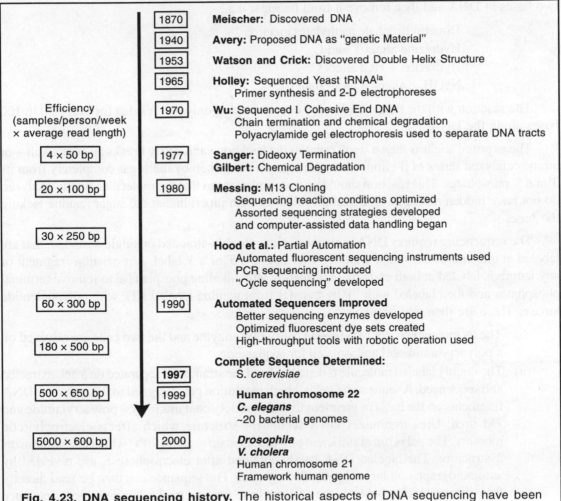

Fig. 4.23. DNA sequencing history. The historical aspects of DNA sequencing have been summarized.

4.6.1 DNA Sequencing by the Maxam and Gilbert Method

This method for DNA sequencing makes use of chemical reagents to bring base-specific cleavage of the DNA.

This method determines the nucleotide sequence of a terminally labeled DNA molecule by breaking it at adenine, guanine, cytosine or thymine with chemical agents. Partial cleavage at each base, produces nested set of radioactive fragments extending from the labeled end to each of the positions of that base. Polyacrylamide gel electrophoresis resolves these single-stranded fragments; their sizes reveal, in order, the points of breakage. The autoradiograph of a gel produced from four different chemical cleavages each specific for a base, then shows a pattern of bands from which the sequence can be directly read. The method is limited only by the resolving power of the polyacrylamide gel.

The base specific cleavage depends upon the chemical agents that selectively alter one or two bases in DNA and then remove it from its sugar *e.g.*,

Dimethylsulphate methylates guanine
Hydrazine alters T and C
Hydrazine + HCl alters C
NaOH   alters A more than C.

The reaction with the bases is a limited one, damaging only one residue for every 50 to 100 bases along the DNA.

The exposed sugar is then a weak point in the backbone and easily breaks an acid alkali – or amine catalyzed series of β-elimination reactions cleaving thereby the sugar completely from its 3' and 5' phosphates. This reaction should go to completion, so that the molecules finally analyzed do not have hidden damages. The strand is cleaved with piperidine at the sugar residue lacking the base.

The sequencing requires DNA molecules, either double-stranded or single-stranded, that are labeled at one end of the strand with ^{32}P. This can be a 5' or a 3' label. A restriction fragment of any length is labeled at both ends-by first treating with alkaline phosphatase to remove terminal phosphates and then labeled with ^{32}P by transfer from gamma-labeled ATP with polynucleotide kinase. There are then two strategies :

(i) The ds molecule is cut by a second restriction enzyme and the two ends are resolved on a polyacrylamide gel and isolated for sequencing.

(ii) The doubly labeled molecule is denatured and the strands are separated on a gel, extracted and sequenced. A sequencing gel is a high resolution gel, designed to fractionate ssDNA fragments on the basis of their length and routinely contains 6-20% polyacrylamide and 7M urea. Urea minimizes DNA secondary structure which affects electrophoretic mobility. The gel is run at sufficient power to heat up to about 70°C which also minimizes 2°structure. The labeled DNA bands obtained after electrophoresis are revealed by autoradiography on large sheets of X-ray film. The sequence can then be read directly from the sequencing ladder in the adjacent base-specific tracks. It is possible to sequence about 100 bases.

General Principles

The principle of this technique involved that a synthetic oligonucleotide primer is annealed to a single standed DNA template. Four different sequencing reactions are set up each containing a DNA polymerase and the four normal dNTPs. One of the precursor or, in some case, the premier is labeled radioactively with ^{32}P, ^{33}P, or ^{33}S or with a nonradioactive fluorescent tag. A small proportion of a 2', 3' – ddNTP is also present on the four reactions. It that carries a 3' – H atom on the deoxyribose moiety, rather than the conventional 3'–OH group. The absence of a 3'–OH group prevents formation of a phosphodiester bond with the succeeding dNTP if a ddNTP molecule is incorporated into a growing DNA chain. Further, extension of the growing chain is impossible. Thus, in a reaction mixture for DNA synthesis, when a small amount of one of the ddNTPs is included with the four conventional dNTPs there is competition between extension of the chain and infrequent, but base-specific, termination. The products of the reaction are a population of oligonucleotide chains whose lengths are determined by the distance between the 5' terminus of the primer used to initiate DNA synthesis and the sites of chain termination. In a sequencing reaction containing ddA, for example, the termination points correspond to all positions normally occupied by a deoxyadenosyl residue. By using the four different dd NTPs in four separate enzymatic reactions, populations of oligonucleotides are generated that terminate at positions occupied by every A, C, G, or T in the template strand. These populations of oligonucleotides can be separated by electrophoresis, and the locations of each band can be ascertained by autoradiography or the emission of fluorescence. When the four populations are loaded into adjacent lanes of a sequencing gel, the sequence of the newly synthesized strand can be read in a 5'–3' orientation by calling the order of bands from the bottom to the top of the gel.

4.6.2 Sequencing by the Chain Inhibitor or Dideoxy Method
(Fig. 4.24)

Atkinson et al. (1969) showed that the inhibitory activity of ddTTP on DNA polymerase I depends on its being incorporated into the growing oligonucleotide chain in the place of dT. Because the ddT contains no 3-hydroxyl group, the chain cannot be extended further, so that termination occurs specifically at positions where dT should be incorporated. If a primer and template are incubated with DNA polymerase in the presence of a mixture of ddTTP and dTTP, as well as the other 3 dNTPs (one of which is labeled with ^{32}P), a mixture of fragments all having the same 3' end with ddT residues at the 3' ends is obtained. When this mixture is fractionated by electrophoresis on denaturing acrylamide gel, the pattern of bands shows the distribution of dTs in the newly synthesized DNA. By using analogous terminators for the other nucleotides in separate incubations and running the samples in parallel on the gel, a pattern of bands is obtained from which the sequence can be read off (Sanger et al., 1977).

Two types of terminating triphosphates have been used, the dideoxy derivatives and the arabinonucleosides. Arabinose is a stereoisomer of ribose in which the 3-hydroxyl group is oriented in trans position with respect to the 2-OH group. The arabinosyl (ara) nucleotides act as chain ending in 3 araC can be furthur extended by some mammalian DNA polymerases. Ratio of terminating triphosphates to normal triphosphates should be such that only partial incorporation of the terminator occurs. For the dideoxy derivatives, this ratio is about 100 and for the arabinosyl derivatives, it is about 5000.

In general, sequences of 1-200 nucleotides for the priming site can be determined with reasonable accurary.

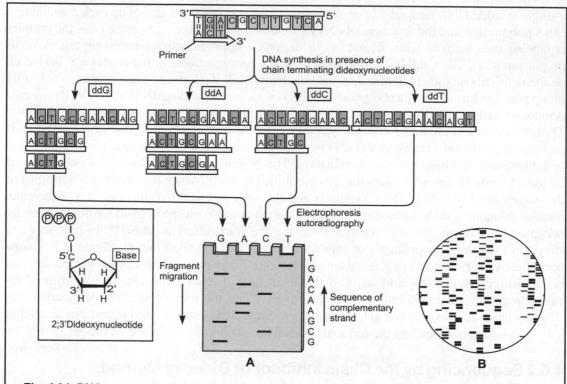

Fig. 4.24. DNA sequencing with DNA polymerase and dideoxynucleotides (ddG, ddA, ddC, and ddT). **(A)** Four reactions are performed, each one containing a separate dideoxynucleotide. The products of each reaction are separated in adjacent lanes of a gel. **(B)** DNA sequencing gel. Each set of four lanes represents one sample (four reactions).

The dideoxy chain termination technique of DNA sequencing is considered to be the easy and quick method of sequencing large DNA fragments by using bacteriophage M13 as a cloning vector.

M13, and single-stranded DNA templates are prepared and used in a primer extension sequencing reaction. About 500 bases can be read from a single sequencing reaction. To determine the sequence of a DNA fragment more than 500 nucleotides in length, different strategies can be used:

(i) Synthetic oligonucleotide primers, complementary to the template DNA near to the point from where to extend the known sequence, can be used. This approach may be good for sequencing small DNA fragments. However, it is costly and time-consuming.

(ii) Progressive deletions of the insert can be made from one end, in such a way as to bring different regions of DNA close to the primer extension site (Henikoff, 1984).

In both of these approaches, the entire DNA fragment is to be cloned into M13 or plasmid. Inserts much greater than 2 kb are usually unstable in the M13 vector. There are problems of secondary structure in DNA sequencing, therefore sequencing of the template in both orientations in usually necessary. There are some other problems with ExoIII, such as dependence upon specific restriction enzyme sites.

To overcome the above problems, the DNA fragment to be sequenced is cut down into fragments of reasonable size that can be easily cloned into M13 by using restricted enzymes. Sequencing of these subfragments leads us to determine the sequence of the entire fragment. To generate subfragment from a larger DNA fragments. There are two methods for generating random fragments.

(a) Enzymatically with DNaseI (Anderson, 1981)

(b) Mechanically by sonication (Deininger, 1983)

A sequencing procedure in which random subfragments are generated mechanically by sonication is described by bankier *et.al.*, 1987. A fragment of up to 20 kb can be sequenced by this procedure. In the sonication method, the sequence of each base in a DNA fragment is read an average of six to eight times before the sequence of the entire fragment is known. This will take care of sequencing errors. The rate of sequencing can be greatly enhanced by using a microtiter plate in place of microfuge tubes for sequencing reaction.

Three Basic Reagents are utilized by Enzymatic Sequencing Reactions

(i) Template : Utilizes a single-stranded DNA or double-stranded DNA that has been usually denatured, by alkali. However, single-stranded DNA templates isolated from recombinant M13 give the best results.

(ii) Primer : For priming of DNA synthesis synthetic oligonucleotide complementary to a specific sequence on the template is used. For DNA synthesis usually the "universal" primer with is complementary to the vector sequences, that flank the target DNA, is used for DNA. Universal primers used for the sequencing of M13 clones are usually 17 to 20 nucleotides long, and complementary to the seuqences immediately adjacent to the HindIII site in the polyclonal region of M13mp 18.

(iii) DNA polymerases: The commonly used enzymes for the dideoxynucleotide-mediated sequencing, are as follows:

- Klenow fragment of *E. coli* DNA polymerase (Sanger *et.al., 1977*).
- Bacteriophage T7 DNA polymerases that have been chemically modified to stop 3'-5' exonuclease activitiy (*e.g.,* Taq DNA polymerase) (Innis *et.al., 1988*).
- Reverse transcriptase (Mierendorf and Pfeffer, 1987).

The international race to sequence the human genome turned gene sequencing into a high speed and high profile endeavour. Now we have machines that create thousands of copies of DNA fragments as a first step towards decoding the sequence, one nucleotide base pair at a time. Groups around the world are working to steel a bit of limelight in sequencing with new approach for sequencing (decoding) having potential for speeding up the process. The push for the developments come from the fact that current DNA decoding schemes, sequence relatively short stretches of DNA, each about 1000 bp long. Researchers wanting to sequence a gene containing about 100,000 bp must sequence overlapping fragments of the gene and then use complex computer programs to reassemble the pieces in the right order. Single molecule sequencers on the other hand hope to sequence DNA segments as long as 50,000 bp, which would, "simplify and quicken up the task of putting the puzzle back together."

In this technique, exonuclease degrades DNA by munching its way through tens of thousands of individual bases, one by one and each base is identified with a laser-based detector. Initially each of the 4 types of bases in DNA are tagged with its own fluorescent compound. One end of DNA strand is then fixed to a high plastic bead that is then held steady with a laser inside a tiny flow chamber. An exonuclease on the other end of the DNA clips off bases from one of DNA's two strands, which then flow past another laser, revealing their presence with bright flash. This technique may yield false-positive results and work for its improvement is still going on. Technique also has potential for DNA fingerprinting (Servic, 1990). Now a days many automatic sequencers have been come up (Fig. 4.25).

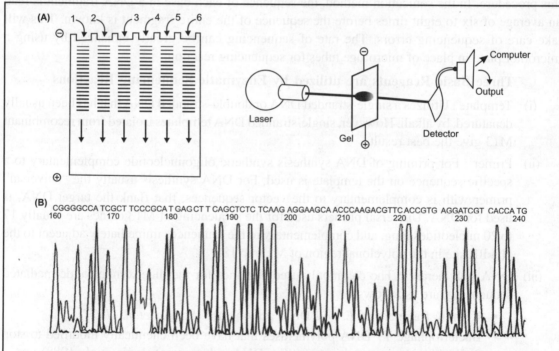

Fig. 4.25. Automated DNA sequencing using fluorescent primers. (A) Principles of automated DNA sequencing. In this all four reaction products are loaded into single lanes of the electrophoresis gel, data is captured during the electrophoresis run. For the base-specific reactions four separate fluorescent dyes are used as labels. During the electrophoresis run, a laser beam is focused at a specific constant position on the gel. The laser causes the dyes to fluoresce as the individual DNA fragments migrate past this position. Maximum fluorescence occurs at different wavelengths for the four dyes, and the information is electronically recorded and the interpreted sequence is stored in a computer database. **(B)** Example of DNA sequence output. This shows a typical output of sequence data as a succession of dye-specific (and therefore base-specific) intensity profiles. The example illustrated represents part of exon 1 of the neurofibromatosis type 2 (*NF2*) gene (Data from Susan Mason, University of Newcastle Upon Tyne). (Adapted from Strachan and Read, 2003)

Microfabricated Electrophoresis: In the early 1990s, the concept of micromachined devices for electrophoretic separation was proved. This sets the stage for DNA microelectrophoresis assays to separate oligonucleotides, restriction fragments, sequencing, genotyping samples and STRs.

Among the most important factors are high cost, long run times large sample volumes and manual operation of gel based electrophoresis devices. They limited the pace of Human Genome Project. Hence, slab-gel and capillary method were optimized to near their theoretical limits, as a result microdevices were able to provide evenly better services, to be of any use.

The device is a typical 8-lane microelectrophoresis device, series of 8 pairs of intersecting enclosed channels are micromachined into fused silica and each has an injector and separation channel. This is housed in a cassette that contains microfluidic reservoirs, electrode and integrated heater. Samples are pipetted into reservoirs connected to the input ends of the injector channels. A voltage is applied to electrophorse the DNA past the channel- intersection points and detection is via laser-induced fluorescence, single base resolution can be achieved for 200 bases in 8 min, for 300 bases in < 11min. one of the strengths of microdevices is the increase in assay speed (Ehrlich and Matsudaira, 1999).

Nanopore sequencers: In 1994, John Karianowicz of NIST, US, with Dan Branton of Harvard University and David Deamer of University of California used an ion channel called a-hemolysin, plucked from the bacterium *S. aureus* and inserted it on a stretched cell membrane across a hole separating 2 compartments containing a K^+ ion solution. Then they added ssDNA to one side and applied a voltage, which dragged the negatively charged DNA through the channel. If the 4 bases all reduce the current to a different degree, one can read the genetic code just by measuring the current.

If it works, the sequencing rates will be off the map, perhaps as high as 1000 bases per second, all 3 billion bps of human genome could be read in a single day. Nanopore sequencers could churn out genetic libraries for other species too and the medical benefits would be huge.

4.7 DNA CHIP TECHNOLOGY

It is the combination of the knowledge of semiconductor and molecular biology with probes. It has allowed us to develop DNA chips, that have applications in understanding various life phenomena. It is possible to analyze tens of thousands of genes on a single "microchip". The microchips and DNA chips are manufactured on similar principle. The difference is in the fact that instead of using shining light through a series of masks to etch circuits into silicon, here a series of masks are used to create a sequence of DNA probes on a glass slide (Fig. 4.26).

DNA is removed from the cells. It is tagged with fluorescent markers and placed on the chip. Hybridized sequences attach to the probes, and unmatched bits of DNA are then washed away. A laser reader, computer and high-powered microscopes are used to analyze thousands of sequences at one time and determine where the tagged DNA has found a match with a chip-mounted DNA probe.

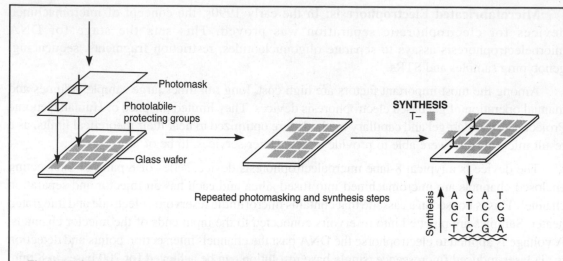

Fig. 4.26. Construction of DNA and oligonucleotide microarrays. The steps involved in the protocol have been highlighted. (Modified from Strachan and Read, 2003)

DNA chip technology has following applications.

(a) Detection of mutations in disease-causing genes
(b) Monitor gene expression in yeast and cancer cell lines
(c) Diagnosing infectious diseases
(d) Knowing whether a pathogen is resistant to certain drugs

DNA chips may contribute to the following areas.

(a) Crop biotechnology
(b) Improved screening for microbes used in bioremediation
(c) Hasten drug discovery
(d) Provide answers to questions about gene function and the significance and clinical manifestations, if any, of polymorphisms in the human genome.

4.8 ANTISENSE TECHNOLOGY

To block or decrease the production of certain proteins, antisense technology is being used. This is accomplished by using small nucleic acids (oligonucleotides) that prevent translation of the information encoded in DNA into a protein.

This technology has many applications. In any situation in which blocking a gene would be beneficial, antisense technology is useful. Attempts are being made to use this technology for the following purposes :

(a) To slow down the food spoilage and control viral diseases
(b) To inhibit the inflammatory response
(c) To treat asthma, cancers and thalasssemia, a hereditary form of anaemia common in many parts of the world

4.9 METABOLOMICS (METABOLIC ENGINEERING)

Antisense technology can be used in metabolonomics (metabolic engineering). Many compounds in nature that have great commercial applicability are proteins. Most compounds produced by plants to deter insect feeding, could be useful as crop protectants, but are not proteins. Their production can be increased in crop plants by using antisense technology to block the production of enzymes in certain pathways. There is, thus, rerouting of the metabolism to favour production of these compounds.

This technology can be used to decrease the production of a substance that is not a protein *e.g.*, cholesterol. With the right antisense molecule, there is a possibility of significantly decreasing production of cholesterol by blocking key enzymatic steps in its synthesis.

This technology has long term benefits as it will allow us to have metabolic products including forms of drugs of our choice.

Interference RNA : This is silencing of gene expression with RNA interference (RNAi).

RNAi is mediated by small interfering RNAs (siRNAs) that are generated from long dsRNAs of exogenous or endogenous origin. Long dsRNAs are cleaved by a ribonuclease III (RNase III) type protein Dicer. Dicer homologues can be found in *S. pombe, C. elegans, Drosophila,* plants and mammals, suggesting that the small RNA-mediated regulation is evolutionarily ancient and may have critical biological roles. SiRNA generated by Dicer is a short (~ 22-nt) RNA duplex with 2-nt overhang at each 3'-end. Each strand contains a 5'-phosphate group and a 3'-hydroxyl group. SiRNA is incorporated into a nuclease complex called RISC (RNA induced silencing complex) that targets and cleaves mRNA that is complementary to the SiRNA. The initial RISC containing a siRNA duplex is still inactive until it is transformed into an active form (RISC*), which involves loss of one strand of the duplex by an RNA helicase activity. The identity of RNA helicase is currently unknown. Dicer has a conserved helicase domain, but it is still not clear whether Dicer actually catalyzes this reaction.

Biochemical studies using *Drosophila* S2 cell extracts and human HeLa cell extracts revealed the presence of argonaute family proteins in the RISC. Argonaute-2 (AGO2) was found in *Drosophila* and two isoforms of eIF2c (eIF2C1 and eIF2C2), in human. Argonaute family proteins are ~ 100-kDa highly basic proteins that contain two common domains, PAZ and PIWI domains. PAZ domain consisting of ~ 130 amino acids is usually located at the centre of the proteins. The C-terminal PIWI domain containing ~ 300 amino acids is highly conserved. The functions of these domains are largely unknown, but the PIWI domain of human eIF2C has recently been shown to be essential for its interaction with Dicer. Depletion experiment of the eIF2C proteins by RNAi have shown that they are required for RNAi. The biochemical functions of argonaute family proteins are still not clear.

The identity of the nuclease that executes the cleavage of mRNA is not clear. Partially purified human RISC is estimated to be between 90 and 160 kDa leaving little room for an additional protein except for eIF2C. Genetic studies of *C. elegans, Drosophila, Neurospora crassa* and plants have revealed several other genes that may be involved in RNA silencing although their biochemical roles remain to be determined.

Persistent RNAi has been observed in *C. elegans* and *N. crassa,* but not in *D. melanogaster* and mammals. RNAi in *C. elegans* can be transmitted to the progeny (F1) although the effect gradually diminishes. RNAi in human cells is transient and usually lasts less than five doubling times. It was reported that SiRNAs are amplified by RNA-dependent RNA polymerase in nematode and fungi, while flies and mammals seem to lack this enzyme.

4.10 TRANSGENICS

Transgenics or transgenic organisms are also called Genetically Modified Organisms (GMOs). In this, gene or genes are introduced using rDNA technology. The organisms (plants and animals) obtained through genetic engineering contain a gene usually from unrelated organism, such genes are called **transgenes** and the plants and animals containing transgenes are known as **transgenic plants** and **animals.** The gene coding for insecticidal protein from *Bacillus thuringiensis* has been transferred to the cotton plant. This transgenic of cotton plant is known as genetically modified cotton *e.g.,* Bt cotton which is resistant to bollworm.

4.10.1 Production of Transgenic Plants

These can be produced by introducing genes in two ways :
   (a)   Vector Mediated
   (b)   Direct Gene Transfer

4.10.1.1 Vector Mediated Transformation
Historical

Agrobacterium is a soil-dwelling bacterium. It infects wound sites on a wide range of plant species and induces the development of crown gall tumours or hairy roots. The cells of crown-galls have acquired the properties of independent, unregulated growth. Although the host range is broad, it does not include monocotyledonous plants. However, with an increase in the knowledge of process of crown gall formation, monocots can also be transformed now.

The disease represents the first documented example of a parasitic interaction between pro-karyotes and eu-karyotes, whereby pathogen (*Agrobacterium*) genetically determines the infected host to express functions that benefit the survival of the parasite. Although the bacteria are required for tumour induction and are mostly found in association with crown gall harbouring plants in nature, they are not necessarily required for tumour maintenance and growth. In fact, sterile crown gall tissues can be readily cultured indefinitely on simple media lacking any added growth hormones.

Taxonomic Classification of *Agrobacterium*

The family Rhizobiaceae comprises only two genera, *Rhizobium* and *Agrobacterium*. The latter is considered to contain four species, *A. tumefaciens, A. radiobacter, A. rhizogenes* and *A. rubi.*

Agrobacterium tumefaciens : It is a flagellate gram-negative phytopathogen that as a normal part of its life cycle genetically transforms plant cells. This genetic transformation leads to the formation of crown gall tumours. These interfere with the normal growth of infected plant. This agronomically important disease affects only dicotyledonous plants, including grapes, stone-fruit trees and roses.

A. rhizogenes: It is quite similar to *A. tumefaciens*. It, however, contains Ri plasmid and induces hairy roots instead of gall tumours in the host plants.

Braun and Mandle (1948) were the first to conclude from their observations that a factor (Tumour inducing principle or TIP factor) is transmitted from the inciting bacteria to the host cells. This results into a stable transformation of the plant cells into crown gall tumour cells in a relatively short time (about 36 hrs). The cells acquire a capacity for autonomous growth, possibly because they persistently synthesize or otherwise become independent of the growth-regulating substances and metabolites that are externally required for cellular growth and division of normal, untransformed cells.

The observation of Morel and his collaborators led to a more fundamental implication of these inciting bacteria. They demonstrated that there are atleast two different types of crown gall inducing bacteria that induce tumours with different properties.

(i) *Agrobacterium* strain induces crown gall tumours in which N-a-(D-1-carboxyethyl)-L-arginine (Octopine) is synthesized.

(ii) Another strain of *Agrobacterium* induces tumours containing N-a-(1, 3-dicarboxypropyl)-L-arginine (Nopaline).

The type of arginine derivative synthesized in the tumour was found to be specified by the particular *Agrobacterium* strain used to incite the tumour and to be independent of the host plant on which the tumours were induced. It was demonstrated that *Agrobacterium* strains that induce the synthesis of octopine in crown gall can selectively use this product, but not nopaline, as the sole carbon and/or nitrogen energy source.

Bacteria that utilize octopine, induce tumours that synthesize octopine and those that utilize nopaline induce tumours that synthesize nopaline. Thus, a genetic linkage between oncogenicity and opine metabolism was established. On the basis of this, a hypothesis was proposed that genetic information could be somehow transferred from bacterium to plant.

In 1973, it was reported that the tumour inducing principle of *Agrobacterium* was carried by extrachromosomal DNA elements of the plasmid type and not of the lysogenic type as had been previously reported.

The discovery of Ti plasmids and of their major role in the crown gall phenomenon, led to the formulation of a general concept about the neoplastic transformation of plant cells. It was observed that both the genes determining opine catabolism in *Agrobacterium* and the ones determining opine synthesis in transformed plant cells are localized on Ti plasmids. This explained the linkage between opine metabolism and oncogenicity. It also provided genetic evidence for the involvement of the Ti plasmid in a mechanism of DNA transfer from bacterium to plant. To explain the role played by Ti plasmids in oncogenicity and in opine synthesis, it was assumed that part or all of the Ti plasmid was somehow transferred to and maintained and expressed in the transformed plant cells. The experiments to demonstrate this were based on a detailed genetic analysis of the Ti plasmid. The transfer DNA or TDNA, which is a part of the Ti plasmid, contains genetic information for plant regulatory functions. These are directly or indirectly responsible for the neoplastic growth pattern. It also encodes genetic information for synthesis of opines (Fig. 4.27).

Depending upon from which bacterial strain the Ti plasmid comes, the length of T-DNA region can vary between approximately 12-24 kilobase pairs (kb). Strains of *A. tumefaciens* that do not possess a Ti-plasmid cannot induce crown gall tumours.

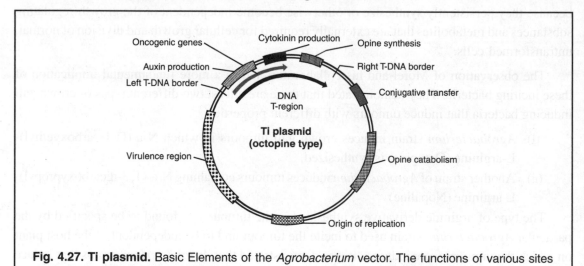

Fig. 4.27. Ti plasmid. Basic Elements of the *Agrobacterium* vector. The functions of various sites on the vectors have been mentioned.

Due to gene transfer mechanism determined by Ti plasmids, agrobacteria have the capacity to force plant cells to genetically divert some of their metabolism to produce compounds (opines) that these bacteria are selectively equipped to catabolize. The transferred plasmid genes coding for "opine synthase" enzyme are linked and shift the cells to a tumourous mode of growth. This establishes relatively large tissue (crown galls) of opine producing cells. To describe this type of parasitic interaction, the term Genetic Colonization was proposed.

Different chromosomally-determined genetic elements have shown their functional role in the attachment of *A. tumefaciens* to the plant cell and bacterial colonization.

(i) *The loci chvA and chvB.* These are involved in the synthesis and excretion of the b-1, 2 glucan

(ii) *The chvE.* It is required for the sugar enhancement for *vir* gene induction and bacterial chemotaxis

(iii) *The cell locus.* It is responsible for the synthesis of cellulose fibrils.

(iv) *The pscA (exoC) locus.* It plays its role in the synthesis of both cyclic glucan and acid succinoglycan

(v) *The att locus.* It is involved in the production of cell surface proteins.

The initial results of the studies on T-DNA transfer process to plant cells demonstrated three important facts for the practical use of this process of transformation of plant cells.

(1) The tumour formation is a transformation process of plant cells resulting from transfer and integration of T-DNA and the subsequent expression of its genes

(2) The T-DNA genes are transcribed only in plant cells and do not play any role during the transfer process

(3) Any foreign DNA placed between the T-DNA borders can be transferred to plant cell, no matter where it comes from. These facts, allowed the construction of the first vector and bacterial strain systems for transformation of plant.

The T-DNA of the Ti plasmid of *Agrobacterium tumefaciens* can be used to introduce modified genes into plants. The Ti plasmid is too large to manipulate easily *in vitro*.

Essential Features of T-DNA

Several essential features are required for *Agrobacterium* mediated transformation of plants:

- **vir genes:** Approximately 35 *vir* genes map outside the T-DNA region called virulence (*vir*) region. It is a regulon organized in six operons that are essential for the

  (i) transfer of T-DNA (*virA, virB, virD,* and *virG*) or

  (ii) for increasing the transfer efficiency (*virC* and *virE*). This encodes products required for excision, transfer and integration of T-DNA into a plant genome.

Vir genes act in trans, meaning they do not need to be physically attached to the T-DNA to cause integration into the plant genome,

- **T-DNA border sequences:** The T-DNA is flanked by right and left T-DNA borders, which are physically attached to the genes to be transferred into the plant genome (Fig. 4.28). These sequences of 25 bp imperfect repeats flank the T-DNA and are required for its transfer. Border sequences encompass the recognition sites for a site-specific endonuclease, which is encoded by the *virD* operon, part of the *vir* genes. The endonuclease cleaves the lower DNA strand of the T-DNA marking the starting point of the transfer.

- **cis-regulatory regions:** Other cis-regulatory regions include promoters and terminators that flank the transgenes and regulate their expression. Commonly used promoters and terminators are the nopaline synthesis gene (*NOS*) and Cauliflower Mosaic Virus (*CaMV*) 35S promoter.

- **selective marker genes:** In plants, they allow the identification and selection of cells with the gene of interest incorporated in their genome. Bacterial selectable markers permit the identification of bacteria.

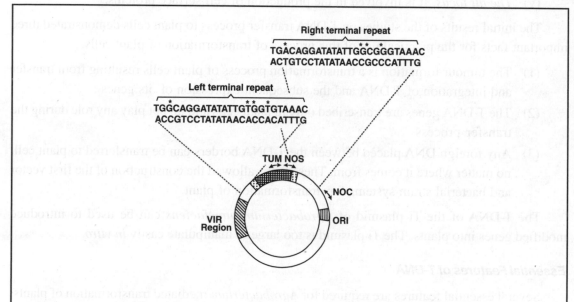

Fig 4.28. Ti plasmid sequences : The figure depicts sequences of right and left borders of Ti plasmid.

Ti Plasmid-Derived Vector Systems

Two basic types of vectors are used to transform a wide range of plants via *Agrobacterium*. These have been described below :

(a) **Binary vector :** In this system, the T-DNA and the *vir* region reside in separate plasmids within the same *Agrobacterium* strain. The *vir* genes are located in a disarmed (without tumour genes) region. Ti plasmid and the T-DNA with the gene of interest is located in a small vector molecule.

(b) **Co-integrated vector :** It is formed by the recombination of a small vector plasmid, *e.g.,* combination of an *E. coli* vector and a Ti plasmid harboured in *A. tumefaciens*. The recombination takes place through a homologous region present in both the plasmids. An engineered T-DNA containing the gene of interest can be in either of the plasmids.

Binary Vectors

It was found out that the *vir* genes do not need to be in the same plasmid with a T-DNA region for allowing it to transfer and insertion of T-DNA into the plant genome. This allowed the construction of a system for plant transformation where the T-DNA region and the *vir* region are on separate plasmids (Fig. 4.29).

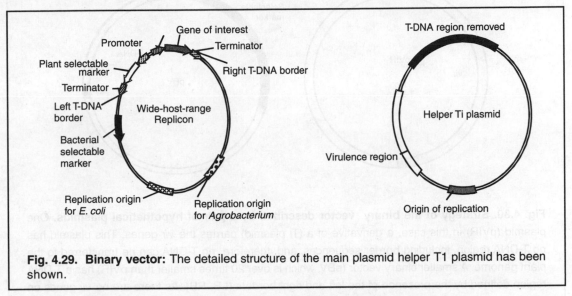

Fig. 4.29. Binary vector: The detailed structure of the main plasmid helper T1 plasmid has been shown.

The two different plasmids employed in the binary vector system are :

- **A wide host range small replicon :** It has an origin of replication (ori) that permits the maintenance of the plasmid in a wide range of bacteria including *E. coli* and *Agrobacterium.* Typically the plasmid contains the following :

  (i)   Foreign DNA in place of T-DNA,

  (ii) The left and right T-DNA borders (or at least the right T-DNA border),

  (iii) Markers for selection and maintenance in both *E. coli* and *A. tumefaciens,*

  (iv) A selectable marker for plants.

- As the tumour-inducing genes located in the T-DNA have been removed, the plasmid is said to be **"disarmed".**

- A helper Ti plasmid, harboured in *A. tumefaciens* lacks the entire T-DNA region, but contains an intact *vir* region. (Fig. 4.30)

In general, the following transformation procedure is followed :

- The recombinant small replicon is transferred via bacterial conjugation or direct transfer to *A. tumefaciens* harbouring a helper Ti plasmid

- The plant cells are co-cultivated with the *Agrobacterium*. It allows transfer of recombinant T-DNA into the plant genome
- Under appropriate conditions, transformed plant cells are selected (Fig. 4.30).

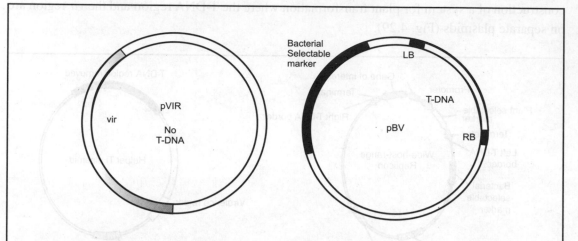

Fig. 4.30. Strategy of the binary vector described by a pair of hypothetical plasmids. One plasmid (pVIR) in this case, a derivative of a (Ti plasmid) carries the *vir* genes. This plasmid has no T-DNA region, including border sequences, and, therefore, no T-DNA can be transferred to the plant genome. A smaller binary vector (pBV, which is over 20 times smaller than pVIR) has a T-DNA region defined by the presence of the left and right borders (LB, RB). As there are no *vir* genes on this plasmid (making the binary vector pBV relatively small), their function is supplied in *trans* by a pVIR. pVIR would normally be maintained in suitable strains of *Agrobacterium* and is not generally isolated or manipulated. Binary vectors are designed to be easily manipulated using standard *in vitro* techniques and contain selectable markers for selection in bacteria (*E. coli* and *Agrobacterium*) outside the T-DNA region. The T-DNA region will, in all probability, contain a selectable marker for use in plants. pBV is normally maintained in *E. coli* the gene of interest (the transgene) being cloned into it. Recombinant *E. coli* colonies are identified after bacterial transformation, and the pBV vector is checked to ensure that the transgenic correctly inserted and has not been mutated. pBVcan then be transferred, by electroporation, into the *Agrobacterium* strain used for plant transformation (which also harbours pVIR) (Adopted and Modified from Plant Biotechnology by Slater *et al.*, 2003).

Co-integrated Vectors

These were among the first types of modified and engineered Ti plasmids devised for *Agrobacterium* mediated transformation. These, however, are not widely used today.

These vectors are constructed by homologous recombination of bacterial plasmids with the T-DNA region of an endogenous Ti plasmid in *Agrobacterium*. For the integration of the two plasmids, a region of homology must be present in both (Fig. 4.31).

In this system, three vectors are necessary:

- **Disarmed Agrobacterium Ti plasmids :** In these, the oncogenes located in the T-DNA region have been replaced by exogenous DNA. The following vectors are included:

  (i) **SEV series:** The right border of the T-DNA together with the phytohormone genes coding for cytokinin and auxin are removed. These are replaced by a bacterial kanamycin resistance gene. The left border and a small part of the left segment (T_L) of the original T-DNA referred to as Left Inside Homology (LIH) are left intact.

  (ii) **pGV series :** The phytohormone genes are excised. These are substituted by part of pBR322 vector sequence. The left and right border sequences as well as the nopaline synthase gene of the Ti plasmid are conserved.

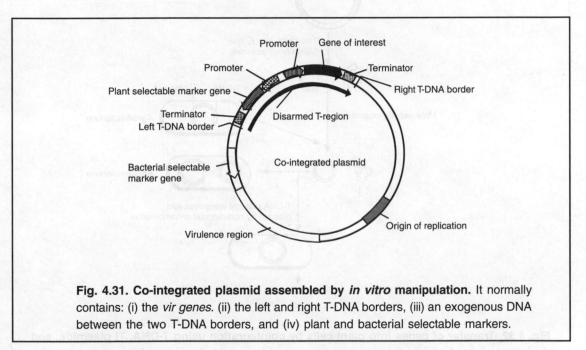

Fig. 4.31. Co-integrated plasmid assembled by *in vitro* manipulation. It normally contains: (i) the *vir genes*. (ii) the left and right T-DNA borders, (iii) an exogenous DNA between the two T-DNA borders, and (iv) plant and bacterial selectable markers.

- **Intermediate vectors :** These are small pBR322 based plasmids (*E. coli* vectors). These contain a T-DNA region. These help to overcome the problems derived from the large size of disarmed Ti plasmids and their lack of unique restriction sites. These vectors are replicated in *E. coli*. They are transferred into *Agrobacterium* by conjugation. They cannot replicate in *A. tumefaciens*. They carry DNA segments homologous to the disarmed T-DNA to permit recombination to form a co-integrated T-DNA structure (Fig. 4.32).

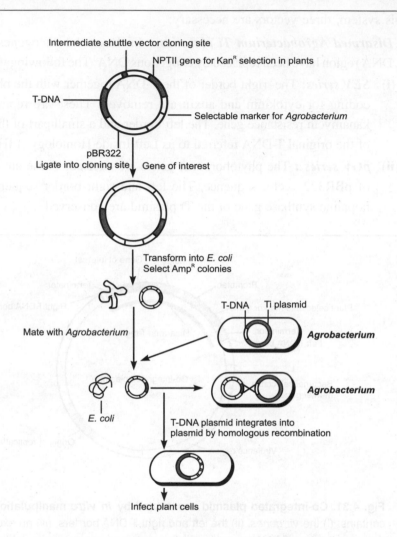

Fig. 4.32. Transfer of genes into plant cells by cointegration using T-DNA, Ti plasmids, and Agrobacterium. A cloned gene can be introduced into plant cells by first inserting it into the cloning site of a plasmid that can replicate in *E. coli* and contains a segment of T-DNA. The resulting intermediate shuttle vector is introduced into *E. coli* cells, and transformants are selected by resistance to ampicillin, encoded within the pBR322 sequences. The plasmid is transferred from the *E. coli* cell to an *Agrobacterium* cell by mating. Once inside the *Agrobacterium*, the plasmids integrate into the Ti plasmid by means of homologous recombination of the T-DNA sequences on the two plasmids. This process places the entire integrative plasmid (the plasmid integrated into the Ti plasmid) between the left and right boundaries of the T-DNA. Plasmids that fail to integrate do not accumulate because they lack an origin of replication for *Agrobacterium*. Agrobacteria containing the recombinant Ti plasmid are selected and used to infect plant cells. Plant cells that have taken up the T-DNA are identified by the plant selectable marker NPTII, which confers resistance to kanamycin. These cells also contain the cloned gene of interest.

Process of Transfer of Genes

Several essential steps are implied in the process of gene transfer from *Agrobacterium tumefaciens* to plant cells (Fig. 4.33). These are :

(1) Colonisation of bacteria
(2) Induction of bacterial virulence system
(3) Generation of T-DNA transfer complex
(4) T-DNA transfer
(5) Integration of T-DNA into plant genome

(a) Colonisation of Bacteria

When *A. tumefaciens* is attached to the plant cell surface, bacterical colonisation takes place. Non-attaching mutants do not have tumour-inducing capacity. An important role in the colonising process is played by the polysaccharides of the *A. tumefaciens* (lipopolysaccharides LPS), and capsular polysaccharides (K-antigens). It has been evidenced that capsular polysaccharides may be playing specific role during interaction with host plant.

Bacterial attachment to the plant cell is facilitated by the products of the genes present in the chromosomal 20kb *att* locus.

(a) The genes, placed at *att* left side, are involved in molecular signalling events
(b) The right side genes are probably responsible for the synthesis of fundamental components.

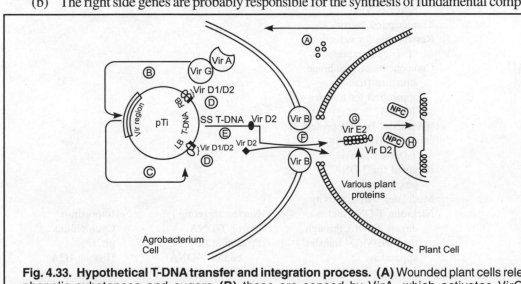

Fig. 4.33. Hypothetical T-DNA transfer and integration process. (A) Wounded plant cells release phenolic substances and sugars **(B)** these are sensed by VirA, which activates VirG by phosphorylation. VirG induces the expression of all the genes in the *vir* region of the Ti plasmid. **(C)** gene products of the *vir* genes **(D)** are involved in a variety of processes.VirD1 and VirD2 are involved in single-stranded T-DNA production, protection and export **(E)** VirB products form the transfer apparatus. **(F)** The single-stranded T-DNA (associated with VirD2) and VirE2 are exported through the transfer apparatus (VirF may also interact with either the VirD2 or VirE2, these are attached to the T-DNA and influence transport and integration). The T-DNA/VirD2/VirE2/plant protein complex enters the nucleus through the nuclear pore complex. Integration into the plant chromosome **(G)** occurs via illegitimate recombination (LB, left border, Rb, Right border, pTi, Ti plasmid, ss, single stranded, NPC, nuclear pore complex) (Adopted and Modified from Plant Biotechnology by Slater, 2003).

(b) Induction of Bacterial Virulence System

The products encoded by the 30-40 kb *vir* region of the Ti plasmid mediate the transfer of T-DNA. This region has at least six essential (*virA, virB, virC, virD, virE, virG*) and two nonessential (*virF, virH*) operons (Table 4.3). The number of genes per operon, however, differs

    (a) *virA, virG* and *virF* have only one gene each

    (b) *virE, virC, virH* have two genes each

    (c) *virD* and *virB* have eleven genes each

Operons *virA* and *virG* express constitutively. They code for a two-component (*virA-virG*) system activating the transcription of the other *vir* genes. Slatter (2003) has summarized the role of various virulent protects gene transfer.

Table 4.3. *Agrobacterium* **Virulence Protein Function** (Source: Slatter, 2003)

Virulence protein	Function in *Agrobacterium* spp.	Function in plant	Plant proteins that interact
VirA	Phenolic sensor kinase Part of two-component system with VirG; phosphorylates and activates VirG		
VirG	Transcription factor Responsible for induction of *vir* gene expression		
VirB1–B11	Components of membrane structure (transfer apparatus) for transfer of T-DNA		
VirC1	'Overdrive' binding protein. It enhances efficiency of T-DNA transfer		
VirD1	Required for T-DNA processing Modulates VirD2 activity		
VirD2	Nicks the T-DNA and directs T-DNA through the VirB/VirD4 transfer apparatus	Nuclear targeting of T-DNA Protection of 5' end of T-DNA from nucleases Possibly involved in integration	Importin-α[a] Cyclophilins[b] pp2C[c] Histone H2A
VirD4	Component of transfer apparatus		
VirE1	VirE2 chaperone Required for VirE2 export from *Agrobacterium spp.*		
VirE2		Single-stranded DNA-binding protein Prevents T-DNA	VIP1[d] VIP2[c]

	degradation by nucleases	
	Involved in nuclear targeting and passage through NPC	
VirF	Cell-cycle regulation Elongation of S-phase	Homologue Plant Skp1[f]
VirJ	Export of T-DNA	

[a] Importin (enables transport through the plant nuclear-pore complex (NPC).

[b] Cyclophilins may cause conformational changes in VirD2 or may be involved in targeting to chromatin and integration of the T-DNA.

[c] Serine/threonine protein phosphatase type 2C (pp2C) negatively affects nuclear import of the T-DNA.

[d] VIP1 facilitates nuclear import and may be involved in directing the T-DNA to chromatin.

[e] VIP2 may target the T-DNA to transcriptionally active chromatin.

[f] Skp1 (an F-box protein) homologue involved in protein degradation and cell-cycle control. May prolong the S-phase and improve the efficiency of integration.

VirA. It is a transmembrane dimeric sensor protein 'that detects signal molecules, released from wounded plants'. These signals include acidic pH, phenolic compounds, such as acetosyringone and certain class of monosaccharides. These synergically act with phenolic compounds. VirA protein has the following domains :

(i)   The periplasmic or input domain

(ii) Two transmembrane domains (TM1 and TM2).

The TM1 and TM2 domains act as a transmitter (signaling) and receiver (sensor).

The periplasmic domain is important for monosaccharide detection. Within the periplasmic domain, an amphipathic helix is present adjacent to the TM2 domain. It has strong hydrophilic and hydrophobic regions. This structure is characteristic for other transmembrane sensor proteins and folds the protein to be simultaneously aligned with the inner membrane and anchored in the membrane.

The TM2 is the kinase domain. It plays a crucial role in the activation of virA, phosphorylating itself on a conserved His-474 residue in response to signalling molecules from wounded plant sites.

Monosaccharide detection by VirA is important amplification system to respond to low levels of phenolic compounds. This system is induced through the periplasmic sugar (glucose/galactose) binding protein ChvE. The latter interacts with VirA. Recent studies for the determination of VirA regions, important for its sensing activity suggested the position which may be involved in TM1-TM2 interaction. This interaction causes the exposure of the amphipathic helix to small phenolic compounds and suggests a putative model for the VirA-ChvE interaction.

Activated VirA transfers its phosphate to a conserved aspartate residue of the cytoplasmic DNA binding protein VirG.

VirG. It functions as transcriptional factor. It regulates the expression of *vir* genes when it is phosphorylated by VirA. The C-terminal region is responsible for the DNA binding activity, while the N-terminal is the phosphorylation domain and shows homology with the VirA receiver (sensor) domain.

External factors like temperature and pH also regulate the activation of *vir* system. The *vir* genes are not expressed at temperatures greater than 32°C, because conformational change in the folding of VirA induces the inactivation of its properties. The effect of temperature on VirA is suppressed by a mutant form of *VirG* (VirGc). It activates the constitutive expression of the *vir* genes. However, this mutant cannot confer the virulence capacity at that temperature to *Agrobacterium*, probably because the folding of other proteins actively participating in the T-DNA transfer process are also affected at high temperature.

(c) Formation of T-DNA Transfer Complex

The activation of *vir* genes carries out the generation of single-stranded (ss) molecules representing the copy of the bottom T-DNA strand. Any DNA placed between T-DNA borders will be transferred to the plant cell, as single stranded DNA, and integrated into plant genome. These are the only *cis* acting elements of the T-DNA transfer system. For this purpose, VirD1 and VirD2 proteins recognize the T-DNA border sequences and nick (endonuclease activity) the bottom strand at each border. The nick sites are considered as the initiation and termination sites for T-strand recovery. Even after endonucleotidic cleavage, VirD2 remains covalently attached to the 5'-end of the ss-T-strand. Thereby, the exonucleolytic attack to the 5'-end of the ss-T-strand is prevented. It also allows to distinguish the 5'-end as the leading end of the T-DNA transfer complex.

VirD1 interacts with the region where the ss-T-strand will be originated. It has been evidenced that for the cleavage of supercoiled stranded substrate by VirD2, VirD1 should be present. The simultaneous restoration of the excised ss-T-strand is evolutionarily related to other bacterial conjugative DNA transfer processes. This includes the generation of single stranded DNA.

The loss of right border either by mutation or deletion completely inhibits the T-DNA transfer. However, the loss of left border only reduces the efficiency of transfer. This reveals that T-strand synthesis is in 5' to 3' direction, and it is initiated at the right border. Even when the left border is mutated or completely absent, the transformation process can take place. The efficiency, however, will be low. An enhancer or "overdrive" sequence is present next to the right border. It is specifically recognized by VirC1 protein. This makes the difference between two T-DNA borders.

(d) Translocation of T-DNA-Complex

Two models have been proposed for this purpose.

First Model : The DNA is transferred to the plant nucleus as ssT-DNA-protein complex. The ssT-DNA-VirD2 complex is coated by the 69-kDa VirE2 protein. VirE2 is a single stranded DNA binding protein. In this way attack of nucleases is prevented. It also extends the ssT-DNA strand. This leads to reduction in the diameter of complex to approximately 2 nm. Thus the translocation through membrane channels becomes easier. VirE2 and VirD2 respectively have two and one plant nuclear location signals (NLS). These proteins play an important role once the complex is in the plant cell mediating the complex uptake to the nucleus.

VirE1 is essential for the export of VirE2 to the plant cell.

Second Model: According to this model, the transfer complex is a single-stranded DNA covalently bound at its 5'-end with VirD2, but uncoated by VirE2. The independent export of VirE2 to plant cell is presented as natural process, and once the naked ssT-DNA-VirD2 complex is inside the plant cell, it is coated by VirE2.

A 9.5 kb virB operon generates a suitable cell surface structure for the transfer of ssT-DNA complex from bacterium to plant. For this transfer, the VirD4 protein is also required. The ATP-dependent linkage of protein complex necessary for T-DNA translocation is provided by VirD4. VirD4 protein is a transmembrane protein. It, however, predominantly located at the cytoplasmic side of the cytoplasmic membrane.

VirB proteins present the hydropathy characteristics similar to other membrane-associated proteins. They are assembled as a membrane-spanning protein channel and involve both membranes. But for VirB11, they have multiple periplasmic domains. The extracellular milieu contains VirB 1of the VirB proteins. It is, however, possible that some of the other VirB proteins may be redistributed during the process of biogenesis and functioning of the transcellular conjugal channel. VirB2 may be translated as a 12 kDa proprotein. It is then proteolytically processed to its mature 7 kDa functional form.

The octopine Ti-plasmid contains two accessory operons. These are *virF* and *virH*.

The virF operon. It encodes for a 23-kDa protein. It functions once the T-DNA complex is inside the plant cells via the conjugal channel or independently, as it was assumed for VirE2 export. The role of VirF is probably to aid in the nuclear targeting of the ssT-DNA complex.

The virH operon. It consists of two genes that code for VirH1 and VirH2 proteins. These Vir proteins are not essential. They, however, can enhance the efficiency of transfer, detoxifying certain plant compounds that can affect the growth of bacteria. The proteins provide host range specificity of bacterial strains for different plant species.

(e) Integration of T-DNA into Plant Genome

As soon as ssT-DNA is inside the plant cell, the complex is targeted to the nucleus crossing the nuclear membrane. VirD2 and VirE2 proteins are involved in this step. A minor role in this process is also played by VirF.

The integration occurs by illegitimate recombination. There is a pairing of a few bases. It provides a minimum specificity for the recombination process. VirD2 is required for the ligation. Some low homologies with plant DNA are found by the 3'-end or adjacent sequences of T-DNA.

It results in the first contact (synapses) between the T-strand and plant DNA.

VirD2 plays an active role in the integration of T-strand in the plant chromosome.

Tobacco was the first plant transformed by *Agrobacterium tumefaciens*. Since then, there has been great progress in understanding the *Agrobacterium*-mediated gene transfer. Genetic manipulation of more than 150 species belonging to diversified families has been carried out by using *Agrobacterium*-mediated or direct transformation method. These include the most major economic crops, vegetables, ornamental, medicinal, fruit, tree and pasture plants.

Agrobacterium tumefaciens naturally infects only dicotyledonous plants and many economically important plants, including the cereals, remained unaccessible for genetic manipulation during long time. For these cases alternative direct transformation methods have been developed. The *Agrobacterium*-mediated transformation, however, is advantageous over direct transformation methods in reducing the copy number of the transgene, potentially leading to fewer problems with transgene cosuppression and instability. In addition, it is a single-cell transformation system and normally do not allow the formation of mosaic plants, which are more frequently produced with the direct transformation method.

Other gene-transfer methods have been developed for monocotyledonous plants. To develop these methodologies for a monocot plant, the critical points are being taken into consideration. These include the cellular and tissue culture methodologies developed for that species. The suitable genetic materials (bacterial strains, binary vectors, reporter and marker genes, promoters) and molecular biology techniques available in the laboratory are necessary for selection of DNA to be introduced. This DNA must be expressible in plant. They make the identification of transformed plants possible in selectable medium by using molecular biology techniques and characterize the transformation events.

4.10.1.2 Non-Vector Gene Transfer in Plants

(i) Microprojectile Bombardment (Biolistics)

Genetic transformation process involves the uptake of naked DNA (gene of interest) by competent cells. It is followed by impregnation into a cell through the cell wall and the plasma membrane. The DNA eventually penetrates into the nucleus. This can be done by delivering foreign genes into plant cells by microprojectile bombardment (Fig. 4.34).

In this technique, a 0.22-Caliber 'gene gun' is used to bombard cells with tiny metallic projectiles coated with the new gene. This method has been used to genetically modify everything from yeast to algae to plants and is also called the 'bioblaster'. In this technique, plant cell wall is not removed. This technique was developed in the Biochemistry Laboratories of Cornell University.

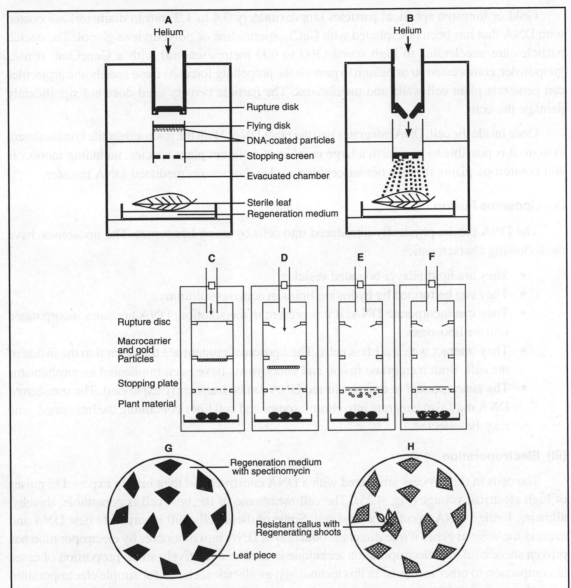

Fig. 4.34. Microprojectile bombardment. (A) A microprojectile system **(B)** Exposure of leaf to bombardment, (C,D,E,F) Details of working of PDS-1000/He particle-bombardment system. The plant tissue is placed into a vacuum chamber (chamber pressure 27 mmHg) 13 cm below the microcarrier stopping plate. The vector DNA-coated particles (the microcarriers) on the macrocarrier membrane are inserted into the apparatus **(C)** Once the vacuum in the lower part of the apparatus is established, the helium pressure above the rupture disc is increased until at 1100 psi (or whatever pressure the rupture disc is designed to rupture at) the rupture disc bursts **(D)**. This propels fragments of the microcarrier and the projectiles down the chamber. The macrocarrier is stopped at the stopping plate **(E)** This allows the microcarriers to pass through and hit the plant material **(F)**, **(G)** Leaf pieces exposed to microprojectiles are plated on the regeneration medium with spectinomycin. **(H)** Resistant callus growing from the regenerating shoot (Modified from Microbial Biotechnology by Loo, 2006).

Gold or tungsten spherical particles (approximately 0.4 to 1.2 mm in diameter) are coated with DNA that has been precipitated with $CaCl_2$, spermidine or polyethylene glycol. The coated particles are accelerated to high speed (300 to 600 metres/second) with a GeneGun. It uses gunpowder, compressed air or helium to provide the propelling force. At these speeds, the projectiles can penetrate plant cell walls and membranes. The particle density used does not significantly damage the cells.

Once inside the cell, DNA integrates into the plant DNA. With the microprojectile bombardment system, it is possible to transform a large number of different plant species, including monocots and coniferous plants that are not susceptible to *Agrobacterium* mediated DNA transfer.

(ii) Liposome Fusion

The DNA can be physically introduced into cells by using liposomes. The liposomes have the following characteristics :

- They are lipid-bilayer bounded vesicles
- They can be formed by hydrating lipids in aqueous solutions
- They can incorporate DNA, if it is present in the solution, DNA becomes incorporated into the liposomes
- They interact with wall-less cells. The liposomal content are transferred to the inside of the cell. Both membrane fusion and endocytosis have been implicated as mechanisms
- The genes present in the transferred DNA can be transiently expressed. The transferred DNA may also integrate into chromosomes and cell lines containing the integrated gene may be selected.

(iii) Electroporation

The cells in this process are mixed with a DNA construct and then briefly exposed to pulses of high electrical voltage (Fig. 4.35). The cell membrane of the host cell is penetrable, thereby, allowing foreign DNA to enter the host cell. Some of these cells will incorporate new DNA and express the desired gene. While direct introduction of DNA into monocots by electroporation has proven successful, the electroporation technique is used in a relatively small proportion of cases in comparison to other methods. In this technology, an electrical circuit for simple electroporation is used.

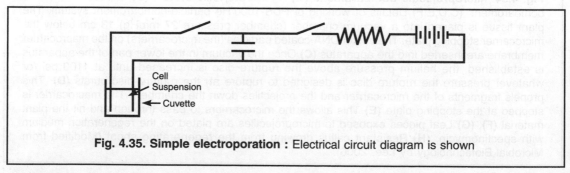

Fig. 4.35. Simple electroporation : Electrical circuit diagram is shown

This process involves the following steps :

- A cell suspension, such as of plant protoplasts, is placed in the cuvette.

- A solution of DNA fragments containing the gene of interest is added. For example, the DNA could include a reporter gene such as that for chloramphenicol acetyltransferase (CAT).

- The capacitor is charged by closing the right-hand switch. When the capacitor has been charged, the direct current pulse is discharged in the cuvette suspension by closing the left-hand switch.

- The DC pulse probably temporarily disrupts the membrane and electrophoresis DNA into cells.

- The cells are put in culture and assayed after various times (24 to 48 h) for the amount of CAT activity.

Gene Constructs for Electroporation

An advantage of DGT techniques is that genes for transfer need not be engineered between *Agrobacterium* T-DNA border sequences, or into a Ti or broad-host-range plasmid. Almost any small high-copy-number *E. coli* cloning vector containing an *E. coli* origin of replication and a bacterial selectable marker gene can be used for delivery (Draper *et al.,* 1988). However, expression of any introduced gene requires appropriate control sequences upstream (5') and downstream (3') of the coding region (Fig. 4.36).

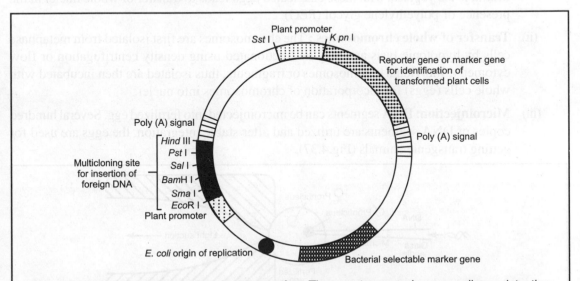

Fig. 4.36. Generalized vector for electroporation. The reporter or marker gene allows detection or selection of transformed cells. The multicloning site allows insertion of a foreign gene for expression in plant cells. The top plant promoter can be replaced with other sequences to be tested as promoters, based on reporter gene expression (Adopted from Draper *et al.,* 1088).

(iv) Microinjection

Micromanipulators fitted with tiny pipettes (tubes) are used to inject DNA directly into plant cells. However, plant cells have a tough cell wall made of cellulose, which accounts for the stiffness. These cell walls are stripped away by the use of enzymes. The new gene can, therefore, be injected into the wall-less cell. These cells are then cultured into whole plants.

By examination under a microscope, a cell is held in place with gentle suction, while being manipulated with the use of a blunt capillary. A fine pipette is then used to insert the DNA into cytoplasm or nucleus.

4.10.2 Production of Transgenic Animals

Following methods can be used for the development of transgenic animals :

- Transfection of fertilized eggs
- Use of stem cells
- Virus vector
- Direct uptake of DNA stimulated by calcium or an electric current ('transfection')
- Use of liposomes

9.10.2.1 Transfection of Fertilized Egg

It may involve the following steps :

(i) **Transfer of whole nucleus :** For the transfer of whole nuclei, the egg cells are treated with cytochalasin-B and subjected to centrifugation causing enucleation. Incubation of desirable karyoplasts with these enucleated eggs leads to transfer of whole nuclei in the presence of polyethylene glycol (PEG).

(ii) **Transfer of whole chromosomes :** The chromosomes are first isolated from metaphase cells by hypotonic lysis and may be fractionated using density centrifugation or flow cytometry. Individual chromosomes or fragments, thus isolated are then incubated with whole cells (eggs) for incorporation of chromosomes into nuclei.

(iii) **Microinjection:** DNA segments can be microinjected into fertilized egg. Several hundred copies of DNA segments are utilized and after stable integration, the eggs are used for getting transgenic animals (Fig.4.37).

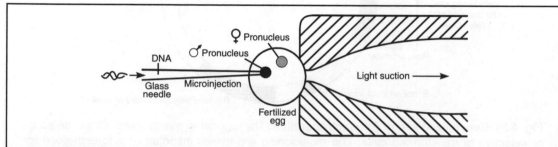

Fig. 4.37. DNA microinjection method. The way the DNA is introduced in male pronucleus is highlighted.

This procedure involves the following steps :

(i)   The number of available fertilized eggs that are to be inoculated by microinjection is increased by stimulating donor females to superovulate. They are given an initial injection of pregnant mare's serum and another injection about 48 hrs later of hCG. A superovulated mouse produces about 35 eggs instead of the normal number of 5 to 10.

(ii) These females are mated and then sacrificed. The fertilized eggs are flushed out from their oviducts.

(iii) The eggs are immediately inoculated. (Fig. 4.38)

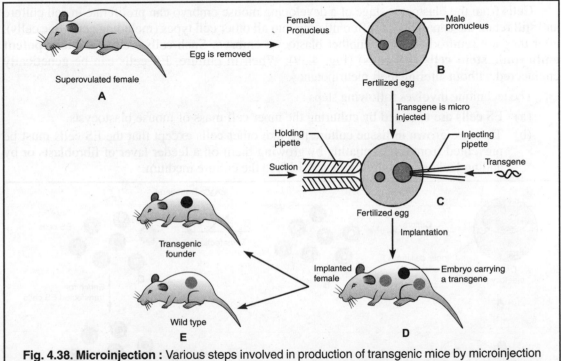

Fig. 4.38. Microinjection : Various steps involved in production of transgenic mice by microinjection method (Modified from Molecular Biotechnology by Glick and Pasternak, 2003)

Quite often, the microinjected transgene construct is in a linear form. In mammals, after entry of sperm into the egg, both the sperm nucleus (male pronucleus) and female nucleus exist separately. After the female nucleus completes its meiotic division to become a female pronucleus, nuclear fusion can occur. The male pronucleus, which tends to be larger than the female pronucleus, can be located by using a dissecting microscope. The egg can then be maneuvered, oriented and held in place, while the DNA is microinjected.

After inoculation, 25 to 40 eggs are microsurgically implanted into a foster mother, which has been made pseudopregnant by being mated to a vasectomized male. In this case, because the male mate lacks sperms, none of the eggs of the foster mother are fertilized. The foster mother will deliver pups from the inoculated eggs about 3 weeks after implantation (in mouse).

To identify transgenic animals, DNA from a small piece of tail can be assayed by either Southern blot hybridization or PCR for the presence of a transgene. A transgenic mouse can be mated to another mouse to determine if the transgene is in the germline of the founder animal. Subsequently, pregnancy can be bred to generate pure (homozygous) transgenic lines.

None of the steps in the procedure is 100% efficient. It is because of this a large number of microinjected fertilized eggs must be used. Furthermore, with this method, the injected DNA integrates at random sites within the genome, and often multiple copies of injected DNA are incorporated at one site. Not all of the transgenic pups will have the appropriate characteristics. In some individuals, the transgene may not be expressed because of the site of integration and in others, the copy number may be excessive and lead to overexpression, which disrupts the normal physiology of the animal.

(iv) Engineered Embryonic Stem Cell Method

Cells from the blastocyst stage of a developing mouse embryo can proliferate in cell culture and still retain the capability of differentiating into all other cell types (including germ line cells), after they are reintroduced into another blastocyst embryo. Such cells are called **pleuripotent embryonic stem cells** (ES cells) (Fig. 4.39). When in culture, ES cells can be genetically engineered without altering their pleuripotency.

The technique involves following steps :

(a) ES cells are obtained by culturing the inner cell mass of mouse blastocysts.

(b) They are grown in tissue culture just like other cells except that the ES cells must be prevented from differentiating by growing them on a feeder layer of fibroblasts or by adding leukemia inhibitory factor (LIF) to the culture medium.

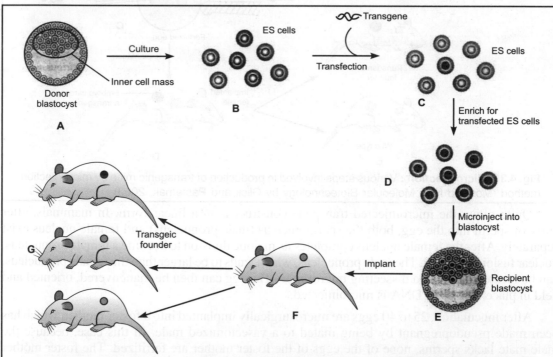

Fig. 4.39. Creation of transgenic mice with genetically engineered embryonic stem (ES) cells.
A,B. From the inner cell mass of a mouse blastocyst an ES cell culture is initiated. **C.** The ES cells are transfected with a transgene. **D.** The transfected cells are then made to grow in culture and identified by either the positive-negative selection procedure or PCR analysis. **E.** Populations of transfected cells are inserted into blastocysts. **F.** These are then implanted into foster mothers. **G.** From founder mice that carry the transgene in their germ lines, transgenic lines can be established by crosses.

(c)   DNA can be introduced into ES cells by transfection, retroviral infection or electroporation. The most important advantage for gene transfer into mice is that cells carrying the transgene can be selected before injecting into the blastocyst. In early experiments, ES cells were infected with retroviral vectors or transfected with plasmids, carrying the *'neo'* marker gene. This gene confers resistance to the antibiotic G 418. Only ES cells that have taken up the *neo* gene grow in medium containing G 418. These G 418– resistant cells are introduced into mouse blastocyst. Not only the resulting mice have *neo* gene integrated into their genome, but also the gene is transmitted to the offspring of the mice, and cell lines from F2 generation are G418-resistant.

(v) Calcium Phosphate Precipitation Technique

Graham and Van der Eb (1973) devised the calcium phosphate precipitation technique for introducing DNA into cells. The DNA and $CaCl_2$ solution are mixed with buffer and a DNA - $Ca_3(PO)_4$ precipitate forms. It can be applied to relatively large number of cells in a culture dish, but it is limited by the variable and usually rather low proportion of cells that take up exogenous DNA. Only a subfraction of these cells will be stably transfected.

(vi) DEAE-Dextran Method

An alternative transfection system employs DEAE-Dextran in the transfection medium, in which the DNA is also present. The DEAE-Dexran is soluble and no precipitate is involved. It is polycationic and possibly acts by mediating, in some unknown way, the productive interaction between negatively charged DNA and components of the cell surface in endocytosis. This method does not appear to be convenient for stable transfectants.

(vii) Electroporation

High voltage electric discharges have been shown to induce cells to fuse via their plasma membranes by creating holes or pores in the cell membrane. This is called **electroporation**. A suspension of plasmid DNA and cells is placed in a chamber and subjected to a high voltage, 2.0 to 4.0 KV, shock at zero degree Celcius. After 10 min., the cells are transferred to fresh medium and grown for two days before imposing selection. This technique is advantageous over other transfection methods as it is easier to use and it can be used for a broad range of cell types.

(viii) Viral Vectors

Many animal viruses have been used as vectors. Virtually every virus that has been studied in any detail and that has a DNA genome or a DNA stage in its replication cycle, has been manipulated in this way.

(a) *SV40* : It was the first animal virus whose genome was completely sequenced. The virus particle contains circular duplex DNA genome of about 5.2 kb. It causes lytic infection of permissive monkey cells. To develop SV40 vectors, following two strategies have been employed.

(i) A region of the viral genome is replaced with an equivalently sized fragment of foreign DNA, and hence produce a recombinant DNA that can replicate and be packaged into virions in permissive cells. In order to supply genetic functions lost by replacement of virus sequences, a helper SV40 virus must genetically complement the recombinant virus.

(ii) In this, the size limitation is to be avoided. Recombinants are constructed which are never packaged into virions and give no lytic infection. These are transiently maintained in host cells as high copy number, unintegrated, plasmid like DNA molecules.

SV40 vectors have the following basic properties :

• They have low to moderate levels of expression in a wide variety of transfected mammalian cells and high levels of expression in transfected COS cells,

• Both genomic DNA and cDNA sequences can be expressed, and there are no constraints on the size of foreign DNA inserted,

• They are generally used as a transient system,

• They contain the SV40 origin of replication, a promoter with a broad host range, and a polyadenylation signal,

• Many vectors of this class also carry splice donor and acceptor signals.

Vectors of this type have been effective in obtaining expression of many foreign genes in monkey cells : mouse beta-globin; rat preproinsulin; the P21 protein of Harvey murine sarcoma virus; influenza virus haemagglutinin and Hepatitis B surface antigen. Late region replacement vectors have also been used to study post translational RNA processing and stability.

(b) *Bovine papilloma virus (BPV)* : Papilloma viruses are small, double stranded DNA viruses that commonly infect higher vertebrates, including man. These viruses cause warts. Most studies at the molecular level have been carried out with a bovine papilloma virus, BPV-1. When rodent cells in culture are morphologically transformed by BPV DNA, 20 to 100 copies of viral DNA persist as extrachromosomal DNA. Not all of the viral genome is required for transformation for establishment of episomal state. The observation that both of these functions can be carried out by a viral fragment of the viral DNA, directly led to the use of BPV DNA as a vector to express rat preproinsulin murine c127 cells.

Following are the basic principles of these vectors :

• They have low to moderate levels of expression in a wide variety of mammalian cells

- Both genomic DNA and cDNA can be expressed
- They are never used as transient expression systems, instead they are used to establish cell lines that contain multiple copies of the gene
- They have segment of BPV DNA, a promoter with a broad host range, and a polyadenylation signal, splice donor/acceptor signals, poisonless plasmid sequence that allows the vector to be propagated in *E. coli*
- They may also carry a selectable marker.

(c) *Recombinant vaccinia viruses* : Vaccinia virus is closely related to variola virus, which causes smallpox. Vaccinia virus recombinants which express antigens of unrelated pathogens have been constructed and attempts are being made to use them as live vaccines against those pathogens.

In an experiment, fragments of vaccinia DNA were cloned in an *E. coli* plasmid vector that contained a non-functional vaccinia thymidine kinase gene. This gene had been rendered inactive owing to the insertion of a vaccinia DNA fragment containing a promoter in the correct orientation. This chimaeric HBsAg was then inserted into vaccinia DNA by homologous recombination.

Subsequently, recombinant vaccinia viruses expressing other important genes have been constructed, using an AIDS virus envelope gene, *HTLV-III* envelope gene and Hepatitis B surface Ag gene.

(d) *Retroviruses* : These have following useful properties :

- They cover a wide host range
- Infection does not lead to cell death
- Viral gene expression is driven by strong promoters
- In the case of murine mammary tumour virus, the promoter fraction can be experimentally switched on and off. Transcription is induced by glucocorticoid hormones.

Retroviruses contain RNA genomes. The virus particle actually contains two copies of viral RNA (Fig. 4.40). The following are the stages of infection of the host cells by retroviruses:

(i) When the viral RNA enters the cell, it is accompanied by reverse transcriptase and integrase which are packaged into the virion.

(ii) The reverse transcriptase then engages in a complex series of cDNA synthesis reactions which lead to the production of a double stranded DNA copy of the viral RNA.

(iii) This DNA copy, which is called the proviral DNA, is slightly longer than the RNA from which it was derived because terminal sequences are duplicated in the process of converting it to the double stranded form.

(iv) The proviral DNA circularizes and through the action of the integrase protein, inserts into the host genome.

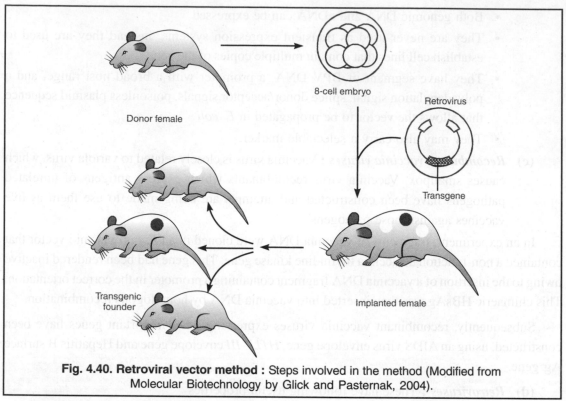

Fig. 4.40. Retroviral vector method : Steps involved in the method (Modified from Molecular Biotechnology by Glick and Pasternak, 2004).

Several retroviral vectors have been developed in which regions of the proviral DNA have been replaced by the foreign DNA.

This method has advantage of being an effective means of integrating the transgene into the genome of a recipient cell. However, these viruses can only transfer small pieces of DNA (8 kb). These, because of the size constraint, may lack essential adjacent sequences for regulating gene expression. There is a further major drawback: although these vectors are designed to be replication defective, they can be produced in the transgenic organism causing contamination.

(e) *Adenovirus vectors*: Adenoviruses have a linear double-stranded DNA genome of about 36 kb. Recombinant viruses can be created by deleting genes *EIA/EIB* and replacing them with the foreign DNA. For viral replication, the deleted genes and transcriptional regulators are required. The recombinant virus is defective. For propagating the defective recombinant virus, a transfected cell line that constitutively expresses the EIA / EIB functions is used. In addition the *E3* gene may be deleted from the adenovirus genome. This gene downregulates the immune response of the host in virus. It, however, is not necessary for replication *in vitro*. With the space created by these deletions, and the limited flexibility in the size of DNA that can be packaged into virions, the maximum insert size is 6-8 kb. When defective recombinant virus is applied to normal susceptible cells, the recombinant DNA episomally persists within the cell for relatively long periods.

Adenoviruses naturally infect the respiratory tract and for this and other reasons, recombinant adenoviruses have been applied to transfect airway cells for somatic gene therapy in cystic fibrosis. Recombinant adenoviruses have also been used for the transfer of gene to skeletal muscle. Transgenesis by adenovirus - mediated gene transfer into mouse zona - free eggs has also been successful. Zona-free mouse eggs at the pronucleus stage were infected with a replication-defective adenovirus vector containing a nuclear targeted *lac Z* gene. In almost all eggs, exogenous β-galactosidase activity was detected at the two cell stage of 27 mice that developed from infected eggs. Three of them carried the integrated exogenous gene mediated by the adenovirus. Two of the three expressed the *lac Z* gene, and all the three mice transmitted the adenovirus - mediated transgene to the F1 progeny (Tsukui *et al.*, 1996).

4.11 CLONING

The term cloning means the production of identical copies of something. In biology, it specifically means genetically identical copies. Asexual reproduction is a natural cloning process, since it generates offspring that are identical to the parent.

4.11.1 Cloning Procedures

The following procedures have been referred to as 'cloning' :

(i) **Adult DNA Cloning (Cell nuclear replacement) :** This involves removing the nucleus from an ova to replace it with that from a cell removed from an individual. Then, the embryo is implanted in a woman's womb and allowed to develop into a new human whose DNA is identical to that of the original individual.

(ii) **Therapeutic Cloning (somatic cell nuclear transfer) :** This starts with the same procedure as is used in adult DNA cloning. The resulting embryo is allowed to grow for perhaps 14 days. Its stem cells can be extracted and encouraged to grow into a piece of human tissue : a complete human organ transplant. This helps in a replacement of organ or a piece of nerve tissue or quantity of skin. The first successful therapeutic cloning was achieved in November 2001 by the Advanced Cell Technology, a Biotech Company in Worcester.

(iii) **Embryo Cloning:** It can be called 'artificial twinning' because it simulates the mechanism by which twins naturally develop. It involves removing one/more cells from an embryo and making the cell to develop into a separate embryo with same DNA as the original. It has been successfully carried out on many species of animals for years.

4.11.1.1 Adult DNA Cloning

It is also called **reproductive cloning** and generates an animal with same nuclear DNA as currently or previously existing animal.

In a process called **Somatic cell nuclear transfer (SCNT)** genetic material from nucleus of donor adult cell is transferred to an egg whose nucleus has been removed. It has been used to clone sheep and other animals.

Following steps are involved in cloning process:

(a) The somatic nucleus (containing full set of paired chromosomes) is transferred to an unfertilized egg from which the nucleus has been removed.
(b) The nucleus itself can be transferred or the intact cells can be injected into the oocyte.
(c) The oocyte with the nucleus from normal cell can be activated by a short electrical pulse.
(d) In sheep, the embryos are then cultured for 5-6 days. Those that appear to be developing normally (about 10%) are implanted into foster mothers.

Using this technology, Roslin institute in Roslin, Scotland, created 'Dolly'. In 1996, a cell was taken from the mammary tissue of a mature 6 year old sheep, while its DNA was in dormant state. It was fused with a sheep ovum which had its nucleus removed. The fertilized cell was then stimulated with an electrical pulse. Out of 277 attempts at cell fusion, only 29 began to divide. These were all implanted in ewes, 13 became pregnant, but only one lamb, Dolly was born. Later, seven more sheep of three breeds were successfully cloned in same institute.

On 1998-July-22, Dr. Ryuzo Yanagimachi of the University of Hawaii announced the cloning of mice.

On 1998-DEC-14, reserachers of the Infertility Clinic at Kyeonghee University in Korea announced that they had successfully cloned a human. Scientists Kim Seung-bo and LeeBo-yeon took an ovum from a woman, removed its DNA and inserted a somatic cell from the same 30 year old woman.

By the end of the year 2000, eight species of mammals had been cloned, including mice, cows, rhesus monkeys, sheep, goats, pigs and rats.

In such experiments, the success rate is quite low i.e.,

(a) average of only 1% of reconstructed embryos lead to live birth,
(b) because of respiratory or cardiovascular dysfunctions, the embryos die late in pregnancy or soon after birth. Abnormal development of placenta is common and this is probably a major cause of foetal loss earlier in pregnancy.

The low success rate is because of the fact that normal development of embryo is dependent on proper methylation state of DNA contributed by the sperm or egg and on appropriate configuration of chromatin structure after fertilization. Somatic cells have very different chromatin structure than that of sperm and reprogramming of the transferred nuclei must occur within a few hours of activation of reconstructed embryo. Incomplete or improper reprogramming will lead to dysregulation of gene expression and failure of the embryo or foetus to develop normally or to non-foetal developmental abnormalities in those that survive.

In addition to nuclear DNA of donor, the clone's genetic materials come from the mitochondria in the cytoplasm of the enucleated egg. Mitochondria, which are organelles that serve as power sources to the cell, contain their own short segments of DNA. Acquired mutations in mitochondrial DNA are believed to play an important role in the aging process.

Dolly's success is truly remarkable because it proved that the genetic material from a specialized adult cell, such as an udder cell programmed to express only those genes needed by udder cells, could be reprogrammed to generate an entire new organism. Before this demonstration, it was believed that once a cell became specialized as a liver, heart, udder, bone or any other type of cell, the change was permanent and other unneeded genes in the cell would become inactive (Fig. 4.41).

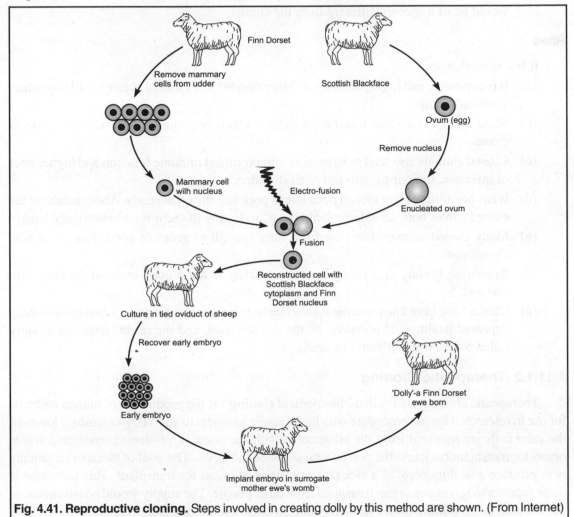

Fig. 4.41. Reproductive cloning. Steps involved in creating dolly by this method are shown. (From Internet)

Advantages of using reproductive cloning

(i) Reproductive cloning can be used to develop efficient ways to reliably reproduce animals with special qualities. For example, drug-producing animals or animals that have been genetically altered to serve as models for studying human disease, could be mass produced.

(ii) Endangered animals or animals that are difficult to breed can be repopulated by reproductive cloning. In 2001, the first clone of an endangered wild animal gaur was born. The young gaur died from an infection about 48 hours after its birth. In 2001, cloning of a healthy baby mouflon, an endangered wild sheep was made in Italy. The cloned mouflon is living at a wildlife centre in Sardinia. Other endangered species that are potential candidates for cloning include the African bongo antelope, the Sumatran tiger, and the giant panda. There is a much greater challenge to scientists for cloning extinct animals because the egg and the surrogate needed to create the cloned embryo would be of a species different from the clone.

Risks

It has several risks.

(a) It is expensive and highly inefficient. More than 90% of cloning attempts fail to produce viable offspring.

(b) More than 100 nuclear transfer procedures could be required to produce one viable clone.

(c) Cloned animals also tend to have more compromised immune function and higher rates of infection, tumour growth and other disorders.

(d) It has been shown that cloned mice live in poor health and die early. About a third of the cloned calves born alive have died young, and many of them were abnormally large.

(e) Many cloned animals have not lived long enough to generate good data about how clones age.

(f) Appearing healthy at a young age unfortunately is not a good indicator of long-term survival.

(g) Clones have been known to die mysteriously. For example, Australia's first cloned sheep appeared healthy and energetic on the day she died, and the results from her autopsy failed to determine a cause of death.

4.11.1.2 Therapeutic Cloning

Therapeutic cloning, also called **"biomedical cloning"** is the production of human embryos for use in research. This is a procedure with initial stages identical to adult DNA cloning. However, the stem cells are removed from the pre-embryo with the intent of producing tissue or a whole organ for transplant back into the person who supplied the DNA. The goal of therapeutic cloning is to produce a healthy copy of a sick person's tissue or organ for transplant. This technique is quite superior to relying on organ transplant from other people. The supply would be unlimited, so there would be no waiting lists. The therapeutic cloning involves the following steps :

(i) A woman's ovum is taken and its nucleus is removed.

(ii) The nucleus is removed from a cell taken from the human and is inserted into the enucleated ovum.

(iii) The resulting ovum is given an electrical shock to start up its embryo making operation. In a small percentage of cases, a pre-embryo or blastocyst will be formed.

(iv) The pre-embryo is allowed to develop to produce many stem cells. The procedure up to this stage is quite similar to that used in adult cloning. However, the pre-embryo is not implanted in a woman's womb to produce a pregnancy.

(v) Stem cells are removed from the pre-embryo. This results in its death.

(vi)   Stem cells are encouraged to grow into whatever tissue/organ is needed to treat the patient.

(vii) The tissue/organ is transplanted into the patient.

Therapeutic cloning has following hurdles :

(a)   Successful isolation and growth of the stem cells in laboratory

(b)   Their differentiation into special cell types (this has been done for most of the 220 cell types in the human body)

(c)   They have to be proven usable in treating patients with diseases, injuries/disorders

(d)   The transplanted tissue must develop normally and must not represent significant 'risk to the patients'

(e)   It has been shown that stem cells from embryos have greater flexibility than from adult cells

(f)   It has also been felt that it involves killing of an embryo *i.e.*, an individual itself

(g)   There is every chance of mutations in stem cells. They are, thereby, rejected by the recipient's body. In certain cases, they have produced tumours also

(h)   The availability of eggs. Where would the eggs come from ? Even the extraction of eggs from women is extremely painful, costly and unreliable. It has been estimated that it takes about 100 eggs if we are lucky to produce a suitable stem cell line. Until the production of stem cells become efficient, very few cures could be made for economic reasons.

Stem cells can be used to serve as replacement cells to treat heart disease, Alzheimer's, cancer, and other diseases (Fig. 4.42).

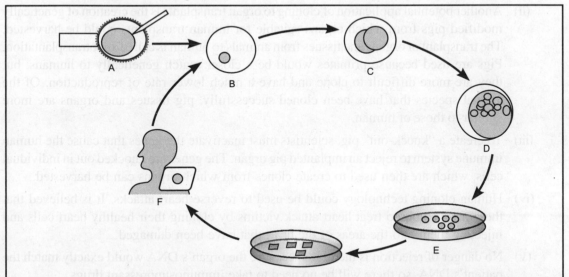

Fig. 4.42. Therapeutic cloning. General strategy for therapeutic cloning. The figure highlights the way the somatic cell nucleus is transferred. **A.** Nuclear material is extracted from the donor egg **B.** The nucleus is harvested from one of the patient's normal somatic cells and inserted into the enucleated egg. **C.** Egg is stimulated to divide to form a blastocyst **D.** Stem cells are harvested from the blastocyst for culturing **E.** Cells are directed to differentiate to form precursors for tissue to be replaced **F.** Cells are transplanted into patient to replace diseased tissue

The therapeutical cloning is quite useful in the sense that with the success of this technology perfectly matched replacement organs would become freely available to the sick or dying people *e.g.,* insulin-secreting cells for diabetes, nerve cells in stroke/Parkinson's disease or liver cells to repair a damaged organ can be grown in this way.

   (i) Therapeutic cloning can be used to generate tissues and organs for transplants (Fig. 4.43).

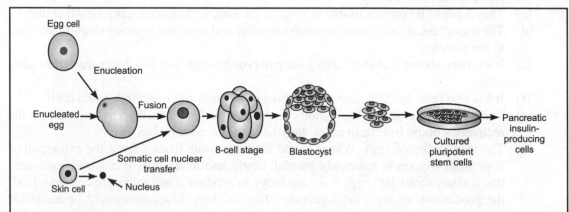

Fig. 4.43. Therapeutic cloning. Diagram showing therapeutic cloning for production of pancreatic cells (From Internet).

   (ii) Another potential application of cloning to organ transplants is the creation of genetically modified pigs from which organs suitable for human transplants could be harvested. The transplant of organs and tissues from animals to human is called xenotransplantation. Pigs are used because primates would be a closer match genetically to humans, but they are more difficult to clone and have a much lower rate of reproduction. Of the animal species that have been cloned successfully, pig tissues and organs are more similar to those of human.

  (iii) To create a "knock-out" pig, scientists must inactivate the genes that cause the human immune system to reject an implanted pig organ. The genes are knocked out in individual cells, which are then used to create clones from which organs can be harvested.

  (iv) Human cloning technology could be used to reverse heart attacks. It is believed that they may be able to treat heart attack victims by cloning their healthy heart cells and injecting them into the areas of the heart that have been damaged.

   (v) No danger of rejection is there. It is because the organ's DNA would exactly match the patient's DNA, so there will be no need to take immunosuppressant drugs.

  (vi) No other individual as a donor is required for organ transplant (like kidney),

 (vii) There is no need of donor and immediate availability of organ will be possible,

(viii) An old organ which has reduced functionality can be replaced by the new one,

  (ix) There is a potential to cure or at least handle disease for which there is no cure as yet.

(x) **Solving infertility problem.** Human cloning could make it possible for many more infertile couples to have children.

(xi) **Defective genes.** The average person carries 8 defective genes inside them. These defective genes allow people to become sick when they would otherwise remain healthy. With human cloning and its technology it may be possible to ensure that we no longer suffer because of our defective genes.

(xii) Cloning the bone marrow for children and adults suffering from leukemia. This is expected to be one of the first benefits to come from cloning technology.

(xiii) **Cancer.** It may be possible to learn how to switch cells on and off through cloning and thus be able to cure cancer. Scientists still do not know exactly how cells differentiate into specific kinds of tissue, nor do they understand why cancerous cells lose their differentiation. Cloning, may serve as a the key to understanding differentiation and cancer.

4.11.1.3 Embryo Cloning

4.11.1.3.1 Historical

Since the late 1970s, cloning of embryos has been used in mice experiments. In animal breeding it is in use since the late 1980s. The procedure splits a single fertilized ovum into two or more clones, each of which is then implanted into the womb of a receptive female.

However, research into cloning of human embryos has been restricted in the United States and in some other countries.

The methods used have been understood for many years and actually used to clone embryos in cattle and sheep. It is possible that someone had successfully used the method on a human embryo in secret. The first publicly announced human cloning was done by Robert J. Stillman and his team at the George Washington Medical Center in Washington D.C. They took 17 genetically flawed human embryos which would have died within days no matter how they were treated. They were derived from an ovum that had been fertilized by two sperms. This resulted in an extra set of chromosomes which doomed the ovum's future. None could have developed into a foetus. These ova were successfully split in 1994-Oct, each producing one or more clones. The main motive of the experiment seems to have been to trigger public debate on the ethics of human cloning.

It is felt that studies be normally limited to the use of preexisting, spare embryos - those that developed during *in vitro* fertilization procedures, that had been performed to assist couples in conceiving. During these procedures generally about 20 or 24 ova are fertilized. Only three or four are implanted in the woman. The extra zygotes are either discarded or frozen for possible future use. New embryos would only be prepared and used if needed for "compelling research." It was recommended that any study be normally terminated within 14 days of conception. Some experiments might be authorized to continue until the 18th day, but no further.

Cloning of human embryo starts with a standard *in vitro* fertilization procedure. Sperm and egg cells are mixed together on a glass dish. After fertilization, the zygote develops into a blastula. The zygote first divides into two, four and then eight cells. To remove the *zona pellucida* covering, a chemical is added to the dish.

(i)   The nutrients to promote cell division are provided by media.

(ii) With the covering removed, the blastula is divided into individual cells which are deposited on individual dishes.

(iii) They are then coated with an artificial *zona pellucida*, and allowed to divide and develop. The experiment by Stillman *et al.* (2001) showed that the best results could be obtained by interrupting the zygote at two cell stage.

(iv) Many of these pairs of zygotes were able to develop to the 32 cell stage, but stalled at that point. This might well had the potential to develop further and even mature into a viable foetus except that the original ovum was defective and would have died anyway. For ethical reasons, the embryos which had no possibility of maturing into foetuses were selected for experimentation.

4.11.2 Advantages of Cloning

The cloning is favoured because of the following reasons :

(i)   Using the DNA from the cell of an adult with the desired traits, an infant with the same potential may be produced.

(ii)   A heterosexual couple in which the husband is completely sterile could use adult DNA cloning to have a child. This technique might be more satisfactory for them than to use sperm of another man.

(iii) A child can be elected by two homosexual females by adult DNA cloning, rather than by artificial insemination. By this technique homosexual, couples can have technicaly their own children.

(iv)   It might make upto understand in a greater way of the causes of miscarriages; this might lead to a treatment to prevent spontaneous abortions. This would be of immense help for women who cannot bring a foetus to term.

(v)   It might lead to an understanding of the mechanisms by which implantation of a morula (a mass of cells that has developed from a blastula) to the wall of the uterus occurs. This might generate new, effective contraceptives that exhibit very few side effects.

(vi)   The rapid growth of the human morula is similar to the rate at which cancer cells propagate. It is believed that if a method is found to stop the division of a human ovum then a technique for terminating the growth of a cancer might be found.

(vii) Parents who are known to be at risk of passing a genetic defect to a child could make use of cloning. A fertilized ovum could be cloned, and the duplicate tested for the disease or disorder. If the clone was free of genetic defects, then the other clone would be as well. The latter could be implanted in the woman and allowed to mature to term.

(viii) In conventional *in vitro* fertilization, attempts are started with many ova, fertilize each with sperm and implant all of them in the woman's womb in the hope that one will result in pregnancy. But some women can only supply a single egg; her chances of becoming pregnant are slim. Through the use of embryo cloning, that egg might be divisible into say 8 zygotes for implanting, the chance of those women becoming pregnant would be much greater.

(ix) Cloning could produce a reservoir of "spare parts". Fertilized ova could be cloned into multiple zygotes; one could be implanted in the woman and allowed to develop into a normal baby; the other zygotes could be frozen for future use. In the event that the child required a bone marrow transplant, one of the zygotes could be taken out of storage, implanted, allowed to mature to a baby and then contribute some of its spare bone marrow to its (earlier) identical twin. Bone marrow can be harvested from a person without injuring them.

(x) A woman could prefer to have one set of identical twins, rather than go through two separate pregnancies. She might prefer this for a number of reasons.

(xi) It is being tried to create transgenic pigs which have human genes. Their heart, liver/ kidneys might be useable as organ transplants in humans.

(xii) Cloning can lead to better understanding of genetics.

(xiii) Transgenic cattle have been produced which produce human hormones/proteins and secrete it along with the milk.

(xiv) Zygotes of a particular gender could be eliminated by the genetic screening test

(xv) A program similar to that of Nazi Germany, whereby, humans were bred to maximize certain traits may be financed by a country. Once the perfect human is developed, embryo cloning could be used to replicate that individual. Similar approach can be used to generate a genetic underclass for exploitation (e.g., individuals with sub normal intelligence).

(xvi) There is always a possibility of killing/insuring the embryos which is protested by pro- lifers.

The mechanism by which twins naturally develop is used for this purpose. It is known that in 10 out of 75 conceptions, the fertilized ovum splits for some unknown reasons and produces monozygotic (identical) twins. This same operation is intentionally done in a laboratory for cloning.

4.11.3 Disadvantages of Embryo Cloning

One of many concerns with human cloning is that cloning of animals sometimes cause foetal overgrowth. The foetus grows unusually large and generally dies. They have underdeveloped lungs and reduced immunity to infection. There are, certain other factors that can also contribute to abnormal development of clones.

It is opposed because of the following reasons :

(i) There is no surety that first cloned humans would be normal. They may be born disabled or disorders may set later in life,

(ii) Cells seem to have an inbuilt defined life span. The clones are, therefore, bound to have reduced life expectancy,

(iii) No semen was required in the creation of Dolly. This excludes the genetic need of a male in future,

(iv) Genetic diversity may be depleted by large scale cloning,

(v) Cloning may have long-term extensive impact on relationship.

Due to these reasons, in most of the countries cloning of humans is banned, but it is largely believed that some scientists are secretely working on it.

It is because of these, Dr Steven Muller headed a panel in the US to produce preliminary cloning guidelines. They suggested that studies be limited to the spare embryos produced during IVF. They also recommended to terminate the studies within 14 days of conception. Some may continue till 18th day, but no further. They recommended to ban implanting human embryos in other species, cloned embryo to another human being and the use of embryos for sex selection.

4.11.4 Risks of Cloning

Reproductive cloning is expensive and highly inefficient.

Cloning animals shows us what might happen if we try to clone humans. What have these animals taught us about the risks of cloning?

a) High failure rate: The success rate of cloning experiments is very low. A large number of cloning experiments fail. Cloning animals through somatic cell nuclear transfer is simply inefficient. The success rate ranges from 0.1 per cent to 3 per cent, which means that for every 1000 trials, only one to 30 clones are made. There may be 970 to 999 failures in 1000 trials. More than 90% of cloning attempts fail to produce viable offspring. More than 100 nuclear transfer procedures could be required to produce one viable clone. Even the successful clones face problem later during the animal's development to adulthood. cloned animals tend to have more compromized immune function and higher rates of infection, tumour growth, and other disorders. It has been shown that cloned mice live in poor health and die early. About a third of the cloned calves born alive have died young, and many of them were abnormally large. Many cloned animals have not lived long enough to generate good data about how clones age. Appearing healthy at a young age unfortunately is not a good indicator of long-term survival. Clones have been known to die mysteriously. For example, Australia's first cloned sheep appeared healthy and energetic on the day she died, and the results from her autopsy failed to determine a cause of death.

Why is this? Here are some reasons:

- The enucleated egg and the transferred nucleus may not be compatible
- An egg with a newly transferred nucleus may not begin to divide or develop properly
- Implantation of the embryo into the surrogate mother might fail
- The pregnancy itself might fail

b) Problems during later development: Cloned animals that do survive tend to be much bigger at birth than their natural counterparts. It is called "Large Offspring Syndrome" (LOS). Clones with LOS have abnormally large organs. This can lead to breathing, blood flow and other problems.

Because LOS doesn't always occur, it cannot reliably be predicted whether it will happen in any given clone. Even some clones without LOS have developed kidney or brain malformations and impaired, which can cause problems later in life.

c) Abnormal gene expression patterns: Are the surviving clones really clones? The clones look like the originals, and their DNA sequences are identical. But will the clone express the right genes at the right time?

Whitehead Institute of Biomedical Research in Cambridge, Massachusetts in 2002 reported that the genomes of cloned mice are compromised. In analyzing more than 10,000 liver and placenta cells of cloned mice, it was discovered that about 4% of genes function abnormally. The abnormalities do not arise from mutations in the genes, but from changes in the normal activation or expression of certain genes.

The programming errors in the genetic material from a donor cell may also cause problems. When an embryo is created from the union of a sperm and an egg, the embryo receives copies of most genes from both the parents. The DNA from the mother and father is imprinted chemically so that only one copy of a gene (either the maternal or paternal gene) is turned on. Defects in the genetic imprint of DNA from a single donor cell may lead to some of the developmental abnormalities of cloned embryos.

It is therefore needed to re-program the transferred nucleus to behave as though it belongs in a very early embryonic cell. This mimics natural development, which starts when a sperm an egg. This program is different for every type of differentiated cell - skin, blood, bone or nerve. Complete reprogramming is needed for normal or near-normal development. Incomplete programming will cause the embryo to develop abnormally or fail.

In a naturally-created embryo, the DNA is programmed to express a certain set of genes. Later on, as the embryonic cells begin to differentiate, there is a change in the program.

d) Telomeric differences: Their chromosomes get shorter with every cell division. This is because the DNA sequences at both ends of a chromosome, *i.e.,* telomeres, shrink in length every time the DNA is copied. The older the animal is, the shorter its telomeres will be, because the cells have divided many, many times. This is a natural part of aging.

The question therefore is what will happen to the clone if its transferred nucleus is already pretty old? Will the shortened telomeres affect its development or lifespan?

No clear answers this questions have been obtained. It was observed that chromosomes from cloned cattle or mice had longer telomeres than normal. These cells showed other signs of youth and seemed to have an extended lifespan compared with cells from a naturally conceived cow. This suggests clones could live longer life spans although many died young after excessive growth. It is being thought that this could eventually be developed to reverse aging in humans, provided that this is based chiefly on the shortening of telomerases. Although some work has been performed on telomerases and aging in nuclear transfer clones, the evidence is at an early stage.

4.11.5 Ethical Issues of Human Cloning

Human cloning and the cloning of human embryos has been opposed by many conservative groups. According to them life begins at the moment of fertilization. Other Christian denominations such as the united chart of Christ do not believe a fertilized egg constitutes a living being, but still they oppose the cloning of embryonic cells. The World Council of Churches, representing nearly 400 Christian denominations worldwide, opposed cloning of both human embryos and whole humans in February 2006. Research and reproductive cloning was apposed by the United Methodist Church in May 2000 and again in May 2004.

The federal government of the United States according to libertarian views on the subject suggest that does not have the power to regulate cloning, as it is not given any such authority by the US.

The main non-religious objection to human cloning is that the cloned individual may be biologically damaged, due to the inherent unreliability of its origin. It is because of the fact that we have been unable to safely and reliably clone non-human primates.

It is believed that with the improvement in cloning research and methods, concerns of safety and reliability will no longer be an issue. It must, however, be pointed out that this has yet to occur. It has been pointed by Rudolph Jaenisch, a Professor at Harvard, that we have become more efficient at producing clones which are still defective. Other arguments against cloning come from various religious orders (believing cloning violates God's will or the natural order of life), and a general discomfort some have with the idea of "meddling" with the creation and basic function of life. This unlease is manifested in contemporary novels, movies, and popular culture, as it did with numerous prior scientific discoveries and inventions. Various fictional scenarios portray clones being unhappy, soulless, or unable to integrate into society. Furthermore, clones are often depicted not as unique individuals but as "spare parts," providing organs for the clone's original (or any non-clone that requires replacement organs).

What so every may be, cloning is a poignant and important topic, reflected by its frequent discussion and debate among politicians, scientists, the media, religions, and the general public.

On December 28, 2006, the FDA approved eating meat from cloned animals. It was said to be virtually indistinguishable from the non-cloned animals. Furthermore, companies would not be required to provide informing the consumer that the meat comes from a cloned animal. All this information has been collected from Internet.

4.11.6 Conclusion

As mentioned earlier cloning has many advantages. It also has is team disadvantages and risks. There are however, many hopes in cloning of plants, animals & humans.

Plants : There are many applications of cloning of plants. These portion not only to agriculture but to medicine as well. Cloning as made it possible to have herbicide resistant crops, earlier/ delayed fruit ripening, longer lasting fruit and higher yields of crops. There are however certain disadvantages too. These mainly concerned with the cost and time needed for the process or consumer resistance. Another disadvantage is that the genetically altered offspring might not be able to reproduce naturally.

Animals: The cloning of animals may lead to :
   i)   improved transgenic animals for biotechnology
   ii) reduced numbers of animals necessary for medical studies
   iii) creation of large numbers of custom-designed transgenic animals that secrete medically useful human proteins in their milk *e.g.,* human clotting factor and fibrinogen.
   iv) creation of model animals for research *e.g.* sheep designed to replicate the effects of human cystic fibrosis

Humans: Following are the application of cloning to human.

i) It might reveal how the environment within the cells of the early embryo regulates gene function.

ii) It might help combat genetic diseaes by allowing researches to turn off bad genes and turn on good genes.

iii) It allows parents of a young child fatally injured to be cloned should enough cells be recovered from the body.

iv) It allows growth of organs intended for transplant therapy.

v) It aids the infertile people.

vi) It can assess how much environment and genetic makeup interact *e.g.,* how much does environment plays in governing intelligence.

4.12 *IN VITRO* FERTILIZATION AND EMBRYO TRANSFER (IVF-ET)

4.12.1 History

Infertility is common, approximately 10% couples have difficulty conceiving a child. In young healthy couples, the probability of conception in one reproductive cycle is typically 20-25% and in one year, it is appropriately 90%. An evaluation is commonly recommended after one year of unprotected intercourse without conception, the standard clinical definition of infertility (Van Vorrhis, 2007).

IVF-ET is a fertility procedure. In 1978 Dr Edwards (an Embryologist) and Dr Streptoe (a Gynaecologist) in England successfully performed it. Since then, the technology has been further refined and has become popular world over. Approximately 1 in 50 births in Sweden, 1 in 60 in Australia and in the United States now result from IVF. In 2003, more than 100,000 IVF cycles were reported from 399 cities in the US, resulting in the birth of more than 48,000 babies. Infact, IVF and similar techniques have resulted in more than 200,000 babies.

The IVF has increased the possibility of continuing pregnancy from practically nil to one chance in 4-6 at IVF centres worldwide. The probability of pregnancy in any given situation can be predicted through evaluation. It depends on many variables such as age and the reproductive health of both the wife and the husband. The chance of success varies from case to case. IVF is never the first step in treatment of infertility. Instead it is reserved for cases in which other methods such as fertility drugs, surgery and artificial insemination have not worked.

4.12.2 Definition

Fertilization involves the union of egg (released from ovary) and sperm (from semen) in the fallopian tube. In IVF, this union is made possible in a lab after eggs and sperms have been collected. It is followed by the transfer of embryos to the uterus to continue growth.

The following procedure is used :

(i) Ova are collected from the woman at the time of her ovulation. Depending upon the cause of infertility, the eggs used may be the woman's own or those from any other donor woman.

(ii)   These are put in a petridish (or test tube).

(iii) The spermatozoa (obtained from a male) and ovum remain in the dish/test tube for a few hours to fuse (this is how the misnomer 'TEST TUBE BABY' came about. It is misleading because individuals don't grow in test tubes).

(iv)   The fertilized ovum is transferred into the woman's uterus.

(v)   If the fertilized ovum gets implanted inside the uterus, the woman becomes pregnant.

4.12.3 Eligibility for IVF

The individuals with following conditions are eligible for IVF.

(i)   *Blocked fallopian tubes :* A women whose both fallopian tubes are absent/blocked due to surgery or tubal infection (STD or tuberculosis)

(ii)   *Endometriosis :* Sometimes due to pelvic anatomy distortions, the endometrium gets adhered to the pelvic region and there is a problem in the movement of spermatozoa and eggs.

(iii) *Reduced sperm counts or motility :* IVF can be performed when sperm counts are not more than 5 million per ml. If the counts are less than that, ICSI (intra cytoplasmic sperm injection) is done.

(iv)   *Patients with unexplained infertility :* A man may have immunological infertility that occurs when antibodies to sperm prevent normal motility and function. Sometimes all the investigations performed on the couple are normal, but still do not normally conceive with routine treatments. In such cases IVF can be done.

Before performing IVF-ET, patient's past medical and infertility history is evaluated. It is followed by consultation in the IVF-ET office, where the medical and scientific aspects of the problem are reviewed.

4.12.4 Steps Involved in IVF

The following steps are involved in IVF :

4.12.4.1 Precycle Evaluation

The following factors determine the success rates of IVF :

(a)   **Response of women to fertility drugs :** This parameter is affected by age. The women with high FSH are more resistant to ovarian stimulation.

(b)   **Presence of endometriosis :** The women undergo hysteroscopy so as to see inside of the uterus by the physician to determine that there are no fibroids (produced by endometriosis).

(c)   **Uterine lining :** Prior to ovulation, the uterine lining is evaluated. It is done by using a sonogram. Certain patterns of uterine lining development are associated with poor pregnancy rate. To take care of any problem, estradiol treatments are given.

4.12.4.2 Induction of Ovulation and Maturation

In order to prepare a proper environment in the woman and to increase the chances of recovering several healthy and mature eggs, the woman undergoes intensive preparation for about two weeks. This includes hormonal therapy with fertility drugs. The production of optimal number of eggs and even the egg retrieval can be timed. It will stimulate the ovarian follicles.

The ovarian follicles can be stimulated by giving the following drugs :

(a) **Leuprolide acetate (lupron)** : It is injected to block secretions of pituitary gland, thereby, optimizing the number of oocytes retrieved.

(b) **Human menopausal gonadotropin (pergonal or HMG) or FSH (or Metrodin) hormones :** This is daily injected for about 6-10 days prior to the procedure.

(c) **Minimizing the chances of cancellation of the patient's cycle.** For this purpose agents that selectively block the endogeneous LH surge (GnRH agonists) are used. These agents act by inducting down regulation and desensitization of pituitary receptors to GnRH (this eliminates any endogenous FSH/LH secretion). These are usually administered daily beginning in the luteal phase of cycle preceding ovarian stimulation.

(d) **Human chorionic gonadotropin (HCG) :** This hormone mimics the action of hormone which naturally induces ovulation. It is injected 34-36 hours before retrieval. It may also be used after retrieval to supplement natural progesterone production. Stimulation is continued with lupron and gonadotropins, until two follicles have reached a mean diameter of 1.5 cm after the hCG is administered.

(e) **Promotion of egg development :** For this purpose serophene pill is used.

In order to monitor patient's response to these drugs, daily sonograms are done and serum estradiol levels are measured.

4.12.4.3 Retrieval of an Ovum

Ovum was used to be recovered by laparoscopy. It has, however, been observed that patients who have undergone a screening laparoscopy, have ovaries inaccessible for ovum retrieval. A conservative ovarian procedure was initially used for patients with extensive periovarian, bowel/omental adhesions, which made the ovaries totally inaccessible for the recovery of an ovum. It made the ovaries more accessible for future ovum retrieval.

The improvements in ultrasound imaging have made the procedure easier and surgery is no more required. The retrieval procedure to obtain the eggs is performed transvaginally using a hollow needle guided by the ultrasound imaging (this is done under adequate sedation and local anaesthesia). Using the needle through vaginal wall, eggs are gently removed from the ovaries. This is called **'follicular aspiration'**. Its timing is crucial because the egg will not develop if it is collected :

   (i) too early or

  (ii) too late,

The egg may also develops poorly or may have already been released by the ovary and lost. The follicles, thus, isolated are immediately brought to the laboratory and examined for the presence of oocytes. Since oocytes are not present in fair percentage of initial follicular aspirates, the follicle is immediately reflushed with heparinized solution. Morphological criteria are used to classify the oocytes that are obtained. These criteria determine the timing of the eggs insemination.

(i) The immature oocytes display compact corona layer with a very small amount of expanded cumulus. The granulosa cells enveloping the immature oocytes display little intercellular expansion and are aggregated in a compact fashion. Such eggs are inseminated only after polar body exclusion or more routinely, the following day.

(ii) Expansion of the corona radiata and that of the extracoronal cumulus display preovulatory oocytes. The granulosa cells are loosely aggregated. They are inseminated at about 6-8 hours after retrieval.

(iii) The oocytes, sometimes have markedly expanded corona layer. Granulosa cells display marked cellular expansion and intercellular spacing. They are inseminated within 1-2 hours after retrieval.

4.12.4.4 Processing of Sperm

It is carried out as follows:

(i) Sperms used for insemination are collected by masturbation.

(ii) Semen is allowed to liquify at room temperature for 30 min.

(iii) It is washed in a modified culture medium.

(iv) The sperm suspension is centrifuged at 20 X G for 5 min. and the supernatant discarded.

(v) The intact pellet is then layered with 1 ml of sperm washing media and incubated for 1 hour in 5% CO_2 with the sperm allowed to swim out of the pellet, theoretically being capacitated during the process.

(vi) The uppermost layer of spermatozoa (motile) is then placed in a separate test tube, counted and the eggs inseminated.

4.12.4.5 *In vitro* Fertilization

(i) The sperm and egg are joined in one of the two ways.

a) They can be mixed in a laboratory dish where egg is inseminated with about 5,00,000 motile sperms

b) Single sperm is injected into each egg using special microscopes, needles and other equipment.

(ii) The inseminated egg is incubated at 37°C in humid atmosphere of 5% CO_2 and air for 18-21 hours.

(iii) After that the oocytes are observed under dissection and inverted stereoscopic microscopes for the presence of pronuclei (and their number) and polar bodies.

(iv) Usually, the cumulus mass surrounding the egg disperses and coronal layer gets loosened.

(v) Once the coronal cells are removed, the presence of pronuclei is then ascertained.

(vi) Eggs containing >3 pronuclei (polypronuclear) will not be transferred for fear of precipitation of an aberrant pregnancy.

(vii) Normal healthy embryos are then incubated at 37°C in a humid atmosphere of 5% CO_2 for another 22-24 hours, after which they are again observed, generally 2-8 cell embryo with even sized blastomeres are seen.

(viii) The pre-embryos are cultured and graded according to the quality of the blastomeres, the number of blastomeres, and the thickness of zona pellucida.

(ix) Since embryo must first hatch from the zona pellucida prior to implantation, those embryos whose zonae are considered too thick undergo 'Assisted Hatching' in the laboratory. In this process, a small dent in the zona pellucida is induced by either dissolving a portion of the zona pellucida with an acid solution or using laser.

4.12.4.6 Transfer of Embryos

When the embryos have reached the four to ten cell stage on the second or third day following follicular aspiration. It is transferred into the uterus. With the development of new types of media, embryos can be cultured for longer periods of time in the laboratory. In *blastocyst* after about five days in culture the embryo development stage just prior to implantation is reached. By extending the culture period, *blastocyst stage culture* it is possible to select best embryos. This improve the pregnancy rates theoretically. This technique may be of greater importance in PCOS patients who produce more embryos, rather than good quality embryos. The transfer of few embryos results in a reduced risk of multiple pregnancy and the serious consequences that can result.

It takes only a few minutes to transfer the embryo. It involves placing a small plastic tube through the cervix into the uterine cavity. The procedure does not require any anesthesia. or sedation. Some centres use ultrasound guidance for the transfer.

Procedure. The selected zygotes are transferred into the uterus approximately 0.5 cm from the uterine fundus (as premeasured ultrasonographically) in 50 ml of a viscous transfer medium using a Teflon catheter. The progesterone treatments are given to the patient daily after egg retrieval to prepare the uterus to support pregnancy and is continued until 10 weeks of pregnancy.

After the pre-embryo placement in the uterus, the patient will lie quietly in the bed for about an hour and will then return home.

The transfer of embryo is followed by a resting period, blood tests and possible ultrasound examination to verify, if pregnancy has been established.

Other points related to IVF

- The total time of procedure is about 3 weeks. Fertility drugs are administrated to stimulate the ovaries. Then during 4-6 days prior to ovulation, patient is monitored by ultrasound and hormonal levels.
- 25% pregnancies of IVF are twins. Extra pre-embryos may be disposed or cryopreserved for future use.

Success rate is higher if donor eggs from young women are used.

Implantation Support and Monitoring

It is unknown when implantation takes place or what can be done to ensure the best chance of implantation. Because of the manipulation of the ovaries that has taken place, additional supplements of hCG and/or progesterone are given to help ensure the optimum environment for implantation.

4.12.4.7 Procedures Associated with IVF/ET

Cryopreservation - IVF/ET involves certain other processes that should be part of the system laboratory. Some of these have been given below: When the number of viable embryos available is more than are needed to successfully establish a pregnancy, they can be frozen. To determine the best quality embryos for transfer, the embryos are allowed to remain in culture long enough. The longer the embryo remains in culture, the less likely a subsequent pregnancy from the frozen embryo will occur. Embryos can be frozen at any time after fertilization. That is embryos frozen at the 2-pronuclear stage, the first day after aspiration, may be better than those frozen on day 3. Those frozen on day 3, may be better than those frozen on day 5. Cryopreservation allows the possibility of an additional attempt at pregnancy without the necessity of the fertility injections or aspiration procedure. It is not known for how long an embryo can remain frozen, but successful pregnancies have occurred ten years after the initial transfer of embryos. As few as 50% of the embryos survive the thaw process, but there is no evidence that cryopreservation is harmful to children born from the technique.

Blastocyst culture. It is an important scientific advance. The technique, however, has been the subject of some major media attention. In some cases, the uterus may still be a better incubator than the laboratory. Some pregnancies may be lost by extending the culture period and delaying transfer. There will be fewer embryos for cryopreservation because of the need to keep more embryos in culture to see which will continue to develop. An additional attempt at success using frozen and thawed embryos may be lost. An extended culture can also mean an extended cost.

Assisted hatching. The egg and later the embryo is surrounded by *zona pellucida* (ZP) a protein halo and later the embryo. As it travels down to the uterus, the ZP prevents attachment of the embryo to the wall of the tube. Once in the uterus, the ZP dissolves and the embryo "hatches." It has been suggested that the ZP abnormally thickens or "hardens". It reduces the success with IVF. In fact, older women and those with PCOS are at a higher risk for this hardening. In assisted hatching, the ZP is thinned either

   a)   mechanically by physically puncturing it, or

   b)   chemically by using an acid solution to partially dissolve it.

While the procedure may improve the chances for pregnancy, it is more costly and could destroy the embryo.

4.12.4.8 Gift and other types of assisted reproduction procedure

There are many variations on the IVF/ET procedures.

GIFT procedure. In this follicles are aspirated during laparoscopy and the unfertilized oocytes placed in the fallopian tube with a sample of prepared sperm. The GIFT procedure is a more "natural approach". It avoids IVF and embryo culture.

The IVF-ET procedure, however, has its own advantages such as avoidance of laparoscopy with its required general anesthesia, postoperative recovery, and higher cost. Knowing that fertilization has occurred and that transport through a possibly abnormal tube is not necessary are additional advantages. IVF-ET may have important diagnostic, prognostic, and therapeutic consequences. The GIFT procedure can only be limitedly used. This procedure has more or less been abandoned in favour of IVF-ET.

Procedures like PROST, ZIFT, and TET are modifications of GIFT/IVF protocols. In these fertilization is determined before transfer into the Fallopian tube. The disadvantage of this group of therapies is that two surgical procedures are required—the follicle aspiration and laparoscopy for transfer.

4.12.4.9 Treatment of severe male infertility

There has been a major breakthrough in treatment of severe forms of male infertility over the last couple of years.

Intracytoplasmic sperm injection (ICSI) A single sperm can now be directly injected into the egg. ICSI bypasses the outer coverings of the egg and thus some of the barriers to fertilization. Since a very small number of sperm are needed for this procedure, it has a very important advantage for men with extremely low sperm numbers, or sperm motility.

ICSI can be used together with aspiration of sperm directly from the testis or the epididymis (TESA). It can restore fertility after vasectomy, or failed reversal of vasectomy. With these two techniques, many men, previously believed sterile can have children. Because of the low sperm number and motility, TESA must be performed in conjunction with ICSI.

4.12.4.10 Out come of IVF/ET pregnancies

Spontaneous abortion (miscarriage) may occur following an IVF pregnancy. There is about 8-15% chance of aborting a pregnancy achieved without any form of fertility therapy. Most of these miscarriages occur in the first 8 weeks after conception. In case of IVF-ET the abortion rate may be slightly higher and it varies from 10 to 20%. This increased probably is more related to egg, sperm, and uterine factors —infertility itself— rather than from the IVF/ET procedure. In both IVF-ET and non-IVF-ET pregnancies after the first 12 weeks pregnancy loss is equally uncommon.

Ectopic pregnancy occurs in about 1% of all pregnancies in the USA. Patients with abnormalities of the Fallopian tubes may have a 10-40% risk of an ectopic pregnancy with a pregnancy achieved without IVF-ET and a small increase in chance of ectopic pregnancy with IVF compared with individuals with no tubal disease. After IVF-ET about 1% of all patients have the risk for ectopic pregnancy.

There is a higher incidence of twins and multiple pregnancies following the transfer of multiple embryos. Occasionally, when multiple pregnancies are noted on an early ultrasound, one or more pregnancies are later absorbed. It leaves a single pregnancy. Multiple pregnancies are diagnosed by ultrasound scan. To reevaluate the progress of pregnancy there is a need of a repeat ultrasound by obstetrician at 16 to 20 weeks.

Just because a pregnancy is achieved by IVF-ET genetic counseling and amniocentesis may be carried out. This should be performed

i)   if there is any family history of genetic diseases or congenital anomalies,

ii) if the age is over 35, or

iii) if there are any other indications for amniocentesis, and/or

iv) genetic counseling, then this should be performed.

Congenital anomalies and/or genetic diseases have been reported in IVF-ET children. The incidence of these problems after IVF-ET has not been found to be higher than in non-IVF-ET pregnancies. It appears that children born after IVF-ET have no greater risk of developmental or learning defects.

4.12.4.11 Risks of IVF and Infertility Therapies

IVF may pose following risks :

(i)   Overstimulation of ovaries causes body fluid to collect in abdomen in some patients undergoing IVF-ET (Ovarian hyperstimulation syndome-OHSS).

(ii) As two embryos are generally placed in the uterus, 20 to 25 % of births result in twins (Multiple gestations).

(iii) The retrieval of eggs is minor surgical procedure and might sometimes result in complications like infection, bleeding and injury to surrounding tissues (Pregnancy complications).

(iv) Long term risks of therapy.

A women's age is a major factor in the success of IVF for any couple. For instance, a women who is 30 and undergoes IVF has a 34.7% chance of having a baby, while a 40-year-old woman has a 15.2% chance. However, the it has recently been found that the success rate is increasing in every age group as the techniques are refined and doctors become more experienced.

Ovarian Hyperstimulation Syndrome (OHSS) - It is difficult to determine in advance whether an individual will over-or under-stimulate with gonadotropin injections. There can be very narrow threshold level between too much and too less. Fertility agents are used to affect some degree of hyperstimulation of the ovaries. Residual cystic change of the ovaries and mild discomfort are quite common. This is of short duration and little risk.

There is a difference between ovarian hyperstimulation and ovarian hyperstimulation syndrome (OHSS). The risk factors for OHSS include younger age and polycystic ovary syndrome. While the cause of OHSS is unknown, it appears to be a discrete disease process associated with altered permeability and leakage of protein-rich fluid from the small vessels of the ovary into the pelvis, abdomen and possibly even around the lungs.

When there is abdominal swelling, ovarian enlargement with cysts up to 5 cm OHSS is mild. Mild OHSS is relatively common after gonadotropin injections. It may be indistinguishable from the natural effects of the ovaries to gonadotropin injections.

In severe OHSS there is marked fluid accumulation and moderate to severe pelvic pain. Sometimes, the amount of fluid is enough to warrant removal. This is usually accomplished with vaginal ultrasound.

Because there is so much loss of fluid from the blood vessels, there is a possibility of blood clot development. In some cases, hospitalization and intravenous fluid is needed. Although rare and now much less common, than in the past, deaths have been reported from severe OHSS.

Prevention is the primary strategy for OHSS. Since risk is related to amount of gonadotropin used for stimulation, use of less gonadotropin therapy translates into less risk. When estradiol levels are very high, these can be controlled with the injection of hCG. Withholding therapy for several days once gonadotropin therapy is started (coasting), will allow the smaller follicles to regress and larger ones to continue to grow before the hCG injection, is another option. There is a decrease in the success with coasting. It is sometimes reasonable to proceed with aspiration in cases of IVF, but all embryos can be frozen for later transfer rather than transfer in the stimulation cycle.

As a precaution, all patients should be weighed at the start of stimulation cycle and after hCG. During gonadotropin stimulation, it is common to have some water retention. Daily weight should be taken if there is the feeling of excessive abdominal fullness or weight gain. There is a need to consult physicians in the following conditions.

a) If the weight gain is over 5 pounds,
b) There is difficulty in breathing,
c) Severe abdominal pain,
d) Vomiting,
e) Fever,
f) Reduced output of urine.

Strenuous exercise and sexual intercourse should be avoided. There should be adequate intake of fluids containing salt (electrolytes). Juices are better than water or soft drinks.

Multiple pregnancies - A multi-fetal pregnancy is accompanied by considerable dangers and the emotional risks and problems. Many infertility patients initially feel very happy on the possibility of twins, or even triplets. But multi-foetal pregnancies are associated with significantly higher physical risks for both mothers and babies. Often multiple pregnancies are born prematurely. Prematurely born infants may experience serious, or life threatening complications, or permanent medical disability.

The improvement of IVF technology is accompanied by the chance of multiple gestations. If one embryo is transferred and the implantation rate is 20%, there is a 20 % chance of a pregnancy and no chance of fraternal (non-identical) twins or triplets. When 2 embryos are replaced the pregnancy chances increase to 40% with 20% chance of twins and 0% chance of triplets. With 3 embryos, the chances of a single pregnancy increase to 60%, twins 40% and triplets 20%. We have limited capacity to select an embryo that will result in a pregnancy. Additional embryos can be frozen for use in other cycles with good success. It is strongly suggested for women under age 35 that no more than 2 embryos are transferred. In women over age 35, each patient should be assessed individually.

Pregnancy complications - After IVF-ET, tubal pregnancy may occur. It often requires surgery. Such surgery may result in the loss of the tube, which, in turn, can further impair fertility

In rare instances, a tubal pregnancy may present a medical/surgical emergency due to shock from blood loss and require transfusion and/or other treatment.

There is about 1-3% risk of congenital malformation in spontaneous pregnancies. Congenital abnormalities and/or genetic disease have been reported in IVF-ET children. There is an increased evidence that a pregnancy after IVF-ET may carry a slightly higher risk of obstetric complications and birth defects. It is thought that therapy itself poses no risk to the pregnancy, but the risk may be due to the same factors that cause infertility. These risks have not been judged high enough to alter decision-making about IVF. The risk that a child will be born with major birth defects increases as its parents' ages increase.

Amniocentesis and/or chorionic villous sampling, each of which can aid in the recognition of genetic defects, should take care of these and other defects can often be detected by ultrasound screening.

Even an apparently normal ongoing pregnancy presents risks to both the mother and the baby. It does not guarantee a normal delivery at term of a normal infant. In pregnancies resulting from intercourse, the rate of serious obstetrical complications is approximately 10-15%. This is no different after IVF-ET.

Long-term risks of infertility therapies – There have been several unconfirmed reports that ovarian cancer is linked with the use of "fertility drugs". These reports are unconfirmed and have been subjected to much criticism.

4.12.4.12 Uses of IVF-ETT

IVF is an elective medical treatment. IVF may provide a couple who has been otherwise unable to conceive with a chance to establish a pregnancy. ETT is mostly used in cattle and buffalo. The procedure includes optimization of *in vitro* fertilization of oocytes, culture of embryos, micro-manipulation and embryo cloning using nuclear transfer, embryo sexing through PCR, reproductive ultrasonography and endocrine profiles for augmenting fertility at various institutes like NDRI, NII and at IVRI. ET technique in camel was standardized at NRC on camel and ET camel calves were produced. Goat kids were also produced through ET technology. Emrbyo sexing method in cattle and buffalo has been developed using bovine Y chromosome and buffalo Y chromosome probe. A programme on Ovum Pick Up technology and cloning has been generated with the aim to produce large number of offspring in a short period. The effort will help reduce the cost of ETT.

ETT demonstration activities were also undertaken at farmer's level through 14 regional centres and a number of ET cattle and buffalo calves were produced. Under open nucleus breeding system (ONBS) programme implemented by NDDB, 145 genetically evaluated cross bred Sahiwal male calves were produced. These bulls are being used in National artificial insemination programme.

Recombinant bovine growth hormone has been expressed in two vectors and a high level of yield has been obtained through fermentor. Biologicals viz., buffalo FSH, buffalo LH, prolectin *etc.* of bovine origin have been produced through biochemical procedures which have great potential in augmenting Embryo Transfer Technology.

The CDC compiles national statistics for all assisted reproductive technology procedures performed in the U.S. The statisticians combine all procedures that constitute assisted reproduction technology (ART) together. These include IVF, GIFT and ZIFT, although IVF is by far the most common. A report on ETT in 2000 found successful pregnancy in 30.7% of all cycles. About 69% of the cycles carried out did not produce a pregnancy. Less than 1% of all cycles resulted in an ectopic pregnancy. About 11% of these pregnancies involved multiple foetuses. About 83% of pregnancies resulted in a live birth. About 17% of pregnancies resulted in miscarriage, induced abortion or a stillbirth.

Other issues: The embryos that are not used in first IVF attempt, can be frozen for later use. This will save money if the IVF is to be undergone a second or third time. If the leftover embryos are not required, they may be donated to another infertile couple, or the clinic can be asked to destroy the embryos. Both the partners must agree before the clinic will destroy or donate the embryo.

The cost: The average cost of an IVF cycle in the U.S. is $ 12,400 according to the American Society of Reproductive Medicine. This price will vary depending on

    a) where one lives,

    b) the amount of medications required to take,

    c) the number of IVF cycles one undergoes,

    d) the amount the insurance company will pay toward the procedure.

The insurance company's coverage of IVF must be thoroughly investigated. A written statement of the benefits should also be procured. Although some states have enacted laws requiring insurance companies to cover at least some of the costs of infertility treatment, many states have not.

4.13 GENOMICS AND BIOINFORMATICS

4.13.1 Genomics

All the DNA in the cells of an organism constitutes genome. Its detailed study is known as **genomics.** Our body contains 100 million cells of over 260 different kinds. In all, there are 23 different chromosomes containing packaged DNA in a haploid set of human genome.

4.13.1.1 Human Genome Project

The human genome project began in 1990 at USA, in collaboration with 18 countries. The goals of this project were to:

    (i)   develop ways of mapping the human genome,

    (ii) store the information in databases and develop tools for data analysis,

    (iii) address the ethical, legal and social issues that may arise from this project

This project aimed to de-code more than 3 billion nucleotides contained in a haploid reference human genome and to identify all genes present in it.

The HGPs ultimate goal was to generate a high quality reference DNA sequence for human genome's 3 billion base pairs and to identify all human genes. Other important goals included sequencing the genomes of model organisms to interpret human DNA, enhancing computational resources to support future research and commercial applications, exploring gene function through mouse-human comparisons, studying human variation, and training future scientists in genomics. The powerful analytical technology and data arising from the HGP present complex ethical and policy issues for individuals and society. These challenges include privacy, fairness in use and access of genomic information, reproductive and clinical issues and commercialization programs to identify and address these implications have been an integral part of the HGP and have become model for bioethics programs worldwide.

- Human genome contains 3.2 billion chemical nucleotide base pairs. The number of genes in humans is about 30-35 thousands. Nine-tenths of our genes are similar to those of a mouse. We have more than twice as many genes as fruit fly, *Drosophila melanogaster*, but only six times more genes than bacterium, *E.coli*.

  We are incredibly 99.9% identical with each other at DNA level. Different human genes vary widely in the length, often over thousands of base pairs.

- The average gene consists of 3000 bp.

- Functions are unknown for more than 50% of discovered genes. The human genome sequence is almost exactly the same in (99.9%) in all people.

- About 2% of genome encodes instructions for the synthesis of proteins. Repeat sequences that do not code for proteins make up atleast 50% of the human genome.

- Chromosome 1 (the largest human chromosome) has the most genes (2968), and the Y has the fewest (231).

Organism	Genomic Size (Base Pairs)	Estimated Genes
Human (*Homo sapiens*)	3.2 billion	25,000
Laboratory Mouse (*M. musculus*)	2.6 billion	25,000
Mustard Weed (*A. thaliana*)	100 million	25,000
Roundworm (*C. elegans*)	97 million	19,000
Fruitfly (*D. melanogaster*)	137 million	13,000
Yeast (*S. cerevisiae*)	12.1 million	6,000
Human immunodeficiency virus (HIV)	9700	9

The human genome project report has given rise to various new areas :

Chemogenomics. Genomic response to chemical compounds.

The goal is the rapid identification of novel drugs and drug targets embracing multiple early phase drug discovery technologies ranging from target identification and validation, over compound design and chemical synthesis to biological testing and ADME profiling.

Metagenomics. It is also knows as environmental genomics, Ecogenomics or community genomics and is the study of genomes recovered from environmental samples as opposed to form clonal cultures.

Nitrogenomics. It is the branch of the study of genomics pertaining to nitrogen assimilation in organisms. Genomics of nitrogen assimilation lie at different levels of organisms, from microbes to higher organisms, where different genetic controls regulate the actual assimilation.

Nutritional Genomics. Nutritional genomics is a science studying the relationship between human genome, nutrition and health.

Nutrigenomics is study of the effect of nutrients on health through altering genome, protocome, metabolome and the resulting changes in physiology.

Nutrigenetics involves the studies of the effect of genetic variations on the interaction between diet and health with implications to susceptible subgroups.

Proteomics. It is the study of the full set of proteins in a cell type or tissue, and the changes during various condition is

Toxicogenomics. It is a form of analysis by which the activity of a particular toxin or chemical substance on living tissue can be identified based upon a profiling of its known effects on genetic material. Once viable, the technique should serve for toxicology and toxin determination, a role analogous to DNA testing in the forensic identification of individuals.

It may also be of use as a preventive measure to predict adverse "side" *i.e.,* toxic effects of pharmaceutical drugs on susceptible individuals. This involves using genomic techniques such as gene expression level profiling and single nucleotide polymorphism analysis of the genetic variation of individuals. Studies of those types are then considered and correlated to adverse toxicological effects in clinical trials so that suitable diagnostic markers for these adverse effects can be developed.

Using such methods, it would then theoretically possible to test an individual patient for his/her susceptibility to these adverse effects before administering the drug. Patient that would show the marker for an adverse effect would be switched to a different drug. While this approach is currently theoretical, it has great potential.

Pharmacogenomics. It is the branch of pharmaceutics which deals with the influence of genetic variation on drug response in patients by correlating gene expression or single-nucleotide polymorphisms with a drug's efficacy or toxicity. By doing so, pharmacogenomics aims to develop rational means to optimize drug therapy with respect to the patients genotype, to ensure maximum efficacy with minimal adverse effects. Such approaches promise the advent of "personalized medicine", in which drugs and drug combinations are optimized for each individual's unique genetic make-up.

Pharmacogenomics is the whole genome application of pharmacogenetics, which examines the single gene interactions with drugs.

Genomic data and technologies are expected to make drug development faster, cheaper, and more effective. Most drugs today are based on about 500 molecular targets, but genomic knowledge of genes involved in diseases, disease pathways, and the drug-response sites will lead to the discovery of thousands of additional targets. New drugs, aimed at specific sites in the body and at particular biochemical events leading to disease, probably will cause fewer side effects than many current medicines. Ideally, genomic drugs could be given earlier in the disease process. As knowledge becomes available to select patients most likely to benefit from a potential drug, pharmacogenomics will speed the design of clinical trials to market the drugs sooner.

Energy and Environmental Applications. GTL program of the U.S. Department of Energy (DOE) is using the Human Genome Project's technological achievements to help solve our growing energy and environmental challenges.

Plants for Biomass, Carbon Storage. Understanding the genes and regulatory mechanisms controlling growth and other traits in recently sequenced popular tree may lead to its use for bioethanol production and for sequestration of carbon.

Microbes Living in Termites. These have been considered a potential source of enzymes for bioenergy production.

GTL researchers are investigating bacteria that live in termite hindguts and churn out wood-digesting enzymes. These proteins may be usable for breaking down plant cellulose into sugars needed for ethanol production. Termites also produce hydrogen as a by-product, a process that may be amenable to scaleup.

Synthetic Nanostructures: Harnessing Microbial Enzyme Functions. Enzymes incorporated into synthetic membranes can carry out some of functions of living cells and may be useful for generating energy, inactivating contaminants, and sequestering atmospheric carbon.

Microbial Genome Analysis for Designing Vaccines and Drugs. The microbial genome analysis will also be extensively utilized for designing diagnostic kits, vaccines and drugs against the diseases caused by these pathogens. This use of microbial genomics has achieved special significance, since it is now known that the pathogens often have the ability to alter their antigenic potential, which enable them to evade human immune system. This has made it more difficult to design vaccines and drugs by conventional methods, thus making the genome analysis more rewarding.

4.13.1.2 Current and Potential Applications of Genome Research

Molecular Medicine

- Improve diagnosis of disease.
- Detect genetic predispositions to disease.
- Create drugs based on molecular information.
- Use gene therapy and controlled systems as drugs.
- Design "custom drugs" based on individual genetic profiles.

Microbial Genomics

- Rapidly detect and treat pathogens in clinical practice.
- Develop new energy source (biofuels).
- Monitor environments to detect pollutants.
- Protects citizenry for biological and chemical warfare.
- Clean up toxic waste safely and efficiently.

Risk Assessment

- Evaluation the health risks faced by individuals who may be exposed to radiation (including low levels in industrial areas) and to cancer causing chemicals and toxins.

Bioarchaeology, Anthropology, Evolution and Human Migration

- Study evolution through germline mutations in lineages.
- Study migration of different population groups based on maternal genetic inheritance.
- Study mutations on Y chromosome to trace lineage and migration of males.
- Compare breakpoints in the evolution of mutations with population ages and historical events.

DNA Identification

- Identify potential suspects whose DNA may match evidence left at crime scenes.
- Exonerate persons wrongly accused of crimes.
- Identify crime, catastrophe, and other victims.
- Establish paternity and other family relationships.
- Identify endangered and protected species as an aid to wildlife officials (*e.g.,* to prosecute poachers).
- Detect bacteria and other organisms that may spoil food and pollute air and soil.
- Match organ donors with recipients in transplant programs.
- Determine pedigree for speed or livestock breeds.
- Authenticate consumables such as caviar and wine.

Agriculture, Livestock Breeding and Bioprocessing

- Grow disease, insect, and drought resistant crops.
- Optimize crops for bioenergy production.
- Breed healthier, more productive, disease resistant farm animals.
- Develop biopesticides.
- Incorporate edible vaccines into food products.
- Develop new environmental clean up uses for plants like tobacoo.

DNA sequences generated in hundreds of genome projects now provide scientists with the lists containing instructions for how an organism builds, operates, maintains, and reproduces itself while responding to various environmental conditions. We do not know much how cells use this information to come alive, however, and functions of most genes remain unknown. We also do not understand how genes and proteins they encode interact with each other and with the environment. If we are to realize the potential of genome projects, with for ranging applications to such diverse fields as medicine, energy and the environment, we must obtain this new level of knowledge.

One of the greatest impacts of whole genome sequences and powerful new genomic technologies may be an entirely new approach to conducting biological research. In the past, researchers studied one or a few genes or proteins at a time. Because biological processes are interwined, these strategies provide incomplete and often inaccurate views.

Attempts are being now made to approach questions systematically and on a much bigger scale. They can study all the genes expressed in a particular environment or all the genes products in a specific tissue, organ or tumour.

4.13.1.3 Prospects and Implications of Human Genome

The genome project is being compared to the discovery of antibiotics. More than 1200 genes are responsible for common cardiovascular ailments, endocrine diseases like diabetes, neurological disorders like Alzheimer's disease, deadly cancers and many more. Even single gene defect is responsible for several inherited diseases. The real impact of human genome is going to be understood by applying bioinformatics tools as so far what we know is structural genomics and for applied purposes, we have to understand functional genomics that is :

   (i)   what makes the genes to transcribe ?

  (ii)   how they get switched on in one type of cell and switched off in the other ?

 (iii)   why there is a differential expression level in different cells and many other questions ?

The Human Genome Project has been a massive, government-subsidized effort. It was designed to sequence the entire DNA of a human.

The completion of the human DNA sequence in the spring of 2003 coincided with the 50th anniversary of Watson and Crick's description of the fundamental structure of DNA.

The Human Genome Project was marked by accelerated progress. In June 2000, the rough draft of the human genome was completed, a year ahead of schedule. In February 2001, special issues of Science and Nature contained the working draft sequence and analyses.

As many as 18 countries have participated in the worldwide effort with significant contributions from the Sanger Centre in the United Kingdom and Research Centres in Germany, France and Japan. Table 4.4 summarizes the goals and achievements of human genome project.

Table 4.4 : Human Genome Project : Goals and Completion Dates

Area	HGP Goal	Standard Achieved	Date Achieved
Genetic Map	2- to 5-cM resolution map (600–1,500 markers)	1-cM resolution map (3,000 markers)	September, 1994
Physical Map	30,000 STSs	52,000 STSs	October, 1998
DNA Sequence	95% of gene-containing part of human sequence finished	99% of gene-containing part of human sequence finished to 99.99% accuracy	April, 2003
Capacity and Cost of Finished Sequence	Sequence 500 Mb/year at <$0.25 per finished base	Sequence>1,400Mb/year at <$0.09 per finished base	November, 2002
Human Sequence Variation	100,000 mapped human SNPs	3.7 million mapped human SNPs	February, 2003
Gene Identification	Full-length human cDNAs	15,000 full-length human cDNAs	March, 2003
Model Organisms	Complete genome sequences of *E. coli, S. cerevisiae, C. elegans, D. melanogaster* drafts of several others, including *C. briggsae, D. pseudoobscura,* mouse and rat	Finished genome sequences of *E. coli, S. cerevisiae, C. elegans, D. melanogaster.*	April, 2003
Functional Analysis	Develop genomic-scale technologies	High-throughput oligonucleotide synthesis	1994
		DNA microarrays	1996
		Eukaryotic, whole-genome knockouts (yeast)	1999
		Scale-up of two-hybrid system for protein-protein interaction	2002

Source: *Science* **300**, 286 (2003) 10.1126/Science.1084564

The human genome project also involved several sub-projects.

(i) **Sequencing of genome of other organisms:** The eventual sequencing of all of the human DNA required some advances in technology. It was because of this, smaller genomes were sequenced – not simply for practice and technological improvement, but also because they are of tremendous value in their own right. These involved the entire DNA sequence of the bacterium (*Escherichia coli*), *Drosophila,* Yeast (*Saccharomyces cerevisiae*), a member of the mustard family (*Arabidopsis thaliana*), and a roundworm or nematode (*Caenorhabditis elegans*). Each of these organisms is a very important experimental system, and knowledge of the entire DNA sequence of each organism is found to help future biological studies. The architecture of the genome of bacterium *Haemophilus influenzae* is given in Fig. 4.43.

(ii) **Detection of genetic marker project :** These are suitably spaced throughout the genome, to allow DNA fragments to be easily cloned and arranged.

(iii) **Detailed restriction maps of the human DNA project :** It is because of the fact that restriction maps make it easier to clone specific fragments of DNA of interest.

(iv) **The DNA sequencing project :** The feasibility of this project was enhanced by the development of automated systems capable of sequencing tremendous amounts of DNA. The technological improvements in fact enhanced the rapidity with which the project could be accomplished.

4.13.1.4 Genome Projects of Model Organisms

As mentioned above, mapping the human genome was not the only scientific focus of human genome project. Right from the beginning, the value of sequencing genomes of five key model organisms was clearly recognized. Since then, however, the list has become quite a substantial one. It includes a wide variety of single-celled microbial organisms as well as various multicellular model organisms. Many of them are particularly suited for genetic analysis. In part, the sequencing of smaller genomes was also considered as a pilot for large-scale sequencing of the human genome.

A. The diversity of prokaryotic genome sequencing projects. A variety of prokaryotic models have long been established. Prokaryotic genomes are typically small (often only one or a few megabases), they are particularly amenable to comparatively rapid sequencing, resulting in the rapid development of many prokaryotic genome projects.

The first prokaryotic genome to be completed in 1995 and was the 1.83-Mb *Haemophilus influenzae* genome (Fig 4.44). It was for the first time that the genome of a free-living organism had been sequenced. It was followed by successes in other prokaryotic genome sequencing. These included the sequencing of genome of the smallest autonomous self-replicating entity (*Mycoplasma genitalium* in 1995), the first archaeal genome (*Methanococcus jannaschii* in 1996). It was followed by the complete sequencing of the 4.6 Mb *E. coli* genome.

Different priorities have been revealed by the list of prokaryotic organisms whose genomes have been sequenced. In some cases, the driving force was to understand evolutionary relationships between different organisms, as in the case of the archaeal genome. *Mycoplasma genitalium* was sequenced to understand the minimal genome (it has only 470 genes). *E. coli* and *Bacillus subtilis,* were sequenced to further basic research on popular experimental organisms.

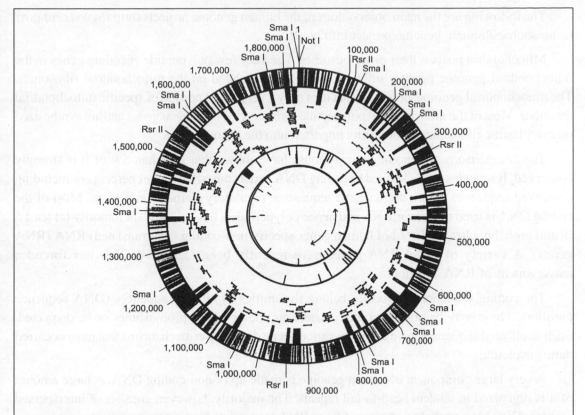

Fig. 4.44. The architecture of the genome of the bacterium *Haemophilus influenzae*. It is based on the complete genomic sequence reported in 1995 (Illustration by Dr. Anthony R. Kerlavage; Institue for Genomic Research. From R.D. Fleischmann et al., "Whole-Genome Random Sequencing and Assembly of *Haemophilus influenzae* Rd." *Science* 1995, 269, 496–512.)

B. Disease related prokaryotic genome projects. In some cases, prokaryotes were selected for genome sequencing because of their association with chronic diseases or because they were known to be causative agents of disease. In addition to have a more complete understanding of these organisms, the new information can be expected to lead to more sensitive diagnostic tools and new targets for establishing drugs/vaccines.

4.13.1.5 General Organization of the Human Genome

The human genome describes the total genetic information (DNA content) in human cells. It comprises two genomes :

(i) A very simple mitochondrial genome with 37 genes

(ii) A complex nuclear genome with about 30,000 or more genes

The nuclear genome provides the great bulk of essential genetic information, most of which specifies polypeptide synthesis on cytoplasmic ribosomes.

The following are the main observations of the human genome projects (http://www.ornl.gov/techresources/human_genome/project/info.html).

Mitochondria possess their own ribosomes. The very few polypeptide encoding genes in the mitochondrial genome produce mRNAs which are translated on the mitochondrial ribosomes. The mitochondrial genome, however, specifies only a very small portion of specific mitochondrial functions. Most of the mitochondrial polypeptides are encoded by nuclear genes and are synthesized on cytoplasmic ribosomes before being imported into the mitochondria.

The comparisons of genome human-mouse have shown that less than 5% of it is strongly conserved. It includes 1.5% devoted to coding DNA and a somewhat higher percentage including conserved sequences within untranslated sequences, regulatory elements and so on. Most of the coding DNA is used to make mRNA and hence polypeptides, but a significant minority (at least 5 % and probably close to 10 %) of human genes specify non-coding (=untranslated) RNA (RNA genes). A variety of novel RNA genes have recently been identified. This has forced a reassessment of RNA function.

The coding sequences frequently belong to families of related sequences (DNA sequence families). These may be organized into clusters on one or more chromosomes or be dispersed. Such duplicated sequences have arisen by various gene duplication mechanisms that have occurred during evolution.

A very large component of human genome is made up of non-coding DNA. A large amount of it is organized in tandem head-to-tail repeats. The majority, however, consists of interspersed repeats. They have probably originated from RNA transcripts by retrotransposition.

4.13.1.5.1 The Nuclear Genome

(i) Size and structure of human chromosomes

More than 99% of the cellular DNA is present in the nucleus of a human cell. The nuclear genome is distributed between 24 different types of linear double-stranded DNA molecules or chromosomes, each of which has histones and other non-histone proteins bound to it.

The 24 different chromosomes (22 autosomes and two sex chromosomes X and Y) can easily be differentiated by chromosome banding techniques and have been classified into groups largely according to size and to some extent, position of centromere.

The DNA selected for sequencing in the Human Genome Project was not the total nuclear genome, but the euchromatic portion. It comprises close to 3000 Mb. Euchromatin represents the true chromatin responsible for encoding of mRNA.

There is also over 200 kb of constitutive heterochromatin, permanently condensed and transcriptionally inactive. This gives a total genome size of the order of 3200 Mb. The average size of a human chromosome is, therefore, about 140 Mb, but with considerable size variations between chromosomes and variable amounts of constitutive heterochromatin, the latter comprises approximately 3 Mb segments at each centromere plus large components on several chromosomes. These include the short arms of the acrocentric chromosomes 13, 14, 15, 21 and 22, the long arm of the Y chromosome and large regions of the long arms of chromosomes 1, 9 and 16.

(ii) Base composition in the human nuclear genome

The draft of human genome sequences has suggested a genome-wide average of 41% GC for the euchromatic component. The base composition, however, considerably varies between chromosomes from 85% GC for chromosomes 4 and 13, up to 49% for chromosome 19. It also considerably varies along the lengths of chromosomes. For example, the average GC content on chromosome 17q is 50% for the distal 10.3 Mb, but it drops to 38% for the adjacent 3.9 Mb. There are regions of less than 300 kb with even wider swings in GC content, for example, from 33.1 to 59.3%.

During chromosome banding, there is a clear correlation between GC composition and the extent of Giemsa staining. For example, 98% of largest-insert clones mapping to the darkest G-bands are in 200 kb regions of low GC content (average 37%). More than 80% of clones, however, map to the lightest G-bands. They are in the regions of high GC content (average 45%). Analysis of the data does not, however, support the existence of **strict isochors** *i.e.,* compositionally homogenous large-scale regions.

(iii) Number of genes in humans and some other organisms

The total number of genes in the human genome is now thought to be in the 30,000-35,000 range, but 37 of these genes are located in the mitochondrial genome. This means approximately on an average, 1400 genes per chromosome are there. The majority of genes are polypeptide-encoding, but a significant minority (at least 5% and probably about 10%) specify untranslated RNA molecules.

The very simple 1 mm long roundworm, *Caenorhabditis elegans* (which consists of only 959 somatic cells and has a genome 1/30 of the size of the human genome), contains 19,099 polypeptide-encoding genes and over 1000 RNA genes. The complexity of genome might not always parallel biological complexity. The *Drosophila melanogaster* has substantially less number of genes than the simpler *C. elegans*.

The genomes of invertebrates (*e.g.,* insects, roundworms, sea squirts) have between 14,000-20,000 genes.

In vertebrates (human, mouse, pufferfish *etc.*), the tendency is towards 30,000-35,000 genes. The unexpected low gene number has also been rationalized both on the basis of the very large increase in transcriptional complexity one might expect as gene number moves from, say, 20,000 to 30,000 and because of the increased frequency of alternative splicing in complex genomes.

(iv) Distribution of human genes

Genes on the chromosomes are not evenly distributed. The constitutive heterochromatin regions are devoid of genes. Even within the euchromatic portion of the genome, however, the gene density can substantially vary between chromosomal regions and also between whole chromosomes. The first generalized insight into whole genome gene distribution was obtained after hybridizing purified CpG island fractions of the genome to metaphase chromosome. On this basis, it was concluded that gene density must be high in subtelomeric regions, and that some chromosomes are gene-rich, while others are gene-poor. The predictions of differential CpG island density and differential gene density were subsequently confirmed when draft sequences encompassing about 90% of the genome were reported.

The difference in percentage GC between Giemsa pale and dark bands also reflects differential gene densities because GC-rich chromosomes *e.g.,* in chromosome 19 pale G bands are comparatively rich in genes. The gene-rich human leukocyte antigen (HLA) complex (180 genes in a span of 4 Mb) is located within the pale 6p21.3 band, while a full 2.4 Mb of DNA which appears to be devoted almost exclusively to a single mammoth gene, the dystrophin gene, lies within a dark G band.

(v) Application of genome analysis

Genomics is likely to make an intensive impact in the near future (Singer and Daar, 2002). The pace of target evaluation and product/service development is bound to increase with the human and some animal genomes almost sequenced and tools for analysis of sequences developed. It may soon be possible to treat inheritable diseases by correcting genetic defects (manipulation of genetic information of the cells in the laboratory, with their subsequent sending back into the patient) and develop personalized treatment regimes (Coutelle and Rodeck, 2002).

The identification of genes resistance to environmental stress and diseases will help in activating them. In some cases, it is also possible to predict the likelihood of populations or individuals susceptible to known diseases, such as cancers (Hsing *et al.,* 2001). This could lead to personalized treatment or designing drugs for populations of known genetic make-up.

4.13.1.5.2 Mitochondrial Genome

The mitochondrial genome consists of a small circular DNA duplex which is densely packed with genetic information.

The human mitochondrial genome is defined by a single type of circular double-stranded DNA. Its complete nucleotide sequence has been established.

   (i)   It is 16,569 bp in length and is 44% G + C.
   (ii) The two DNA strands have significantly different base compositions : the heavy (H) strand is rich in guanines, the light (L) strand is rich in cytosines.
   (iii) The mitochondrial DNA is principally double stranded, but a small section shows a triple-DNA strand structure due to the repetitive synthesis of short segment of heavy strand DNA, 7S DNA.

In human cells typically thousands of copies of the double-stranded mitochondrial DNA molecule are present. The number can considerably vary in different cell types.

The mitochondrial genome of the zygote is usually exclusively determined by that originally found in the unfertilized egg as the sperm is contributing only nuclear genome. The mitochondrial genome is, therefore, maternally inherited: males and females both inherit their mitochondria from their mother. During mitotic cell division, the mitochondrial DNA molecules of the dividing cell segregate in a purely random way to the two daughter cells.

4.13.1.5.2.1 Mitochondrial Genes

(i)   The human mitochondrial genome contains 37 genes (Fig. 4.45).

(ii) The heavy strand for 28 of the genes is the sense strand, for the other nine, the light strand is the sense strand.

(iii) Of the 37 genes, a total of 24 specify a mature RNA product : 22 mitochondrial tRNA molecules and two mitochondrial rRNA molecules, a 23S rRNA (a component of the large subunit of mitochondrial ribosomes) and a 16S rRNA (a component of the small subunit of the mitochondrial ribosomes). (Table-4.5)

(iv) The remaining 13 genes encode polypeptides which are synthesized on mitochondrial ribosomes. Each of the 13 polypeptides encoded by the mitochondrial genome is a subunit of one of the mitochondrial respiratory complexes, the multichain enzymes of oxidative phosphorylation which are engaged in the production of ATP.

(v) In the mitochondrial oxidative phosphorylation system, there is, however, a total of about 100 different polypeptide subunits. The vast majority are, therefore, encoded by nuclear genes.

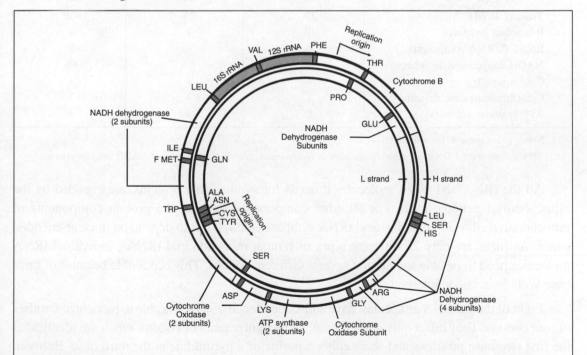

Fig. 4.45. Map of human mitochondrial DNA. The tRNA genes are designed by three-letter abbreviations corresponding to the amino acid to which each tRNA binds. Human mitochondrial DNA codes for 22 tRNAs, 2 rRNAs, and 13 polypeptides.

4.13.1.5.2.2 *The Mitochondrial Genetic Code*

The mitochondrial genetic code is used to decode the heavy and light chain transcripts to give a total of only 13 polypeptides. This very small functional load has allowed the mitochondrial genetic code to drift from the 'universal' genetic code (which is retained for nuclear genes because of the need to conserve the functions of 30,000 or so genes).

(i) There are 60 mitochondrial sense codons. It is one less than in the nuclear genetic code and four stop codons.

(ii) Two of the four stop codons, UAA and UAG, also serve as stop codons in the nuclear genetic code, but the other two are AGA and AGG which specify arginine in the nuclear genetic code.

(iii) UGA encodes tryptophan rather than serving as a stop codon and AUA specifies methionine not isoleucine.

Table 4.5. Genes Identified in Mitochondrial DNA

Gene	Human	Yeast	Plants
Ribosomal RNAs			
Large subunit	16S	21S	26S
Small subunit	12S	15S	18S
5S RNA	—	—	—
Transfer RNAs	22	24	–30
Ribosomal proteins	0	1	+
RNase P (RNA component)	—	+.	?
NADH dehydrogenase subunits	7	0*	6?
Cytochrome b	+	+	+
Cytochrome oxidase subunits	3	3	3
ATP synthase subunits	2	3	4

Note : (+) = present, (–) = absent

*The mitochondrial DNAs of other fungi, such as *Neurospora*, encode six subunits of NADH dehydrogenase.

All the rRNA and tRNA molecules it needs for synthesizing proteins, are encoded by the mitochondrial genome itself. For all other components (such as the protein components of mitochondrial ribosomes, amino acyl tRNA synthases *etc.*), it has to depend on nuclear-encoded genes. As there are only 22 different types of human mitochondrial tRNAs, individual tRNA molecules need to be able to interpret several different codons. This is possible because of third base Wobble in codon interpretation.

Eight of the 22 tRNA molecules, have anti-codons each of which is able to recognize families of four codons. They differ only at third base, 14 recognize pairs of codons which are identical at the first two base positions and share either a purine or a pyrimidine at the third base. Between them, a total of 60 codons [(8 × 4) + (14 × 2)] are recognized by 22 mitochondrial tRNA molecules.

Besides the presence of differences in genetic capacity and different genetic codons, the mitochondrial and nuclear genomes differ in many other aspects of their organization and expression.

4.13.1.5.2.3 *Distinguishing Characteristics of Mitochondrial Genome*

(i)   The human mitochondrial genome is extremely compact : approximately 93% of the DNA sequence represents coding sequence.

(ii)   All 37 mitochondrial genes lack introns and they are tightly packed (on an average one per 0.45 kb).

(iii) The coding sequences of some genes (notably those encoding the sixth and eigth subunits of the mitochondrial ATPase) show some overlap. In most other cases, the coding sequences of neighbouring genes are contiguous or separated by one or two non-coding bases.

(iv)   Some genes even lack termination codons; to overcome this deficiency, UAA codons have to be introduced at the post-transcriptional level.

The significant region without any known coding DNA is the displacement (D) loop region. In this region, a triple-stranded DNA structure is generated by duplicated synthesis of a short piece of the H-strand DNA, known as 7S DNA. The mitochondrial genome probably originated following endocytosis of an aerobic prokaryote and subsequent large scale gene transfer to the nuclear genome.

4.13.1.5.3 Chloroplast Genome

Most chloroplast genomes are in the size range of 120-160kb. Based on plant species, they are closed circular molecules (Fig. 4.46).

Chloroplast DNA molecules are usually present in groups of several molecules in nucleoids. The supercoiling and folding of the cpDNA to form nucleoids is based on basic histone-like proteins. The nucleoids are bound to chloroplast membranes at precise locations. It depends on species and stage of plastid development. Such an arrangement helps ensure segregation of cpDNA at plastid division.

Coding The chloroplast genome encodes many of the components of its own protein synthesizing apparatus which include:

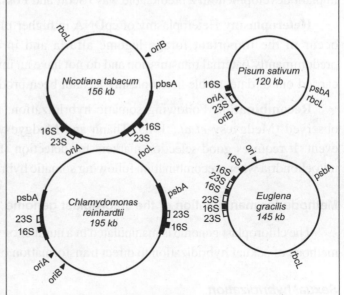

Fig. 4.46. Gene maps of cpDNA from *Nicotiana tabacum, Pisum sativum, Chlamydomonas reinhardtii* and *Euglena gracilis*. The figure show the location of 16S (closed blocks) and 23S (open blocks) rRNA genes, inverted repeats (thick lines), *psbA* and *rbcL* genes, and putative origins of replication (ori). The map diagrams of *C.reinhardtii* and *E.gracilis* cpDNA are modified from Palmer (1985). The map diagram of *P. sativum* is also modified from Palmer (1985b), with the ori regions from Meeker *et al.* (1988). The map diagram of *N. tabacum* is modified from Shinozaki *et al.*1986 (Adopted from Slater, 2003).

(a) the ribosomal and transfer RNAs,

(b) a number of ribosomal proteins and a protein synthesis initiation factor.

It also appears that the cpDNA encodes chloroplast RNA polymerase subunits. The chloroplast protein synthesizing system translates the cpDNA encoded photosynthetic proteins. Four of the five major thylakoid protein complexes contain a number of cpDNA-encoded proteins including the psbA protein of PSII. The large subunit of ribulose bisphosphate carboxylase is also encoded by cpDNA. The role of cpDNA is the photosynthetic process and hence plant productivity is quite evident.

Copy number In most cases every plastid of a plant contains multiple copies of the same type of cpDNA. In spinach, chloroplasts in mesophyll cells contain 57 to 353 copies of cpDNA (depending on leaf age). The root proplastids may, however, contain as few as ten (Scott and Possingham, 1980). Even in this latter case, root cells with about ten plastids (Possingham, 1980) will contain 100 cpDNA molecules. Spinach mesophyll cells each contain up to 200 chloroplasts, thus, giving a huge cpDNA copy number per cell. The cpDNA copy number per plastid changes in plant development in a predictable way (Scott and Possingham, 1983).

Heteroplasmy Heteroplasmy of cpDNA in higher plants is not common, however, it does occur in the important forage legume alfalfa and in conifers. These latter species have predominantly paternal transmission and do not have an inverted repeat. In somatic hybridization carried out to date, stable heteroplasmy has not been produced.

Recombination Following somatic hybridization, recombination of cpDNAs has been observed (Medgyesy *et al.,* 1985; Thanh and Medgyesy, 1989). It must, however, be a rare event. It requires good selection markers for detection (Rose *et al.,* 1999). It is in contrast to mitochondria where recombination following somatic hybridization is quite common.

Methods for manipulation of the chloroplast genome

The chloroplast genome is manipulated in a number of ways. These range from the established methods of sexual hybridization to direct transformation of the chloroplast genome.

Sexual hybridization

Species with maternal plastid inheritance: In plants having maternal inheritance, it is possible to transfer a new chloroplast type by hybridization and extensive backcrossing. Kaul in 1988 demonstrated the principle of this methods. Hybridization of the female parent transferred the chloroplast to the male parent with the required nuclear genotype. The hybrid plant produced is repeatedly backcrossed as a female to the male parent with the required nuclear genes. This plant breeding strategy is valuable where a plastid character has been identified that warrants complete replacement of the existing plastid. This approach, however, is time-consuming and results in at least some change in the nuclear genotype.

Species with biparental plastid inheritance: Cytoplasmic hybrids can be produced in those species having biparental inheritance of plastid genomes. It is not possible in those species with uniparental maternal plastid inheritance. An example of an agricultural species showing a high paternal plastid transmission is alfalfa, where mixed cytoplasms are present, recombination is possible. These are based on somatic hybridization studies (Medgyesy *et al.*, 1985; Thanh and Medgyesy, 1989) and observations on sexual hybrids of *Oenothera* (Chiu and Sears, 1985).

In the case of biparental inheritance, where a mixed population of plastids is present, sorting out would normally be expected to occur, but in some species, heteroplasmy is possible (Rose *et al.*, 1986). This allows the potential for organelle complementation. Much more flexibility to a breeding programme for improved cytoplasmic characters is given by biparental inheritance. However, unidirectional plastid transfer without backcrossing steps requires organelle transfer without nuclear transfer by one of the parents.

Production of mutations in chloroplast DNA: A large number of cpDNA mutations have arisen spontaneously or have been induced by mutagens or nuclear 'mutator' genes. They were originally identified as cpDNA mutations by their non-Mendelian inheritance. These mutations represent valuable genetic resources for the dissection of the chloroplast genome as well as for research on the manipulation and transformation of the chloroplast genome, where they have particular value as markers. These mutations have been shown to affect the chloroplast translational apparatus, photosystem I, photosystem II, the cytochrome f/b6 complex, carbon fixation or the ATP synthase.

Somatic hybridization

This involves the transfer of chloroplast genomes by cybrid formation (there is no nuclear hybridization in this case). Cybrid plants contain the nucleus of one species and part or all of the cytoplasm derived from another. Cybrid production provides following advantages to plant breeding.

(i)   Transfer of chloroplast can be carried out in single step. Traditionally alloplasmic plants have been produced by recurrent backcrossing after hybridization (Feldman and Sears, 1981). Flowering plants which have uniparental maternal inheritance have benefitted from it.

(ii) Chloroplasts may also be transferred beyond the normal boundaries defined by sexual hybridization. Cybrid production can occur across wider genetic distances than nuclear hybridization. Studies in *Nicotiana,* clearly show the importance of this phenomenon.

(iii) The cytoplasmic contribution of the male parent in sexual reproduction in sexual reproduction is usually very small. Somatic hybridization provides an advantage of producing a heterokaryon with approximately equal contributions.

4.13.1.6 Applications of Genomics in Drug Development

The completion of human genome project is revolutionizing biomedical research. The dramatic increase in the amount of genetic and structural information about individual genes and their expression profiles have opened the door to a new paradigm for practising medicines that promises to transform healthcare. Research in human genome is leading to the identification of a range of molecular markers for predisposition testing, disease screening and prognostic assessment and markers used to predict and monitor drug response.

Microarray technology is a means to determine the expression of potentially all human genes at the level of messenger RNA. It has wide field applicability including

(i)   development of more global understanding of the gene expression abnormalities

(ii) discovery of new diagnostic and prognostic indicators and biomarkers of therapeutic response

(iii) identification of genes involved in conferring drug sensitivity and resistance

(iv) prediction of patients most likely to benefit from the drug and use in general pharmacogenomic studies.

An insight to efficacy and adverse drug reactions at early stage of drug development through toxicogenomics studies can be used as a guide to define specific biomarkers of toxicity and efficacy of drug candidate.

Genome wide linkage analysis and positional cloning have identified unique genetic differences between individuals for disease susceptibility. Most human sequence variation is attributable to SNPs (single nucleotide polymorphisms), it is single base substitution of one nucleotide with another. SNPs have a low rate of recurrent mutations, making them stable indicators of disease susceptibility.

The number of polymorphisms identified in genes encoding drug metabolizing enzymes, drug transporters and receptors is rapidly increasing. In many cases, these genetic factors have a major impact on pharmacokinetics and pharmacodynamics of a particular drug and thus, influence the sensitivity of such a drug in an individual patient with a certain genotype, thereby, suggesting the importance of pharmacogenetics and pharmacogenomics.

Thus, the potential of all these integrated technologies is to develop personalized or tailored medicines best suited for an individual's genotype : the right drug in the right dose, to the right person at the right time.

4.13.1.6.1 Foundation of Personalized Therapy

Personalized medicine simply means the prescription of specific treatments and therapeutics best suited for an individual. The basic foundations for developing personalized systems of medicines are pharmacogenetics, pharmacogenomics and pharmacoproteomics.

4.13.1.6.2 Pharmacogenetics and personalized medicines

The term pharmacogenetics implies the interactions between drug and characteristics of the individuals. It is based on observations of clinical efficacy or safety and tolerability profile of a drug in individuals – the phenotype. It takes into account the interindividual differences in the observed response that may be associated with the presence or absence of individual-specific biological markers allowing prediction of individual drug response. As mentioned above, such markers are most commonly polymorphic at the level of nuclear DNA.

Many drug metabolizing enzymes are polymorphic such as cytocrome P450 2D6 (CYP 2D6), cytochrome P450 2C9 (CYP 2C9) and thiopurine-S-methyl transferase (TPMT) and have differential impact on drugs for treatment with anti-depressants, oral anticoagulants and cytostatics, depending on allelic variant in the genotype.

One of the pharmacogenetic based drug is Herceptin. It is a humanized monoclonal antibody directed against the *her-2*-oncogene. It is possible to detect HER 2 protein receptor or copies of *her-2* gene sequence, which are prerequisite for choosing the appropriate therapy. Since, this drug inhibits 'gain of function' variant of the oncogene, it is ineffective in the 2/3 of patients who do not 'overexpress' the drug's target. The food and drug administration's (FDA) approval of the pharmacogenomic marker *her-2* linked to herceptin represents an important precedent for regulatory approval of personalized medicine product.

4.13.1.6.3 Pharmacogenomics: Translation of functional genomics into rational therapeutics

Genetic polymorphisms in drug-metabolizing enzymes, transporters, receptors and other drug targets have been linked to interindividual differences in the efficacy and toxicity of many medications. Pharmacogenomic studies are rapidly elucidating the inherited nature of these differences in drug disposition and effects, thereby, enhancing drug discovery and providing a stronger scientific basis for optimizing drug therapy on the basis of each patient's genetic constitution.

It is mainly concerned with a comprehensive, genome-wide assessment of the effects of certain interventions, mainly drugs or toxicants. It is the systematic assessment of how chemical compounds modify the overall expression pattern in certain tissues of interest.

Overall pharmacologic effects of medications are typically not monogenic traits, rather they are determined by the interplay of several genes encoding proteins involved in multiple pathways of drug metabolism, disposition and effects. It is seen that the individuals with homozygous wild-type drug metabolizing enzymes and drug receptors would have a high probability of therapeutic efficacy and low probability of toxicity, in contrast to those with homozygous mutant genotypes for the drug-metabolizing enzymes and the drug receptors, in which the likelihood of efficacy is low and toxicity is high.

New tools for elucidating polygenic components of human health and disease are provided by human genome project, coupled with functional genomics and high throughout screening methods, is proving powerful.

4.14 PROTEOMICS

Earlier it was thought that the study of product of transcription *i.e.,* transcriptosome is sufficient to have an idea of gene expression. According to Strachan and Read (2003) it is not so. The authors have ascribed this to the following facts :

(i) All the mRNAs in the cell are not translated so the transcriptosome may include gene products that are not found in the proteome.

(ii) The rates of protein synthesis and protein turnover also differ among transcripts. The abundance of a transcript, however, does not necessarily correspond to the abundance of the encoded protein. It is because of this that the transcriptosome may not accurately represent the proteome *i.e.,* total protein content either qualitatively or quantitatively.

(iii) Besides these, the post translational modifications also regulate the proteome. Generally the protein activity depends on post-translational modifications. These can not be predicted from the level of the corresponding transcript. Many proteins are present in the cell as inert molecules. These get activated by processes such as proteolytic cleavage or phosphorylation. In fact variations in the abundance of a specific post-translation variants are quite significant.

These facts, therefore, necessitate the study of proteins in detail *i.e.,* Proteomics.

Proteomics provides the information required to establish a link between gene expression and function. Stathmin, a signalling protein is found at high levels in various cancers, including childhood leukemias, but only the phosphorylated form of the protein is a useful marker of the disease. Therefore, in order to fully understand the functional molecules present in the cell, it is necessary to study the proteome directly.

Uses

Proteomics has the following uses :

(i) It can be used to monitor the abundance of different gene products.

(ii) The expression of all the proteins in the cell can be compared among related samples. This allows the identification of proteins with similar expression patterns.

(iii) It also highlights important changes in the proteome that occur, for example, during disease or the response to particular external stimuli. This is also termed as expression proteomics.

(iv) By investigating protein interactions, functions of proteins can often be established.

Proteins carry out their activities in the cell by interacting with other molecules. The establishment of specific interactions between proteins can help to assign individual functions to each one of them.

Proteins can also be linked into pathways and networks. Further information can be derived from protein interactions with small molecules (which may act as ligands, cofactors, substrates, allosteric modulators *etc.*) and with nucleic acids. The analysis of protein interactions overlaps with the analysis of protein structure. It is possible to predict the interaction of proteins with other proteins and with smaller molecules from the knowledge of a three dimensional structure of the protein, which can be very useful in the development of effective drugs. The comparison of protein structures can help to determine evolutionary links between genes and investigate their function. The following is an overview of proteomics and methodologies used to study it.

Genomics has provided information about the gene activity which links with diseases. However, gene sequence information does not provide the complete information about the structure and activities of a protein. It is because of this, now a days, for the systematic analysis of protein profiles of tissues, the term **proteomics** is used. Proteome refers to all proteins produced by a species.

With time, however, proteome shows variations. The proteome is defined as, the proteins present in one sample (like tissue, organism, cell culture) at certain point of time.

As we know, genomics deals with one genome per organism, the proteomics may be large in number in an organism and varies with the nature of sample, development state, disease, drug treatment *etc*. It provides help in the discovery of new protein markers for diagnostic purposes and of novel molecular targets for drug discovery.

Proteomics, therefore, is a study of proteins on a large-scale using new technologies.

Proteomics also refers to characterize proteins at broad level *viz.*, study their interactions and identify their functional rates.

Study of proteomics at different segments can be categorized into four different levels:

(a) Expression
(b) Cell map/interaction
(c) Functional
(d) Structural

(a) **Expression proteomics:** It is quiet similar to DNA array for transcription profiling. In this approach, the expressed protein from cell or tissue extracts is quantitatively analyzed for protein profiling or the protein maping.

> This can be done by using two-dimensional (2-D) gel electrophoresis for separating proteins and identifying them by mass spectrometry. However, these approaches don't provide functional information for differentially expressed proteins.

(b) **Cell map proteomics :** Subcellular location of proteins and protein-protein interactions can be determined by the cell map proteomics. In this approach, the protein is first purified and then subjected to mass spectrometric identification for protein-protein interaction. One of the most popular techniques used for this purpose yeast two cell hybrid system.

> Protein-protein interaction may provide enough information about the protein function and may elucidate functional pathways relevant to disease. In context of protein function, we are unable to identify the exact physiological rate of the proteins. This is one of the limitations of this technique.

(c) **Functional proteomics :** It is useful in the drug development or to find novel inhibitor to perturb the function of protein or enzyme that leads to disease or malfunction in body. It includes the use of specific blocking molecules such as neutralizing antibodies and aptamers, pharmacological inhibitors, transdominant mutations and chromophore-assisted laser inactivation. Another strategy for defining protein function is arraying functional proteins on chip.

(d) **Structural prediction :** It helps in the deduction of the 3-D structure of proteins, *viz.,* the protein folds, its basic frame and amino acid sequence, its pattern or arrangement in a protein molecule, active site determination *etc.* It is still not possible to deduce all possible functions from a protein's 3D structure with high confidence because of the lack of understanding of protein structure, function relationships (Fig. 4.47).

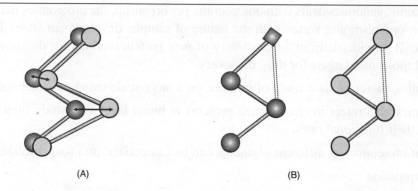

(A)   (B)

Fig. 4.47. Comparison of protein structures, with circles representing Carbon atoms of each amino acid residue and lines representing the path of the polypetide backbone in space. (A) Intermolecular comparison involves the super imposition of protein structures and the calculation of distances between equivalent atoms in the superimposed structures. These have been shown as bi-directional arrows. These distances are used to calculate the root mean square deviation (RMSD). A small RMSD value computed over many residues is an evidence of significantly conserved tertiary structure. **(B)** Intramolecular comparison involves side-by-side analysis based on comparative distances between equivalent atoms within each structure (Modified from Strachan and Read, 2003).

Structural and computational technology is far more useful to predict the exact structure of the proteins, to design protein ligands *etc.*

Techniques and Tools used in Proteomics

(a) General : Fundamental steps involved in proteome analysis are:

(a) **Sample collection, handling and storage :** Sample may be body fluid or tissue samples, from patients of transplant rejection, cancer *etc.* Sample quality does matter.

(b) **Preparation of sample :** First the pure cell sample is isolated.

(c) **Separation of protein :** It should be done by electrophoresis.

(d) **Identification of protein :** Protein patterns on 2-D electrophoresis gels are analyzed using image analysis techniques and software tools.

(e) **Characterization :** Proteins are excised for identification and characterization. These include post-translational modifications using PMF (peptide mass fingerprinting), mass spectrometry methods and bioinfotools. Other techniques may include, protein chips/microarrays and Y2H *etc.*

(b) Specialized : There are so many techniques evolved in proteomics.

(i) 2-D gel electrophoresis

It is the method of choice to separate and identify proteins in a sample by displacement in 2 dimensions.

It is primary and most important technique to separate proteins for further characterization by mass spectroscopy. Comparison of two or more samples to find the differences in their protein expressions simultaneously can be done by 2-D gel electrophoresis (Fig. 4.48).

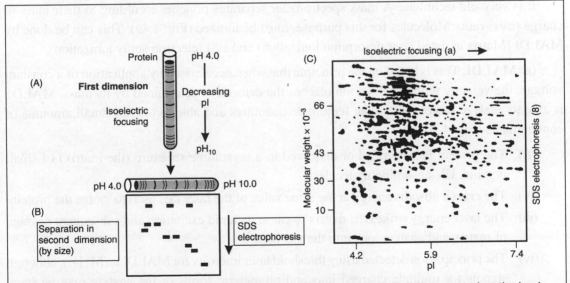

Fig. 4.48.: Two dimensional gel electrophoresis (A) The proteins are first separated on the basis of their charges by isoelectric focusing **(B)**. The resulting gel strip thus produced is applied to an SDS-polyacrylamide gel **(C)**. The proteins are separated into bands by mass (Adopted from Molecular Cell Biology by Lodish *et al.*, 2003).

Basic principle of this technique is that, in first dimension, the proteins are first separated according to the respective pI characteristic values in a thin gel. After that the gel is incubated in detergent solution for short time. The proteins are then separated in a second dimension according to their size.

SDS-PAGE (sodium dodecyl SO_4^- polyacrylamide gel electrophoresis) is also used to separate the proteins from mixtures of protein samples and determine the molecular weight of the proteins.

(ii) ICAT (Isotope Coded Affinity Tags)

To compare relative protein abundance between two samples, this is an alternative technique to SDS-PAGE. The comparison may establish between healthy and diseased tissue, but may be one of any number of comparisons. For this purpose, different affinity tags are used for two different samples. Tags are structurally similar except that one has a linker containing eight hydrogen atoms, the other a linker with 8 deuterium atoms. Thus, one state of the cell is labeled with light tag and the other with the heavy tag. They are mixed and subjected to digestion and separation by ion-exchange chromatography. The final elute is analysed by mass spectrometer and their relative abundance is quantified. This technique is quicker than gel electrophoresis.

(iii) Mass Spectroscopy

It is very old technique. A mass spectrometer separates proteins according to their mass to charge (m/z) ratio. Molecules for this purpose, must be ionized (Fig. 4.49). This can be done by MALDI (Matrix assisted laser desorption ionization) and ESI (electron spray ionization).

(a) MALDI. This is based on the principle that when accelerated by application of a constant voltage, the velocity with which an ion reaches the detector is determined by its mass. MALD1 is able to analyze proteins less than fentomole quantities and able to tolerate small amounts of contaminants (Fig. 4.50).

   (i)   The protein is suspended or dissolved in a crystalline structure (the matrix) of small organic, UV-absorbing molecules.

   (ii) The crystal absorbs energy at the same value of the laser *i.e.,* used to ionize the protein.

   (iii) The laser energy strikes the matrix to cause its rapid excitation and subsequent passage of matrix and analyte ions into the gas phase.

   (iv) The principal ion detected using threshold laser intensity for MALDI is ($M^+ H^+$), although signals for multiple charged ions and oligomeric forms of the analyte may be seen, especially for large proteins.

   (v)   The ionized protein is accelerated by an electrostatic field and expelled into a flight tube.

   (vi) As it exits the flight tube, the mass analyzer is encountered.

   (vii) The analyzer is often a time of flight (TOF).

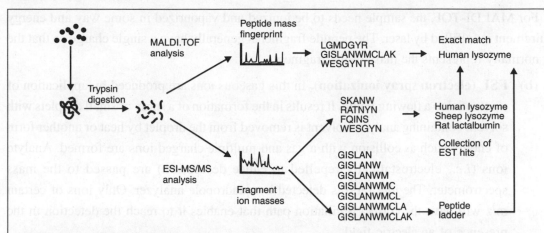

Fig. 4.49. Protein annotation by mass spectrometry : Individual protein samples (*e.g.,* spots from 2D-gels) are digested with trypsin. It cleaves on the C-terminal side of lysine (K) or arginine (R) residues as long as the next residue is not proline. MALD1-TOF can be used to analyze the tryptic peptides as intact molecules. The masses are, thus, used as search queries against protein databases. Algorithms are used to analyze protein sequences. They cut them with the same cleavage specificity as trypsin, and compare the theoretical masses of these peptides with the experimental masses obtained by MS. Normally, the masses of several peptides should identify the same parent protein *e.g.,* lysozyme. If the protein is not in the database, there may be no hits. It might, however, has been subject to post-translational modification or artifactual modification during the experiment. In such case ESI-tandem mass spectrometry can be used to fragment the ions. The fragment ion masses can be used to search EST databases and obtain partial matches. It may lead eventually to the correct annotation (Modified from Strachan and Read, 2003).

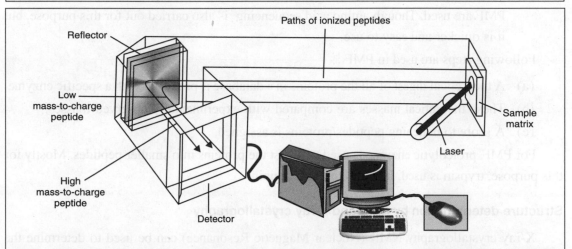

Fig. 4.50. Principal of maldi-tof MS : The analyte (usually a collection of tryptic peptide fragments) is mixed with a matrix compound. It is then placed near the source of a laser. The analyte/matrix crystals are heated up by laser heats. It causes the analyte to expand into the gas phase without significant fragmentation. Then the ions travel down a flight tube to a reflector. The later focuses the ions onto a detector. The time of flight (the time taken for ions to reach the detector) is dependent on the charge/mass ratio and allows the mass of each molecule in the analyte to be recorded (Adopted from Strachan and Read, 2003).

For MALDI-TOF, the sample needs to be ionized and vapourized in some way and energy requirement is fulfilled by laser. The peptide fragments generally have a single charge so that the m/z normally represents the mass of the fragment.

(b) **ESI (electron spray ionization).** In this gaseous ions are produced by application of a potential to a flowing liquid. It results in the formation of a spray of small droplets with solvent containing analyte. Solvent is removed from the droplet by heat or another form of energy such as collision with a gas and multiple charged ions are formed. Analyte ions (*i.e.,* electrostatically repelled to cause desorption) are passed to the mass spectrometer. The m/z ratio is detected by quadropole analyzer. Only ions of certain m/z will have the correct oscillation path that enables it to reach the detection in the presence of an electric field.

(c) **MS/MS/Tandem mass spectrometry analysis.** In this technique, a mixture of peptides is separated out into individual peptides that are energized and susceptible to breakage at peptide bonds. Then the mass of the resulting peptide fragments is determined. We can finally deduce the protein sequence by calculating the masses of fragments differing by one amino acid.

MS/MS has certain limitations in its use. Large ions are produced *i.e.,* dependent on analytic procedure used that interferes in calculations.

(d) **PMF (Peptide mass fingerprinting).** To identify protein, spots obtained from 2D-gel PMF are used. Though amino acid sequencing is also carried out for this purpose, but it is quicker and easy to use.

Following steps are used in PMF :

(a) A theoretical digest of all the proteins in a database is produced with a specific enzyme.

(b) These theoretical masses are compared with experimentally observed masses.

(c) A score to matching peptides/proteins is assigned.

For PMF, proteolytic enzyme is used to digest the proteins into smaller peptides. Mostly for this purpose, trypsin is used. It cuts lysine and arginine.

Structure determination by NMR and X-ray crystallography

X-ray crystallography NMR (Nuclear Magnetic Resonance) can be used to determine the secondary and tertiary protein structure. In either case, the protein should be better than 95% pure for optimal results. For purification, gel or column separations, dialysis, differential centrifugation, salting out or HPLC are the choice of methods.

X-RC (X-ray Crystallography) is a powerful technique. When X-rays are directed at a crystal of protein or a derivative of the protein, they get diffracted from the crystals. The rays are scattered in a pattern dependent on the electron densities in different portions of the proteins and detected by the detector. It also analyzes the X-ray pattern to determine the secondary or tertiary structure proteins. There are certain limitations of X-RC.

It, however, has so many advantages. It reveals very precise and critical structural data about amino acid orientation which is then used to understand protein-interactions and design drugs in structure based pattern.

NMR (Nuclear Magnetic Resonance) is also used for structure determination of proteins. It has a distinct advantage over XRC. NMR can be useful for the compounds that will not be crystallized or sometime show difficulty in crystallization.

It can resolve protein structures upto 100,000 MW. NMR facilitates to reveal details about specific sites of molecules without having to solve their entire structure. Modern NMR spectroscopy is particularly adapt at revealing how active sites of enzymes work. NMR has now been upgraded to TOESY (Transfer Overhauser Spectroscopy). It facilitates shape determination of small molecules bound to very large ones and helps define the binding pocket of the macromolecule.

Protein Microarrays

These have been developed to analyze protein expression profiling or protein interaction profiling. Surface of the array is coated with cationic, anionic, hydrophobic, hydrophilic sub-stances *e.g.,* antibodies, receptors, ligands, nucleic acids, carbohydrates *etc.* Some surfaces have broad specificity and bind only a few proteins from a complex sample. The sample is brought in contact with specific surface coated array/chip, it is then washed to reduce non-specific binding and analyzed by MALDI-TOF.

By this technique, we can find the ligand which binds to specific receptors, and binding domains for protein-protein interactions.

Phage Display Method

This is used to make protein libraries or viral surfaces that are screened for activity as a group (Fig. 4.51).

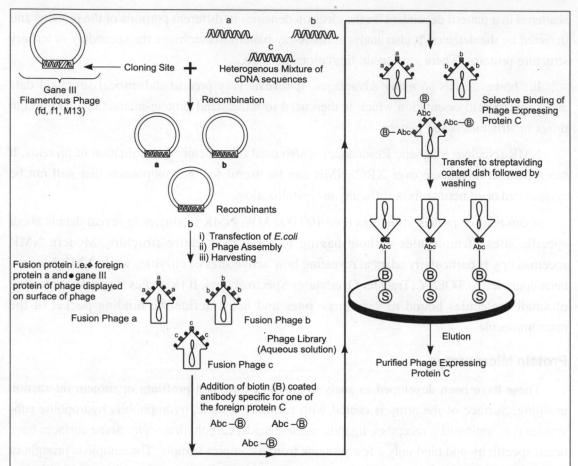

Fig. 4.51. The principle of phage display: Bait protein can be immobilized on the surface of microtiter wells or membranes. All other proteins in the proteome are then expressed on the surface of bacteriophage, by cloning within a phage coat protein gene. It creates a phage display library. The wells or membranes are then flooded with the phage library for any given bait protein (X). Phage carrying interacting proteins will be retained. The phage displaying non-interacting proteins will be washed away. The bound phage can be eluted in a high-salt buffer and used to infect *E. coli*. It produces a large amount of phage particles containing the DNA sequence of the interacting protein (Modified from Strachan and Read, 2003).

Protein-protein interactions

Proteins at every cellular process interact with other proteins to proceed the cellular reactions *e.g.*, carbohydrate and lipid metabolism, cell-cycle regulation protein nucleic acid metabolism, signal transduction *etc.* All cellular functions, therefore, depend on the complex network of cellular protein-protein associations.

Experimentally this can be proved by yeast two hybrid (Y2H) system affinity chromatography and some useful tools used in computational biology (Fig. 4.52).

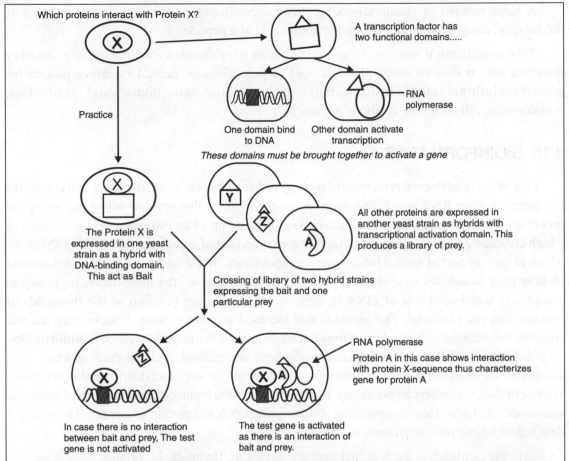

Fig. 4.52. Principle and practice of yeast two-hybrid system : Two functionally independent domains are present on transcription factors. One of these is for DNA binding and the other for transcriptional activation. Although these do not have to be covalently joined together can be assembled to form a dimeric protein. Bait proteins are expressed in one yeast strain as a fusion with a DNA-binding domain. The candidate prey are expressed in another strain as fusions with a transactivation domain. On mating the two strains, functional transcription factors are assembled only if the bait and prey interact. The inclusion of a reporter gene being activated by the hybrid transcription factor helps in detecting this. Though the principle is simple, the technique is prone to the problem of reliability and reproducibility. Due to spontaneous autoactivation (the bait or prey can activate the reporter gene on their own), sticky baits and prey (proteins that interact non-specifically with many others) and irrelevant interactions (chance interactions between proteins that would never encounter each other under normal circumstances, such as those usually found in separate compartments) may lead to false positives. False negatives may arise due to the errors in PCR in the construct and due to non-physiological conditions (the assay takes place in the nucleus, so proteins from other compartments may not fold or assemble properly may cause false negative) (Adopted from Human Molecular Genetics Strachan and Read, 2003).

Post translational modification

For important cellular processes, some modifications in protein structure are necessary steps for its functions. Many proteins are activated by phosphorylation or some are expressed in an inactive form.

A large number of chemical modifications occur in protein structure *e.g.,* oxidation of methionine, acetylation of the N-terminal amino acid of a peptide *etc.*

This modification supresses the ionization in peptide/amino acid sequence. Another modification is cyclization of glutamic acid to pyro-glutamic acid. To analyze protein for post-translational modifications, mass spectrometry and monoclonal antibodies, immunoassays/fluorescence methods are used.

4.15 BIOINFORMATICS

One of the fundamental principles of biology is that within each cell, the DNA that comprises the genes encodes RNA which in turn produces the proteins that regulate all of the biological processes within the organism. The human body is made up of an estimated 10^{12} cells, each of which contains 23 pairs of chromosomes that are comprised of approximately 3,000-3500 genes which in turn are part of some 3 billion pairs of DNA bases. While we have a basic understanding of how gene sequences code specific proteins, we do not have the information necessary to completely understand role of DNA in specific diseases or the function of the thousands of proteins that are produced. The methods that are used to collect, store, retrieve, analyze and correlate this mountain of complex information are grouped into a discipline called **bioinformatics**. Vast amount of DNA sequences have already been determined, and the pace at which new sequences are characterized is continuously accelerating. To store and distribute this enormous volume of data, computers are necessary. Bioinformatics is the rapidly developing area of computer science devoted in collecting, organizing and analyzing DNA and protein sequences. The principal data banks where such sequences stored are :

(i) the GenBank at the National Institute of Health, Bethesda, Meryland,

(ii) the EMBL Sequence Data Base at the European Molecular Biology Laboratory in Heidelberg.

These databases continuously exchange newly reported sequences and make them available to molecular cell biologists throughout the world on the Internet. Newly derived sequences can be compared with previously determined sequences to search for similarities. These are called **homologous sequences.**

Protein-coding regions can be translated into amino acid sequences, which can also be compared. Because of degeneracy in the genetic code, more homology is shown by related proteins than the genes encoding them. Databases of partial cDNA sequences (EST database) are particularly useful in designing probes for screening libraries.

The goal of bioinformatics is thus, to provide scientists with a means to explain :

- normal biological processes
- malfunction in these processes which leads to diseases
- approaches to improve drug design and discovery

Bioinformatics encompasses the use of tools and techniques from three separate disciplines;

(i) **Molecular biology.** It is the source of data to be analyzed

(ii) **Computer science.** It supplies the hardware for running analysis and the networks to communicate the results

(iii) **The data analysis algorithms.** It strictly defines bioinformatics.

4.15.1 A Brief History of the Field

1949 – A technique of electrophoresis was introduced by Tselius for separating proteins in solution.

1951 – Pauling and Corey proposed the structure for the alpha-helix and beta-sheet

1954 – Perutz's group developed heavy atom methods to solve the phase problem in protein crystallography.

1955 – The sequence of the first protein bovine insulin was announced by F.Sanger.

1958 – The first integrated circuit was constructed by Jack Kilby at Texas Instruments.

1969 – The ARPANET was created by linking computers at Standford, UCSB, The University of Utah and UCLA.

1970 – The details of the Needleman-Wunsch algorithm for sequence comparison were published.

1971 – Ray Tomlinson (BBN) invented the email program.

1972 – The first recombinant DNA molecule was created by Paul Berg and his group.

1973 – The Brookhaven Protein Data Bank was announced.

1974 – Vint Cerf and Robert Kahn developed the concept of connecting networks of computers into an 'internet" and developed the Transmission Control Protocol (TCP).

1975 – Two dimensional electrophoresis, where separation of proteins on SDS polyacrylamide gel was combined with separation according to isoelectric points, was announced by P.H. O'Farrell.

E.M. Southern published the experimental details for the Southern Blot technique of specific sequences of DNA.

1976 – The Unix-To-Unix Copy Protocol (UUCP) was developed at Bell Labs.

1977 – Full description of Brookhaven PDB (http://www.pdb.bnl.gov) was published

Allan Maxam and Walter Gilbert (Harvard) and Frederick Sanger (U.K. Medical Research Council), reported methods for sequencing DNA.

1980 – First complete gene sequence for an organism (QX 174) was published. The gene consists of 5,386 base pairs which code nine proteins.

1981 – Smith Waterman algorithm for sequence alignment was published.

1985 – FASTP algorithm was published.

The PCR reaction was described by Kary Mullis and co-workers.

1986 – "Genomics" appeared for the first time to describe the scientific discipline of mapping, sequencing and analyzing genes. The term was coined by Thomas Roderick as a name for the new journal.

The SWISS-PROT database was created by the Department of Medical Biochemistry of the University of Geneva and the European Molecular Biology Laboratory (EMBL)

1987 – Use of yeast artificial chromosomes (YAC) was described by David T.Bruke and his colleagues.

The physical map of *E. coli* was published (Y.Kohara *et al.*).

Perl (Practical Extraction Report Language) was released by Larry Wall.

1988 – National Centre for Biotechnology Information (NCBI) was established at the National Cancer Institute.

The Human Genome Initiative was started (Commission on Life Sciences, National Research Council. Mapping and Sequencing the Human Genome, National Academy Press : Washington, D.C.).

The FASTA algorithm for sequence comparison was published by Pearson and Lupman.

A new program, an Internet computer virus designed by a student, infected 6,000 military computers in the US.

1990 – BLAST program (Altschul *et al.*) was implemented.

1992 – Genome Systems, Gaithersburg Maryland, was formed by William Haseltine.

The Institute for Genomic Research (TIGR) was established by Craig Venter.

Genome Therapeutics announced its incorporation.

Mel Simon and coworkers announced the use of BACs for cloning.

1995 – The *Haemophilus influenzae* genome (18 Mb) was sequenced.

The *Mycoplasma genitalium* genome was sequenced.

The Prosita database was reported by Bairoch *et al.*

Affymetrix produced the first commercial DNA chips.

Structural Bioinformatics, Inc. founded in San Diego, CA.

1997 – Genome for *E. coli* (4.7 bp) was published.

Oxford Molecular Group acquired the Genetics Computer Group.

LION Bioscience AG founded as an integrated genomics company with strong focus on bioinformatics. The company is built from IP out of the European Molecular Biology (EMBL), the European Bioinformatics Institute (EBI), the German Cancer Research Center (DKFZ), and the University of Heidelberg.

Paradigm Genetics Inc., a company focused on the application of genomic technologies to enhance worldwide food and fibre production, was founded in Research Triangle Park, NC.

1998 – Genomes for *C. elegans* and baker's yeast were published.

2000 – The genome for *Pseudomonas aeruginosa* (6.3 Mbp) was published.

The *A. thaliana* genome (100 Mb) was sequenced.

The *D. melanogaster* genome (180 Mb) was sequenced.

2001 – The human genome (3,000 Mbp) was published.

2002 – Structural Bioinformatics and Gene Formatics emerged.

An international sequencing consortium published the full genome sequence of the common house mouse (2.5 Gb).

2004 – The draft genome sequence of the brown Norway laboratory rat, *Rattus norvegicus,* was completed by the Rat Genome Sequencing Project Consortium. The paper appeared in the April 1 edition of Nature.

4.15.2 Sequences and Nomenclature

The IUPAC Symbols

Many times the nomenclature system adopted in bioinformatics work is based on the International Union of Pure and Applied Chemistry (IUPAC) recommendations. It is useful to follow this nomenclature system so that data sets from different laboratories situated around the world can be easily and uniformly compared. The database institutions and the journals that publish research reports strictly follow these recommendations to ensure uniformity and to aid rapid reproducibility. Details are available on the IUPAC Website. For routine work using the nucleic acid and protein sequence data, the following system of IUPAC nomenclature can be used.

Symbol	Meaning
A	Adenine
G	Guanine
C	Cytosine
T	Thymine
U	Uracil
Y	Pyrimidine (C to T)
R	Purine (A to G)
W	Weak (A to T)
S	Strong (C to G)
K	Keto (T to G)
M	Amino (C to A)
B	not A (C or G or T)
D	not C (A or G or T)
H	not G (A or C or T)
V	not T (A or C or G)
X, N, ?	unknown (A or C or G or T)
O	deletion
—	deletion

DNA and Protein Sequences

There are many different representations (or, in other words, formats) of DNA and protein sequences (joined with ancillary annotations). Almost every program introduces its "native" format optimized for a particular program needs, and some programs include converters from one format to another. However, some programs read sequences only in one or the other formats. For successful data anlaysis, it is important to recognize formats and to be able to interconvert files in different formats.

There are sets of symbols used to abbreviate nucleotides and amino acids. The characters can either be upper or lower case. The characters enable input of nucleic acid sequences taking full account of ambiguities if any, in the sequences.

The protein sequences are given by the one-letter code used by the Margaret Dayhoff's group. They are as follows :

Amino acid	One-letter code
ala	A
arg	R
asn	N
asp	D
asx	B
cys	C
gln	Q
glu	E
gly	G
glx	Z
his	H
ileu	I
lys	K
leu	L
met	M
phe	F
pro	P
ser	S
thr	T
trp	W
tyr	Y
val	V
deletion	—
nonsense (stop)	*
unknown amino acid	X
unknown (incl. deletion)	?

The simplest format for DNA/protein sequences is a FASTA format (or Pearson format). It is used in a variety of molecular biology software. Every sequence in the file starts with "greater than" character (>). That character is followed by an identifier of a sequence (*e.g.,* name, description, gi-number) and a carriage return (it is also called a **paragraph sign**). This line is called a **definition line.** Everything after carriage return is considered as a sequence.

4.15.3 The Concept of Directionality

In biological systems, the usual direction in which the DNA and RNA are synthesized is in the 5'-3' direction. This is universal and, therefore, is helpful to adopt this fact as a way to collect and store data in the sequence databases. The nucleotide sequence data are generally present in the database as they have been submitted or published, subject to some conventions which have been adopted for the database as a whole. The sequences are always listed in the direction 5'- to 3'- regardless of the published order. Bases are numbered sequentially beginning with 1 at the 5'- end of the sequence. The complementary sequence is described with a 'c' indicated next to the position of the sequence. Complementary sequence also runs 5'-3' but in the opposite direction to the given strand. Only one strand of the DNA sequence is given in a database entry. The complementary strand will be inferred using programs available in various packages or from various websites. In the case of proteins, they are synthesized in the cell from N-terminus to the C-terminus. It is useful to adopt this convention in database entry for protein sequences. Thus, the concept of directionality used in biological systems is useful in describing the conventions to be adopted by the Database institutions. The advantage here is the universality of these fundamental biological processes in almost all living organisms.

For example, if we have given a sequence without any label, how we will find out whether it is a DNA sequence or a RNA sequence or a protein sequence. For this purpose the usual approach taken by standard computer programs like sequence search programs is to use the first 20 symbols. If the symbols encountered switch between any of the 4 bases only, then the sequence at hand is taken as a DNA sequence. Instead of T if U is encountered, then it is a RNA sequence. But if the symbols switch between any of the 20 (greater than 4), then it is taken as protein sequence.

4.15.4 Different Types of Sequences

ESTs. These are small pieces of DNA sequences (usually 200 to 500 nucleotides long) that are generated by sequencing either one or both ends of an expressed gene. The idea is to sequence bits of DNA that represent genes expressed in certain cells, tissues or organs from different organisms and use these "tags" to fish a gene out of a portion of chromosomal DNA by matching base pairs. The challenge associated with identifying genes from genomic sequences varies among organisms and is dependent upon genome size as well as the presence or absence of introns, the intervening DNA sequences interrupting the protein coding sequence of a gene. Millions of ESTs have been deposited in a special database called dbEST.

GSTs. The genome sequence tag (GST) approach was pioneered by Dunn *et al.* Genomic sequence tags (GSTs) are short (*e.g.,* 21 base) sequence fragments sampled more or less at random from microbial genomes in the given population. Such tags are inexpensive to assay, yet long enough to allow for straightforward species identification against sequence databases.

cDNA. It is a form of DNA prepared in the laboratory using an enzyme called **reverse transcriptase.** cDNA production is the reverse of the usual process of transcription in cells because the procedure uses mRNA as a template rather than DNA. Unlike genomic DNA, cDNA contains only expressed DNA sequences, or exons.

Organelle DNA : Eukaryotic cells have organelles such as mitochondria and chloroplasts. These organelles have their own store house of information in the form of organelle DNA. Organelle DNA codes for a few genes. The coding information for the rest of the genes resides in the nuclear DNA of the same cell.

(a) **Chloroplast DNA :** It is a circular molecule ranging from 120 to 217 kilobase pairs in photosynthetic land plants. Chloroplast DNA sequences have become a useful tool for the study of plant taxonomy, phylogeny and population genetics. Chloroplasts are sub-cellular organelles with their own DNA which is unlike the genomic DNA of a plant because it is inherited in a non-Mendelian mode. Chloroplast DNA is inherited from one parent only and because of this, does not undergo recombination and has a highly conserved gene order.

(b) **Mitochondrial DNA :** It consists of a circular DNA molecule, with 5 to 10 copies per organelle. Mitochondria are organelles found in eukaryotic cells that serve as powerhouses; harnessing the majority of usable energy from a simple sugar molecule. Individual mtDNA molecules are small and usually circular. mtDNA is present in 50 to 100 copies per diploid cell. The mitochondrial genome can undergo many different types of gross chromosomal rearrangements. Mitochondrial genomes are also characterized by a very high mutation rate. The proteins encoded by mtDNA are synthesized within the mitochondria on small ribosomes known as **mitoribosomes**. Mitochondrial genes can contain introns though mammalian mtDNA apparently does not. The mammalian mitochondrial genome contains 16,500 base pairs.

GOBASE is a taxonomically broad organelle genome database that organizes and integrates diverse data related to mitochondria and chloroplasts.

4.15.5 Main Databases

NCBI has a multi-disciplinary research group composed of computer scientists, molecular biologists, mathematicians, biochemists, research physicians, and structural biologists concentrating on basic and applied research in computational molecular biology. Fundamental biomedical problems are being understood at the molecular level using mathematical and computational methods. These problems include gene organization, sequence analysis and structure predictions. A sampling of current research projects include :

(a) detection and analysis of gene organization,

(b) repeating sequence patterns,

(c) protein domains and structural elements,

(d) creation of a gene map of the human genome,

(e) mathematical modeling of the kinetics of HIV infection,

(f) analysis of effects of sequencing errors for database searching,

(g) development of new algorithms for database searching and multiple sequence alignment,

(h) construction of non-redundant sequence databases,

(i) mathematical models for estimation of statistical significance of sequence similarity and vector models for text retrieval.

NCBI assumed responsibility for the GenBank DNA sequence database in October 1992. NCBI staff with advanced training in molecular biology build the database from sequences submitted by individual laboratories and by data exchange with the international nucleotide sequence databases, European Molecular Biology Laboratory (EMBL) and the DNA Database of Japan (DDBJ). In addition to GenBank, NCBI supports and distributes a variety of databases for the medical and scientific communities. These include the Online Mendelian Inheritance in Man (OMIM), the molecular modeling database (MMDB) of 3D protein structures, the Unique Human Gene Sequence Collection (UniGene), A Gene Map of the Human Genome, the Taxonomy Browser and the Cancer Genome Anatomy Project (CGAP) in collaboration with the National Cancer Institute.

Additional software tools provided by NCBI include: Open Reading Frame Finder (ORF Finder) and the sequence submission tools, Sequin and Banklt.

4.15.6 Database Retrieval Tools

Various data retrieval and submission tools may be found on the NCBI web site, including Text term searching

Enterz : Enterz is NCBIs search and retrieval system that provides users with integrated access to sequence, mapping taxonomy and structural data. Enterz also provides graphical views of sequences and chromosome maps. A powerful and unique feature of Enterz is the ability to retrieve related sequences, structures and references. Enterz, therefore, integrates the scientific literature, DNA and protein sequence databases, 3D protein structure and protein domain data, population study datasets, expression data, assemblies of complete genomes and taxonomic information into a tightly interlinked system.

It provides integrated access to nucleotide and protein sequence data from over 100,000 organisms along with three-dimensional protein structures, genomic mapping information, and PubMed Medline.

Taxonomy database : The taxonomy database contains the names of all organisms that are represented in the NCBI genetic database by at least one nucleotide or protein sequence.

Citation matcher : The journal literature is available through PubMed, a Web search interface that provides access to over 11 million journal citations in MEDLINE and contains links to full-text articles at participating publishers' websites.

4.15.7 Sequence Similarity Searching

Powerful computer programs have been devised to permit searching of nucleic acid and protein sequence databases for significant sequence matching (sequence homology) with a test sequence under investigation. Different BLAST and FASTA are the popularly used programs (Ginsburg, 1994 and Table below).

Program	Compares
FASTA	A nucleotide sequence against a nucleotide sequence database or an amino acid sequence against a protein sequence database
TFASTA	An amino acid sequence against a nucleotide sequence database translated in all six reading frames
BLASTIN	A nucleotide sequence against a nucleotide sequence database
BLASTX	A nucleotide sequence translated in all six reading frames against a protein sequence database
EST BLAST	A cDNA/EST sequence against cDNA/EST sequence databases
BLASTP	An amino acid sequence against a protein sequence database
TBLASTN	An amino acid sequence against a nucleotide sequence database translated in all six reading frames.

Note: As the design of comparable programs such as FASTA and BLAST is different, they may give different results (Ginsburg, 1994). All of the above programs are accessible through the Internet from various centres, such as the US National Centre for Biotechnology Information (http ://www.ncbi.nih.gov/) and the European Bioinformatics Institute (http://www.ebi.ac.uk/).

Programs such as BLAST and FASTA use algorithms to identify optimal sequence alignments and typically display the output as a series of pair-wise comparisons between the test sequence (query sequence) and each related sequence which the program identifies in the database (subject sequences).

Different approaches can be made to calculate the optimal sequence alignments. For example, in nucleotide sequence alignments, the algorithm devised by Needleman and Wunsch (1970) seeks to maximize the number of matched nucleotides. In contrast, other programs such as that of Waterman *et al.* (1976) have an object to minimize the number of mismatches. Pair-wise comparisons of sequence alignments are comparatively simple when the test sequences are very closely matched and have similar, preferably identical, lengths. When the two sequences that are being matched are significantly different from each other, and especially when there are clear differences in length due to deletions/insertions, considerable effort may be necessary to calculate the optimal alignment.

If the nucleotide sequence under investigation is a coding sequence, then nucleotide sequence alignments can be aided by parallel amino acid sequence alignments using the assumed translational reading frame for the coding sequence. This is so because there are 20 different amino acids, but only four different nucleotides. Pair-wise alignments of amino acid sequences may also be aided by taking into account the chemical subclasses of amino acids. Conservative substitutions are nucleotide changes which result in an amino acid change but where the new amino acid is chemically related to the replaced amino acid, typically belongs to the same subclass. As a result, algorithms used to compare amino acid sequences typically use a scoring matrix in which pairs of scores are arranged in 20 × 20 matrix where higher scores are accorded to identical amino acids and to ones which are of similar character (*e.g.,* isoleucine and leucine) and lower scores are given to amino acids that are of different character (*e.g.,* isoleucine and aspartate; Henikoff and Henikoff, 1992). The typical output gives two overall results for percent sequence relatedness, often termed % sequence identity (matching of identical residues only) and % sequence similarity (matching of both identical residues and ones that are chemically related.

4.15.8 Analyses using Bioinformatics Tools

Many kinds of analyses can be made using various bioinformatics tools.

1. Molecular Medicine

The human genome will have profound effects on the fields of biomedical research and clinical medicine.

(a) **Diagnosis and molecular mechanism of disease :** We can search for the genes directly associated with different diseases and begin to understand the molecular basis of these diseases more clearly.

(b) **More drug targets :** With an improved understanding of disease mechanism and using computational tools to identify and validate new drug targets, more specific medicines can be developed.

(c) **Personalized medicine :** Clinical medicine will become more personalized with the development of the field of pharmacogenomics.

(d) **Gene therapy :** The potential for using genes themselves to treat disease may become a reality.

2. Microbial Genome Application

The arrival of the complete genome sequence and their potential to provide a greater insight into the microbial world and its capacities could have far reaching implications for environment, health, energy and industrial applications.

In 1994, US DOE initiated microbial genome project to be ultimately helpful in

(a) waste clean up

(b) climate change

    (c)   antibiotic resistance

    (d)   evolutionary studies.

3. Agriculture

The sequencing of genomes of plants and animals should have enormous benefits for the agricultural community.

Bioinformatics tools can be used to search for the genes, their role and pattern of expression.

Genetic knowledge could then be used to produce stronger, more drought-disease and insect-resistant crops having improved nutritional quality. Success has been obtained in transferring genes into rice to increase level of vitamin A, iron and other micronutrients and growing in poorer soils and in drought conditions.

Animals : Sequencing projects of many farm animals are underway for improving the production and health of livestock.

4. Comparative Studies

Analyzing and comparing the genetic material of different species is useful for understanding the functions of genes, the mechanisms of inherited diseases and species evolutions.

4.15.9 Information from Stored Data

The following are the examples of use of database in obtaining useful information :

Homologous amino acid sequences are found in proteins with similar functions. These correspond to important functional domains in the three-dimensional structure of proteins. It has been observed that protein encoded by a newly cloned gene exhibits such homologies with proteins of known function. This can provide revealing insights into the function of the cloned gene.

The example is of *BRCA*-1 gene. There is a high probability of developing breast cancer before the age of 50 in women who inherit a mutant form of this gene. cDNA of the *BRCA*-1 mRNA was cloned and *BRCA*-1 protein was sequenced. Sequence comparison revealed that *BRCA*-1 protein is distantly, but significantly related to the *S. cerevisiae* Rad 9 protein, a cell-cycle checkpoint protein, which functions to control cell division in yeast. From this clue, several laboratories have initiated experiments to determine if the *BRCA*-1 protein functions similarly to the *S. cerevisiae* Rad 9 protein.

This indicates that the molecular basis of inherited human diseases can be gained by identifying and cloning the associated mutant gene and then comparing the sequence of the encoded protein with the sequences of other proteins stored in data banks. This general approach for revealing the molecular function of various proteins simply by sequencing the DNA that encodes them are bound to increase as the sequences of more proteins with known functions are determined. The development of new computer methods for identifying potentially significant relationship between sequences are also going to help in this direction.

Comparative Analysis of Genomes

The comparative analysis of genomes has been made possible by the availability of complete genome sequences of several organisms. It is now possible to know about the biology of organisms that are difficult to culture, even when few, if any, of their proteins have been isolated and studied directly. This can be understood in detail by taking the example of organisms, *Mycoplasma genitalium*, *Methanococcus jannaschii* and *Haemophilus influenzae*. Most proteins from these organisms have not been studied directly. Their functions in fact are inferred from their sequences by comparison with previously studied proteins of model organisms like *E. coli* and *S. cerevisiae*.

Smallest known genome of any cell is present in *M. genitalium*. It contains long open reading frames (ORFs), sequences that can encode polypeptides containing 100 or more amino acids. Most of the proteins encoded by these long ORFs have been assigned functions based on their sequence homology with proteins of known function from *E. coli* and other bacteria. It is difficult to culture *M. genitalium,* which lives as a parasite within epithelial cells lining primate urogenital and respiratory tracts. But we can know much about *M. genitalium* from an analysis of its genome sequence.

The *M. genitalium* genome does not encode many of the enzymes of intermediary metabolism found in free living organisms (*e.g., E. coli*) that can synthesize all their constituents from simple molecules such as acetate and ammonia plus inorganic ions. It is suggested that *M. genitalium* produces its macromolecules primarily from preformed precursors that are transported into its cytoplasm from the cytoplasm of its eukaryotic host cell. The *M. genitalium* genome does not recognize genes encoding the enzymes of the tricarboxylic acid (TCA) cycle and the cytochromes required for oxidative phosphorylation, it does encode the enzymes required for glycolysis of glucose to lactate and acetate. Thus this organism most likely produces ATP principally from the comparatively simple and low-yield glycolytic pathway. Infact the *M. genitalium* genome almost completely lacks the proteins that regulate transcription in other bacteria. This indicates that for cell growth and division, only minimal regulation of gene expression is required. Alternatively, *M. genitalium* may regulate the expression of its genes by mechanisms that are quite different from that in other prokaryotes.

The genome from an archaean *M. jannaschii* has also been completely sequenced. This is a strict anaerobe. It lives in an oxygen-free environment and derive chemical energy by the reduction of CO_2 with H_2 to produce methane. Genes encoding all the required enzymes for this reductive pathway have been identified in the *M. jannaschii* genome. Comparison of the sequences of *M. jannaschii* proteins with proteins performing equivalent functions from bacteria and eukaryotes has provided insights about the evolution of these three major classes of life on earth.

One of the first important human pathogens whose genome has been sequenced is *H. influenzae*. It causes life-threatening bacterial meningitis, an inflammation of the meninges that cover the central nervous system. As a pathogen, it normally lives in close association with eukaryotic host cells. *H. influenzae* does not have many of the enzymes of intermediary metabolism that are present in *E. coli*. It is because of this, *H. influenzae*, like *M. genitalium*, cannot grow on minimal medium. An understanding of the complete set of *H. influenzae* genes may provide clues for designing more effective therapies against *H. influenzae* infection. Because of the potential of this approach, scientists are determining the complete genome sequences of a number of pathogenic bacteria.

Questions

1. What is recombinant DNA technology ? Describe in detail.
2. What do you mean by the following ?
    (a) Gene Cloning
    (b) Gene Library
    (c) cDNA
    (d) Shotgun strategy
    (e) Mitochondrial genome
3. Explain the following
    (a) Genomics
    (b) Proteomics
    (c) The nuclear genome
4. What is Bioinformatics ? Write an essay on bioinformatics.
5. What is cloning ? Give its different types and describe them in short.
6. What are transgenics ? How we can produce transgenic plants ?
7. Give various strategies to produce transgenic animals.

Chapter 5

Specialized Tools in Biotechnology-III : Protein Based Technologies

5.1 PROTEIN ENGINEERING

Proteins are the important molecules of life. Protein engineering is the generation of novel proteins with new functions. Attempts are being made to engineer them by various ways (Fig. 5.1).

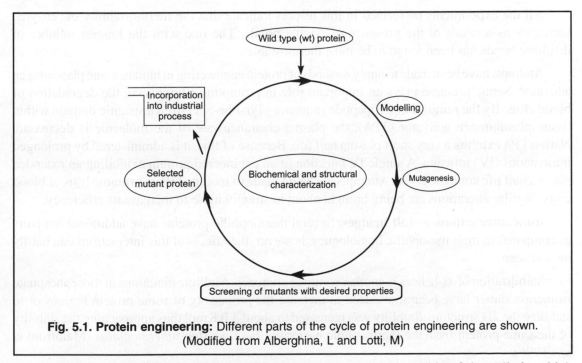

Fig. 5.1. Protein engineering: Different parts of the cycle of protein engineering are shown. (Modified from Alberghina, L and Lotti, M)

Candidates for protein engineering are proteins that are used outside of the cells in which they were originally made. These include proteins that are:

(a) administered to people as therapeutic agents

(b) used in manufacturing

(c) added to products to enhance them.

They are engineered for the following purposes:

(i) Enhancement of Protein Stability

Stability of protein is important in biotechnology. Stability include:

- temperature stability
- pH stability
- stability towards detergents
- resistance to proteolytic enzymes

Stability can be introduced by several methods. It can be done by the introduction of cystein residues which allow formation of disulfide bond. The amino acid residues that are targetted to become cystein residues are spatially close to each other in the active enzyme. This ensures that the overall conformation of molecule would remain essentially unaffected by the formation of the new disulfide linkage. This is evident from the example of lysozyme. It has a cystein residue at position 7. This in turn enhances the stability of protein. The targetted amino acids are not involved in the active site of the enzyme.

All the experiments performed in this respect indicate that the thermostability of enzyme increases as a result of the presence of double bonds. The one with the longest number of disulfide bonds has been found to be most thermostable.

Attempts have been made to apply methods of protein engineering in human tissue plasminogen activator. Serine protease plays an important role in promoting fibrinolysis, the degradation of blood clots. By the removal of a tripeptide sequence (Tyr-Phe-Ser) from a specific domain within tissue plasminogen activator (TPA), the plasma clearance rate of the molecule is decreased. Native TPA exhibits a very short plasma half life. Because of this, it is administered by prolonged intravenous (IV) infusion. A single IV injection of an engineered protein exhibiting an extended plasma half life could be given. Attempts are being made to modify TPA to promote lysis of blood clots. Similar alterations are being brought about in streptokinase to increase its efficiency.

Ionic interactions or salt bridges. Several thermophilic proteins have additional ion pairs as compared to their mesophilic homologues; however, the effects of this interactions can hardly be foreseen.

Stabilization of α-helices is another common approach. Multiple mutations in triosephosphate isomerase dimer have been introduced to increase the propensity of some protein regions or to stabilize the 2D structure. Stability was increased to about 3.0 Kmol thus approaching the stability of the same protein from the thermophilic bacterium *Bacillus stearothermophilus* (Mainfroid *et al.*, 1996).

Hydrophobic effect is the well known effect by which hydrophobic regions are sequestered in the protein interior. Even the tight and highly organic packing of the protein core is important and in fact the native state of protein shows packing densities similar to those of organic crystals. Substitutions of hydrophobic residues in the protein cores with smaller amino acids has been carried out.

Binding of ion metals or cofactors can increase structure stability when they bind with folded state with high affinity as compared to unfolded one, *e.g.*, Ca^{2+}, Zn^+, Cu^{2+}.

(ii) Improvement in Specificity of Proteins

Protein engineering techniques are also being used to improve the properties of proteins. Such a specificity includes fine tuning by the addition of specific property or regulatory system. This generally requires gene for the engineered protein to be transferred into the genome of the plant or animal that is to be utilized as a source of protein. The expression of the new gene should support in a way comparable to the wild-type protein. It may be necessary to remove or otherwise inactivate the wild-type gene of the protein. It is now possible to use random mutagenesis to improve the gene for a poorly-behaved lysine-enriched protein.

(iii) Use of Chemically Synthesized or Modified DNA Sequences to Have Proteins of Choice

Chemical synthesis of gene fragments may help to have engineered proteins. The nucleotide sequence of any DNA can be translated in bacterial and mammalian cells into a linear sequence of amino acids making up a polypeptide chain. Completely synthetic genes of lengths up to 100 or more nucleotides specifying any desired polypeptide sequence can be made. Hybrid genes coding for newly designed protein sequences can be created by joining synthetic gene segments to parts of natural genes. The synthetic fragment can be used to replace or extend a part of a natural gene by oligonucleotide mismatch mutagenesis or the polymerase chain reaction (PCR). The synthetic gene is incorporated into a plasmid of high copy number to direct the synthesis of the designed polypeptide chain either in bacteria or cultured eukaryotic cells. These new materials can be very cheap.

To follow genetic route to new proteins is sometimes problematic. It is because of the non-availability of efficient ways of incorporating non-biological amino acids. Natural enzymes often contain prosthetic groups which are necessary for their catalytic activities. But organic conductors and transition state metal chelators can easily be inserted into short, chemically synthesized peptides, and the development of a protein ligase which specifically links a cyanoethyl ester-activated peptide to a much larger section of protein has been made by the genetic route. It has opened the way to much more versatile synthetic engineered proteins, that can be constructed from the 20 amino acids.

(iv) Mutagenesis

It is possible to engineer the proteins by causing mutations. Mutations can now be caused at the site of choice in the gene and then have new proteins (Fig. 5.2). Various techniques are available for this purpose (Table 5.1).

Table 5.1: Experimental Approaches for Protein Engineering

Rational design: It allows for the introduction of mutations targeted to specific protein sites. It requires a detailed know how of the protein structure and of structure-function relationships.

Molecular evolution: It does not require any knowledge of the protein structure and mechanism of action. It is based on random generation of a vast number of mutants followed by screening for the desired function.

Generation of random libraries: It is the production of large collection of proteins, peptides of region thereof. It is often coupled with such display to ease screening of the mutants.

De novo **protein design:** It generates novel structural scaffolds able to accommodate active sites or other protein functions.

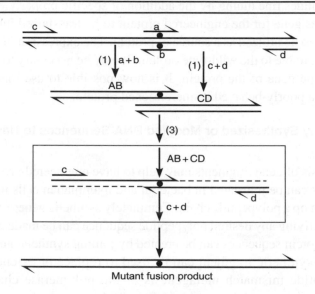

Fig. 5.2. Schematic diagram of site-directed mutagenesis by overlap extension. The dsDNA and synthetic oligos are represented by lines with arrows that indicate the 5'-3' orientation. The site of mutagenesis is indicated by small black **circle**. Oligos are denoted by lower-case letters corresponding to the oligo primers used to generate that product. The proposed intermediate steps taking place during the course of reaction are represented by boxed portion of the figure. The denatured fragments anneal at the overlap and are extended 3' by DNA polymerase (dotted line) to form the mutant fusion product. The mutant fusion product is further amplified by PCR by adding additional primers 'a' and 'd'.

Now computer graphic analyses are employed for site-directed mutagenesis. These are based on the knowledge of three dimensional structure and primary sequence of protein. For proper modification, a model for the relation between protein structure and functionality is needed. Based on the sensitivity of the amino acid sequence to oxidation or hydrolysis, it is possible to develop a desired protein structure model. This can be done for obtaining a protein of required functionality. Then computer graphic analysis is used to design amino acid replacements. We can carry out deletions or insertions of the sequences by altering the DNA sequence of the cloned gene by site-directed mutagenesis. However structure and sequence analyses still remain bottlenecks in protein engineering (Table 5.2).

Table 5.2: Hiccups in the Field of Protein Engineering

(i)   Scarce information on the second genetic code, *i.e.,* rules relating sequence to structure in protein

(ii)   Limited number of known 3D structures

(iii) Limited or not obvious information of the structure-function relations in proteins

(iv)   Difficulties in the management of information.

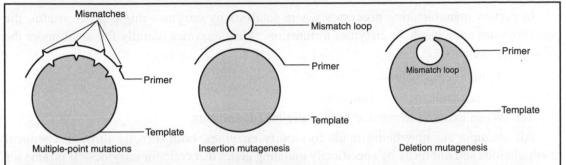

Fig. 5.3. The use of oligonucleotide-directed mutagenesis. It can generate multi-point mutations, insertions and deletions.

The site-directed mutagenicity modifies the genetic sequence so as to make a different protein. To produce modified protein, the altered DNA is cloned and expressed in a fermentation process. By predicting how specific changes in amino acid sequence can alter three-dimensional structures, new proteins are designed by computer graphics.

Oligonucleotide or oligodeoxynucleotide site directed mutagenesis helps in having substitution, insertion or deletion of single or small number of bases. For more changes, whole sequences are deleted from proteins by specific digestion with restriction endonucleases. This is followed by joining of the ends. Using oligonucleotide-directed mutagenesis, specific restriction sites should be inserted at the correct positions (Fig. 5.3).

Although most of the protein engineering work has been directed at changing the catalytic properties of existing enzymes, methods have been found out to synthesize novel catalysts : antibodies with catalytic ability or enzymes. Antibodies resemble enzymes because both are proteins that bind to specific molecules. A lot of information is available in databases and tools for the analysis for persons working in field of protein engineering (Table 5.3).

Table 5.3: Some Websites of Interest to Protein Engineers

Structures and structure analysis

Protein Data Bank, Brookhaven National Lab.	http://www.pdb.bnl.gov
SCOP-Structural classification of proteins	http://scop.mrc-lmb.cam.ac.uk/scop
RasMol	http://www.umass.edu/microbio/rasmol

Sequences and sequence analysis

European Bioinformatic Institute, UK	http://www2.ebi.ac.uk
European Molecular Biology Lab., G	http://www.embl-heidelberg.de
Expasy Molecular Biology Server	http://www.expasy.ch
NCBI Genebank	http://www.ncbi.nlm.nih.gov
Pedro's Biomolecular Research Tools	http://www.public.iastate.edu/

5.2 ENZYME ENGINEERING

Enzymes are beautifully designed for the role they play in nature as catalysts of the biochemical reactions on which living organisms depend. Such enzymes function best under conditions that are compatible with life : neutral pH, mild temperature and pressure, and an aqueous (water-based) environment.

In certain manufacturing processes where catalysis by enzymes might prove useful, the conditions are quite harsh for enzymes to function. Most enzymes literally fall apart under the following conditions :

(a) At high temperatures

(b) In very acidic or basic solutions

(c) When exposed to organic (non-water-based) solvents.

All attempts are now being made to modify enzymes, both directly through chemical manipulations and indirectly by specifically mutating genes that code for enzymes. It is done for the following purposes :

(a) To increase their stability under harsh manufacturing conditions,

(b) To broaden their substrate specificity,

(c) To improve their catalytic power.

It is evident from all this that bioinformatics is bound to help us in understanding intricacies in the life system and apply this understanding for the benefit of man. Enzymes can be engineered to have enhanced properties. These are optimized by natural selection for action on natural chemicals under natural conditions. Attempts are now being made to improve enzymes directed against unnatural substrates or for use under unnatural conditions.

Engineered enzymes are also used to produce required stereoisomers of a compound for medical use. Stereoisomers often occur in chemicals containing an asymmetric carbon atom. Many modern drugs have such asymmetric centres and the different isomers can have very different effects. Since the tragedies that resulted when the stereoisomer of a drug that was active against morning sickness turned out to impair foetal development, the pharmaceutical industry is increasingly introducing drugs which are single compounds, not the naturally occurring mixture of two stereoisomers. Often highly selective enzymes that can discriminate between the two stereoisomers of a compound are used in the production process, and their usefulness is being extended by protein engineering.

Attempts are being made to custom design enzymes to catalyze reactions for which there are no known enzymes. This has opened a new world of exciting possibilities. It is possible that we might be able to create antibodies that can function as proteases, the enzymes that break down proteins.

To improve existing proteins and to create proteins not found in nature, protein engineering technology will often be used in conjunction with genetic engineering. In theory, it should eventually be possible to create any protein from scratch. Research efforts are primarily aimed at modifying existing proteins (Table 5.4).

Table 5.4: Industrial Enzymes: Targets of Protein Engineering

> **Starch processing enzymes:** α-amylase, glucoamylase, glucose isomerase
>
> **Detergent enzymes:** proteases, lipases, amylases
>
> **Texiles:** cellulases, amylases, lipases, catalases, peroxidases
>
> **Leather industry:** proteases, lipases
>
> **Pulp and paper:** xylanases, cellulases

5.3 BIOCATALYST IMMOBILIZATION

It is the conversion of enzymes from a water-soluble, mobile state to a water-insoluble, immobile state. It prevents enzyme diffusion in the reaction mixtures and facilitates their recovery from the product stream by solid/liquid separation techniques. The immobilization has the following advantages:

(i)   Multiple and repetitive uses of a single batch of enzymes.

(ii)   Creation of a buffer by the support against changes in pH, temperature and ionic strength in the bulk solvent, as well as protection from shear forces.

(iii)   No contamination of processed solution with the enzyme.

(iv)   Analytical considerations, especially with respect to long half-life for activity and predictable decay rates.

An enzyme in general is attached to a solid support material so that substrate can be continually converted to product. Enzymes can, thus, be recycled and used many times. The purpose is to:

(i)   increase the stability of enzyme

(ii)   increase the ability to recycle the enzyme

(iii)   separate the enzyme easily from the product

5.3.1 Principle

The enzymes can be confined to a restricted space by various ways leading to a heterogenous reaction system. The immobilization is carried out in such a way that the substrate solution can easily pass through this region and efficiently interacts with the catalyst. It can be achieved by having substrate and catalyst in different phases, the latter being retained in some type of reaction vessel.

As most of the biological materials are handled in aqueous solutions, the catalyst is best attached to a solid phase which can be separated from the liquid by simple physical method. If the substrate is in a gas phase or in water, immersible phase is applied. The latter makes it possible to consider catalyst confined in a discontinuous aqueous phase. The catalyst does not even have to be fully attached to support material so long as the complex is retained in the reaction vessel. The procedure, therefore, in which enzymes and cells are held back in a reactor by means of semi-permeable membrane, fall into the immobilized system. A number of biocatalysts can now be immobilized.

5.3.2 Techniques

The following techniques are used for immobilization of enzymes :

(a) Adsorption   (b) Covalent attachment

(c) Cross linking   (d) Entrapment and

(e) Encapsulation.

5.3.2.1 Adsorption

One of the earliest methods used for immobilizing enzyme is physical adsorption to an insoluble matrix. It relies on non-specific physical interaction (Fig. 5.4) between protein and the surface of the matrix by mixing a concentrated solution of enzyme with the solid. Two categories of solids are used for immobilization. These are :

(i) **Natural :** These include charcoal, alumina, calcium phosphate, hydroxylapatite, collodion, collagen and wood chips.

(ii) **Synthetic :** The exchange resins such as DEAE cellulose, DEAE sephadex, Dowex 50 and a variety of phenolic resins belong to this category. The process is dependent on the pH, solvent type, ionic strength, quality of enzyme, adsorbance time and temperature. Some commonly used adsorbants include alumina for acylase and amylase, cellulose for cellulase, clays for catalase, glass for urease *etc.*

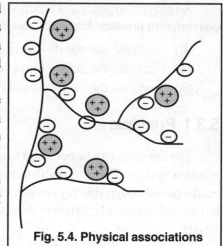

Fig. 5.4. Physical associations

An enzyme solution is added to the support, mixed and surplus enzymes are then removed by washing. The process is carried out by contacting enzyme and supporting in an agitated reactor or by passing the enzyme through a packed bed column or membrane formed from the support material. The attachment of cofactors such as pyridoxal phosphate or the use of hydrophilic side chains can improve adsorption on the support. Enzymes can be attached to ion-exchange resins by electrostatic attractions.

The method of immobilization can be very weak and the enzyme may be easily desorbed and lost. The following factors are involved in adsorption by electrostatic attractions:

(i) The surface area to volume ratio

(ii) The particle size

(iii) The ratio of hydrophilic to hydrophobic groups

The disadvantages are:

(i) the change in pH, temperature *etc.* can lead to desorption,

(ii) it is not specific for the enzyme,

(iii) it requires stabilization,

(iv) it is not useful in case of multi-enzyme complex.

5.3.2.2 Covalent Bonding

Enzymes are composed of chains of amino acids which have a number of reactive side chains that can be utilized for the formation of chemical bonds and activated supports (Fig. 5.5). Some of these reactive amino groups are carboxyl, sulphydryl, hydroxyl, imidazole and phenolic groups. Since covalent bonds are strong, very little leakage of enzyme from the support occurs. This is based on the covalent attachment of enzymes to water-insoluble matrices. Attachment must involve functional groups on the enzyme not responsible for catalytic action and thus attachment reagent must be carefully selected.

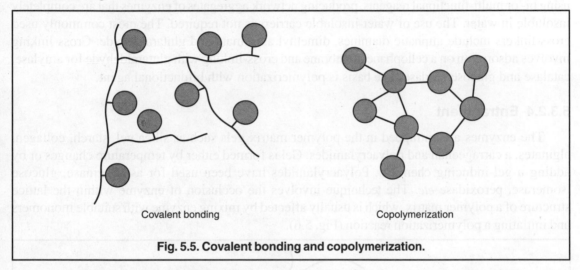

Covalent bonding   Copolymerization

Fig. 5.5. Covalent bonding and copolymerization.

The followiong supports are commonly used:

(i)   Natural insoluble polymers (such as dextrose, agarose and collagen)

(ii) Synthetic polymers such as polyvinyl alcohol, xylan, polystyrene *etc,*

(iii) Inorganic supports include silica, glass, magnetic ion particles *etc*.

The following methods have been developed:

(i)   Covalent attachment in the presence of a competitive inhibitor or substrate.

(ii) A reversible covalently linked enzyme inhibitor complex.

(iii) A chemically modified soluble enzyme whose covalent linkage involves new active residues.

(iv) A zymogen precursor.

The support used includes porous glass and ceramics for amyloglucosidase, cellulose for amylase, nylon for urease, alumina for glucose oxidase and polymers for trypsin. This method involves support activation and enzyme attachment. Amino, hydroxyl, carboxyl groups form the chemical bond. Cyanogen bromide (CnBr) activation of Sephadex and Sepharose is one of the most widely used covalent attachment techniques. It is because of the fact that it gives an enzyme immobilization with good activity, coupling yield, stability, filterability and economical cost.

The technique of covalent coupling is advantageous. The enzymes have increased stability as a result of which their properties are not altered by temperature, pH, solvent *etc*.

The disadvantage of this method is that the conditions of immobilization are harsh and may inactivate or change the properties of enzymes.

5.3.2.3 Cross-Linking

Cross linking of proteins to other proteins or to functional groups on an insoluble support matrix also provide immobilization. This is based on covalent bonding between enzyme molecules using bi- or multi-functional reagents, producing network aggregates of enzymes that are completely insoluble in water. The use of water-insoluble carriers is not required. The most commonly used cross-linkers include aliphatic diamines, dimethyl adipimate and glutaraldehyde. Cross-linking involves adsorption on a cellophane membrane and cross-linking with glutaraldehyde for amylase, catalase and glucose oxidase. The basis is polymerization with bifunctional agent.

5.3.2.4 Entrapment

The enzymes are entrapped in the polymer matrix gels such as silica gel, starch, collagen, alginates, a carrageenan and polyacrylamides. Gel is formed either by temperature changes or by adding a gel-inducing chemical. Polyacrylamides have been used for asparaginase, glucose isomerase, peroxidase *etc*. The technique involves the occlusion of enzyme within the lattice structure of a polymer matrix, which is usually affected by mixing enzyme with suitable monomers and initiating a polymerization reaction (Fig. 5.6).

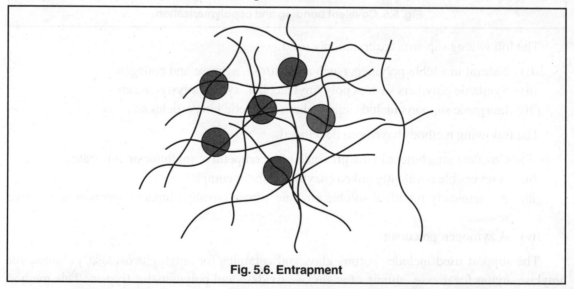

Fig. 5.6. Entrapment

The enzyme in this technique is subjected to limited interference as it is simply conferred in the entrapping system. This method is particularly useful for whole cell immobilization.

The disadvantages of the method include the fact that the monomers react with the catalyst. A diffusional barrier for the substrate is imposed by the presence of gel.

5.3.2.5 Encapsulation or Membrane Confinement

A number of methods can be used for this purpose. All the methods utilize the semipermeable membranes that are impermeable to enzymes, but permeable to low molecular weight substrates and products. Colloidion membranes have been used for catalase and L-asparaginase, cellulose derivatives for lipase and nylon for urease and trypsin.

Semipermeable membranes confine the enzymes, but allow free passage of substrates and the reaction product. Microencapsulation is preferred by depositing the polymer around emulsified aqueous droplets either by interfacial coactivation or interfacial polycondensation.

A polymer such as cellulose nitrate can separate out around enzyme microdroplets formed by agitating an aqueous dispersion of an enzyme, in a water-immiscible solvent *e.g.,* ethyl cellulose, nitrocellulose, polystyrene, polyethylene, polyvinyl acetate, polymethyl-metacrylate and polyisobutylene. This can be accomplished by interfacial polymerization, in which an aqueous solution of an enzyme and monomer can be dispersed in a water-immiscible solvent. The second hydrophobic monomer is then added, dissolved in the solvent and polymer is then formed by chemical reaction at the interface of the microdroplets.

The disadvantage of using microencapsulation within thin walled spheres is that many interfacial polymerization procedures can cause inactivation of an enzyme.

This problem can be overcome by producing hollow fibres instead of spheres by placing the enzyme in the fibre and sealing the end. The added advantage of hollow fibres is that it is easy to fill them with a variety of catalysts. It also allows recovery of trapped enzyme.

In general, the entrapment methods fall into two categories :

(a) **Single membranes :** In this, single membrane or barrier encloses a defined area containing the catalyst.

(b) **Three dimensional gel :** In this, the catalyst is dispersed within a three dimensional gel.

(a) **Single membrane reactor :** The simplest form of membrane reactor is formed by the membrane filtration units used to concentrate biochemicals. There is no doubt that reverse osmosis, ultrafiltration and microfiltration membranes can be employed. The ultrafiltration molecules provide the most convenient compromise between the low pressure drop to get a good flux, and size discrimination to give good catalyst retention.

(b) **Three dimensional gels :** For the immobilization, a number of hydrogels have been used. These techniques involve gelation of an aqueous suspension of the cell so that the biocatalyst is distributed throughout the solid mass.

Selection of Technique

While selecting the technique, one must keep in mind the following points :

(i) It should be mild, cheap, safe, versatile and suitable for scale-up

(ii) It should provide easy control of the type and amount of enzymes

(iii) There is minimum enzyme leakage from the support

(iv) It prevents partitioning between substrate, product and support.

The purity of enzyme prior to immobilization affects the stability of the immobilized complex. A crude enzyme preparation can withstand more denaturing and inactivation mechanisms inherent in immobilization techniques. Economic consideration plays a large role in determining feasibility and benefits of immobilization. Immobilization is generally not required for reactions that can be performed satisfactorily with inexpensive, crude enzyme preparations, or where large quantities of products are not required.

There is no doubt that all the enzymes can be immobilized, but their use, desired end result economics and substrate constraints should be considered. Each enzyme has a unique surface chemistry and operational stability. Consequently, no one immobilization method is generally applicable to all the enzymes. Substrate diffusion is critical in immobilized enzymes, since it must penetrate the immobilized matrix to get to the enzyme.

Applications of Immobilized Enzymes

Immobilized enzymes find applications in the following areas:

Analysis: The use of immobilized enzymes allows continuous and reagentless assays to be performed *e.g.,* immobilized urease is used to determine urea in serum and immobilized peroxidase to determine hydrogen peroxide.

Preparative use: Immobilized enzymes have been widely used in the field of affinity chromatography to isolate enzyme inhibitor and labeled peptides.

Food industry: Immobilization is used to produce glucose syrups using immobilized amylases. Hydrolysis of sucrose using immobilized invertase has also been achieved. Using immobilized papain, beer haze can be overcome. They are used for meat processing *etc.*

Medical uses: Immobilized enzymes are used in the detection of diseases and for estimation of glucose, urea *etc.* in blood using glucose oxidase and uricase, respectively.

Other uses: Immobilized enzymes are used in the production of ethanol by fermentation.

5.4 BIOSENSOR TECHNOLOGY

Chemical engineers, biotechnologists and medical researchers all have a common problem. They deal with processes involving complicated chemical reactions, which they need to monitor and control before the specific chemical reaction that interests them, proceeds too far. Often, the traditional techniques of analytical chemistry are not suitable. By the time a technician has removed a sample from the process, taken it to laboratory and carried out analysis, the results have taken too long to be of any use to an engineer running the process. In other cases such as medical research, it may not be possible to remove samples without interfering with process that is being measured.

It is because of this, many industries, not just biotechnology, medicine and agriculture need "Molecular Sensors" that detect specific molecules. There is worldwide demand for precise sensors that can carry on working within chemical and biological processes.

This realization in recent years has prompted the development of sensor technologies suitable for monitoring and controling the biological systems and processes having commercial potential. At the same time, it must be kept in mind that sensing devices or biosensors could revolutionize analytical biotechnology in areas such as health care, agriculture and pollution monitoring.

A biosensor is an analytical device which converts biological responses into electrical signals. The term biosensor is often used to cover sensor devices used to determine the concentration of substances and other parameters of biological interest even where they do not directly utilize a biological system.

The unique feature of a biosensor is that the device incorporates a biological sensing element in close proximity or integrated with a transducer to give a reagentless sensing system for the target analyte.

5.4.1 Components of a Biosensor

A biosensor is made up of following components (Figs. 5.7, 5.8):

(a) Biocatalyst

The biological response of the biosensor is determined by the biocatalytic membrane which accomplishes the conversion of reactant to products. The biocatalysts used can be enzymes, cells or antibodies. It is advantageous to use immobilized biocatalyst as it provides reusability, long half-life of biocatalyst and ensures that same catalytic activity is present for a series of analyses. This is an important factor for securing reproducible results. The following different methods are used for immobilization :

    (i)   Adsorption

    (ii) Covalent binding

    (iii) Entrapment

    (iv) Microencapsulation

    (v) Cross binding

(b) Transducer

It is the key part of biosensor which makes use of a physical change accompanying the reaction.

    (*i*)   *Calorimetric transducer :* This detects change in the heat output or adsorption during reaction.

    (*ii*) *Potentiometric transducer :* It detects changes in distribution of charges causing an electric potential.

    (*iii*) *Amperometric transducer :* It detects movement of electrons produced in a redox reaction.

    (*iv*) *Optical transducer :* It detects light output during the reaction.

    (*v*) *Piezo-electric transducer :* It detects the effect due to mass of reactant or product.

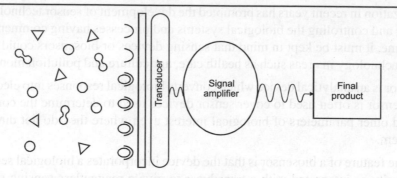

Fig. 5.7. Principle of a biosensor (Modified from Kreuzer, 2003).

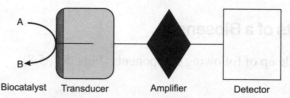

Biocatalyst   Transducer   Amplifier   Detector

Fig. 5.8: General structure of biosensors. The substrate is converted by the biocatalyst into product. There is a concurrent change in a physicochemical parameter (*e.g.,* heat, electron transfer, light, ion or proton flow *etc.*). It is converted into an electrical signal by the transducer, amplified and processed by a detector.

(c) Amplifier

It is an electronic circuit which amplifies the electrical signals produced by the transducer.

(d) Processor

It processes the signal by subtracting the base line signal which is taken by an electrode without a biocatalyst and converts the resultant signal to digital output.

(e) Monitor or Display

It displays the processed digital reading.

There are certain problems in assemblage of different components of a biosensor. These are:

(*i*) It requires immobilization of the biorecognition molecule
(*ii*) The biorecognition molecule should be specific for a particular analyte
(*iii*) Transduction, the biorecognition event should be efficient
(*iv*) Calibrations in data processing have to be done

5.4.2 Types of Biosensors (According to Biocatalyst used)

(a) Enzyme Biosensor

In this case the catalytic property is exploited. The enzyme is immobilized on a membrane which covers the transducer. A particular enzyme is used for a particular analyte and the product formed or other changes occurred during the enzyme analysis are sensed by a transducer lying nearby.

(b) Bacterial Biosensor

These sensors directly utilize a chain of bacterial enzyme reactions to achieve detection of one or more species. They are based on the same principle as the enzymatic device in that ultimately a current or a potential difference is used as an index of chemical concentration.

(c) Immuno-Biosensor

In case of immuno-biosensor, the antigen-antibody specific reaction properties are exploited. Labeled antibodies or antigens may be used in biosensors based on enzyme linked immunoassay (EIA or ELISA) principles.

There is competition between the antigen and the enzyme labeled antigen for binding with the immobilized antibody. Alternatively, the free antigen may be used to displace the enzyme labeled antigen from its complex with the immobilized antibody.

The enzyme-label is measured by adding the substrate specific for the enzyme and monitoring the product. The signal is then combined with a biosensor to form an immuno-biosensor.

5.4.3 Types of Biosensors (According to transducer used)

(a) Calorimetric Biosensor

Many enzyme catalyzed reactions are exothermic, generating heat, which may be used as a basis for measuring the rate of reaction and hence analyte concentration. This represents the most general type of biosensor. The temperature changes are usually determined by means of thermistors at the entrance and exit of small packed bed columns containing immobilized enzyme within a constant temperature environment. Under such controlled conditions, up to 80% of the heat generated in the reaction may be registered as a temperature change in the sample stream.

The sensitivity of thermistor biosensors is quite low for a majority of applications. Although greater sensitivity is possible using more exothermic reactions, the low sensitivity of system can be increased by increasing the heat output by the reaction. In the simplest case, this can be achieved by linking together several reactions in a reaction pathway, all of which contribute to heat output. Thus the sensitivity of glucose analysis using glucose oxidase can be more than double by using catalyst within the column reactor to utilize the hydrogen peroxide product.

(b) Potentiometric Biosensor

In order to transduce the biological response into an electrical signal, they make use of ion-selective electrodes. This consists of a membrane containing immobilized enzyme, surrounding the probe from a pH meter, where the catalyzed reaction generator absorbs hydrogen ions. The reaction occurring next to this sensing glass electrode causes a change in pH which may be directly read from the pH meter display.

(c) Amperometric Biosensor

These biosensors function by the production of current when a potential is applied between two electrodes (Fig. 5.9). The most common amperometric biosensor is glucose biosensor which is based on a Clark oxygen electrode. In this case, the platinum cathode is held next to an oxygen permeable membrane on which glucose oxidase enzyme is immobilized. This membrane is also permeable to substrate glucose.

$$Ag \text{ anode} \quad 4 Ag^+ + 4 Cl^- \longrightarrow 4AgCl + 4 e^-$$

$$Pt \text{ cathode} \quad O_2 + 4 H^+ + 4 e^- \longrightarrow 2H_2O$$

When a potential of $- 0.6$ V is applied to the platinum cathode relative to Ag/AgCl electrode, a current proportional to oxygen concentration is generated. As enzyme glucose oxidase changes glucose to gluconic acid by consuming O_2 and produces H_2O_2, the glucose biosensor records the reduced O_2 concentration and this decreased amount of O_2 can be calibrated with the amount of glucose present.

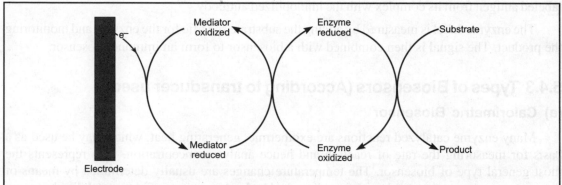

Fig. 5.9. An Amperometric biosensor arrangement. A mediator is used to transfer electrons from an electrode to an enzyme-catalyzed redox reaction.

In this case, measurement of hydrogen peroxide present which is directly proportional to the glucose concentration can be second approach. This can be done by applying a potential of 0.68 to the Pt anode relative to Ag/AgCl electrode.

$$Ag \text{ anode} \quad H_2O_2 \longrightarrow O_2 + 2H^+ + 2 e^-$$

$$Pt \text{ cathode} \quad 2AgCl + 2e^- \longrightarrow 2Ag^+ + 2Cl^-$$

A major disadvantage of this biosensor is its dependence on the dissolved oxygen concentration.

(d) Optical Biosensor

There are two main areas of development of optical biosensors. These involve determining changes in light absorption between the reactants and products of a reaction or measuring the light output, the luminescent process. The former usually involves the widely established use of colorimetric strips. These are single use disposable cellulose pads impregnated with enzymes and reagents. The most common use of this technology is for whole blood monitoring in diabetes control. The most promising biosensor involving luminescence uses firefly luciferase to detect the bacteria in food or clinical samples. Bacteria are specifically lysed, ATP released reacts with luciferin and oxygen to produce light.

$$\text{ATP + D-Luciferin + O}_2 \xrightarrow{\text{(Luciferase)}} \text{Oxyluciferin + AMP + PPi + CO}_2 \text{ + Light (562 nm)}$$

The light produced may be photometrically detected with the use of high voltage photomultiplier tubes or photodiode systems of low voltage (Fig. 5.10).

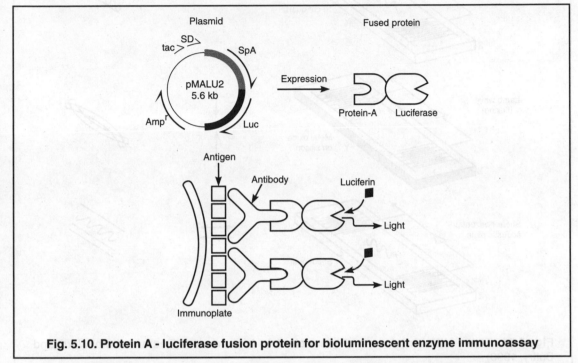

Fig. 5.10. Protein A - luciferase fusion protein for bioluminescent enzyme immunoassay

The most important field application of its sensor type is, thus, cell fluorescence. The advantages offered by optical biosensor are that they are not susceptible to disturbances caused by electric fields and are suitable for continuous indication. The sample also remains chemically unchanged during measurement.

The disadvantage is that they may be operated only in the dark as day light disturbs the measuring procedure.

(e) Piezoelectric or Acoustic Biosensor

Piezoelectric materials due to their ability to generate and transmit acoustic waves in a frequency dependent manner can be used as sensor transducer. The optimal resonant frequency for acoustic wave transmission is highly dependent on the physical dimensions and properties of the piezoelectric crystal. Changes in the mass of material at the surface of the crystal will cause quantifiable changes in the resonant frequency of that crystal.

There are three classes of mass balance acoustic transducer (Ward and Buttry, 1990). Some of these transmit an acoustic wave from one crystal face to another *i.e.*, bulkwave (BW) devices. There are others that transmit the acoustic wave along a single crystal face from one location to another *i.e.*, surface acoustic wave (SAW) devices (Fig. 5.11).

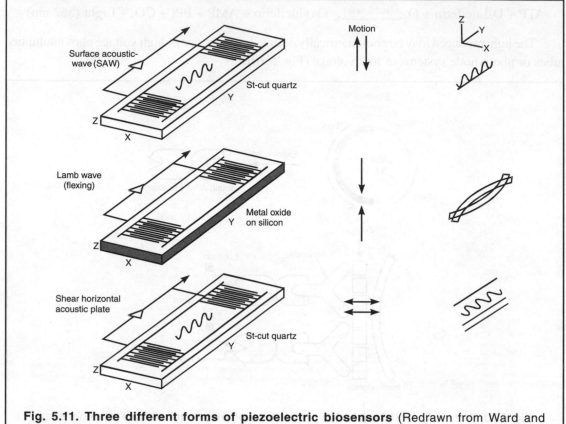

Fig. 5.11. Three different forms of piezoelectric biosensors (Redrawn from Ward and Buttry, 1990)

The operation of these devices in the gas phase is well established, but is not fully understood in liquid media. Piezoelectric technology has been available for several years and is relatively cheap. When the problems of non-specific binding and poor sensitivity are solved, then this technology may provide an option for transduction in affinity biosensors.

5.4.4 Applications of Biosensors

Biosensors have been hailed as the solution to many analytical problems, from toxin detection in industrial effluent to drug monitoring *in vivo*. The extent and diversity of activity in the research laboratories, indicated by the patents and research papers published each year, suggest that so called "biosensor revolution" is already here. There are two broad areas of applications in which biosensors seem particularly suited.

A. Medical Applications

The feature of biosensors fit very well into various applications within the field of medical diagnostics. Medical diagnostics can be conveniently divided into *in vivo* and *in vitro* applications.

In vivo **diagnostics :** This area of medical diagnostics has received a great deal of attention in recent years. For example, the idea of continuous *in vivo* monitoring of blood glucose, coupled with an implanted "artificial pancreas" has been posed as potential cure for diabetes. Other *in vivo* monitoring concepts have included therapeutic drugs, as well as a variety of endogenous components, as markers for various disease states.

In vitro **diagnostics :** There are three main application areas in *in vitro* diagnostics .

(*i*) *Centralized testing :* It is carried out in a hospital or other specialized medical laboratories.

(*ii*) *Decentralized testing :* It is carried out in the doctor's office, accident or emergency ward operation theatre, ambulance or sports medicine clinic.

(*iii*) *Consumer testing :* These are tests purchased and used by the general public.

B. Non-Medical Applications

Biosensors have applications in many non-medical fields. Indeed, a great deal of effort has been directed towards selected areas of analysis within a number of industries. These include the food, environmental monitoring, agriculture and related industries, as well as defense and general industrial processing. Some biosensor devices, however, do not fall clearly into any of these groups.

The Food Industry : The food industry is a major producing industry which is fragmented into a number of different companies, products and services worldwide. Although the industry is very large, when compared with the medical product industry, profit margins are low.

Biosensors can be used for the quality control of raw materials and in monitoring and control of manufacturing processes, either on-line and off-line *e.g.,* monitoring product's shelf-life.

The most viable opening in the food industry can arise where a biosensor can fulfill a common need for analysis. The examples include rapid detection of total microbial contamination, presence of toxins and sugar quantification in drink products besides improvements in production and quality control. These two specific applications offer the best market opportunities for biosensors in the food industry.

Environmental Monitoring : Biosensors are useful for monitoring air, land and water. There is an intense interest in monitoring the environment and biosensors can be used in the form of small portable analyzers for use in the field. Biosensors are useful for monitoring pollutants, chemical residues, pesticides, toxins or microorganisms in sea water, waste water, rivers, reservoirs and supplies.

Defense : The military has long had an interest in portable sensors for personal, tank or field unit use. The sensor could be used by the infantry and other front line units to detect and identify agents of chemical warfare, such as mustard and nerve gases.

Agriculture and Related Industries : The number of possible applications of biosensors in agriculture is enormous. In analysis, such as the detection of viral, bacterial and fungal diseases of crops and livestock, plant nutrients, fertilizers and pesticides can be envisaged.

Conclusions

The markets for biosensor based products have been slow to evolve, mainly because the problems in developing biosensors have been underestimated. In addition to an improved understanding of fundamental biomolecular interactions, one of the key factors in the success of biosensor is improved manufacturing processes to allow the production of reproducible and quality products.

5.5 BIOPROCESS TECHNOLOGY

Bioprocess technology includes the following stages (Fig. 5.12):

Stage I   : **Upstream processing :** It involves preparation of liquid medium, separation of particulate and inhibitory chemicals from the medium, sterilization, air purification *etc.*

Stage II : **Fermentation :** It involves the conversion of substrates to desired product with the help of biological agents such as microorganisms.

Stage III : **Downstream processing :** It involves separation of cells from the fermentation broth, purification and concentration of desired product and waste disposal or recycle.

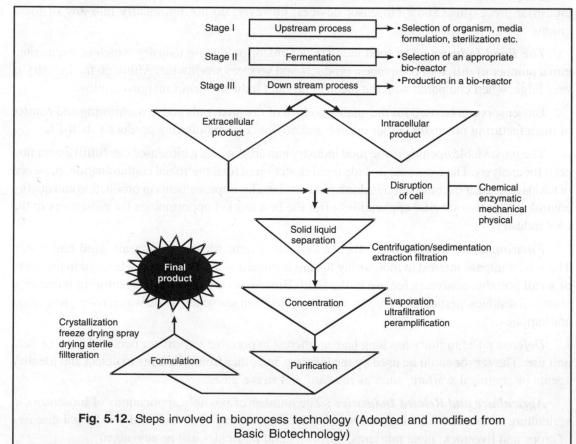

Fig. 5.12. Steps involved in bioprocess technology (Adopted and modified from Basic Biotechnology)

Depending on the type of product, the concentration levels it was to produce and the purity desired, the fermentation stage might constitute anywhere between 5-50% of the total fixed and operating costs of the process. Therefore, optimal design and operation of bioreactor frequently dominates the overall technological and economic performance of the process.

5.5.1 Upstream Process

5.5.1.1 Selection of Microorganisms

Screening: It includes selective procedures that allow the detection and isolation of microorganisms of interest from among a large microbial population.

(*i*) **Primary screening**: It allows the detection and isolation of microorganisms that possess potentially interesting industrial applications.

(*ii*) **Secondary screening**: It further tests the capabilities of and gain information about these organisms. It is conducted on agar plates, in flasks or small fermenters containing liquid media or as a combination of these approaches. For initial secondary screening, the use of agar plates, although not as sensitive as liquid culture, is advantageous.

The secondary screening can be qualitative or quantitative in its approach. The secondary screening helps us to:

(*a*) have organisms which should yield the type of information that is needed in order to evaluate the true potential of a microorganism for industrial usage

(*b*) determine whether the microorganisms are actually producing new chemical compounds not previously described or alternatively for fermentation processes that are already known, secondary screening should determine whether a more economical process is possible

(*c*) detect optimal requirements (pH, aeration *etc*.) associated with particular microorganisms, both for growth of the organism and for the formation of products

(*d*) to know the gross animal, plant or human toxicity attributed to the fermentation product

(*e*) reveal whether a product resulting from a microbial fermentation occurs in the culture broth in more than one chemical form, and whether it is an optically or biologically active material

(*f*) reveal whether microorganisms are able to chemically alter or even destroy their own fermentation products.

5.5.1.2 Media Formulation

For an individual fermentation process, detailed investigation is needed to establish the most suitable medium.

(i) The media should produce maximum yield of product per gram of substrate used.

(ii) It should be available throughout the year

(iii) It should allow maximum rate of product formation.

(iv) It should cause minimal problems during sterilization of media.

The media, however, must meet certain basic requirements. It is relatively simple to devise medium containing pure compounds for small scale fermentation, but for large scale process this is not necessary.

The elemental requirement of cell biomass and metabolic production must be satisfied by the constituents of a medium. There must be an adequate supply of energy for biosynthesis and cell maintenance.

The main constituents of media include :

(a) Water

(b) Energy source

(c) Carbon source

(d) Nitrogen source

(e) Vitamin source

(f) Buffers

(g) Precursors and Metabolite regulators.

5.5.1.3 Sterilization

If fermentation is attacked by an unwanted foreign microbe, then it would lead to loss of productivity. The containment may also outgrow the production organism or it may degrade the desired product. Thus, contamination has to be avoided. It may be achieved through the removal or destruction of any organism that will adversely affect the process or products. It is difficult, however, to remove selectively those organisms that may adversely affect specific process. For this reason, medium sterilization is carried out to remove or destroy all microorganisms present in the material.

The choice of sterilization method depends on several factors. These include :

(i) effectiveness in achieving an acceptable level of sterility,

(ii) reliability, effect (positive or negative) on medium quality and cost, including operating and capital expense to achieve sterilization.

Sterilization can be carried out by air sterilization using thermal destruction by dry heat, moist heat, pasteurization, filtration etc. Besides these, chemicals and ionization radiation can be used for sterilization.

This can be achieved by:

(i) using a pure inoculum

(ii) sterilization of the medium

(iii) sterilization of the fermenter

(iv) sterilization of the air used for aeration

Medium Sterilization

The media can be sterilized by many ways, but because of practical reasons two of them are universally used.

(a) Steam sterilization

(b) Filter sterilization

Steam sterilization: Steam sterilization can be achieved in two ways:

 (i) **Batch sterilization process:** The batch sterilization of the medium can be done in the fermentation vessel or in a separate vessel. The highest temperature which appears to be feasible for batch sterilization is 121°C.

(ii) **Continuous sterilization process:** There are two types of continuous sterilizers which may be used for the treatment of media:

    (a) the indirect heat exchanger and

    (b) direct heat exchanger.

In the double spiral type indirect heat exchangers, there are two sheets of stainless steel curved around an axis to form a double spiral. Steam is passed through one spiral and medium through the other in counter current streams. Incoming unsterile medium is used as a cooling agent.

In the direct heat exchangers or steam injectors, steam is directly injected into the unsterile broth.

Filter sterilization of medium: The major exception is the use of filteration for the sterilization of media for animal cell culture because it contains heat-labile components.

Two types of filters are available:

(a) fixed pore filters or absolute filters and

(b) non-fixed pore filters or depth filters.

Usually filter sterilization of medium is done by absolute filters which consist of membrane cartridges that are filled into stainless steel modules.

Sterilization of the Fermenter

Fermenter can be sterilized by heating the jackets of the fermenter with steam and sparging steam into the vessel through all entry points. Steam pressure is held at 15 psi for 20 minutes in the vessel.

Air Sterilization : Air sterilization can be carried out by the following agents :

(*a*) Heat

(*b*) Ultraviolet rays and other electromagnetic waves

(*c*) Germicidal sprays

(*d*) Filtration

For air sterilization, filtration is commonly used. The filtration devices used include depth type filters such as granulated carbon beds and vessels containing packed fibre glass or packed cotton. These filtration devices will provide a fair statistical probability of organism retention as long as they are properly maintained and care is taken to control the quality and velocity of the inlet air. Such air filters, however, fail to perform with high efficiency, if the influent air is not dried or if the air velocity is not optimum. Because organism retention is dependent on the depth of the filter medium used, these filters have high pressure drops which restrict air flow to the fermenter and increase compressor costs. In addition, depth filters trap moisture and organic contaminants and provide an environment conducive to microbial growth.

Membrane air or gas filter cartridges typically contain a pleated or hydrophobic filter medium having a fixed submicron pore structure or uniform size distribution. Use of a hydrophobic filter medium prevents transmembrane pressure drop increase due to wetting of filter. This also ensures reliable organism retention by excluding moisture which can decrease filtration efficiency. The submicron structure allows quantitative microbial retention and the pleated filter medium results in a high filter area in small unit volume which can be easily handled for servicing and installation.

5.5.2 Fermentation

Fermentation (latin verb perverse to boil) by using the metabolic activity of microorganisms, leads to complex transformation upon organic material. Originally, fermentation referred to the bubbling observed when sugar and starch materials underwent a transformation to yield alcoholic beverages. This term was later on applied to the process in which alcohol was formed from sugar, regardless of whether the causative agent is biological or non-biological. Pasteur, however, considered fermentation to apply to those anaerobic reactions through which microorganisms obtained energy for growth in the absence of oxygen. Fermentation today has a much broader meaning. It applies to both the aerobic and the anaerobic metabolic activities of microorganisms in which specific chemical changes are brought about in an organic substrate.

It concerns itself with the isolation and description of microorganisms from natural environment, such as soil or water and with the culture conditions required for obtaining rapid and massive growth of these organisms in the laboratory and in large scale culture vessels known as **fermenters** or **bioreactors** (Fig. 5.13).

5.5.2.1 Bioreactors

A fermenter is a bioreactor. It is used to provide optimum conditions for the controlled growth of microorganisms by regulating agitation, temperature and aeration of the fermentation. Broth parameters such as pH and dissolved oxygen concentration are needed to be measured to monitor and control fermentation process. For a successful fermentation, the maintenance of aseptic operating conditions is important. The fermenter vessels are constructed of stainless steel and are jacketed and pressure coded so that the equipment can be safely steam-sterilized *in situ*.

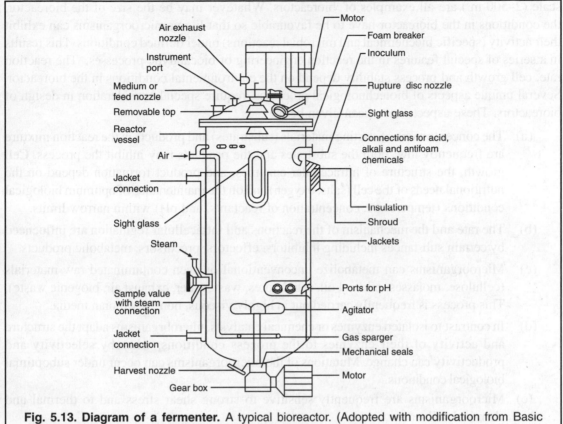

Fig. 5.13. Diagram of a fermenter. A typical bioreactor. (Adopted with modification from Basic Biotechnology by Colin Retledge and Bjorn Kristiansen, 2001).

Fermenters can be operated on a batch, semi-continuous, or continuous basis. Fermentation can occur in :

   (*i*) static or agitated cultures,

   (*ii*) the presence or absence of oxygen, and

   (*iii*) aqueous or low moisture conditions

The biocatalysts (microorganisms, enzymes, cells) can be free or attached to surfaces by immobilization. In general, fermentation industries require a bioreactor that can meet a number of different operating conditions including varying consistency, aeration rate, agitation intensity and fermentation volume.

All the biotechnology-based production processes depend upon bioreactors or fermenters. These processes include production of vaccines, proteins, organic acids, amino acids, antibiotics, or carrying out of microbial transformation, bioremediation and biodegradation and microbial inoculants for use as biofertilizers.

At every step of the development of a biotechnological process, bioreactor is invariably used. The sizes of the bioreactor can vary over several orders of magnitudes. The microbial cell (few μm^3), shake flask (100-1000 ml), laboratory fermenter (1 – 50 L), pilot scale (0.3 – 10 m^3) to plant scale (2–500 m³) are all examples of bioreactors. Whatever may be the size of the bioreactor, the conditions in the bioreactor have to be favourable so that living microorganisms can exhibit their activity (specific biochemical and microbial reactions) under defined conditions. This results in a series of special features in the reaction engineering of biocatalytic processes. The reaction rate, cell growth and process stability depend on the environmental conditions in the bioreactor. Several unique aspects of biotechnological processes, require special consideration in design of bioreactors. These aspects have been given below :

(a) The concentrations of starting materials (substrates) and products in the reaction mixture are frequently low; both the substrates and the products may inhibit the process. Cell growth, the structure of intracellular enzymes, and product formation depend on the nutritional needs of the cell (salts, oxygen) and on the maintenance of optimum biological conditions (temperature, concentration of reactants, and pH) within narrow limits.

(b) The rate and the mechanism of the reactions and intracellular regulation are influenced by certain substances including inhibitors, effectors, precursors, metabolic products.

(c) Microorganisms can metabolize unconventional or even contaminated raw materials (cellulose, molasses, mineral oil, starch, ores, wastewater, exhaust air, biogenic waste). This process is frequently carried out in highly viscous, non-Newtonian media.

(d) In contrast to isolated enzymes or chemical catalysts, microorganisms adapt the structure and activity of their enzymes to the process conditions, whereby selectivity and productivity can change. Mutations of the microorganisms can occur under suboptimal biological conditions.

(e) Microorganisms are frequently sensitive to strong shear stress and to thermal and chemical influences.

(f) Reactions generally occur in gas-liquid-solid systems, the liquid phase usually being aqueous.

(g) As biochemical conversion progresses, the microbial mass can increase. During the reaction, there may be a growth on the walls, flocculation or autolysis of microorganisms.

(h) Continuous bioreactors often exhibit complicated dynamic behaviour.

5.5.2.2 Requirements of Bioreactors

Due to above mentioned demands made by biological systems on their environment, there is no universal bioreactor. However, the general requirements of the bioreactor are as follows:

(a) The design and construction of biochemical reactors must preclude foreign contamination (sterility). Furthermore, during fermentation, monoseptic conditions should be maintained and ensure containment.

(b) Optimal mixing with low, uniform shear

(c) Adequate mass transfer (oxygen)

(d) Clearly defined flow conditions

(e) Feeding of substrate with prevention of under or over-dosing

(f) Suspension of solids

(g) Gentle heat transfer

(h) Compliance with design requirements such as ability to be sterilized, simple construction, simple measuring control, regulating techniques, scale up, flexibility, long-term stability, compatibility with up-down stream processes; antifoaming measures.

Features of Bioreactor Design

Irrespective of the specific bioreactor configuration used, the vessel must be provided with certain common features:

(i) The reactor vessel is provided with a vertical sight glass and side ports for pH, temperature and dissolved O_2 sensors as minimum requirements.

(ii) Retractable sensors that can be replaced during operation are increasingly used.

(iii) Connections for acid and alkali (for pH control), antifoam agents and inoculum are located above the liquid level in the reactor vessel.

(iv) For the introduction of air (or other gases, such as CO_2 or ammonia for pH control) a sparger is provided near the bottom of the vessel.

(v) The agitator shaft has a steam-sterilizable single or double mechanical seals.

(vi) Double seals are preferred. They, however, require lubrication with cooled, clean steam condensate. Alternatively, when torque limitations allow, magnetically coupled agitators may be used. This eliminates the mechanical seals.

(vii) Bioreactor is so designed that the following conditions are met with :

(a) **Control of foaming:** Foam is inevitably caused by aeration and agitation. This foam is controlled with a combination of chemical antifoam agents and mechanical foam breakers. Foam breakers are used only when the presence of antifoam in the product is not acceptable or if the antifoam interferes with downstream processing operations such as membrane-based separations or chromatography. The shaft of the high-speed mechanical foam breaker must also be sealed using double mechanical seals.

(b) **Maintenance of sterilization:** Although the sterilization temperature generally does not exceed 121°C, the vessel is designed for a higher temperature, typically 150-180°C. The vessel is designed to withstand full vacuum, or it could collapse, while cooling after sterilization. Using saturated clean steam at a minimum absolute pressure of 212 kPa, the reactor can be sterilized in place.

(c) **Protection against overpressure:** A rupture disc located on top of the bioreactor provides overpressure protection. Usually this is a graphite burst disc because it does not crack or develop pinholes without falling completely. The rupture disc is piped to a contained drain.

(d) **Supply of media etc.:** The head plate of the vessel has nozzles for media or feed addition and for sensors (e.g., the foam electrode), and instruments (e.g., the pressure gauge).

(viii) Few internals: There should be very few internals in the vessel

The design should take into account the needs of clean-in-place and sterilization in place procedures:

(a) There should be no crevices and stagnant areas in the vessel, because these may provide pockets for the accumulation of liquids and solids, *i.e.,* leading to chances of contamination.

(b) The design of minor items such as the gasket grooves is important. It is preferred to use the channels with rounded edges which can be easily cleaned.

(c) The joints should be welded and couplings should be avoided.

(d) Complete displacement of all air pockets in the vessel and associated pipework should be allowed by steam connection.

(e) The design of the exterior of a bioprocess plant should be such that it has smooth contours and minimum bare threads.

(*ix*) The reactor vessel is invariably jacketed. Normally, the design of jacket is with the same specifications as the vessel. The jacket should be covered with a chloride-free-fibreglass insulation that is fully enclosed in protective sheet. A cover valve located on the jacket or its associated piping provides a relief over pressure protection to the jacket. Most of the bioreactors including jacket are made up of stainless steel.

5.5.2.3 Bioreactor Systems

A microbial fermentation can be viewed as a **three-phase system.** It involves liquid-solid, gas-solid, and gas-liquid reactions.

The liquid phase: It contains dissolved nutrients, dissolved substrates and dissolved metabolites.

The solid phase: It consists of individual cells, pellets, insoluble substrates, or precipitated metabolic products.

The gaseous phase: It provides a reservoir for oxygen supply and for removal of CO_2.

5.5.2.4 Agitation (Stirring and Mixing)

A suitable mixing device brings about the transfer of energy, nutrients, substrate and metabolite within the bioreactor. The efficiency of any one nutrient may be crucial to the efficiency of the whole fermentation.

The stirring of a bioreactor brings about the following :

- Dispersion of air in the nutrient solution
- Homogenization to equalize the temperature and concentration of nutrients throughout the fermentor
- Suspension of microorganisms and solid nutrients
- Dispersion of immiscible liquids

The relative velocity between the nutrient solution and the individual cell should be about 0.5 m/sec. According to the way they behave when stirred, nutrient solutions can be subdivided into two groups:

(i) **Viscous** solutions with Newtonian and non-Newtonian properties

(ii) **Viscoelastic** solutions, in which normal liquid-state properties are not observed in stirred vessels.

The **viscosity**, the ability of a material to resist deformation, is the most significant property that affects the flow behaviour of a fluid. Such behaviour has a marked effect on pumping, mixing, heat transfer, mass transfer and aeration.

This second group is represented by only a few examples. These include polysaccharides and certain antibiotic fermentations. Most fermentation solutions fall into the first category. Un-inoculated solutions and bacterial cultures often behave as simple Newtonian liquids.

With many mycelial organisms, changes occur during the fermentation not only in the amount of mycelium, but in the characteristics of the nutrient solution. During metabolism, substrates are taken up. The proportion of undissolved substrates is, thus, reduced. At the same time, metabolites are excreted. It, thus, affects the viscosity of the solution.

Agitators: The agitator is used for the mechanical stirring inside a fermenter vessel. Agitators can be classified as disc turbines, vaned discs, open turbines of variable pitch and marine propellers (Fig. 5.14).

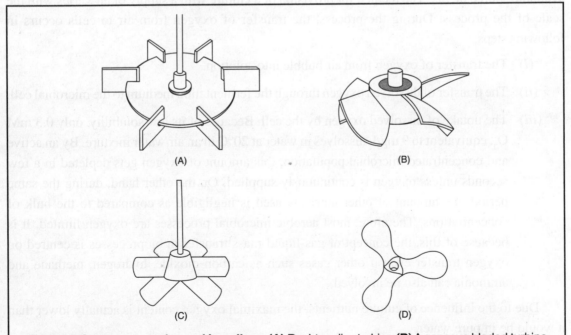

Fig. 5.14. Some commonly used impellers: (A) Rushton disc turbine **(B)** A concave bladed turbine **(C)** A hydrofoil impeller and **(D)** A marine propeller. (Redrawn & modified from Basic Biotechnology, by Colin Retledge and Bjorn Kristiansen, 2001)

The disc turbine consists of a disc with a series of rectangular vanes set in a vertical plane around the circumference and vaned disc has a series of rectangular vanes attached vertically to the underside.

The vanes of a variable pitch open turbine and the blades of a marine propeller are attached directly to a boss on the agitator shaft.

5.5.2.5 Aeration

One of the most critical factors in the operation of fermenter is the provision of adequate gas exchange. The majority of fermentation processes is aerobic and therefore, requires the provision of oxygen. It is the most important gaseous substrate for microbial metabolism, and carbon dioxide is the most important gaseous metabolic produce.

The oxygen demand of an industrial fermentation process is normally satisfied by aerating and agitating the fermentation broth. In many aerobic fermentation systems, the rate of oxygen transfer to the cells is the limiting factor which determines the rate of biological conversion. The availability of oxygen for microbial use depends upon the solubility and mass transfer rate of oxygen in the fermentation broth and the rate of microbial utilization of the dissolved oxygen.

Oxygen is normally supplied to microbial cultures in the form of air, as this is the cheapest available source of gas. The method for provision of a culture with a supply of air varies with the scale of the process. During the process, the transfer of oxygen from air to cells occurs in following steps:

(*i*)   The transfer of oxygen from air bubble into solution.

(*ii*) The transfer of dissolved oxygen through the fermentation medium to the microbial cell.

(*iii*) The uptake of dissolved oxygen by the cell. Because of its low solubility, only 0.3 mM O_2, equivalent to 9 mg/l, dissolves in water at 20°C in an air/water mixture. By an active and concentrated microbial population, this amount of oxygen gets depleted in a few seconds unless oxygen is continuously supplied. On the other hand, during the same period, the amount of other nutrients used is negligible as compared to the bulk of concentrations. Therefore, most aerobic microbial processes are oxygen limited. It is because of this, the concept of gas-liquid mass transfer in bioprocesses is centred on oxygen transfer even if other gases such as carbon dioxide, hydrogen, methane and ammonia can also be involved.

Due to the influence of culture nutrients, the maximal oxygen content is actually lower than it would be in pure water.

The solubility of gases follows **Henry's law** in the gas pressure range over which fermenters are operated. This means that if the oxygen concentration in the gas phase increases, there is an increase in the O_2 proportion of the nutrient solution. Consequently, during aeration with pure oxygen, the highest O_2 partial pressures are attained. Compared to the value in air (9 mg O_2/l), 43 mg O_2/l dissolves in water when pure oxygen is considered.

Several independent partial resistances must be overcome for oxygen to be transferred from a gas bubble to an individual cell (Fig. 5.15).

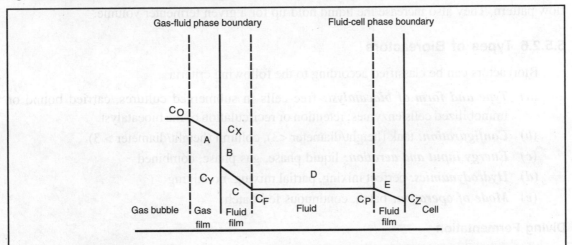

Fig. 5.15: Resistances for Oxygen Transfer from Air Bubble to the Microbial Cell (A) Resistance within the gas film to the phase boundary **(B)** Penetration of the phase boundary between gas bubble and liquid **(C)** Transfer from the phase boundary to the liquid **(D)** Movement within the nutrient solution **(E)** Transfer to the surface of the cell.

For fermentations carried out with single celled organisms such as bacteria and yeasts, the resistance in the phase boundary between the gas bubble and the liquid is the most important factor controlling the rate of transfer.

Microbial cells near gas bubbles may absorb oxygen directly through the phase boundary and there is an increase in the rate of gas transfer to such cells.

The O_2 transfer within the agglomerate can become the limiting factor in cell agglomerates or pellets.

The mass transfer of oxygen into liquid can be characterized by the **oxygen transfer rate (OTR)** or by the **volumetric oxygen transfer coefficient (K_La)**. These values have been thoroughly examined as a critical parameter for bioreactor function. The oxygen transfer rate and the volumetric oxygen transfer coefficient are dependent on the following parameters:

- Vessel geometry: diameter, capacity *etc.*
- Mixing properties: power, impeller configuration and size, baffles
- Aeration system: sparger rate, geometry, location
- Nutrient solution: composition, density, viscosity

- Microorganism: morphology, concentration
- Antifoam agent
- Temperature

The baffles serve to disrupt the vortex pattern that develops around a single-shaft impeller rotating in an unconstrained fluid. The baffles produce a large planar liquid surface and a uniform flow pattern. They also increase the liquid hold-up for a given fermenter volume.

5.5.2.6 Types of Bioreactors

Bioreactors can be classified according to the following criteria

(a) *Type and form of biocatalyst:* free cells in submerged cultures, carried bound or immobilized cells/enzymes; retention or recirculation of the biocatalyst

(b) *Configuration:* tank (height/diameter <3), column (height/diameter > 3)

(c) *Energy input and aeration:* liquid phase, gas phase, combined

(d) *Hydrodynamics:* perfect mixing, partial mixing, no mixing

(e) *Mode of operation:* batch, continuous fed-batch.

Diving Fermentation

(i) The microorganisms or cells are first cultured in the smallest bioreactors

(ii) Then the contents of this reactor are transferred to a larger, pre-sterilized, medium filled reactor

(iii) It is finally done in the production fermenter, the largest reactor.

Commercial processing is generally carried out as submerged culture with the biocatalyst suspended in a nutrient medium in a suitable reactor. Following types of bioreactors are commonly used:

1. Stirred Tank Bioreactors (STB)

Stirred tank reactor is the choice for many (more than 70%) processes, though it is not the best. STBs have the following functions:

(a) Homogenization

(b) Suspension of solids

(c) Dispersion of gas-liquid mixtures

(d) Aeration of liquid

(e) Heat exchange

In the stirred tank, air is introduced under the agitator via a sparger. It can be readily adapted for multi-purpose use, and is suitable for low-viscosity broths. To improve the mixing efficiency, hydrodynamically shaped agitators are provided (Fig. 5.16).

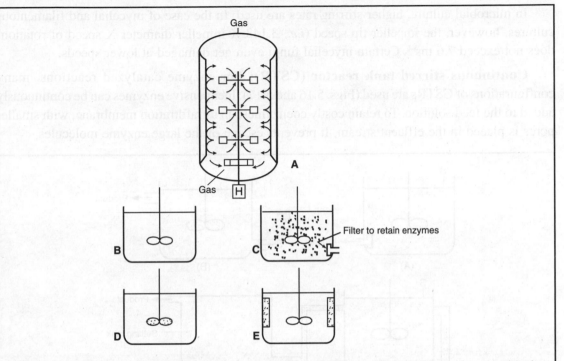

Fig. 5.16. Different types of stirred tank bio-reactors. **(A)** Generalized structure of a stirred tank reactor **(B)** Stirred tank for soluble enzymes (To immobilization) **(C)** Stirred tank for immobilized enzymes **(D)** Stirred tank with immobilized enzyme on paddles **(E)** Stirred tank with immobilized enzymes on basket baffles (Modified from "Handbook of Enzyme Biotechnology". 1985. A. Wiseman, ed., Chichester, U.K. : Ellis Harwood, Ltd.)

It consists of a cylindrical vessel with motor-driven central shaft. It supports one or more agitators. The shaft may enter through the top or the bottom of the reactor vessel.

For microbial culture, the vessels have four baffles. They project into the vessel from the walls. This prevents swirling and vortexing of the fluid. The width of baffle is 1/10 or 1/12 of the diameter of tank diameter. The aspect ratio (*i.e.*, height-to-diameter ratio) of the vessel is 3-5.

In animal cell culture applications, aspect ratios do not normally exceed 2. The vessels for animal cell culture generally do not have baffle (especially small-scale reactors) to reduce turbulence that may damage the cells. However, there may be single, large diameter, low-shear impeller such as a marine propeller. A perforated pipe ring sparger with a ring diameter is also present. It is slightly smaller than that of the impeller. Gas is sparged into the reactor liquid below the bottom impeller.

The number of impellers depends on the aspect ratio.

The bottom impeller is located at a distance about 1/3 of the tank diameter above the bottom of the tank. Additional impellers may be presented. These are spaced approximately 1.2 impeller diameter distance apart. The diameter of impeller is about 1/3 of the vessel. For highly viscous mycelial broths, bulk mixing is done by using larger hydrofoil impellers with diameters of 0.5 to 0.6 times the tank diameter.

In microbial culture, higher stirring rates are used. In the case of mycelial and filamentous cultures, however, the impeller tip speed (*i.e.*, 3.143 X impeller diameter X speed of rotation) does not exceed 7.6 ms^{-1}. Certain mycelial fungi even get damaged at lower speeds.

Continuous stirred tank reactor (CSTR): For enzyme catalyzed reactions, many configurations of CSTRs are used (Figs. 5.16 and 5.17). Inexpensive enzymes can be continuously added to the feed solution. To retain costly coenzymes, an ultrafiltration membrane, with smaller pores is placed in the effluent stream. It prevents escape of the large enzyme molecules.

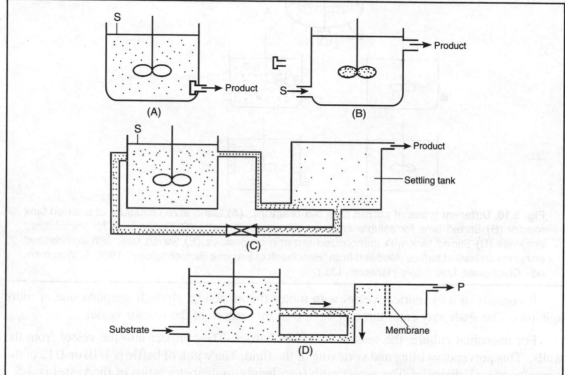

Fig. 5.17. Different types of continuous flow stirred tank reactors.(A) Stirred tank reactor with filtration recovery **(B)** Stirred tank reactor with immobilized enzyme on paddles **(C)** Stirred tank reactor with settling tank recovery **(D)** Stirred tank reactor with ultrafiltration recovery. (Adopted from "Handbook of Enzyme Biotechnology". 1985)

If enzymes are immobilized on insoluble particles, then a screen in the effluent line will be sufficient. The enzymes can also be attached to the agitator shaft using screen baskets. The same effect can be achieved by circulating reaction mixture from a well-mixed reservoir through a short-packed column of immobilized enzyme.

Reasonably good mass transfer rates and mixing are provided by the STBs. The cost of operation is lower and the reactors can be used with a variety of microbial species. The mixing concepts are well developed because stirred tank reactor is commonly used in chemical industry.

2. Airlift Bioreactors

To agitate the fluid by means of draft tubes or external recycle, air is used in air lift fermenters. It eliminates the need for expensive mechanical mixture (Fig. 5.18).

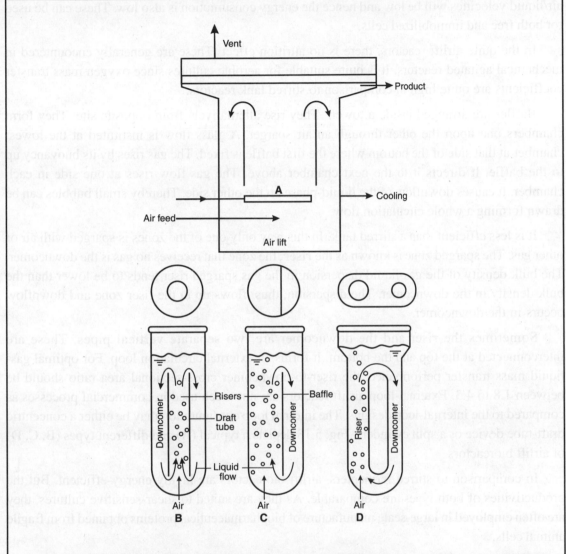

Fig. 5.18. Air-lift fermenter (A) Typical air lift bioreactor (B) Internal loop configuration (C) Draft tube (D) Split cylinder device (adopted from Basic Biotechnology, 2002).

(a) A rapid, low-shear movement is created by aeration.

(b) The liquid movement is initiated by injection of air at the foot of the riser column. This has pilot scale application in culture of plant and animal cells. The shear effects of the stirring gear are avoided.

These are generally classified as pneumatic reactors without any mechanical stirring arrangements for mixing. Adequate mixing of the liquid is ensured by turbulence caused by the fluid flow. In the central section of the reactor the draft tube is provided. The introduction of the fluid (air/liquid) causes upward motion and results in circulatory flow in the entire reactor. The air/liquid velocities will be low and hence the energy consumption is also low. These can be used for both free and immobilized cells.

In the quite airlift reactors, there is no attrition effect. These are generally encountered in mechanical agitated reactors. It is quite suitable for aerobic cultures since oxygen mass transfer coefficients are quite high in comparison to stirred tank reactors.

Baffles are arranged inside a tower. They rise alternatively from opposite site. They form chambers one upon the other through an air sparger. A glass flow is instituted at the lowest chamber at that side of the bottom where the first baffle is fixed. The gas rises by its buoyancy up to the baffle. It directs it to the next chamber above. The gas flow rises at one side in each chamber. It causes downflow of the liquid-phase on the other side. Thereby small bubbles can be drawn forming a whole circulation flow.

It is less efficient than a stirred tank. In this way only one of the zones is sparged with air or other gas. The sparged zone is known as the **riser**; the zone that receives no gas is the downcomer. The bulk density of the gas-liquid dispersion in the gas sparged riser tends to be lower than the bulk density in the downcomer. The dispersion, thus, flows up in the riser zone and downflow occurs in the downcomer.

Sometimes the riser and the downcomer are two separate vertical pipes. These are interconnected at the top and the bottom. It forms an external circulation loop. For optimal gas-liquid mass transfer performance, the riser-to-downcomer cross-sectional area ratio should be between 1.8 to 4.3. External-loop airlift reactors are not much used in commercial processes as compared to the internal-loop designs. The internal loop configuration may be either a concentric draft-tube device or a split cylinder. Fig. 5.18 depicts a typical (A) and different types (B, C, D) of airlift bioreactors.

In comparison to stirred fermenters, airlift bioreactors are highly energy-efficient. But the productivities of both types are comparable. As they are suited to shear-sensitive cultures, they are often employed in large-scale manufacture of biopharmaceutical proteins obtained from fragile animal cells.

Heat and mass transfer capabilities of airlift reactors are atleast as good as those of other systems, and airlift reactors are more effective in suspending solids than are bubble columns.

The performance characteristics of airlift bioreactors are determined by the rate of gas injection and the resulting rate of liquid circulation. In general, the rate of liquid circulation increases with the square root of the height of the airlift device. As a result, the reactors are designed with high aspect ratios.

3. Fluidized Bed Reactors

Fluidized bed bioreactors (FBB) (Fig. 5.19) have received increased attention in the recent years. It is because of their advantages over other types of reactors. Most of the FBBs developed for biological systems involve cells as biocatalysts. These are three phase (solid, liquid and gas) systems. The FBBs are generally operated in co-current upflow with liquid as continuous phase. Usually fluidization is obtained either by external liquid re-circulation or by gas fed to the reactor. In the case of immobilized enzymes, the usual situation is of two-phase systems involving solid and liquid, but the use of aerobic biocatalyst necessitates introduction of gas (air) as the third phase.

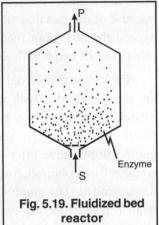

Fig. 5.19. Fluidized bed reactor

A differentiation between the three phase fluidized bed and the airlift bioreactor would be made on the basis that the latter have a physical internal arrangement (draft tube). It provides aerating and non-aerating zones. The circulatory motion of the liquid is induced due to the draft tube. Basically the particles used in FBBs can be of three different types:

(i) Inert core on which the biomass is created by cell attachment.

(ii) Porous particles in which the biocatalyst is entrapped.

(iii) Cell aggregates/flocs (self-immobilization).

In comparison to conventional mechanically stirred reactors, FBBs provide a much lower attrition of solid particles. The biocatalyst concentration can significantly be higher and washout limitations of free cell systems can be overcome.

In comparison to packed bed reactors, FBBs can be operated with smaller size particles. Its use is limited because of the following drawbacks :

(i) Clogging

(ii) High liquid pressure drop

(iii) Channeling

(iv) Bed compaction.

The smaller particle size facilitates higher mass transfer rates and better mixing. The volumetric productivity attained in FBBs is usually higher than in stirred tank and packed bed bioreactors.

4. Packed Bed Columns

A packed bed is constituted by a bed of solid particles, usually with confining walls. The biocatalyst is supported on or within the matrix of solids that may be porous or a homogenous non-porous gel. The solids may be particles of compressible polymeric or more rigid material.

There is a continuous flow of a fluid containing nutrients through the bed to provide the need of the immobilized biocatalyst. Metabolites and products are released into the fluid and removed in the outflow.

The flow may be upward or downward, but normally there is a downflow under gravity. If the fluid flows up the bed, the maximum flow velocity is limited. It is because of the fact that the velocity cannot exceed the minimum fluidization velocity otherwise the bed will fluidize.

Packed bed or fixed bed bioreactors are commonly used with attached biofilms especially in wastewater engineering (Fig. 5.20). The use of packed bed reactors gained importance after the potential of whole cell immobilization technique has been

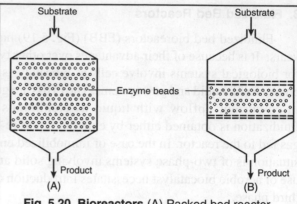

Fig. 5.20. Bioreactors (A) Packed bed reactor and (B) Flat bed Reactor. (Modified from Food Biotechnology Technomic publication. 1992).

demonstrated. The immobilized biocatalyst is packed in the column and fed with nutrients either from top or from bottom. One of the disadvantages of packed beds is the changed flow characteristic due to alterations in the bed porosity during operation. When the soft gels like alginates, carrageenan *etc.* are used, the bed compaction which generally occurs during fermentation results in high pressure drop across the bed. In many cases, the bed compaction is so severe that the gel integrity is severely hampered. In addition channeling may occur due to turbulence in the bed. Though packed beds belong to the class of plug flow reactors in which back mixing is absent, in many of the packed beds slight amount of back mixing occurs which changes the characteristics of fermentation. Packed beds are generally used where the rate of reaction is governed by substrate inhibition. To reduce the pressure drop across the length of the reactor, inclined bed, horizontal bed, rotary horizontal reactors several modifications such as tapered beds and flat beds have been tried, but the success had been limited.

The depth of the bed is limited by several factors. These include:

(*i*) the density and the compressibility of the solids

(*ii*) the need to maintain a certain level of a critical nutrients, such as O_2 through the entire depth

(*iii*) the flow rate that is needed for a given pressure drop.

With the increase in the depth of bed, there is a decrease in the gravity-driven flow rate through the bed as the fluid moves down the bed nutrients. In this way substrates are depleted. On the other hand, there is an increase in the concentration of metabolites and products. The environment of a packed bed is non-homogeneous. The concentration variations along the depth can, however, be decreased by increasing the flow rate.

If the reaction consumes or produces H^+ or OH^-, gradients of pH may occur. The control of pH by addition of acid and alkali is quite difficult. Beds with greater void volume permit greater flow velocities through them. The concentration of the biocatalyst in a given bed volume, however, declines as the voidage (void volume) is increased.

If the packing *i.e.,* the biocatalyst-supporting solid is compressible, its weight may compress the bed unless the packing height is kept low.

Because of a reduced voidage, flow is difficult through a compressed bed. Packed beds are extensively used as immobilized enzyme reactors. Such reactors are very useful for product inhibited reactions. The product concentration varies from a low value at the inlet of the bed to a high value at the exit. In this way, only a part of the biocatalyst is exposed to high inhibitory levels of the product.

5. Deep Shaft Bioreactor

It is based on air-lift principle, but by using hydrostatic pressure, large aeration is provided. There is looping of a long thin vessel far below the sparger. As the medium flows downward, to regions of great hydrostatic pressure, air is compressed, and the vent is kept higher than the inlet of feed. The net effect on air is an expansion, which causes the circulation in the broth (Fig. 5.21).

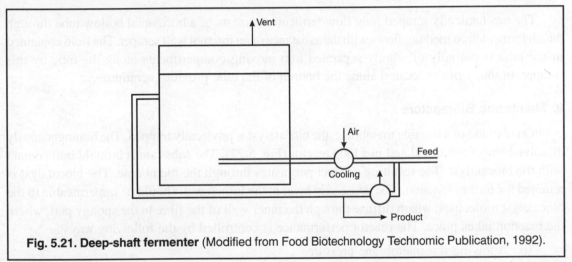

Fig. 5.21. Deep-shaft fermenter (Modified from Food Biotechnology Technomic Publication, 1992).

6. Roto-Fermenter

It is made up of

   (i)   a fermentation vessel

   (ii) a rotating microporous membrane

   (iii) a filter chamber.

At the bottom of the fermenter, fresh medium and air are injected. The cells grow in the annular space around the rotating membrane. They can be retained inside the fermenter. The rotating microporous membrane allows to continuously remove the cell free filtrate. It, thus, performs the dual function of cell growth and concentration with the simultaneous removal of metabolic products. The roto-fermenter can, therefore, serves both as a fermenter and a cell separator.

7. Solid-State Fermenter

To culture on solid nutrients such as bran through which air can diffuse, solid-state fermentation is used *e.g.,* in compositing of organic wastes such as garbage to produce fertilizer.

The substrates used in solid-state fermenters are cereals, legume bran, wood straw and other animal and plant materials. The design is controlled by the characteristics of heat and mass transfer. The limiting factors are inter-particle and intra-particle diffusions. By proper mixing and aeration, satisfactory inter-particle oxygen transfer is achieved. CO_2 build-up in the void spaces should be avoided. Intra-particle mass transfer is the transfer of nutrients and enzymes within the fermentation substrates and concerned with the role of enzymes in hydrolyzing the water-insoluble polymers into soluble substrate. Microbial heat generation is much greater per unit volume than for liquid fermentation. By increasing the aeration rate, heat can be removed.

8. Scraped Tubular Fermenter

The mechanically scraped plug flow fermenter consists of a horizontal hollow tube through which fermentation medium flows with the assistance of an internal wall scraper. The fluid contained in the tube is partially or wholly separated into moving compartments along the tube by this scraper. In this, orifices located along the bottom of the tube provides aeration.

9. Membrane Bioreactors

On one side of a suitable membrane, the biocatalyst is physically trapped. The homogeneously dissolved enzyme is circulated in a loop reactor (Fig. 5.22). The substrate is brought into contact with the biocatalyst. The resulting product permeates through the membrane. The biocatalyst is retained for further reaction. The dense skin layer at the lumen wall should be impermeable to the biocatalyst molecules, which diffuse through the inner wall of the fibre to the spongy part, where the reaction takes place. The reactor performance is controlled by the following ways :

(i) Applying transmembrane pressure

(ii) Axial flow rate control.

Design of continuous flow reactors where reaction could be achieved without biocatalyst loss in the effluent stream has been permitted by the availability of improved immobilization technique. Biocatalysts are more protected and less exposed to denaturation. A sterile filter is used to introduce the homogeneously soluble enzyme to the reactor. It is easily kept sterile. If there is deactivation of enzyme by the supply of additional biocatalysts, it is possible to maintain a constant activity in the reactor. This reactor has the following characteristics:

(*i*)   Continuous operation

(*ii*) No back-mixing

(*iii*) Can be used as a multi-enzyme reactor

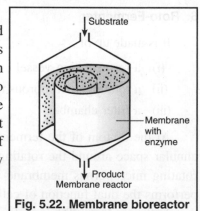

Fig. 5.22. Membrane bioreactor
(Adopted from Biotechnology :
Technomic Publication)

(iv) Can be used for very diluted solution *e.g.,* for the treatment of wastewater

(v) Low residence time of substrates and products.

The successful operation of reactor requires good control of concentration polarization and minimization of fouling effect. The operating parameters such as recirculation rate, temperature and transmembrane pressure have impact on the system.

10. Hollow Fibre Bioreactor

The hollow fibre membrane forms the basis of this bioreactor (Fig. 5.23).

(a) A bundle of thousands of these hollow fibres is formed and contained in a housing.

(b) Cells in hollow-fibre bioreactors are cultured on the exterior surface or extra capillary space (ECS) of the fibres.

(c) Medium is circulated through the inner lumen or intra-capillary space (ICS).

(d) Nutrients and oxygen are supplied by diffusion of medium across the porous membranes from the ICS to the ECS.

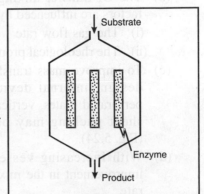

Fig. 5.23. Hollow fibre reactor (Adopted from food biotechnology Technomic Publication, 1992)

(e) The waste products may, however, exchange in the opposite direction, away from the cells.

(f) The flow of medium through the hollow fibres in the conventional systems is unidirectional.

(g) Low molecular weight species such as glucose, vitamins and amino acids are readily transported across the membrane and nourish the cells.

11. Other Types of Bioreactors

Many variations of the above-mentioned reactors have been developed for various specified functions *e.g.,* for producing yeast based flavour enhancers, Provesta Corp used 25,000 L continuous fermenter employing high-cell-density technology. This allowed high oxygen and heat transfer rates. It resulted in high cell density and increased yeast production. To produce very high oxygen transfer rate, this design utilizes:

(i) high gas velocity

(ii) high operating pressure.

It has the following features:

(i) An agitator at the bottom of the fermenter provides a high shear rate.

(ii) The foam level in the fermenter is controlled by mechanical foam brakes at the top of the vessel.

(iii) An efficient heat exchanger system is used.

Bubble Column Reactors

In these, usually the column is cylindrical with an aspect ratio of 4-6 (height/diameter).

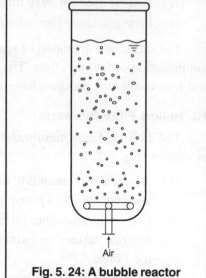

(a) At the base of the column, gas is sparged through perforated pipes, perforated plates or sintered glass or metal microporous spargers.

(b) The O_2 transfer mixing and other performance factors are influenced by the following factors:

    (i) The gas flow rate

    (ii) The rheological properties of the fluid.

(c) To improve mass transfer and modify the basic design, internal devices such as horizontal perforated plates, vertical baffles and corrugated sheet packing may be placed in the vessel (Fig. 5.24).

Fig. 5. 24: A bubble reactor (adopted from Basic Biotechnology).

(d) With increasing vessel diameter, there is an improvement in the mixing for a given gas flow rate.

(e) With an increase in gas flow rate, mass and heat transfer, there is an increase in prevailing shear rate.

5.5.2.6 Selection of Bioreactor

While selecting fermenter, the following factors are to be kept in mind :

(i) Flexibility

(ii) Minimum cost.

(iii) Thermal control

(iv) Product selectivity

(v) Biological constrains of the organism

(vi) Production scale

(vii) Level of technology available

5.5.2.7 Photobioreactor

Important chemicals, such as astaxanthin and β-carotene are provided by some micro-algae and cyanobacteria. Cyanobacteria, such as *Spirulina,* are also grown as human health-foods. Sunlight or artificial illumination are required for photosynthetic culture. For heterotrophic culture of certain algae, instead of sunlight organic energy source such as glucose is used. Such cultures, however, lack photosynthetic pigments and, thus, are different in their productive behaviour. Attempts are being made to develop heterotrophic strains for production of commercially valuable products.

Two types of photobioreactors are being used for monoculture of such organisms:

    (a) Open photobioreactors   (b) Closed photobioreactors

(a) **Open bioreactors.** Open ponds and raceways are often used to culture micro-algae for that large scale production. Photosynthesis can occur only at relatively shallow depths because it needs light. Algal ponds are typically no deeper than 0.15m. Photoinhibition is caused by too much light. In such cases, a slight reduction in light intensity can accelerate the rate of photosynthesis. Light penetration is decreased by increasing cell population and also from the self-shading effect of cells. Photosynthesising algal cells also need a source of carbon, usually carbon dioxide. Dissolved bicarbonate species may provide a part of it. Cells convert CO_2 to carbohydrate and other cellular components. The photosynthetic productivity can be reduced by high concentration of CO_2.

(b) **Closed photobioreactor.** The monocultures consists of arrays of transparent tubes. These may be made of glass, or more commonly, a clear plastic. The tubes may be laid horizontally, or arranged as long rungs on an upright ladder (Fig. 5.25). A continuous single run tubular loop configuration is also used or may be wound helically around a vertical cylindrical support. Sometimes in addition to the tubes, flat or inclined thin panels may be used. In fact a solar receiver is constituted by an array of tubes on a flat panel in relatively small-scale operations. A variety of methods are used to circulate the culture through the solar receiver. These include centrifugal pumps, positive displacement mono pumps, Archimedean crews and airlift devices. Airlift pumps perform well. They have no mechanical parts and can be easily operated. These suited to shear sensitive applications.

To aid periodic movement of cells from the deeper poorly lit interior to the regions nearer the walls. The flow in a solar receiver tube or panel should be turbulent enough. Generally, a minimum Reynolds number value of 10^4 is good. Turbulence is needed to improve radial mixing but too much of it can be harmful. The velocity everywhere should be sufficient to prevent sedimentation of cells. Typical linear velocity through receiver tubes tend to be 0.3-0.5 ms^{-1}.

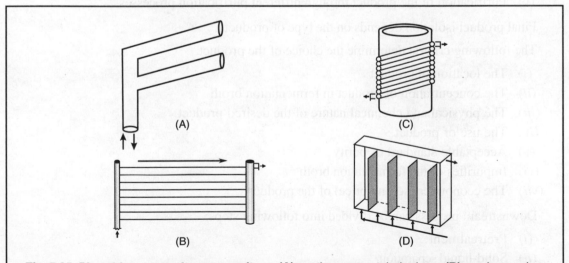

Fig. 5.25. Photobioreactors for monoculture. (A) continuous run tubular loop; **(B)** a solar receiver made of multiple parallel tubes **(C)** helical wound tubular loop **(D)** flat panel configuration. Configuration (A) and (B) may be mounted vertically or parallel to the ground (Modified from Microbial Biotechnology, Lee Yuan Kunn, World Scientific, NJ 2006).

Because of the need to maintain adequate sunlight penetration, a tubular solar receiver cannot be scaled up by simply increasing the true diameter. The diameter should not exceed 6 cm, although this constraint may be relaxed somewhat by deploying specially designed static mixers that improve radial mixing inside the tube. The performance of culture can be improved by reducing the tube diameter from 6 to 5 cm and further to 2 cm without using the mixer. Light penetration depends on biomass density, cellular morphology and pigmentation, and absorption characteristics of the cell-free culture medium.

5.5.3 Downstream Processing

It involves the separation of biomolecules produced through fermentation technology. As soon as the fermentation is over, it is desirable to get a high quality of product with minimum cost. The overall yield decreases with the increase in number of involved steps. It also increases the capital investment.

The choice of downstream process depends on trial and error method. There is a need to optimize the process for the particular product. In general, the downstream process involves the following steps :

- (*i*) The separation of solids (cell mass) from the liquid. This is **solid-liquid** separation,
- (*ii*) If the products are extracellular in nature, the primary isolation of the product from the supernatant is required
- (*iii*) If the products are intracellular, there is a need to disintegrate the cells
- (*iv*) Primary isolation step is the concentration. In this, maximum amount of water is removed. This is also called **"volume reduction step"**.
- (*v*) Purification of the product through different purification processes.

Final product isolation depends on the type of products.

The following criteria determine the choice of the product.

- (*i*) The location of product
- (*ii*) The concentration of product in fermentation broth
- (*iii*) The physical and chemical nature of the desired product
- (*iv*) The use of product
- (*v*) Acceptable standard of purity
- (*vi*) Impurities of the fermentation broth
- (*vii*) The economics (selling price) of the product

Downstream process can be divided into following steps:

- (*i*) Pretreatment
- (*ii*) Solid-liquid separation
- (*iii*) Concentration
- (*iv*) Purification
- (*v*) Formulation

5.5.3.1 Pretreatment

Pretreatment is given by heating the harvested fermenter broth. The proteins during this process are coagulated and these are easy to filter. This simple method improves the handling characteristics. Generally, to enhance the rate of filtration, filter aids are added. Different types of filter aids like Kiselghur (diatomaceous earth) *etc.* are used. These improve the porosity of the resultant filter cake. It leads to faster flow rate. The compressibility character of the cake is decreased by filter and it also reduces the penetration of small particles such as bacteria or fragmented mycelia into the precoat. It also pasteurizes the whole broth. The addition of electrolytes to the fermentation broth may promote coagulation and flocculation in initial solution. The electrolytes may be :

(i) simple (*e.g.,* ferric chloride and alum, acids and base and synthetic polyelectrolytes),

(ii) cationic, anionic and non-ionic such as polyacrylamide, polyethylamines and polyamine derivatives.

The electrolytes reduce the electrostatic repulsion in colloidals. The particles get coagulated and hence easier to filter. The pH is changed by acids and bases and hence reduces the charge.

5.5.3.2 Solid-Liquid Separation

It is a primary recovery operation. It is used for the following purposes :

(*i*) For the separation of whole cells from the culture broth

(*ii*) Removal of cell debris

(*iii*) Collection of protein precipitate

(*iv*) Collection of inclusion bodies *etc.*

Many methods are used for the separation of cells. Centrifugation and filtration are the commonly used unit operations .

Cells can be separated or concentrated by following methods :

(*a*) **Sedimentation or flocculation :** It is useful when the density difference between cells and suspension medium is large.

(*b*) **Centrifugation-decanter, solid-injecting and nozzle separators :** These are used when the density difference is not too small.

(*c*) **Filtration for diminishing density differences**

(*d*) **Floatation**

(*e*) **Coagulation**

Yeasts and bacteria are usually homogeneously suspended in the fermentation broth. Some bacteria may form slime layers depending on the strain and fermentation conditions and this will lead to separation problems.

Filamentous fungi are frequently characterized by a network of intertwined filaments. They produce viscous fermentation broths that may be difficult to dehydrate. These fungi under certain conditions will form agglomerates called **'pellets',** which due to their large size (100-4000 mm), are relatively easy to recover.

The initial recovery phase of the processing includes the gross separation of components such as cells and broth.

The recovery of intracellular products involves the following steps :

(*i*)   Concentration of cells to a 50% solution,

(*ii*) Rupture the cells to release the product,

(*iii*) Partial purification to yield an intermediate product

Extracellular products are small molecules or macromolecules that are transported through the cell membrane. Products are concentrated, usually from 0.1 to 100g/L of broth, because they are often present in low concentration.

5.5.3.2.1 Flocculation

To ensure efficient solid-liquid separation, pretreatment or conditioning of the broth is required. It changes the following:

(*i*)   The biomass particle size

(*ii*) The fermentation liquor viscosity,

(*iii*) The interactions between the biomass particles

This indicates the use of a filter aid (body-feed) in the broth and/or for precoating the filter medium. Filter aids are incompressible, discrete particles of high permeability with size ranging from 2-20 mm. Filter aids should be inert to broth being treated. Most frequently Diatomite (skeletal remains of aquatic plants), Perlite (volcanic rock processed to give expanded structure) and inactive carbon are used.

Flocculating agents such as polycations, either cellulosic or based on synthetic polymers, inorganic salts or mineral hydrocolloids are added to cause agglomeration of individual cells or cell-particles into large flocks. These can be easily separated at low centrifugal forces. The flocculation is affected by the following factors :

(*i*)   Flocculating agent itself,

(*ii*) Physiological stage of the cells,

(*iii*) The ionic environment,

(*iv*) Temperature and nature of the organism

5.5.3.2.2 Centrifugation

The removal of solids in centrifugation is based on the density difference between particles to be separated and the surrounding medium.

Generally at a laboratory scale, batch centrifugation is applied. Although these provide high centrifugal forces, the processing capacity is low. The use of such centrifuges is limited on an industrial scale. At industrial scale normally continuous flow centrifuges are used. There is a continuous feeding of the slurry and collection of the clarified solution in these. The deposited solids can either be continuously or intermittently removed from the centrifuge.

Microbial cells and other discrete large particles are removed. The supernatant obtained by centrifugation is not free of cells (10^3-10^5 cells ml^{-1}) and costs of maintenance and power consumption are quite high. It is quite difficult to separate particulate debris from cell homogenate by centrifugation.

5.5.3.2.3 Filtration

A filter medium constitutes the separating agent. It retains the particles according to size and allows the passage of liquid through the filter. In cake filtration, the particles are retained as a cake on the filter medium.

The flow through the filter layers is dependent on the area of the filter and flow resistance provided by the filter medium and the cake. If the particles do not penetrate the filter medium, the flow resistance of the latter is not changed. As the cake-layer gets thicker, the resistance is increased.

The filter cakes obtained in many cases of biomass separation are compressible. The changing effective pressure difference, thus, influences the flow through the filter. Some of the filter media used are the perforated sintered metals, cloths, synthetic fibres, cellulose, glass wool, ceramics and synthetic membranes.

Many types of filtration equipments are available. Because of a contained operation, the filter press is formed of a sequence of perforated plates. These alternate with hollow frames mounted on suitable supports.

(i) The plates are covered with a filter medium (cloth). This results in the formation of a series of cloth-walled chambers.

(ii) The slurry can be forced into these under pressure. The solids are retained within the chambers.

(iii) The filtrate discharges into hollows on the plate surface and hence to drain points.

(iv) The hydraulic pressure is released at the end of the filtration cycle.

(v) The cake is manually removed from the cloth.

(vi) Filter sizes range from 120 cm^2 from laboratory scale to 14400 cm^2 (per plate) for a production scale equipment.

The cake chamber in membrane filter process is covered with a rubber membrane. It is inflated using air or water. It allows *in situ* compacting of the cake. In this way the yield is increased and the cake is dry. The use of microporous membrane operated under pressure has provided a viable alternative to centrifugation as a means to remove suspended particles from process fluids.

5.5.3.2.4 Floatation

Floatation involves the following steps:

(*i*) Particles are adsorbed on gas bubbles

(*ii*) They are trapped in a foam layer and can be collected.

The gas may either be sparged into the particulate feed or very fine bubbles can be generated from the dissolved gases by releasing the overpressure or by electrolysis.

In some cases, floatation is a feasible alternative for cell separation. This process has been used in waste water treatment and for the recovery of single cell proteins (SCPs).

5.5.3.2.5 Coagulation

The effluent stream with microbial cells is withdrawn from the broth and pH is lowered. The acidified effluent stream is heated at 45-90°C before passing through a separator.

The formation of stable foam is supported by the presence of 'collector substances' such as long-chain fatty acids or amines.

The decanter consists of a rotating bowl. It is supported at both the ends and an internally arranged screw conveyer rotates at a slightly lower speed in the same direction as the bowl. Solid-ejecting separators are disc-bowl machines. These eject the solids intermittently during operation. The concentrate in the nozzle separators is continuously discharged through nozzles. This takes place under pressure at the top of the machine.

5.5.3.3 Release of Intracellular Components

Attempts are to be made to liberate maximum amount of product in an active state. The presence of inactivating effects of shear, temperature and proteases is, however, to be kept in mind. The choice of disruption method has to be empirically made at the same time.

Cell Disruption

It is required to release the intracellular enzymes or other biologically active compounds, and to facilitate their recovery. It is easier to disrupt mammalian cells/tissues, because they don't have protective cell wall. This involves the mechanical, physical or chemical breakage of cell wall or membrane. The majority of proteins currently commercially available are produced by animal cell culture. The proteins may be intracellular or secreted into the medium. Generally, if the protein is intracellular, its subsequent purification needs more exhaustive downstream processing procedures than those required for a protein secreted directly into the culture medium.

1. Disruption of Microbial Cells

The disruption of microbial cells is complex and requires a more violent method. Typically following cell disruption techniques are used (Table 5.5):

(*i*) Breaking the cells by mechanical forces

(*ii*) Preferentially damaging the cell wall *e.g.*, by drying

(*iii*) Enzymatic lysis or lysing primarily the membranes *e.g.*, by treatment with chemicals.

Table 5.5: Methods used to Disrupt Microbial Cells

Mechanical	Physical	Chemical
Liquid shear/high pressure	Ultrasonic/sonication	Alkalies
Homogenizer	Osmotic shocks	Enzymes
Solid shear/beads	Freezing/thawing	Detergents

A. Mechanical disruption of cells

It is the most common means of releasing intracellular products both at the laboratory and industrial scales.

(*a*) **Homogenization** : Industrial scale disruption of microbial cells is achieved by high-pressure homogenization. The cell suspension is forced at high pressure through an orifice of narrow internal diameter to emerge at atmospheric pressure. The sudden release of pressure creates a liquid shear capable of disrupting the cells.

(*b*) **Solid shear/beads** : A technique of microfluidization has also been introduced for cell disintegration. Here the cell suspension is fed under higher pressure through a chamber where it is split into two streams to be directed at each other at high velocity before emerging at atmospheric pressure. The cells are disrupted by agitation with glass in bead mills by a combination of high shear and impact with the cells. The size of beads varies from 0.2-0.5mm for bacteria to 0.4-0.7 mm for yeast.

(*c*) **Grinding with abrasives** : In this case cell pastes are grinded in a mortar with an abrasive powder. This includes glass, alumina or Kieselguhr. This technique using machines was originally developed for wet grinding and dispension of pigments in the printing and paint industries. To release proteins from a wide variety of microorganisms, a typical product, the Dynamill can be used.

B. Non-mechanical disruption of cells (Physical and Chemical Methods)

It is possible by physical, chemical or enzymatic means. Drying is widely applied method of cell lysing. This causes changes in the structure of the cell wall and makes it possible to subsequently extract the cell membrane. It is particularly effective for animal cells that lack a cell wall. The presence of cell wall in the case of microbial cells may necessitate application of additional measures.

(*a*) **Induction of lysis** : This involves non-mechanical means for the hydrolysis of the cell wall. By physical stresses, such as a sudden depression, osmotic shock or rupture with ice crystals, cell membrane can be ruptured. To damage the lipoprotein of cell membranes, certain lytic agents such as cationic and anionic detergents are used. Such chemicals may destroy the intracellular materials, thereby, limiting their use. For the release of hydrolytic enzymes and binding proteins from a periplasmic space of a number of Gram negative bacteria including *Salmonella typhimurium* and *E. coli*, osmotic shock has been used.

The technique involves suspending the cells in buffer solution to free them from growth medium and then suspending in 20% buffered sucrose. The cells, after being allowed to equilibrate, are harvested and rapidly suspended in water at about 4°C. After the first purification step, only about 4-8% of total bacterial protein is released by osmotic shock in 90% pure state. The cytoplasmic protein on the other hand was only 10% pure after some purification step.

(b) **Ultrasonics :** Liquid processors of about 500 W power, working on ultrasonics are commonly used for research, analytical and light industrial applications. The sonicator may be consisting of a tapped titanium disruptor and a cool-running 20 KHz ultrasonic converter. Ultrasonication disrupts the cells by cavitation, and is commonly used at laboratory scale. On a larger scale, removal of the heat generated is difficult. Heating of sample is prevented by alternative on and off cycles.

(c) **Cell breakage by heat shock (thermolysis) :** It is quite easy and cheap. It however, is useful for heat-stable products only. Treatment with organic solvents and detergents to render cells more permeable is often used in the laboratory scale. Frequently toluene is used. It acts by dissolving membrane phospholipids and creating pores in the cell membrane. It is possible to selectively liberate enzymes from the periplasmic space by treatment with water-miscible solvents such as methanol. It, however, requires spark-proof equipment and special precautions with regard to fire safety.

(d) **Mini bomb cell disruptor :** The sudden depressurization of cell suspension disrupts the cells.

   (i) The mini-bomb is pressurized with air, nitrogen or carbon dioxide.

  (ii) The pressure is released after the cells have equilibrated.

 (iii) A short treatment at moderate pressure will rupture the outer cell membranes leaving the nuclei intact. After separating these fractions, a second treatment can be used to rupture the nuclei.

 (iv) It results in a uniform homogenate of whole cells and sub-cellular components.

(e) **Use of enzymes for cell lysis:** It is advantageous as it provides selectivity during the release of product. It can be used under mild conditions. The lytic enzyme lysozyme is extensively used. This enzyme damages the peptidoglycan layer, thereby, the internal osmotic pressure of the cells bursts the periplasmic membrane. Gram-positive bacteria are more susceptible to attack by lysozyme. The lysis of Gram-negative bacteria on the other hand requires passage of lysozyme through the outer membrane. It is aided by the addition of EDTA. For the degradation of yeast cell walls, glucanase and mannanase, in combination with proteases, are used. Application of lytic enzymes is limited by their availability, costs and the need for removal in subsequent purification steps.

A combination of enzymatic/chemical lysis with mechanical disintegration has been suggested in enhancing the efficiencies of the respective methods. It saves time and energy and facilitates subsequent processing.

For the large scale disruption, it is preferred to use mechanical methods that are simple, effective and readily scalable. In most of the large-scale installations, the homogenizer is used.

The bead mill is used in about 10% and the remaining use a variety of chemical and physical methods. Disruption by ultrasonics produces heat and product activity may be decreased by overheating. Other physical methods require relatively more time and energy.

Microbial enzymes can digest away the outer wall of microbial and plant cells, thus allowing easier disruption of the cell membrane. However, these enzymes are relatively expensive. The chemical treatment involves the main problems of cost, possible toxicity and recovery of chemicals.

The choice of method for large-scale cell disruption is determined by the following factors :

(*a*)   Characteristics of the microbes *i.e.,* susceptibility of cells to breakage and ease of extraction from cell

(*b*)   Cost

(*c*)   Speed

(*d*)   Sensitivity of the product to heat and shear

(*e*)   Location of the desired product within the cell.

It is required to optimize operating parameters such as capacity, pressure, temperature and residence time *etc.*

2. Homogenization of Animal/Plant Tissue

It is normally easy to break the animal cells because they do not have cell walls. In order to homogenize the animal cells the following steps are followed:

(*i*)   Animal tissue is cut into small pieces

(*ii*)   There are suspended in ice-cold homogenization buffer

(*iii*)   They are grinded in a blender

Plant cells have tough cell walls. They are more resistant.

(*i*)   The extract of a fleshy or non-fibrous plant tissue is prepared by rapid homogenization of the material.

(*ii*)   It is suspended in a partially frozen suitable buffer in a Waring blender pre-cooled at −20° C.

The more fibrous material, which is difficult to macerate, is frozen and ground to a dry powder before adding the extraction buffer for homogenization. Concentrated buffers with pH values around 6.5-7.2 are used.

(*i*)   In order to neutralize the acidic materials including the phenols, and also certain phenol scavengers such as polyvinylpyrrolidone (PVP) and/or Amerlite, a hydrophobic polystyrene-based adsorbent is used.

(*ii*)   Reducing agents such as ascorbate and thiols are used to prevent the accumulation of quinines. This in turn prevents the inactivation of enzymes during extraction.

Cellulases and pectinases are also used for the digestion of plant cell walls to release the protoplasmic material and not the vacuole contents. It, however, has limitation in the large scale use because of high enzyme concentration.

5.5.3.4 Concentration of Biological Products

Once the cells are separated from the whole broth, 85-98% of the filtrate contains water. The product forms only a minor constituent. The removal of water is very costly. The following processes are used to have a water free product :

(i)   Evaporation

(ii) Membrane filtration

(iii) Liquid-liquid extraction

(iv) Precipitation

(v)   Adsorption

The technique is chosen normally on the basis of the nature of the product. Care is taken that there is minimum loss of product activity.

5.5.3.4.1 Evaporation

It is a simple process. In many cases, however, the process consumes energy. It is quite reliable and simple. It is often applied on a large scale. Steam is the source of heat. In biotechnology, the evaporator must often serves as a multi-purpose equipment. It should be able to

(i)   handle a broad range of product viscosity (1-10000 mPa)

(ii) heat sensitive products

(iii) give a minimum of scale formation, fouling and foaming

The basic unit of evaporators comprises the following:

(i)   A heating section to which the steam is fed

(ii) A section where the concentrate and vapours are separated

(iii) A condenser for condensing the vapour

(iv) The required vacuum and product pumps, control equipment *etc*.

Several different types of equipments have been developed. They range from laboratory scale (0.5-1.01 h^{-1} water evaporation capacity) to very large industrial scale (150 m^3 h^{-1}). To decrease the steam consumption, multiple-effect (stage) evaporators have been designed. In these the liquid flows through a number of stages (maximum seven) using the vapour of one stage as a heating source for the next.

(a) **The falling film evaporators.** In these the liquid to be concentrated flows down long tubes. It is uniformly distributed over the heating surface as a thin film. The vapours flowing in the same direction increase the linear velocity of the liquid. This improves the heat transfer. Residence times in the evaporator are in the order of minutes. These evaporators are used for concentrating viscous products (up to 200 mPa). These are frequently used in fermentation industry.

(b) **Plate evaporators.** In these the heating surface is a plate as opposed to a tube. These are used for higher viscosities. These have a relatively large evaporation area in a small volume, but the possibility to treat viscous and solid-containing fluids is limited.

(c) **Forced film evaporators.** These are used for higher viscosities. These have mechanicaly driven liquid film. In some cases it produces a dry product. The resistance time ranges from a few seconds to a few minutes.

(d) **Centrifugal force-film evaporators.** These permit a further reduction in the residence time so that even the heat labile substances can be concentrated under gentle conditions. Evaporation takes place on a heated conical surface or plates over which the transport of the liquid takes place through the centrifugal force produced by the rotating bowl.

5.5.3.4.2 Liquid-Liquid Extraction

Liquid-liquid extraction is applied on a large scale in biotechnology. It is used both for concentration and purification. In this, there is a transfer of solute from one liquid phase to another. The efficiency of an extraction process is governed by the distribution of substances between two phases. These are defined by the partition coefficient kL (=concentration of substance in extract phase/concentration of substance in raffinate phase). The demands on a liquid-liquid extraction process are influenced by the physio-chemical properties of the product.

(a) Extraction of low molecular weight products

For extracting small lipophilic target molecules, an organic solvent is used. It is difficult to design an efficient extraction process for hydrophilic compounds. Any one of the following methods can be used for extraction in organic solvent :

Physical extraction: According to its physical preference, the compound distributes itself between the two phases. By screening for the solvents that would lead to a high k value and also show a maximal difference in k for the different components present in the crude mixture, the extraction is optimized for non-ionizing compounds.

Dissociative extraction: To achieve separation, difference in the dissociation constant of the ionizable components is exploited. To overcome an adverse ratio of partition coefficients, these differences are quite large. This technique is used to extract penicillin and some other antibiotics.

Reactive extraction: For this purpose, a carrier, such as an aliphatic amine or a phosphorous compound, is added to the organic solvent. It forms selective solvation bonds or stoichiometric complexes. These are also insoluble in the aqueous phase. Thus, the compound is carried from the aqueous to the organic phase. This type of extraction is useful for compounds that have a high solubility in aqueous medium, such as organic acids.

The formation of emulsions at the interface is avoided in most cases. It is done by removing cells and other particulates prior to extraction. Once the extraction is completed, the product is recovered from the solvent. It is done by distilling of the following :

(*i*) The product in case of a high-boiling solvent, or

(*ii*) The solvent when this is low boiling

If the product is heat sensitive, it is recovered by back-extraction into a new aqueous phase under conditions different from the first extraction *e.g.*, penicillin is extracted into butyl acetate or amyl acetate from the fermentation medium at pH 2.5-3.0 and back-extracted into aqueous phosphate buffer at pH 5-7.5.

When it is desired to have a high extraction yields, multi-step extraction in a countercurrent mode is used. This saves both the solvent and time. Different kinds of extraction equipments are available.

Supercritical fluid (SCF) extraction: Sometimes the conventional extraction causes toxicity and flammability of organic solvents. In such cases, supercritical fluid (SCF) extraction is used. SCFs are materials that exist as fluids above their critical temperature and pressure. Many properties of SCFs are intermediate between those of gases and liquids. Their diffusivity is higher than those of liquids, but viscosity is lower. SCFs are preferred extractants because their solvent properties are highly suitable for extraction proteins.

Aqueous two-phase systems (ATPS) : These are prepared by mixing two different polymers, or a polymer and a salt above certain concentrations. The water is the major component (80-95%). The two phases formed are enriched in the respective phase components:

(a) *Top phase* : Its main component is polyethylene glycol (PEG).

(b) *The bottom phase* : It is composed of dextran or salt. The interfacial tension between the two phases is significantly lower than that in water-organic solvent systems. The separation of phase is also slower. It varies between a few minutes and 1-2 hours. It is generally speeded up by the centrifugation at low g values. Partitioning of a component in aqueous two-phase systems is based on the following factors :

(*i*)   Its surface characteristics

(*ii*) Nature of the phase component

(*iii*) The ionic composition

It has been seen that:

(*i*)   small molecules are more or less equally distributed between the two phases

(*ii*) the partition of particulate matter is invariably one-sided

(*iii*) macromolecules cover a wide range

When large scale isolation of protein from crude homogenates having high viscosity and heterogeneous distribution of particle sizes is to be carried out, the extraction process is useful over centrifugation and filtration. These techniques have high capacity (biomass to volume ratio) and straightforward scale-up.

PEG/Salt systems are being used for industrial scale separation in ATPs system. It is because of their relatively low cost. It has been observed that expensive salt in the bottom phase can be replaced by the following:

(i) Biodegradable polymer

(ii) The incorporation of a suitable ligand molecule into the extraction phase (by being coupled to the polymer forming the phase)

The selectivity of protein extraction in ATP system may be increased. It selectively pulls up the target protein, the latter is subsequently released into a new bottom phase made up of either salt or fresh polymer supplemented with free ligand.

Reverse micelles: The extraction system made of these is used in the downstream processing of proteins. These are thermodynamically stable aggregates of surfactant molecules and water in organic solvents. The polar surfactant head groups point towards the interior of the aggregates.

These form a polar core in which water can be solubilized to generate water pools. The aliphatic chains of the surfactant protrude into the surrounding organic phase. These systems can then be used to solubilize enzyme and there is no loss of activity. These can then be used to catalyze many organic transformations. Reverse micellar systems can also be used for the separation and recovery of proteins by extraction from aqueous feed. There is probably an involvement of electrostatic interactions between the protein surface and charged surfactant during extraction. It is dependent on pH and ionic strength.

5.5.3.4.3 Membrane Filtration

For separation of biomolecules and particles and concentration of process fluids, membrane technology is quite useful. The membrane function has also been integrating with other separation principles.

Microfiltration and ultrafiltration: Semi-permeable membrane during separation acts as a selective barrier. The molecules/particles bigger than the pore sizes are retained by these and the smaller molecules are allowed to permeate through the pores, according to the type of force driving the transport through the membrane. These are named by the size of pores in the filter such as

(i) *Microfiltration:* It is used for separation of particles, typically 0.02-10 μm in diameter.

(ii) *Ultrafiltration:* It separates polymeric solutes in the 0.001-0.02 μm range.

(iii) *Reverse osmosis or hyperfiltration:* It separates ionic solutes typically less than 0.001 μm.

The primary recovery stages of downstream processing involve microfiltration and ultrafiltration. The sieve action of pores determines the selectivity of membranes. It is generally expressed in terms of molecular weight cut off. This can also be done by hydrophilic/hydrophobic nature and charge interactions.

Dead end filtration : Earlier separation with membranes was done by dead-end filtration, in which the feed flows onto the membrane. There is a deposition of cake on the surface and in the pores of filters by settling of particle depositions of colloidal species, adsorption of macromolecular solutes and precipitation of small solutes. It grows in thickness with time and it results in the slowing down of the flow through the filter.

Cross-flow or tangential flow filtration : In this a flow of feed stream is maintained parallel to the separation surface. The purpose is to provide sufficient shear force close to the membrane surface. This do not allow the settling of particulate matter on or within the membrane structure. The membranes used in cross-flow filtration, are also subject to fouling. The cake thickness in this system remains limited to a thin layer.

5.5.3.4.4 Adsorption

It involves interaction between molecules of a fluid mixture and the surface of a solid *e.g.*, batch adsorption in stirred tanks for industrial production. Molecules and particulates in a fluid distribute themselves between a solid and the fluid. Strong binding facilitates adsorption even from crude mixtures. It minimizes the amount of adsorbent required and there is minimum loss of resolution. Adsorption can be used for purification as well as concentration at the early stages of a process.

These are made of micro/macro-porous membrane matrices with ion exchange groups and affinity ligands. These have been developed to bind proteins from the clarified feed pumped over them. Desorption of the protein is later on carried out by using solutions as in chromatography. A total surface area for adsorption equivalent to chromatography gels is provided by a stack of membranes. Liquid transport gives similar high resolution separation. It tremendously increases the speed of separation.

Adsorption is advantageous because it is cheaper and easier to monitor and control than chromatrography. In addition it has following other advantages :

(*i*) Reduced binding site occupancy by unwanted materials

(*ii*) Reduced mass action effects

(*iii*) An uptake that is much less dependent on input concentration

It has following disadvantages:

(*i*) Inactivation of ligand

(*ii*) Lower ligand density than in ion exchange

(*iii*) Less resolution.

A particular molecule can be concentrated from a crude extract by capturing on high-capacity solid adsorbent particles. Activated charcoal has been quite commonly used as adsorbent material for product concentration. Ion exchange resins have also been used for initial capturing of low molecular weight products and proteins from crude extracts. It is because of the following characteristics :

(i) High binding capacity

(ii) Applicability to harsh cleaning in place (CIP) protocols

(iii) Relatively low cost

Synthetic adsorbents with hydrophobic adsorption characteristics have also been developed. These can be used for the extraction of organic compounds.

This procedure is advantageous over conventional solvent extraction, because the adsorbent processes require much smaller amounts of the toxic and flammable organic solvents. The adsorbents based on their pore structure also have a molecular sieving function so that fractionation by molecular size can takes place during the process.

5.5.3.4.5 Prevaporation

It is a membrane based process. This is used for recovery and concentration of volatile products. The technique, however, is not suitable for large-scale work because it involves the following :

(*i*)   High energy consumption

(*ii*)   Insufficient selectivity of the membrane and

(*iii*)   Difficult process design due to a temperature drop across the membrane

5.5.3.4.6 Perstraction

This technique is used for product concentration and recovery. It combines membrane processes and solvent extraction. Molecules that partition into the liquid filling the membrane pores can be transferred by using the membrane as a barrier between an aqueous feed and an organic solvent. It is useful for separating the hydrophobic substances.

(i)   The membrane protects the cells from the toxic or inhibitory effects of the extraction solvent.

(ii)   The membranes could be hydrophobic or hydrophilic.

(iii)   It depends on the direction of mass transfer.

The hydrophobic membranes have pores filled with the organic phase used. On the other hand, the pores in hydrophilic membranes help to extract the non-polar product from the aqueous medium. These are filled with a suitable aqueous buffer. It facilitates the removal of product from the solvent.

5.5.3.4.7 Precipitation

It is very commonly used for the concentration of proteins and polysaccharides. It is generally used for precipitation of proteins. It is usually based on a decrease in solubility induced by external factors. Commonly salts or organic solvents are used as precipitating agents in industry. These precipitation processes are non-specific in the sense that they exploit the ionic and hydrophobic interactions, which are common to all proteins.

There are a few examples of semi-specific precipitations :

(i)   Protein-carbohydrate complexes have been precipitated by means of borax additions.

(ii)   The use of affinity interactions has provided selectivity of precipitation.

In this the affinity interaction creates large complexes such as between antigen and antibody. This is used in immunoprecipitation.

Selective precipitation of multimeric proteins is done by this principle (having more than one binding site for a ligand). Both homo-and hetero-functional ligands can be used.

Homobifunctional ligands. These are synthesized by coupling two ligand molecules by a spacer. The modified ligand bridges different protein molecules. They, thus, form aggregates. The precipitation of the affinity complex occurs only at a definite ratio of the ligand and the protein. Transition metal ions are able to precipitate proteins by bonding with the surface histidine residues. The more the residue, the easier is the precipitation.

Heterobifunctional ligands. In these one functionality is responsible for the affinity binding and the other for the precipitation. It facilitates affinity precipitation in a more general mode. The precipitating component of the heterobifunctional ligand is a smart polymer which responds to minor changes in an environmental parameter *e.g.,* pH, temperature, ionic strength, *etc.,* by a visible change in solubility. Chitosan, alginate, poly-(N-isopropylacrylamide), hydroxypropyl-methyl cellulose *etc.* are the examples of such polymers.

5.5.4 Purification

This step is applied for high-solution purification. The following techniques can be used for this purpose.

5.5.4.1 Immunoaffinity Purifications

For the antigens that stimulate their production, immobilized antibodies may be used as affinity adsorbents. A variety of chemical coupling procedures are used to immobilize on a suitable support matrix. Mammalian polyclonal antibodies raised against purified preparation of antigens are commonly used for initial immunoaffinity columns.

5.5.4.2 Chromatography

It is very common separation technique for analytical and preparative work. This technique has been a standard laboratory practice for many years for the purification° of proteins. The chromatographic techniques are also be used for the isolation of much larger quantities of proteins, although the order in which they are used, must be carefully considered. This mechanism of the separation depends in different cases on :

    (*i*)   adsorption

   (*ii*)   ion-exchange

  (*iii*)   affinity to immobilize ligands

  (*iv*)   size exclusion

   (*v*)   molecular sieving effects

The systems are based on operating a packed column in a batch mode with the appropriate selection of packing and operating conditions for optimum resolution.

The efficiency of liquid chromatography is determined by:

    (*a*)   length of the column

    (*b*)   size of the particle and packing quality

    (*c*)   mobile phase velocity and viscosity.

Although it is not generally considered applicable to the purification of low value high volume products, chromatography has been used even on the industrial scale for the purification of high value low-volume products, typically therapeutic or diagnostic proteins. This method allows to purify a single protein with the required selectivity from a complex mixture of proteins upto a final purity of greater than 99.8%. However, typical recovery values might range from 75-95%. It is, therefore, must to design a purification scheme, which yields the desired level of purity of protein using as few chromatographic steps as possible. Each step needs to be optimized for maximizing yields.

The components to be separated are distributed between stationary and mobile phases.

The stationary phase is usually composed of uniformly sized particles. It is packed in a column and equilibrated with a suitable solvent.

(i)   The mixture to be separated is loaded on the column.

(ii) This is followed by the pouring of the mobile phase.

(iii) Elution of the components is achieved either in an isocratic mode *i.e.,* the same mobile phase is maintained throughout separation of the components depending on differences in retention time in the column or by gradient elution where the mobile phase is continuously changed to facilitate the release of the components bound to the stationary phase by non-covalent forces.

(iv) The eluate from the column can be continuously monitored *e.g.,* by monitoring the ultraviolet absorption at 280 nm for proteins and is collected in fractions of definite volume using a fraction collector.

For protein separation, various modes of chromatography are used. These separations may be based on the following factors:

(i)   Size (size exclusion chromatography)

(ii) Characteristics of surface groups or recognition properties of protein molecules (adsorption chromatography).

The chromatographic matrix should be:

(i)   Inert to prevent any non-specific adsorption

(ii) Rigid, thereby, resist compression at high flow rates

(iii) Chemically stable to withstand harsh cleansing procedures

(iv) Bead shaped for good flow properties

(v)   Porous to provide a high surface area and to allow free passage of molecules.

The chromatographic materials can be:

(i)   organic and inorganic

(ii) synthetic and natural

Fast protein liquid chromatography (FPLC) and high performance liquid chromatography (HPLC) utilizing rigid matrices for operation with relatively high pressure, are now common tools for purification of commercial proteins. All the normal chromatographic techniques are commercially available in the HPLC-mode.

(i) *Flow-through or perfusion chromatography*

It uses a chromatography matrix based on poly-(styrene-divinylbenzene) or agarose- or methyl-methacrylate containing two types of pores:

(a)   The big transecting through pores. These allow convective flow through the particle

(b)   Smaller diffusive pores lining the pores. These provide a large adsorptive surface area.

This combination provides a rapid mass transport of sample molecules and there is no sacrifice of resolution and capacity. In conventional chromatography, however, the sample transport is diffusion-limited.

(ii) *Radial flow chromatography*

In this the innovative design of columns allows the application of sample to the outer wall and move radially by sideways flow of the eluant. The flow rate depends on length and radius and the capacity of the column. The radius influences the resolution also. Scale-up to larger volumes does not need to be done by increasing the bed height, and thus the problem of bed compression of soft gels obtained in conventional downward flow of eluant is overcome.

(iii) *Size-exclusion chromatography*

This involves partitioning of proteins between the stationary liquid held by pores of the gel particles and the mobile liquid in the void volume between the particles. The gel matrices used for this technique have a defined pore size range. It allows small molecules into the pores. The larger molecules, however, are excluded and pass through the column with the mobile liquid.

Size-exclusion chromatography is useful for only small sample volumes (equivalent to 2-5% of the total bed volume). It is used as a final polishing step in a purification protocol to separate the protein from aggregates of degradation products *etc.* and even for buffer exchange.

(iv) *Gel-filtration chromatography*

This method allows the selection of an eluate fraction containing molecules of a specific molecular size. This is also termed as **gel permeation chromatography**. This depends on the ability to produce chromatographic gels of precisely controlled pore size using materials such as:

(*a*)   cross-linked dextrans

(*b*)   agarose

(*c*)   porous glass

(*d*)   polyacrylamide

For the fractionation of proteins:

(*i*)   the protein containing solution is percolated through a column packed with porous gel matrix in bead form

(*ii*)   the small molecules are partially retained in the matrix pores, whereas larger ones pass more quickly through the bed

(*iii*) by increasing the bed width, but not the bed height, scale up is achieved. The progress of smaller proteins through the column is retarded as such as molecules are capable of entering gel beads

(v) HPLC (High performance liquid chromatography)

The use of microparticulate stationary phase media of very narrow diameter increases the chromatographic flow rate. By this the time required for molecules to diffuse in and out of porous particles is reduced. Thus, resolution is not lost. Reduction in particle diameter leads to an increase in the pressure required to maintain a given flow rate. To provide fraction of multi component feed streams, small rigid beads (10 to 100 mm diameter) are used.

This technique can be used for the separation of molecules of similar size and composition. This provides better resolution, speed and ease of quantitative sample recovery.

(vi) Centrifugal partition chromatography (CPC)

To perform separation of complex mixture of chemical substances, CPC in the absence of a solid support utilizes liquid-liquid partition, counter current distribution of solute mixture between two immiscible liquid phases.

The stationary phase: The solvent is retained in the column by centrifugal force.

The liquid mobile phase: As in conventional chromatography, it flows through the stationary phase. Solvent can be reclaimed for reuse by either fractional or azeotropic distillation.

Within column cartridges, separation columns are connected in series. They are arranged around the rotor of a centrifuge. Their longitudinal axes are parallel to the direction of the applied centrifugal force. To separate low-molecular weight substances, organic solvent partition systems are used. The commonly used organic solvents include butanol, chloroform and ethyl acetate. Generally, water-miscible organics are used as modifier to adjust the partition coefficients of the sample components. Combination of hydrocarbons with methanol or acetonitrile was used to separate fat-soluble vitamins, fatty acids, sterols and hydrophobic substances. CPC can easily be adapted for large-scale continuous separation.

(vii) High-performance centrifugal partition chromatography (HPCPC)

It is a process-scale separation technology. The processing cost is minimized by a solid matrix. It contains covalently attached side by side groups of opposite charge.

(viii) Adsorption chromatography

In adsorption chromatography, resolution of the macromolecules is a surface-mediated process *i.e.,* there is differential adsorption of molecules at the surface of the matrix. The matrices employed are derivatized to contain covalently attached functional groups, which adsorb and separate proteins by different mechanisms.

(ix) Ion-exchange chromatography

This is based upon the principle of reversible electrostatic attraction of a charged molecule. This method is based on the following :

(a) Net electrostatic charge

(b) Charge density

(c) Molecular size of the protein

(d) pH and ionic strength of the solution

Proteins may subsequently be eluted by altering the pH or by increasing the salt concentration of the irrigating buffer.

The most commonly used ion-exchange materials include resins and celluloses.

The protein selection of the matrix depends upon the following factors:

(i) Molecular weight

(ii) Flow rate

(iii) Ionic strength

(iv) Resolution

The scale-up is achieved by increasing column cross-sectional area. It has following stages:

Stage I : In this the sample is applied and adsorbed.

Stage II : In this unbound substances can be washed out from the exchanger.

Stage III : Enzymes of interest are eluted from the column and collected in separate fractions.

Since, different proteins have different affinities for the ion exchanger, due to differences in their charges, the fractionation is achieved.

Traditionally ion-exchange media for the fractionation of proteins have been based on cellulose substituted with various charged groups. Cellulose ion exchangers are ideally suited to batch type operations at the early stages of a process. Ion exchange chromatography is most widely used because it has:

(i) general applicability

(ii) good resolution and

(iii) high capacity

It is also insensitive to sample volumes and is often used in the initial phase of downstream processing to provide both product purification and volume reduction of the process fluid.

Compounds are separated according to the difference in their surface charges. Hence pH of medium is one of the most important parameters for binding of the target molecule, as it determines the effective charge on both the target molecule and the ion exchanger. Ionic bound molecules are eluted from the matrix, either by increasing the concentration of salt ions which compete for the same binding sites on the ion exchanger, or by changing the pH of the eluant so that the molecules lose their charges.

Displacement chromatography is an alternative mode, wherein a more heavily charged molecule is used to displace the bound material.

Ion exchangers are grouped as follows :

(i) Anion exchangers with positively charged groups like diethylaminoethyl (DEAE) and quaternary amino (Q)

(ii) Cation exchangers having negatively charged groups like carboxymethyl (CM) and sulphonate (S).

In the purification of L-asparaginase from *Erwinia chrysanthemia,* a six fold purification and 100 fold reduction in volume can be achieved by batch adsorption and elution from CM-cellulose. For routine use, cellulose is not useful in large-scale columns because the volume changes with a change in pH or ionic strength. It is difficult to regenerate without unpacking the column, which is quite difficult. This volume change is less marked with cross linked, beaded form. Ion exchange Sephadex gels are prepared by introducing ion exchange groups into sephadex G-25 or G-50.

(x) *Reversed phase chromatography or hydrophobic interaction chromatography*

This method uses the following:

(i) A non-polar stationary phase

(ii) A porous microparticulate chemically bonded alkyl silica

(iii) A polar mobile phase

In this, hydrophobic contacts between the molecule of interest and the stationary phase take place. Most proteins are folded in such a way that the majority of their hydrophobic amino acid residues are internally buried in the molecule. In this technique protein molecules are fractionated by exploiting differing degrees of such surface hydrophobicity. It depends on the occurrence of hydrophobic interaction between the hydrophobic patches on the protein surface and hydrophobic groups covalently attached to a suitable matrix.

(xi) *Hydrophobic interaction chromatography*

This is high capacity method. It may be used early with dilute product streams. It leads to concentration and purification. It is analogous to reverse phase chromatography (RPC). It relies on the following factors :

(i) Comparatively weak hydrophobic interactions between hydrophobic ligands, alkyl or aryl side chains on the gel matrix

(ii) The accessible hydrophobic amino acids on protein surface

For separation of proteins, differences in the content of these amino acid residues can be used. A medium favouring hydrophobic interactions is used for binding *e.g.,* a solution of high salt concentration. By reducing the hydrophobic interactions, elution of bound material is achieved. It is done either by :

(*i*) lowering the salt concentration

(*ii*) the temperature

(*iii*) decreasing the polarity of the medium (by inclusion of solvents like ethylene glycol or ethanol in the buffer)

Matrices with different hydrophobic groups like butyl-, octyl- and phenyl- are commercially available.

(xii) *Affinity chromatography*

Molecular recognition forms the basis of adsorption and separation by affinity chromatography. In this, one of the reactants is an 'affinity pair', the ligand is immobilized on a solid matrix and is used under suitable conditions to fish out the complementary structure (the ligate). The binding is reversible and by changing the buffer conditions, it can be broken. It relies on the interaction of a protein with an immobilized ligand (Fig. 5.26).

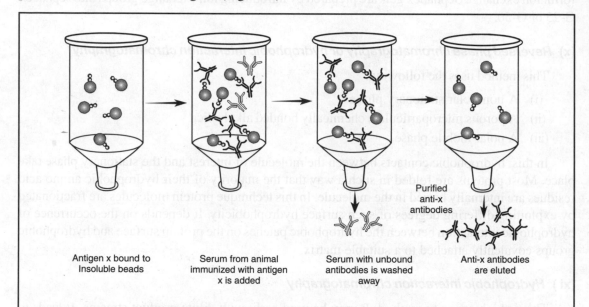

Antigen x bound to Insoluble beads

Serum from animal immunized with antigen x is added

Serum with unbound antibodies is washed away

Purified anti-x antibodies

Anti-x antibodies are eluted

Fig. 5.26. Affinity chromatography : By passing the mixture through a matrix of insoluble beads to which the antigen is attached, antibodies specific for a particular antigen can be purified from a mixture of antibodies in serum or other solutions. The antigen-specific antibodies will bind to the beads, whereas the nonbonding antibodies are washed away. The bound antibodies can then be recovered (eluted) by changing the pH or ionic strength of the solution so that the affinity of antigen binding is lowered.

A ligand, can either be specific for a protein *e.g.,* a substrate-substrate analogue, inhibitor or antibody. It may also be able to interact with a variety of molecules *e.g.,* AMP, ADP, NAD dyes or hydrocarbon chains. When a protein with impurities is passed on this column, the enzyme is covalently bound to the immobilized protein. The desired protein is then recovered by washing the column with an appropriate solvent. The affinity of the interaction is such that only the substance to be purified is adsorbed. A suitable support matrix and a ligand specific for the molecule of interest are required. A number of matrices are commercially available. These are supplied in activated form for coupling to specific chemical groups on the ligand of interest.

The choice of matrix is based on the following characteristics :

(i) Rigidity

(ii) Insolubility

(iii) Stability

(iv) Porosity

(v) Ability to couple ligand

(vi) Cost

A variety of ligands, with specificity for one or a group of proteins, can be used for affinity chromatography.

The ligand must chemically attach to the matrix and have an interaction of appropriate affinity for the molecule to be purified. The interaction must be reversible to permit recovery of product. Dye ligand affinity chromatography has been developed for large scale use. Now a large number of ligands is available. Attempts are being made to replace very expensive natural ligands by synthetics.

These techniques can be used to concentrate and stabilize the protein when absorbed onto the process by a single affinity stage.

Probably the affinity chromatography provides an elegant method for the purification of protein from the complex mixture. It is extensively used on a laboratory scale, however, it is also now used in industrial scale purifications.

Affinity chromatography can be used for the following purposes :

(i) To obtain extremely high purification. In some cases it can be upto several thousand folds in a single step

(ii) To separate active from inactive form of proteins. It can be done by using an antibody or pseudo-substrate

(iii) This technique also allows amount of an impurity from an otherwise pure product

The preparation of gel matrices involves chemical cross linking polymeric molecules. These include dextran, agarose, acrylamide and vinyl polymers. The degree of cross linking in fact controls the average pore size of the gel prepared. Most gels are then available in a variety of pore sizes. The higher the degree of cross linking introduced, the smaller the average pore size and more rigid the resultant gel beads. Various highly cross linked gel matrices such as G-25 or Bio Gel P2 have pore sizes that exclude all protein polymers from entering the gel matrix. Such gels may be used to separate proteins from other molecules used to remove low molecular weight buffer component and salts from protein solution.

To separate proteins from each other, gels of larger pore sizes are used. The application of chromatography largely depends on the following factors:

(a) Molecular weight of the protein of interest

(b) Molecular weight of the major proteinaceous containment

Potentially, such methods possess very high resolving power. Downstream processing infact exploits the specificity of affinity interactions earlier in the separation train so as to reduce the number of purification steps. This, however, puts extra demands on the ligands with respect to chemical and biological stability. It is because of these, pseudo biospecific ligands like dyes and metal ions are used.

(xiii) *Chromatofocusing*

It separates proteins on the basis of their isoelectric points. In this a buffer of one pH is percolated. A continuous pH gradient may be set up along the length of the column due to natural buffering capacity of the exchanger. In order to achieve maximum resolution, along the length of the column, a linear pH gradient may be set up.

It is because of this, element buffer and exchanger which exhibit even buffering capacity over a wide range of pH values are used. The range of pH gradient achieved depends on the pH at which the ion exchanger is pre-equilibrated and on the pH of the element buffer.

A weak anion exchanger which exhibits a high buffering capacity is used on the matrix in chromatofocussing. The technique involves following steps :

- (*i*) The anion exchanger is re-equilibrated at a high pH value
- (*ii*) The sample is then usually applied in the running buffer. Its pH is lower than that of the pre-equilibrated column
- (*iii*) The column is constantly percolated due to its lower pH value relative to the initial column pH. It results in the establishment of an increasing pH gradient down the length of the columns.

5.5.5 Formulation of Product

The commercial viability of a biotechnological product is dependent on the maintenance of its activity and stability during distribution and storage.

Low molecular weight products such as bulk solvents, bulk organic acids, *etc.* are formulated as concentrated solutions after removing most of the water.

When high purity is required, small molecules such as antibiotics, citric acid, sodium glutamate *etc.* are crystallized from solution by the addition of salts once they have reached the required degree of purity. During downstream processing and subsequently during storage, proteins are particularly sensitive to loss of biological activity. The following factors are playing role in this process :

- (*i*) Oxidation
- (*ii*) Temperature
- (*iii*) Presence of proteases or dry powders

In order to prolong the shelf life of a product, a variety of stabilizing additives are included in the formulations. Suitable stabilizer is empirically chosen. The following non-specific chemical additives are regularly used as stabilizers in protein formulations :

    (*i*)   Salts (ammonium sulphate or sodium chloride)

    (*ii*) Sugars (sucrose, lactose *etc.*)

    (*iii*) Polyhydric alcohol (sorbitol, glycerol *etc.*) or polymers (polyethylene glycol, bovine serum albumin *etc.*)

Bulk enzymes are commonly sold as concentrated liquid formulations. It is, however, often preferred to dry the product to decrease the volume as well as the denaturing reactions that are enhanced in aqueous solution. Gentle drying methods are required for dryers, as byproducts are often sensitive to heat. These, depending on the mechanism of heat transfer, can be classified as follows :

    (*i*)   Contact

    (*ii*) Convection

    (*iii*) Radiation-dryers

Batchwise drying in many contact dryers is facilitated using mechanically moved layers. It has following advantages :

    (i)   The uniform thermal stresses exerted on the material being dried

    (ii) High throughput

    (iii) Possibility for development of continuous processes.

A common feature of convection dryers is that the movement of the material to be dried is promoted by a flow of gas. Drying of large streams of product is achieved in a very short time.

Spray drying: In this, aerosol of tiny droplets is generated by passing the product containing liquid through a nozzle or a rotating atomizing disc and directing it into a stream of hot gas. The water in the droplets evaporates. Solid product particles are thus left behind.

Spray-dryers are used for drying large volumes of liquids.

Chamber dryer: It may be used for smaller quantities. In this the product is placed on shelves and the transfer of heat takes place partly by contact and partly by convection. The material is exposed to severe non-homogeneity in thermal stress. This method is useful only at relatively low temperature.

Freeze-drying or lyophilization: It is one of the least harsh methods of protein drying. It is used for the drying of pharmaceutical products, diagnostics, foodstuffs, viruses, bacteria *etc*. The drying principle is based on sublimation of the liquid from a frozen material. The liquid containing the product is frozen, ideally to a temperature below its glass transition temperature and subjected to vacuum in a freeze-dryer. While maintaining the internal vial temperature still below the glass transition value, the shelf temperature is increased to a temperature above zero to promote efficient sublimation of the crystallised water.

After the primary drying, the protein cake still retains a significant amount of water which is removed by sublimation during secondary drying by increasing the internal vial temperature.

Subsequently, the vacuum is released and the vials are sealed under a high vacuum. The heat transfer is solely via contact and not by convection, therefore one needs to operate at low vacuum for a efficient drying process. Normally, prior to freeze-drying, additives are included in the solution. These may be compounds

    (a) For maintaining the activity of the protein in the finished product

    (b) To prevent product blow out during lyophilisation (*e.g.*, mannitol) and

    (c) To enhance product solubility

All these additives influence the glass transition value of solution and hence the freezing temperature prior to drying.

Questions

1.   What is downstream processing ? Explain in detail.

2.   What is bioprocess engineering ? Describe

3.   What do you know about the following ?

    (a) Biosensors

    (b) Immobilization of enzymes

    (c) Chromato-focusing

4.   What is protein engineering ? Discuss various methods of protein engineering.

5.   What is enzyme immobilization ? Discuss various methods involve in immobilization and their application.

6.   What are biosensors ? Discuss their components, types and applications.

7.   Discuss methods for screening of industrial important microorganism.

8.   What is media formation ?

9.   What is sterilization ? Discuss various methods of sterilization.

10. What are fermenters ? Discuss various types and their applications.

11. Discuss various methods of separation of solid-liquid during downstream processing.

12. What is cell disruption ? Discuss various methods of cell disruption.

13. What is column chromatography ? Discuss various chromatographic methods used to separate proteins.

Chapter 6

Immunological and Molecular Diagnostic Techniques

6.1 DIAGNOSTIC TECHNIQUES

Living things have a precarious existence. They are constantly threatened by changes in the environment, such as alterations in climate or competition by new kinds of neighbours. Such changes like these have eliminated many species. These conditions develop over many generations and allow time for evolutionary adaptation by mutation and selection of offspring that are better suited to the new surroundings than their ancestors.

A vertebrate is an attractive culture ground for many kinds of virus, bacteria, fungi, protozoan and metazoan parasites. With their capacity for extremely rapid multiplication, viruses, bacteria and protozoan parasites may cause an epidemic which can sweep through population within weeks. It is, therefore, neccesary to develop diagnostic tests with greater sensitivity and specificity.

One of the most crucial aspect of modern medical treatment is the accurate and timely diagnosis of the disease.

Diagnosis is a technique to identify the nature of any illness or problem by examination of the symptoms, and to find out the reason behind that illness or problem. Once we know the exact reason then it is possible to focus on either to eliminate or kill the cause. The various techniques used for diagnosis are :

- Immunological diagnostic techniques
- Molecular diagnostic techniques
- Pharmaceutical diagnostics

6.2 IMMUNOLOGICAL DIAGNOSTIC TECHNIQUES

Diagnostic techniques depend on the following four main features of immune response :

- An immune response is specific,
- The response is potentially diverse,
- It is adaptive,
- It has the ability to respond to unexpected stimuli.

These are based on the ability of antibodies to recognize specific molecules (antigens) in a sample.

Antigen recognition by the antibodies, which is important for the body's defense against disease, has been exploited in the development of diagnostic tests for diseases. They have an innate ability to bind to specific substances. The extent of their use in diagnostics, however, is determined by the methods by which that binding is converted into a measurable signal – the label. At present, most immuno assays use antibodies tagged with radioisotopes, enzymes or fluorescent dyes.

In enzyme immunoassay (EIA), a colour producing enzyme is coupled to the antibody and results can be observed either by eye or spectrophotometrically. This technique is modified by using the enzyme cascade in which several enzyme reactions are coupled to produce a significant amplification of the original binding signal. In fluorescence immunoassay and luminescence immunoassay, the label emits fluorescence or light respectively. In radioimmunoassay a radioisotope is tagged to the antibody.

Most of the antibodies used are the monoclonal antibodies developed through the hybridoma technology in given 1975 by Kohler and Milstein in which antibody producing cells were fused with tumour cells to produce a hybrid which could be grown as culture in laboratory and which produced antibodies of a single specificity – monoclonal antibodies. Monoclonal antibodies are also being used to image tumour by conjugating them to radioactive isotopes. Following injection of the conjugates, body scanning techniques can be used to localize and quantify cancerous tissue. This allows us to give an initial diagnosis or assessment whether or not the disease is responding to the conventional treatments. Following is the account of such techniques used in diagnosis.

6.2.1 Radio Immuno Assay (RIA)

It is one of the most sensitive techniques for detecting antigen or antibody. It was first developed by Berson and Rosalyn Yalow in 1960 to determine levels of insulin-anti-insulin complexes in diabetics. The technique is now used for measuring hormones, serum proteins, drugs and vitamins at concentrations of 0.001 µg/ml or less. In 1977, some years after Berson's death, the significance of the technique was acknowledged by the award of a Nobel Prize to Yalow.

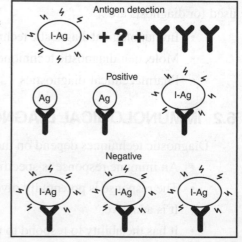

Fig. 6.1. Radio immuno assay : This is normally carried out as competition assay. A known amount of antigen radiolabeled by iodination ($^{131 \ or \ 125}$Ag) is mixed with a known amount of antibody and the specimen is tested for that antigen. If the test specimen contains no antigen, all the radioactive antigen will combine with the antibody. This can be detected by precipitation with ammonium sulfate or a second antibody. Conversely, if antigen is present in the specimen, it will compete for antibody with the radioactive antigen and the number of radioactive counts precipitated will be reduced.

It involves competitive binding of radiolabeled antigen and unlabeled antigen to a high-affinity antibody. The labeled antigen is mixed with antibody at a concentration that saturates the Ag binding sites of the Ab molecule, and then increasing amounts of the test sample containing unlabeled Ag of unknown concentration is added. The Ab does not distinguish labeled from unlabeled Ag and so the two kinds of Ag's compete for available binding sites on the Ab so that with increasing concentrations of unlabeled Ag, more labeled Ag will be displaced from the binding sites.

This test uses radioactive material to detect small amount of antigens and to quantitate amount of sametigen in tissues, body fluids or blood. These assays are based on competition between a labeled known antigen and the same unlabeled antigen in the test sample. How is this done? Radiolabeled antigen (labeled with ^{131}I or ^{125}I) is mixed with antibody and with the test sample and complexes are precipitated either with 58% ammonium sulfate or by adding a second antibody directed against the first one. If the specimen contains the antigen, all the radioactivity will be precipitated. If significant amounts of antigen are present in the sample, they will compete with radiolabeled antigen and reduce the proportion of radioactive count in the precipitate. There are many modifications of this test (Fig. 6.1).

6.2.2 RIST (Radio Immuno Sorbent Test)

The antibody content of a patient's serum can be assessed by the ability of antibody to bind to antiglobulin, which has been immobilized on a solid surface which may be a nitrocellulose paper disc or a polycarbonate tube.

This test is usually done for the detection of IgE antibodies in severely allergic patients. Here, anti-IgE raised in rabbits is used in labelled and unlabelled form.

   (i)   Coat plate or couple disc with anti IgE.
   (ii) Block to decrease non-specific binding.
   (iii) Add test serum and incubate to detect IgE.
   (iv) Wash with saline to remove excess serum and serum proteins.
   (v)   Add labelled anti-human IgE.
   (vi) Wash
   (vii) Count: Geiger Muller counter measures radioactivity and quantity of bound antibody is
         detected.

6.2.3 RAST (Radio Allergo Sorbent Test)

It is used for measuring the amount of IgE to specific allergen in the patient's serum. Specific allergen is absorbed on to a solid surface and treated with patient's serum. The mixture is incubated for some time and later, excess of serum is washed with saline. Specific IgE from the patient's serum will bind to specific allergen, which is immobilized. Bound IgE is detected by adding radio-labelled anti-IgE, as is done in case of RIST.

Limitation

This method can not be used to detect antigen concentrations lower than 5 µg/ml.

6.2.4 Precipitation Reactions

The interaction between an antibody and a soluble antigen in an aqueous solution forms a lattice. This eventually develops into a visible precipitation. It occurs more slowly and often takes a day or two to get completed.

Formation of an Ag-Ab lattice depends on the valency of both the antibody and the antigen:

- The antibody must be bivalent; a precipitate will not form with monovalent Fab fragments.

- The antigen must be either bivalent or polyvalent; that is, it must have at least two copies of the same epitope, or have different epitopes that react with different antibodies present in polyclonal antisera.

The requirement that protein antigens be bivalent or polyvalent for a precipitin reaction to occur, is illustrated by myoglobin. Myoglobin precipitates well with specific polyclonal antisera, but fails to do so with a specific monoclonal antibody because it contains multiple, distinct epitopes, but only a single copy of each epitope. Myoglobin, thus, forms a cross-linked lattice structure with polyclonal antisera, but not with monoclonal antibody.

6.2.4.1 Precipitation Reaction in Fluids

The basic type of antigen-antibody reaction is the precipitation reaction. This test allows the free diffusion of antigen and antibody.

The interaction of proper ratios of these antigens with specific antibodies results in a precipitation reaction. At the interface of the solution of antibody and antigen, a faint line appears. These tests are carried out in capillary tubes. The antigen is overlayered on the antiserum. A ring of precipitation will be seen at the interface of the two solutions, if specific antibody and antigens are present. These tests are called **ring tests** and **capillary tube tests** (Fig. 6.2).

6.2.4.2 Precipitation Reaction in Gels

(a) Radial immunodiffusion (Mancini Method)

In this, antigen or antibody is incorporated in the gel. If the gel contains the suitable dilution of antibody, antigen is added to the punched wells and allowed to diffuse through

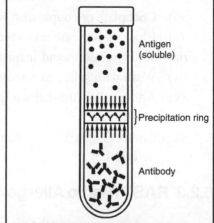

Fig. 6.2. A precipitation ring test. Antibody and antigen diffuse toward each other in a test tube. A precipitation ring is formed at the zone of equivalence.

the agarose, overnight. As this happens, a precipitin ring is formed around the well. The diameter of the precipitin ring is proportional to the antigen concentration.

By preparing a standard graph after measurement of the diameter of rings formed against known concentration of antigen against a known dilution of antibody, the concentration of unknown antigen can be found.

(b) Double Immunodiffusion (Ouchterlony Method)

These reactions when performed in an agarose gel are most stable. Antibody-antigen reactions are commonly performed in the clinical laboratory in agarose by a technique called double immunodiffusion or Ouchterlony method.

In the Ouchterlony method, both antigen and antibody diffuse radially from wells toward each other. They, thus, establish a concentration gradient. When an equilibrium is reached, there is formation of a visible line of precipitation, a **precipitin line.** This technique is quite useful for determining the relationship between antigens and the number of different Ag-Ab systems present.

The technique involves the following steps :

(i) Wells of appropriate diameter are formed in an agarose gel in a small petri dish or glass slide and spaced appropriately.

(ii) In one well, the specimen containing the unknown soluble antigen is placed.

(iii) In an adjacent well, a known antibody-containing solution is placed.

(iv) The antigen and antibody molecules in solution diffuse out of the wells and through the porous agarose.

(v) If antigen specific for the known antibody is present, the two components combine and produce a visible precipitin band or line of precipitation at a point of optimal concentration of each component (Fig. 6.3A).

Diffusion is a slow process and is generally not amenable to rapid diagnosis. It is most commonly used to detect fungal exoantigens or serum antibodies.

The pattern of precipitin lines that form when two different antigen preparations are placed in adjacent wells, indicates whether or not they share epitopes.

- Identity occurs when two antigens share identical epitopes. As the antiserum forms a single precipitin line with each antigen, the two lines grow towards each other and fuse to give a single curved line (Fig. 6.3B).

- When two antigens are unrelated (*i.e.,* share no common epitopes, the antiserum forms an independent precipitin line with each antigen and the two lines cross (Fig. 6.3C).

- There is partial identity when two antigens share some epitopes, but one or the other has a unique epitope (s). The antiserum forms a line of identity with the common epitope (s) and a curved spur with the unique epitope (s) (Fig. 6.3D).

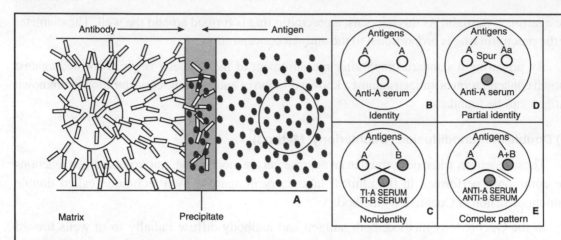

Fig. 6.3. Double immuno diffusion assay. (A) Mode of movement of antibody and antigen from their respective wells towards each other to form a precipitate. **(B)** Characteristics of identity, **(C)** Reaction of nonidentity **(D)** Partial identity, **(E)** A complex pattern.

(c) Immunoelectrophoresis

This technique combines separation by electrophoresis with identification by double immunodiffusion. An antigen mixture is first electrophoresed to separate its components by charge. Then the troughs are cut into the agar gel parallel to the direction of the electric field. It is followed by the addition of antiserum to the troughs (Fig.6.4). Antibody and antigen then diffuse towards each other and produce lines of precipitation where they meet in appropriate proportions.

Immunoelectrophoresis is used in clinical laboratories to detect the presence or absence of proteins in the serum. A sample of serum is electrophoresed and the individual serum components are identified with antisera specific for a given protein or immunoglobulin class. This technique is useful in determining whether a patient produces abnormally low amounts of one or more isotypes.

It is a strictly qualitative technique. It can detect relatively high antibody concentrations. It is useful for detecting quantitative anomalies only when there is much difference in the diseased and normal states. The examples are immunodeficiency and immunoproliferative disorders.

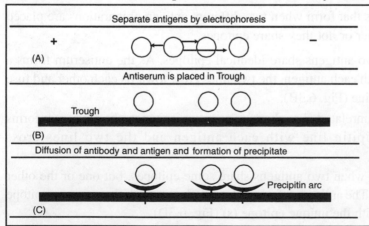

Fig. 6.4. Immunoelectrophoresis. (A) An electrical charge is used to separate antigens in an agar gel. **(B)** Antibody (antiserum) is then placed in a trough cut parallel to the direction of the antigen migration. **(C)** The antigens and antibodies diffuse through the agar and form precipitin arcs.

Rocket electrophoresis: It is also a quantitative technique. It permits measurement of antigen levels. Hence, a negatively charged antigen is electrophoresed in a gel containing antibody. The precipitate formed between antigen and antibody has the shape of a rocket. Its height is proportional to the concentration of antigen in the well.

This technique has a limitation. The antigen in this technique has to be negatively charged for electrophoretic movement within the agar matrix. Some proteins, such as immunoglobulins are not sufficiently charged and hence cannot be quantitatively analyzed by this technique. It is also not possible to measure the amounts of several antigens in a mixture at the same time.

(d) Counter Immunoelectrophoresis (CIEP)

It is a modification of the principle of immunodiffusion. By applying an electric current, the migration of soluble antigen and antibody can be quickened. For this purpose :

   (i)   antigen and antibody solutions are placed in adjacent wells cut in the agarose supported on a glass or plastic surface,

  (ii)   the gel is placed in an alkaline buffer-containing electrophoresis chamber and the electric field is applied,

 (iii)   under the buffer conditions chosen, most antigens have a net negative charge (anions) and thus migrate in the gel towards the positively charged electrode (anode),

 (iv)   antibody molecules, on the other hand have very weak negative charge or are neutral under alkaline buffer conditions. In the charged field, they do not migrate significantly. By the effect of buffer ions, they are, however, carried towards the negatively charged electrode (cathode).

This phenomenon is known as **electroendosmosis.** If specific antigen and antibody are present, the antigen and the antibody will meet in the gel at some point of optimum proportion or equilibrium and form a visible precipitin band (Figs. 6.5, 6.6).

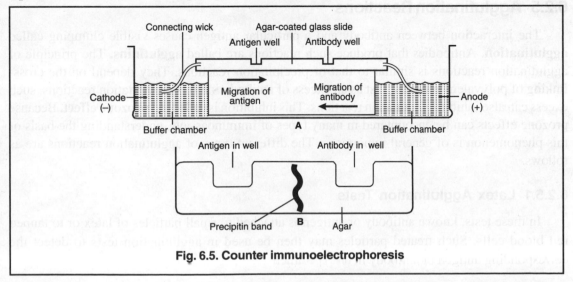

Fig. 6.5. Counter immunoelectrophoresis

The process is completed in an hour or less by CIEP, whereas in gels by passive diffusion, 24 hours are required for precipitin bands to form. Counter immunoelectrophoresis plates are generally ready immediately after electrophoresis. For this purpose non-specific precipitin band is removed after washing. The gel is subjected to overnight refrigeration. It allows the specific

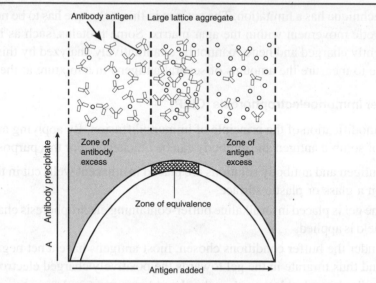

Fig. 6.6. How the precipitation occurs during the combination of antigen and antibody: Graph showing the formation of precipitation curve on the basis of ratio of antigen to antibody. The zone of equivalence represents the optimal ratio of antigen and antibody.

bands to intensify. The test was very popular in the 1970s for a number of applications, but now it has been replaced by particle agglutination tests and immunoassays for rapid microbial antigen detection.

6.2.5 Agglutination Reactions

The interaction between antibody and a particular antigen causes visible clumping called **agglutination.** Antibodies that produce such reactions are called **agglutinins.** The principle of agglutination reactions is similar to that of precipitation reactions. They depend on the cross-linking of polyvalent antigens. Just as an excess of antibody inhibits precipitation reactions, such excess can also inhibit agglutination reactions. This inhibition is called the **prozone effect.** Because prozone effects can be encountered in many types of immunoassays, understanding the basis of this phenomenon is of general importance. The different types of agglutination reactions are as follows.

6.2.5.1 Latex Agglutination Tests

In these tests, known antibody or antigen is attached to small particles of latex or to tanned red blood cells. Such treated particles may then be used in agglutination tests to detect the corresponding antigen or antibody in test specimen.

(A) For antigen detection, latex beads coated with a known amount of antibody are mixed with a specimen of patient-serum which may contain antigen 'Ag'. If antigen is present, the antibody coated latex particles form a lattice connected by the antigen molecules (as long as the antigen is multivalent). The lattice is macroscopically visible as agglutination of the particles (Fig. 6.7).

(B) For antibody detection, the patient's serum which is being tested for the content of antibody molecules is mixed with a known amount of antigen (Ag) attached to latex spheres or RBCs. If antibody is present, the particles will agglutinate.

Some important controls must accompany the test. These controls include the following:

(i)   A positive antigen control (a solution containing the known antigen of interest)

(ii) A negative antigen control (a solution not containing the antigen)

(iii) A control latex suspension to detect the presence of non-specific agglutination reactions.

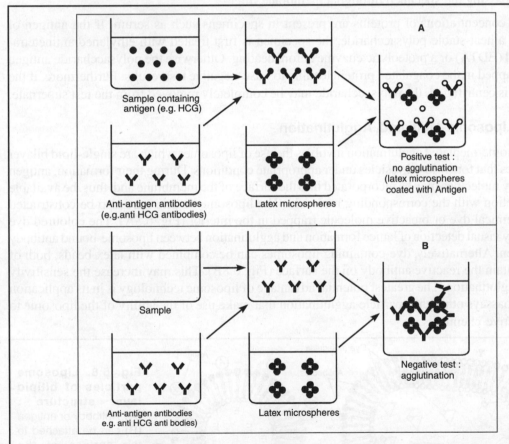

Fig. 6.7. Latex agglutination test. (A) In a positive test, the sample containing antigen is mixed with a solution of antibody specific for that antigen. In the second step, latex microspheres coated with antigen are added. If antigen is present, it binds to antigen-specific antibodies. This prevents them from agglutinating the microspheres. (B) In a negative test, the microspheres coated with antigen are agglutinated by antigen-specific antibody and hence agglutination takes place (adopted with modifications from Diagnostic Microbiology, Mahon et.al., 2000).

The control latex involves testing the patient specimen with latex beads coated with an immunoglobulin whose specificity is not directed to the test antigen. This non-immune serum is generally obtained from the same animal species in which the specific antibody was made. A non-specific agglutination reaction occurs when the patient's specimen reacts with both the test and the control latex. When such reactions occur, the test is uninterpretable. A positive test result requires that the test latex, but not the control latex, agglutinates the patient specimen.

To eliminate or minimize non-specific agglutinations, a number of specimen pretreatment procedures can be used. It can be done by removing or inactivating factors in the specimen responsible for these reactions. These procedures include :

(i) specimen centrifugation to remove particulate material,

(ii) boiling to inactivate protein constituents (acceptable when test antigen is a heat-stable polysaccharide),

(iii) passing the specimen through a membrane filter.

High concentrations of proteins are present in specimens such as serum. If the antigen of interest is a heat-stable polysaccharide, the specimen is first treated with ethylenediaminetetra-acetic acid (EDTA) or a proteolytic enzyme before heating. Otherwise the polysaccharide antigen may be trapped in the coagulated protein leading to false-negative test results. Furthermore, if the specimen is centrifuged, the polysaccharide may be completely removed from the test supernate.

6.2.5.2 Liposome-Mediated Agglutination

Liposome-mediated agglutination involves the use of liposomes which are single-lipid bilayer membranes that form closed vesicles under appropriate conditions. During their formation, antigen or antibody molecules may be incorporated into the surface of the membrane and, thus, be available for interaction with the corresponding molecule. The liposome vesicle may also be constructed with a chemical dye or bioactive molecule trapped in the interior (Fig. 6.8A). The coloured dye allows easy visual detection of lattice formation and agglutination between liposome-bound antibody and antigen. Alternatively, dye-containing liposomes can be combined with latex beads, both of which contain the reactive antibody on the surface (Fig. 6.8B). This may increase the sensitivity of latex agglutination. The greatest potential advantage of liposome technology is in its application to immunoassays other than particle agglutination that make use of the ability of the liposome to carry reactive chemicals.

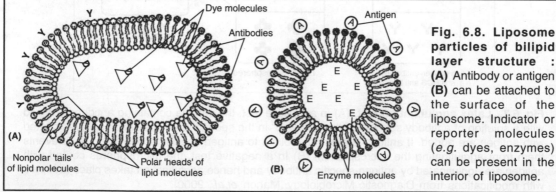

Fig. 6.8. Liposome particles of bilipid layer structure : **(A)** Antibody or antigen **(B)** can be attached to the surface of the liposome. Indicator or reporter molecules (*e.g.* dyes, enzymes) can be present in the interior of liposome.

6.2.5.3 Coomb's Test

This anti-globulin test was devised by Coombs Mourant and Race in 1945 for the detection of anti-Rh antibodies that do not agglutinate Rh positive RBCs in saline. When sera containing incomplete anti-Rh antibodies are mixed with Rh positive red cells, the antibody globulin coats the surface of RBCs, though they are not agglutinated. When such RBCs are washed free of all unattached protein and treated with a rabbit antiserum against a human gammaglobulin (Coomb's serum), the cells get agglutinated. This is the principle behind this test.

(a) **Direct Coomb's Test:** Here, the sensitisation of the erythrocytes with incomplete antibodies takes place *in vivo,* as in the case of haemolytic disease of the new born due to Rh incompatibility.

When red cells of erythroblastotic infants are washed free of unattached protein and then mixed with a drop of Coombs serum, agglutination results. This test comes often negative in haemolytic disease, due to ABO incompatibility.

(b) **Indirect Coomb's Test:** *In vitro* sensitisation of red cells with the antibody globulin is performed.

Now-a-days, Coomb's test is also being used to detect incomplete or non-agglutinating antibody, as in case of brucellosis.

6.2.5.4 Haemagglutination

Agglutination reactions are routinely performed to type red blood cells (RBCs). In typing for the ABO antigens, RBCs are mixed on a slide with antisera to the A or B blood group antigens. If the antigen is present on the cells, they agglutinate and form a visible clump on the slide. The basis for matching blood types for transfusions is determination of which antigen is present on donor and recipient RBCs. Red blood cell glycoproteins and plasma antibodies are found in individuals with each blood type. Type A individuals have the A glycoprotein on their red blood cell surfaces. Their plasma contains antibodies only against B glycoproteins (anti-B antibodies). Therefore, their plasma antibodies do not bind to their own red blood cell glycoproteins and their own cells do not clump. Type B individuals have B glycoproteins and anti-A antibodies. The AB blood contains red blood cells with both the A and B glycoproteins, but neither of the antibodies. Type O individuals have both anti-A and anti-B antibodies, but no reactive glycoproteins on their red blood cells. In the reaction between type A blood cells transfused into type B blood, type B blood contains many anti-A antibodies. Because each antibody has two sites that can bind the A glycoprotien, the A cells become clumped, held together by the anti-A antibodies. These clumps can become large clots, with serious medical consequences.These glycoproteins on red blood cells may react with antibodies in the blood plasma. If a patient with type B blood receives a transfusion of type A blood, the anti-A antibodies in the patient's serum cause the type A blood cells to clump. On the basis of these studies, we can say O type is universal donor and type AB is universal recipient (Fig. 6.9; Table 6.1).

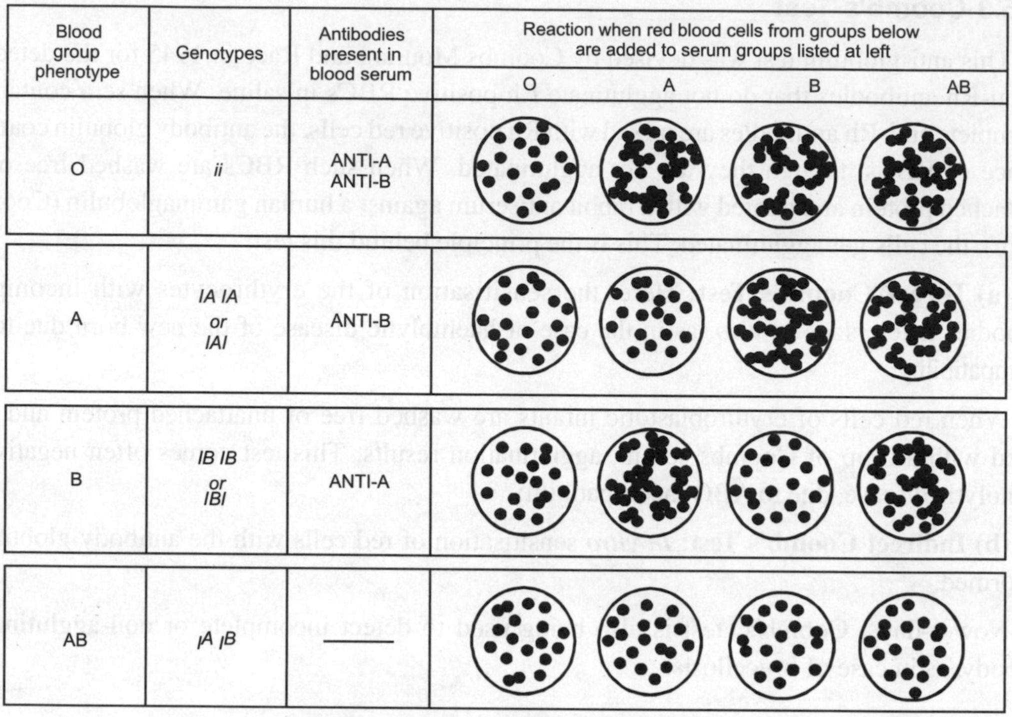

Blood group phenotype	Genotypes	Antibodies present in blood serum	Reaction when red blood cells from groups below are added to serum groups listed at left			
			O	A	B	AB
O	ii	ANTI-A ANTI-B				
A	IA IA or IAI	ANTI-B				
B	IB IB or IBI	ANTI-A				
AB	IA IB	—				

Fig. 6.9. Agglutination Pattern with Different Blood Groups.

Table 6.1. Effects of Blood Transfusions

Donor	Recepient	Effect on Blood Donation	Permissible Type
A	A	—	Yes
	B	Clumping	No
	AB	—	Yes
	O	Clumping	No
B	A	Clumping	No
	B	—	Yes
	AB	—	Yes
	O	Clumping	No
AB (Universal recepient)	A	Clumping	No
	B	Clumping	No
	AB	—	Yes
	O	Clumping	No
O (Universal donor)	A	—	Yes
	B	—	Yes
	AB	—	Yes
	O	—	Yes

6.2.5.4 (A) Bacterial Agglutination

A bacterial infection often elicits the production of serum antibodies specific for surface antigens on the bacterial cells. Bacterial agglutination reactions help to detect the presence of such antibodies. In this method

   (i)   serum from a patient thought to be infected with a given bacterium, is serially diluted in a series of tubes to which the bacteria are added.

   (ii) the test tube showing visible agglutination will reflect the serum antibody titre of the patient.

   (iii) the agglutinin titre is defined as the reciprocal of the greatest serum dilution that elicits a positive agglutination reaction.

For example, if serial two fold dilutions of serum are prepared and if the dilution of 1/640 shows agglutination, but the dilution of 1/1280 does not, then the agglutination titre of the patient's serum is 640. In some cases serum can be diluted up to 1/50,000 and still shows agglutination of bacteria (Fig. 6.10).

The agglutinin titre of an antiserum can be used to diagnose a bacterial infection. Patients with typhoid fever for example, show a significant rise in the agglutination titre to *Salmonella typhi*. Agglutination reactions also provide a way to type bacteria. For instance, different species of the bacterium *Salmonella* can be distinguished by agglutination reactions with a panel of typing antisera.

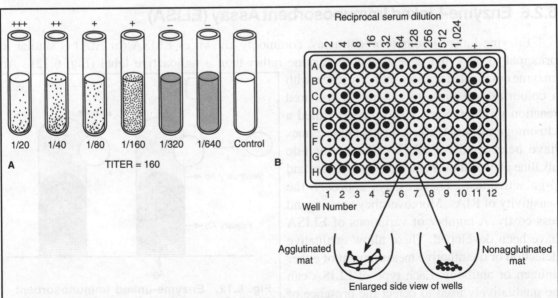

Fig. 6.10. Agglutination tests. (A) Tube agglutination test for determining antibody titer. The titre in this example is 160 since there is no agglutination in the next tube in the dilution series 1/320). **(B)** A microtitre plate shows haemagglutination. The antibody is placed in the wells (1–10). Positive controls (row 11) and negative controls (row 12) are included. Red blood cells are added to each well. If sufficient antibody is present to agglutinate the cells, they sink as a mat to the bottom of the well. If insufficient antibody is present, they form a pellet at the bottom.

6.2.5.4 (B) Viral Agglutination

Like bacteria certain viruses can also agglutinate a variety of cells *e.g.,* RBCs (Fig. 6.11).

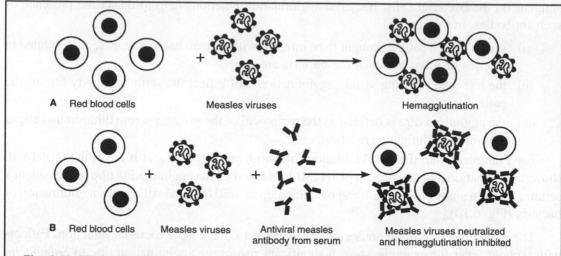

Fig. 6.11. Viral haemagglutination. (A) Certain viruses can bind to red blood cells and cause haemagglutination, **(B)** if serum containing specific antibodies to the virus is mixed with the red blood cells, the antibodies will neutralize the virus and inhibit haemagglutination (a positive test).

6.2.6 Enzyme-Linked Immunosorbent Assay (ELISA)

Enzyme-linked immunosorbent assay, commonly known as ELISA (or EIA) is similar in principal to RIA, but depends on an enzyme rather than a radioactive label (Fig. 6.12). An enzyme conjugated with an antibody reacts with a colourless substrate to generate a coloured reaction product. Such a substrate is called a **chromogenic substrate.** A number of enzymes have been used for ELISA. These include alkaline phosphatase, horseradish peroxidase and β-galactosidase. These assays approach the sensitivity of RIAs. Moreover they are safe and less costly. A number of variations of ELISA have been developed. These allow qualitative detection or quantitative measurement of either antigen or antibody. Each type of ELISA can be qualitatively used to detect the presence of antibody or antigen. Alternatively, a standard curve based on known concentrations of antibody or antigen is prepared, from which the unknown concentration of a sample can be determined.

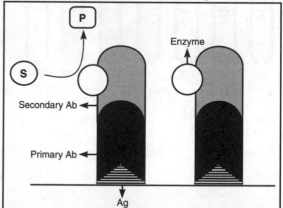

Fig. 6.12.   Enzyme-linked immunosorbent (ELISA) assay. The primary antibody (Ab) recognizes the immobilized antigen (Ag) and the secondary Ab will specifically interact with the primary Ab. These interactions can be detected using an appropriate substrate (S) for the enzyme attached to the secondary Ab, the product (P).

Indirect ELISA

Antibody can be detected or quantitatively determined with an indirect ELISA. Serum or some other sample containing primary antibody (Ab_1) is added to an antigen-coated microtitre well and allowed to react with the antigen attached to the well. After any free Ab_1 is washed away, the presence of antibody bound to the antigen is detected by adding an enzyme-conjugated secondary anti-isotype antibody (Ab_2), which binds to the primary antibody. Any free Ab_2 is then washed away, and a substrate for the enzyme is added. The amount of coloured reaction product that forms is measured by specialized spectrophotometric plate readers called ELISA reader, which can measure the absorbance of all of the wells of a 96 well plate in less than a few seconds (Fig. 6.13).

To detect the presence of serum antibodies against human immunodeficiency virus (HIV), the causative agent of AIDS, indirect ELISA is used. In this assay recombinant envelope and core proteins of HIV are adsorbed to solid-phase antigens on microtitre wells. Individual infected with HIV will produce serum antibodies to epitopes on these viral proteins. Generally, serum antibodies to HIV can be detected by indirect ELISA within 6 weeks of infection.

Sandwich ELISA

Antigen can be detected or measured by a sandwich ELISA. In this technique, the following steps are involved (Fig. 6.13):

(i) The antibody (rather than the antigen) is immobilized on a microtitre well.

(ii) A sample containing antigen is added and allowed to react with the immobilized antibody.

(iii) After the well is washed, a second enzyme-linked antibody specific for a different epitope on the antigen is added and allowed to react with the bound antigen.

(iv) After any free second antibody is removed by washing, substrate is added, and the coloured reaction product is measured.

One of the rapid chromatographic tests was developed by Rockeby Biomed, Ltd. (Shaw House, Singapore; www.rockeby.com). Rockeby presented on its proprietary rapid bird flu test, which was prepared by Western Australia's Murdoch University School of Veterinary and Biomedical Sciences, at the Second International Anti-Avian Influenza Conference in Paris at the Pasteur Institute on May 31 and June 1, 2007.

Antigen detection tests could be used for preliminary investigation of avian flu (H5N1) outbreaks as low cost, simple flock tests in sick and dead birds for the rapid detection of H5N1 infection. In the March 2007 issue of the Journal of Avian Diseases, investigators from Murdoch University (Pearth, Australia, www.murdoch.eu.au) reported the results of their evaluation of five influenza antigen detection tests. The tests included two rapid chromatographic immunoassay, an H5HA antigen detection enzyme linked immunosorbent assay (ELISA), an influenza A antigen detection ELISA and H5 rapid immunoblot assay. Avian influenza H5N1 virus positive swab samples were used to estimate sensitivity of the diagnostic tests. This included H5N1 infected swabs from various bird species such as chickens, various other land-based poultry, ducks and geese and various wild birds.

Competitive ELISA

In this technique, antibody is first incubated in solution with a sample containing antigen. The antigen-antibody mixture is then added to an antigen coated microtitre well. The more the antigen present in the sample, the less free antibody will be available to bind to the antigen-coated well. Addition of an enzyme-conjugated secondary antibody (Ab$_2$) specific for the isotype of the primary antibody can be used to determine the amount of primary antibody bound to the well as in an indirect ELISA. In the competitive assay, however, the higher the concentration of antigen in the original sample, the lower is the absorbance.

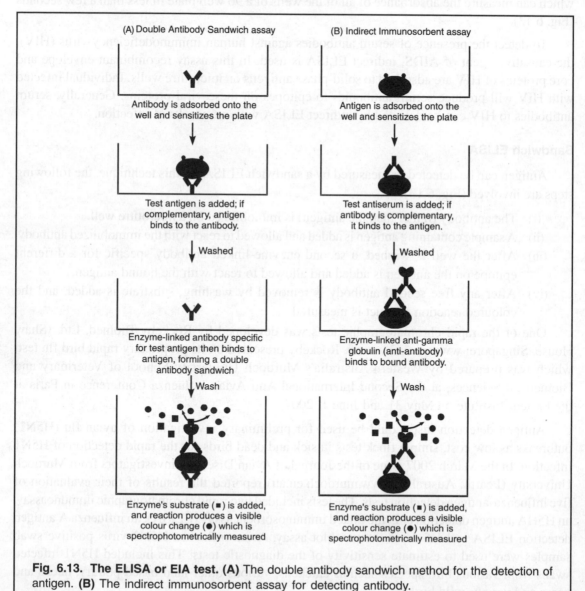

(A) Double Antibody Sandwich assay

Antibody is adsorbed onto the well and sensitizes the plate

Test antigen is added; if complementary, antigen binds to the antibody.

↓ Washed

Enzyme-linked antibody specific for test antigen then binds to antigen, forming a double antibody sandwich

↓ Wash

Enzyme's substrate (■) is added, and reaction produces a visible colour change (●) which is spectrophotometrically measured

(B) Indirect Immunosorbent assay

Antigen is adsorbed onto the well and sensitizes the plate

Test antiserum is added; if antibody is complementary, it binds to the antigen.

↓ Washed

Enzyme-linked anti-gamma globulin (anti-antibody) binds to bound antibody.

↓ Wash

Enzyme's substrate (■) is added, and reaction produces a visible colour change (●) which is spectrophotometrically measured

Fig. 6.13. The ELISA or EIA test. (A) The double antibody sandwich method for the detection of antigen. **(B)** The indirect immunosorbent assay for detecting antibody.

Chemiluminescence

Measurement of light produced by chemiluminescence during certain chemical reactions is a convenient and highly sensitive alternative to absorbance measurements in ELISA assays. In this a luxogenic (light-generating) substrate is used in place of chromogenic substrate in conventional ELISA reactions. For example, oxidation of the compound luminol by H_2O_2 and the enzyme horseradish peroxidase (HRP) produces light:

$$\text{Ab-HRP} + \text{Ag} \longrightarrow \text{Ab-HRP- Ag} \xrightarrow[H_2O_2]{\text{Luminol}} \text{Light}$$

The chemiluminescence assays are more sensitive than the chromogenic ones. By switching from a chromogenic to luxogenic substrate in general, the detection limit can be increased to more than 200 folds. The addition of enhancing agents increases it further in fact, under ideal conditions, as little as 5×10^{-18} moles of target antigen have been detected.

6.2.7 Complement Fixation Test

Complement enters into a complex with certain antigen-antibody reactions. Once complement has entered into this complex, it is not free to enter into a second reaction, that is, it is fixed. By finding out, if the complement has become fixed, it is possible to determine, if an antigen antibody reaction has taken place. To destroy any residual complement that might be present, serum sample is first heated to 56°C for 30 minutes (Fig. 6.14).

The known antigen and a standard amount of complement are then added to the serum. If the serum contains specific antibodies to the antigen, a complex of complement, antigen and antibody will form and complement will be fixed. An indicator system is used to determine if the reaction has occurred. A known antigen and a known antiserum are used in the **indicator system.** The antigen is sheep red blood cells (SRBCs) and the antiserum contains antibodies against it (SRBC's). The sheep red blood cells and antiserum are added to the original test system; if a specific reaction does not occur in the original test, the

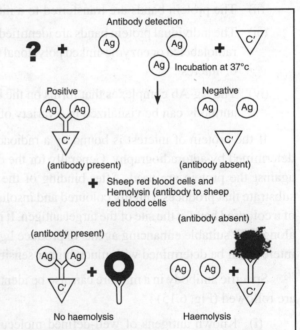

Fig. 6.14. Complement fixation test : The patient's serum is tested for the presence of antibody molecules (Y shape). Serum is mixed with the antigen (Ag within circles) and complement (C within triangle). If the serum contains complement-interacting antibodies (top), complement will be fixed, and become unavailable. Haemolysin (antibody) coated sheep red blood cells are added later. On the other hand, if the patient serum is negative for antibodies (bottom) complement will be available to lyse the haemolysin-coated sheep RBCs. Thus, haemolysis indicates the absence of complement fixing antibodies.

complement is still available to complex with the indicator system which results in the lysis (dissolving) of the red blood cells. This reaction is readily visible.

If the sheep red blood cells do not lyse, it is assumed that the original reaction fixed the complement and so specific antibodies must have been present in the original serum. This is an indirect test for antibodies.

6.2.8 Western Blotting

Western blotting helps in the identification of specific protein in a complex mixture of proteins. The following steps are involved :

(i) A protein mixture is electrophoretically separated on an SDS-polyacrylamide gel (SDS-PAGE), a polyacrylamide slab gel infused with sodium dodecyl sulfate (SDS), a dissociating agent.

(ii) The protein bands are transferred to a nitrocellulose membrane by electrophoresis

(iii) The individual protein bands are identified by flooding the nitrocellulose membrane with radiolabeled or enzyme-linked polyclonal or monoclonal antibody specific for the protein of interest.

(iv) The Ag-Ab complexes that forms on the band containing the protein recognized by the antibody can be visualized in a variety of ways.

If the protein of interest is bound by a radioactive antibody, its position on the blot can be determined by autoradiography. Generally for the detection of protein enzyme-linked antibodies against the protein are used. After binding of the enzyme-antibody conjugate, a chromogenic substrate that produces a highly coloured and insoluble product is added. It causes the appearance of a coloured band at the site of the target antigen. It is possible to use chemiluminescent compound along with suitable enhancing agents to produce light at the antigen site, the site of the protein of interest can be determined with much higher sensitivity.

Specific antibody in a mixture can also be identified by Western blotting. The following steps are followed (Fig. 6.15) :

(i) Known antigens of well-defined molecular weight are separated by SDS-PAGE and blotted onto nitrocellulose.

(ii) The separated bands of known antigens are then probed with the sample suspected of containing antibody specific for one or more of these antigens.

(iii) Reaction of an antibody with a band is detected by using either radiolabeled or enzyme-linked secondary antibody that is specific for the species of the antibodies in the test sample.

This procedure is used as confirmatory test for HIV. It is used to determine whether the patient has antibodies that react with one or more viral proteins.

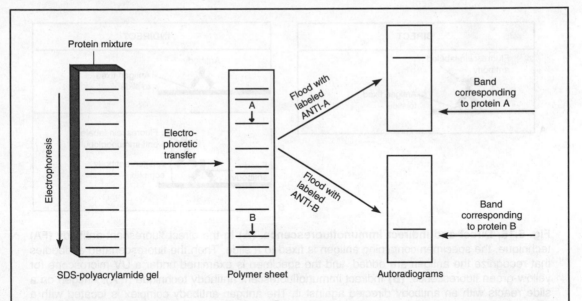

Fig. 6.15. Western blotting. In this a protein mixture is separated by electrophoresis. Then the protein bands are transferred onto a nitrocellulose or other PVDF membrane. After the sheet is flooded with radiolabeled specific antibodies, the various protein bands can be visualized by autoradiography.

6.2.9 Immunofluorescence

Albert Coons in 1944 showed that antibodies could be labeled with molecules that have the property of fluorescence. Fluorescent molecules absorb light of one wavelength (excitation) and emit that of another (emission). If antibody molecules are tagged with a fluorescent dye, or fluorochrome, immune complexes containing these fluorescently labeled antibodies (FA) can be detected by coloured light emission when excited by light of the appropriate wavelength. It is possible to see the antibody molecules bound to antigens in cells or tissue sections. The emitted light can be seen with a fluorescence microscope equipped with a UV light source.

- The most widely used label for immunofluorescence studies is fluorescein, which is an organic dye. It absorbs blue light (490 nm) and emits an intense yellow-green fluorescence (517 nm).

- Another organic dye, rhodamine absorbs in the yellow-green range (515 nm) and emits a deep red fluorescence (546 nm). It emits fluorescence at a longer wavelength than fluorescein. It, therefore, can be used in two-colour immunofluorescence assays.

- The light-gathering proteins are phycoerythrin and other phycobiliproteins. In some species of algae these play important roles in photosynthesis. Because they are efficient absorbers of light (~30-fold greater than fluorescein) and brilliant emitters of red fluorescence, these proteins have become widely used labels for immunofluorescence.

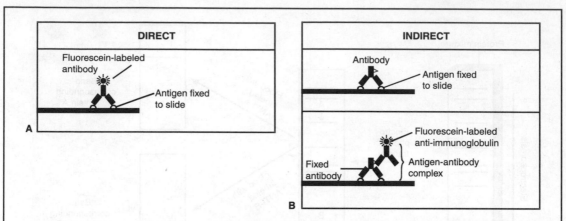

Fig. 6.16. Direct and indirect immunofluorescence. (A) In the direct fluorescent-antibody (FA) technique, the specimen containing antigen is fixed to a slide. Then the fluorescenated antibodies that recognize the antigen are added, and the specimen is examined under a UV microscope for yellow-green fluorescence. **(B)** Indirect immunofluorescent antibody technique (IFA). Antigen on a slide reacts with an antibody directed against it. The antigen-antibody complex is located with a fluorescent labelled secondary antibody that recognizes the primary antibody.

The cell membrane molecules or tissue sections can be directly or indirectly stained with fluorescent-antibody (Fig. 6.16).

(*i*) *Direct immunofluorescence:* In this the specific antibody (the primary antibody) is directly conjugated with fluorescein.

(*ii*) *Indirect immunofluorescence:* The primary antibody is unlabeled and is detected with an additional fluorochrome-labeled reagent.For indirect staining, a number of reagents have been developed. The most common is a secondary antibody, a fluorochrome-labeled anti-isotype antibody, such as fluorescein-labeled goat anti-mouse immunoglobulin. Another reagent is fluorochrome labeled protein A from *Staphylococcus aureus*. It binds with high affinity to the Fc region of IgG antibody molecules. Another indirect approach uses a secondary biotin-conjugated anti-isotype antibody. It is followed by fluorochrome-conjugated avidin, a protein that binds to biotin with extremely high affinity.

6.2.10 Flow Cytometry

The fluorescent antibody techniques are extremely valuable qualitative tools, but they do not provide quantitative data. This drawback, has been overcome by the development of flow cytometry (Fig. 6.17). A version of this is called the FACS (fluorescence-activated cell sorter). It was designed to automate the analysis and separation of cells stained with fluorescent antibody. To count single intact cells in suspension, a laser beam and light detectors are used in FACS. The laser light is deflected from the detector when a cell passes the laser beam in the FACS and this interruption of the laser signal is recorded.

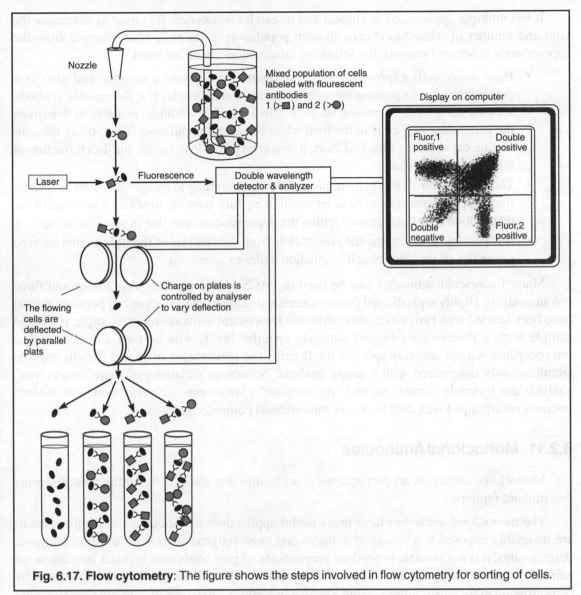

Fig. 6.17. Flow cytometry: The figure shows the steps involved in flow cytometry for sorting of cells.

The cells with a fluorescent tagged antibody bound to their cell surface antigens are excited by the laser. The emitted light is recorded by a second detector system located at a right angle to the laser beam. The simplest form of the instrument called FAC Scan counts each cell as it passes the laser beam. It records the level of fluorescence emitted by the cells. Plots of the number of cells are generated as the ordinate and their fluorescence intensity as the abscissa by the attached computer. A FAC Scan cannot sort or separate the cells.

The FACSort is used to determine which and how many members of a cell population bind fluorescent labeled antibodies in cell sorting and the instrument is used to place cells having different patterns of reactivity into different containers.

It has multiple applications in clinical and research laboratories. It is used to determine the kind and number of white blood cells in each population in patient's blood sample from the appropriately processed analysis, the following information can be obtained.

- How many cells express the target antigen as an absolute number and also as a percentage of cells passing through the beam. For example, if a fluorescent antibody specific for an antigen present on all T cells is used, it would be possible to determine the percentage of T cells in the total white blood cell population. Then, using the cell-sorting capabilities of the FACSort, it would be possible to isolate the T-cell fraction of the leucocyte population.

- The distribution of cells in a sample population according to antigen densities as seen by fluorescence intensity can be determined. It is, thus, possible to obtain a measure of the distribution of antigen density within the population of cells that possess the antigen.

- It is possible to determine the size of cells from the analysis of the light-scattering type properties of the types of cell population under examination.

Many fluorescent antibodies can be used in FACS at one time for two colour and three colour analysis. Highly sophisticated flow cytometers simultaneously analyze cell populations that have been labeled with two or even three different fluorescent antibodies. For example, if a blood sample with a fluorescence tagged antibody specific for T cells is used and also with a phycoerythrin-tagged antibody specific for B cells, the percentages of B and T cells may be simultaneously determined with a single analysis. Numerous variations of such "two-colour" analyses are routinely carried out, and "three-colour" experiments are performed with modern instruments equipped with dual laser and sophisticated computer analysis.

6.2.11 Monoclonal Antibodies

Monoclonal antibodies are preparations of antibodies that are specific for the same antigenic determinant (epitope).

The monoclonal antibodies have many useful applications in various areas. As the animals are invariably exposed to a variety of antigens and most antigens contain a variety of antigenic determinants, it is not possible to produce preparations of pure antibodies in intact animals as the animal serum will always contain a mixture of antibodies. Monoclonal antibodies are produced by cells growing in (*in vitro*) cultures, using a unique procedure. Two types of cells are fused together into a single cell called hybridoma. The hybridoma cells are able to express properties of both the parent cells. The two cells used to make monoclonal antibodies are mouse tumour cells, called **myeloma cell** and an activated B cell from the spleen of an immunized mouse. The myeloma cell has ability to grow indefinitely in a culture, but is not able to produce specific antibodies. The activated B cell can produce specific antibodies against a single antigenic determinant, but cannot grow in culture. Once a hybridoma that is secreting antibodies against a known antigen is produced, clones of this cell can be propagated which in turn will produce large amounts of the pure or monoclonal antibodies. It is necessary to screen many hybridoma cells to find the one that is producing the desired specific antibody.

6.3 DNA DIAGNOSTIC METHODS

6.3.1 Nucleic Acid Hybridization Techniques

The process of nucleic acid hybridization involves the formation of stable double-stranded nucleic acid from complementary single-stranded molecules. The single-stranded molecules can be RNA or DNA, and the resultant hybrids formed can be DNA-DNA, RNA-RNA or DNA-RNA.

A probe is a labeled single-strand sequence of nucleic acid that is complementary to the nucleic acid sequence to be detected and can either be DNA or RNA.

The targeted nucleic acid sequence is referred to as the target and it too can be RNA or DNA. This target can be located within the specimen or in a colony, either from an agar plate or broth culture.

6.3.1.1 Preparation of Nucleic Acid Probes

In standard nucleic acid hybridization assays, the probe is labeled in some way. Nucleic acid probes may be made as single-stranded or double-stranded molecules, but the working probe must be in the form of a single-strand (Fig. 6.18).

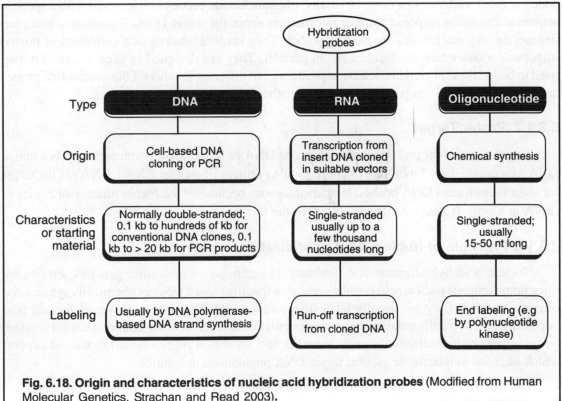

Fig. 6.18. Origin and characteristics of nucleic acid hybridization probes (Modified from Human Molecular Genetics, Strachan and Read 2003).

6.3.1.1.1 Conventional DNA Probes

These are isolated by cell-based DNA cloning or via PCR. In both the cases, probes to begin with are usually double-stranded. DNA cloned within the cells may range in size from 0.1 kb to hundreds of kilobases, but DNA cloned by PCR is usually less than 1 kb in length. The probes are usually labeled by incorporating labeled dNTPs during an *in vitro* DNA synthesis reaction.

6.3.1.1.2 RNA Probes

They are derived from single-stranded RNA molecules typically a few hundred bp to several kilobases long. They can be conveniently generated from DNA which has been cloned in a specialized plasmid vector containing a phage promoter sequence immediately adjacent to the multiple cloning site. The RNA synthesis reaction is performed using the relevant phage RNA polymerase and the four rNTPs, at least one of which is labeled. Specific labeled RNA transcripts can then be generated from the cloned insert.

6.3.1.1.3 Oligonucleotide Probes

These are single-stranded and very short (typically 15-50 nucleotides long). These are made by chemical synthesis (unlike other probes which originate from DNA cloning). To a starting mononucleotide which is initially bound to a solid support, mononucleotides are conventionally added at the 3' end, one at a time. Generally, oligonucleotide probes are designed with a specific sequence chosen in response to prior information about the target DNA. Sometimes, however, degenerate oligonucleotides are used as probes. They involve labeling of a collection of related oligonucleotides which are synthesized in parallel. They are designed in such a way that they need to be identical at certain nucleotide positions, but different at others. Oligonucleotide probes are often labeled by incorporating ^{32}P atom or other labeled group at the 5' end.

6.3.1.2 Probe Target

Target material for probe technology can be DNA or RNA. Most commercial probes utilize DNA as a target. Gen-Probe Inc. (San Diego, CA) utilizes ribosomal RNA (rRNA) as the target for chemiluminiscent DNA probes. It is advantageous because of the higher number of copies of rRNA in a cell. It increases the sensitivity of the test.

6.3.1.3 Principle of Nucleic Acid Hybridization

Nucleic acid hybridization is a fundamental technique in molecular genetics. Probes are often homogeneous nucleic acid population (*e.g.,* a specific cloned DNA or chemically synthesized oligonucleotide). They are usually labeled in solution. On the other hand, target nucleic acid populations are typically unlabeled complex populations and are usually bound to a solid support. Some important hybridization assays, however, use unlabeled probes bound to a solid support which are used to interrogate labeled target DNA populations in solution.

The specificity of the interaction between probe and target comes from base complementarity, because both populations are treated in such a way as to ensure that all the nucleic acid sequences present are single stranded. Thus, if either the probe or the target is initially double-stranded, the individual strands must be separated (denatured), generally by heating or by alkaline treatment.

After mixing single strands of probe with single strands of target, strands with complementary base sequences are allowed to reassociate (= reanneal) to form double-stranded nucleic acids. This leads to the formation of two types of products (Fig. 6.19) :

- **Homoduplexes :** Complementary strands within the probe, or within the target reanneal to regenerate double-stranded molecules originally found in the probe or target populations;
- **Heteroduplexes :** A single-stranded nucleic acid within the probe base-pairs with a complementary strand in the target population. The double-stranded nucleic acid formed represents a new combination of nucleic acid strands. Heteroduplexes define the usefulness of a nucleic acid hybridization assay because the whole purpose of the hybridization assay is to use the known probe to identify related nucleic acid fragments in the target.

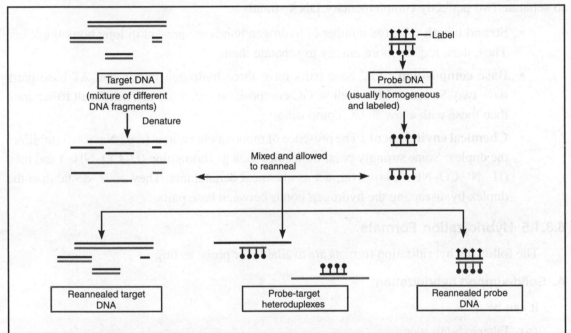

Fig. 6.19. A nucleic acid hybridization assay. It requires the formation of heteroduplexes between labeled single-stranded nucleic acid probes and complementary sequences within a target nucleic acid. (Adopted and modified from Human Molecular Genetics, Strachan and Read, 2003)

6.3.1.4 Melting Temperature and Hybridization Stringency

Stringency can be regarded as the specificity with which a particular target sequence is detected by hybridization probes. Stringency is most commonly controlled by the temperature and salt concentration in the post hybridization wash. The melting temperature (Tm) of a probe-target hybrid can be calculated to provide a starting point for the determination of correct stringency. The Tm is the temperature at which the probe and target are 50% dissociated. For probe longer than 100 base pairs.

$$Tm = 81.5°C + 16.6 \log M + 0.41 \, (\% \, G + C)$$

The Tm can be determined experimentally, but is usually calculated from the following formula called "Wallac Rule"

$$Tm = (4 \times [G + C] + 2 \times [A + T])$$

Where, [G + C] is the number of G and C nucleotide in the primer sequence and [A + T] is the number of A and T nucleotides.

Denaturation of double-stranded probe DNA is generally carried out by heating a solution of the labeled DNA to a temperature that is high enough to break the hydrogen bonds holding the two complementary DNA strands together. The following factors determine the energy required to separate two perfectly complementary DNA strands :

- **Strand length :** A large number of hydrogen bonds are present in long homoduplexes. They, thus, require more energy to separate them

- **Base composition :** GC base pairs have three hydrogen bonds and AT base pairs have two. Strands with a high % GC composition are, thus, more difficult to separate than those with a low % GC composition

- **Chemical environment :** The presence of monovalent cations (*e.g.*, Na^+ ions) stabilizes the duplex. Some strongly polar molecules such as formamide ($H–CO–NH_3^+$) and urea ($H–_3N^+–CO–NH_3^+$), however, act as chemical denaturants. They, thus, destabilize the duplex by disrupting the hydrogen bonds between base pairs.

6.3.1.5 Hybridization Formats

The following hybridization formats are available for probe testing :

A. Solid-support hybridization

It can be :

  (a) Filter hybridization

  (b) Modified filter : microtitre tray or test tube hybridization

  (c) Sandwich hybridization

B. In solution hybridization

    In situ hybridization

C. Southern hybridization

A. Solid–support hybridization

(a) **Filter hybridization :** In filter hybridization with oligonucleotide probes, the hybridization step is usually performed at 5°C below Tm for perfect matched sequences. For every mismatched base pair, a further 5°C reduction is necessary to maintain stability. In this, the target nucleic acid is secured to a membrane (nitrocellulose or nylon fibre filter paper) (Fig. 6.20). The probe is then overlaid to attach to target, if present after a series of washes, detected as a dot, spot or blot (Fig. 6.21). It depends upon the manner in which nucleic acid is added to the membrane.

Immobilized target   Probe   Hybrid

Fig. 6.20. A, Solid-state hybridization. The target is immobilized on the substrate. The labeled probe is made to hybridize with this fixed probe. (Modified from Diagnostic Microbiology by Mahon *et.al.*, 2000)

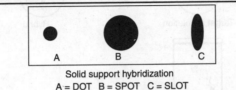

Solid support hybridization
A = DOT   B = SPOT   C = SLOT

Fig. 6.21. Resultant reactions of solid support hybridization: It can be seen as a dot, spot or slot.

(b) **Test tube/tray hybridization :** This method may be modified to use microtitre tray, well or inside of a test tube as the solid support.

(c) **Sandwich hybridization :** Two probes are utilized by the solid support.

(i) Unlabeled probe is attached to the solid support

(ii) The other labeled probe is added after the target. If present, it allows to hybridize to the first probe (Fig. 6.22).

This "sandwich" may help to reduce background non-specific binding at the expense of some reduced sensitivity.

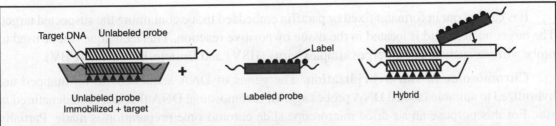

Target DNA   Unlabeled probe   Label

Unlabeled probe   Labeled probe   Hybrid
Immobilized + target

Fig. 6.22. Sandwich hybridization : The target DNA is fixed on the substratum. The unlabeled probe is hybridized with the DNA. Then a labeled probe is added to it, and if complementary sequences are present, hybridization takes place. (Modified from Diagnostic Microbiology by Mahon *et.al., 2000*)

Reverse hybridization assays : In these the probe population is unlabeled and fixed to the solid support. The target nucleic acid, however, is labeled and present in aqueous solution.

B. In solution hybridization

It involves movement of both target and probe in a liquid environment. It provides kinetics of interaction. The technique may be 5 to 10 times faster than solid-support hybridization. The use of hydroxylapatite, magnetic beads, or differential chemical hydrolysis has facilitated the required differential separation and removal of the unbound probe. It depends upon the commercial application of probe technology (Fig. 6.23).

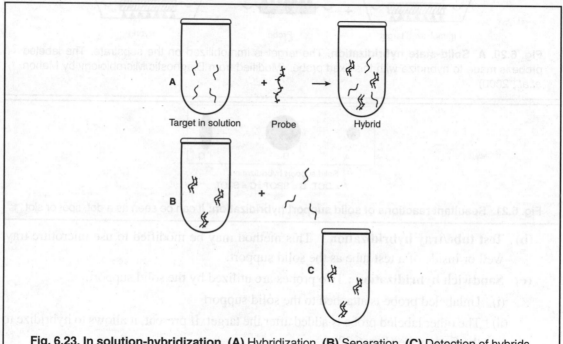

Target in solution   Probe   Hybrid

Fig. 6.23. In solution-hybridization. (A) Hybridization, **(B)** Separation. **(C)** Detection of hybrids.
(Modified from Diagnostic Microbiology by Mahon *et.al.,2000*)

(a) *In situ* hybridization

It is carried out in formalin fixed or paraffin embedded tissue containing the suspected target. The target nucleic acid is located in the tissue by positive reaction. This technique can be used to probe viruses such as HPV, herpes simplex virus (HSV) and Epstein-Barr virus (EBV).

Chromosome *in situ* hybridization. The genes of DNA sequences to be mapped are hybridized to suitable labeled DNA probes against chromosomal DNA that has been denatured *in situ*. For this purpose an air-dried microscope slide chromosome preparation is made. Partially purified chromosomal DNA is obtained by treatment with ribonuclease (RNase) and proteinase K. It is denatured by exposure to formamide. The denatured DNA is then available for *in situ* hybridization with an added solution containing a labeled nucleic acid probe. It is overlaid with a coverslip. Chromosome banding of the chromosomes can be arranged either before or after the hybridization step. It, however, depends on:

(a) the techniques used

(b) the signal obtained after removal of excess probe with the chromosome band pattern

Tissue *in situ* hybridization. A labeled probe is hybridized against RNA in tissue section (Wilkinson, 1998).

(i) Tissue sections are made from either paraffin-embedded or frozen tissue using a cryostat, and then mounted on to glass slide.

(ii) A hybridization mix including the probe is applied to the section on the and slide covered with a glass coverslip.

(iii) Formamide at a concentration of 50% is used to reduce the hybridization temperature and minimize evaporation problems.

Although double-stranded cDNAs have been used as probes, single-stranded complementary RNA probes (ribo-probes) are preferred. The sensitivity of initially single-stranded probes is generally higher than that of double-stranded probes, presumably because a proportion of the denatured double stranded probe renatures to form probe homoduplexes. cRNA riboprobes that are complementary to the mRNA of a gene are known as antisense riboprobes. They can be obtained by cloning a gene in the reverse orientation in a suitable vector such as pSP64. The phage polymerase in such cases will synthesize labeled transcripts from the opposite DNA strand to that which is normally transcribed *in vivo*. Useful controls for such reactions include sense riboprobes which should not hybridize to mRNA except in rare occurrences where both DNA strands of a gene are transcribed.

Using either selected radioisotopes, notably ^{35}S, or by non isotopic labeling, labeling of probes is performed. The hybridized probe in the former case is visualized using autoradiographic procedures. The localization of the silver grains is often visualized using only dark field microscopy (direct light is not allowed to reach the objective, instead the illuminating rays of light are directed from the side so that only scattered light enters the microscopic lenses and the signal appears as an illuminated object against a black background). However, bright field microscopy (where the image is obtained by direct transmission of light through the sample) provides better signal detection.

Hybridization assays using cloned target DNA and microarrays

Colony blot and plaque lift hybridization. Colonies of bacteria or other suitable host cells which contain recombinant DNA can generally be selected or identified by the ability of the insert to inactivate a marker vector gene *(e.g.,* β galactosidase, or an antibiotic resistance gene). If, however the desired recombinant DNA contains a DNA sequence that is closely related to an available nucleic acid probe, it can be specifically detected by hybridization. In the case of bacterial cells used to propagate plasmid recombinants, the cell colonies are allowed to grow on an agar surface and then transferred by surface contact to a nitrocellulose or nylon membrane. This process is known as **colony blotting.**

Alternatively the cell mixture is spread out on a nitrocellulose or nylon membrane placed on top of a nutrient agar surface. In both the cases colonies are allowed to form directly on top of the membrane. The membrane is then exposed to alkali to denature the DNA prior to hybridizing with a labeled nucleic acid probe.

The probe solution is removed after hybridization and the filter is washed extensively dried and submitted to autoradiography using X-ray film. The position of strong radioactive signals is related back to a master plate containing the original pattern of colonies, in order to identify colonies containing DNA related to the probe. These can be individually picked and amplified in culture prior to DNA extraction and purification of the recombinant DNA.

When using phage vectors similar procedure can be followed. The plaques which are formed following lysis of bacterial cells by phage will contain residual phage particles. A nitrocellulose or nylon membrane is placed on top of the agar plate in the same way as above and when removed from the plate will constitute a faithful copy of the phage material in the plaques, a so called plaque lift.

Gridded high density arrays of transformed cell clones or DNA. After making it possible to create complex DNA libraries, attempts were made to develop more efficient methods of clone screening. It was made possible to pick individual colonies and transfer them onto large membranes in the format of a high-density gridded array.

The application of robotic gridding devices simplified the process of generating arrays. These could perform the necessary spotting automatically by pipetting from clones arranged in microtitre dishes into pre-determined linear co-ordinates on a membrane. The rapid and efficient library screening was permitted by using high density clone filters. These could be copied and distributed to numerous laboratories throughout the world.

DNA microarray technology. Because of their huge capacity for miniaturization and automation (Schena *et al.*, 1998) DNA microarrays have provided a scale-up in hybridization assay technology. Microarray construction involves quite different procedures. Here, instead of porous membranes, the surface involved has chemically treated glass microscopic slides. There has now been a move towards more porous substrates such as nitrocellulose-coated glass surfaces in an effort to bind more DNA and increase sensitivity. According to differences in how the nucleic acid samples are generated and delivered to the microarray, there are two quite distinct types of microarray technologies.

Microarray of pre-synthesized nucleic acid. The nucleic acids present in the microarrays have previously been synthesized (often they consist of collections of different DNA clones but they could in principle be collections of previously synthesized oligonucleotides). The construction of the microarray in this case means that individual DNA clones or oligonucleotides are plotted at individual locations on the surface of a microscope slide, specified by precise x, y coordinates in a miniaturized grid.

Microarray of oligonucleotides synthesized *in situ*. The afymetrix company has pioneered this. It typically involves a combination of photolithography technology from the semi-conductor industry with the chemistry of oligonucleotide synthesis. In this case, many thousands of different oligonucleotides are assembled *in situ* on the surface of a glass slide, in a series of sequential synthesis steps involving, adding one nucleotide at a time. The process requires covalent coupling of mononucleotides to a linker molecule which terminates with a photolabile protecting group.

(b) Southern hybridization

The techinque of Southern hybridization is widely used in a clinical or forensic setting to identify individuals, determine relatedness, and to detect genes associated with genetic abnormalities or viral infections. Southern blot analysis is also used in basic scientific research to confirm the presence of an exogenous gene, evaluate gene copy number or to identify genetic aberrations in models of disease.

The first step in successful Southern hybridization is to obtain DNA that is reasonably intact. The test DNA must be fragmented with restriction enzymes, which cut the double strands of DNA at multiple sequence-specific sites. This creates a set of fragments of specific sizes that represent the region of DNA between restriction sites. The fragmented DNAs are size-fractionated via agarose gel electrophoresis and are subsequently denatured, which enables them to hybridize to complementary nucleic acid probes. The DNAs are transferred to a solid support such as nylon or nitrocellulose membrane via capillary action or electrophoretic transfer and are permanently bound to the membrane by brief ultraviolent (UV) crosslinking or by prolonged exposure to temperature of 80°C or so. Blots at this stage may be stored for later use or may be probed immediately (Fig. 6.24). For separating very large fragments of DNA an alternative electrophoresis method is required.

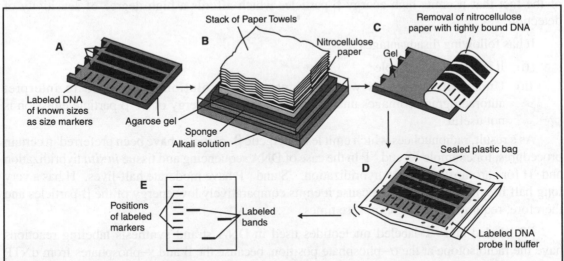

Fig. 6.24. Southern hybridization. (A) DNA fragments separated by agarose gel electrophoresis. **(B)** Separated DNA fragments are blotted onto nitrocellulose paper. **(C)** Labeled DNA probe is hybridized to separated DNA. **(D)** Labeled DNA probe is hybridized to complementary DNA bands visualized by autoradiography. **(E)** Autoradiograph showing the bands. (From Internet)

Northern blot hybridization

It is a variant of Southern blotting. In this the target nucleic acid is undigested RNA. This method is based on the principle to obtain information on the expression pattern of specific genes.

The cloned gene can be used as a probe and hybridized against a Northern blot containing samples of RNA isolated from a variety of different tissues in different lanes. This can provide information on the range of cell types in which the gene is expressed and the relative abundance of transcripts. Evidence for different isoforms (*e.g.*, resulting from alternative promoters, splice sites, polyadenylation sites *etc.*) can be provided by revealing transcripts of different sizes.

6.3.1.6 Preparation of Probe and its Labeling

A label or marker is needed to detect the hybridization reaction that may have occurred between probe and target. In most cases, this involves labeling of the probe with any of the markers.

6.3.1.6.1 Isotopic labeling and detection

Traditionally, nucleic acids have been labeled by incorporating nucleotides, containing radioisotopes. Such radiolabeled probes contain nucleotides with a radioisotope generally ^{32}P, ^{33}P, ^{35}S or ^{3}H. These can be specifically detected in solution or much more commonly within a solid specimen.

The intensity of an autoradiographic signal is dependent on the intensity of radiation emitted by the radioisotope, and the time of exposure, which may often be long (1 or more days, or even weeks in some applications). ^{32}P has been widely used in nucleic acid hybridization assays because of the fact that it emits high-energy β-particles which affords a high degree of sensitivity of detection.

It has following disadvantages :

(i) It is relatively unstable

(ii) Under circumstances when fine physical resolution is required to interpret autoradiographic images unambiguously, the high energy of ^{32}P β particle emission is not useful.

As a result, radionuclides which emit less energetic β–particles have been preferred in certain procedures, for example ^{35}S and ^{33}P in the case of DNA sequencing and tissue *in situ* hybridization and ^{3}H for chromosome *in situ* hybridization. ^{35}S and ^{33}P have moderate half-lives. ^{3}H has a very long half life, but is not useful because it emits comparatively low energy of the β-particles and therefore, requires very long exposure times.

^{32}P-labeled and ^{33}P-labeled nucleotides used in DNA strand synthesis labeling reactions have the radioisotope at the α–phosphate position, because the β and γ–phosphates from dNTP precursors are not incorporated into the growing DNA chain. Kinase-mediated end-labeling, however, uses [γ-^{32}P] ATP.

In the case of ^{35}S-labeled nucleotides which are incorporated during the synthesis of DNA or RNA strands, the NTP or dNTP carries a ^{35}S isotope in place of the O⁻ of the α-phosphate group.

^{3}H-labeled nucleotides carry the radioisotope at several positions. Specific detection of molecules carrying a radioisotope is most often performed by autoradiography.

6.3.1.6.2 Non-isotopic labeling systems

These have widespread applications in a variety of different areas. Two major types of non-radioactive labelings are used :

- **Direct non-isotopic labeling :** In this method nucleotide containing an attached labeled group is incorporated. Such systems generally, involve incorporation of modified nucleotides containing a fluorophore. The latter is a chemical group which can fluoresce when exposed to light of a certain wavelength (Fig. 6.25).

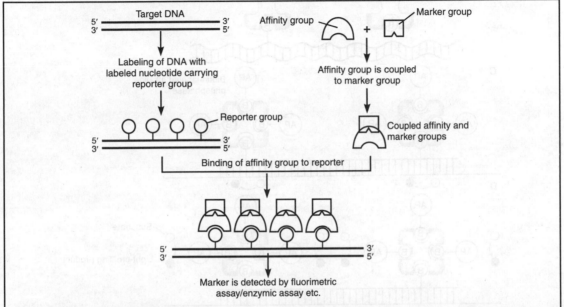

Fig. 6.25. General principles of indirect non-isotopic labeling. The reporter groups recognized by the protein (Often it is a specific antibody or any other ligand which has a very high affinity for specific group). In the case of biotin as a reporter, it is streptavidin. Detection of marker can be carried out in various ways. If it carries a specific fluorescent dye, its detection can be carried out by fluorimetric assay. If it is an enzyme (*e.g.*, alkaline phosphatase), it can be coupled to an enzyme assay. Its product can be colorimetrically measured. (Modified from Molecular Human Genetics, Vol 3, 2004).

- **Indirect non-isotopic labeling :** It involves chemical coupling of a modified reporter molecule to a nucleotide precursor. After incorporation into DNA, affinity groups can be specifically bound by an affinity molecule, a protein or other ligand which has a very high affinity for the reporter group. A marker molecule group which can be detected in a conjugated form to the affinity molecule is a marker molecule or group suitable assay. The reporter molecules on modified nucleotides should sufficiently protrude far from the nucleic acid backbone. It facilitates their detection by the affinity molecule and so to separate the nucleotide from the reporter group, a long carbon atom spacer is required.

There are two indirect non-isotopic labeling systems :

- **The biotin-streptavidin system :** It utilizes the extremely high affinity of two ligands:

(i) Biotin (a naturally occurring vitamin) which acts as the reporter,

(ii) It is a streptavidin bacterial protein, that is used as an affinity molecule.

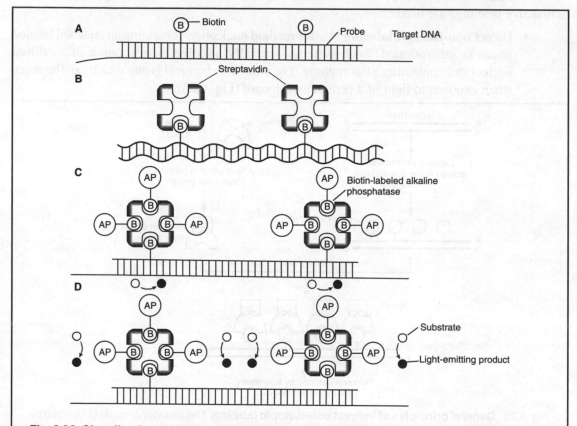

Fig. 6.26. Chemiluminescent detection of target DNA. (B, Biotin, AP, Alkaline phosphatase). **(A)** A biotin-labeled probe is bound to the target DNA. **(B)** Streptavidin is bound to the biotin molecules. **(C)** Biotin-labeled alkaline phosphatase binds to streptavidin, **(D)** Alkaline phosphatase converts the substrate into a light-emitting product (Adopted and modified from Molecular Biotechnology by Glick, B.R. and Pasternak, J.J. ASM Press, 1998).

Biotin and streptavidin bind together extremely tightly with an affinity constant of 10^{-14}, one of the strongest known in biology.

Biotinylated probes can be easily made by including a suitable biotinylated nucleotide in the labeling reaction (Fig. 6.26).

- **Digoxigenin :** It is a plant steroid (obtained from *Digitalis* plants). A specific antibody has been raised against it. The digoxigenin-specific antibody permits detection of nucleic acid molecules which have incorporated nucleotides containing the digoxigenin reporter group.

A variety of different marker groups or molecules can be conjugated to affinity molecules such as **streptavidin** or the **digoxigenin-specific antibody**. They include various fluorophores or enzymes such as **alkaline phosphatase** and **peroxidase** which can permit detection via colorimetric assays or chemiluminiscence assays *etc.*

Protocol

(*a*) A biotin labeled probe is hybridized to the target DNA.

(*b*) Streptavidin is bound to the biotin molecules. Alternatively, avidin, a chicken egg white protein can be added.

(*c*) Biotin-labeled alkaline phosphatase (or peroxidase) binds to streptavidin.

(*d*) A chromogenic or chemiluminiscent substrate is added depending on the biotin labeled enzyme used. The conversion of substrate into product is accompanied either by colour change or production of light. This indicates the hybridization of target with probe DNA.

This system has following advantages :

(*a*) The harmful substances like X-rays are not involved.

(*b*) Biotin-labeled DNA is stable for atleast one year at room temperature.

(*c*) Sensitivity of chemiluminiscence detectors is as high as radioactivity detectors.

(*d*) Signal detection can be completed within few hours.

6.3.1.6.3 Limitations of Usages of Probe

(**i**) **Radioactivity :** The use of radioactive probes is limited to the laboratories having sophisticated facilities and probe license. But as mentioned above, non radioactive probes can be used.

(**ii**) **Small quantity of target material :** Sometimes the material to be detected is present in extremely small quantities that

(*i*) cannot be seen under a light microscope,

(*ii*) cannot generate a detectable immune response and

(*iii*) in the quantities that often lie below the limit of detection of standard immunoassays, which is of the order of 10^8 molecules of target protein.

The problems have been solved by the advent of the technique of PCR.

6.3.2 Pulsed Field Gel Electrophoresis (PFGE)

Agarose gel electrophoresis can resolve DNA fragments in a limited size range, from 100 bp to about 30 kb. The DNA molecules pass through pores in the agarose gel and small molecules are able to migrate more quickly through the pores. The sieving effect is no longer effective above a certain size of DNA fragment, however, the resolution of DNA fragments above 40 kb is quite limited.

PFGE can resolve DNA fragments in a size range from about 20 kb to several Mb in length. The very large DNA molecules contained in mammalian chromosomes-typically hundreds of Mb in length-cannot be size-separated by this method. Specialized restriction nucleases, however, can cleave vertebrate DNA rather infrequently producing large restriction fragments which can be size separated by PFGE. These *rare-cutter restriction endonucleases* often recognize GC-rich recognition sequences which contain one or more CpG dinucleotides. Human DNA and other vertebrate DNAs have comparatively few recognition sequences for restriction enzymes which cut at sequence CpGs because the CpG dinucleotide occurs at low frequencies in vertebrate DNA.

6.3.3 Polymerase Chain Reaction (PCR)

The exponential amplification methods that produce a gain of 10^8 fold are based on either DNA or RNA replication. This methodology of PCR is used in the field of diagnostic assays for the detection of infectious agents or genetic defects. PCR primers can be designed which will specifically amplify the required (may be infectious agent) DNA present within a DNA sample (Fig. 6.27). After amplification, hybridization should detect many copies of that DNA now present. If the patient is not infected, then nothing will be amplified by this. There will, thus, be no hybridization signal. PCR kits are already available for commercial diagnostic systems.

Clinical microbiology and molecular pathology laboratories have more recently adapted amplification methodologies for the detection of pathogens and/or their products in infectious material. The methods potentially can approach the sensitivities of culture and are in general higher than those of probes. Amplification is a molecular procedure that increases the number of nucleic acid copies in a specimen to millions in a very short period of time (often less than 5 hours). With its inception, it first appeared that amplification would take over all conventional methods in the clinical laboratory. However, it has become obvious that the technique, although excellent in some areas, may not always be the method of choice. Its successes include identification of organisms for which there are no culture methods available or for diseases in which the need to know the answer very rapidly overrides the costs.

The first method of amplification was Polymerase Chain Reaction (PCR). It is a major mode of this molecular technique and leads the way in many applications. Originally described by Mullis in 1990, it involves repeating three-step process

    (a)   Denaturation of target DNA,

    (b)   Annealing of primers (oligonucleotides) to the single-stranded DNA and

    (c)   Enzymatic extension to synthesize complementary strands of the DNA.

A typical reaction involves a total of 25 to 40 cycles. Each cycle is a sequential series of three different temperature settings. Table 6.2 identifies the action of each step in the amplification process. The methodology is similar to the *in vivo* mode of DNA replication. The following components are required :

    1.   A short single-stranded DNA (ssDNA) molecule, known as an oligonucleotide primer that initiates the DNA for amplification.

    2.   A DNA polymerase enzyme that catalyzes the formation of additional molecules of double-stranded DNA (dsDNA).

    3.   The individual building blocks of DNA, the deoxynucleotide bases (dATP, dCTP, dGTP and dTTP or dUTP).

Table 6.2: Reaction Steps Required for Amplification of DNA by PCR.

Step	Temperature°C*	Action
Denaturation of strands (dsDNA)	94	Double-stranded DNA separated into single (ssDNA)
Primer annealing	55	Attachment of oligonucleotide primers to complimentary regions on ssDNA
Extension	72	Synthesis of dsDNA (5'→ 3'); catalyzed by *Taq* DNA polymerase

* Temperature varies from reaction to reaction.

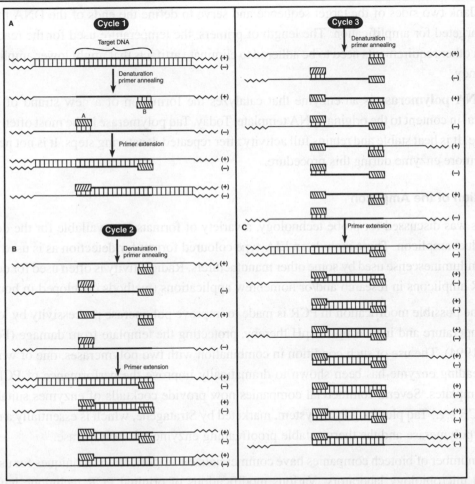

Fig. 6.27. Polymerase chain reaction. Schematic diagram of PCR **(A)** As a result of the first cycle, two long products are formed. These are defined at one end by a primer sequence. **(B)** The second cycle results in the generation of two long products and two intermediate products (defined at three ends by primer sequences). **(C)** The third cycle results in the generation of two short products defined at all the four ends by primer sequences. (Modified from New England Journal of Medicine 322 : 179, 1990).

Nucleic acid sequence information is available for most organisms of clinical importance. Careful evaluation of this information determines the target for amplification. Microorganisms typically contain unique region in their genetic material that is/are characteristic of a particular strain. Primers designed to recognize these unique regions permit highly specific identification of a single strain or species. Conversely, the amplification may be intentionally non-specific. If primers are chosen that flank a region shared by a diverse group of organisms, the resulting amplification would yield the detection of more than one type of organisms.

Proper selection of a pair of oligonucleotide primers is essential to the ultimate success of amplification. Primers are short pieces (18 to 28 nucleotides in length) of single-stranded DNA. They flank two sides of the target sequence and serve to define the ends of the DNA that have been targeted for amplification. The length of primers, the temperature used for the reaction, as well as other requirements need to be adhered to eliminate artifacts that might lower amplification efficiency.

DNA polymerase is an enzyme that catalyzes the formation of a new strand of dsDNA identical in content to the original DNA template. Today, Taq polymerase is the most often utilized enzyme. It is heat stable and retains full activity after repeated denaturing steps. It is not necessary to add more enzyme during this procedure.

Detection of the Amplicon

As was discussed for probe technology, a variety of formats are available for the detection of product amplicon. These include an EIA-type coloured format for detection as is used in PCR or chemiluminescense used by some other manufacturers. Radioactivity is often used for detection of PCR amplicons in research and/or homebrew applications (methods developed in house).

One possible modification in PCR is made to improve polymerase processivity by lowering the temperature and increasing the pH thereby, protecting the template from damage (Ford and Rose, 1994). The use of such condition in combination with two polymerases, one of which is a proofreading enzyme has been shown to dramatically improve the performance of PCR using long templates. Several commercial companies now provide cocktails of enzymes suitable for long PCR *e.g.,* Taq plus long PCR system, marketed by Stratagene, which is essentially a mixture of Taq polymerase and the thermostable proofreading enzyme Pfu polymerase.

A number of biotech companies have commercialized amplification techniques for use in the clinical microbiology laboratory. Various modifications of original PCR technique have been developed. The principle of Transcription Mediated Amplification (TMA) is the method of Gen-Probe Inc. (San Diego, CA). In this procedure, rRNA is the target material and a transcriptase enzyme is used to produce amplicons in a thermally stable environment. Because ssRNA is the target, no denaturation of the nucleic acid is needed.

Ligase Chain Reaction (LCR): It consists of three steps, as in PCR

(a) Denaturation of the target nucleic acid

(b) Annealing of primers

(c) An enzymatic ligation of the primers by a thermostable DNA ligase.

Two pairs of primers are used in the LCR process and these are designed to hybridize to adjacent target sequences. After annealing, these primers are linked by the ligase enzyme. Two chimeric single-standard oligonucleotides act as synthetic target molecules for further rounds of primer annealing and ligation. The amplicon is a short sequence of dsDNA 40 to 50 nucleotides long, containing one molecule each of the original four primers. The LCR procedure is very similar to PCR. It can, however, also be used in genetic screening, for example, for detection of a mutation that may be responsible for a specific disease or for detection of a mutation that might confer resistance to antibiotics.

Nucleic acid based sequence amplification (NASBA): It is a system developed by Organon Teknika (Raleigh-Durham, NC). The acronym *NucliSense* has recently been introduced in place of NASBA.This amplification procedure utilizes enzymes : reverse transcriptase, a RNase, and RNA polymerase. An RNA target is first reverse transcribed to produce a cDNA copy; these cDNA copies are transcribed into ssRNA molecules by the RNA polymerase. The recognition site for the polymerase is specified so that only correctly primed DNA sequences are transcribed. The amplicon is RNA.

Strand displacement amplification (SDA): It is the amplification product of Becton Dickinson Microbiology Systems (Sparks, MD). A PCR process occurs first, in which dsDNA target material is denatured, primers are annealed and complementary DNA is synthesized by a DNA polymerase. The newly formed strands are phosphorothiolated because of the incorporation of an alpha-thio-modified nucleotide triphosphate. The specific primers that anneal to specific target sites also contain a specific restriction endonuclease recognition site (Hind III). When Hind III is added to the reaction mixture, the second phase of SDA, different from PCR, is initiated. Hind III-produces nicks in the non-phosphorothiolated DNA polymerase. As the polymerase proceeds along the template, it "displaces" the nicked strand as an intact molecule. It is then capable of acting as a template for further reactions. Simultaneously, an exact replica of it is synthesized by the polymerase to replace the nicked strand. The thermally stable process of nicking, cDNA synthesis and strand displacement results in millions of copies of amplicons.

In the branched chain cDNA amplification, signal amplification has been utilized by the Chiron Crop (Emeryville, CA). By using a series of probes, one of which is a capture probe and another a detector probe, the specific RNA target is captured and attached to a substrate in a microtiter plate. The signal or detector probe is specific for the captured target. This latter probe has attached to it many "branches" that can themselves attach to enzyme molecules. These can then be detected with a chemiluminescent substrate. The number of branches per detector molecule is specific and consistent. Thus, it is possible to quantify the number of these molecules that are attached to the appropriate target. This method of amplification is also known as the **"Christmas tree method"**. It is because the branches are said to light up with the detector when positive for correct target and look like a Christmas tree. There are not much chances of contamination of subsequent reactions because the signal and not the target is amplified. It is in contrast to amplification systems in which the target itself is amplified.

Hybrid capture technique: Digene Diagnostics (Silver Springs, Md.) has formulated this method, which does not require any amplification. It can, however, provide a very sensitive method for the detection of target material in clinical specimens. The target material (specimen) is applied in solution, to a probe-labeled tube and if complementary, the probe and target will hybridize on the inside wall of the tube. Subsequently, a capture probe, specific for the hybrid only, is added. It captures any hybrids that have been formed. The capture probe is then detected.

Cycling probe technology : It is being employed by ID Biomedical Labs (Burnaby, Canada). This is a gene-based method and is used for the detection of specific target sequences. It utilizes a chimeric DNA-RNA-DNA probe containing an RNase H sensitive sessile link that is cleaved by RNase H when hybridized to a complementary DNA sequence. The cleaved probe fragments dissociate from the target. It releases the target for hybridization with another probe molecule. The cycle of hybridization, RNase H-mediated probe cleavage and probe dissociation repeats multiple times. This results in an accumulation of cleaved probe fragments. This technique has been used for the detection of the *mecA* gene of *S. aureus* and the resistance genes of vancomycin-resistant enterococcal strains.

Clinical Application of Amplification

The PCR is being used for the detection of many pathogens. In addition, any of the other amplication methods may be employed for specific detections. A combination of "home-brew" is used in clinical microbiology laboratories. It involves the PCR procedures that are developed by the laboratory itself. Commercially available purchased kits in which amplification can be performed are also used.

PCR based systems with fluorescent tags : In these systems, fluorescent dyes are used. A fluorescent compound emits light of a longer wavelength after it absorbs light of a shorter wavelength (Fig. 6.28).

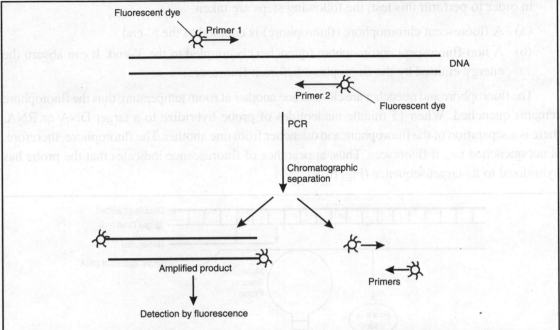

Fig. 6.28. Fluorophore based PCR. Primers are labeled with a fluorescent dye. The target DNA is amplified. The amplified DNA is separated by chromogenic assay and detected by fluorescence measuring techniques (Modified from Molecular Biotechnology by Glick, B.R. and Pasternale, J.J. ASM Press 1998, Glick).

Commonly used dyes are

- *Fluorescein*: It appears green under certain wavelengths.
- *Rhodamine*: It appears red.

General protocol of the system is given below :

(a) Fluorescent dye is bound to the 5′-end of each primer (primers are specific to the pathogen).

(b) The DNA is amplified by PCR using these fluorescent labeled primers.

(c) After amplification, the amplified fragments are separated from primers.

(d) The presence of label is then detected.

If target DNA is not present in the sample, then no fluorescent product will be observed.

6.3.4 Molecular Beacon System

A molecular beacon is a single-strand probe which is 25 nucleotide in length:

- 15 nucleotides in middle are complementary to the target,
- 5 nucleotides at each end are complementary to each other.

In order to perform this test, the following steps are taken.

(a) A fluorescent chromophore (fluorophore) is coupled to the 5′-end

(b) A non-fluorescent chromophore (quencher) is coupled to the 3′-end. It can absorb the energy emitted by the fluorophore before it fluoresces.

The fluorophore and quencher are close to one another at room temperature thus the fluorophore remains quenched. When 15 middle nucleotides of probe hybridize to a target DNA or RNA, there is a separation of the fluorophore and quencher from one another. The fluorophore, therefore, is not quenched *i.e.,* it fluoresces. Thus, appearance of fluorescence indicates that the probe has hybridized to its target sequence (Fig. 6.29).

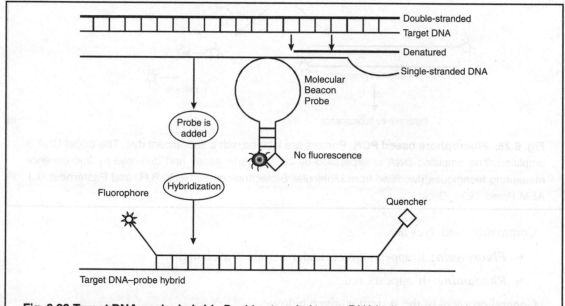

Fig. 6.29.Target DNA-probe hybrid : Double-stranded target DNA is denatured and hybridized with Molecular beacon probe (Modified from Molecular Biotechnology by Glick and Pasternak, 2003).

The temperature in this system is normally maintained at near-ambient. It is because the high temperature can also cause the nucleotides to become unpaired. This will result in separation of fluorophore and quencher and thus will give us false-positive results. The 15 nucleotides should be exactly complementary to target DNA.

Detection of Infection

If an individual is infected with a particular virus, then genetic material from that virus will be present and it will be different from the individual's DNA. Hybridization is quite suitable for the detection of unique DNA sequence. DNA can be isolated from the patient's blood. This DNA sample will include the viral DNA along with the normal components, assuming the individual is infected.

Questions

1. What do you know about the following:
   (a) ELISA
   (b) RIA
   (c) Probing of DNA
2. Write short notes on:
   (a) Molecular Beacon
   (b) PCR
   (c) Hybridization stringency
   (d) Hybridization formats
3. Write an essay on immunological diagnostic techniques.
4. (a) What are SNPs? Explain in detail.
   (b) What do you mean by mini and micro satellites ? Give their importance in genetic diagnostics.

Chapter 7

Applications of Biotechnology in Health

7.1 INTRODUCTION

Health care is the top priority of all the nations. Anything that may improve it is more than welcome, especially if it affects individuals directly and independently. For many life-threatening illnesses, access to drugs, diagnostics and vaccines remains limited. Medical biotechnology is likely to help in the following :

   (i) Improvement in the accuracy and speed of diagnosis of diseases and identification of pathogens.

  (ii) Prevention (through efficient, cheap and safe vaccines) of diseases.

 (iii) Management (new drugs and genetic profiles) of illnesses.

Attempts have been made to improve diagnosis, prognosis and therapeusis of the diseases by developing and using new and better agents. There have been dramatic changes in this area especially after the completion of human genome project and introduction of stem cell research.

This field has received significant attention from industry, government and international bodies. Biotechnology in this area has made significant inroads both in developed and developing countries. The field, with the exception of animal cloning and stem cell research, is generally less controversial. Developing countries, most of which are in the tropics, have a high disease burden. Some of the major health diseases include malaria and tuberculosis, which spread fast in countries with poor housing and sanitation, as well as HIV/AIDS. The economic development of nations, where millions of able-bodied people spend a significant proportion of their time bedridden, is seriously affected. Thousands of people in these countries die from preventable and treatable diseases. There are few alternative drugs for some of the most devastating diseases (*e.g.,* malaria) in poor countries as the economic cost of producing drugs may exceed the profits from their sales. Biotechnologically produced drugs may cut down the cost of drug development. Biotechnology and nanotechnology tools promise to increase the pace of target molecule validation, which will in turn shorten the time required to develop a number of drugs for the same disease. But the rate at which disease-causing organisms are developing resistance to the few available drugs, threatens to outpace drug development (Newton and White, 1999).

Biotechnology has provided the knowledge and tools to understand the processes that underlie many diseases and to design therapies that specifically interfere with them. The main areas are :

1. **Diagnostics:** These involve immuno-diagnostics, which make use of monoclonal antibodies, DNA probes *etc.*

2. **Therapeutics:** These include anti-sense oligonucleotides, biopharmaceuticals and therapeutic monoclonal antibodies.

3. **Targeted drugs:** These involve the use of liposomes, monoclonal antibodies, nasal delivery systems *etc.*

4. **Vaccines**

7.2 RECOMBINANT DNA TECHNOLOGY FOR THE DIAGNOSIS OF DISEASES

With the completion of the human genome project, DNA probes can be powerful tools for the diagnosis of human diseases, both genetic and acquired. Molecular diagnostic techniques have found applications in virtually all areas of medicine. These include the following:

• Detection of inherited mutations that underlie the development of genetic diseases either prenatally or after birth.

• Detection of acquired mutations that underlie the development of neoplasms, their accurate diagnosis and classification.

• Identification of individuals predisposed to conditions such as diabetes or coronary heart diseases.

• Diagnosis of infectious diseases, including AIDS disease.

• Determination of relatedness and identity in transplantation, paternity testing, and forensic medicine.

In addition, future uses of molecular diagnosis will include detection of polymorphisms that influence disease susceptibility, as occurs, for example, in lung cancer. Certain polymorphisms in the P-450 monooxygenase system predispose to cigarette smoke-induced lung cancer. In the field of pharmacogenomics, certain polymorphisms dictate susceptibility to actions and adverse effects of drugs.It is possible that with high throughput analysis, complete genetic profiles of individuals may allow presymptomatic diagnosis of all possible genetic diseases and the risk for environmentally induced disorders.

Most of these involve the use of hybridization techniques *i.e.* hybridization of target DNA with the probe DNA. The details of hybridization have been given in the chapter on technology.

7.2.1 Diagnosis of Infectious Diseases

Hybridization is quite suitable for the detection of unique DNA sequences. DNA can be isolated from the patient's blood. If the individual is infected, this DNA sample will include the infectious organism's (*e.g.*, virus, bacteria) DNA along with the normal components.

If we know what organism we are looking for and if the DNA sequence of at least a portion of that organism is already known, then a small probe that will specifically hybridize to that DNA can be constructed.

After the DNA sample has been isolated, it is separated by gel electrophoresis and transferred onto a nitrocellulose filter, hybridization with the viral probe can determine whether the organisms (virus, bacteria *etc.*) DNA is present in the DNA sample. If it is, then the specific infection can be diagnosed.

Probes are available from Gen-Probe Inc. (San Diego, CA) for detection of *Legionella sp.* and *Mycoplasma pneumoniae*. However, use of a radioactive label on the probe and lack of much interest among clinicians and laboratories for their use have made them somewhat obsolete. Direct specimen probes for viral respiratory pathogens are available from a variety of manufacturers, including ENZO Diagnostics, Inc. (Syosset, NY) and Digene (Silver Springs, MD). *In situ* probes for EBV, cytomegalovirus (CMV) and adenovirus are available, but in general, for research use only.

Bacteria. A variety of microbes can be confirmed by several acridinium ester-labeled probes (*e.g.*, Accuprobe, Gen-Probe, San Diego, CA).

Some probes can detect gastrointestinal pathogens, such as to detect enterotoxigenic (ETEC), enteroinvasive (EIEC), enteropathogenic (EPEC), enterohaemorrhagic (EHEC), and diffuse adherence (DAEC) *E. coli* strains causing diarrhoea. DNA probe can detect anaerobes and other bacteria associated with periodontal disease.

***Probes for Mycobacterium* sp.** A number of commercial probes are available for the culture confirmation of *Mycobacterium sp.* Accuprobe (Gene-Probe Inc., San Diego, CA) helps to detect *Mycobacterium tuberculosis* complex (*Mycobacterium tuberculosis, Mycobacterium bovis, Mycobacterium africanum* and *Mycobacterium microti*) and *Mycobacterium avium* complex (*M. avium* and *Mycobacterium intracellulare*), *Mycobacterium gordonae* and *Mycobacterium ransasii*. For all, isolates from solid medium or broth (BACTEC, MGIT or from one of the automated mycobacterial detection systems) can be probed in 1.5 to 2 hours. Many reports in the literature have verified the specificity of the probe. Sensitivity is not in question with solid media and testing from broth has a good sensitivity provided the amount of growth in the broth is sufficient. Probes are available for the direct detection of *Plasmodium sp., Borrelia burgdorferi,* hepatitis B and C viruses, and a variety of other infectious diseases.

Fungi. Accuprobes (Gen-Probe Inc., San Diego, CA) for fungal identification are available for *Histoplasma capsulatum, Blastomyces dermatitidis, Cryptococcus neoformans* and *Coccidioides immitis*. The same format for hybridization (*e.g.*, in solution hybridization using an acridinium ester label and chemiluminescent detection) is employed. A DNA probe for *Candida albicans* was reported by Cheung and Hudson (1988), but it was found to be genus specific.

Recent advances in the detection of infectious diseases

Scientists at University of Texas (UT) Southwestern, Children's Medical Center Dallas (TX, USA, www.utsouthwestern.edu) and the Baylor Institute for Immunology Research came up with a new approach in the detection of infections diseases.

Different viruses and bacteria trigger the activation of very specific genes that code for proteins called **receptors** in leukocytes, that help the body fight infections. Scientists surmised that if they looked at the leucocytes, they could detect the specific pattern of receptors similar to a disease fingerprint and be able to identify which infection was present. The process identifying such biosignatures is called **gene expression profiling,** which is done using microarray analysis.

RNA was extracted from a drop of blood and placed on a special gene chip (microarray) which contains probes for the whole human genome and measures which gene turned on or off.

Gene expression patterns were examined in leucocytes from 29 children known to have one of four common infections : flu (*influenza* A), staph (*Staphylococcus aureus*), strep (*Streptococcus pneumoniae*) or *E. coli* (*Escherichia coli*). Thirty five genes were analyzed that help distinguish infections and identified infectious agents with better than average success rate. Doctors were able to distinguish between the influenza, *E. coli* and strep infections in 95% of cases. A different set of genes distinguished *E. coli* from staph infections with 85% accuracy. Further investigation demonstrated clear distinction between viral and bacterial pneumonias.

The next step is to study whether the microarray analysis can be applied in a more challenging clinical setting, such as an emergency room (ER). When a child comes in with a fever to the ER, it is to be seen if we can predict who just has a virus and can go home and who has to be admitted and put into the intensive care unit and treated with antibiotics.

According to Dr. Ramilo, pediatricians understand that this could change the way we do things. It could also prove useful for identifying previously unknown illnesses or biologic weapons. Even if we don't know which pathogen it is, we still can tell which family or which group it's in, so if someone engineers a virus that has never been seen, we will have hints that it's close to something that is known, he said.

PNA FISH for *Staphylococcus aureus*: Clinical data have demonstrated significantly reduced mortality and hospital costs associated with Staphylococcal bloodstream infections (BSIs) due to a new nucleic acid test.

The 2.5 hour fluorescence *in situ* hybridization (FISH) assay using fluorescence-labeled peptide nucleic acid (PNA) probes that target the species specific ribosomal RNA (rRNA) in bacteria and yeast has been developed. PNA FISH is a qualitative nucleic acid hybridization assay intended for identification of organisms from blood cultures. Results are visualized using fluorescence microscopy. Fluorescing cells indicate the target species, while no fluorescence indicates that another species is present in the positive blood culture.

A group of infectious disease specialists, headed by Shmuel Shoham M.D. presented data from study of AdvanDx, Inc.'s (Woburn, MA, USA; www.advandx.com) PNA FISH diagnostic test at the 2007 annual scientific meeting of the society for healthcare epidemiology of America. The study took place at the Washington Hospital Centre.

During the study, 202 patients whose blood culture tested positive for Gram-positive cocci (indicating a Staphylococcal bloodstream infection) were alternately assigned to a control, or intervention group. In the intervention group, PNA FISH results and general organism information were relayed to the treating clinician via a call from a hospital liason, whereas control group patients did not receive a call with PNA FISH results.

According to the study, reporting of PNA FISH results led to an 80% reduction in intensive care unit (ICU) related mortality due to *S. aureus* BSIs; median hospital cost savings of $19,441 per patient; and a 61% reduction in patients receiving antibiotics for coagulase-negative staphylococci (CNS), which is often a blood culture contaminant that leads to unnecessary antibiotic therapy even though the patient does not have a true bloodstream infection.

PACE system for STDs : Commercial probes are available for the diagnosis of *Neisseria gonorrhoeae* and *Chlamydia trachomatis*. The PACE system (Gen-Probe Inc., San Diego CA) utilizes a chemiluminescent detection method. The same specimen collection tube can be used to collect one specimen for both the procedures. After collection of the cervical or urethral specimen, cells are lysed, releasing the target nucleic acid, if present. At this point, all organisms are killed and, thus, if culture is desired in conjunction with the probe assay, separate specimen would have to be collected.

The PACE system for direct detection of pathogens differs from the cultural confirmation application from Gen-Probe. In the PACE procedure, after hybridization with specific probe, the unbound probe is removed via a magnetic bead separation step. The added magnetic beads will attract the bound probe (probe-target complex, if present), and when tubes are placed on a magnetic separation rack and then decanted, the unbound probe is eliminated in a supernatant. The remaining material is resuspended and read in the luminometer as with the Accuprobe products.

For the identification of *Trichomonas vaginalis, Gardnerella vaginalis* (as a marker for bacterial vaginosis), and *Candida albicans,* a probe called AFFIRM (Becton Dickinson, Sparks, MD) is marketed. All three agents are detected in about 45 minutes, the assay requires a small piece of equipment.

For the diagnosis of HSV and HIV, probes are commercially available from a variety of manufacturers, including PathoGene (ENZO, Syosset, NY) and HSV Disk (Diagnostic Hybrids, Inc., Athens, OH). HPV can be detected by probes manufactured by Digene (Silver Springs,

MD), BioPap (ENZO Diagnostics, Inc., Syosset NY) and DAKO Biotinylated HPV Probes (DAKO Corporation, Carpentera, CA).

7.2.2 Diagnosis of Genetic Diseases

DNA hybridization can also be used for the diagnosis of genetic diseases. Suppose a specific genetic mutation causes a particular disease, *e.g.,* Alzheimer's disease, Duchennes muscular dystrophy, cystic fibrosis, sickle cell anaemia and Huntington's, DNA probes or complementary DNA to that mutation can be constructed. It will allow hybridization to detect the presence of mutations. It is then possible to prepare DNA samples from patients and determine whether they possess the mutated or wild type gene.

Because DNA probes home in on the genetic material itself, they are quite suited for locating defects in human DNA that are associated with heritable diseases.

Concurrently, many newer and more sensitive diagnostics are being developed which will reduce the time needed to diagnose a disease and will be cheaper. These include use of microchips *etc.* If a genetic defect can be detected, it may be possible to reverse or cure it by what we call **gene therapy.** If a specific gene is identified as defective, it may be possible to replace or supplement it with a normal one. Haemophiliacs have a defective gene for one of the proteins involved in blood clotting. Such patients are generally treated by administering the proper quantities of the missing proteins at regular intervals. This is just an effective preventive method.

Diagnosis of genetic disease and gene therapy now offer hope for curing many diseases, but raises a number of ethical and moral questions.

The use of recombinant DNA technology for the diagnosis of inherited diseases has several distinct advantages over other techniques.

   (i)   It is remarkably sensitive.

   (ii) The amount of DNA required for diagnosis by molecular hybridization techniques can be readily obtained from 100,000 cells.

   (iii) Furthermore, the use of PCR allows several million fold amplification of DNA or RNA, making it possible to use as few as 1 cell for analysis.

   (iv) Tiny amounts of whole blood or even dried blood can supply sufficient DNA for PCR amplification.

   (v)   DNA-based tests are not dependent on a gene product that may be produced only in certain specialized cells (*e.g.,* brain) or expression of a gene that may occur late in life.

   (vi) Because virtually all cells of the body of an affected individual contain the same DNA, each postzygotic cell carries the mutant gene.

These features have profound implications for the prenatal diagnosis of genetic diseases because a sufficient number of cells can be obtained from a few milliliters of amniotic fluid or from a biopsy of chorionic villus that can be performed as early as the first trimester.

Diagnosis of genetic diseases requires examination of genetic material (*i.e.,* chromosomes and genes). Hence, two general methods are employed:

   (a) Cytogenetic analysis

   (b) Molecular analysis.

Cytogenetic analysis It requires karyotyping.

Prenatal chromosome analysis. It should be offered to all patients who are at risk of cytogenetically abnormal progeny. It can be performed on cells obtained by amniocentesis, on chorionic villus biopsy, or on umbilical cord blood. This can be performed in the following conditions:

   (i) Advanced maternal age (>34 years) because of greater risk of trisomies.

   (ii) A parent who is a carrier of a balanced reciprocal translocation, Robertsonian translocation, or inversion (in these cases the gametes may be unbalanced, and hence the progeny would be at risk for chromosomal disorders).

   (iii) A parent with a previous child with a chromosomal abnormality.

   (iv) A parent who is a carrier of an X-linked genetic disorder (to determine fetal sex).

Postnatal chromosome analysis. It is usually performed on peripheral blood lymphocytes with the following situation :

   (i) Multiple congenital anomalies.

   (ii) Unexplained mental retardation or developmental delay.

   (iii) Suspected aneuploidy (*e.g.,* features of Down's syndrome).

   (iv) Suspected unbalanced autosome (*e.g.,* Prader-Willi syndrome).

   (v) Suspected sex chromosomal abnormality (*e.g.,* Turner syndrome).

   (vi) Suspected fragile-X syndrome

   (vii) Infertility (to rule out sex chromosomal abnormality) if multiple spontaneous abortions (to rule out the parents as carriers of balanced translocation; both partners should be evaluated).

The karyotyping cannot detect diseases caused by subtle changes in individual genes. Traditionally the diagnosis of single-gene disorders has depended on the identification of abnormal gene products (*e.g.,* mutant haemoglobin or enzymes) or their clinical effects, such as anaemia or mental retardation (*e.g.,* phenylketonuria). It is now possible to identify mutations at the level of DNA and offer gene diagnosis for several mendelian disorders.

Two distinct approaches can be used in the diagnosis of single gene diseases by recombinant DNA technology:

   (a) Direct detection of mutations

   (b) Indirect detection based on linkage of the disease gene with a harmless "marker gene".

A. Direct gene diagnosis

According to McKusick, direct gene diagnosis is the diagnostic biopsy of the human genome. It depends on the detection of an important qualitative change in the DNA. Several methods are in use for the direct gene diagnosis. Almost all of these are based on PCR analysis. If RNA is used as a substrate, it is first reverse transcribed to obtain cDNA and then amplified by PCR (RT PCR). This method is often abbreviated as RT-PCR.

(a) **RFLP :** This technique is based on the fact that some mutations alter or destroy certain restriction sites on DNA. The example is the gene encoding factor V. This protein is involved in the coagulation pathway, and a mutation affecting the factor V gene is the most common cause of inherited predisposition to thrombosis. Exon 10 of the factor V gene and the adjacent intron have two Mnl1 restriction sites. A G-to-A mutation within the exon destgoys one of the two Mnl1 sites restriction. Two primers that bind to the 3' and 5' prime ends of the normal sequence are designed to detect the mutant gene. The PCR is performed with these primers. The amplified normal DNA and patient's DNA are then digested with the Mnl1 enzyme. Under these conditions, the normal three DNA fragments (67 base pairs, 37 base pairs, and 163 base pairs long) are provided. On the other hand, the patient's DNA yields only two products, an abnormal fragment that is 200 base pairs and a normal fragment that is 67 base pairs long (Fig. 7.1).

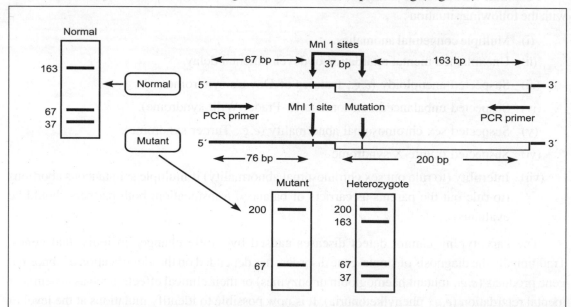

Fig. 7.1. Direct gene diagnosis. Polymerase chain reaction analysis is used for the detection of coagulation factor V mutation. One of the two Mnl1restriction sites is destroyed by G-to-A substitution in an axon. The mutant allele, therefore, gives rise to two, rather than three, fragments by PCR analysis (Adopted and modified from Kumar *et. al.*, 2004, Saunders Publ).

(b) **Detection of mutations that affect the length of DNA :** These include deletions, expansions *etc.* Several diseases, such as the fragile-X syndrome, are associated with trinucleotide repeats. This mutation can also be detected with PCR. To amplify the intervening sequences, two primers that flank the region affected by trinucleotide repeats are used (Fig. 7.2). The size of the PCR products obtained from the DNA of normal individuals, or those with premutation, is quite different. It is because of the large differences in the number of repeats. These size differences are revealed by differential migration of the amplified DNA products on a gel. Since the affected segment of DNA is too large for conventional PCR, at present the full mutation cannot be detected by PCR analysis. In such cases, a Southern blot analysis of genomic DNA has to be performed.

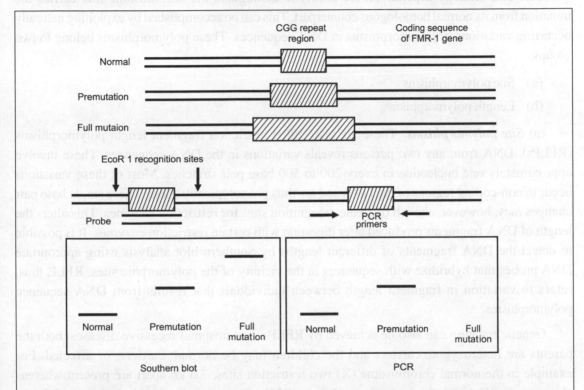

Fig. 7.2. Application of PCR and Southern blot analysis in the diagnosis of fragile-X syndrome. The differences obtained by PCR in the size of CGG repeat between normal and premutation yield products of different sizes and mobility. With a full mutation, the region between the primers is too large to be amplified by conventional PCR. In southern blot analysis the DNA is cut by enzymes that flank the CGG repeat region and is then probed with a complementary DNA that binds to the affected part of the gene. A single small band is seen in normal males, a higher-molecular-weight band in males with premutation, and a very large (usually diffuse) band in those with the full mutation. (Adopted and modified from Kumar *et. al.*, 2004, Saunders Publ)

(c) Detection of presence and absence of mutations by the use of fluorophores: This can be done in "real time" (*i.e.,* during the exponential phase of DNA amplification) PCR. In this no time is wasted in mutation detection by removing the restriction digestion and electrophoresis steps. As mentioend earlier for such molecular beacon's are used. These have already been described in Chapter 4.

B. Indirect DNA diagnosis : Linkage analysis

In a large number of genetic diseases, including those that are relatively common, information about the gene sequence is not available and thus direct gene diagnosis cannot be carried out. Alternative strategies are employed to track the mutant gene on the basis of its linkage to detectable genetic markers. For this purpose, it is required to determine whether a given foetus or family member has inherited the same relevant chromosomal region(s) as a previously affected family member. This strategy depends on the ability to distinguish the chromosome that carries the mutation from its normal homologous counterpart. This can be accomplished by exploiting naturally occurring variations or polymorphisms in DNA sequences. These polymorphisms belong to two groups:

(a)   Site polymorphisms

(b)   Length polymorphisms.

(a) *Site polymorphisms.* These are also called restriction fragment length polymorphisms (RFLPs). DNA from any two persons reveals variations in the DNA sequences. These involve approximately one nucleotide in every 200 to 500 base pair stretches. Most of these variations occur in non-coding regions of the DNA and are thus phenotypically silent. These single base pair changes may, however, abolish or create recognition sites for restriction enzymes. This alters the length of DNA fragments produced after digestion with certain restriction enzymes. It is possible to detect the DNA fragments of different lengths by Southern blot analysis using appropriate DNA probes that hybridize with sequences in the vicinity of the polymorphic sites. RFLP, thus, refers to variation in fragment length between individuals that results from DNA sequence polymorphisms.

Genetic tracking can also be achieved by RFLPs. In autosomal recessive diseases, both the parents are heterozygote carriers and the children may be normal, carriers, or affected. For example in the normal chromosome (X) two restriction sites, 7.6 kb apart are present whereas chromosome W carries the mutant gene. It has a DNA sequence polymorphism that result in the creation of an additional (third) restriction site for the same enzyme. The additional restriction site in fact does not resulted from the mutation, but from a naturally occurring polymorphism. When DNA from such an individual is digested with the appropriate restriction enzyme and probed with a cloned DNA fragment that hybridizes with a stretch of sequences between the restriction sites 7.6 kb band is yielded the normal chromosome, whereas the other chromosome (carrying the mutant gene) produces a smaller, 6.8 kb, band. Thus it produces 2 bands. It is possible by this technique to distinguish family members who have inherited both normal chromosomes from those who are heterozygous or homozygous for the mutant gene. RFLP can also be detected by PCR followed by digestion with the appropriate restriction enzyme and gel electrophoresis. In this case however, target DNA should be of the size that can be amplified by conventional PCR (Fig.7.3).

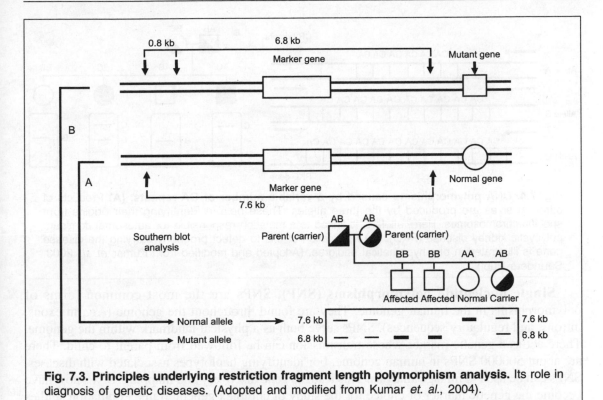

Fig. 7.3. Principles underlying restriction fragment length polymorphism analysis. Its role in diagnosis of genetic diseases. (Adopted and modified from Kumar *et. al.*, 2004).

(b) *Length polymorphisms:* Short repetitive sequences of non-coding DNA are present in human DNA. The number of repeats affecting such sequences varies greatly between different individuals. Thus, the resulting length polymorphisms are quite useful for linkage analysis. These polymorphisms are often subdivided on the basis of their length into following groups:

(i) Microsatellite repeats

(ii) Minisatellite repeats.

Microsatellites. These are usually less than 1 kb. These are characterized by a repeat size of 2 to 6 base pairs.

Minisatellites. These comparatively, are larger (1 to 3 kbL) and the repeat motifis are usually 15 to 70 base pairs. Within a given population, the number of repeats, both in microsatellites and minisatellites, is extremely variable. These stretches of DNA, therefore, can effectively be used to distinguish different chromosomes Microsatellite polymorphisms can be used to track the inheritance of autosomal dominant diseases such as polycystic kidney disease (PKD). In this case, allele C, which produces a larger PCR product than alleles A or B, carries the disease-related gene. All individuals who carry the C allele are affected. In linkage studies, microsatellites have assumed great importance and hence in the development of the human genome map. Now-a-days linkage to all human chromosomes can be identified by microsatellite polymorphisms (Fig. 7.4).

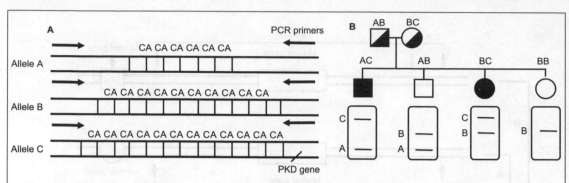

Fig. 7.4. DNA polymorphisms caused by a variable number of CA repeats. (A) Products of different sizes are produced by the three alleles. These help in identifying their origins from specific chromosomes. Here, allele C is linked to a mutation responsible for autosomal dominant polycystic kidney disease (PKD). **(B)** Application of this to detect progneny carrying the disease gene is illustrated in one hypothetical pedigree. (Adopted and modified from Kumar *et. al.*, 2003, Saunders Publ)

Single nucleotide polymorphisms (SNP). SNPs are the most common forms of polymorphisms in the human genome. They are found throughout the genome (*e.g.*, in exons, introns, and regulatory sequences). SNPs serve both as a physical landmark within the genome. There act as a genetic marker whose transmission can be followed from parent to child. There are about 500000 SNPs in human genome. For identifying haplotypes associated with diseases, SNPs can be used in linkage analysis. This leads to gene discovery and mapping. SNPs now have become the genetic marker of choice for the study of complex genetic traits. Population studies have found associations between specific SNPs and multifactorial diseases such as hypertension, heart disease, or diabetes. For example, certain polymorphisms within the angiotensinogen gene are associated with variations in resting blood pressures and a predisposition to hypertension. Attempts are being made to map all SNPs in the human genome. These would facilitate the eventual construction of "SNP chips" for genetic risk profiling of individuals.

In linkage studies, the mutant gene itself is not identified because there are certain limitations

(a) For diagnosis, several relevant family members must be available for testing. In the case of autosomal recessive disease, a DNA sample from a previously affected child is necessary to determine the polymorphism pattern that is associated with the homozygous genotype.

(b) Key family members must be heterozygous for the polymorphism (*i.e.*, the two homologous chromosomes must be distinguishable for the polymorphic site). There can be only two variations of restriction sites (*i.e.*, presence or absence of the restriction site). This is, therefore, an important limitation of RFLPs. Microsatellite polymorphisms have multiple alleles and hence much greater chances of heterozygosity. These are, therefore, much more useful than restriction site polymorphisms.

(c) Normal exchange of chromosomal material between homologous chromosomes (recombination) during gametogenesis may lead to "separation" of the mutant gene from the polymorphism pattern with which it had been previously coinherited. This may lead to wrong genetic prediction in a subsequent pregnancy. In fact the closer the linkage, the lower the degree of recombination and the lower the risk of a false test.

Molecular diagnosis by linkage analysis is quite useful in the antenatal or presymptomatic diagnosis of disorders such as Huntingtons disease, cystic fibrosis, and adult polycystic kidney disease. Direct gene diagnosis becomes the method of choice when a disease gene is identified and cloned. However, the disease is caused by several different mutations in a given gene (*e.g.,* fibrillin-l), direct gene diagnosis is not feasible, and linkage analysis remains the preferred method.

7.2.3 Other Advances in Diagnosis

Detection of Human Papillomavirus

A molecular diagnostic test enables the detection of human papillomavirus (HPV) in cervical cancer surveillance. Cervical cancer is caused in 99.7% of cases by persistent infection with HPV. Most HPV infections are transient and do not induce cervical cancer. However, a small, but significant number of HPV infections progress to cervical cancer. It is the expression of viral oncoproteins E6 and E7, which initiates the cervical cancer process, that makes them early cancer markers. It has also been demonstrated that genotypes 16 and 18 cause approximately 70% of cervical cancer worldwide.

Most cervical cancer surveillance methods currently in use detect viral DNA. Such testing, however, does not make it possible to most effectively assess the virus' oncogenicity. Developed by bioMerieux (Marc l'Etoile, France; www.biomerieux.com), NucliSens EasyQ HPV is based on a new concept that benefits from all the latest discoveries in the field, directly determining the expression of oncogenic risk factors by detecting the mRNA of the E6 and E7 proteins.

NucliSENS HPV EasyQ is the first real-time amplification/detection test with this degree of automation. It is currently available in France, the Netherlands, Belgium, Switzerland, Spain and Austria and will be progressively marketed in other countries according to local regulations.

The test enhances the quality of women's lives by reducing the need for invasive examinations and treatments. The test contributes to the fight against cancer and complements current methods for the detection and monitoring of cervical cancer.

Detection of ovarian cancer

Ciphergen's diagnostic program is related for the detection of high-risk disease at early stage. Commonly known as the "silent killer," ovarian cancer according to the American Cancer Society leads to approximately 15,000 deaths each year in the United States. Approximately 20,000 new cases are diagnosed each year, with the majority of patients diagnosed with late stage disease where the cancer has spread beyond the ovary. The prognosis is poor in these patients, leading to the high mortality from this disease. Ovarian cancer has up to a 90% cure rate following surgery and/or chemotherapy if detected in stage 1.

Several protein biomarkers have been discovered that may be potential diagnostic markers in detection of early stage ovarian cancer. The study's data were presented at the American Association for Cancer Research annual meeting (Lost Angeles, CA, USA) held in April 2007. Ciphergen (Freemont, CA, USA; www.ciphergen.com) researchers and collaborators discovered four proteins in urine that separate women with early cancer from healthy individuals with a sensitivity of 56% and a specificity of 95%. These results suggest the possibility that these proteins in combination with other biomarkers, could aid in the diagnosis of early-stage ovarian cancer. To elucidate these findings, Ciphergen employed protein expression profiling methods to analyze urine samples from 400 women, including 288 women with epithelial ovarian cancer, 52 with early stage disease and 176 with late stage disease, 74 women with benign ovarian disease and 98 normal, healthy controls.

"By demonstrating that these findings are reproducible, Ciphergen hopes to advance these markers beyond the clinical validation process and into clinical trials."

Ciphergen has multiple ovarian cancer diagnostic tests in development, including an ovarian cancer triage test designed to distinguish between benign and malignant pelvic masses, one to predict recurrence of ovarian cancer and a test to aid physicians in identifying women considered at high risk for ovarian cancer.

Discovery of Four New Breast Cancer Genes

More than 10% of women in the United States and the United Kingdom are affected by breast cancer, which is believed to have a significant genetic component. But until this new study, scientists could only account for about 25% of the genetic part of breast cancer risk. They suspected that the residual genetic variance was probably due to a variety of different gene mutations.

To look for further mutations that might be associated with breast cancer, Dr. Easton and colleagues from 15 countries conducted a two-stage Genome-Wide Association Study (GWAS) in nearly 4,400 women with breast cancer and a similar number of control subjects. They then confirmed 30 suspected single nucleotide polymorphisms (SNPs) in about 22,000 women with breast cancer and a similar number of controls spanning 22 studies.

Overall, over 225,000 SNPs were looked at and five until now unknown loci in the genome have demonstrated strong and consistent links with breast cancer incidence. Four of these have been found to be the most plausible : *FGFR2, TNRC9, MAP3K1,* and *LSP1*. It has been reported that there might well be many other mutations that contribute to a higher risk of breast cancer that could be found using their approach.

According to Dr. Douglas Easton, from the University of Cambridge (UK; www.cam.ac.uk), who was involved in the research, this discovery could help physicians improve their prediction of breast cancer risk and choose better treatments to prevent and cure the disease. But more significantly is how these findings help researchers find out how the disease works.

The gene whose variants appear to present the greatest risk of the four is fibroblast growth factor receptor 2 (FGFR2). Women, who have two copies of the high-risk variants of this gene, estimated that about 16% of the female population, have a 60% greater risk of getting breast cancer compared with women who have none.

These findings alone are not enough to enable clinicians to scan a woman's genes and inform her what is likely to be her breast cancer risk. But as more of these studies are performed and more risk factor revealed, the more realistic such a scenario will become. According to Dr. Easton as more genes are identified, tests will become more predictive.

Dr. Easton believes a similar approach will work for other cancers, and he is now evaluating prostrate cancer. Two other studies published in the May 27, 2007, issue of the journal Nature Genetics have also discovered genetic variants that appear to increase the risk of breast cancer.

One study conducted in the United States and led by Dr. David J. Hunter, from the department of medicine, Brigham and Women's Hospital and Harvard Medical School (Boston, MA, USA), is also a Genome Wide Association Study. This GWAS evaluated about 2,000 postmenopausal women and 2,000 control subjects, and found several genetic variants in the FGFR2 gene associated with increased risk of breast cancer.

In the other study, led by Dr. Simon N. Stacey from decode genetics (Reykjavik, Iceland), an international team of scientists found mutations on chromosomes 2 and 16, which appear to increase the risk of estrogen receptor-positive breast cancer. Overall, they studies over 4,500 affected Icelandic women and 17,000 controls.

7.3 THERAPEUTICS

Today biotechnologically produced drugs have started hitting the market. These include the following:

7.3.1 Antisense Oligonucleotides

New approaches to the development of therapeutic drugs have been facilitated by the advances in synthetic chemistry and molecular genetics. The traditional drugs work to inhibit an enzyme or protein. The new approaches, however, involve blocking genetic messages to turn off production of disease causing proteins at the source. Malfunctioning of deleterious genes is inhibited by blocking the code and hence it is used as a mean to control diseases.

In this approach any of the following are targetted:

(i) A nucleotide sequence on a single-stranded messenger RNA (mRNA), which encodes for disease causing proteins

(ii) Double-stranded DNA from which mRNA is transcribed.

Once a target sequence is determined, a complementary or an antisense DNA sequence can be made. This will bind and inactivate the genetic message.

A 15-25 nucleotides long antisense oligonucleotide is expected to be highly selective in its ability to recognize and bind to its target sequence. The affinity for hybridization can be reduced by several orders of magnitude by a single mismatch in complementarity. This expected high specificity will probably minimize the side effects. As a significant number of diseases have genetic origin, there is a potential for wide range of applications and sizeable markets.

For putting antisense oligonucleotide principle into practice, the following requirements must be met:

- Under physiological conditions (*e.g.,* resist nucleases), the complex formed between the oligonucleotide and the complementary target sequence should be stable.
- The interaction between oligonucleotides and the target sequence should be specific.
- Under *in vivo* conditions, half-life should be sufficiently long.
- They should be able to pass through the cell membrane.

Thus, development of antisense oligonucleotides requires that all the four conditions must be met at all times.

Gene silencing has been tried in the treatment of various diseases. The malarial parasite *Plasmodium falciparum* uses the *var* gene responsible for the regulation of variable surface protein PFEMP1, which hides the parasite from the host's immune response. While the *P. falciparum* genome contains at least 50 *var* genes, only one is expressed at any given time, thereby, giving rise to a single version of the PFEMP1.

The region around the *var gene* was analyzed at the Walter ELIZAHall institute of Medical Research, Melbourne, Australia (www.wehi.edu.an) by inserting a gene for drug resistance adjacent to it. In some parasites, the DNA region was active and the parasites showed resistance to the drug, in others, the drug killed the parasites.

Another potential control mechanism for gene silencing is compartmentalization. FISH technique was used to determine the position of *var* genes in ON and OFF states. It was found that there appeared to be discrete nuclear compartment that allowed gene expression to occur when a *var* gene moved into this compartment, it became activated and its version of PFEMP1 was expressed.

The gene silencing suggested the involvement of SIR2 (Silent informate regulator 2) protein which is known to play a role in gene silencing in yeast by modifying gene packaging. In order to know the functioning of SIR2, a parasitic line was developed that lacked the *sir2* gene. Silencing probably occurs by packaging up the DNA into a light form and preventing it from being expressed. The tight packaging involves SIR2 protein.

7.3.2 Human Therapeutics Based on Triple Helix Technology

A new class of drugs-initially applicable for the effective and specific treatment of viral infections-eventually perhaps cancer and immunological disorders, is being developed by the application of triple helix technology to rationale drug delivery. Attempts are on to chemically modify the oligonucleotides to enhance their affinity and efficiency. There are attempts to make comprehensive studies on pharmacokinetics, toxicity, mutagenicity and *in vivo* efficacy of these compounds.

7.3.3 Developing rDNA Products for Treatment of Various Ailments

For haemophilia A, frequent infusion of plasma derived human factor VIII is required. There is, however, an associated drawback of potential viral contamination, high cost and limited plasma availability. Now factor VIII replacement therapy has been improved by the increased knowledge of molecular mechanisms regulating blood coagulation. Homogeneous pure preparation of factor VIII has been produced through r-DNA technology. It can now be used for successful treatment of haemophilia A.

7.3.4 RNAi as Therapeutics

The idea of using RNAi for therapeutic purpose has been extensively tested for last couple of years since Tuschl's pioneering work on siRNAs. Candidate diseases for such treatment include viral infections. Kim (2003) has summarized the role of RNAi as therapeutic and dominantly inherited genetic disorders.

The first obvious target for such application was the human immunodeficiency virus (HIV). Viral genes including *tat, rev, nef,* and *gag* have been silenced. This resulted in successful inhibition of viral replication in cultured cells. Cellular genes such as *CD4, CCR5* and *CXCR4* required for viral infection have also been targeted. A major cause of chronic liver disease, Hepatitis C virus (HCV) has a genome of a single-stranded RNA, which has been made an attractive target for RNAi. Expression of RNAs from HPV replicon was inhibited in cell culture. This may lead to a new therapy for this virus. Human papilloma virus (HPV) is believed to contribute to tumourigenesis. Attempts have been made to silence *E6* and *E7* genes of HPV type 16 by siRNA. It resulted in reduced cell growth and induced apoptosis in cervical carcinoma cells. Reduction in hepatitis B virus (HBV) RNAs and proteins has been induced by siRNA-producing vectors in cell culture and in mouse liver. Influenza virus was also challenged with siRNA specific for nucleocapsid (NP) or a component of the RNA transcriptase which abolished the accumulation of viral mRNAs.

By using small interfering RNA molecule (sRNA), it has been demonstrated that inhibition of Survivin causes upto 70 % reduction in the growth of cancer cells due to significant increase in apoptosis. *Survivin* is a human gene encoding a structurally unique apoptosis inhibitor. It is not detectable in terminally differentiated adult tissues. However, survivin becomes prominently expressed in transformed cell lines and in all the most common human cancers of lung, colon, pancreas, prostate and breast. Survivin is also found in high grade non Hodgkin's lymphoma. It was observed that when injected directly into the tumour or intravenously, the siRNA interfered with survivin's functions. In this way a 70% reduction of the tumour growth was seen. It was observed that there was no apparent adverse impact on normal cells by the survivin siRNA, encoding plasmid (Biotech News International, 10(5) : 9-10-2005).

Exquisite sequence specificity of RNAi enables specific knockdown of mutated genes. Such possibility was first tested on an oncogene, *k-ras,*(v12) whose loss of expression led to loss of anchorage-independent growth and tumourigenicity .This approach was particularly encouraging because it was successful not only in tissue culture, but also in an animal model (mouse). Similar studies quickly followed using various forms of siRNA. Oncogenes can be activated by chromosomal translocation fusing two parts of unrelated genes. *M-bcr/abl* fusion leads to leukemic cells with such a rearrangement. Transfection of dsRNA specific for the M-bcr/abl mRNA has been shown to downregulate the fusion protein in K562 cells.

Another cause of tumourigenesis is overexpression of oncogenes. Overexpression of P-glycoprotein (*P-gp*), the *mdr1* gene product, confers multidrug resistance (MDR) to cancer cells. RNAi successfully reduced P-gp expression and, thereby, drug resistance. Expression of endogenous *erbB1* can be suppressed by RNAi in A431 human epidermoid carcinoma cells. Combined RNAi to reduced expression of *c-raf* and *bcl-2* genes may also represent a novel approach to leukemia. Blocking angiogenesis is another important anti-cancer strategy. There are atleast five isoforms of vascular endothelial growth factor (VEGF). These are thought to perform different functions in tumour angiogenesis. Specific knock-down is possible by using RNAi. This provides a new tool to study isoform-specific VEGF function as well as to treat cancer.

Mutations on one allele whose gene product acts transdominantly cause dominantly inherited genetic disorders. Specific abrogation of the mutated gene would leave the unaffected allele to restore the normal cellular function. At least eight human neurodegenerative disorders including Huntington's disease and spinobulbar muscular atrophy (Kenney's disease) are caused by the expansion of trinucleotide (CAG) repeats encoding an increased polyglutamine tract. Although the mechanism underlying neurodegeneration is not clear, aggregation of mutant polyglutamine proteins is related to the toxic gain-of-function phenotype. SiRNA targeting the 5'-end or 3'-end of the CAG repeat rescued the polyglutamine toxicity in cultured cells. This has opened up the possibility for new approaches.

Other diseases considered for RNAi-based therapy include Fas-induced fulminant hepatitis. Intravenous injection of siRNA targeting Fas, reduced Fas expression in mouse hepatocytes. This led to resistance to apoptosis and protection of mice from liver fibrosis.

Before RNAi becomes a realistic tool in clinics many hurdles need to be circumvented.

(i) Enough amount of siRNA should be delivered into enough number of target cells, efficiently and stably. This delivery problem may be solved by chemically modifying siRNA to make it more stable, penetrable and cost-effective. Alternatively, siRNAs can be delivered by way of viral vectors. Viral vectors such as lentiviral vectors would have unique advantages over synthetic siRNAs in terms of persistency. Developing optimal vectors will greatly accelerate siRNA-mediated gene therapy. A related issue is "targeted" administration of siRNA. This is hard to be achieved with synthetic siRNAs. For DNA-based RNAi, however, inducible/repressible promoters can be used to regulate siRNA expression in a tissue specific manner.

(ii) Another problem has a origin from the technique's own merit; sequence specificity. Frequent mutations of target genes may allow escape from specific inhibition of disease genes, especially in viral infection. A "combination" strategy using several different siRNAs may minimize the escape.

7.3.5 Biotherapy and Pharmaceutical Products

A large number of pharmaceutical products are compounds that are derived from :

(*i*)   synthetic chemical processes

(*ii*) naturally occurring sources (plants, microorganisms)

(*iii*) combination of both.

Such compounds are used to regulate essential body functions or to combat disease-causing microorganisms.

Much of the focus of the drug research has been on artificially engineered versions of natural biological regulators, hormones, immunomodulators, growth factors and key physiological enzymes. Biopharmaceuticals attempt to mimic the natural disease fighting mechanisms of the body. Sometimes there is difficulty to implement this strategy in practice. It is because of the fact that most of these regulatory molecules have multiple functions. The administration of large doses has created unexpected side effects. However, many biotechnologically produced drugs, have shown strong therapeutic promise.

In treating cancer there has been mixed success of biopharmaceuticals. In a few rare cancers clear positive results have been demonstrated by interferons. Powerful antitumour activity has been observed in trials with interleukin – 2(IL-2). Positive results have also been obtained with combination therapies, where several drugs administered together appear to act synergistically, but working out the ideal combination of drugs may, however, take many years.

Attention is now being directed to the body's own regulatory molecules, which normally occur only in very small concentrations and have predominantly been subjected to modern methods of extraction or synthesis. Limited quantities of some of these compounds have historically been derived from organs of cadavers and from blood banks. Genetic engineering is now used to have some of these scarce molecules in unrestricted quantities. This technology is being used to produce many molecules such as hGH. Many of these protein products are used to treat or cure diseases. Biopharmaceuticals are the protein drugs, vaccines *etc*.

It is now possible to produce these biopharmaceuticals in a form similar to that normally occurring in the human body. In fact attentions are on to design meaningful improvements in activity, stability or bioavailability. These products will be free from the dangerous contaminants that have occasionally arisen from extraction of cadavers, *e.g.,* the degenerative brain disease Creutzfeldt-Jakob disease has been associated with the administration of human growth hormone from early cadaver extractions.

The following requirements should be met for the successful development of biopharmaceuticals:

(*i*)   Identification and characterization of the native compounds by advanced biochemical/ biomedical research (*ii*) Identification of the relevant gene sequences and their insertion into a mammalian or microbial host cell

(*iii*) Availability of bioprocess technology to grow the organisms and to isolate, concentrate and purify the chosen compounds

(*iv*) Expertise skilled in molecular biology and cloning technology, clinical and marketing.

About 100 to 150 new biopharmaceuticals are currently undergoing final clinical trials.

There are, however, several restrictions of these protein like pharmaceuticals. These limit the use and size of market. One of them is their low stability, but now the attempts are on to get rid of this and other related problems.

Biopharmaceuticals are now becoming increasingly relevant in biological applications, but their share in the market is still only a small part of the pharmaceutical industry. Biotechnology has accelerated screening, bioassaying and the production of new drugs, and also explains more accurately how drugs act in the human system. Biotechnology is also expected to reduce the huge cost of new drug development.

7.3.5.1 Endogenous Therapeutic Agents

The human body produces many of its own therapeutic compounds and a large number of them are proteins. As proteins, they are prime candidates for possible production by genetically engineered bacteria. Such production would provide quantities that would allow us to analyze their functions in a better way and make their commercialization economically feasible. As we increase our understanding of these and other endogenous therapeutic agents, we will be able to capitalize more on the body's innate healing ability. Following are the examples of some of the endogenous therapeutic compounds :

- Interleukin-2 activates T-cell responses.
- Erythropoietin regulates red blood cell production
- Tissue plasminogen activator dissolves blood clots

Some of the biopharmaceuticals which are being produced have been discussed below:

7.3.5.1.1 Insulin

It is a protein hormone made in the pancreas. It plays a vital role in the regulation of blood sugar levels. Its deficiency is one of the causes of the disease, diabetes mellitus (sugar diabetes). In this, blood sugar levels are raised with harmful consequences. At least 3% of the population is affected by diabetes mellitus. Worldwide more than 2 million people use insulin and the world market is worth several hundred million pounds a year.

Insulin is generally isolated from the pancreas of slaughtered pigs and cattle and is injected into the patients. However, due to minor differences in the amino acid composition of insulin from species different from us and traces of impurities, some patients are allergic to animal insulin and show the damaging side effects as a result of the injections. The introduction of genetic engineering, however, has provided a solution to this. The gene for human insulin is inserted into a bacterium which is grown in a fermenter to make large quantities of the protein (Fig. 7.5).

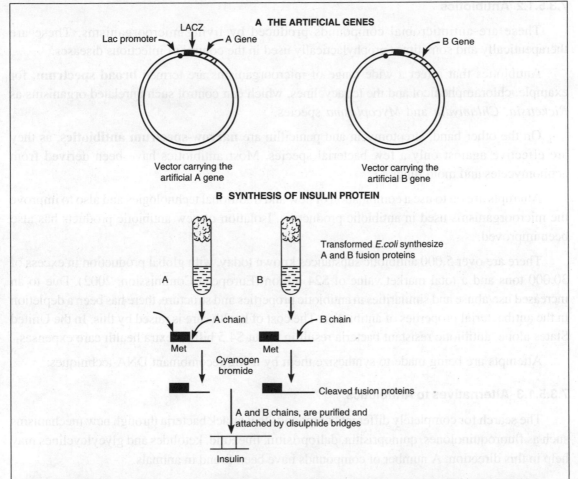

A THE ARTIFICIAL GENES

LACZ
Lac promoter — A Gene

B Gene

Vector carrying the
artificial A gene

Vector carrying the
artificial B gene

B SYNTHESIS OF INSULIN PROTEIN

Transformed *E.coli* synthesize
A and B fusion proteins

A   B

A chain   B chain

Met   Met
Cyanogen
bromide

Cleaved fusion proteins

A and B chains, are purified and
attached by disulphide bridges

Insulin

Fig. 7.5. Synthesis of insulin through recombinant DNA technology. (A) Genes for A and B chains are added to the vectors, which are inserted in *E.coli*. **(B)** *E.coli* are made to grow and produce A and B chains of proteins which are then treated with cyanogen bromide to have cleaved fusion proteins. These in turn are joined by disulfide bridges.

The question is how to switch the gene on in the bacterium. Not all the genes in a cell are switched on at any one time. The promoter region present next to the gene has to be activated before it is expressed. If the new gene is inserted in the middle of an existing one, the switch for that gene may be used. The gene used in *E. coli* was for β-galactosidase. Its promoter is switched on if *E. coli* is grown in a medium containing lactose.

The original technique was developed by Eli Lilly and Company and in 1982, human insulin, marketed as 'humulin', became the first genetically engineered pharmaceutical product to be approved for use.

Now for its production, genetically engineered eukaryotic cells in culture or transgenic animals can be used.

7.3.5.1.2 Antibiotics

These are antimicrobial compounds produced by living microorganisms. These are therapeutically and sometimes prophylactically used in the control of infectious diseases.

Antibiotics that affect a wide range of microorganisms are termed **broad spectrum**, for example, chloramphenicol and the tetracyclines, which can control such unrelated organisms as *Rickettsia*, *Chlamydia* and *Mycoplasma* species.

On the other hand, streptomycin and penicillin are **narrow-spectrum antibiotics,** as they are effective against only a few bacterial species. Most antibiotics have been derived from actinomycetes and mold fungi.

Attempts are on to use a combination of new and traditional technologies and also to improve the microorganisms used in antibiotic production. Isolation of new antibiotic products has also been improved.

There are over 5,000 antibiotic substances known today, with global production in excess of 30,000 tons and a total market value of $24 billion (European Commission, 2002). Due to an increased use/abuse and similarities in antibiotic properties and structure, there has been a depletion in the antibacterial properties of antibiotics. The cost of health care is raised by this. In the United States alone, antibiotic resistant bacteria result in about $4.5 billion extra health care expenses.

Attempts are being made to synthesize them by using recombinant DNA techniques.

7.3.5.1.3 Alternatives to Antibiotics

The search for completely different compounds that attack bacteria through new mechanisms such as fluoroquinolones, quinoprisitin, dalfoprisitin, linezolid, ketolides and glycylcyclines may help in this direction. A number of compounds have been found in animals.

Attempts are being made to isolate anti-microbial peptides such as magnainin from frogs.

Specific genes in the sequenced genomes of major pathogenic bacteria are scanned to identify gene(s) coding for key, but unique metabolic processes in the pathogenic microorganisms. To attack such a process, an inhibitor molecule can then be engineered. The bacterial genomics have already given about 500 to 1,000 new broad-spectrum antibacterial targets. In addition, there is now stress on bacteriophages. They are thought to be of significant help for a few specific applications (European Commission, 2002). The antibiotics are being combined with compounds, **"guardian-angel,"** that neutralize antibiotic-resistant bacteria *e.g.,* Clavulanic acid. It is ineffective in protecting the cephalosporins, but is frequently used in hospitals. The β- lactamase inhibitors have been developed (European Commission, 2002).

7.3.5.1.4 Somatostatin

It is difficult to isolate the growth hormone somatostatin from animals. Half a million sheep brains are required to yield 0.005g of pure somatostatin. By cloning the human gene for somatostatin into a bacterium, the same amount of hormone can be produced from 9 litres of a transgenic bacterial fermentation (Fig. 7.6). One child in 5000 suffers from hypopituitary dwarfism resulting from growth hormone deficiency. The easy availability of this biopharmaceutical will be of immense benefit to the child sufferers. The annual world market is estimated at $US 100 million.

However, a potential massive market could arise from the increasing evidence that this growth hormone can increase muscle formation in normal individuals and is now being exploited by some athletes. It is being claimed that regular administration of the hormone can improve quality of life in the aged.

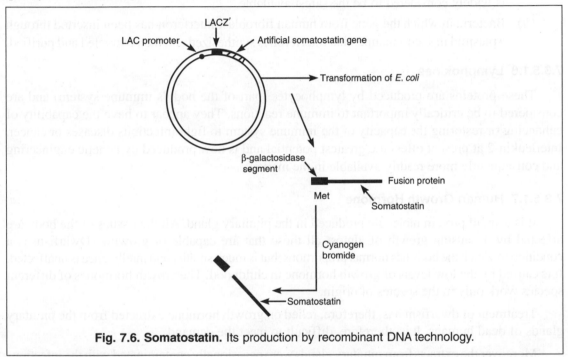

Fig. 7.6. Somatostatin. Its production by recombinant DNA technology.

7.3.5.1.5 Interferons

In 1957, two British researchers discovered substances produced within the body that could act against viruses by making cells resistant to viral attack. Most vertebrate animals can produce interferons, and many animal viruses can induce their *in vitro* synthesis and become sensitive to them. Only minute amounts of interferon are produced within cells, and it is quite complicated to extract and separate them from other cellular proteins.

Human interferons are glycoproteins. They are believed to play a part in controlling many types of viral infections, including the common cold. They have the potential to control cancer.

Different types of interferons are characteristic of individual species of animals. Mouse interferon will respond to mouse cells, but not to human cells and *vice versa*. Different tissues from the same species also produce different interferons. Thus, interferon for human studies must be derived from human cells. Most early human interferon production was carried out in

Finland using leucocytes from blood, and the small amounts of interferon produced, this way was used for limited clinical tests throughout the world.

The current interest in interferons is with regard to their ability to inhibit cancer in experimental animals. Interferons attack the cancer cells by inhibiting their growth, and that of any virus involved in the cancer processes. They can also stimulate the body's natural immune defences against the cancer cells.

There are two sources of interferons :

(*i*) Growing human diploid fibroblasts attached to a suitable surface. The interferon produced is widely considered to be the safest available.

(*ii*) Bacteria in which the gene from human fibroblast interferon has been inserted through a plasmid in such a manner that interferon is synthesized and then extracted and purified.

7.3.5.1.6 Lymphokines

These proteins are produced by lymphocytes (part of the body's immune system) and are considered to be critically important to immune reactions. They appear to have the capability of enhancing or restoring the capacity of the immune system to fight infectious diseases or cancer. Interleukin-2 at present offers the greatest potential and is now produced by genetic engineering and consequently more readily available in the market.

7.3.5.1.7 Human Growth Hormone

It is a small protein molecule produced in the pituitary gland. All the tissues of the body are affected by it causing growth of almost all those that are capable of growing. Dwarfism is a condition in which the body has normal proportions, but is much smaller and intelligence is unaffected. It is caused by the low levels of growth hormone in childhood. The growth hormones of different species work only in the species of origin.

Treatment of dwarfism has, therefore, relied on growth hormone extracted from the pituitary glands of dead humans. It is, therefore, difficult to meet the demand.

Moreover the extracts from pituitary glands were occasionally contaminated with the infectious protein that caused Creutzfeldt-Jakob disease (the same protein that may cause mad cow disease). In the 1970s, after several fatalities among people the treatment was withdrawn.

Genentech, a California-based company, has produced human growth hormone (hGH) from genetically engineered bacteria which contains human gene for the hormone. It can be produced in much larger quantities and in a pure form. Near-normal height in children suffering from growth hormone deficiency is restored by the regular injections of the hormone.

The cDNA is added to the bacterium. It is made from mRNA using reverse transcriptase. Before adding it to the vector, the cDNA is joined to the signal sequence from the bacterium.

7.3.5.2 Natural Products as Pharmaceuticals

Many plants produce compounds that have human therapeutic value. For years, we have used a chemical derived from foxglove (*Digitalis*) for treating heart conditions.

(i) A newly discovered chemical extracted from yew trees is being used to treat breast and ovarian cancers.

(ii) Ticks and leeches are being investigated for the source of secretions to be used as anticoagulant compounds.

(iii) Poison arrow frogs are looked for source of painkillers.

(iv) Researchers studying osteoporosis used a rat model to show that a compound isolated from white onions decreased bone loss by inhibiting osteoclast bone absorption actively. The active ingradient for this purpose was found to be gamma L-glutamyl-trans-S-1 propenyl-L-cysteine sulfoxide (GPCS). A peptide with a mass of 306 daltons demonstrates dose dependent inhibition of osteoclast resorption activity with a minimal effective dose being about 2mM.

(v) Attention is paid to the extraordinarily diverse ecosystems found in the sea to look for compounds that heal wounds, destroy tumours, prevent inflammation, relieve pain, and kill microorganisms, 2,400 kg of sponges have been found to yield 1 mg of an anticancer drug. The use of sponges as source of a pharmaceutical, would not be economically feasible and it would be an ecological disaster. Attempts are, therefore, being made to culture sponge cells in laboratory. Genes required to produce the compounds are being identified and attempts are to be made to move them into organisms that do well under culture conditions.

By developing sophisticated cell culture and bioprocessing technologies, we will broaden our ability to use many more compounds from nature.

7.3.5.3 Biopolymers as Medical Devices

Nature has provided us with substances that are useful medical devices. Some are superior to inorganic, man made substances, because, being biological materials, they are more compatible with our tissues and are degraded and absorbed when they have performed their function. The manufacture of biological polymers is also environmentally more benign. Some of the naturally occurring biopolymers used as medical devices are:

- **The carbohydrate hyaluronate :** It is a viscous, elastic, plastic like, water-soluble substance that is used to treat arthritis, to prevent postsurgical scarring in cataract surgery and for drug delivery.

- **Adhesive protein polymers :** These are derived from living organisms and are replacing sutures and staples in wound healing. They set quickly, produce strong bonds and are absorbed.

- **Chitin :** This carbohydrate is found in the exoskeletons of insects and crustaceans, combined with a natural fibre polynosic, creates a material that limits bacterial and fungal growth.

7.3.5.4 Designer Drugs

Using principles of protein engineering and computer molecular modeling, we may be able to design effective therapeutic compounds before stepping into a laboratory.

7.3.5.5 Vaccines

A vaccine is a living or a non-living preparation that can be fool the body's immune system into thinking that it is a pathogenic organism. The body is unable to distinguish a harmless invader from a pathogen. It can only determine that it is harbouring foreign material (an antigen). The presence of an antigen initiates complex sequence of actions, the aim of which is to contain, neutralize and destroy the invader. The body produces antibodies, which specifically bind to the antigen.

Antigenic material also stimulates the immune "memory" which allows the body to act more quickly and effectively in combating subsequent invasion by the same organism or organisms, which to the body's immune system appear similar.

Vaccination has eradicated small pox and drastically reduced the incidence of diphtheria, tetanus, T.B., polio, cholera and whooping cough in developed countries. Many infections, however, remain as challenges for bio-technology. These include viral diseases such as influenza, hepatitis and AIDS, parasitic diseases like malaria, leishmaniasis, bacterial diseases like leprosy *etc.* and attempts are being made in this direction.

The production of human vaccines by recombinant methods (Fig. 7.7) has been quite successful and is going to allow for new approaches to disease control. Recombinant hepatitis B vaccine has gained regulatory approval and has high market sales.

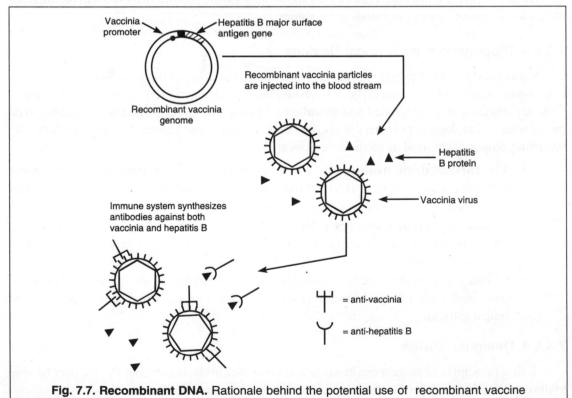

Fig. 7.7. Recombinant DNA. Rationale behind the potential use of recombinant vaccine

To develop vaccine against malaria is a major priority of World Health Organization. A series of potentially protective vaccines have been developed using molecular techniques, but none has been found to be encouraging because of the complexity of the parasite's life cycle and the antigenic diversity exhibited by each stage.

Progress towards developing a malaria vaccine is being made by several approaches. The question is whether a malaria vaccine will fulfill the criteria for vaccines in general, namely safety, efficacy, stability, the ease of administration and cost. Any vaccine produced will be of very restricted use because of the variations that exist between different geographical isolates.

Third generation whooping cough vaccines : R-DNA methods have been used to remove the enzymatic activity of Pertuses toxin to obtain a new molecule, which is devoid of toxicity and can be used for safer vaccination against the disease.

Modern vaccine technologies
Genetically improved live vaccines

Genetically attenuated microorganisms: The strategy of attenuation by genetic engineering of the live, intact pathogen has some potential drawbacks. The virulence and the life cycle of the pathogen have to be known in detail. It is also obvious that the protective antigens must be known: attenuation must not result in reduced immunogenicity. It has also to seen be that reversion of the attenuated microorganism occurs during its production or its presence in the host. This means that subtle changes in the genome are not desirable. The homologous engineering is mainly restricted to pathogens that are used as starting materials for the production of subunit vaccines.

An example of an improved live vaccine obtained by homologous genetic engineering is an experimental, oral cholera vaccine. An effective cholera vaccine should induce a local, humoral response in order to prevent colonization of the small intestine. Initial trials with *Vibrio cholerae,* cholera toxin (CT) mutants caused mild diarrhoea, which was thought to be caused by the expression of accessory toxins. A natural mutant was isolated that was negative for these toxins. Next, CT was detoxified by rDNA technology. The resulting vaccine strain, called CVD IO3 is well tolerated by children (Suharyono *et al.,* 1992) and challenge experiments with adult volunteers showed protection (Tacket *et al.,* 1992).

Live vectors

A way to improve the safety or efficacy of vaccines is using live harmless (*i.e.,* non-pathogenic or attenuated) viruses or bacteria as carriers for antigen from other pathogen.

Advantages of vaccinia virus as vector include the following:

(i)   Its proven safety in humans as a smallpox vaccine

(ii) The possibility for multiple immunogen expression

(iii) The ease to produce

(iv) Its relative heat-resistance

(v)   Its various possible administration routes

A multitude of live recombinant vaccinia vaccines with viral and tumour antigens have been constructed, several of which are being tested in the clinic. It has been demonstrated that the products of genes coding for viral envelope proteins can be correctly processed and inserted into the plasma membrane of infected cells. Problems related with the side effects or immunogenicity of vaccinia may be circumvented by the use of attenuated strains or poxviruses with a non-human natural host.

Genetically improved subunit vaccines
Genetically detoxified proteins

A biotechnological improvement of the acellular pertussis vaccine has been the switch from chemically to genetically inactivated pertussis toxin. The principle of both chemical and genetic inactivation is schematically illustrated. Chemical treatment with formaldehyde results in a cripple protein molecule with partial loss of conformational and antigenic properties. This reduces its immunogenicity, whereas potential reversal to a biologically active toxin is a major concern. Variations in the extent of detoxification can affect both the immunogenicity and toxicity of the product. In contrast, genetic detoxification by site-directed mutagenesis warrants the reproducible production of a non-toxic mutant protein that is highly immunogenic because the integrity of immunogenic sites is fully retained. In the example, of pertussis toxin codons for two amino acids were mutated in the cloned pertussis gene, which abolished the toxicity of the protein without changing its immunological properties. The altered gene was then substituted in Bordetella pertussis for the native gene (Nencioni *et al.*, 1990). Other candidates for genetic detoxification are diphtheria, tetanus and cholera toxins. Alternatively, proteins can be detoxified by genetic deletion of active sites or subunits.

Proteins expressed in host cells

To improve the yield, facilitate the production and/or improve the safety of protein-based vaccines, protein antigens are sometimes expressed by host cells of the same (homologous) species or of different (heterologous) species that are safe to handle and/or allow high expression levels.

Heterologous hosts used for the expression of immunogenic proteins include yeasts, bacteria and mammalian cell lines. Hepatitis B surface antigen (HBsAg) which previously was obtained from plasma of infected individuals, has been expressed in Baker's yeast (*Saccharomyces cerevisae*; Valenzuela *et al.*, 1982) and in mammalian cells (Chinese Hamster Ovary cells, Burnette *et al.*, 1985) by transforming the host cell with a plasmid containing the HBsAg-encoding gene. Both expression systems yield 22-nm HBsAg particles that are identical to those excreted by the native virus. Its advantages are:

    (i)   safety
   (ii)   consistency
  (iii)   quality
  (iv)   high yield

The yeast-derived vaccine has become available worldwide and appears to be as safe and efficacious as the classical plasma-derived vaccine.

The experimental multivalent meningococcal vesicle vaccine is an example of the expression of multiple antigens in homologous host cells (Van der Ley *et al.*, 1995). The vaccine is prepared by extraction of vesicles from the meningococcal outer membrane. These vesicles serve as a natural carrier for immunogenic outer membrane proteins (OMPs), which are incorporated into the vesicle membrane. Each wild-type meningococcus strain expresses strain-specific OMPs. Taking a wild-type strain as starting point, mutant strains expressing OMPs specific for three strains have been made through transformation with plasmid constructs in *E. coli* and their recombination into the meningococcal chromosome. Outer membrane vesicles of two trivalent strains have been prepared and combined to a hexavalent vaccine, which is presently being evaluated in infants for its safety and efficacy.

Recombinant peptide vaccines

After identification of a protective epitope, it is possible to incorporate the corresponding peptide sequence into a carrier protein containing Th-epitopes through genetic fusion (Francis, 1991). The peptide-encoding DNA sequence is synthesized and inserted into the carrier protein gene. Such fusion proteins comprise HBsAg, hepatitis B core antigen, and β-galactosidase. An example of the recombinant peptide approach is a malaria vaccine based on a 16-fold repeat of the Asn-Ala-Asn-Pro sequence of a *Plasmodium falciparum* with the HBsAg gene and the fusion product was expressed by yeast cells (Vreden *et al.*, 1991). Genetic fusion of peptides with proteins offers the possibility to produce protective epitopes of toxic antigens derived from pathogenic species as part of non-toxic proteins expressed by harmless species. Furthermore, a uniform product is obtained in comparison with the variability of chemical conjugates.

Anti-Idiotype antibody vaccines

Antibodies can be elicited against any antigenic structure on any molecule, including antibodies themselves. The concept of anti-iodiotype vaccines is eliciting antibodies against the antigen-binding site of protective antibodies.

(i) A monoclonal antibody (MAb-1) that recognizes a protective epitope of a particular immunogen is selected.

(ii) A monoclonal antibody (MAb-2) is generated against the idiotype *i.e.*, the three-dimensional structure of the antigen-binding site of MAb-1.

Hence, MAb-2 imunologically mimics the protective epitope of the immunogen and may, thus, be used as a vaccine component. The original epitope is not necessarily of protein origin, it shows that immunological mimicry is not always present at the atomic level (Pan *et al.*, 1995). This makes the approach especially attractive for non-protein epitopes. For instance, anti-idiotype antibodies carrying the 'internal image' of carbohydrates that are difficult to produce and/or isolate in large quantities or of immunogenic carbohydrate residues of toxic lipopolysaccharides or glycoproteins may serve as vaccine components. Large quantities of monoclonal antibodies are easy to produce using modern hybridoma technology. Experimental anti-idiotype vaccines that have been studied in animals include, amongst others, a *Streptococcus pneumoniae* vaccine based on phosphorylcholine mimicking antibodies (McNamara *et al.*, 1984) and a vaccine consisting of anti-idiotype antibodies resembling lipopolysaccharide from *Pseudomonas aeruginosa* (Schreiber *et al.,* 1991). The clinical applicability of anti-idiotype antibodies, however, remains to be established. A drawback is that the major structural part of the anti-idiotype antibody molecule does not have any relationship with the structure of the original antigen and may give rise to unwanted (immunological) reactions, unless human, humanized and/or single-chain MAbs are used.

Synthetic peptide-based vaccines

Another form of molecule mimicry are synthetic peptides, which are vigorously being explored for immunization. Through recent improvements in solid-phase peptide synthesis, large quantities of oligopeptides that are capable of eliciting an immune response towards the native protein can be prepared nowadays. Primarily, peptide-based vaccines based on antibody recognition have been designed. Two approaches can be discerned, depending on whether the epitope is continuous or discontinuous.

1. Immunogenic epitopes are determined by DNA cloning and nucleotide sequencing of protein antigen and serology studies. The small linear peptide sequence is chemically synthesized and can be used as a vaccine component. A limitation of this concept is that it is only applicable to continuous epitopes that are solely determined by the primary amino acid sequence and not by the conformation of the epitope. Many B-cell epitopes, however, are conformationally determined and/or discontinuous. For continuous conformational epitopes, synthetic peptides can be forced to adopt the proper conformation by cyclization.

2. It is particularly useful for discontinuous epitopes. In this case, the optimal sequence of a synthetic peptide is not easy to determine. With current technology, however, thousands of peptides can be rapidly synthesized at random and screened for optimal binding to protective antibodies (Geysen *et al.,* 1986, 1987). The sequence of a selected peptide can, if necessary, be optimized for antibody binding by selectively substituting one or more amino acid residues. Such peptides approximating the native epitope (but not necessarily containing the exact (linear or non-linear) sequence of the epitope) are referred to as mimotopes. In theory, analogous with anti-idiotype antibodies, mimotopes may be useful as internal image not only of peptide epitopes, but also of non-protein structure.

Similar to B-cell epitope peptides, T-cell epitope peptide vaccines can also be designed. T-cell epitopes usually have a continuous, non-conformational nature and are, therefore, relatively easy to mimic after their sequence has been identified, in analogy with the approach for continuous B-cell epitopes.

Synthetic peptide vaccines have the following advantages :

  (i)   They can be prepared in unlimited quantities using solid-phase technology

  (ii) They are easily purified by HPLC methods

  (iii) They do not contain infectious or toxic material

The use of synthetic peptides as vaccine has two main complications regarding their immunogenicity.

(a) Short peptide antigens are usually poorly immunogenic. This can be alleviated by :

    (i)   synthesizing them as multiple antigen peptides (MAPs; Tam, 1988)

    (ii) coupling them to a carrier protein (Francis, 1991).

MAPs consist of branched multimers with a small oligolysine core at the centre. Apart from MAPs containing multiple copies of a single epitope, multivalent peptides consisting of different covalently linked epitopes can be constructed, including combination of B-and T-cell epitopes. Examples of increased immunogenicity of synthetic MAPs are experimental malaria vaccines consisting of combined B- and T-cell epitopes (Tam *et al.*, 1990) or a multimeric tetrapeptide with the sequence of a repetitive *Plasmodium falciparum* surface antigen epitope (Pessi *et al.*, 1991). A convincing success of a synthetic peptide-carrier protein vaccine *in vivo* was reported by Langeveld *et al.* (1994). Peptides with the sequence of the amino-terminal region of protein VP2 of canine parvovirus were synthesized and chemically coupled to a protein (Keyhole limpet haemocyanin). This vaccine induced full protection against virulent virus in dogs.

(b) They can adopt various conformations, which upon immunization may give rise to antibodies that recognize the peptide, but not the native antigen. This is especially true for conformational epitopes. This problem may be overcome by the cyclization of peptides by using chemical linkers (usually oligopeptides). Thus, the conformation of the peptide is constrained to that of the native epitope. The success of cyclization is determined by the nature of the peptide as well as length and conformation of the cyclic construct. One of the first examples of the successful induction of the proper conformation through cyclization has been reported by Muller *et al.* (1990). Antibodies raised to ovalbumin conjugates of cyclic peptide analogs of influenza virus haemagglutinin reacted with native haemagglutinin. The immunogenicity of the peptides was strongly dependent on the loop conformation and on the orientation of the peptide on the carrier protein. Hoogerhout *et al.* (1995) have shown that the ring size and the cyclization chemistry are of crucial importance for the imunogenicity of cyclic peptide analogs (Coupled to tetanus toxoid) of meningococcal OMP epitope.

Nucleic acid vaccines

A revolutionary application of rDNA technology in vaccinology has been the introduction of nucleic acid vaccines (Davis and Whalen, 1995). In this approach plasmid DNA or messenger RNA encoding the desired antigen is directly administrated into the vaccine. The foreign protein is then expressed by the host cells and generates an immune response.

Plasmid DNA is produced by replication in *E. coli* or other bacterial cells and purification by established methods (*e.g.*, density gradient centrifugation, ion-exchange chromatography). Only parenteral administration has proven to be effective in animals. Intramuscular injection seems to be the preferred administration route. The favourable properties of muscle cells for DNA expression are probably due to their relatively low turn-over rate. It prevents the plasmid DNA to be rapidly dispersed in dividing cells. After intracellular uptake of the DNA, the encoded protein is expressed on the surface of host cells. After a single injection, the expression can last for more than one year.

Nucleic acid vaccines offer the safety of subunit vaccines and the advantages of live recombinant vaccines. Possible disadvantages of nucleic acid immunization concern acceptability issues. The main pros and cons of nucleic acid vaccines are listed in Table-7.1. An advantage of RNA over DNA is that it is not able to incorporate into host DNA. A drawback of RNA, however, is that it is less stable than DNA.

Table 7.1 : Advantages and Disadvantages of Nucleic Acid Vaccines

Advantages	Disadvantages
Low intrinsic immunogenicity of nucleic acids	Effects of long-term expression unknown
Induction of long-term immune response possible	Formation of anti-nucleic acid antibodies
Induction of both humoral and cellular immune responses	Possible integration of the vaccine DNA into the host genome
Possibility of constructing multiple epitope plasmids	Concept restricted to peptide and protein antigens
Heat-stability	Ease of large-scale production

The concept is still in its infancy. In particular, the long-term safety of nucleic acid vaccines remains to be established. On the other hand, nucleic acids coding for a variety of antigens have shown to induce protective, long-lived humoral and cellular immune responses in various animal species. Examples are hepatitis B vaccine (Davis *et al.*, 1994), influenza vaccine (Webster *et al.*, 1994) and HIV vaccine (Coney *et al.*, 1994).

Production of vaccines in cells in culture

Except for synthetic peptides, vaccines are derived from microorganisms or from animal cells. For optimal expression of the required vaccine component (s), these microorganisms or animal cells can be genetically modified. Animal cells are used for the cultivation of viruses and for the production of some subunit vaccine components, and have the advantage that the vaccine components are released into the culture medium.

Three stages can be discerned in the manufacture of cell-derived vaccines.

(i) Cultivation

(ii) Downstream processing

(iii) Formulation

The development of the seed strain is a crucial part in the development of vaccines. The strain has to be characterized well in order to insure its genetic stability (*e.g.*, with regard to the synthesis of the antigens) during cultivation. Then the master and working seed lots are prepared. The development of the strain as well as the production and control of the seed lots have to be performed under 'good manufacturing practice' (GMP) conditions.

Bacteria and yeasts are relatively easily cultivated in bioreactors. The cultivation of animal cells is more complicated, because they are very sensitive to environmental factors like shear and oxygen concentration, and the composition of the culture media is complex. The seed culture, the medium composition, the cultivation conditions (such as pH, dissolved oxygen), and the criteria for harvesting should be well defined. The cultivation conditions have to be chosen in such a way that scaling up to production scale does not affect the quality of the vaccine component.

After cultivation, the vaccine component has to be separated from the bacteria, yeast or animal cells and from other unwanted cell suspension components. The applied downstream processing procedures depend on several factors like the cell type, the localization (cellular or released) of the vaccine component, and the physico-chemical characteristics of the component. If the component is linked to a microorganism or a cell, the microorganisms or cells have to be collected. If the component is secreted, the cell free culture liquid is collected. For the separation of cells, filtration and centrifugation techniques are most commonly applied and the cell-free culture liquid for the release of cell-associated vaccine components.

The infection of naked DNA into muscles or skin cells also elicits an immune response. The immune response against the therapeutic protein in early gene therapy trials was too strong for the gene therapy to be effective. This results were though disappointing provided new hope to the DNA-based vaccines. These are considered improved vaccines with fewer side effects.

7.3.5.6 Transgenic Animals as a Source of Biopharmaceuticals

Many biopharmaceuticals are produced by using genetically engineered mammalian cells or microbial fermentations. However, with the development of transgenic animals, it has become possible to produce certain human proteins of biopharmaceutical potential including tissue plasminogen activator, blood clotting factors, *etc.* in the lactating glands of several animal species, such as mouse, sheep, cow and pig. These proteins are secreted in milk and more easily extracted and purified.

It is from this, the term **Pharming** was coined to convey the idea that milk from transgenic farm animals (Pharm) can be used as a source of authentic human protein drugs or pharmaceuticals. There are a number of reasons why the mammary gland should be used in this way. Milk is a renewable, secreted body fluid that is produced in substantial quantities and can be frequently collected without harm to the animal. A novel drug protein that is confined to the mammary gland and secreted into milk should have no side effects on the normal physiological processes of the transgenic animals and should undergo post-translational modifications, that at least, match closely to those in humans. Finally, purification of protein from milk, which contains only a small number of different proteins should be relatively straightforward.

One of the earliest successes in creating transgenic animals was a mouse. A growth hormone gene from a rat was inserted into the genome of a mouse. Attached to the growth hormone gene was a powerful promoter which was stimulated by the presence of heavy metals in the mouse's diet. When these heavy metals were included in the mouse's food, the growth hormone gene was almost continually 'switched on'. This made the mouse grow 2-3 times faster than mice without the gene. The mouse with the growth hormone gene also finished growth at about twice as large as normal. This was achieved through genetic engineering.

Production of human pharmaceuticals in transgenic animals – is rapidly becoming a reality. The animals can be considered in biotechnology terms as bioreactors operating on a continuous basis. Overall this could be a massive global market. At present the major limiting factor is achieving the number of transgenic animals.

An American company is producing human haemoglobin in the blood of transgenic pigs that can serve as a human blood substitute. Such transgenic haemoglobin could capture a massive market. Each year world wide, 70 million units of human blood are transfused at a cost of $US10 billion. This transgenic haemoglobin would be free from human pathogens such as HIV and would not need typing or matching before transfusion, because it is not composed of red blood cells. Much needs to be done before this becomes a reality.

Once manufacturing biopharmaceuticals in animals becomes efficient, the flow of more than a hundred protein-based drugs currently in advanced phases of clinical trials and many more that are in development in the laboratory, would increase. For example, 14 varieties of therapeutic proteins have been produced in milk of goats engineered by GTC Biotherapeutics, United States. A flock of transgenic goats can be created for about $100 million, a third of the cost of building a protein-production facility. In addition, when a drug maker needs to double the production, the solution is to breed more animals, instead of spending $300 million on a new factory. This could decrease the cost of purified therapeutic protein from $150 to between $1 and $2 a gram (*The Economist*, 2003).

Chicken multiply and mature early. The desired proteins can be recovered from their eggs. An ideal storage medium for compounds is egg white. In July 2002, two antibodies (one human and one murine) were produced by TranXenoGen, United States, in the albumin of transgenic chicken. Insulin and human serum albumin have also been produced in the eggs of transgenic chicken (*The Economist*, 2003).

7.3.5.7 Genetically Engineered Plants as a Source of Pharmaceuticals and Edible Vaccines

Pharmacological products are also produced in genetically engineered plants. About 300 trials of genetically engineered crops to produce various therapeutic products have been initiated. These include :

  (i)   modified tobacco plants that produce Interleukin-10 for the treatment of Crohn's disease,
  (ii) GM potatoes that produce antibodies for reducing the risk of rejection in kidney transplants,
  (iii) GM tobacco that produces vaccines against hepatitis B and drugs against HIV/AIDS,
  (iv) potatoes that produce human insulin.

Other GM plant-produced substances include enkephalins, alpha-interferon, serum albumin and glucocerebrosidase. Clinical trials have begun on crop-grown drugs to treat cystic fibrosis, non-Hodgkin's lymphoma and hepatitis B (*The Economist*, 2003).

The various firms and research centres have shown the effectiveness of plants as bioreactors. These include SemBioSys Genetics, Inc., Canada, Plant Biotechnology, Inc., United States. The ProdiGene, Inc. United States have demonstrated that mice fed on potatoes expressing the beta-subunit of cholera toxin (CTB) were resistant to the cholera toxin (Langridge, 2000). Greater protection against cholera to humans than to mice is provided by CTB.

Transgenic tomatoes containing a gene from *Escherichia coli* that can protect against diarrhoeal diseases have been produced (Lemonick, 2003).

Tomatoes, bananas and potatoes among other crops are being used in several laboratories around the world to develop vaccines what we call edible vaccines.

Biopharming is mainly driven by a cost advantage. For example, medicinal products could be synthesized in plants at less than one tenth of the cost of conventionally manufactured drugs and vaccines. By the end of the current decade, biopharmaceuticals are projected to grow into a $20 billion industry. By this, the cost of treating some diseases may be brought down (Roosevelt, 2003).

GM crops are being used for the production of food, drugs *etc*. Pharmagedden's crop products carrying drugs, vaccines and industrial chemicals will be used on the dinner tables. But there are many fears, which will be discussed in the last chapter.

The production of transgenic animals and plants would require relatively small investments and minimal cost of maintenance. It is because of this they may provide the economically viable option for independent production of therapeutic proteins in developing countries.

7.3.5.8 Regenerative Medicine : Stem Cells and their Relevance Today

There is a remarkable capacity in human body to repair and maintain itself. The prospects of using the body's natural healing processes are now thought to be used not to treat debilitating diseases, but perhaps to cure them.

Scientists and clinicians have been depending in the background on stem cell research for years, but it took two breakthroughs – animal cloning by Ian Wilmut, Keith Campbell and colleagues and the derivation of pluripotent human embryonic stem (ES) cells by Jamie Thomson and Co-workers to really shake up the thing. They first showed that an adult cell nucleus can be reprogrammed to produce an entire animal, a dramatic demonstration of hidden potential and the second provided a possible source of cells for cell based therapies for many human diseases. Because pluripotent stem cells are able to form virtually any cell type present in the body, many new disease treatments become possible. In theory, neurons and glia could be produced to treat neurodegenerative diseases such as Parkinson's and Alzheimer's. Muscle cells could be produced to treat muscular dystrophies and heart disease, haematopoietic stem cells could be produced to treat leukaemias and AIDS.

As mentioned earlier in Chapter 3 stem cells are defined as cells that have the ability to perpetuate themselves through self-renewal and to generate mature cells of a particular tissue through differentiation. In most tissues, stem cells are rare. As a result, stem cells must be identified prospectively and purified carefully in order to study their properties.

Pathways regulating stem cell, self-renewal and oncogenesis

Because normal stem cells and cancer cells share the ability to self-renew, it seems reasonable to propose that newly arising cancer cells appropriate the machinery for self-renewing cell division that is normally expressed in stem cells. Evidence shows that many pathways that are classically associated with cancer may also regulate normal stem cell development *e.g.,* the prevention of apoptosis by enforced expression of the oncogene *bcl-2* results in an increased number of HSCs *in vivo.* This suggests that cell death has a role in regulating the homeostasis of HSCs.

Stem cells in tissue engineering

The list of tissues with potential to be engineered is growing steadily. Tissues that can now be engineered using stem cells comprise a diverse range from epithelial surfaces to skeletal tissues. These systems are inherently different in their rate of self-renewal and physical structure, the two important determinants of any attempt to reconstruct tissues using stem cells.

Stem cell cloning

The word 'clone' derives from the Greek term Klon, meaning a 'sprout' or 'twig'. It refers to a method of reproduction apart from the parental, sexual mating process that is characteristic of most organisms. Attempts are being made to develop a clone of cells from stem cells that develop very early in life *e.g.,* Wisconsin Biotechnology Company unveiled a Holstein bull calf (appropriately named Gene) that was developed by stem cell cloning. Gene was a transgenic calf because it carried foreign genes. Before the fusion process, researchers had inserted genetic markers into the stem cells and now those markers could be located in the cell of the calf. The calf demonstrated that stem cells could be manipulated by inserting new genes.

In November 1998, researchers reported that they had established human embryonic stem cell (ESC) lines capable of existing indefinitely in lab dishes and giving rise to any human cell type. One group isolated and cultured certain cells from a frozen blastocyst, which is the 5 day old ball of 100 to 150 cells that eventually develops into an embryo. The other group isolated and cultured ES cells from progenitor germ cells, isolated from foetusus that had been aborted. Unlike adult cells (ASc), embryonic stem (ES) cells are pluripotent, they can differentiate into any kind of cell in the body. In addition to their total development plasticity, it is believed that hES cells in culture may be able to reproduce without limit. The availability of a source of undifferentiated human cells opens up new avenues for treating diseases with cell therapy, studying human development, discovery of new drugs, testing drug safety and engineering replacement tissues. However, many people object to any research that involves growing human embryos in labs, even if it is limited to the earliest stages of development.

There are three potential sources of human pluripotent ES cells.

1. Isolation of ES cells from surplus blastocysts created during *in vitro* fertilization (**IVF**) treatment, which would otherwise be destroyed.
2. Extraction of cells destined to form eggs or sperm from aborted or miscarried foetuses.
3. Production of tailor-made ES cells from patient's own differentiated adult cells using cloning techniques.

American scientists have used the first two methods to isolate human pluripotent stem cells (**hPSCs**), and have successfully grown them in the laboratory. Their hPSC cultures can be multiplied indefinitely and are, therefore, a valuable resource for stem cell research. The work raises the possibility that collections of specific stem cell types could be generated for transplant therapy. UK research using human embroys is governed by the Human Fertilization and Embryology Act (1990), which restricts research on embryos on more than 14 days old for certain permitted purposes only. Government of US widened the Act's scope to include research into stem cell therapies from January 31, 2001, in the light of recommendations of the August, 2000 Donaldson Report and following free votes in parliament and in the House of Lords. The stem cells used for therapeusis must be compatible whit immune system of the recipient (Fig.7.8).

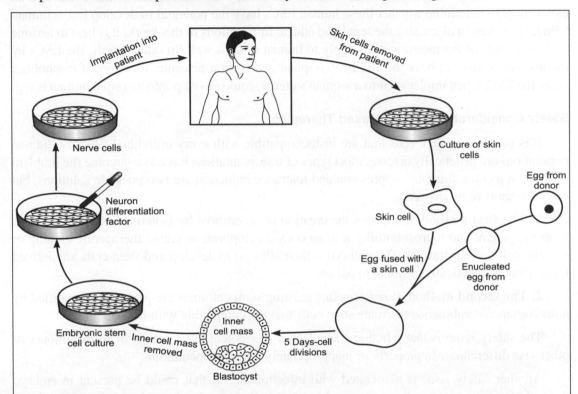

Fig. 7.8. Stem cells compatible to immune system: The therapeutic potential of stem cells can be maximized if they are not seen as foreign by the immune system. For this purpose patient's own cells can be used *e.g.,* in the case of Parkinson's disease nuclei are removed from the patient's skin cells. They are implanted in a donated, enucleated egg. After the cells are cultured for approximately 4 to 5 days, a blastocyst is produced. The inner cell mass of the blastocyst is removed and cultured. Thereby a line of hES cells is created. These are genetically identical to the (except for the mitochondrial DNA from the egg donor) cells of patient. The ES cells are converted by treatment with appropriate differentiation factors into neural stem cells and then into fully differentiated nerve cells. These are implanted in the patient in hope of replacing the diseased or injured cells with healthy nerve cells (Modified from Kreuzer and Massey, 2003).

To assess the implications of ESC research, we need to understand the methodologies used in human ESC cultures. After fertilization, cells of the embryo divide until they form a blastocyst, hollow ball of approximately 140 cells in approximately 5 days. Blastocytes consist of an inner cell mass. The trophoblast cells differentiate into placental tissue, while the inner cell mass is made up of the cells that will divide and develop into the individual. The inner cell mass gives rise to the ESC lines.

Working on human ESCs, researchers obtained 14 blastocysts that were produced through *in vitro* fertilization and donated with informed consent, by the couples. Scientists removed the inner cell masses and successfully established five ESC lines that have shown remarkable plasticity. When implanted in mice, these human ESCs have differentiated into tissues resembling neural epithelium, bone cartilage, gut, striated muscle and kidney. Of course, no one has conducted an experiment to determine whether these human ESCs have the potential to develop into a human being. However, in discussing the legal and ethical implications of this work, it is best to assume that the results of the mouse studies apply to human ESC as well. In other words, the ESCs by themselves would not have the ability to implant and form a placenta, however, if trophoblast cells, the ESC, when implanted into a woman's uterus, could develop into a normal human being.

Safety Considerations in Cell Based Therapies

It is possible to have cells that are histocompatible with every individual. Because human populations are genetically diverse, most types of transplantations have to overcome the problem of tissue rejection. Immune suppression and tolerance induction are two possible solutions, but both are short-term answers.

1. The first method. It requires the creation of an embryo by isolating a somatic nucleus from the patient and reprogramming it in an oocyte cytoplasm so called therapeutic cloning or somatic cell nuclear transfer. The embryo is then allowed to develop and stem cells are derived from it that is genetically identical to patient.

2. The second method. It requires that existing stem cell lines are genetically modified by homologous recombination to create stem cells that are compatible with the patient.

The safety issue is that whether transplanted pluripotent stem cells will form tumours or otherwise differentiate improperly or inappropriately after transplantation.

Another safety issue is associated with infectious agent that could be present in embryo derived pluripotent stem cells or acquired by stem cells in feeder dependent culture containing bovine serum.

Issues associated with ESC research

A number of people are concerned about the source of ESCs. If we decide to, we should not use blastocysts as sources of ESC in the future. The question arises that is it acceptable to use the immortal ESC lines researches have already established? or should they be discarded? All these need thorough thinking.

The Future of Stem Cell Therapy

It is being thought that stem cell therapies could become a clinical reality in 5-10 years time, but a huge research effort will be needed to achieve this goal.

The combined characteristic of developmental versatility and unlimited capacity for self-renewal make stem cells excellent therapeutic tools. For example, medical researchers in Germany have used cultured AS cells to repair tissue severely damaged after heart attacks. They injected the patient's own heart stem cells into their coronary arteries and saw improvements within a few weeks. On an average, the area of damaged heart tissue decreased by 36% and heart function improved 10%.

Stem cell research has following priorities

- Understanding mechanisms of differentiation and development.
- Identification, isolation and purification of different adult stem cell types.
- Controlling differentiation of stem cells to target cell types needed to treat disease.
- Learning to make stem cell transplants compatible.
- Demonstrating normal cell development and function and appropriate growth control in stem cell transplants.
- Confirming the results of successful animal experiments in humans.

7.3.5.9 Targeted Drugs

For quite sometime, attempts have been made on targeted drug delivery. The vast majority of new generation therapeutic drugs are peptides or protein molecules that the body will readily break down unless they are protected in some way. The new therapeutic technologies require new delivery technologies designed to :

(*i*)   deliver these powerful, but delicate molecules to their sites of action and

(*ii*) control their release to maximize their therapeutic benefit.

The first boom in delivery technologies came in 1950s. Encapsulation made the oral administration of many drugs possible. In the 1970s transdermal patches appeared. This allowed timed absorption of drugs through the skin. The developments of minute infusion pumps, liposomes and nasal sprays, monoclonal antibodies and bio-erodable polymers have further helped.

7.3.5.9.1 Liposomes

Liposomes are readily adsorbed by lipoprotein coated cells. Attempts have been made to adopt them as drug delivery vehicles (though it can be difficult to direct these miniscule capsules to the target site). Liposomes consist of two layers of highly polar phospholipids, either natural or synthetic in origin. A coating of phospholipids on an aqueous droplet (containing a water soluble drug for example) quickly orient themselves with their hydrophilic portions towards water and their hydrophobic portions "tucked in" to form a double membrane.

Liposomes are preferentially adsorbed by the reticuloendothelial system. Because the droplets are too large to pass through capillary walls, injected liposomes accumulate in the liver and spleen where they can persist for days or weeks. They also accumulate at the sites of inflammation.

Various workers have proposed the use of liposomes to deliver drugs via oral inhalation (bronchodilators, for example), occular droplets (for dry eye and glaucoma), injection (for metastatic carcinoma and systemic fungal infections) and topical administration (Minoxidil for male pattern baldness).

7.3.5.9.2 Nasal sprays and transmembrane carrier

A drug can carry a carrier molecule as a sort of passport helping it penetrate some of the body's key barrier membranes such as mucous-membranes of the nasal passages. The latter is particularly useful for delivering smaller peptide drugs (especially hormones such as insulin).

7.3.5.9.3 Biodegradable polymers

The drugs are being made to be delivered to the target by encasing them in a polymer matrix that dissolves slowly in the body's aqueous environment. The precise kinetics of the dissolution is important. Water attacked many early candidate polymers. This altered their permeability causing them to "dump" their proteins rather than releasing them at a constant, controlled rate.

Attempts are on to develop polymers of highly hydrophilic molecules, such as anhydrides. Water binds tightly only to the exposed surfaces, lifting the material off in monomolecular layer. This maintains the shape of the delivery matrix. The embedded drug is delivered at a constant rate.

7.3.5.9.4 Osmotics

The osmotic pumps of polymers are being developed to deliver peptide drugs. It is a tablet coated with a permeable (but insoluble) polymer membrane. Water diffuses through the polymer, dissolves the drug at a controlled rate from the surface of tablets. The dissolved protein (too large to diffuse through the membrane) under osmotic pressure is then expelled through a minute laser drilled hole. The delivery rate can be controlled by varying the size of the laser hole and the thickness and composition of the membrane.

Now an improved system with two chambers has been made. One is totally enclosed in a permeable membrane containing an osmotic agent, the other is bonded side by side with first. It contains the drug. As the osmotic compartment swells in an aqueous environment, it presses against the wall it shares with the drug containing compartment. This pressure forces dissolved drug out through another laser bored channel.

7.3.5.9.5 Monoclonal Antibodies

The cancerous cells are identical to normal cells in almost every respect and this hampers the treatment of cancer, because the therapeutic agents, directed against cancer cells, are also likely to kill normal cells. It has, however, been found that the surfaces of cancer cells in the body differ in a few respects from those of normal cells. Since monoclonal antibodies recognize specific antigens on cells, they are being used to image tumours and in therapy against melanomas, lymphomas, and breast and colon cancer.

Several approaches can either be used alone or linked to toxic drugs. The theory is that when the antibodies bind to the tumour, they attract the cells of the immune system to act against the cancerous tissue. Monoclonal antibodies may be used to target toxic drugs to the interior of cancer cells.

7.3.5.10 Modified Proteins

The modification of the activity of proteins is another research area of interest. Tumour-destroying proteins as well as proteins that will protect the immune system against viruses and cancers, just as vaccines do, are being developed by Genencor designing. Similarly, effective versions of interferons alpha and gamma, yet to be tested in people, are being developed by Maxygen. They are also developing proteins that would behave as vaccines against bowel cancer and dengue fever. Others include Viracept, a protease inhibitor for HIV (by Agouron) and Relenza™ (by Biota Holdings), an inhibitor of neuraminidase of the influenza virus (The Economist, 2003).

7.3.6 Pharmacogenomics

7.3.6.1 Introduction

Interindividual variability in drug response, ranging from no therapeutic benefit to life threatening adverse reactions, is influenced by variation in genes that control absorption, distribution, metabolism and excretion of drugs. Differences in DNA sequences altering the expression or function of proteins that are targeted by drugs can significantly contribute to variation in the responses of individuals. Many of the genes examined in early studies were linked to highly penetrant, single-gene traits, but future advances hinge on the more difficult challenge of elucidating multigene determinants of drug response. This interaction of genomics and medicine has the potential to yield a new set of molecular diagnostic tools that can be used to individualize and optimize drug therapy.

Pharmacogenomics, the study of the genetic basis of therapeutics, identifies discrete genetic differences among individuals that play a critical role in drug response. This term in fact describes a polygenic or genome-wide approach to identify genetic determinants of drug response, capitalizing on information from the Human Genome Project and on advances in technology (for example, high-throughout sequencing, DNA and protein microarrays and bioinformatics). The integration of such sophisticated genomic tests with extensive phenotypic characterization of uniformly treated patients is essential if we are to define the inherited nature of most drug effects. DNA tests based on these genetic variations can predict how a patient will respond to a particular medicine. Clinicians will use them to select optimal therapy and tailor dosing regimens; the benefits will include reduced incidence of adverse drug events, improved clinical outcomes, and reduced costs.

7.3.6.2 History and Background

The idea that genes control some drug responses was suggested in the 1950s because of the link between inheritance or ethnicity and aberrant drug responses (Evans *et al.*, 1999). This notion was strengthened by family and twin studies in the 1960s and 70s (Vesell *et al.*, 1989), extended by biochemical studies in the 1970s and 80s (Guengerich *et al.*, 1998), and solidified by molecular genetics in the 1980s and 90s (Meyer *et al.*, 1997). Cloning and characterization of the first human gene containing DNA sequence variations (genetic polymorphisms) that influence drug metabolism did not take place until the late 1980s (Gonzalez *et al.*, 1988). In the decade that followed, genes responsible for many such inherited differences in drug metabolism and dispositions were isolated, characterized and linked to differences in drug effects (Weinshibourn *et al.*, 2003). During this period, polymorphisms in genes encoding drug transporters and targets were also identified and shown to alter drug response (Evans *et al.*, 2003).

There are many challenges that we have to overcome if we are to understand fully the contribution of genetic polymorphisms to inter-individual differences in drug effects and to translate this new knowledge into clinical practice. It is a new, but rapidly growing field and has a great potential in the near future. It will evolve from its current state of one-drug one-gene, to one in which multiple genetic (polygenic) determinants of drug effects will be defined and used to individualize drug therapy.

7.3.6.3 Genes and Polymorphisms Governing Drug Responses

Short of being able to sequence the entire genome of every patient with an aberrant drug response, how does one rationally go about identifying genetic polymorphisms that influence drug effects? The 'candidate gene' or 'candidate pathway' approach has been used to make an informed prediction of the genes in which polymorphisms might affect the disposition or response to a given drug. In many cases, the known phenotypic variability in the gene product guided the search for a genetic basis for such differences.

These can be several reasons for the failure of the candidate-gene approach one of these could be that other mechanisms alter protein function (for example, post-translational modifications). Even if the selected gene is important, it may be technically difficult to identify the functionally important polymorphisms. These can include promoter or enhancer polymorphisms, gene duplications, synonymous coding SNPs that affect transcript stability, like 3' untranslated region (3' UTR) or intronic SNPs that cause splice variants that create early stop codons. SNPs can also cause significant amino-acid substitutions (Evans *et al.*, 1999).

'Pathways' of genes may be more important than individual genes, with the effects of polymorphisms in networks of genes acting together to create a single phenotype.

(i) Genetic Polymorphism in Drug Transporters

Although passive diffusion accounts for cellular uptake of some drugs and metabolites, increased emphasis is being placed on

   (i)   the role of membrane transporters in absorption of oral medications across the gastrointestinal tract;
   (ii) excretion into the bile and urine;
   (iii) distribution into "therapeutic sanctuaries," such as the brain and testes;
   (iv) transport into sites of action, such as cardiovascular tissue, tumour cells, and
   (v)   infectious microorganisms (Schuelz *et al.*, 1995).

It has been proposed that some of these transporters, such as P-glycoprotein, may not be essential for viability, because knockout mice appear normal until challenged with xenobiotics. However, other transporters are likely to play critical roles in transport of endogenous substances. Although polymorphisms in P-glycoprotein have been reported (Schinkel *et al.,* 1989), and such variation may have functional importance for drug absorption and elimination, the clinical relevance of polymorphisms in drug transporters has not yet been fully elucidated.

(ii) Genetic Polymorphisms in Drug Metabolism and Disposition

Polymorphisms in drug-metabolizing enzymes have more subtle, yet clinically important consequences for interindividual variability in drug response. Such polymorphisms may or may not have clear clinical importance for affected medications. It depends on

   (i)   the molecular basis of the polymorphism,

   (ii) the expression of other drug-metabolizing enzymes in the patient,

   (iii) the presence of concurrent medications or illnesses, and

   (iv) other polygenic clinical features that impact upon drug response.

The landmark approval of Bidil, a drug used to treat heart failure specifically in blacks, has raised questions about the role of race-specific drugs in medical practice. Hearing early reports that the drug is ready being prescribed to people outside the prescription guidelines, experts are finding out a biological marker that can better predict who will respond to the drug. The US Food and drug administration (FDA) approved Bidil after studies showing that it improves mortality after heart failure by 43% among blacks, but the drug was found to be ineffective in whites (N. Engl. J. Med. 2004). It is not yet clear why the drugs should be more effective in blacks, but one theory holds that Bidil compensates for a nitric oxide deficiency common in this population.

In addition to detoxifying and eliminating drugs and metabolites, drug-metabolizing enzymes are often required for activation of prodrugs. Many opioid analgesics are activated by *CYP2D6* (Desmeules *et al.,* 1996). It renders the 2 to 10% of the population who are homozygous for nonfunctional *CYP2D6* mutant alleles relatively resistant to opioid analgesic effects. It is, thus, not surprising that there is remarkable interindividual variability in the adequacy of pain relief when uniform doses of codeine are widely prescribed.

At present, the best-recognized and completely developed examples of genetic polymorphisms that alter drug response in humans are monogenic (single gene) traits that affect drug metabolism. Thiopurine S-methyltransferase (TPMT) is one such example. It has influence on the metabolism of thiopurine drugs *e.g.,* mercaptopurine and azathioprine in patients who inherit non-functional *TPMT* alleles (Evans *et al.,* 2001). These drugs are clinically used as immunosuppressants and to treat neoplasias. Because TPMT is the predominant inactivation pathway for these medications in haematopoietic tissues, patients who inherit a TPMT deficiency accumulate excessive concentrations of the active thioguanine nucleotides in blood cells when treated with mercaptopurine or azathioprine. This can lead to severe and potentially life-threatening haematopoietic toxicity (Evans *et al.,* 2001).

TPMT-deficient patients (that is, patients with two nonfunctional alleles) can be treated successfully using much lower doses of thiopurines (around 5–10% of the conventional dose) (Evans *et al.*, 1991). We can now clinically detect the inactivating single-nucleotide polymorphisms (SNPs) in the human *TPMT* gene. It is possible to use *TPMT* genotyping to make treatment decisions (Abott *et al.*, 2003; Marshal *et al.*, 2003). To some extent, this is related to the perceived high cost of genotyping, even though this process has been shown to be cost-effective (Phillips *et al.*, 2001).

(iii) Genetic Polymorphisms in Drug Targets

Most drugs interact with specific target proteins to exert their pharmacological effects, such as receptors, enzymes, or proteins involved in signal transduction, cell cycle control, or many other cellular events. Molecular studies have revealed that many of the genes encoding these drug targets exhibit genetic polymorphism, which in many cases alters their sensitivity to specific medications. The examples are polymorphisms in β-adrenergic receptors and their sensitivity to β-agonists in asthmatics (Martinez *et al.*, 1999), angiotensin converting enzyme (ACE) and its sensitivity to ACE inhibitors (Vanderkleij *et al.*, 1997), angiotensin II T1 receptor and vascular reactivity to phenylephrine (Henrion *et al.*, 1998) or response to ACE inhibitors (Benetos *et al.*, 1996), sulfonylurea receptor and responsiveness to sulfonylurea hypoglycemic agents (Essen *et al.*, 1996), and 5-hydroxtryptamine receptor and response to neuroleptics such as clozapine (Arraz *et al.*, 1995).

In addition, genetic polymorphisms that underlie disease pathogenesis can also be major determinants of drug efficacy, such as mutations in the apolipoprotein E (*apoE*) gene and responsiveness of patients with Alzheimer's disease to tacrine therapy (Poirier *et al.*, 1995) or cholesteryl ester transfer protein polymorphisms and efficacy of pravastatin therapy in patients with coronary atherosclerosis (Kuivenhover *et al.*, 1998). The risk of adverse drug effects has also been linked to genetic polymorphisms that predispose to toxicity, such as dopamine D3 receptor polymorphism and the risk of drug-induced tardive dyskinesia (Steer *et al.*, 1997), potassium channel mutations and drug-induced dysrhythmias (Abott *et al.*, 1999), and polymorphism in the ryanodine receptor and anesthesia-induced malignant hyperthermia (Gillard *et al.*, 1992). Another important source of genetic variation in drug sensitivity is polymorphisms in genes of pathogenic agents (human immunodeficiency virus, bacteria, tuberculosis, and others).

7.3.6.4 Pharmacogenomics to Clinical Uses : Limitations

Despite all the advances made in molecular genetics, pharmacogenomics is still not commonly used in clinical practice. There are several well-established examples of common genetic polymorphisms such as *CYP2D6* and codeine activation, *TPMT* and mercaptopurine inactivation, *CYP2C9* and warfarin inactivation that have a greater impact on drug effects than do laboratory parameters that are currently used to adjust drug therapy (for example, serum creatinine reflects renal function and is used to adjust drug doses in kidney failure). Furthermore, for each of these polymorphisms, robust molecular genotyping methods have been established. But transfer of pharmacogenetics into clinical practice is still not being followed.

Clinicians normally start treatment with the default 'average dose'. Individualizing dosages, even based on easily assessed characteristics of patient (such as age or renal function), have not been widely embraced by the medical or pharmaceutical communities. There is resistance to relying on tests for every medical decision, and a 'trial and error' approach to drug dosing has become widely accepted. The pharmacogenomics in fact require a laboratory test to look for polymorphic type of a particular gene involved in drug response. It also requires an interpretation of genotypes which makes clinicians to receive further training in molecular biology or genetics.

It is at the moment difficult to conduct definitive clinical pharmacogenomic studies to prove that individualization of drug therapy on the basis of genetics improves clinical outcomes. The limitations to provide evidence to catalyze a change in clinical practice include the factors like the multigenic nature of most drug effects, the difficulty in controlling for non-genetic confounders (for example, drug interactions, diet and smoking) and the lack of funding for large-scale pharmacogenomic studies with adequate follow-up.

In fact there are certain examples of genetic technology that have been validated to the point where they can be used in clinically licensed and regulated laboratories. The diverse range of mechanisms that account for genetic polymorphisms (SNPs, insertions/deletions, splice variants and so on) means that providing definitive results for even a single gene is a tremendous challenge, as has been demonstrated for *Brca1* and *Brca2,* which are implicated in breast cancer, and *Cftr*, which has a role in cystic fibrosis (Eccles *et al.,* 2003).

Ultimately, testing for pharmacogenetic polymorphisms must overcome the same technical hurdles that govern other molecular diagnostics. The positive point is that a given genotype needs to be determined only once, unlike a measure of renal function.

7.3.6.5 Conclusion and Future Direction

There is a great potential for the human genome to yield new insights into the pathogenesis of human diseases and to reveal new strategies for their prevention or treatment. Pharmacogenomics may enhance drug discovery and development in two ways:

(i) By the identification of drug targets

(ii) By subpopulation-specific drug development.

Genomics can be used to identify new targets through the discovery of genes that are under- or over-expressed in cancer cells that are sensitive to anticancer agents compared with those that are resistant. The products of such over-expressed genes represent possible targets for inhibitors that could reverse the drug resistance phenotype.

Pharmacogenomics enables to identify genetic polymorphisms that predispose patients to adverse drug effects that, although may occur in only a small subset of the people treated with a new medication, are sufficiently toxic to jeopardize further development of the drug for all patients. There are differences in the genotypic subgroup frequencies of many pharmacogenetic polymorphisms (for example CYP3A5, nitric oxide synthetase, angiotensin converting enzyme) amongst different races (Kuehl *et al.,* 2001; Holden *et al.,* 2003).

Genetic differences between racial groups may result in the preferential development of drugs that would benefit one group more than another, whether international or not.

The information about potential value of all the possible factors that influence the effects of new agents, will help in determining the important role in drug discovery and development.

Advances in technology are likely to bring down cost of genotyping and establishing definitive polygenic models for optimizing drug therapy. But as pharmacogenomic strategy becomes a routine part of drug discovery and development, such models should emerge from large scale clinical trials and follow-up of new medications. Pharmacogenetics research network (http://www.pharma.GKb.org) funded by the National Institute of Health coupled with other academic and industry initiatives, is a driving force behind this science (Altman *et al.,* 2003). As there are millions of human genome polymorphisms, the best way forward is to replicate large-scale clinical trials of uniformly treated and evaluated patients. These trials should incorporate comprehensive and rigorous pharmacogenomic studies. These should be coupled with preclinical experimental models that reinforce genotype-phenotype clinical associations. Incorporation of genetics into clinical trials may not be deemed "innovative" by conventional peer review, but will be necessary for clinical progress.

If the use of medication is to evolve from a 'trial and error' approach to individualized therapy using genetics, at least two aspects of health care need to change.

(a) We must introduce protection against the misuse of genetic information and accept the added costs that may be increased during the transition to genetically guided decisions about drug therapy.

(b) In the long run, decreasing the frequency of adverse drug effects and increasing the probability of successful therapy will probably lower the cost of health care.

Pharmacogenomics has the potential to facilitate this process by translating knowledge of human genome variability into better therapeutics.

7.4 GENE THERAPY

7.4.1 Introduction

The most far reaching and controversial area of genetic engineering of humans is gene therapy. This is the treatment of disease by the transfer and expression of genetic material in a patient's cells in order to restore normal cellular function. It is of two types :

(i) Germ cell gene therapy

(ii) Somatic cell gene therapy

Germline gene therapy is the permanent introduction of genetic material to germ cells, which allow generational passage of genes to offspring. Transfer of DNA to somatic cells that cannot be transmitted to new generations is called somatic cell gene therapy. Germline gene therapy would result in the permanent introduction of new or altered traits into the desirable, to "correct" conception of individuals. Germline gene therapy has had no place in the current model of medicine and is not being considered or conducted so far, as it has raised a number of ethical concerns.

Most diseases under consideration for gene therapy result from single genes or interactions among multiple genes, and the affected individual is the target of somatic gene therapy. Treatment is directed to individuals who are affected by or predisposed to the development of significant life-threatening diseases.

Gene therapy has been focusing on correcting single-gene defects (mutations), such as cystic fibrosis and haemophilia that have been observed in families by their Mendelian pattern of inheritance. It is thought that hundreds of such diseases could be treated by this process. Many diseases at present have no effective treatment. Gene therapy, therefore, could offer hope to so many people. Somatic cell gene therapy for complex multifactor diseases *e.g.,* Parkinson's disease, cancer and other such diseases is being experimented. In many of these diseases, there can be involvement of a number of genes as well as an interaction with environmental factors.

Gene therapy is a complex series of events depending heavily on new biotechnological techniques. Therapy will require a full understanding of the mechanism which will help in following:

(i)   The defective or unusual gene exerts its effect on the individual

(ii) The defective gene gets switched off

(iii) The substitution of a healthy gene copy can be attained

Each year, a few children are born with severe combined immune deficiencies, in which the affected person has virtually no defense against disease. In about 25% of these children, the failure of the immune system can be traced to abnormal alleles of a single gene. In the late 1980s, R. Michael Blaese of the National Cancer Institute and W. French Anderson of the National Institute of Health, USA inserted normal alleles of this gene into white blood cells and obtained reasonably normal gene expression. At 12:52 p.m., September 14, 1990, the first clinical trial of human gene therapy began: A four – year-old girl received a transfusion of her white blood cells that were genetically engineered to contain the normal allele. The treatment worked.

In 1994, experimental trials for cystic fibrosis, a usually fatal inherited lung disease, were begun with a "nasal spray" containing viruses engineered to infect lung cells and provide them with a crucial protein that helps to prevent fluid buildup in the lungs.

Despite the risks it has, gene therapy is highly supported. For the first time on 18th August 2003, treatment of Parkinson's disease via gene therapy was tested on humans. In this, virus carrying the gene for the synthesis of dopamine (whose absence causes the disease) was injected in the brain. The scientific community was divided about this approach. Some neurologists considered this to be highly risky. The results were expected within two months after the initial injection.

The most likely candidates for genetic engineering in humans are, therefore, defects involving discrete structures, such as lungs or bone marrow, that are normally the only, or the most important, active sites of transcription of certain genes. These therapies would, of course, offer effective treatments only for the affected individual. The patient's reproductive cells would not be "fixed," so he or she could still pass the genetic defect on to future generations. Some of the protocols are now proceeding through regulatory processes. The patent coverage for gene construction, delivery systems and supporting technology has been granted. Private medical systems are internationally seeing gene therapy as a very, lucrative market in the affluent nations.

There is currently, ns available cure for the lysosomal storage disease (LSD). However, the scientist at St Jude successfully treated a laboratory model of the LSD called GM1-gangliosidosis using bone marrow cells (BMCs) into which the gene for an enzyme that degrades a fat molecule called GM1 has been introduced. GM1 is a normal component of normal brain cells. But in GMI gangliosidosis brain cells lack the enzyme β-galactosidase and GM1 accrues to such an extent that, it breaks down the normal function of cell and causes it to self distinct. BMC includes a population of so called pluripotent stem cells.

After the genetically adapted BHCs were infused into the laboratory model, the resulting monocytes migrated to the deteriorating brain cells that lacked the gene for β-galactosidase. These cells gathered the enzyme released by the monocytes and utilized it to degrade excess GM1, thereby, fixing the potentially fatal buildup of the molecule (Biotech News International, Vollo, 2005).

There have been unprecedented advances in understanding the molecular biology of clinical medicine particularly of inherited disorders. It is thought that upto 4000 such disorders are existing. Although fortunately some of these disorders are rare, others are prevalent. There have been attempts to have techniques to identify genes that cause or contribute to diseases in humans. It is all because of the leaps made in molecular genetics and recombinant DNA technology. The list of diseases for which candidate gene is identified grows daily and the human genome initiative has been rapidly determining the location of these genes and will no doubt identify many new genes involved in diseases. The ability to manipulate DNA has made it possible to transfer corrected or correcting genes for therapy of genetic and gene influenced diseases. With the localization and cloning of the defected gene in cystic fibrosis in 1987, the possibility of curing it and other such disorders by the introduction of the correct, non-defective gene into the patients, the gene therapy has been postulated.

The term gene therapy is a broad one : it includes many different strategies. All these strategies are designed to overcome or alleviate disease by a procedure in which genes, gene segments or oligonucleotides are introduced into the cells of an affected individual. The genetic material may be transferred as follows:

    (i) Directly into cells within a patient (*in vivo* gene therapy)

    (ii) Cells may be removed from the patient and the genetic material inserted into them *in vivo*, prior to replacing the cells in the patient (*ex vivo* gene therapy)

Vectors or gene transfer vehicles transfer plasmid DNA, RNA or oligonucleotides into target cells altering the expression of specific mRNA that directs the synthesis of therapeutic protein by the transfected cells. Because the molecular basis of diseases can widely, vary some gene therapy strategies are particularly suited to certain types of disorders, and some to others. Following are the major disease classes :

    (i) Infectious diseases

    (ii) Cancers

    (iii) Inherited disorders

    (iv) Immune system disorders

The basic need for gene therapy has been to treat diseases for which there is no effective treatment. Gene therapy has the potential to treat all the above classes of disorders. Depending on the basis of pathogenesis, different gene therapy strategies can be considered.

Current gene therapy is exclusively somatic gene therapy, the introduction of genes into somatic cells of an affected individual.

7.4.2 Strategies of Gene Therapy

Strachan and Read (2003) have pointed out following strategies of gene therapy.

(a) Gene augmentation therapy (GAT): In the diseases caused by loss of function of a gene it can be helpful in increasing the amount of normal gene product to the required level, by introduction of extra copies of the normal gene. This may restore the normal phenotype. As a result, GAT is targeted at clinical disorders where the pathogenesis is reversible. It also helps to have no precise requirement for expression levels of the introduced gene and a clinical response at low expression levels. This has been particularly applied to autosomal recessive disorders where even modest expression levels of an introduced gene may make a substantial difference. This approach is not suitable in dominantly inherited disorders as gain of function mutations are not treatable and, even if there is a loss of function mutation, high expression efficiency of the introduced gene is required. Individuals with 50% of normal gene product are normally affected. There is a challenge to increase the amount of gene product towards normal levels.

(b) Targeted killing of specific cells : This approach is commonly tried in the cancer gene therapy. Genes are directed to the target cells and then expressed so as to cause cell killing.

Cells can be directly killed, if the inserted genes are expressed to produce a lethal toxin (suicide genes), or a gene encoding a prodrug is inserted, concerning susceptibility to killing by a subsequently administered drug.

Immunostimulatory genes are used to provoke or enhance an immune response against the target cell *i.e.,* indirect cell killing.

(c) Targeted mutation correction : There may not be any use of gene augmentation, if an inherited mutation produces a dominant-negative effect. It is required to correct the resident mutation. This approach has not yet been applied because of practical difficulties. In principle it can be done at different levels: at the gene level (*e.g.,* by gene targeting methods based on homologous recombination or at the RNA transcript levels (*e.g.,* by using particular types of therapeutic ribozymes or therapeutic RNA editing.

(d) Targeted inhibition of gene expression : If diseased cells display a novel gene product or inappropriate expression of a gene (as in the case of many cancers, infectious diseases, *etc.*) a variety of different systems can be used specifically to block the expression of a single gene at the DNA, RNA or protein levels. Allele specific inhibition of expression may be possible in some cases. This permits therapies for some disorders resulting from dominant negative effects.

7.4.3 Techniques of Gene Therapy

The gene therapy approaches which simply rely on gene transfer and expression *i.e.,* classical gene therapy are different from other approaches (targeted inhibition of gene expression and targeted gene mutation correction).

In any gene therapy protocol, the basic steps are as follows :

1. Isolation of gene to be used for gene therapy
2. Delivery and targeting to the appropriate tissue
3. Controlled expression of the therapeutic gene

7.4.3.1 Insertion of Gene into the Cells of Patients

In order to overcome the disease, cloned gene have to be introduced and expressed in the cells of a patient. Usually this involves targeting the cells of diseased tissues. In some cases, unaffected tissues are deliberately targeted as it is sometimes useful to target genes to healthy immune system cells in order to enhance immune responses to certain cancer cells and infectious agents. Genes may be initially targeted to one type of tissue, while their gene products may be delivered to a remote location. In muscle fibres the myonuclei have the advantage of being very long lived. Therefore, the genetically engineered myoblasts, have potential to ameliorate some nonmuscle diseases through long term expression of exogenous genes which encode a product secreted into the blood stream.

In the transfer of genes for gene therapy, there are two major general approaches :

(i) *Ex vivo* **gene transfer.** In many cases cells can be transfected with the gene *in vitro* and transferred into a patient. This is referred to as the autogene *ex vivo* approach. This initially involves transfer of cloned genes into cells grown in culture. Those cells which have been transformed successfully are selected, expanded by cell culture *in vitro*, then introduced into the patient. Normally the autologous cells are used to avoid rejection of the introduced cells by immune system. The cells are initially collected from the patient to be treated and grown in culture before being reintroduced into the same individual. This process is applicable to tissues that can be removed from the body, altered genetically and returned to the patient, where they will engraft and survive for a long period of time (*e.g.,* cells of the haematopoietic system, skin cell *etc.*).

It has therefore to be ensured that

(a) cells from patients will not be confronted and rejected by the immune system.

(b) transfecting the cells *in vitro* is more efficient. In some cases, with the use of selection markers on the vectors, it may be possible to isolate specifically the transfected cells. The cell provides an ideal milieu for DNA/vector containing the therapeutic gene and will protect it from degradation.

(ii) *In vivo* **gene transfer.** In this the cloned genes are directly transferred into the tissues of the patient. This may be the only possible option in tissues where individual cells cannot be cultured *in vitro* in sufficient numbers (*e.g.,* brain cells) and/or where cultured cells cannot be re-implanted efficiently in patients. For this purpose, liposomes and certain viral vectors are increasingly being employed. In the latter case, it is often convenient to implant vector-producing cells (VPCs), cultured cells which have been infected by the recombinant retrovirus *in vitro*. In this case, the VPCs transfer the gene to surrounding disease cells. As there is no way of selecting and amplifying cells that have taken up and expressed the foreign gene, the success of this approach is crucially dependent on the general efficiency of gene transfer and expression (Fig. 7.9 and 7.10).

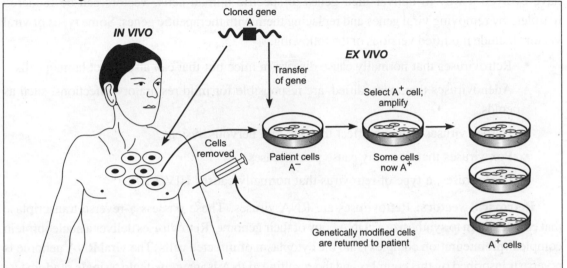

Fig. 7.9. *In vivo* and *ex vivo* gene therapy. In *ex vivo* cells are removed from the patient modified in laboratory and returned to patient. In *in vivo* the cells are modified within the patient's body by introducing cloned gene. (Adopted and modified from Human Molecular Genetics, Strachan and Read, 2004).

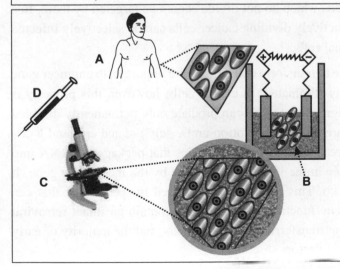

Fig. 7.10. Details of *in vivo* gene therapy. (A) Cells obtained from outpartient skin biopsy are isolated in TKT's facility. **(B)** Gene of therapeutic interest is introduced into cells by electroporation. **(C)** Genetically engineered cells are propogated and characterized **(D)** Genetically engineered cells are returned to physician for injection. (bioweb. wku. edu Transcriptsome 17/review html)

7.4.3.2 Vehicles of Gene Transfer

Generally gene therapies require efficient transfer of cloned genes into diseased cells so that the introduced genes are expressed at suitably high levels *i.e.*, it involves targeting a defective gene for replacement by a normal one.

(a) Viral vector mediated gene transfer

Viral vectors used in gene therapy trials: These are most commonly used, as viruses have evolved a way to deliver their genes to specific human cells. Vectors can be rendered harmless by removing viral genes and replacing them with therapeutic genes. Some types of viral vectors include modified versions of the following:

- Retroviruses that normally cause disease in mice but that can also infect human cells.
- Adenoviruses that, unmodified, are responsible for mild respiratory infections, such as colds.
- Herpes viruses that can affect the skin and nervous system.
- Pox viruses that normally cause pox diseases.
- Lentiviruses, a type of retrovirus that normally causes HIV/AIDS.

Retroviral vectors: Retroviruses are RNA viruses. These possess a reverse transcriptase that enables them to synthesize a cDNA copy of their genome. Retroviruses deliver a nucleoprotein complex (preintegration complex) into the cytoplasm of infected cells. The viral RNA genome is reverse transcribed by this complex and the resulting cDNA is integrated into a single random site in a host cell chromosome. Integration requires the retroviral cDNA to gain access to the host chromosomes, and it is only able to do this when the nuclear membrane dissolves during cell division. These retroviruses, therefore, can only infect dividing cells, thereby, the potential target cells are limited. The cells such as neurons which do not divide cannot be targeted by them. In a normally nondividing tissue like brain, actively dividing cancer cells can be selectively infected and killed without major risk to the normal cells.

Similarly native retroviruses are able to transform cells. It is quite important to engineer gene therapy vectors so that this possibility is eliminated. In cancer cells, however, this property is useful. Attempts are being made to design systems that can produce only permanently disabled viruses. In retroviruses normally there are three transcription units, *gag, pol* and *env,* and a cis-acting RNA element Ψ in retroviruses recognized by viral proteins that package the RNA into infectious particles. The *gag, pol* and *env* in the vector are replaced by the therapeutic gene. It has a maximum cloning capacity of 8 kb. This construct is packaged in special cell that can contribute the necessary *gag, pol* and *env* functions, but does not contain an intact retroviral genome. Retroviruses are very efficient at transferring DNA into cells, and the majority of early trials of gene therapy have used retroviral vectors (Fig. 7.11).

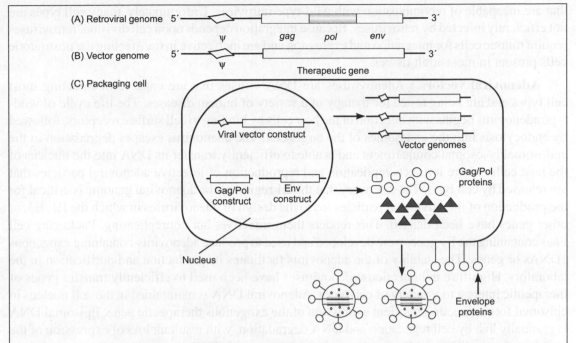

Fig. 7.11. Construction and packaging of a retroviral gene therapy vector. The gag/pol and env genes are replaced by therapeutic gene in retroviral genome. The recombinant viral genomes are packaged into defective but replication deficient virus particles which bud off from the cell and are recovered from supernatant (Adopted and modified from Strachan and Read, 2005).

Limitations of retroviruses for gene transfer : Retroviral genomes can only incorporate genes of approximately 4 kb. Most cDNAs for human genes are about 2.5-3 kb in length; however, many genes, including haemoglobin, require genome enhancer sequences on the 5' and 3' ends to direct their expression. This effectively limits the use of retroviruses to the transfer of relatively small genes or cDNAs that lack desirable regulatory sequences. A major advantage of retroviruses is the ability to permanently integrate exogenous genes into the human genome. However, integration is predominantly random. It makes the integration into normal genes, possible thereby leading to disrupted expression of essential genes. Majar problem is that the integrase enzyme can insert to genetic material of the virus in any arbitrary position in the genomes. This genetic material happens to be inserted is the middle of one of the original genes of the host cell, this gene will be disrupted. It this gene happens to be one regulating cell, division, uncontrolled cell division can occur. An insertion near a tumour suppressor gene could alter its expression and cause tumours. Insertional mutagenesis is an important theoretical and practical concern. Gene therapy trials to treat sever combined immunodeficiency (SCID) were halted or restricted when leukemia was reported in several of the potent. Such events however appear to be rare. A more serious limitation to the safety of retroviruses for gene therapy is the potential recombination of the therapeutic virus with endogenous retroviruses that can complement the defective virus and allow production of infective retroviral particles. Such recombinations may occur in the packaging cell lines and major efforts have been directed to ensure the production of "minimal" retroviruses that are incapable of recombination with wild-type retrovirus. Unfortunately, many cell types are not efficiently infected by retroviruses. Because integration depends upon cell division, retroviruses require mitotic cells for integration and expression and are ineffective in transfecting the postmitotic cells present in most adult tissues.

Adenoviral vectors : Adenoviruses are DNA viruses that are capable of infecting most cell types and are being tested for therapy of a variety of human diseases. The life cycle of wild-type adenovirus begins with infection of the host cells by binding to cell surface receptors, followed by endocytosis into the endosomes of the target cell. The adenovirus escapes degradation in the endosomal/lysosomal compartment and is able to efficiently transfer its DNA into the nucleus of the host cells, where it directs replication and reproduction of infective adenoviral particles that are released by cell lysis. The discovery that the E1 region of the adenoviral genome is critical for the production of infective viral particles led to the design of adenoviruses in which the E1, E3, or other genes have been deleted. This renders them infective, but nonreplicating. Packaging cell lines containing the E1 genes were developed and used to produce adenovirus-containing exogenous cDNAs or genes. The stability of the adenovirus facilitates its production and purification in the laboratory. High-titre noninfectious adenoviruses have been used to efficiently transfer genes of therapeutic interest to a variety of cell types. Adenoviral DNA is maintained in the cell nucleus in episomal form, directing transient expression of the exogenous therapeutic gene. Episomal DNA is gradually lost by cell replication and DNA degradation, with resultant loss of expression of the transfected gene (Fig. 7.12).

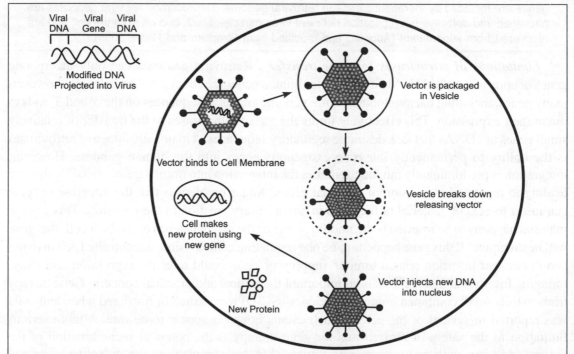

Fig. 7.12. Gene therapy using an Adenovirus vector. A new gene is inserted into an adenovirus vector, which is used to introduce the modified DNA into a human cell. If the treatment is successful, the new gene will make a functional protein (Website).

Gene inserts of up to 7-8 kb can be accommodated in this large genome of adenovirus (75kb). Highly efficient packaging cell lines are available and powerful cell selective promoters can be incorporated into the constructs. In contrast to retroviral vectors, which require substantial *ex vivo* manipulation for infection of pluripotent stem cells, the adenoviral vectors are highly efficient *in vivo* for delivery of exogenous genes in a variety of cell types.

(a) Direct instillation of recombinant adenovirus into the respiratory tract is being studied for gene transfer of the cystic fibrosis transmembrane conductance regulator (CFTR) gene for therapy of cystic fibrosis.

(b) Administration of adenovirus into the portal vein in experimental animals directs high levels of gene expression in hepatocytes that could be used to transfer α-1-antitrypsin or blood clotting factors.

Because adenoviruses are highly efficient, they may have advantages in the therapy of some diseases. For example, specific antitumour or highly toxic gene sequences could be used to treat malignancies.

The DNA molecule is left free in the nucleus of the host cell, and the instructions in this extra DNA molecule are transcribed just like any other gene. The only difference is that these extra genes are not replicated when in the cell is about to undergo cell division. So the descendants of that cell will not have the extra gene. This means that treatment with the adenovirus will require regular doses to add the missing gene every time new cells are produced without the gene.

Herpes simplex virus vectors : These vectors are tropic for the central nervous system (CNS). There can establish life-long latent infections in neurons. They are non-integrated and so long-term expression of transferred genes is not possible. Their major applications are expected to be in delivering genes into neurons for the treatment of neurological diseases, such as Parkinson's disease and for treating CNS tumours. They have comparatively large insert size capacity (> 20 kb).

Adeno associated vectors (AAVs) : AAV is a small (4.7 kb) parvovirus. Its life cycle requires coinfection with adenovirus. AAV is :

(a) non-pathogenic,

(b) infects dividing and nondividing cells,

(c) integrates into the human genome,

(d) does not express its own genome following transfection of the target cells

Wild type AAV is capable of site-specific integration on chromosome 19. Production of AAV requires a packaging cell line for replication of recombinant vectors containing the genes of interest. To achieve viral replication, adenovirus must be added *in vitro*. The need for the adenovirus derives from the life cycle of the AAV. Infective viral particles enter cells via receptor-mediated endocytosis, and their DNA becomes integrated into the genome of cells. During this phase, the virus is non-replicative and no additional viral particles are produced. Upon exposure of these cells to adenovirus, AAV replication and protein synthesis is activated and the life cycle of the adenovirus is terminated. The AAV genome contains genes for the entire assembly and production of infective viruses. The release of the integrated virus requires the presence of adenovirus genome and the production of adenoviral proteins E1, E1A and S4. Following production of capsid and envelope proteins, defective AAV particles are secreted and purified for use in transfer. Because the AAV viral cycle includes termination of the adenovirus life cycle, the pathologic consequences of adenoviral infection may be blunted. Similar to other viral vectors, incorporation of specific human genes into AAV requires the removal of DNA sequences necessary for the life cycle of the AAV. Packaging occurs only in cells that are transfected or infected with adenovirus to produce recombinant AAV viruses.

Current recombination AAVs lack the genetic signals that allow selective integration of the AAV genome on human chromosome 19. The recombinant AAVs used for gene transfer are normally maintained as episomes. They do not integrate into the host genome *in vivo*. The use of AAV is limited by the lack of a highly productive packaging cell line, limitations in the size of DNA that can be inserted (> 4-5 kb) and some restrictions in cell target. Even with these limitations, AAVs are promising reagents for gene therapy and are being tested for treatment of cystic fibrosis and other diseases.

A parvovirus vector is also being used to introduce the CFTR into the sinuses of cystic fibrosis patient at the Standford Univ. Medical Centre. The virus vector is administered to the maxillary sinuses which are small easily accessible and have the same surface tissue as the lungs. This vector is reported to cause less respiratory inflammation and a lower immune response than the adenoviral vectors.

(b) Non-viral methods of gene delivery

A variety of non-viral methods have been developed for gene transfer *in vivo* and *in vitro*. These methods use recombinant, plasmid DNA produced in bacteria. A variety of formulations including naked DNA and liposomal and protein-DNA conjugates, have been utilized for gene transfer. Incorporation of plasmid DNA into small lipid vesicles or "liposomes". When formulated with cationic lipids, the efficiency of DNA delivery to target cells is markedly enhanced. The organ distribution and cellular specificity of the expression of the transferred genes is changed by altering the composition of liposomes. The lipid DNA complex binds to the plasma membrane of target cells. In this process the electrostatic interactions might be playing some role. The vesicle is internalized into the endosomes and lysosomes of the target cell. Some DNA that escapes degradation in the lysosomes, is released to the cytosol, and is transported to the nucleus. The DNA is maintained as an episome in the nucleus of the target cells. Gene expression is transient because DNA is not integrated into the genome. To correct most genetic disorders, repeated administration of DNA complexes is required.

Plasmid DNA vectors are likely to be applicable for gene therapy of cancer for which toxic genes would be required for relatively short period. Plasmid DNA is readily produced in large quantities *in vitro* and there are no major limitations on the size of plasmid DNA that can be transferred into cells. Thus large cDNAs or genes, including complex regulatory elements, can be incorporated into plasmid-based gene transfer systems. For gene transfer artificial chromosomes are also being designed.

Direct injection/particle bombardment: In some cases, DNA can be injected directly with a syringe and needle into a specific tissue, such as muscle. In the case of DMD, where early studies investigated intramuscular injection of a dystrophin minigene into a mouse model, *mdx* the approach is made by particle bombardment techniques. In this case DNA is coated on to metal pellets and fired from a special gun into cells. This approach was successfully used for gene transfer into a number of different tissues. There is, however, poor efficiency of gene transfer, and a low level of stable integration of the injected DNA. The latter property, however, is particularly disadvantageous in the case of proliferating cells. There is, thus, a need of repeated injections.

Receptor-mediated endocytosis: DNA in this case is coupled to a targeting molecule that can bind to a specific cell surface receptor. Then endocytosis and transfer of the DNA into cells tarkes place. The sialoglycoprotein receptors present there get induced. The transfer of exogenous DNA into liver cells can be done by coupling of DNA to a sialoglycoprotein via a polycation such as polylysine. The complexes can be infused into the liver either via the biliary tract or vascular bed. They are then taken up by hepatocytes. In a general approach, the transferring receptor, which is expressed in many cell types, but is relatively enriched in proliferating cells and haemopoietic cells, is utilized. Gene transfer efficiency may be high, but the method is not designed to allow integration of the transferred genes. A further problem has been that the protein-DNA complexes are not particularly stable in serum. Moreover, the DNA conjugates may be entrapped in endosomes and degraded in lysosomes, unless previously co-transferred with or physically linked to an adenovirus molecule (Fig. 7.13).

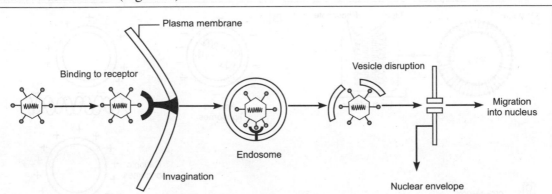

Fig. 7.13. Gene transfer via receptor-mediated endocytosis. After binding the ligand, the plasma membrane invaginates then pinches off, leaving the receptor-ligand complex in an intracellular vesicle, the endosome. The ligand might be an adenovirus carrying a therapeutic gene, as shown here, or it could be the therapeutic DNA bound to some other molecule for which the target cell has a specific receptor. Endosomes are normally targeted to lysosomes for degradation. In order to be expressed, the foreign DNA must somehow escape from the endosome before this happens, and reach the nucleus. Adenoviruses specifically disrupt endosomes, allowing efficient escape. (Adopted and modified from Strachan and Read 2005)

Liposomes: These are spherical vesicles made to synthetic lipid bilayers which mimic the structure of biological membranes. The DNA to be transferred is packaged *in vitro* within the liposomes and used directly for transferring it to a suitable target tissue *in vivo*. Sonication of preparation of equal amounts of the synthetic cationic lipid (DOTMA) and a fusogenic lipid

(DOPE) results in a liposome preparation commercially sold as Lipofoetus. Other mixtures can be combined to obtain liposomes or cytosomes. These liposomes are composed of small unilamellar vesicles with a size range of 50-200 nm in diameter. Mixing of these cationic liposomes with an aqueous solution of DNA results in interaction of the two components to form a positively charged complex, containing some encased DNA that binds and then fuses with the negatively charged cell surface and is subsequently taken by the cell with high efficiency. Liposomes have become popular vehicles for gene transfer in *in vivo* gene therapy because of the safety concerns when using recombinant viruses. This system can also be utilized to deliver recombinant proteins and antisense oligonucleotides into the cell. However, efficiency of gene transfer is low, and the introduced DNA is not designed to integrate into chromosomal DNA, the expression of the inserted genes, therefore, is transient (Fig. 7.14). The following factors have made this system quite attractive for gene transfer :

(a)   Commercial availability and stability of liposomes.

(b)   Ease of bulk synthesis of novel liposomes.

(c)   High binding efficiency for naked DNA or RNA with no limitations on size.

(d)   Ability to transfect most cells.

(e)   Lack of immunogenicity or biohazardous activity making repeated usage a possibility.

(f)   Replication of the recipient cell not necessary for DNA uptake.

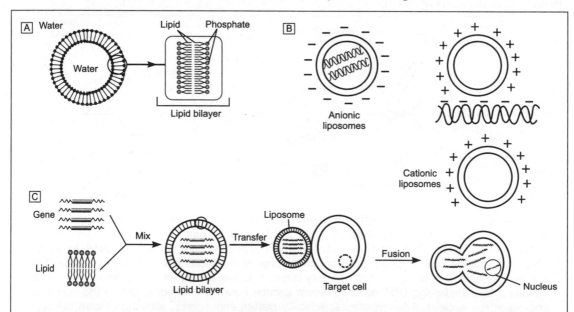

Fig. 7.14. Use of liposomes for gene delivery *in vivo*. (A) Liposomes are synthetic vesicles that form spontaneously in aqueous solution when certain lipids are mixed. They can carry a positive (cationic) or negative (anionic) surface charge, depending on the chemistry of the lipids used. **(B)** The DNA cargo is transported inside anionic liposomes or bound to the surface of cationic liposomes. **(C)** DNA delivery occurs when a liposome fuses with the plasma membrane of a cell. (Adopted and modified from Strachan and Read 2005).

Host immunity and gene transfer

A major hurdle in the application of gene transfer technology for therapy of genetic and acquired diseases is the host immune response, both humoral and cellular. Immune responses to the vectors, as well as to the recombinant therapeutic protein, results in clearance of cells transfected by the vector.

C. Alternative strategies for gene therapy

Another approach to gene therapy involves the development of "organoids". Cells that are genetically altered to produce high levels of a secretable gene product are expanded *ex vivo*. These cells are transplanted back into the recipient so that the gene product is secreted or delivered locally or systemically to a variety of cell types. Organoids are useful for disorders that can be treated by a secreted protein that functions at distant cellular sites. This approach has been used for the expression of factor VIII for haemophilia, α-glucuronidase for correction of mucopolysaccharidosis (MPS) type VII, and various cytokines for treatment of cancer.

(a) The factor VIII cDNA can be permanently transferred to fibroblasts or other cells that are capable of post-translationally modifying and producing factor VIII zymogen for secretion into the circulation. *Ex vivo* transfected cells are subdermally placed in blood vessels, in the peritoneal cavity or are attached to synthetic membrane supports and transplanted into the recipient. Following vascularization of the graft, factor VIII is secreted into the circulation.

(b) Treatment of mucopolysaccharidosis type VII, a α-glucuronidase deficiency, might be achieved by organoids capable of attaching the specific ligand, mannose-6-phosphate, to the α-glucuronidase polypeptide. The α-glucuronidase-ligand complex may then be delivered to cells expressing mannose 6-phosphate receptors on their surface. The therapeutic enzymes would be delivered to the lysosomes of the target cell via the endosomal pathway, correcting the α-glucuronidase deficiency in cells bearing this receptor. The pathology of the visceral organs in animal models of MPS-type VII is reversible by this approach.

(c) Cytokine or cytotoxic genes can also be transferred to somatic cells that have been transplanted directly into sites of tumour involvement. The expression of the cytokines, acting as immuno-attractants, stimulates an immune response to the tumour cells. Producer cells containing a retrovirus that expresses herpes thymidine kinase gene have been injected into brain tumours, causing infection of adjacent tumour cells and rendering them susceptible to killing by nucleotide analogs (ganciclovir) that are metabolized to toxic compounds by the viral thymidine kinase (TK), but not by the host cell. Systemic administration of ganciclovir to the patient kills tumour cells expressing retroviral-TK.

Gene therapy for targeted inhibition of gene expression *in vivo*

By selectively inhibiting the expression of a predetermined gene *in vivo*, certain human disorders can be treated. This general approach, in principle, is particularly suited to treating cancers and infectious diseases and some immunological disorders. In this case the expression of a specific gene that allows the cancerous cells, infection, allergy, inflammation *etc.,* to flourish, by interfering with normal cell function, is knocked out. For example, selectively inhibiting the expression of a particular viral gene that is necessary for viral replication, or an inappropriately activated oncogene *etc.*

Certain dominantly inherited disorders can be treated by targeted inhibition of gene expression. If a dominantly inherited disorder is due to the loss-of-function mutation, treatment may be possible by conventional gene augmentation therapy. However, since heterozygotes with 50% of normal gene product can be severely affected, successful gene therapy for heterozygotes requires efficient expression of the introduced genes. Dominantly inherited disorders which arise because of a gain-of-function mutation may not be amenable to simple addition of normal gene. In some cases instead, it may be possible, to specifically inhibit the expression of the mutant gene, but the expression of the normal allele must be maintained. Such specific inhibition of gene expression is facilitated, if pathogenic mutation results in a significant sequence difference between the alleles.

To inhibit the expression of selected gene, a variety of different strategies can be used.

The specific *in vivo* mutagenesis of that gene, alters it to a form that is no longer functional. Gene targeting by homologous recombination offers the possibility of site-specific mutagenesis to inactivate a gene. This technique has only very recently become feasible with normal diploid somatic cells. It, however, is still very inefficient. Currently, methods of blocking the expression of a gene without mutating it are preferred. This can be done at the different levels,

(i) At the DNA level (by blocking transcription)

(ii) At the RNA level (by blocking post-transcriptional processing, mRNA transport or engagement of the mRNA with the ribosome)

(iii) At the protein level (by blocking post-translation processing, protein export or other steps that are crucial to the function of the protein).

Therapy by selective inhibition of gene expression is technically possible at three expression levels.

(i) **Triple helix therapeutics :** These involve binding of gene-specific oligonucleotides to double-stranded DNA in order to inhibit transcription of a gene.

(ii) **Antisense therapeutics :** These involve binding of gene-specific oligonucleotides or polynucleotides to the RNA, in some cases the binding agent may be a specifically engineered biozyme (a catalytic RNA molecule that can cleave the RNA transcript).

(iii) **Intracellular antibodies (intrabodies) and oligonuleotide aptamers :** It involves the construction of antibodies that can be directed to specific locations within cells in order to bind a specific protein or oligonucleotide aptamers that can bind specifically to a selected polypeptide.

(i) Triple helix therapeutics

Synthetic short oligonucleotides (15-27 nucleotides long) are capable of specifically binding to a sequence of double-stranded DNA. They form a triple helix. The oligonucleotide binds by

Hoogsteen hydrogen bonds to the double-stranded DNA, without disrupting the original Watson-Crick hydrogen bonding. The most stable Hoogsteen-bonded structures are G bound to a GC base pair and T bound to an AT base pair. Although such structures can inhibit DNA replication *in vitro*, helicases can unwind triple strand structures *in vivo*. However, triplex formation has been shown to block binding of transcription factors *in vitro* and also, at least in some cases, there is an evidence of gene-specific inhibition of transcription in intact cell.

Oligonucleotides are large polyanionic hydrophilic structures. They are, thus, not ideally suited to diffusing across the highly hydrophobic plasma membrane. They are, therefore, directly delivered into the cytoplasm using cell permeabilization techniques. This provides the most efficient approach to enable subsequent transfer into the nucleus and delivery is carried out using liposome method. Thereafter, the oligonucleotides can rapidly migrate to the nucleus by passive diffusion through the pores of the nuclear envelope. The oligonucleotides inside the cell are exposed to nuclease attack, notably from exonucleases, and the half-life of conventional oligonucleotides with phosphodiester bonds is typically about 20 min. In order to protect the oligonucleotides from nuclease attack, it is usual for the 3' and 5' ends to be chemically modified. These modifications involve incorporation of sulfur containing phosphorothioate bonds to generate the S-oligonucleotides.

There are, however, some general difficulties. Comparatively large amounts of oligonucleotide are required for inhibition of gene expression. The limitation imposed by Hoogsteen hydrogen bonding is also a problem. The target sequences need to carry virtually all their purine bases on one DNA strand. This problem is solved by replacement of the phosphate groups by different chemical groupings that allow triplex-forming oligonucleotides to 'hop' from one strand of the bound DNA duplex to the other.

(ii) Antisense oligonucleotides

During transcription, only one of the two DNA strands in a DNA duplex, the template strand, serves as a template for making a complementary RNA molecule. As a result, the base sequence of the single-stranded RNA transcript is essentially identical (except that U replaces T) to the other DNA strand, commonly called the **sense strand.** Any oligonucleotide or polynucleotide which is complementary in sequence to an mRNA sequence; including the template strand of the gene, can, therefore, be considered to be an antisense sequence.

Binding of an antisense sequence to the corresponding mRNA sequence would be expected to interfere with translation and, thereby, inhibit polypeptide synthesis. Naturally occurring antisense RNA is known to provide a way of regulating the expression of genes in some plant and animal cells, as well as in some microbes. Synthetic oligonucleotides can be designed to be complementary in sequence to a specific mRNA and when transferred into cells, show evidence of inhibition of the corresponding gene. The concept of antisense therapeutics was, therefore, developed. By this, unwanted expression of a specific gene in disease tissues could be selectively inhibited using an artificially gene-specific antisense sequence. A variety of different types of antisense sequences can be used.

Antisense genes : Even when chemically modified, antisense oligonuleotides are not stable indefinitely, continuous supply of antisense sequence can be assured by using a form of expression cloning. In this a specifically designed antisense gene is transferred into the relevant cells. It is possible to engineer such a gene by constructing a minigene in which an inverted coding sequence is placed downstream of a powerful promoter. The DNA strand that normally serves as the sense strand is now transcribed to give an antisene RNA.

Antisense oligodeoxynucleotides: The artificial antisense oligonucleotides are quite commonly used as they can be synthesized easily. They can be transferred efficiently into the cytoplasm of cells using liposomes and their intracellular stability is improved by using chemically modified oligonucleotides, notably S-oligonucleotides. Although antisense oligonucleotides migrate to the nucleus, they do not bind the double-stranded DNA. It is because of the fact that they are not designed to participate in Hoogsteen hydrogen bonding. There is a preference for antisense oligodeoxynucleotides (ODNs) to oligoribonucleotides, because they are generally less vulnerable to nuclease attack. Moreover, they have the additional advantage of inducing the destruction of an mRNA to which they bind. It is known that an ODN-mRNA hybrid, like all DNA-RNA hybrids, is vulnerable to attack and selective cleavage of the RNA strand by a specific class of intracellular ribonuclease, RNAse H. There is a great therapeutic potential of the antisense ODNs.

Ribozymes : The RNA molecules are functionally different from DNA molecules. They collectively can serve diverse functions, rather than simply being involved in transfer of genetic information. The activation energy for specific biochemical reaction can be lowered by some RNA molecules. In this way they effectively function as enzymes (ribozymes). The transcripts of group I introns are autocatalytic and self-splicing. Other ribozymes which cleave RNA, are trans-acting, that is they cleave an RNA sequence on a different molecule. They contain two essential components : target recognition sequence (which base-pair with complementary sequences on target RNA molecules) and a catalytic component, much like the active site of an enzyme responsible for cleaving the target RNA molecule, while it is held at place by base pairing. The cleavage leads to inactivation of the RNA. It is probably because of subsequent recognition of the two unnatural ends by intracellular nucleases. The examples include human ribonuclease P and various ribozymes obtained from plant viroids (virus like particles).

Genetic engineering can be employed to custom design the recognition sequences so that it contains antisense sequences that can base-pair to a specific mRNA molecule, but the catalytic site is kept intact. It is then possible to transfect the engineered genes which can be transcribed to produce the desired ribozyme into suitable cells. Ribozymes have been designed against specific oncoproteins.

RNA editing can also be used as alternative therapeutics. In this, complementary RNA oligonucleotide specifically binds to a mutant transcript at the sequence containing the pathogenic point mutation, and an RNA editing enzyme, such as double-stranded RNA adenosine deaminase is used to direct the desired base modification.

(iii) Inhibition of polypeptide function

The function of a specific polypeptide can be inhibited by artificially designed intracellular antibodies (intrabodies), oligonucleotides (aptamers) and mutant proteins.

***Intracellular antibodies (intrabodies)* :** The antibody normally functions extracellularly as after synthesis they are either secreted into the extracellular fluid or remain membrane bound on the B-cell surface as antigen receptors. Antibodies are being engineered to design the genes encoding intracellular antibodies, or intrabodies. This may lead to help in using antibodies within cells to block the construction of viruses of harmful proteins, such as oncoproteins. It is being tried to engineer the antibody F105 which binds to gp 120, a crucial human immunodeficiency virus (HIV) envelope protein that the AIDS virus uses to attach to and infect its target cells. This envelope protein is derived from a larger precursor gp 160 which is synthesized in the endoplasmic reticulum. Marasco and co-workers designed a novel F105 gene which encoded an antibody that was stably expressed and retained in the endoplasmic reticulum without being toxic to the cells. The engineered antibody binds to the HIV envelope protein within the cell and inhibits processing of the gp 160 precursor, thus, reducing the infectivity of the HIV-1 particles produced by the cell.

***Oligonucleotide aptamers* :** By delivering 25% each of the four bases, A, C, G and T at each base position during oligonucleotide synthesis, fully degenerate oligonucleotides can be synthesized. In this way enormous number of sequence permutations can be generated. To screen for the ability to bind to a selected target protein (protein epitope targeting), the resulting mixture of oligonucleotides can be used. In fact the use of partially degenerate oligonucleotides is preferred so that the concentration of individual oligonucleotides is not too low. This in fact means simultaneous screening of many thousands of oligonuleotides and so the chance of at least one epitope of the target protein being specifically bound by an oligonucleotide can be high. The bound oligonucleotide sometimes known as an **adaptamer** or **aptamer** can be eluted from the protein and sequenced to identify the specific recognition sequence. Transfer of large amounts of a chemically stabilized aptamer into cells can result in specific binding to a predetermined polypeptide. This leads to blockage of its function. In this way, protease thrombin was inhibited.

The future use of oligonucleotide aptamers, however, to inhibit specific intracellular protein targets will inevitably involve genetic modification of cells and can, therefore, be considered as a form of gene therapy.

Mutant proteins : The production of a mutant polypeptide that binds to the wild type protein, inhibiting its function can be involved in naturally occurring gain of function mutation. The wild-type polypeptides in such cases naturally associate to form multimers and the process is inhibited by the incorporation of a mutant protein. By designing genes to encode a mutant protein that can specifically bind to and inhibit a predetermined protein, such as a protein essential for the life-cycle of a pathogen, gene therapy is possible. An attempt has been made to artificially produce a mutant HIV-I protein to inhibit multimerization of the viral core proteins.

Though in principle, artificial correction of a pathogenic mutation *in vivo* is possible, it is however, not efficient and not readily amenable to clinical applications.

Gene therapy may not be easily carried out for certain disorders. For example, dominantly inherited disorders where a simple mutation results in a pathogenic gain of function cannot be treated by gene augmentation therapy and targeted inhibition of gene expression may be difficult to achieve. To inhibit novel or inappropriate gene expression in human cells, target inhibition is best suited, for example expression of viral genes, oncogenes *etc.* Expression of a gain-of-function mutant allele may need to be inhibited, but expression of a very similar wild-type allele has to be retained. If a significant change in sequence at the site of the mutation is carried by the mutant allele, it may be possible to achieve selective inhibition. If, however, the change is a simple mutation, *e.g.,* single nucleotide substitution, then the need of other approaches may be required. Targeted mutation correction can be carried out by inserting some reagents into cells in order to change the mutant sequence back to a form that is compatible with normal function.

A different way can be used to correct a specific mutation selectively *in vivo*, mostly at the DNA level. Gene targeting techniques are based on homologous recombination.

(i) Because this approach offers the ability to make site-specific modification of endogenous genes, it represents a potentially powerful method for gene therapy. By this both **acquired** and **inherited mutations** could be corrected and novel alterations can be engineered into the genome. This technique has so far been largely limited to pluripotent mouse embryonic stem cells. It has also been applied to normal diploid somatic cells. However, the enormous inefficiency of this procedure (even when using the ideal target of cells cultured *in vitro*) and the need to correct the defect in many different cells *in vivo* has meant that clinical applications are a long way off.

(ii) The genetic defect can be repaired at the RNA level. Further a therapeutic ribozyme can be used.

7.4.4 Gene Therapy for Diseases

Successful gene therapy requires that the:

- Genetic malfunction/nature of a disease is understood;
- Therapeutic material can be delivered to the target cells in the affected tissue or organ;
- Therapeutic material is active for the intended duration and delivers the intended benefit to the target cells;
- Harmful side effects, if any, are manageable.

Therapeutic material can be delivered to the target cells in two mainways. **A.** It can be inserted into cells from the affected tissue outside the body, and these cells then returned to the body. **B.** It can be delivered directly into the body at the required site. Either way a 'delivery vehicle' called a vector is used to get the therapeutic material to the patient's target cells. Vectors are most commonly based on modified viruses, because these can target and enter cells efficiently.

(a) Inherited diseases: Different genetic disorders can be targeted by gene therapy. In the common non-mendelian genetic diseases, a complex interplay between different genetic loci and/ or environmental factors may be involved and gene therapy approaches in such cases may not be straightforward. Single gene disorders, where individuals are severely affected and where there is no effective treatment, are more obvious candidates for gene therapy.

Recessively inherited disorders. The recessively inherited disorders result from a simple deficiency of a specific gene product. These are generally the most amenable to treatment. To overcome the genetic deficiency, high level expression of an introduced normal allele should be sufficient. Because the mutations are almost simple loss-of-function, recessively inherited disorders have been of special interest as candidates for gene therapy.

Affected individuals have deficient expression from both alleles and so the disease phenotype is due to complete or almost complete absence of normal gene expression. Heterozygotes, however have about 50% of the normal gene product and are normally asymptomatic. Additionally, there are, at least some cases, with a wide variation in the normal levels of gene expression, so that a comparatively small percentage of the average normal amount of gene product may be sufficient to restore the normal phenotype. It is also often observed that the severity of the phenotype of recessive disorders is inversely related to the amount of product that is expressed. As a result, even if the efficiency of gene transfer is low, modest, expression levels of an introduced gene may make a substantial difference. This is quite unlike dominantly inherited disorders where heterozygotes with loss-of-function mutations have 50% of the normal gene product and may yet be severely affected.

The recessively inherited disorders are generally amenable to gene augmentation therapy, their degree, however, differs. In addition to the question of accessibility of the disease tissue, some disorders may be difficult to treat for other reasons. Â-thalassemia which results from mutations in the α-globin gene, provide a good example. This is a severe disorder affecting hundreds of thousands of people world-wide. The gene is very small and has been extensively, characterized. The disorder is recessively inherited and affects blood cells, but so far no successful attempt for the gene therapy has been obtained.

Adenosine deaminase deficiency (ADA) : It was selected as the prototypic genetic disorder for initial gene therapy trials. Severe combined immune deficiency and loss of both T-cell or B-cell function caused partial or complete deficiencies of ADA. While bone marrow transplantation is done to treat ADA deficiency, many patients lack an appropriately matched bone marrow donor. There is intracellular accumulation of a toxic purine intermediate, deoxyadenosine. Deoxyadenosine is cytotoxic to both T and B cells. It results in the loss of cellular and humoral immunity. It leads to recurrent infections that cause childhood death in patients with ADA deficiency. ADA is a very rare recessively inherited disorder. It is involved in the purine salvage pathway of nucleic acid degradation. It is housekeeping enzyme which is synthesized in many different types of cell. An inherited deficiency of this enzyme has, however, particularly severe consequences in the case of T lymphocytes, one of the major classes of immune system cells. As a result, ADA-patients suffer from severe combined immunodeficiency. This severe disorder was particularly amenable to gene therapy for a variety of reasons :

(a) The ADA gene is small, and had previously been cloned and extensively studied,

(b) The target cells are T cells which are easily accessible and easy to culture enabling *ex vivo* gene therapy,

(c) The disorder is recessively inherited.

The gene expression is not tightly controlled (normal individuals show a huge range in enzyme levels, from 10 to 5000% of the average levels). Engraftment of T cells alone may be sufficient as the allogenic bone marrow transplantation can cure the disorder. The transfer of normal ADA genes into ADA-T cells has been carried out to result in restoration of the normal phenotype.

In order to go for gene therapy, autologous bone marrow of ADA patients was *ex vivo* transfected with a recombinant retrovirus containing the ADA cDNA and returned to cells of the patients. The therapy represents an isogenic transplant that avoids graft-versus-host reactions because autologous marrow is used. Correction of the recombinant cell results in their selective growth advantage over uncorrected cells.

The first apparently successful gene therapy was initiated on 14 September 1990. The patient, Ashanthi DeSilva, suffering from adenosine deaminase (ADA) deficiency was just 4 years old. The gene therapy trials have been carried out in many other inherited genetic disorders (Table 7.2).

Table 7.2: Examples of Current Gene Therapy Trials for Inherited Disorders

Disorder	Cell altered	Gene therapy strategy
ADA deficiency	T cells and haemopoietic stem cells	*Ex vivo* GAT using recombinant retroviruses containing an *ADA* gene
Cystic fibrosis	Respiratory epithelium	*In vivo* GAT using recombinant adenoviruses or liposomes to deliver the *CFTR* gene
Familial hypercholesterolemia	Liver cells	*Ex vivo* GAT using retrovirus to deliver the LDL receptor gene (*LDLR*)
Gauchher's disease	Haemopoietic stem cells	*Ex vivo* GAT retroviruses to glucocerebrosidase GAT, gene augmentation therapy

Familial hypercholesterolemia (FH): This disorder is caused by a dominantly inherited deficiency of low density lipoprotein (LDL) receptors. These are normally synthesized in the liver. It is characterized by premature coronary artery disease. About 50% of heterozygous affected males die by 60 years of age, unless treated. Homozygotes are occasionally seen because FH is such a common single gene disorder. They suffer precocious onset of disease and increased severity, with death from myocardial infarction commonly occurring in late childhood. There was an attempt with gene therapy for FH for a 28 year old woman who was homozygous for a pathogenic missense mutation in the *LDLR* gene. She had a myocardial infarction at the age of 16 and required coronary artery bypass surgery at the age of 26.

The differentiated hepatocyte is refractory to infection with retroviruses. However, hepatocytes can be cultured *in vitro* and, under such conditions are susceptible to retroviral infection. *Ex vivo* gene therapy is possible in cultured hepatocytes. Genes can be injected via the portal venous system. The gene therapy involved:

(a) surgical removal of a sizeable portion of the left lobe of the patient's liver,

(b) disaggregation of the liver cells and plating in cell culture prior to infection with retroviruses containing a normal human *LDLR* gene.

(c) The genetically modified cells were infused back into the patient through a catheter implanted into a branch of the portal venous system.

(d) The patient's LDL/high density lipoprotein (HDL) ratio subsequently declined from 10-13 before gene therapy to 5-8 and such improvement was maintained over a long period.

LDL uptake in target cells is blocked by mutations in the low density lipoprotein (*LDL*) receptor gene. This causes severe, premature atherosclerosis in affected patients.

(a) The normal LDL receptor was transferred by retroviral vectors into hepatocytes isolated from the liver of patients with LDL receptor gene defects *ex vivo*.

(b) The transfected cells were grown in culture and reinjected into the patient portal vein and recolonized in the liver.

(c) Decreased serum cholesterol levels were observed after reimplantation with the autologous, genetically corrected hepatocytes.

A variety of metabolic disorders, including phenylketonuria, clotting disorders, α-1 antitrypsin deficiency, *etc.* can be treated in this way.

Cystic fibrosis (CF) : It is the most common lethal genetic disease in whites. It affects approximately 1 in 2,500 infants in North America.

(a) It is caused by mutations in a membrane protein, the cystic fibrosis transmembrane conductance regulatory of CFTR.

(b) The CFTR is primarily expressed in epithelial cells and acts as a Cl^- transport protein that enhances Cl^- transport across the epithelial surfaces of numerous organs, including the lung, gastrointestinal tract and pancreas.

(c) The most common mutation, a deletion of a phenylalanine codon at position 508 of the polypeptide, produces a CFTR protein that is not properly routed to the apical membrane of the affected cells and, therefore, fails to transport Cl^- following stimulation with 3'-5-cyclic adenosine monophosphate (cAMP).

(d) Lack of Cl^- and fluid secretion results in the accumulation of mucus in the secretory ducts of many organs, including the liver, pancreas, gastrointestinal tract, reproductive organs and lung.

(e) Both *in vivo* and *in vitro*, evidences support the feasibility of transferring the wildtype CFTR to somatic cells of the epithelium of the lung for correction of the lethal pulmonary complications related to mucous plugging and recurrent infection seen in cystic fibrosis.

(f)   A variety of viral and non-viral strategies for transfer of the CFTR cDNA have been actively studied in the laboratory.

(g)   Phase 1 clinical trials, testing the ability to transfer the human CFTR cDNA with liposomes, adenovirus and adeno-associated viruses have been initiated and more than 40 patients have been studied.

(h)   Transfer of the CFTR to CF cells and animals with the CF gene defect, restores Cl$^-$ ion transport and can correct the intestinal obstruction in CFTR-deficient transgenic mice in the absence of discernible toxicity.

(i)   The respiratory tract is uniquely accessible via the tracheal-bronchial tree and the systemic and pulmonary vasculatures.

It has been seen that transfer of the CFTR cDNA to CF epithelial cells can restore normal ion transport properties to the cells. The vectors fail to integrate into stem cells of the pulmonary epithelium, therefore, the present transfer vectors will require repeated administration to patients with cystic fibrosis. To correct the CF deficit throughout the life of patients with cystic fibrosis, vectors that are neither toxic nor immunologic will be required. The retroviruses, however, do not efficiently transfect respiratory epithelial cells *in vivo*.

Duchenne muscular dystrophy (DMD) : It is a severe X-linked recessive disorder. The affected males in this suffer progressive muscle deterioration. They are confined to a wheel chair in their teens and usually by the third decade die. The target tissue is skeletal muscle and initial interest in treatment for this disorder focused on cell therapy because of the unique cell biology of muscle as well as muscle fibres (myofibres-very long, post-mitotic, multinucleate cells). Mononucleate myoblasts are present in skeletal muscle. These are normally quiescent, but can divide and subsequently fuse with myofibres to repair muscle damage. Although implanting normal or genetically modified myoblasts into diseased muscles appeared attractive, but there are many difficulties in the approach. Because adult skeletal muscle fibres are post-mitotic, retroviral vectors cannot be used. They are, thus, not susceptible to retroviral infection. To deliver genes to muscle fibres *in vivo,* adenovirus vectors have been used.

For gene therapy, haemoglobinopathies and other haematologic disorders are also potential candidates. Because haematopoietic stem cells can be permanently transfected by retroviral vectors *ex vivo* and the cells returned to the marrow of the patient, disorders such as sickle cell anaemia, thalassemia, Fanconi anaemia type C, and other disorders of erythropoiesis and myelopoiesis are potentially amenable to gene therapy.

Gaucher's disease and mucopolysaccharidosis VII (MPS-VII) are lysosomal storage diseases. These are targets for gene therapy. Gaucher's disease is caused by mutations in acid β-glucosidase. This is a protein found in the inner lysosomal membrane of the cell. The lack of MPS-VIII enzyme, β-glucoronidase, causes defects in the soluble lysosomal protein secreted from cells. Transfection of bone marrow progenitor cells with retrovirus containing the cDNA encoding these proteins should correct the enzymatic defect in the macrophage/monocyte-derived cells. Because acid β-glucosidase is not secreted, correction of Gaucher's disease by gene therapy requires transfection of a high percentage of these cells with retrovirus capable of inducing the synthesis of the acid β-glucosidase cDNA. Bone marrow ablation or pretreatment with effective enzyme therapy may be required to provide sufficient space for the growth of transfected bone marrow progenitor cells. Nevertheless, human trials using retroviral gene transfer have been initiated.

The major visceral pathology in MPS-VII derives from involvement of bone marrow monocyte/macrophage-derived cell lines and chondrocytes. Central nervous system involvement also occurs in this disease. Gene therapy for MPS-VII, retroviral transfection of bone marrow progenitor cells, may provide functional macrophages that are distributed throughout the body and are capable of secreting the recombinant protein. By expressing high levels of β-glucuronidase, macrophages could also act as organoids producing the enzyme that will be taken up by macrophages in various organs. Visceral organ correction of the MPS-VII mouse indicates that both direct cellular correction and uptake of β-glucuronidase by macrophages may repair the metabolic abnormalities in MPS-VII. In the mouse model of MPS-VII, neither enzyme nor transplanted macrophages entered the brain, and pathologic correction was not achieved in that organ, while visceral manifestations of the disease were eradicated by gene transfer.

Prenatal approaches to gene therapy of Herlitz Junctional Epidermolysis Bullosa

According to (Park *et al.* 2008) herlitz junctional epidermolysis bullosa (H-JEB), a fatal hereditary disease caused by the absence of the epidermal basement membrane protein laminin-5, presents from birth with widespread skin blistering and peeling away of the eqidermis. Usually the oropharyngeal mucosa and the mucous membranes of the trachea are also involved. Chronic infections combined with loss of protein and iron, in addition to poor feeding, contribute to impaired wound healing and failure to thrive. There is no specific therapy for H-JEB, and most patients die early in infancy. Park *et. al.* (2008) have observed that in the majority of H-JEB cases, nonsense mutations in the *LABM*3 gene encoding the β3 chain of laminin-5 are responsible for the phenotype. They have established DNA-based prenatal diagnosis. *In vitro*, full phenotypic correction of H-JEB cells can be achieved by transfer of *LAMB*3 cDNA. They have shown that marker gene vectors injected into the amniotic cavities of mice reach not only the whole surface of the foetal skin including the epidermal stem cells, but also the oropharyngeal mucosa, the upper airways and the digestive tract of the foetus, thus all the target tissues for a treatment of H-JEB. Since gene therapy *in utero* appeared simple to perform and a single, minimally invasive prenatal administration of the corrective gene might suffice to prevent H-JEB manifestation completely, they generated different viral and non-viral vectors carrying *LAMB*3-cDNA. These vectors were tested on patient's keratinocytes, shown to be stable in amniotic fluid and are currently being evaluated *in vivo* by administration into the amniotic cavities of LABM3-deficient mice. If successful, this approach may prove promising for the curative treatment of a variety of heritable extracellular matrix diseases Fig. 7.15.

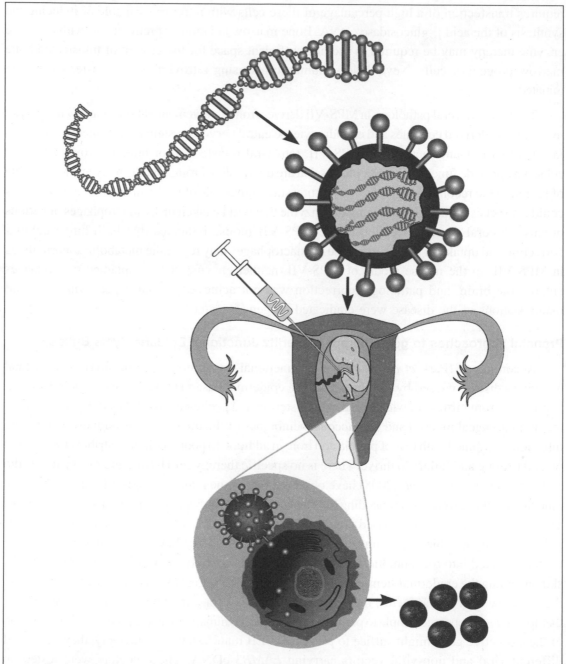

Fig. 7.15. Prenatal approach to gene therapy. Gene therapy of Herlitz Junctional Epidermolysis Bullosa (www.eml.molmed.uni-erlangen.de/AGSchneider/IUGTscheme.)

Use of antibodies that penetrate neurons to destroy Alzheimer's associated amyloid
It is being tried to understand how immune-based therapies can treat Alzheimer's disease (AD)-by studying how antibodies go inside brain cells to reduce levels of Alzheimer's linked amyloid peptides that form plaques between neurons.

According to Dr. Gunnar Gouras, Associate Professor of Neurology and Neuroscience at Weill Cornell Medical College this internalization and activity of the antibody within the cell was a big surprise and something they really haven't appreciated in neurological medicine. This has given new hope for the use of immunotherapy against Alzheimer's, while casting intriguing new light on other disease processes.

Over the past six years, advantage is being taken of breakthroughs in neuroscience research to help answer that question. Special transgenic mice are being developed to closely approximate the progress of human Alzheimer's disease.

According to the Alzheimer's Association there are currently no effective treatments to fight Alzheimer's disease, which now affects over 5 million Americans. It is now predicted that unless new ways are found to prevent or treat the disease, the total could climb to 16 million by 2050.

Dr. Gouras' team exposed amyloid-filled neurons from these mice to immune antibodies similar to those used in clinical trials. They then examined changes in these cells in the lab, using microscopy, immunofluorescence and other high-tech methods.

According to " Dr. Gouras instead of working outside the cell, it was found that these antibodies to beta amyloid bind with a specific part of amyloid precursor protein (APP) — a precursor molecule to beta amyloid — as it lies on the outside of the affected cell. This complex then gets internalized within the cell, where it works to decrease the levels of amyloid peptides, the building block of plaques that are found outside and between cells. In fact, the antibodies cut down on intracellular amyloid accumulation by about one-third.

How might antibodies working inside neurons decrease exterior plaque levels? The researchers aren't still sure, but they have already ruled out some of the most obvious answers.

Dr. Davide Tampellini, a researcher in the Weill Cornell Laboratory of Alzheimer's Disease Neurobiology did not find evidence that the antibody somehow inhibits the activity of either of the two cellular enzymes—secretases—that we know help produce "beta amyloid". It has been observed that the presence of the antibody appears to boost secretase activity.

It is possible that the antibody is affecting key trafficking mechanisms within the cell, thereby, increasing the degradation of existing beta amyloid before it makes its way to the surface.

It is clear from the study that immune-based therapy does work to rid brain cells of amyloid —giving new impetus to the search for a safe, effective Alzheimer's vaccine.

Cancer gene therapies

For cancer gene therapy, many different approaches can be used. In a few cases, the gene therapy approach has focused on targeting single genes, such as TP53 gene augmentation therapy and delivery of antisense *k ras* genes in the case of some forms of non small cell lung cancer. In most cases, however, targeted killing of cancer cells has been conducted without knowing the molecular etiology of the cancer.

Gene transfer into tumour-infiltrating lymphocytes : Earlier a population of immune system cells were used for specifically targeting a foreign protein to a tumour. The therapy could be considered to be a form of adoptive immunotherapy because a gene encoding a cytokine, tumour necrosis factor-α (TNA-α), was transferred into tumour-infiltrating lymphocytes (TILs) in an effort to increase their anti-tumour efficacy. The TIL population of T lymphocytes which can seek out an infiltrate tumour deposits, such as metastatic melanomas. TNF-α is a protein naturally produced by T lymphocytes which, if infused in sufficient amounts in mice, can destroy tumours. However, it is a toxic substance and intravenous infusion of TNF has significant cellular vectors for transferring the toxic protein directly to tumours.

Adoptive immunotherapy by genetic modification of tumour cells : Good results were obtained from animal studies in which murine tumour cells were genetically modified by the insertion of genes encoding various cytokines (several different interleukins (IIs), TNF-α, interferon (IFN-), granulocyte-macrophage colony-stimulating factor (GM-SCF) and then re-implanted in mice. In each case, the genetically altered tumour cells either never grew or grew and then regressed.

Most of these were then systemically immune to re-implantation of nonmodified tumours. When animals with established sizeable tumours were treated, the results, however, were much less satisfactory. This gave impetus to the idea of modifying a patient's own tumour cells for use as a vaccine adoptive immunotherapy. For treating a wide variety of cancers, human gene therapy trials have been approved for the insertion of cytokine genes using retrovirus vectors.

In such cases, the patients are specifically immunized against their own tumours by genetically modifying the tumours with one of a variety of genes that are expected to increase the host immune reactivity to the tumour. Other genes such as foreign HLA antigen genes have also been transferred to tumours for this purpose.

Adoptive immunotherapy by genetic modification of fibroblasts : There is a difficulty with *ex vivo* therapy for tumours in growing tumour cells *in vitro* because not more than 50 per cent cell grow. In some cases, fibroblasts have been targeted. The basis of some clinical trials for treatment of various cancers has been provided by the transfer of genes encoding the cytokines IL-2 and IL-4 into skin fibroblasts grown in culture. The IL-2 and II-4 secreting fibroblasts are mixed with irradiated autologous tumour cells. They are then injected subcutaneously. It is hoped that the local production and secretion of cytokines by the transferred fibroblasts will induce a vigorous immune response to the nearby irradiated tumour cells and thereby, result in a systemic anti-cancer immune response.

It has been tried to use retrovirus mediated transfer of a gene encoding a 'prodrug,' a reagent that confers sensitivity to cell killing following subsequent administration of a suitable drug. The retroviruses were provided in the form of murine fibroblasts. These produce retroviral vectors (retroviral vector-producing cells or VPCs), in this case in brain tumour cells, such as recurrent glioblastoma multiforme. Within growing tumours, the cells were directly implanted into multiple areas. For this purpose, stereotactic injections guided by magnetic resonance imaging were given. Once injected, the VPCs continuously produce retroviral particles within the tumour mass and transfer the genes into surrounding tumour cells. Although retroviruses are not normally used for *in vivo* gene therapy because of their sensitivity to serum complement, they are comparatively stable in this special environment and have the advantage that, since they only infect actively dividing cells, the tumour cells are a target, but not nearby brain cells.

Another prodrug gene that was transferred is a *HSV* gene. It encodes thymidine kinase (HSV-tk). HSV-tk confers sensitivity to the drug gencyclovir by phosphorylating it within the cell to form gancyclovir monophosphate. It is subsequently converted by cellular kinases to gancyclovir triphosphate. DNA is inhibited by this compound. It causes cell death. Such a therapy appears to benefit from a phenomenon known as the by-stander effect : adjacent tumour cells that have not taken up the *HSV-tk* gene, may still be destroyed. This is probably due to diffusion of the gancyclovir triphosphate from cells which have taken up the *HVS-tk* gene. This uptake might be through gap junctions.

Other Immunological approaches : Two other *ex vivo* gene therapy strategies using immunological approaches are used for the destruction of tumour.

(i) An antisense insulin-like growth factor-1 (*IGF*1) gene is transferred into tumour cells in order to block production of IGF-1. It has been shown that when tumour cells modified in this way are re-implanted *in vivo,* an immune response which can lead to destruction of non modified tumours can set in.

(ii) In this costimulatory molecule, such as B7-1 or B7-2, which are normally present on lymphocytes being required for full T lymphocyte activation, are inserted.

MicroRNAs as tumour suppressors: It has been observed that microRNAs, can function as tumour suppressors in laboratory studies. According to Drs. Yong Sun Lee and Anindya Dutta (UVA) microRNAs can function as tumour suppressors *in vitro*. It has been pointed out that overexpression of HMGA2 is an important feature of many medically important tumours like uterine fibroids. It is very exciting to realize that microRNAs have an important role in suppressing the overexpression of HMGA2, and so may have a role in the causation and perhaps the cure of a disease that is responsible for the vast majority of hysterectomies in the Western world.

Studying chromosomal HMGA2 translocations that are associated with human tumours. It has been found out that in normal cells, a microRNA called let-7 binds to the 3' end of the HMGA2 mRNA transcript and suppresses its expression in the cell cytosol. However, chromosomal breaks that shorten the 3' end of the HMGA2 transcript, and prevent let-7 binding, result in aberrantly high levels of HMGA2 expression and tumorigenesis. HMGA2 has been reported to be a target of let-7, and that the let-7 microRNA functions as a tumor suppressor to prevent cancer formation in healthy cells.

Gene therapy for infectious disorders: Somewhat difficult approaches for gene therapy for treating infectious disorders are used. There are certain strategies that intend to affect the life-cycle of the infectious agent or reduce its ability to undergo productive infection. Some infectious agents are genetically comparatively stable. There may, however, be rapid evolution, undergone by other. These present problems for any general therapy, *e.g.,* AIDS. In this the infectious agent, HIV-1 appears to mutate rapidly.

Gene therapy for AIDS : The infectious agent for AIDS is a retrovirus known as HIV-1 which can infect helper T lymphocytes, a crucially important subset of immune system cells. The HIV-1 is deadly because of its two features :

(i)   It eventually kills the helper T cells (thereby rendering patients susceptible to other infections).

(ii) The provirus tends to persist in a latent state, before being suddenly activated (the lack of virus production during the latent state complicates anti-viral drug treatment). The high rate of mutations in HIV genome poses a major problem.

The following strategies can be used :

(a)   A variety of gene therapy strategies can be envisaged for treating AIDS.

The infected cells can be killed directly (by insertion of a gene encoding a toxin or a prodrug) or indirectly, by enhancing an immune response against them. This can involve transferring a gene that encodes an HIV-1 antigen, such as the envelope protein gp 120, and expressing it in the patient in order to provoke an immune response against the HIV-1 virus or the patient's immune system can be boosted by transfer and expression of a gene encoding a cytokine, such as an interferon. An approach, applicable to all disorders caused by infectious agents, is used to find a means of interfering with the life-cycle of the infectious agent.

(b)   **Interference with the HIV-1 life cycle.** For such an inhibition, three major levels can be used.

(i)   **Blocking HIV-1 infection.** The virus normally infects HIV-1 T lymphocytes by binding of the viral gp 120 envelope protein to the CD4 receptor on the cell membrane. Transfer of a gene encoding a soluble form of the sCD4 antigen (sCD4) into lymphocytes or haemopoietic cells and subsequent expression will result in circulating CD4. If the levels of circulating sCD4 are sufficiently high, binding of sCD4 to the gp 120 protein of HIV-1 viruses could be reduced.

(ii) **Inhibition at the RNA level :** The production of HIV-1 RNA can be selectively inhibited by standard antisense/ribozyme approaches, and also by the use of RNA decoys. The latter strategy exploits unique regulatory circuits which operate during HIV replication. Two key HIV regulatory gene products are tat and rev these bind to specific regions of the nascent viral RNA, known as TAR and RRE respectively. Artificial expression of short RNA sequences corresponding to TAR or RRE will generate a source of decoy sequence which can compete for binding of tat and rev, and possibly, thereby, inhibit binding of these proteins to their physiological target sequences.

(iii) **Inhibition at the protein level :** In one strategy for this purpose, intracellular antibodies are designed. In another approach, genes that encode dominant-negative mutant HIV proteins which can bind to and inactivate HIV protein (transdominant proteins) are introduced. For example, trans-dominant mutant forms of the gag proteins have been shown to be effective in limiting HIV 01 replication, possibly by interfering with multimerization and assembly of the viral core.

7.4.5 Technical Considerations of Gene Therapy

These have been explained in postnote brought out by the parliamentary office of sciences and technology on June 2005 (Number 240 appeared in the internet).

7.4.5.1 Delivery of Gene

Even for single-gene disorders, successful gene delivery is not easy or predictable. For example, although the genetic basis of cystic fibrosis is well known, the presence of mucus in lungs makes it physically difficult to deliver genes to the target lung cells. Delivering genes for cancer therapy may also be complicated where the disease may be in several sites. Gene therapy trials for X-chromosome linked severe combined immunodeficiency (X-SCID) however, have been more successful. In this case, the genetic basis of the disease is well-understood and the therapeutic material can be delivered using the established procedure of bone marrow transplant. A viral vector has been used to introduce functioning copies of the gene whose malfunction causes X-SCID into blood producing stem cells from the patient's bone marrow. These are then transplanted back into the patient. In most cases the modified bone marrow successfully supplied the missing gene product, and the patients are able to lead normal lives.

For *in vivo* techniques, the challenge of inserting the genes is even greater. The vector carriers have a difficult task to complete. They must deliver the genes to enough cells for results to be achieved and they have to remain undetected by the body's immune system.

7.4.5.2 Durability and Integration

In some gene therapy approaches there are attempts to achieve long-term effects. In such cases it is must that the therapeutic material must remain functional for the intended duration or gene therapy. It can be achieved by two possible ways to use multiple rounds of gene therapy, or to integrate therapeutic genes so they remain active for some-time.

Integrating therapeutic DNA into the target cells' genetic material has a long-lasting. It may however have concerns over possible undersirable side effects. This approach has been used in trials to treat babies with X-SCID syndrome. In UK trials T babies responded well and remained healthy well to gene therapy and remain healthy. However, 3 of those in a similar French trial went on to develop leukaemia-like symptoms. It is possible that the therapeutic material might have integrated where it could affect another gene to produce rapid growth of cancerous cells. Attempts are being made to achieve long-term therapeutic effects without integration. It is being done by using stable, non-integrating, vectors.

When the more immediate effects are to be obtained integration of therapeutic DNA into the target cells is not the aim. In gene therapy to treat cancer, the aim may be to use 'suicide' genes to kill cancerous cells as quickly as possible.

7.4.5.3 Immune Response

The viral vector used to deliver gene therapy, may be recognized by body as 'foreign'. They may mobilize the immune system to attack it. In the case of cancer, the aim of the gene therapy may be to trigger such an immune response. In other cases, immune responses may reduce the efficacy of gene therapy. It may cause the patient to stop responding after a few applications or inducing serious side-effects. It is also possible that an enhanced response to vectors encountered previously may make it difficult to give repeat applications of gene therapy.

7.4.5.4 Safety of Vectors

Since 1990 using experimental gene therapies more than 3,000 patients with serious diseases have been treated. To date, there have been just two fatalities that have been directly attributable to such treatments. A small minority of cases have raised potential concerns. The use of viral vectors has been suggested as a factor in the death of a US gene therapy patient and the cases of leukaemia like symptoms after X-SCID gene therapy trials in France.

Important information on vector safety can be obtained from research on animals preclinical study, mice developed liver tumours after exposure to one particular vector. On the basis of this GTAC published an open letter drawing attention to potential safety concerns. The UK Health and Safety Executive also issued information and interim advice on containment. These steps were taken in case these observations affected any pre-clinical or clinical studies planned or in progress.

7.4.5.5 Uncertainty

Though GTAC and MHRA make every effort to maximize patient safety, uncertainties remain. The range of explanations that have been advanced to account for leukaemia-like symptoms in the French X-SCID gene therapy trial, illustrate the difficulties faced by regulatory bodies. These explanations include problems with the integration of the therapeutic material, an immune response to the vector, a family history of cancer, and preexisting infections and other problems symptomatic of immune deficiency.

Conclusions

According to Parliamentary Office of Science and Technology, UK gene therapy has following applications:

- Gene therapy can potentially treat diseases such as CF, cancers, heart disease and HIV infection. To date, no gene therapy clinical trial has given rise to the development of a commercial treatment in the UK.

- While industry considers commercial gene therapy likely for some cancers in the next few years, others suggest it may be 10 to 15 years before gene therapies are widely available.

- Potential issues with gene therapy include effective delivery, longevity of the therapy and safety concerns.

- Gene therapy clinical trials in the UK are regulated by GTAC and MHRA. MHRA is also responsible for regulating medicinal gene products in the UK.

- The regulatory system considers all gene therapy research proposals on a case-by case basis, taking ethical and scientific factors into account. Some have called for more open and transparent assessments.

- Technical developments may necessitate changes to the current regulatory framework, which has separate arrangements for therapies derived from human genes, stem cells, cells and tissues.

7.4.6 The Ethics of Human Gene Therapy

All current gene therapy trials involve treatment for somatic tissues. Somatic gene therapy is considered appropriate for therapy of lethal diseases, but not for cells of the human germline. Extensive procedures have been developed to ensure rigorous scientific assessment of safety and efficacy in gene therapy trials. More than 200 individuals have been given recombinant cells or recombinant vectors for gene therapy of lethal diseases, including adenosine deaminase deficiency, cancer, cystic fibrosis and familial hypercholesterolemia. As application of gene transfer is extended to other nonlethal diseases, continued safeguards will be required. In principle somatic gene therapy has not raised many ethical concerns other than its possible application in any treatment involving genetic modification of an individual's cells in order to enhance some trait, such as height, without attempting to treat disease. Care has to be taken to keep in view the safety of the patients, as the methodology for gene therapy is not perfect as yet.

Germline gene therapy, involves the genetic modification of germline cell (*e.g.,* in the early zygote). It is quite different from somatic cell gene therapy. It has been successfully practiced on animals (*e.g.,* to correct-thalassemia in mice). It has, however, not been used for the treatment of human disorders. Human germline gene therapy involves ethical concerns and limitations of the technology for germline manipulation. Germline gene therapy is not practiced for the following reason :

The imperfect technology : It requires modification of the genetic material of chromosomes (most easily by chromosomal integration of an introduced gene). But it is not possible with the vector systems used for this process. In somatic gene therapy, the only major concern about lack of control over the fate of the transferred genes is the prospect that one or more cells undergo neoplastic transformation. In germline gene therapy, however, only one cells involved in genetic manipulation.

Ethical reasons : Genetic modification of human germline cells may have consequences not just for the individual whose cells were originally altered, but for all those who inherit the genetic modification in subsequent generations. Moreover, it would deprive the subsequent generations of choice about whether, the incorporation of gene at inappropriate place may lead to cancer or any other disorder. In addition to the question of the rights of individuals in the future, this technology will inevitably lead to a slippery slope towards genetic enhancement. This would entail a program of positive eugenics, whereby planned genetic modification of the germline could involve artificial selection for genes that are thought to confer advantageous traits. Inevitably, even if this were judged to be acceptable in principle, the question arises who decides what traits are useful ?

Role of germline therapy : Germline genetic modification may be considered as a possible way of avoiding what would otherwise be certain inheritance of a known harmful mutation. However, how often does this situation arise and how easy would it be to intervene ? There are two ways for having a 100% chance of inheriting a harmful mutation.

(i) When an affected woman is homoplasmic for a harmful mutation in the mitochondrial genome and wishes to have a child. However, multiple mitochondrial DNA molecules involved are still far from providing gene therapy for such disorders.

(ii) Inheritance of mutations in the nuclear genome is another problems. To have a 100% risk of inheriting a harmful mutation would require mating between a man and woman both of whom have the same recessively inherited disease. This is a rare occurrence. Instead, the vast majority of mutations in the nuclear genome are inherited with almost a 50% risk (for dominantly inherited disorders) or a 25% risk (for recessively inherited disorders). *In vitro* fertilization provides the most accessible way of modifying the germline. If however, the chance that any one zygote is normal is as high as 50 to 75%, gene transfer into an unscreened fertilized egg which may well be normal would be unacceptable. Even if the safety of the techniques for germline gene transfer improves markedly in the future. These are however some risks in the procedure to identify a fertilized egg with the harmful mutation, thus screening using sensitive PCR-based techniques would be required. The same procedure, however, can be used to identify fertilized eggs that lack the harmful mutation. Since *in vitro* fertilization generally involves the production of several fertilized eggs, it would be easier to screen for normal eggs and select these for implantation, rather than to attempt genetic modification of fertilized eggs identified as carrying the harmful mutation.

Newer advances use of recombinant protein in therapy

Though successes were obtained in the production of recombinant proteins in bacterial hosts, they are generally not capable of fully processing and secreting many potentially therapeutic polypeptides that can only be produced in eukaryotic cells, because the latter involve, specific post-translational processing, including proteolytic processing, oligosaccharide addition and modifications, and polypeptide folding required to produce functional, stable recombinant protein. Culture of genetically altered eukaryotic mammalian cells is used to produce many proteins for human therapeutics. DNAse, tissue plasminogen activator (TPA), erythropoietin, GM<– CSF and others have been produced in this way. Transgenic animals and plants are also used for this purpose. Transgenic goats and cattle have been produced in which the recombinant gene is expressed from promoters that are active only in the mammary gland. Active recombinant proteins are appropriately processed and can be obtained in large quantities for therapeutic use. Similarly, domesticated plants, such as tobacco, have been used to express human transgenes in leaves.

Protein modeling and engineering

By the use of site-directed mutagenesis (to make mutations in polypeptide sequences that can be expressed by recombinant DNA technology), novel therapeutic proteins are produced. The alteration of amino acids or groups of amino acids that compose functional domains in the polypeptide can improve synthesis rates, processing stability or biologic activity. Antigenicity of the proteins can be modified to overcome host cells immunity. Novel proteins can be designed for administration and the improved genes can also be transferred to direct the synthesis of the protein. Some of the limitations of gene transfer systems can be overcome by combining the approaches of genetic therapeutics and the rational design of proteins. A gene encoding an engineered protein with an increased catalytic rate constant could be incorporated into cells. It leads to the production of a protein with an enhanced therapeutic effect.

7.5 DNA FINGERPRINTING, DNA FOOTPRINTING AND BIOMETRICS

7.5.1 DNA Finger Printing

Like the fingerprints that came into use by detectives and police labs during the 1930s, each person has a unique DNA fingerprint (Fig. 7.16). Unlike a conventional fingerprint that occurs only on the fingertips and can be altered by surgery, a **DNA fingerprint** is the same for every cell, tissue, and organ of a person. It cannot be altered by any known treatment. Consequently, DNA fingerprinting is rapidly becoming the primary method for identifying and distinguishing among individual human beings. An additional benefit of DNA fingerprint technology is the diagnosis of **inherited disorders** in adults, children, and unborn babies. Even blood-stained clothing from Abraham Lincoln has been analyzed for evidence of a genetic disorder called Marfan's Syndrome.

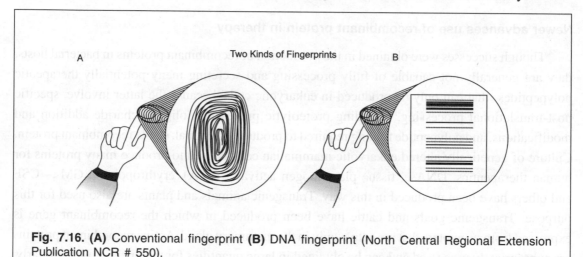

Fig. 7.16. (A) Conventional fingerprint **(B)** DNA fingerprint (North Central Regional Extension Publication NCR # 550).

The techniques of RFLP analysis is used to indicate that though all the individuals possess many common genes, have many differences in their genetic material. Most of these differences occur in non-coding regions of the DNA. It is because of this they do not have effect on the individuals. Many non-coding regions are quite variable. It is, therefore, possible that two individuals will differ in the particular non-coding regions they possess, unless they are related and, therefore, share genetic material. These genetic differences are identified by DNA fingerprinting. This determines whether two different DNA samples are from the same or from different persons.

The basic technique involves the following steps:

(a) The preparation of a DNA sample from biological material found at a crime scene – blood, semen or even hair.

(b) Generally to amplify specific fragments of DNA, PCR is used because little biological material may be found.

(c) Other DNA samples are prepared from suspects.

(d) The DNA samples are digested with restriction enzymes and are separated and displayed by gel electrophoresis.

(e) Following transfer to a membrane, the DNA can then be subjected to hybridization.

(f) Then appropriate DNA probes are selected.

(g) It is attempted to visualize one or more DNA fragments that are known to be highly variable among individuals.

(h) The probes are, therefore, chosen so that they hybridize to these variable regions.

(i) If the identical bands following hybridization are showing DNA in samples from a suspect and from the crime scene, then within some statistical limits, the two DNA samples can be said to be from the same person.

(j) If, on the other hand, different bands are shown by the DNA samples, then the DNA is not from the same person.

Uses of DNA Fingerprints: DNA fingerprints are useful in several areas of society. They are used in human health and the justice system.

Diagnosis of inherited disorders: It is used to diagnose inherited disorders in both prenatal and newborn babies in hospitals around the world. These disorders may include cystic fibrosis, hemophilia, Huntington's disease, familial Alzheimer's, sickle cell anemia, thalassemia, and many others.

Early detection of such disorders enables to make the parents sensitive for proper treatment of the child. In some programs, genetic counsellors use DNA fingerprint information to help prospective parents understand the risk of having an affected child. In other programs, prospective parents use DNA fingerprint information in their decisions concerning affected pregnancies (Fig.7.17).

Development of cures for inherited disorders : Research programs to locate inherited disorders on the **chromosomes** depend on the information contained in DNA fingerprints. By studying the DNA fingerprints of relatives who have a history of some particular disorder, or by comparing large groups of people with and without the disorder, it is possible to identify DNA patterns associated with the disease in question. This work is a necessary first step in designing an eventual genetic cure for these disorders.

Forensic or criminal : The investigating agencies throughout world have begun to use DNA fingerprints to link suspects to biological evidence—blood or semen stains, hair, or items of clothing—found at the scene of a crime.

DNA fingerprints also help in the court system is to establish paternity in custody and child support litigation.

Personal identification : Because same DNA fingerprint is present in every organ or tissue of an individual the armed services in some countries have just begun a program to collect DNA fingerprints from all personnel for use later, in case they are needed to identify casualties or persons missing in action.

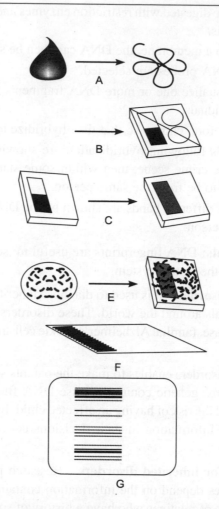

Fig. 7.17. The process of DNA Fingerprinting. (A). The process begins with a blood or cell sample from which the DNA is extracted. **(B).** The DNA is cut into fragments using a restriction enzyme. The fragments are then separated into bands by electrophoresis through an agarose gel. **(C).** The DNA band pattern is transferred to a nylon membrane. **(D).** A radioactive DNA probe is introduced. The DNA probe binds to specific DNA sequences on the nylon membrane. **(E).** The excess probe material is washed away leaving the unique DNA band pattern. **(F).** The radioactive DNA pattern is transferred to X-ray film by direct exposure. When developed, the resultant visible pattern is the DNA FINGERPRINT. (Sources Biotechnology training programme Inc. Ed. by Glenda D. Webbo).

Legal and scientific debate on the use of DNA as an evidence. As erroneous or improper conclusions may be made because of technical difficulties, the reliability of the technique, thus, can be questioned. Another concern is the statistical validity of the results. The statistical analysis is complex and is based on its own set of assumptions which only the person with experience can describe assumptions that are nearly impossible for non-experts to recognize, let alone judge.

7.5.2 DNA Footprinting

It is carried out by using the following techniques as described in the website http://bioweb.wku.edu/courses/biol350/Transcriptome17/Reviw.html.

7.5.2.1 DNase Footprinting

It is possible to determine the location of protein binding sites by allowing the proteins to bind to the DNA followed by the addition of DNase I. The locations where proteins are bound to DNA will be protected from nicking by DNase I and when fragments are separated on a gel, those protected by proteins will be absent in the protected areas.

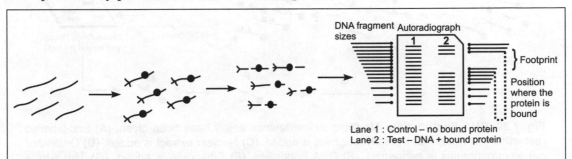

Fig. 7.18. DNase Footprinting. (A) DNA molecule is end labeled and regulatory protein is added. **(B)** It is subjected to limited DNase digestion. **(C)** Gel electrophoresis, autoradiography are performed to detect the fregment. DNA molecules are cut at any phosphodiester bond not protected by the protein.

7.5.2.2 Gel Shift Assays

Proteins that bind to DNA can be eluted and characterized by protein gel electrophoresis. The specificity of protein binding can be tested by gel-shift or gel-retardation assays (Fig. 7.19). When a protein is tightly bound to the DNA, the migration of labeled DNA fragments through a gel will be retarded or shifted. Competition assays combined with gel-shift assays determine the binding affinity of the protein to the DNA.

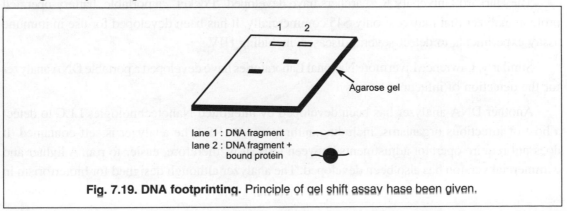

Fig. 7.19. DNA footprinting. Principle of gel shift assay hase been given.

7.5.2.3 Modification Interference Assay

Bases are modified by Dimethyl sulphate. In chemical sequencing of Gs and As is the first step. Methylation of DNA can sometime prevent proteins from binding to the DNA. Typically the DNA regions that bind to proteins are undermethylated. Gel shift assays bands that are not retarded in their migration are then isolated and exposed to piperidine which cleaves the DNA at the modified G or A. The size of the remaining band the can let us know the locaion of the DNA binding site (Fig 7.20).

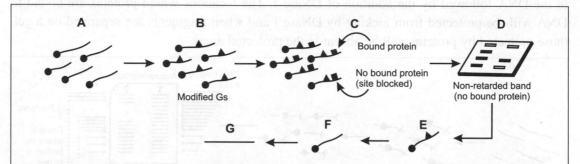

Fig. 7.20. DNA Footprinting: Modification interference assay hase been given. **(A)** End-labelled restriction fragments. **(B)** Dimethyl sulphate is added. **(C)** Nuclear extract is added. **(D)** Organized gel electrophoresis is performed. **(E)** DNA is purified. **(F)** Piperidine is added. **(G)** The size is determined by polyacrylamide gel electrophoresis (http://bioweb.wku.edu/courses/biol350/Transcriptomel7/Review.html).

7.7 BIOMETRICS

It is automated method of recognizing a person on the basis of physical features such as face, fingerprints, hand, iris and voice. This is being deployed to enhance the identification and verification of individuals. **Indian Technology,** 'iris recognition technology' is already being used for identification and verification of persons at some airports and borders. It is more accurate, faster and easy to use than standard methods.

The Harvard University Researchers have developed "Pocket", a portable, battery-operated protein analyzer that may cost only $45 commercially. It has been developed for use in immune assay experiments to detect several diseases, including HIV.

Similarly, Lawrence Livermore National Laboratories have developed a portable DNA analyzer for the detection of infectious agents.

Another DNA analyzer has been developed by integrated Nanotechnologies LLC to detect a host of infectious organisms, including anthrax and SARS. The analyzer is self-contained. It does not require operator adjustments between tests. It is, therefore, easier to run. A lighter and commercial version has also been developed. The analyzer although designed for bioterrorism in developed countries, is a useful tool for developing countries too as it could cut down reagent needs and overcome current lack of trained personnel. The costs of equipping, maintaining and running diagnostic laboratories in developing countries are likely to be reduced by the introduction of these and others and improve health.

A chip-sized version of a common detector used to identify proteins, DNA and other molecules has been developed at Purdue University. This is able to radically reduce the size of detection equipment, in a similar way in which the move from separate transistors to integrated circuitry changed the size and power of computers and similar equipment. It is expected that in the very near future, tools that can analyze a human genome in a day and make it possible to profile entire populations, will be developed.

The Affymetrix Gene Chip array measures the activities of thousands of genes simultaneously.

Cloning, especially of animals, may soon become a routine practice. A cloning chip that could denucleate, one of the first hurdles in cloning more than a hundred cells at once, has now been developed by Aegen Bioscience.

7.7 BIOMATERIALS AND STEM CELLS

Biorubber : It has been developed at MIT and can be safely used in engineering heart valves, blood vessels, livers and other elastic body tissues.

Biogel : It can change its characteristics in response to stimulus. The property of the gel makes it a candidate for construction of controlled drug delivery devices and similar tools.

Tissue based bioreactor for stem cells : A bioreactor containing a matrix of hundreds of membranes has been developed at Berlin's Charité Hospital. In this human adult living stem-cells have been coaxed to grow into complex living tissue like a healthy liver. On feeding a patient's blood through the bioreactor, the cultured liver cells start all the normal, healthy functions of the patient's own diseased organ. The bioreactors are being used in clinics at Berlin and Barcelona to save the lives of patients whose own livers have stopped functioning, but whose donor organs have not yet arrived. Attempts are being made to target the liver's regenerative capacity so that there is no need of transplantation. The patients may be hooked up to the reactor so their own livers can take time off and recuperate. Even the attempts are being made to develop complete liver in tissue culture.

For the first time, a system has been devised in lab mice that duplicates neurogenesis, the process of producing new brain cells in a culture dish. Regenerative scientists from the University of Florida's McKnight Brain Institute (Gainesville, USA) reported in an article published in the June 13, 2005 issue of the Journal Proceedings of the /US/National Academy of Sciences that they developed a cell culture technique that offers the promise of generating a limitless supply of an individual's own brain cells to possibly repair disorders such as epilepsy and Parkinson's disease (Biotech New International, Vello, 2005).

Embryonic stem cells : A firm Geron has made these to turn into different types of normal cell lines. These may be used to repair damaged tissue (heart, muscle, pancreas, bone, brain, spinal injuries and liver). This could solve rejection of organs or tissues derived from stem cells of another organism. It is worth to mention the case of Molly Nash.

Molly Nash (8 years in 2003) was born with a disease Franconi's anaemia. In this disease bone-marrow cells fail to function. New cells from a donor who is genetically exactly similar were needed for Molly. Pre-implantation genetic diagnosis (PGD), a test using embryonic cells was performed to save Molly. With the help of PGD, Molly's parents conceived their son Adam, When he was 2.5 years old in 2003, his umbilical-cord blood was taken to save his sister's life. PGD, however, is transforming reproductive medicine by giving parents unprecedented control over what genes their offspring will have. Cautions must be taken not to misuse PGD. Many clinics in London, Chicago, Tel Aviv and Brussels provide the PGD service.

For this purpose a single human stem cell is plucked from a three-day-old embryo or less. It is developed in a fertility clinic, where would be parents get their eggs harvested, fertilized and grown in the laboratory. By day three, the egg cell may divide, on average, into only six stem cells. To find out if it carries the genes for Tay-Sachs or cystic fibrosis or sickle cell anaemia, the DNA is analyzed in the laboratory. The number of people on lists for organ transplant remains high, but a few viable organs are available to save these lives. Work is going on to :

   (i)   make stem cells to grow into organs,

   (ii) overcome rejection of organs from other animals (e.g., pigs), and

   (iii) target ailing tissues/organ's ability to regenerate.

Despites impressive results, there is no consensus on the use of embryonic cells and the limits to which such use is permitted. In 2002, EU adopted a moratorium on the Commission's funding of research on embryonic stem cells until clear-cut and strict ethical rules are made. The United States has also curtailed Federal funding for embryonic research. A decision on human cloning in 2003 was postponed in the UN.

The United Kingdom, Singapore and China being more liberal in regulation, embryonic researches are persuing the field further.

The development of research on adult stem cells, spurred by therapeutic aims, may raise more formidable problems than those aimed at forbidding destruction or creation of embryos. It is because of lack of a clear definition of an embryo in science. Ethical reflection and debates appear to be based on the representation of life which the continuous discoveries put in question. At the same time, people want therapies without ethical dilemmas, the absence of risk without questioning our representations of life (Renard and Bonniot de Ruisselet, 2003). Throughout history, ethical codes have been revised by mankind to meet new challenges. The continuous creation of knowledge, innovations and technologies require the design of new procedures and the participation of academic, professional associations and legislatures.

Nanoparticles and Human Disease

Scientists have devised a new way to fight certain diseases including cancer, based on the nanoparticles that can both identify and kill cancerous cells. The study published in the April 13, 2005 issue of the Journal Nanoletters reported that current molecular imaging approaches only identify malignancies, but do not provide a treatment method. We can look for a marker that may indicate a significant clinical problem, but we can not do anything about it. Attempts are on to locate the cancerous cells.

A genetic assay screens for 23 mutations that account for approximately 90% of cystic fibrosis cases in the US Caucasian population. A new microarray system is developed by Nanogen, Inc. (San Diego, CA, USA, www.nanogen.com) an automated multiplexing platform that laboratories use to detect genetic sequences.

Its a NanoChip 400 microarray system. The NanoChip 400 instrument is an automated multiplexing platform that laboratories use to detect genetic sequences. Tests can be performed using reagents supplied by Nanogenor laboratories that can develop a variety of "homebrew" assays. The NanoChip instrument employs Nanogens core microarray technology, which utilizes patented microfluidics and electronic technology to automate sample handling and detection of results.

The Cystic Fibrosis Kit is intended to be used for carrier testing in adults of reproductive age, as an aid in newborn screening, and in confirmatory diagnostic testing in newborns and children.

According to Dr. Howard C. Birndorf, Nanogens Chairman, the premarket notification of the NanoChip system and the Cystic Fibrosis assay to the FDA is an important milestone.

The American College of Obstetrics and Gynaecology (ACOG) and the American College of Medical Genetics (ACMG) have recommended that genetic testing for cystic fibrosis be offered to all couples in the U.S. currently planning a pregnancy.

Newer Advances in Cancer Therapy

(A) Silica nanoparticles adapted to deliver water-insoluble cancer drugs

Cancer researchers have found a way to administer hydrophobic chemotherapeutic drugs that do not require dissolving the medication in toxic solvents. Investigators at the Jonsson Cancer Centre (Los Angeles, CA, USA; www.cancer.mednet.ucla.edu) used the pores of fluorescent mesoporous (i.e., containing pores with diameters between 2 and 50 nm) silica nanoparticles to sequester the hydrophobic anticancer drug camptothecin (CPT). This drug and its derivatives are considered to be among the most effective of the currently available anticancer drugs. Although studies have demonstrated their effectiveness against a number of different types of cancer cells in vitro, clinical application of CPT in humans has only been carried out with derivatives that have improved water solubility.

According to Dr. Fuyuhiko Tamanoi, Professor of Microbiology Immunology and Molecular Genetics at the Jonsson Cancer Center, Silica nanoparticles containing CPT were taken up by cancer cells, which then began to die. In order to be used on humans, current cancer therapies such as CPT or Taxol, which are poorly water-soluble must be mixed with organic solvents in order to be delivered into the body. These elements produce toxic side effects and in fact decrease the potency of the cancer therapy. Silica nanomaterials show promise for delivering camptothecin and other water insoluble drugs. Hydrophobic anticancer drugs have successfully been into mesoporous nanoparticles and delivered into human cancer cells to induce cell death.

(B) Light activated molecules trigger double stranded DNA damage that kills cancer cells

Recently cancer a panel of chemotherapeutic agents that induce apoptosis in tumour cells by causing damage to the cancer cells double-stranded DNA when exposed to light has been developed.

Investigators at Florida State University (Tallahassee, USA; www.fsu.edu) worked with a class of reagents known as simple lysine conjugates. These small molecules were known to be capable of selective DNA damage at sites approximating a variety of naturally occurring DNA damage patterns. This process transformed single-strand DNA cleavage into double-strand DNA cleavage. Cells are incapable of repairing double-strand damage, which triggers apoptotic pathways.

It is known that lysine conjugates get activated by certain wavelengths of lights, thus, the molecules get photoactivated. Cancer killing properties are expressed upon demand only when the reagents are in exactly the right place and when their concentration is high inside the cancer cells (Proceedings of the U.S. National Academy of Sciences August 7, 2007). It has been revealed that while several lysine conjugates demonstrated little effect upon cultured cancer cells, metastatic human kidney cancer cells without light. Upon phototherapy activation they killed more than 90% of the cancer cells with a single treatment.

According to Dr. Igor V. Alabugin, Associate Professor of Chemistry and Biochemistry at Florida State University when one of the two strands of cellular DNA is broken, intricate cell machinery is mobilized to repair the damage. This process functions in an environment full of ultraviolet irradiation, heavy metals, and other factors that constantly damage our cells. Ways are being found out to induce apoptosis in cancer cells by damaging both of their DNA strands. A group of cancer killing molecules known as lysine conjugates can identify a damaged spot in a single strand of DNA and then induce cleavage on the DNA strand opposite the damage site. This double cleavage of the DNA is very difficult for the cell to repair and typically leads to apoptosis.

Subset of cancer stem cells found to suppress chemotherapy agent

It has now been suggested that for chemotherapy to be effective in treating lung cancers, it must be able to target a small subset of cancer stem cells, which have been shown to share the same protective mechanisms as normal lung stem cells.

Current cancer therapies often succeed at initially eliminating the bulk of the diseased, including all rapidly proliferating cells, but are eventually thwarted because they cannot eliminate a small reservoir of multiple-drug-resistant tumour cells, called cancer stem cells, which ultimately become the source of disease recurrence and eventual metastasis.

Scientists from the University of Pittsburgh School of Medicine (PA, USA; www.medschool.pitt.edu) presented this groundbreaking study at the tissue engineering and regenerative medicine international society (TERMIS) North American Chapter meeting.

Vera Donnenberg, Assistant Professor of Surgery and Pharmaceutical Sciences, University of Pittsburgh Schools of Medicine and Pharmacy, used cell surface markers and dyes to identify cancer stem cells as well as normal adult stem cells and their progeny in samples obtained from normal lung and lung cancer tissue samples. A very small, rare set of resting cancer stem cells have been identified in the lung cancer samples that looks and behave much like normal adult lung tissue stem cells. Both the cancer and normal stem cells have been found to be equally protected by multiple drug resistance transporters, even if the bulk of the tumour responds to chemotherapy.

According to Dr. Donnenberg, the very fact that cancers can and do relapse after apparently successful therapy, indicates the survival of a drug-resistant, tumour-initiating population of cells in many types of refractory cancers. "Because of the similarities between the way that normal stem cells and cancer stem cells protect themselves, cancer therapies have to be specifically designed to target cancer stem cells while sparing normal stem cells".

Questions

1. What is the role of biotechnology in medicine ? Explain in detail.
2. Describe in detail the immunological assays for the diagnosis of a disease.
3. Describe the use of molecular probes in the diagnosis of diseases.
4. What do you mean by the following :
   (a) Biopharmaceuticals
   (b) Biomimetics
   (c) Biopharming
   (d) Tissue based bioreactor for stem cells.

Chapter **8**

Biotechnology in Agriculture

8.1 INTRODUCTION

Agriculture is an ancient practice, which in the last century or so, has been moulded into a science through the developments of modern genetics and biotechnology. In the last couple of years, advances in understanding of plant cell and molecular biology have provided the basis for the improvement in agriculture produce. The world population is set to increase by 90 million a year, between now and 2030 (an additional ca. 3.25 billion people), with the majority of this increase occurring in the developing countries. The question is, can the earth's resources sustain such growth at a time when there is a continual decline in agriculturally exploitable land, shifting weather patterns and the associated localized scarcity of water and a build up of environmental pollutants due, in part at least, to the farming techniques associated with high input agriculture.

At the same time, there is a growing realization that good health is dependent upon adequate nutrition and as the middle class grows with increasing economic prosperity in many parts of the world, they will demand dietary upgrading. This in turn leads to a higher standard of living and an associated lowering of the birth rate.

Despite the dramatic increase in yield achieved by the plant breeders over the last 50 years, coupled with the 'green revolution' which led to the development of a wide range-of grain crops capable of responding to high input agriculture, the world grain production per person reached a peak in the mid 1980s and is now declining. Recent reports have shown that droughts across the world in the last few year has led to a decline in the world grain stocks, which cushioned the world against serious scarcity, and they have slumped to their lowest level ever and are now well below the safety level of 60 days supply. Grain prices have escalated and we shall be paying more for our daily bread, while others will starve.

Uptil now the skills of the plant breeder and farmer coupled with technological innovation by the agrochemical industries has allowed us to increase world food production to broadly sustain the increase in population. This state of affairs is unlikely to continue and can only be addressed by increased efforts to control the birth rate, improve the agricultural resource base and applying the emerging technologies to sustain yields and decrease our dependence upon the application of agrochemicals derived from oil-based feedstock.

The challenges which have been met by the plant breeder depended upon the selection and breeding of plants which exhibit desirable traits. These include increased yield, response to applied fertilizers, resistance to pests, pathogens and stress and improved quality of the harvested product. During the last decade, the skills of the plant breeder have been supplemented by what is termed as plant biotechnology.

The revolution which has occurred in plant science over the last 10 years and has been termed plant biotechnology has resulted from an increase in our understanding of how cells and organisms work at the molecular, biochemical and physiological levels and also from the development of techniques which allow the transfer of genes from one plant species to another, or from other organisms such as bacteria. These technologies complement classical plant breeding and potentially reduce the time scale required to produce a new variety or hybrid.

Agriculture or plant biotechnology is an expanding area of research and is the product of several disciplines such as molecular biology, tissue culture, chemical engineering, plant pathology and increasing accountancy. The agricultural practices are now undergoing dramatic changes. It is being done to have sustainable agriculture. It is possible only if the plants are able to face the abiotic and biotic stresses. The concept of sustainable agriculture means different things to different people. Specific details may vary, but all share a common theme, a longer term view than the one dictated by short-term economic considerations and a tripartite set of interdependent dimensions for measure sustainability—ecological, economic and social. Each one of the three dimensions has many facets or levels of impact. Biotechnology is expected to contribute to sustainable agriculture, if farmers maintain the quality of biotic and abiotic resources, they depend upon and decrease the consumption of non renewable resources. This can be achieved if the newer techniques of biotechnology are used.

(i)   The test based on DNA probes of monoclonal antibodies can be used to identify diseased plant much earlier. This will decrease the spread of disease and hence the use of fungicides to make the plants more resistant.

(ii) The detection system can be used to determine the resistance or susceptibility of insect pests to insecticides to be used.

(iii) The yield can be improved by metabolic engineering. It will not need application of more fertilizers.

(iv) RNA interference can be used to block the expression of certain genes and shunt most of the plant's resources to the part of the plant.

All these are in fact concerned with the improvement of quality, nutritional value and yield of agricultural products and decreasing the cost of production. These aims are being targeted through the two fundamental techniques of biotechnology.

1.   Plant Cell and Tissue Culture
2.   Genetic Engineering

8.2 PLANT TISSUE CULTURE AND ITS APPLICATIONS

With the advances in cell and tissue culture, it is now possible to grow whole plants, parts of plants and also single cells in sterile media containing minerals, growth factors, a carbon source and plant growth regulators. Culturing of plant tissues gives us the power to manipulate and exploit their growth for commercial purposes.

Some of the technologies used for this purpose are as follows :

  (a) Micropropagation
  (b) Somatic embryogenesis
  (c) Modification by somaclonal variations
  (d) Protoplast fusion
  (e) Virus elimination
  (f) Embryo rescue
  (g) Haploid production and ploidy manipulation
  (h) Production of chemicals by cultured cells
  (i) Plant transformation

Most of these have been described earlier in the chapter on cell culture technologies.

8.2.1 Micropropagation

It is achieved by maintaining organized tissues by the multiplication of meristems and axillary buds. A range of tissues can be explanted from different species as a source of material for micropropagation like shoot meristems, stem segments with axillary bud tissues that will form adventitious shoots and adventitious embryos.

8.2.2 Somatic Embryogenesis

It is the production of embryo-like structures from somatic cells. A somatic embryo is independent bipolar structure and is physically not attached to tissue of origin. It can germinate to form plantlets on culture medium without hormones, till they reach the size suitable for transfer to soil. The production of 'artificial' seeds by encapsulation of somatic embryo is being done. It is possible to dehydrate encapsulated embryos and to store them until they are required in bulk for plantation in field. Encapsulation of somatic embryos has been applied to crops such as celery, alfalfa and cauliflower. A number of biotechnology companies are actively engaged in research to expand the range of plants and to convert the potentials into a practical agricultural procedure.

8.2.3 Modifications by Somaclonal Variations

Somaclonal variation is a general phenomenon of all plant regeneration systems that involve a callus phase. This includes plants regenerated from protoplasts, cultured explants, microspores, anthers, ovaries *etc*. Spontaneous mutations occur at a rate on an average of 1 in 10^6 and this is generally too low for practical exploitation. Somaclonal variations can produce variants in 15% or more of progeny. The process of regeneration sieves out genotypes with most harmful changes and somaclones can usually be stabilized in a single generation.

For many crop species, potentially useful somaclonal variants have been produced. There is prospect of producing **'designer vegetables'** by upgrading existing varieties and many products are already in the market.

8.2.4 Protoplast Fusion

Removal of cell walls renders protoplasts of different types amenable to fuse together. The initiation of fusion of protoplast can be done by chemicals, electrofusion, heterofusion, direct fusion, macrofusion, microfusion *etc*. By hybridization of somatic cells (Fig. 8.1), we can overcome the cross – incompatibility of unrelated plants.

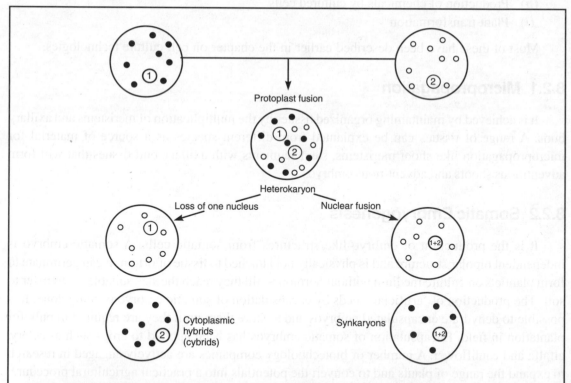

Fig. 8.1. Protoplast fusion. The different kinds of hybrids which can be obtained following fusion of two different protoplasts. The nuclei of the two parental cells are labelled 1 and 2 and the plastids have been shown as dark or hollow circles.

Practical applications are possible with protoplast technology. Work on members of Solanaceae family has led to successful hybridization of genera such as *Nicotiana, Datura, Petunia, Solanum* and *Lycopersicon*. Successful hybridizations have also been obtained in *Umbelliferae, Rotaceae* and *Fabacosae*. By using this technology, interspecific hybrids *e.g.*, hybrids between rice and barnyard grass have been produced. Plant regeneration from cereal protoplast is currently being extensively studied particularly in connection with cereal transformation *e.g.*, albino barley plants, sterile plants of maize *etc.* have been produced.

8.2.5 Virus Elimination

Under normal conditions, plants are susceptible to a range of pathogens including bacteria, fungi, viruses, nematodes and insects. It is possible to eliminate most pathogens by chemical treatments or other means, but viruses pose a problem. Biotechnology has a solution to this problem.

Observations have shown that distribution of viruses in plants is uneven and very few of them are present in apical meristem. Meristem tip culture methods has been developed for virus elimination.

In this way virus free plants have been produced from a large number of species. In potato alone, at least 136 cultivars have been freed from virus infection. Pathogen free plants have also been produced from stocks infected with mycoplasma, fungi and bacteria. This approach has, therefore, become invaluable practical technique.

8.2.6 Embryo Rescue

In sexual crosses where the parents are taxonomically distant, barriers that prevent successful hybrid production may be overcome by *in vitro* culture techniques. Especially in cases where fertilization is successful, but embryo fails to develop, immature zygotic embryos can be excised and cultured and hybrid plants can be regenerated.

8.2.7 Haploid Production

Haploid plants have gametic chromosome numbers. These are very useful in plant breeding, both for rapid production of homozygous lines following chromosome doubling to original ploidy level and for detection and selection of recessive mutants.

Haploid plants have been regenerated in more than 50 species through anther cultures, with the majority in the Gramineae, Solanaceae and Cruciferae. These include crop species such as wheat, barley, maize, rice, rye, grasses, Brassicas, potato, tomato and tobacco.

8.2.8 Production of Chemicals by Cultured Cells

A large number of commercially important chemicals are directly derived from plant material and are almost invariably secondary metabolites *i.e.*, compounds that have not been considered essential for primary metabolic functions of plants.

8.2.9 Plant Transformation

It is possible to introduce one or more genes directly into plant cells and to transform them. The gene transfer may be mediated by *Agrobacterium* or direct *e.g.,* agroinfection, macro- and micro- injection, shotgun transformation, DNA/pollen transfer, chemical mediated DNA uptake, electroporation *etc.*

8.3 GENETIC ENGINEERING OF PLANTS

It involves the movement of one or a very small number of known genes into a plant to introduce specific characteristics. Genetic engineering has been used for a number of reasons *e.g.,*

(a) To make crops resistant to insect pests.

(b) To make crops resistant to viral diseases.

(c) To produce edible vaccine proteins in fruits.

(d) To produce other medicinal proteins and secondary metabolites in plants.

Transgenic organisms provide an alternative to traditional methods of animal and plant breeding and hence they offer an exciting new way forward in agriculture. Improving crops or domestic animals by traditional methods is a slow process as it relies on chance because of crossing over in meiosis and random segregation of chromosomes during sexual reproduction. It takes 7-12 years to develop a new cereal variety. The agriculture scientists are incorporating into our crops through genetic engineering the same characteristics that they had incorporated into crops through selective breeding.

Genetic engineering helps to add new genes directly, without relying on sexual reproduction. It is possible to have 'designer' plants and animals with desirable properties such as disease resistance. Animals and plants can become 'living factories' for useful products, just like bacteria in fermenters. The challenge in agriculture is to improve food production in the developing countries. Some of the new techniques are to be applied to regions where food shortage is greatest.

The transgenics have the following advantages :

- A gene for a desirable characteristic can be identified and cloned.
- All the beneficial characteristics of an existing variety can be kept and just the desired new gene can be added.
- There is no need of sexual reproduction.
- Transgenesis is much faster than conventional breeding.
- Selection of a particular trait is easy.

Plant cells in culture whether normal or transformed can be upstreamed by using fermenter.

8.3.1 Applications of Genetic Engineering of Plants

Prior to genetic engineering, the exchange of DNA material was possible only between individual organisms of the same species. With the advent of genetic engineering in 1972, scientists have been able to identify specific genes associated with desirable traits in one organism, and transfer those genes beyond the boundaries of species into another organism. For example, genes from bacteria, viruses or animals may be transferred into plants to produce genetically modified plants having changed characteristics. This method, therefore, allows mixing of genetic material among species that otherwise cannot naturally breed.

Some current applications of plant biotechnology include developing the plants :

- that are resistant to diseases, pests and stress,
- whose fruits and vegetables remain fresh for longer periods of time which is extremely important in tropical countries,
- that possess healthy fats and oils,
- that have increased nutritive value,
- that have a higher expression of the anti-cancer proteins naturally found in certain plants such as soybeans,
- that have whole range of higher value added feed,
- that have applications such as lignin modification in trees that will make possible much higher fibre extraction rates in the paper and pulp industry,
- that are capable of producing new substances, including biodegradable plastics and small proteins or peptides such as prophylactic and therapeutic vaccines,
- that have properties of sensing pollutants

Some of the applications have been given below :

The first generation of genetically engineered crops to come to market was for better pest management capabilities. These included a number of herbicide tolerant insects and disease resistant crop varieties.

8.3.2 Herbicide Resistance

Selective herbicides are routinely applied to control weeds among crop plants that would otherwise compete for available nutrients, space and light and, thus, reduce crop yield and quality. The introduction of genes which give resistance to certain herbicides is an interesting application of genetic engineering in plants. When the crop is sprayed with that herbicide, only the weeds are killed. This solves the problem that weed killers are not normally very selective. In developed countries where modern agricultural methods are used, weeds can reduce crop yields by over 10%. It is estimated that genetically engineered herbicide resistance could double or even quadruple yields in some parts of Africa where serious weed problems are found. The parasitic weeds broomrape (*Orobanche*), found north of the Sahara, and witchweed (*Striga*) found in sub-Saharan Africa are quite harmful. They affect maize, millet, wheat, sorghum, sunflowers and legumes. Herbicide-resistant corn, wheat, sugar beet and oilseed rape have so far been produced by developed countries for their own use. These are resistant to the herbicide Basta.

A single new gene may be all that is necessary to confer resistance to herbicide. For example, the broad-spectrum weed killer glyphosate (phosphonomethylglycine) acts by inhibiting the enzyme

5-enolpyruvyl-3-phosphoshikimate synthetase that converts phosphoenol pyruvate and 3-phosphoshikimic acid to 5-enolpyruvyl-3-phosphoshikimic acid in the shikimic acid pathway in bacteria (Fig.8.2). Following mutagenesis of *Salmonella typhimurium,* an altered synthetase enzyme resistant to glyphosate has been identified and the altered *aroA* gene that codes for it has been cloned and introduced into *E.coli*, where it confers resistance to glyphosate. The next step is to transfer this gene, with appropriate control signals into crop plants. Similarly, propanil (3, 4-dichloropropionanilide) is a selective photosynthetic herbicide used for post emergence weed control in rice fields. Resistance in rice is conferred by the mitochondrial enzyme arylacylamidase, which rapidly hydrolyzes acylamilide herbicides and its gene could be moved to other crop plants.

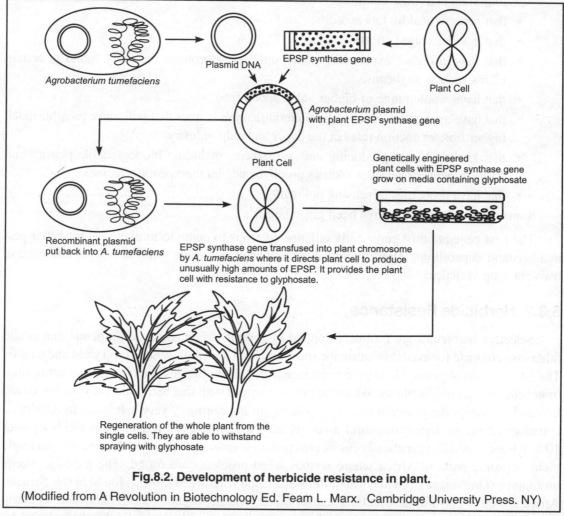

Fig.8.2. Development of herbicide resistance in plant.

(Modified from A Revolution in Biotechnology Ed. Feam L. Marx. Cambridge University Press. NY)

Genes conferring resistance to herbicides may also be identified in weed species that have developed resistant biotypes under long periods in which one herbicide has routinely been applied, as has been found for S-triazines (atrazine and simazine), paraquat and dichloro-rofop-methyl. Resistance genes may also be obtained from algae (*e.g., Chlamydomonas*).

8.3.3 Resistance to Insects

Enormous crop losses are caused by insects. Since the Second World War, many chemicals, starting with DDT, have been used as insecticides. But now we are aware of the ecological damage done by these chemicals and that is why attempts are being made to reduce or avoid the problem.

Insects play a vital role in nature as well as delighting the observer by their beauty and variability. However, due to our heavy reliance on a very few crop species, man has created a monoculture in which the major crop plants are often seriously damaged by insect pests. To counter this, a range of effective chemical insecticides has been developed since the Second World War. The total estimated cost of such agrochemicals is some $8 billion per annum and yet of the order of 13% of crop production is still lost to insects. To reduce dependence upon potentially environmentally damaging pesticides, a range of naturally occurring insecticidal proteins has been identified in bacteria and plants, the genes which encode them have been isolated and transferred to a number of crop plants.

The genetically engineered insect resistant plants are useful because they do not involve the use of pesticides which are :

- expensive and time consuming to apply,
- non selective and, therefore, kill harmless and useful organisms as well, such as pollinators.

As a form of biological control, the bacteria themselves can be applied to the crop. This, however, is expensive because they quickly die. It makes the regular spraying necessary.

Two general strategies have been adopted in creating transgenic plants that are resistant to insects. One takes advantage of the insecticidal protein produced by *Bacillus thuringiensis*, the other exploits genes encoding plant proteins which form part of the plant's normal defensive strategy and have insecticidal activity.

Bacillus thuringiensis synthesizes an insecticidal crystal protein. It resides in inclusion bodies produced during sporulation of the bacteria. *Bacillus thuringiensis* can be used against several species of insect pests. It is 80,000 times more powerful than the organophosphate insecticides commonly sprayed on crops and is fairly selective. It kills only the larvae of certain species. Different insects are killed by different strains of *Bacillus thuringiensis*, mainly the larvae of moths and butterflies (caterpillars) and of some hemipterans, such as white flies (maggots) and the larvae of flies such as mosquitoes (aquatic larvae). Some of them kill nematode worms that can also be pests.

The crystal protein, when ingested by insect larvae, is solubilized in the alkaline conditions of the insect midgut and processed by midgut proteases to produce a protease-resistant polypeptide toxic to the insect. The toxin specifically binds to the inside of the insect's gut and the epithelium is damaged so that digested food is not absorbed. The insect larvae starve to death. Attempts are being made to isolate the toxins and to stabilize them by protein engineering. The gene responsible for the production of toxin is taken and genetically engineered into plants, so that they get permanent protection. Caterpillars or other larvae eating their leaves would then die. This has been successfully achieved for some plants, such as maize. European corn borer attacks maize. The insect larva tunnels into the plant from eggs laid on the undersides of leaves. The toxicity of protein is limited to the insect and this specificity is carried further with different bacterial strains producing proteins toxic to specific insects. For example, the toxin produced by *B.thuringiensis* varieties *berliner* and *kurstaki* are active against Lepidoptera, whereas that produced by *israelensis* is active against Diptera. The toxin has been used for some time as a biological insecticide and although a great deal of work has been carried out on isolating the toxin genes, comparing the differing domains of the protein is necessary for activity, the exact mechanism of toxicity is still not known.

In the case of the *berliner* variety, the protein comprises 1115 amino acids, but the amino terminal region located between amino acids 29 and 607 is essential for activity. In fusing this to a kanamycin-resistant gene encoding NPT II and in turn linking this to a constitutive promoter and transferring this to a novel host plant, individuals resistant to kanamycin as well as insect predilection were formed.

There are more than 500 different strains of Bt, each of which is specifically toxic to a very limited range of insect species and does no harm to animals, birds, fish and beneficial insects. In fact various strains of Bt have been marketed for many years in powder form as highly effective bioinsecticides. The gene encoding the insecticidal protein from specific strains of Bt has been isolated and transferred to a range of crop plants including potato, corn and cotton. Insect pests such as Colorado Beetle and European Corn Borer, which attack the potato and corn respectively, are rapidly killed when they start eating the plants due to the presence very low level of Bt insecticidal protein produced by the plant. The first commercial use of potato and corn protected with the *Bt* gene will occur in the near future and, subject to legislation, will be introduced into Europe within the next few years.

Insecticidal proteins have been identified in plants as diverse as Snowdrop and Cowpea and transgenic plants expressing their proteins are being tested in both the laboratory and field. If a combination of such insecticidal proteins can be introduced into plants, it should be possible to produce crops with durable resistance to major insect pests.

The advantages of insect resistant plants include, season long protection, which is independent of the weather, only crop-eating insects are exposed, the insecticidal material is confined to plant tissues and the active factor is biodegradable and non-toxic to man and animals. The final bonus being a reduction in pesticide use.

A similar transgenic approach is being used to engineer control of a wide range of fungal pathogens. This involves the identification of antifungal proteins and/or the isolation of natural resistance genes which are then transferred into crop species and provide control against target organisms.

In this case the assay involved cutting leaf discs from the transgenic plants and feeding these to young larvae of *Manduca sexta* (tobacco hook worm) with insect mortality and weight loss being scored. In this case, discs from the plants resistant to kanamycin showed 75-100 % mortality of the larvae. Though such results were encouraging, often when using the native toxin gene, it was observed that relatively low levels of protein products were seen. This is a consequence of the differing codon usage between bacteria and plants. Thus, many toxin genes have been 're-engineered' to conform to plant codon usage and this has boosted expression up to 500 fold. Using this approach, potato varieties can be protected against Colorado beetle and corn against the European corn borer.

Now the bacteria may not be used. Some polypeptides that inhibit proteinases in the insect gut are produced by some plants, particularly those of the legume family such as peas and beans.

This reduces the ability of insects to digest protein and prevents or slows down their growth. The genes responsible for this character have been transferred to some other crops that do not have them. This has been successful in giving resistance to seeds against some beetle larvae that feed on them.

8.3.4 Viral Resistance

Plant viruses are serious pests of crop. To minimize crop losses resulting from viral infection, several approaches are available. These include breeding for resistance, propagation of virus-free material, chemical control of viral vectors such as insects or changing agronomic practices such as rotating sensitive host crops with non-host crops or changing planting times. The effects of these measures are, however, limited and there are few effective controls of viral infection so the losses that result, can be substantial.

Initially the concept of cross-protection raised the interest in engineering viral resistance in plants. Viral cross-protection is based on the observation that if a plant is infected with a virus producing mild symptoms those (the symptoms) produced by a related, more virulent virus are reduced when compared with plants infected with the second virus alone. Although the viral coat protein has been implicated in the effect, the precise mechanism of cross-protection is not known.

Coat-protein-mediated protection in transgenic plants was first demonstrated by linking TMV coat protein gene to the 35S RNA promoter and transferring to tobacco. When inoculated with TMV, these plants displayed delayed or no TMV symptoms. It was subsequently shown that the inoculated plants displayed reduced number of lesions as well as reduced levels of virus accumulation. Protection has been demonstrated in transgenic plants using coat protein genes from a variety of differing viruses.

Similar experiments have been performed with potato, tomato and alfa alfa to protect them from virus. The coat protein of soybean mosaic virus, which does not infect tobacco, confers resistance to two unrelated viruses, potato mosaic virus (PMV) and tobacco etch virus (TEV). By engineering plants to express coat proteins from differing viruses, broad resistance can be achieved.

8.3.5 Abiotic Stress Tolerance

The average crop yields do not achieve their full potential. It is because of the fact that their environmental growth conditions are not optimal. They are in fact stressed in some way. The major stresses in the field are :

    (i)   water,

    (ii)   high salt,

    (iii)   high or low temperatures

It is possible to select plants that are more stress-tolerant from populations and to subject individual plants to treatments that can increase stress resistance. It indicates that there are genetic aspects of stress which are amenable to manipulation. The protein synthesis is changed in response to high temperature. This provides information on environmentally regulated gene expression. When soybean seedlings are transferred to 45°C for 2 hours, they die, but survive if they are pretreated for 3 hours at 40°C before being transferred to 45°C. The 40°C treatment induces mRNA that codes for "heat shock" proteins and their presence correlates with heat-stressed plants in the field. Heat-shock genes have been cloned and it has been observed that several multigenic families are involved that are similar in a variety of crop plants.

70% of water of the globe is consumed by irrigation and in this way certain non-renewable resources of water are continuously depleting. Attempts are on to produce transgenic plants with decreased need of water.

Similarly, on the addition of elicitor (cell wall fractions of fungus or UV light treatment of Parsley cell suspensions), proteins involved in defense mechanisms in plants are induced. The proteins are involved in the biosynthesis of related furanoumarins (antimicrobial, phytoalexin-like) and flavonoids that may protect plants from excess UV irradiation. Genetic engineering techniques have lead to an understanding of such stress-related responses that can be beneficially applied.

Toxicity of soil by metals particularly aluminium is affecting its ability to sustain plant life. It has been reported that more than one third of the world's soil suffers from toxicity of alcohol. In the humid tropical climates of many developing countries, the problem is more serious. According to Kreuzer and Maassay (2004) aluminium toxicity is also a consequence of soil acidification because of acid deposition. The plant root cells are injured by aluminium ions. They, thus, interfere with growth of root and uptake of nutrients. Attempts have been made to make crop plants more resistant to aluminium. Corn, rice and papaya have been transformed with a bacterial gene for the enzyme citrate synthase by a maxican group. These plants release citric acid which binds to soil aluminium and prevents it from entering the plant roots. It has been found that genetically modified plants were able to germinate and develop at aluminium concentration that are toxic to non transgenic plants.

8.3.6 Hormone Engineering

Plant growth regulators regulate the physiological properties of many crop plants. It is now possible to manipulate these regulators internally in dicotyledonous crops by altering expression of oncogenic loci of T-DNA. This approach has been used in potato because physiological properties such as tuber induction and photoperiodic response can be readily modified by hormones. Potato plants transformed by *Agrobacterium* have already been regenerated. In this case, the T-DNA of a shoot-inducing strain of *A. tumefaciens* apparently altered internal hormones so that the transformed potato plants readily produced aerial tubers. Carrot plants transformed by *A. rhizogenes* have also been regenerated.

8.3.7 Genetic Manipulation of Flower Pigmentation

Certain plant species have always been appreciated for their ornamental characteristics. The most important near term application of genetic engineering to the ornamental plant industry is likely to be in the area of flower colour. Longer lasting varieties will soon be bred. The large scale production of ornamental plants will also benefit from disease and herbicide resistance genes that are already being tested in field crops. Today flower breeding has become widespread industry and consequently there is a wide scope for improvement. Several enterprises have started molecular flower-breeding programmes, which use recombinant DNA techniques.

The colour of flowers is due to their capabilities of biosynthesizing the pigments. Flowers have two main types of pigments :

- **Flavonoids** – These contribute to a range of colours from yellow to red to blue. These flavonoid molecules are anthocyanins which are glycosylated derivatives of cyanidin (red), pelargonidin (brick red), delphinidin (blue), petunidin and malvidin. They are localized in the vacuole.
- **Carotenoids** – These are commonly the pigments for yellow to orange flowers.

The side chain substitutions of different chemical structures determine the colour of flower. The cyanidin derivatives produce more red and the delphinidin derivatives produce more blue. An active research area in the flower industry now is to develop blue varieties of major cut flower species like rose, chrysanthemum, carnation and gerbera by biotechnology.

Flower colour is also influenced by co-pigmentation with colourless flavonoids, metal complexation, glycosylation, acetylation, methylation and vacuolar pH.

It is being attempted to improve flower appearance and post harvest lifetime. It has been possible by traditional breeding techniques over the years to create thousands of new varieties that differ from one another in colour, shape and plant architecture. However, it is a slow and painstaking procedure. It is limited by the gene pool of a particular species. The rose is the world's most popular flower. However, despite sustained efforts made by rose breeders, amateurs and professionals, the rose displays so many beautiful colours except blue.

The biosynthetic pathway of the flavonoid pigments is quite well known. One of the key enzymes involved in this pathway is flavonoid 3', 5'-hydroxylase. Unfortunately the rose is deficient of this key enzyme and incapable to synthesize the major blue pigment delphinidin. The genetic engineering has made it possible to even produce blue rose. The genes which may lead to the synthesis of blue pigment in flowers are called "blue genes". The natural gene pool limitation can be overcome by isolating the "blue gene" from other beautiful blue flowers and put it into rose to help the biosynthesis of blue pigment. The genes of flavonoid 3'-5'-hydroxylase have been identified and cloned from a number of plants. For example one blue gene from the *Petunia* is used to make a blue rose.

By manipulating the genes for enzymes in the anthocyanin biosynthesis pathway, uniquely coloured flowers can be developed. The most common type of flower pigments, anthocyanins are synthesized from the amino acid phenylalanine by a series of enzyme-catalyzed reactions.

8.3.8 Nitrogen Fixation

Nitrogen fixation is the process by which atmospheric nitrogen gas is reduced to ammonia within cells so that it can be used to make proteins and other organic compounds. The process can be carried out only by certain bacteria. Some nitrogen-fixing species live in the root nodules of plants, particularly peas, beans and alfalfa which are legumes. These benefit the plants, but artificial nitrogen fertilizers have to be added to most crops, if the soil is not to become deficient in nitrogen, especially where the same crop is grown year after year in the same soil as is often the case with cereals. If plants contain their own nitrogen-fixing genes, enormous savings can be made in the time, money and energy used in making, transporting and spraying the fertilizers. Strategies have been developed to improve the nitrogen content of plants grown on poor soils by genetic engineering of nitrogen fixation. This can be done by increasing the host range of symbiotic nitrogen fixing bacteria and improving the efficiency of nitrogen fixation.

Genetic engineering of nitrogen fixation is made difficult by the fact that it is a complex process involving many enzymes. About 17 genes are involved (the *Nif* genes) (Fig. 8.3).

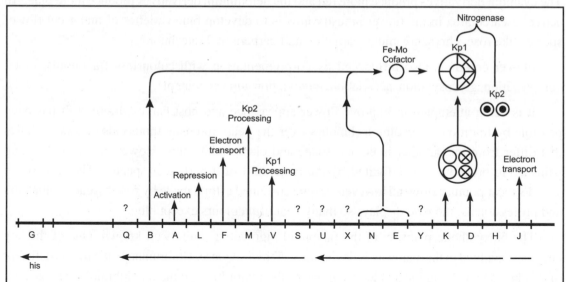

Fig. 8.3. The *nif* gene cluster of *Klebsiella pneumoniae*. The seventeen *nif* genes are arranged in eight transcriptional units, The dimensions and orientations of these genes are indicated by the arrow. The roles of the individual *nif* genes where known definitely, are shown. *nifH* specifies the polypeptide of nitrogenase reductase (Kp2) and *nifD* and *nifK* specify the two polypeptides of nitrogenase (Kp1). *nifQ* is probably to be involved in the uptake of molybdenum and *nifS* and *nifU* may be involved in the processing of nitrogenase. The other genes of the complex. Fe–Mo cofactor, iron-molybdenum cofactor are regulated by *nifA* and *nifL*. (From Ray Dixon, Unit of Nitrogen Fixation, University of Sussex.)

It seems that 85 to 90% of the plant kingdom, includng virtually all of our crops, already has at least some of the genetic machinery they need in order to establish relationships with nitrogen fixing bacteria. The goal of making all crops capable of forming symbiotic relationships with nitrogen fixers may not be as easy as was once thought. Moreover, part of the process is anaerobic, requiring a means of excluding oxygen. A great deal of work has been carried out in this regard.

8.3.9 Increasing Shelf Life

Soft fruits are artificially ripened *e.g.,* tomatoes, bananas and red peppers are normally picked green and are ripened using ethane gas in warehouses. This means when picked, they are still hard. It reduces bruising and the fruit is to be picked mechanically and tipped into containers. It also allows controlled ripening so that the fruit will have maximum appeal to the customer. However, due to biochemical changes during transport and shipping, much of the flavour of the fruit is lost. A company Calgene in the USA and the ICI Seeds in the UK, have produced a genetically engineered tomato in which the ripening process is slowed down by **gene silencing**. The fruit can, therefore, be left on the plant for longer, giving both increased yields and a fuller development of flavour. There is, thus, double advantage for the farmer and customer. The tomatoes first went on sale in the USA in 1995 as 'Flavr Savr' tomatoes. They are expensive, but do taste better. They were first introduced in the UK by Sainsbury in 1996, initially in tomato paste.

8.3.10 Plants as Bioreactors : Production of Chemical Drugs and Other Products

The new techniques make it possible to use crop plants as manufacturing plants to produce new medicines. Even though humans have been using wild plants for centuries, they were not food plants. The pharmaceutical plants of the future will include food crops that are capable of synthesizing medicines. Because they are not intended for food consumption, they will be grown by a few specially trained and tightly supervised growers and harvested, transported and processed under strict confinements.

Use of Plant Cell Cultures in Production

- Plant cell in culture are used for the production of various products of use to mass? The cells are upscaled using various fermenters (Fig. 8.4 and 8.5).
- The most common types of system on the bench is a stirred-jar fermentor which is mainly used for microbial cultivation although some minor alteration is made while upstreaming.
- Since the cultured plant cells are more fragile than bacterial cells and also specific to each of the cell line, the system for their culture should be comparatively soft.
- A company in Germany, DIVERSA, has equipped 5 sophisticated fermentors of up to 75,000 L for plant cell cultures. Although detailed modifications to those fermentors has not been disclosed, the photos show that they are similar to ordinary microbial fermentors. The company has cultivated *Echinacea purpurea* cells for the manufacture of immuno-biologically active polysaccharides.

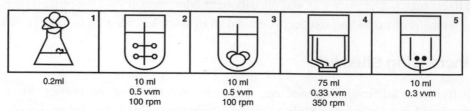

1	2	3	4	5
0.2ml	10 ml 0.5 vvm 100 rpm	10 ml 0.5 vvm 100 rpm	75 ml 0.33 vvm 350 rpm	10 ml 0.3 vvm

Fig. 8.4 Various fermenters used for upscaling the plant cells in culture (Source : Internet)

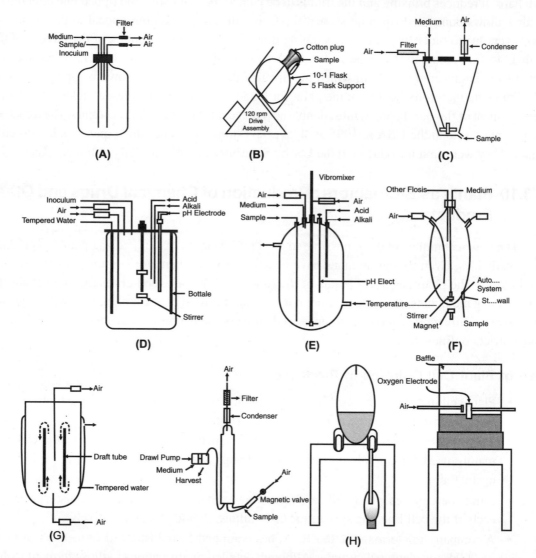

Fig. 8.5 Various fermenters used for upscaling plant cells in culture (A) Carboxy system
(B) Roller Flask system **(C)** V-shape reactor **(D)** Stirrer-jar fermenter **(E)** Vibromixer fermenter
(F) Continuous fermenter of Miller **(G)** Airlift fermenter **(H)** Continuous fermenter of Kurz (Adopted
from 20animal%20and%20plant%20cells%20culture.pdf)

Using these systems attempts are being made to bring quantitative and qualitative changes in plant oils, carbohydrates and proteins for food use. However, because plants are capable of converting carbon dioxide to simple sugars via photosynthesis, they can exploit the ultimate renewable energy source-, the sun - and act as biorefineries for the production of a wide array of non-food materials.

Millions of tons of organic chemicals and polymers are produced annually from crop plants. In addition, plants produce of the order of 100,000 different natural products - some of which are already exploited by the pharmaceutical industry. An increase in our understanding of the complex biosynthetic pathways involved in the synthesis of these compounds coupled with the development of transgenic systems for the production of commercially important compounds raises the potential for metabolic engineering. This will allow the exploitation of plants as biorefineries to produce carbohydrate, lipid and protein based polymers for non-food use and as feedstock for the chemical industry.

The recognition that plants collectively have the capacity to synthesize a very wide spectrum of useful compounds of high value also means the potential exists to produce transgenic plants capable of synthesizing pharmaceutical proteins, enzymes for animal feed and as a source of industrial bulk enzymes.

The non-food products which are likely to be produced by genetic engineering of plants are industrial oils and fatty acids and biodegradable plastics. In California, Calgene have introduced a single gene from the California bay tree into oil seed rape plants which then synthesize an altered pattern of fatty acids. In effect the transgenic plants convert more of their carbon into a shorter, 12 carbon length fatty acid - lauric acid which is used to make soaps and detergents and less into the normal 18- carbon length compounds. In a slightly more complex manipulation, Chris Sommerville from the Carnegie Institute in Stanford, California, has introduced two genes which encode enzymes responsible for synthesizing the naturally occurring biodegradable plastic, polyhydroxybutyrate (PHB), from a bacterium *Alcaligenes eutrophus,* into plants. When the enzyme products from these genes are targeted to plastids, significant yields of PHBs are obtained. PHBs are 100% biodegradable and share many of the characteristics of polyethylene. A closely related biodegradable plastic, a polyhydroxyalkanoate is already produced commercially by bacterial fermentation and is called Biopol™. It has a wide range of uses including packaging containers, as a biodegradable carrier for long-term dosage of drugs, herbicides and insecticides, for making surgical pins, sutures and swabs as well as a wide range of disposable consumer products.

Plants are being genetically manipulated to produce increased levels of anti-oxidants (plant carotenoids) and proven anticancer compounds as well as reduced levels of potential carcinogens. Plants also provide a potential alternative to fermentation-based production systems for the expression of foreign proteins with pharmaceutical or industrial value.

Plants are being used to produce drugs and vaccines. Pharmacological products are also produced in genetically engineered plants. About 300 trials of crops genetically engineered to produce various therapeutic products have been initiated. These include:

(i)   modified tobacco plants that produce Interleukin-10 for the treatment of Crohn's disease,

(ii) GM potatoes that produce antibodies for reducing the risk of rejection in kidney transplants,

(iii) GM tobacco that produces vaccines against hepatitis B and drugs against HIV/AIDS, potatoes that produce human insulin,

(iv) the human enkephalin gene has been expressed in plants. Soybeans with a higher expression of anticancer proteins (normally found in soybean but in low quantity) have been produced.

Other GM plant-produced substances include enkephalins, alpha-interferon, serum albumin and glucocerebrosidase. Clinical trials have begun on crop-grown drugs to treat cystic fibrosis, non-Hodgkin's lymphoma and hepatitis B (*The Economist*, 2003).

Technology is being developed for the production and oral delivery of inexpensive vaccines, especially for use in the developing world. Major targets include vaccines effective against bacteria and viruses that cause diarrhoeal diseases and hepatitis B. Charles Arntzen and his colleagues have transformed potato with the gene encoding the B subunit of the *E. coli,* heat labile enterotoxin which causes diarrhoea, particularly in children in developing countries. When potatoes expressing this enterotoxin were fed to mice, both serum and secretary antibodies were induced. These antibodies were protective in bacterial toxin assays *in vitro* and, thus, provided the first proof of concept for edible vaccines.

Recognizing that vaccine delivery in raw potato is not practical, Arntzen has developed a genetic transformation system for banana which is grown in most developing countries and fed uncooked to children and adults. This opens the way for the production of 'edible vaccines' which, if shown to be orally active, would provide an economically viable production and delivery system. Edible vaccine production in crop plants may also have value as a supplement to animal feed and, thus, improvement of animal health.

Transgenic tomatoes containing a gene from *Escherichia coli* that can protect against diarrhoeal diseases have been produced (Lemonick, 2003).

By expressing industrial enzymes in transgenic seeds using plant genetic engineering techniques, novel products can be created. The seed formulated enzyme provides instant packing, resulting in stable, safe and convenient handling and storage. The transgenic seeds can be applied without purification of the enzyme, in the industrial process where the activity is required and, thus, would be most compatible where seeds are presently used, as for example in the production of feed, food, beer and raw material processing industries.

The Dutch company PlantZyme have, for example, engineered the gene encoding Phytase from the common fungus *Aspergillus niger* into tobacco and oil seed rape seed. This enzyme catalyzes the conversion of phytate, the principal storage form of phosphorous in plant seeds into inorganic phosphate and myo-inositol. When plant seeds are compounded in animal feed, the inorganic phosphates stored as phytate cannot be utilized by monogastric animals such as chicken and pigs and inorganic phosphate has to be added to meet the animals requirements. This leads to an increase in excretion of phosphate in manure of pigs and poultry which in regions of intensive livestock farming causes environmental pollution and eutrophication of surface waters. The inclusion of phytase containing seeds in the feed optimizes phosphate utilization from the stored phytate, removes the need to add phosphate to fodder and significantly reduces the excretion of phosphate.

The various firms and research centres have shown the effectiveness of plants as bioreactors. These include SemBioSys Genetics, Inc., Canada, Plant Biotechnology, Inc., United States. The ProdiGene, Inc. United States, have demonstrated that mice fed on potatoes expressing the beta-subunit of cholera toxin (CTB) were resistant to the cholera toxin (Langridge, 2000). Greater protection against cholera to humans than to mice is provided by CTB.

Biopharming is mainly driven by a cost advantage. For example, medicinal products could be synthesized in plants at less than one tenth of the cost of conventionally manufactured drugs and vaccines. By the end of the current decade, biopharmaceuticals are projected to grow into a $20 billion industry. By this the cost of treating some diseases may be brought down (Roosevelt, 2003).

Pharmagedden crop products carrying drugs, vaccines and industrial chemicals will be used on the dinner tables. But there are many fears, which will be discussed in the last chapter. The production of transgenic animals and plants would require relatively small investments and minimal cost of maintenance. It is because of this they may provide the economically viable option for independent production of therapeutic proteins in developing countries.

The technology to create transgenic plants has formed the basis for a new 'green revolution'. There are opportunities in all areas of crop protection and crop improvement in both the developed and developing countries. In addition, the potential for metabolic engineering should allow the exploitation of plants as biorefineries to produce high-value pharmaceuticals, carbohydrates and lipid based biopolymers for non-food use and as feedstock for the chemical industry. If this potential can be realized, it will go some way towards reducing our dependence upon fossil fuels such as petroleum and natural gas and would have the added advantage of depending upon a renewable energy source, the sun, and involve the use of more environmentally benign materials which result in the production of less toxic waste.

Plant biotechnology may provide solutions to some of the problems outlined in the introduction. However, as in the past with other new technologies, scientific advances have moved ahead of society awareness and acceptance of the technology. It is likely that community confidence will, for a while, be the limiting factor in the adoption of these technologies into our agriculture and food production industries. However, if we are to satisfy the environmental concerns associated with modern high input agriculture and feed the increasing world population, it seems that gene technology has many advantages and will be accepted.

Intended Changes for Future Transgenic Crops

The intended changes may enhance safety and improved nutritional value. Allegenic modification techniques *i.e.,* recombinant techniques, mutagenesis *etc.* can be directed to these dual objectives.

8.3.11 Improvement of Food Quality

Attempts are being made to improve the bread-making quality of the high-yielding British wheats. Improving the quality of protein will improve the flour quality. The nutritional quality of plant foods is increased by increase in the proportion of essential amino acids. Many legumes are deficient in sulfur-containing amino acids. Biotechnology derived cooking oils have already been introduced in the market. In fact concentration of saturated fatty acids has been decreased in certain vegetable oils by genetic engineering. However, the conversion of linoleic acid to alpha linoleic acid in these oils has been increased. Not only does this conversion decreases the saturated fatty acid concentration, it also increases the type of oil, found mainly in fish, that is associated with lowering cholesterol levels.

Another nutritional concern related to edible oils is the negative health effects produced when vegetable oils are hydrogenated to increase their heat stability for cooking or to solidify oils used in making marginin. The soybean oil has been given these same properties, not through hydrogenation, but by using genetic engineering to increase the amount of the naturally occurring fatty acid stearic acid.

Attempts are being made to have nutritionally beneficial products such as grains with a complete protein profile; fruits, vegetables, and grains with improved vitamin and mineral content. Potatoes are being engineered with higher solid content, so that when fried, less oil is absorbed by them.

The health benefits of functional foods are also being improved. Functional foods are the ones that contain significant levels of biologically active components. These impart health benefits or desirable physiological effects beyond our basic need for sufficient calories, the essential amino acids, vitamins and minerals. The example of functional foods include :

(i) compounds in garlic and onions. These lower cholesterol and improve the immune response,

(ii) anti-oxidants found in green tea,

(iii) the glucosinolates in broccoli and cabbage that stimulate anticancer enzymes.

Attempts are on to increase the production of these types of compounds.

An important concern due to the unique capabilities provided by recombinant DNA technology is related to food allergenicity. People allergic to certain foods become aware of their allergies early and avoid those foods. Now crop developers can transfer genes from allergenic foods (*e.g.,* Brazil nuts) to non allergenic food crops. This could pose major health risk to people with allergies to gene donor unless the crop breeders can prove that the transferred gene does not encode the donor's allergenic protein.

8.3.12 Biological Methods to Protect Crops

It has been discovered that like animals, plants have endogenous defense systems :

(i) The hypersensitive response (HR) and

(ii) Systemic acquired resistance (SAR)

Attempts are being made to search for chemicals that can be used to trigger these two means of defense so that plants can better protect themselves against insects and diseases.

Each of these two mechanisms is equipped with a diverse arsenal of molecular weapons for protection against disease and herbivory. The hypersensitive response (HR), is a localized response to infection. Plants respond to molecules from the infecting pathogen through ROS, which in turn trigger at least three biochemical responses.

(i) The ROS cause cross-linking of molecules in the plant cell wall, which reinforces plants physical protection against pathogens.

(ii) The programmed cell death cascade is initiated in the infected plant cell. When the infected cell commits suicide, the pathogen is unable to survive and reproduce thwarting the pathogen's spread to other cells.

(iii) ROS act as warning signals to cells surrounding the dying cell, which stimulate the activation of a second defense system, the systemic acquired response (SAR). In the SAR, uninfected cells begin to biochemically equip themselves for the coming onslaught.

8.4 DANGERS OF GENETIC ENGINEERING OF THE PLANTS

It is important to assess the environmental hazards related to release of transgenic resistant plants. It is necessary that appropriate, containment measures are taken during field trials *e.g.,* in transgenic virus resistant sugar beet, containment measures include preventing the flowering of plants to avoid the escape of transgenic pollen and avoiding resistance tests against the whole virus to guard against the possibility of the selection of strains with increased virulence.

Side effects : We still do not understand living systems completely enough to perform DNA surgery without creating mutations and these could be harmful to the environment and our health. We are experimenting with very delicate, yet powerful forces of nature, without full knowledge of the repercussions.

Widespread failure of crop : We are trying to make profit by patenting genetically engineered seeds. This means that, when a farmer plants genetically engineered seeds, all the seeds will have identical genetic structure. As a result, if a fungus, a virus or a pest develops which can attack this particular crop, there could be widespread crop failure.

Absence of long-term safety testing : Genetic engineering uses materials from organisms that have never been part of the human food supply. To change the fundamental nature of the food we eat, without long-term testing, we do not know the safety of the foods.

Toxins : Unexpected mutations can be caused by genetic engineering in an organism. It can create new and higher levels of toxins in foods.

Allergic reactions : Genetic engineering can also produce unforeseen and unknown allergens in foods.

Antibiotic resistant bacteria : Antibiotic-resistance genes are used to mark genetically engineered cells. This means that genetically engineered crops contain genes, which confer resistance to antibiotics. These genes may be picked up by bacteria, which may infect us.

Untracable Problems : Without labels, our public health agencies are unable to trace problems of any kind, back to their source. The potential for tragedy is staggering.

Increased use of herbicides : It is estimated that genetically engineered herbicide-resistant plants will greatly increase the amount of use of herbicides. By knowing that their crops can tolerate the herbicides, farmers will use them more liberally.

More pesticides : GE crops often manufacture their own pesticides and may be classified as pesticides by the EPA. This strategy will put more pesticides into our food and fields than ever before.

Damage to ecology : The influence of a genetically engineered organism on the food chain may damage the local ecology. The new organism may compete successfully with wild relatives, causing unforeseen changes in the environment.

Recombinant DNA technology makes it possible to introduce a piece of DNA consisting of either single or multiple genes that can be defined in function and even in nucleotide sequence. With classical techniques of gene transfer, a variable number of genes can be transferred, but prediction of precise number of the traits that have been transferred is difficult and we cannot always predict the phenotypic expression that will result.

It is, therefore, must that the sincere cautions must be taken, while constructing and releasing transgenics to field.

Gene flow from crops to other plants

The use of transgenic crops has triggered questions about the possibility that the transgene will move to other, non transgenic plants, both wild and crop plants, via pollination. Gene flow between plants can occur with any croop. There is abundant evidence that it occurred with other crops long before modern biotechnology appeared on the scene.

Gene flow depends on cross pollination. Whether the unintended recipient of the gene is a wild plant or non-transgenic crop, the probability of gene flow from a transgenic crop depends first and foremost on the potential for cross-pollination between the transgene crop (pollen donor) and the other plant. Pollen transfer and successful fertilization are necessary, but not sufficient for gene flow to occur, because there are many postfertilization barriers to reproduction if the two plants do not have a sufficient amount of genetic similarily. Cross-pollination between two different plants that produces fertile, viable offspring is known as **hybridization.** To hybridize with each other (in the absence of human intervention), plants must be very closely related to each other.

(i) *Corn and the teosintes:* Gene flow from corn to some of the teosintes can occur. Cross-pollination can lead to successful hybridization, but the success rate varies according to the direction of hybridization. A number of studies have shown that genes are more likely to move from teosinte to corn than in opposite direction.

(ii) *Soybeans and wild soy relatives.*

8.5 ANIMALS BIOTECHNOLOGY

Animal cells in culture and transjenic animal have great potential in agriculture.

The first reports of the application of modern biotechnology to animals appeared in the 1980s. Genetic engineers inserted novel genes into mice, rats, pigs, and fish to achieve faster growth rates, improved resistance to disease, and other effects. Although some of these traits were attainable through traditional breeding methods, genetic engineering can produce greater (or more dramatic) effects while expanding the range of potential traits. In 1983, the cover of *Science*, one of the most widely read scientific journals in the United States, sported a photo of a huge mouse bearing novel genes that accelerated its growth rate (Palmiter *et al.,* 1983). Shortly after, scientists in China reported the first successful insertion of novel growth hormone genes into fish. These events stirred substantial debate and captured the interest of biochemists, geneticists, aquaculture scientists, and private entrepreneurs, leading to more transgenic research in laboratories around the world, some of it focusing on fish and other aquatic organisms.

8.5.1. Use of Animal Cells

Animal cells in culture are used for following purposes.

- A resource in biotechnological processes.
- Increasingly used in molecular biology developments, fermentation technology, the production of diverse **health care products**, and **cell-based screening systems** in toxicology and pharmacology.
- Large-scale technology began with the use of primary monkey kidney cells for the production of poliomyelitis vaccines in the 1950s.
- The next step was taken in 1964 with the commercial production of foot-and-mouth disease virus (FMDV) from baby hamster kidney (BHK 21) cells for veterinary purposes.
- Interferon was the first licensed product derived from Namalwa cells, a human lymphoblastoid heteroploid cell line, in the late1970s. Cell are cultured in various ways (See Chapter 3). They can be grown in suspension or as monolayer (Fig. 8.6). These cells can be upscaled for commercial use.

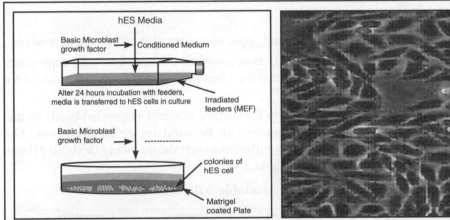

Fig.8.6 Animal cell in culture : Human Embryonic stem cells in culture (Adopted from 20 Animal%20and%20plant%20cells%20culture.pdf)

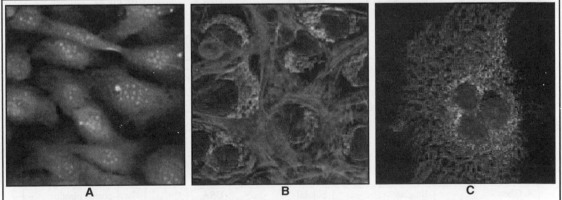

Fig.8.7 Cell Culture (cancer cell) (A) Human bone osteosarcoma Cells, **(B)** Bone cancer (U-2 OS Line) **(C)** Human cervical adenocarcinoma cells (HeLa Line) (From Internet)

Upscaling of Animal Cells in Culture

- Except for the production of activated or inactivated viral vaccines, virtually all mammalian cell culture processes now employ continuous cell lines, Most of these have been adapted to grow in suspension, They do not need a surface support.
- Two types of ideal reactors are used for this purpose :
  - (a) Well-mixed stirred tank and
  - (b) Plug-flow (tubular) reactor

Bioreactor for production by animal cells

A Stirred tank reactor

- In this the substrate's added is through feeding instantaneously. It is distributed throughout the entire reactor.
- The gas sparging may be employed with the agitator that provides intimately mixed gas and liquid

It has the following characteristics:

- Cell Support for Mixing Vessels.
- Many cell types used in tissue engineering applications are also anchorage dependent.
- The cells can be cultivated in a relatively homogeneous environment with appropriate environmental control. The can be infected for virus production, or can be induced for product formation.
  - Microcarriers allow cells to attach to the surface of small suspended beads so that conventional stirred tank bioreactors can be used for cell cultivation. The microcarriers can be made of many different materials, including dextran, gelatin, polystyrene, glass, and cellulose (Table 8.1).

Table 8.1: Commercially Available Microcarriers

Type	Trade Name	Company ?	Country	Composition
1. Dextran	Cytodex-1	Pharmacia ?	Sweden	DEAE-dextran
	Cytodex -2	Pharmacia ?	Sweden	Quaternary amine Coated dextran
	Supperbeads ?	Flow Labs	USA ?	DEAE – dextran
	Microdex ?	Dextran Products	Canada	DEAE dextran
	Domacell	Pfeifer Langer	Germany	DEAE dimmers-dextran
2. Plastic ?	Biosilon	Nunc ?	Denmark	Polystyrene-charged
	Biocarriers	Biorad	USA	Polyacrylamide/ DMAP
	Cytospheres	Lux	USA	Polystyrene charge
3. Gelation	Cytodex-3	Pharmacia	Sweden	Gelatin-coated dextran
	Gelibeads	KC Biologicals? Hazelton Labs	USA ?	Gelatin
4. Glass	Bioglas	Solohil Engs	USA	Glass-coated plastic
	Ventreglas	Ventrca	USA	Glass-coated plastic
5. Cellulose	DE-52/53	Whatman	UK	DEAE-Cellulose

- There is a variant of microcarriers. These have large pores of tens of micrometers in they interior. Their aggregate culture which is promoted by calcium
- Calcium alginate is used for the cell entrapment

(B) The Tubular Reactor (These include hollow fibre and ceramic systems)

- In this the liquid phase moves as a plug flow. There is thus no variation of axial velocity over the cross section.
- It is more difficult to scale up than mixing vessels, because the concentration gradient of essential nutrients, oxygen in particular, inevitably becomes limiting in the downstream region of the reactor. (Fig. 8.8)

Fig. 8.8 Bioreactor used in animal cell culture (Internet)
(Adopted from 20Animal%20and%20plant%20cells%20culture.pdf)

The Membrane material used in the hollow fiber reactor are given in the Table 8.2.

Table 8.2: Membrane Materials Used in Hollow Fiber Reactor and Kidney Dialysis

Membrane material	Cell type
Cellulose nitrate or	
Cellulose acetate	Liver cells
Cellulose acetate	Hybridomia
	Hybridoma
Polysullone	Chinese hamster lung fibroblast
	MDCK (Madin-Darby Canine Kidney)
	Hepatocytes
Polymethyl-methacrylate	Kidney
Cuprophane or	
polyacrylonitrile	Kidney
Acrylic copolymer	SVT 3 cells

(Source Internet)

Operation Modes

- Mammalian cell cultures are operated in batch, fed-batch, and continuous modes, but the batch culture is used most frequently. During the culture the oxygen is provided continuously in batch culture through surface aeration membrane creation, or direct gas sparging. The typical densities achieved are rather low, about $2-4 \times 106$ cells/mL by using a conventional medium.

Biotechnology is providing new ways to improve the health of animals and increase their productivity. The advances in diagnosis and therapeusis are helping to improve the health of animals.

Diagnostics : In order to diagnose the diseases of animals, various monoclonal antibodies and probes have been developed. In fact many diagnostic kits have been developed. The accurate and earlier diagnosis help in choosing appropriate therapy. All these efforts are to help in reducing the spread of diseases such as Brucellosis, Pseudorabies, scours, foot and mouth disease, trichinosis *etc*.

Therapeutics : To produce sufficient quantities of endogenous therapeutic proteins found in animals, genetic engineering techniques are being used.

To fight virus, interferons and interleukin-2 proteins are naturally made by cattle. The genes for these proteins have been cloned and expressed in bacteria so as to produce proteins of choice.

Prevention of disease : Attempts are also being made to prevent diseases of livestock by developing disease resistance in them. Some breeds are naturally resistant to some bacterial diseases such as mastitis and, thus, resistance has a genetic basis. If only one or a few genes are responsible for disease resistance, creating transgenic animals, resistant to bacterial diseases may be possible.

8.5.2 Transgenic Animals and Their Uses

Until recently, selective breeding was the only way to enhance the genetic features of domesticated animals. Many generations of selective matings are required to improve livestock and other domesticated animals genetically for traits such as milk yield, wool characteristic *etc*. Combination of mating and selection, although time consuming and costly, has been experimentally successful. However, once an effective genetic line has been established, it becomes difficult to introduce new genetic traits by selective breeding methods.

During the 1980s, with considerable efforts, the idea of genetically manipulating animals by introducing genes into fertilized eggs was converted into a reality. For this purpose, following generalized steps are followed :

- A cloned gene is injected into the nucleus of a fertilized egg.
- Inoculated fertilized eggs are implanted into a receptive female (because successful completion of mammalian embryonic development is not possible outside of a female).
- Some of the offspring derived from the implanted eggs carry the cloned gene in all of their cells.
- Animals with the cloned gene integrated in their germ line cells are bred to establish new genetic lines.

Gene transfer in animals (Transfection) can be carried out at cellular level and transfected cells may be used for a variety of purposes. These include :

(i)   production of chemicals and pharmaceutical drugs,

(ii) study of structure and function of genes,

(iii) production of transgenic animals for increased milk production, increased growth of livestock, large scale production of valuable proteins and for improvement of wool in sheep,

(iv) for studying fundamental problems of mammalian gene expression and development,

(v)   for establishing animal model systems for human diseases.

Transgenic animals help in the production of relatively large quantities of rare and expensive proteins for use in medicine. This process is sometimes referred to as 'pharming' of drugs.

The term **Pharming** was originally coined to convey the idea that milk from transgenic farm animals (Pharm) can be used as a source of authentic human protein, drugs or pharmaceuticals. These drugs are not produced in bacteria as they do not always have the necessary machinery to process the proteins *i.e.*, proteins have to be precisely folded or modified using mammalian cell machinery *e.g.*, factor IX protein after the production has to have a –COOH group added to some of its amino acids.

There are number of reasons why the mammary gland should be used in this way. Milk is a renewable, secreted body fluid that is produced in substantial quantities and can be frequently collected without harm to the animal. A novel drug protein that is confined to the mammary gland and secreted into milk should have no side effects on the normal physiological processes of the transgenic animals and should undergo post translational modifications, that at least, match closely those in humans. Finally, purification of protein from milk, which contains only a small number of different proteins, should be relatively straight forward.

The mammary glands are also now used to produce the proteins so that these can be harvested by milking the animals. For example AAT (α-1-antitrypsin) has been manufactured by PPL Pharmaceuticals in Edinburgh. AAT is a naturally occurring protein found in human blood. A genetic disease is caused by a mutant form of the gene that codes for it. As a result of uninhibited elastase activity, it leads to **emphysema.** Elastase is an enzyme produced by some white blood cells which destroy elastic fibres in the lungs as part of the normal turnover of elastic tissue. Its activity is normally regulated by AAT, which inhibits the enzyme (Smoking is also thought to inhibit AAT. This explains a link between smoking and emphysema). AAT is made in the liver and can be supplied by the usual means. Wild type (healthy human) gene for AAT has now been added to sheep and the mammary gland of the sheep is used to express the gene.

The gene has a high degree of expression. It is switched on most of the time. This results in nearly 50% of the milk protein being human AAT. The sheep cells carry out the correct modification of the protein.

8.5.2.1 Examples of the Transgenic Animals and their Applications

One of the first reports of transgenic animals published in December 1982, involved transfer of growth hormone (GH) gene (from rat) fused to the promoter for the mouse metallothionein 2(MT) gene. Since then, many animals including those in cattle, sheep, goats, pigs, rabbits, chickens and fish have been produced.

8.5.2.1.1 Transgenic Mice

The first transformation of mouse embryos with foreign DNA was reported by Jaemisch and Mintz (1974). In this original experiment, the SV40 DNA was incorporated into the somatic cells; however, germline transmission was not achieved. Subsequently, germline transmission of foreign

DNA was demonstrated by exposure of zona - free 4 to 8 cell stage mouse embryos to Moloney Leukemia retrovirus by Jaemisch (1976). In 1980, Gordon *et al.* reported the first transformation of mouse embryos by microinjection of recombinant DNA into the pronucleus of fertilized oocytes (Fig. 8.9).

Transgenic mice can be used for the following purposes :

(i)   As model systems for human diseases

(ii) As test cases to determine the worthiness of a proposed production scheme

Whole animal models stimulate both the onset and progress of a human disease. These provide a system for testing potential therapeutic agents. Mouse models for human genetic diseases such as Alzheimer's disease, arthritis, muscular dystrophy, tumourigenesis, hypertension, neurodegenerative disorders, endocrinological dysfunction, coronary disease, and many others have been developed.

Transgenic mice have also been used as a tool for the analysis of eukaryotic gene expression and function *e.g.*, tetracyline regulated gene expression in transgenic mice (Bluethmann *et al.*, 1996).

Transgenic mice have also been used as models for expression systems that are designed for secretion of the product of a transgene into milk. For example, large quantities of authentic cystic fibrosis transmembrane regulator (CFTR) protein is required to study its function and to formulate potential therapies for treating cystic fibrosis. The CF is caused as follows :

(i)   The primary effect of a faulty CF gene is an alteration in the function of a CFTR, which normally acts as a chloride channel.

(ii) There is an accumulation of mucus in the ducts of several organs, especially the lungs and pancreas as a consequence of the disruption of the proper flow of chloride ions in and out of cells.

(iii) This mucus becomes the site of bacterial infection.

(iv) As the bacteria die, their released DNA makes the mucus very thick.

(v)   This thickened mucus blocks the duct and, thus, prevents normal organ function. This exacerbates the effects of CF.

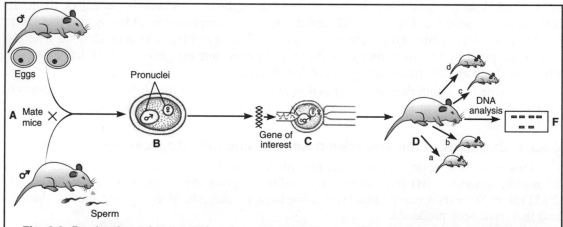

Fig. 8.9. Production of transgenic mice : Transgenic mice can be produced by microinjection method. **(A)** Male and female mice are mated **(B)** Fertilized egg having male and female pronuclei are retrievd **(C)** Foreign DNA is microinjected into pronucleus with gene of interest **(D)** Eggs are implanted into pseudo pregnant mouse **(E)** Offspring are produced **(F)** These are tested for the presence of transgene by PCR.

Transgenic mouse lines carrying the CFTR sequence under the control of beta-casein gene regulatory sequences have been established. Milk of transgenic females contain the CFTR protein bound to the membrane of fat globules. The CFTR protein is glycosylated and readily extracted from the fat-rich fraction of milk.

Protection against neonatal infections of the entire tract can be provided by the generation of transgenic mice producing virus neutralizing antibodies in milk. 18 lines of transgenic mice secreting a recombinant monoclonal antibody (Mab) neutralizing transmissible gastroenteritis coronavirus (TGEV) into the milk were generated to provide passive immunity.

The genes encoding a chimeric Mab with the variable modules of the murine TGEV-specific Mab 6A.C3 and the constant modules of a human IgG isotype Mabs were expressed under the control of regulatory sequences derived from the whey acidic protein, which is an abundant milk protein (Enjranes *et al.*, 1998).

Successes have been reported to have high level expression of recombinant human fibrinogen in the milk of transgenic mice. Fibrinogen is a complex plasma protein. Three expression cassettes, each containing the genomic sequence for one of the three human fibrinogen chains controlled by sheep whey protein β-lactoglobulin promoter sequences, were co-injected into fertile mouse eggs. More than 80 % of the transgenic founders contained all the three fibrinogen genes (Prunkard *et al.*, 1996).

Low lactose milk has been produced by creating a transgenic mice carrying a hybrid gene in which the intestinal lactase - phlorizin hydrolase cDNA was placed under the conrol of mammary - specific alpha-lactalbumin promoter (Freund *et al.*, 1999).

Study of G protein signalling *in vivo* for the purpose of transgenic mice has also been used. Conditional expression and signalling of a specifically designed Gi-coupled receptor has been studied in a transgenic mice (Conklin *et al.*, 1999).

Knockout Animals

For discovering the function (s) of genes for which mutant strains were not previously available, knockout mice are valuable tools. The following two generalizations have emerged by examining knockout mice :

(i) Knockout mice are often surprisingly unaffected by their deficiency. Many genes turn out not to be indispensable. The mouse genome appears to have sufficient redundancy to compensate for a single missing pair of alleles.

(ii) Most genes are pleiotropic. They are expressed in different tissues in different ways and at different times in development.

Transfection of human chromosome in mice : Experiments have been performed to find if fragments of human chromosomes 1,4 or 2 could be transferred into and propagated in mice. These chromosomes have been chosen because they harbour the genes necessary to produce functional heavy lambda, or kappa chains of human immunoglobulins, respectively. It has been found that DNA from all the three human chromosomes could be introduced into mouse chromosomes. The human chromosomes can be propagated without any fusion to the mouse genome. The tissue specific expression of these human genes is preserved in the mouse. Normal DNA rearrangements have been undergone by the genes to form functional immunoglobulin subunits with functions similar to those in human cells. We can now perform large, human DNA experiments in mice (Sikorski and Richard, 1997).

Expression of human growth hormone (hGH): Transgenic mice expressing hGH in their bladder epithelium, resulting in its secretion into the urine at 100-500 ng/ml have been reported. A 3.6 kb promoter of mouse uroplakin II gene was used (Wall *et al.,* 1998).

Site specific recombination systems using transgenic mice as powerful tools for introducing pre-determined modifications into eukaryotic genomes have also been produced. It is possible to allow manipulation of chromosomal DNA in a spatially and temporarily controlled manner in mice. These offer unprecedented possibilities for studying mammalian genome functions and for generating animal models for human diseases (Metzger and Feil, 1999).

8.5.2.2 Transgenic Livestock

For livestock in general, attempts are being made to create animals with inherited resistance to bacterial, viral and parasitic diseases. For these, lines of animals that are resistant to infectious agents providing inherited immunological protection have also been created by transgenesis.

Attempts are being made for the secretion of recombinant proteins into the milk of transgenic animals. Milk contains very high levels of proteins such as casein, β-lactoglobulin and whey acidic protein. Production of these proteins is tightly regulated by promoters. These limit gene expression only to cells of the mammary gland. Transgenic animals are expected to act as bioreactors that continually secrete high levels of a desired protein into their milk. The protein can be harvested simply by milking the animals, and then standard chromatographic procedure is used to purify it. If the mammary gland is to be used as a bioreactor, then dairy cattle, are likely candidates for transgenics.

Following steps are involved in development of transgenic cattle:

Collection of oocytes
↓
In vitro maturation
↓
In vitro fertilization
↓
Centrifugation of eggs
↓
DNA microinjection into male pronuclei
↓
In vitro embryonic development to blastocyst stage
↓
Embryo implantation
↓
Screening of offspring for transgene

The goals of transgenesis of dairy cattle include the change of the constituents of milk. The amount of cheese produced from milk is directly proportional to the k-casein content. Expression of a lactase transgene in the mammary gland could result in milk that is free of lactose. PPL have created 'Rosie'- the first transgenic cow producing functionally active human lactase in the milk (Chawla, 1999; Whitelaw, 1999). Reports of transgenic cow producing lactoferin and human α-lactalbumin are also there.

There has been controversy over the introduction of human growth hormone genes into farm animals. When this was done, the normal controls over production of the hormone were avoided. Transgenic sheep which over produce growth hormone put on weight more quickly, making more efficient use of their food. However, they are more prone to infection, tend to die young and the females are infertile. Similarly, transgenic pigs grow leaner meat more efficiently. However, even more side effects were noted than with sheep, including arthritis, gastric ulcers, heart and kidney disease. Until ways are found to regulate the genes more precisely, the process will not be commercially used.

8.5.2.3 Transgenic Sheep

Sheep provide a very low (0.1 to 0.2 %) rate of transgenesis. If only transgenic viable embryos are transferred to surrogate ewes, this rate can be improved. The first reports of transgenic sheep were published by J.P.Simons (1988) of Edinburgh. Two transgenic ewes were produced, each carrying about 10 copies of human anti-haemophilic factor IX gene (cDNA) fused with the β-lactoglobulin gene. The gene, thus, had a tissue specific expression and ewes secreted human factor IX into their milk; this human factor is active even though the expression of transgene is low.

Ovine primary foetal fibroblasts were co-transfected with a neomycin resistance marker (neo) and a human coagulation factor IX genomic construct. For nuclear transfer to enucleated oocytes, two cloned transfectants and a population of neomycin-resistant cells were used as donors. Six transgenic lambs were live born. Three produced from cloned cells contained factor IX and neotransgenes. It is, thus, possible to subject the somatic cells to genetic manipulation *in vitro* and produce viable animals by nuclear transfer (Schnieke *et al.,* 1997).

Transgenic sheep can be produced by transfer of nuclei from transfected foetal fibroblasts besides by using DNA microinjection technique.

Five transgene sheep involving transgene fusion of the ovine β-lactoglobulin gene promoter fused to the human α-antitrypsin gene have also been produced at Edinburgh. The protein purified from milk had a biological activity similar to that from human plasma derived α-antitrypsin. The deficiency of human α-antitrypsin leads to a lethal disease emphysema.

The first ever mammalian clone, Dolly was produced from an adult sheep (Figs. 8.10, 8.11). Ian Wilmut and Keith Campbell at the Roslin Institute in Edinburgh, Scotland, cloned Dolly by transferring the nucleus from an udder cell into an egg whose DNA had been removed. However, this sent ethical shock waves around the world (Pennisi and William, 1997).

Several flocks of sheep each producing different proteins are there. Tracy was the first transgenic sheep to produce AAT. The offspring of transgenic animals continue to carry the gene, so whole flocks of transgenic sheep can eventually be built up. Transgenic animals are seemingly perfectly normal, and no ill effects have been detected as a result of the treatment.

Same gene for AAT is present in all the cells of sheep. The question is how we ensure that only the cells of the mammary gland makes it? For this, body's own regulatory system is used. Every cell in the body contains the genes for milk proteins. However, for a gene to be expressed, the promoter, must be switched on. By cloning one of the sheep's own milk protein promoters and attaching it to the human DNA, we can ensure that only mammary glands express that DNA. For this purpose, promoter used is of β-lactoglobulin as lactoglobulin protein is present in high concentration in milk.

Ability of sheep for wool growth can also be increased by this methodology. For this purpose, genes essential for synthesis of some important amino acids found in keratin proteins of wool, have been cloned. They were introduced in embryos to produce transgenic sheep.

Growth hormone genes have also been introduced and can be used to promote body weight.

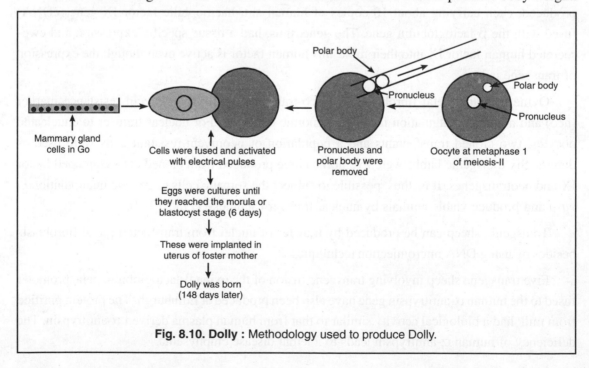

Fig. 8.10. Dolly : Methodology used to produce Dolly.

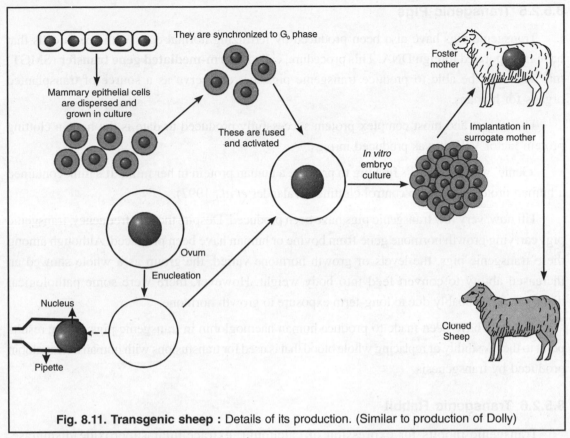

Fig. 8.11. Transgenic sheep : Details of its production. (Similar to production of Dolly)

8.5.2.4 Transgenic Goats

Transgenic goats have been produced by John Mc Pherson and Karl in U.S.A. These goats expressed a heterologous protein, a variant of human tissue type plasminogen activator in their milk. This protein is used for dissolving blood clots *i.e.,* for treatment of coronary thrombosis. A cDNA representing LAtPA was linked with either the murine whey acid promoter (WAP) or α, β casein promoter in an expression vector and injected into early embryos surgically obtained from oviducts of superovulated dairy goats. Five offspring were delivered by the transgenic female. One of these was transgenic, showing expression of LAtPA at a low level of few milligrams per litre of milk. In another case of a transgenic goat, few grams of LAtPA per litre of milk could be obtained.

Transgenic goats could also be produced by nuclear transfer of foetal somatic cells. Donor karyoplasts were obtained from a 40-day transgenic female foetus produced by artificial insemination of a non-transgenic adult female with semen from a transgenic male (Echelard *et al.,* 1999). Anti-thrombin III, α-antitrypsin, growth hormone, monoclonal antibody producing transgenic goats are also on record.

8.5.2.5 Transgenic Pigs

Transgenic pigs have also been produced by fertilizing normal eggs with sperm cells that have incorporated foreign DNA. This procedure, called **sperm-mediated gene transfer** (SMGT) may someday be able to produce transgenic pigs that can serve as a source of transplanted organs for humans.

The largest and most complex protein successfully produced to date is the human clotting protein factor VIII. It was produced in pigs.

'Genie' was the world's first pig to produce a human protein in her milk. It's milk contained a human protein C (acts to control clotting) (Valander *et al.,* 1997).

Till now, very few transgenic pigs have been produced. Despite the low frequency, transgenic pigs carrying growth hormone gene from bovine or human have been produced. Although among these transgenic pigs, the levels of growth hormone varied, the group as a whole showed an increased ability to convert feed into body weight. However, there were some pathological disturbances, probably due to long-term exposure to growth hormone.

Attempts have been made to produce human haemoglobin in transgenic pigs. These results point to the possibility of replacing whole blood that is used for transfusions with human haemoglobin produced by transgenesis.

8.5.2.6 Transgenic Rabbit

Transgenic rabbits for expression of calcitonin, extracellular superoxide dismutase, erythropoietin, growth hormone, insulin like growth factor 1, interleukin 2 have been produced.

8.5.2.7 Transgenic Chicken

Attempts have been made to develop transgenic chicken. It is because of the following facts:

- It grows faster than sheep and large numbers can be grown in close quarter
- It synthesizes several grams of protein in the "white" of their eggs

Two methods have succeeded in producing chicken carrying and expressing foreign genes :

(a) Infecting embryos with a viral vector carrying
  - the human gene for a therapeutic protein
  - promoter sequence that will respond to the signals for making proteins (*e.g.,* lysozyme) in egg white.
(b) Transforming rooster sperm with a human gene and the appropriate promoters and checking for any transgenic offspring.

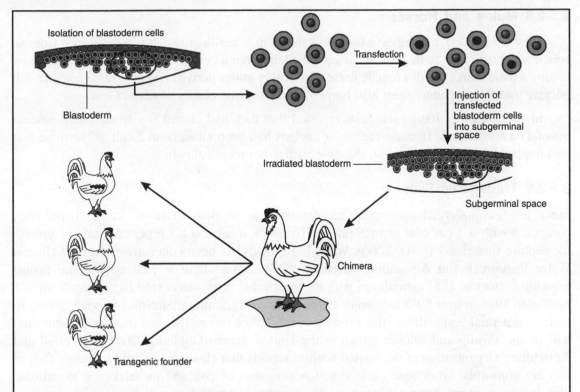

Fig. 8.12. Production of transgenic chicken: Blastoderm cells are separated from the early developing embryo. They are transfected with the construct (gene) of choice. The transfected cells are inserted back into subderminal disc. These cells must grow into transgenic/chimeric chicken. (Adopted and modified from Microbial Biotechnology by Glick and Pasternak, 2004)

A procedure involving use of engineered cell from embryos has provided possibilities to develop transgenic chicken. Blastoderm cells can be removed from a donor chicken, transfected with cationic lipid (liposome) transgene DNA complexes (lipofection) and reintroduced into the subgerminal space of embryos of freshly laid eggs (Fig. 8.12). Some of the progeny will consist of mixture of cells, some of which are from donor, but most of which are from the recipient.

Such a genetic mixture is called a **chimaera.** In some of the chimaeras, cells that were descended from transfected cells may become part of germ line tissue and form germ cells. Transgenic lines can then be established from these chimaeras by rounds of matings. The proportion of donor cells in chimeras can be increased to enhance the possibility of obtaining germ line chimeras if the recipient embryos are irradiated with a dose of 540 to 660 rads for 1 hour before the introduction of the transfected cells. The radiation treatment destroys some, but not all the blastoderm cells, thereby, increasing the final ratio of transfected cells to recipient cells.

It may be possible for chicken to produce as much as 0.1 g of human protein in each egg that they lay. Transgenic chicken are now able to synthesize human proteins in the "white" of the eggs.

8.5.2.8 Mules and Horses

On 4th May 2003, a horse gave birth to Idaho Gem, a healthy mule. It had developed from an enucleated horse oocyte that had received a nucleus from a cell of a mule foetus (produced by mating a male donkey with a female horse). Two other mules derived from the same somatic-cell nuclear transfer procedure were also born. These were true clones of Idaho Gem.

In August 2003, Italian scientists reported that they had cloned (by somatic-cell nuclear transfer) a baby horse. Because the donor nucleus had been taken from a cell of the mare that was implanted with the blastocyst, she gave birth to her identical twin !

8.5.2.9 Transgenic Fish

Since 1984, commercial aquaculture has expanded at an annual rate of almost 10 per cent, compared with a 3 per cent growth rate for livestock meat and a 1.6 per cent rate of growth for capture fisheries-1 (FAO, 2000). While the growth has been concentrated in Asia (Inland Water Resources and Aquaculture Service, 1997), aquaculture is also one of the fastest growing sectors of U.S. agriculture, with the total value of products sold increasing from $45 million in 1974 to over $978 million in 1998 (National Agricultural Statistics Service, 2000). In fact, commercial aquaculture sites produce nearly all of the catfish and trout and about one-half of the shrimp and salmon grown in the United States (Goldburg, 2001).The Food and Agriculture Organization of the United Nations reports that close to 100 million metric tons of fish are consumed worldwide each year. Consumption of fish and its relevance to national diet, however, varies between countries. Developed countries Generally consume more than developing countries. Japan, for instance, has some of the highest per capita consumption levels, while Africa and the Near East have some of the lowest (Westlund, 1995). Meeting the demand for seafood will likely require a number of steps including, but not limited to, continuing increases in aquaculture production, restoring depleted capture fisheries-2 maintaining currently productive ones, and assuring sustainable practices in any newly developed fisheries (Marine Stewardship Council 1998; Peacey, 2000).

Certain basic reasons those attract the aquatic animals, particularly fish grown in aquaculture systems two for research these are:

(i) Fish lay eggs in large quantities   and those eggs are more easily manipulated, making it easier for scientists to insert   novel DNA into the eggs of fish than into the eggs of terrestrial livestock. (In comparison, cows and pigs produce fewer eggs at a time, and once scientists insert novel   DNA, they must re-insert the altered eggs into the animal)

(ii) Aquaculture is one of the fastest growing food-producing sectors globally, suggesting growing demand for more aquaculture products.

A genetically engineered variety of Atlantic Salmon that grows to marked weight in about 18 months, compared to the 24 to 30 months that it normally takes to reach that size, has been created. The raising of these so called transgenic fish could be faster and cheaper for farmer because it takes less feed and about half the time to produce a crop they can send to market.

Scientists in Canada have added a gene from another fish (the ocean trout) to salmon which activates the salmon's own growth hormone gene. The salmons grow up to 30 times their normal weight and at 10 times the normal rate. Scottish fish farmers started breeding them on a trial basis in 1996.

There are several processes used to insert "new" DNA into fish, ranging from inserting genetic material directly into eggs to subjecting fish eggs to electrical pulses, which form pores and allow foreign DNA to access the eggs. The precise location where the new genetic material has attached to the original DNA is unknown and may vary between individual fish so scientists need to check to ensure the inserted gene is present and determine if it functions as expected. Once scientists have determined that the genes have been inserted, the fish are raised like other farmed fish.

Although this discussion is focused on transgenic fish, other transgenic aquatic organisms, including marine and freshwater plants and shellfish, are being fast-tracked for commercialization.

Risks Associated with Transgenic Fish

Many risks are associated with the transgenic fish

Food Safety Issues. Potential human consumption of transgenic fish and shellfish raises an assortment of food safety concerns.

- Genes inserted to promote disease resistance may cause transgenic fish to absorb toxic substances (like mercury) at a higher rate and pass these toxic substances on to consumers.
- Roughly 90 per cent of food allergies can be attributed to consumption of eggs, fish, shellfish, milk, peanuts, soybeans, tree nuts, and wheat. If proteins used in the production of transgenic species originate from one of these eight sources, there may be potential for allergic reactions among consumers.
- The majority of transgenic fish have been inserted with growth genes. Large doses of growth hormones may pose health risks if consumed in raw and uncooked foods like sushi.

Environmental concerns Transgenic fish pose potential threats to natural ecosystems and native species populations that are not fully understood and remain insufficiently studied. However, it is known that:

- Fifty percent of all intentionally introduced fish have had harmful economic or environmental effects;
- Sixty-six percent of all unintentionally introduced fish have had harmful economic or environmental effects;
- Millions of farmed fish escape from open water facilities each year and contaminate native populations; and
- It is inevitable that transgenic fish will escape from aquaculture pens or field trial parameters.

Although these statistics reflect impacts from the releases of non-transgenic fish, the implication is clear–introduction of foreign species to an ecosystem does not occur without negative consequences. The environmental risks of releasing transgenic fish may threaten biodiversity in natural ecosystems and the genetic integrity of those systems.

Genetic alterations in transgenic fish may give them competitive advantages over native species.

- By using growth hormone genes, researchers have been able to increase growth rates 2 to 11 times faster than the normal rate. Faster development leads to earlier sexual maturity and potentially more breeding opportunities than their native counterparts.
- If transgenic fish are genetically enabled to breed earlier and at a faster rate, transgenic genes are more likely to be spread throughout native populations. This would reduce the genetic diversity of the native population.
- It has been indicated that transgenic fish may reach maturity faster, They however also die sooner. This *"Trojan gene"* scenario could have devastating consequences if transgenic fish interbreed with native populations. Additionally, if transgenic genes for rapid growth are correlated with shorter life expectancy, the overall life expectancy for the entire population may be reduced.
- Transgenic fish may have similar effects on natural ecosystems as exotic species. An increased growth rate is often accompanied by a voracious appetite, and transgenic fish may out-compete native species for resources, destroy plants and sensitive habitat, and/or alter the food chain in an ecosystem.

8.5.3 Applications of Transgenic Animals

The potential benefits of transgenic animals, however, do not stop at food production. The first transgenic animals were created to advance basic biomedical research, genetically modifying lab rats, mice, rabbits, and monkeys to give them characteristics that mimic human diseases. These research resources, for example, rapidly advanced the understanding of oncogenes.

Ways are being found out to genetically modify the organs of animals, such as pigs, for possible transplantation into humans.

High production animals are developed as more or less on pure breeding lines.

The product of the injected gene stimulates growth, animals that acquire this gene should grow faster and require less feed. An enhancement of feed efficiency by a few percent would have a profound impact on lowering the cost of production of either beef or pork.

8.6 BIOTECHNOLOGY OF SILKWORM

Sericulture, in India, is one of the important agro-based cottage industries employing six million people. India, now is the second largest silk producer after China. India also produces three other commercial varities: *tassar, eri* and *muga*. Indian sericulture mainly relies on its polyvoltine breeds which are being used as female parents and are crossed with bivoltine strains.

Careful examination reveals that indian silk is qualitatively poor and the quantitative recovery is far from satisfactory. The bulk of indian silk which comes from polyvoltine x bivoltine hybrids. It is qualitatively inferior to the bivoltine silk produced in Japan and China. The changes can only be ushered in with an improvement in the economic standard of the farmers/people. The next alternative is to make the silkworm strains more adapted to our conditions to produce higher yield and better silk. The genetic analysis of Indian strains shows that :

(a) they are endowed with excellent adaptability to tropical conditions

(b) they have high dietary efficiency

(c) the are easy to manage

Further, these features are coupled with distinct biochemical and quantitative differences that are quite different from those of bivoltine strains.

In silkworm, the genes encoding the following proteins have been successfully expressed: human interferon, chloramphenicol acetyl transferse, IGF-II, human interleukin, mouse interleukin, HIV, Adenovirus, Encephalitis virus, bovine papilloma virus, galactosidase, Gag, Pol, gp 41 and gp120 (Dutta and Nagraj, 1993).

The proteins secreted into haemolymph could be easily purified. The foreign proteins so far expressed and purified from *Bombyx* source show that they are almost identical to those in humans.

The domesticated silkworm *"Bombyx mori"* has also attained importance as a convenient model system for the study of gene expression and its control mechanisms. A genomic library of the silkworm *B. mori* has been established in bacteriophage *lambda charon* 4A. From this library, a fibroin (major silk protein) clone has been isolated and characterised. The clone contained the entire coding sequence of fibroin gene, the 3'-flanking region and nearly 4kbp of the 5'-upstream region. Using this cloned fibroin gene in combination with other methods, the synthesis of fibroin mRNA during different stages of larval development has been quantitated. The synthesis of fibroin protein was also quantitated simultaneously using a labelled antibody approach. The transcription of fibroin gene was 'on' at all stages of development, however, the mRNA was not efficiently translated into protein at earlier stages of development. The translational block of the polysome bound fibroin mRNA appeared to be due to non-availability of cognate tRNA's. The expression of various tRNA molecules was transcriptionally regulated in a development stage specific manner. The methylation status of the DNA in the posterior and middle silk glands as a function of development has also been investigated in order to understand the role of DNA base modification in the expression of fibroin gene. The possibility of increasing the fibroin gene dosage to increase the fibroin production has been suggested.

8.6.1 Use of Biotechnology for Improving Silkworm

An excellent but different genetic repertoire provides an opportunity to scientists working an silkworm. The work initiated in this direction in different labs is given below:

(i) *RFLP (Restriction fragment length polymorphism):* For many years now, silk worm geneticists have been identifying, cataloguing and mapping single gene markers in silk worm. This map offers limited scope to understand the genome. Apart from this, nothing is known about genetic factors which contribute to the yield. The ideal starting point for laying the foundation for new breeding technologies is to harness already proven recombinant DNA technology for the potential improvement of this insect. The RFLPs, which allow direct examination of certain DNA sequence differences between individuals, provide a large number of genetic markers representing the coding and non-coding sequences.

(ii) *Use of Baculovirus for the production of proteins of interest:* In the study of gene expression in eukaryotes, two different types of strategies are followed to introduce foreign gene; one approach is to integrate the foreign genetic material into the host chromosomal DNA, while the other is to transport foreign DNA into cell nucleus, which gives transient expression. While the first method is being developed in silk worm, the second is extensively used in it. The *Bombyx mori* nuclear polyhedrosis virus (BMNPV) has become an extremely useful vector in the expression of foreign genes at very high levels to produce biologically active proteins both in *Bombyx* cell lines and under *in vivo* conditions. Silkworm is a very useful host for mass production of useful proteins, because its development is extraordinarily rapid. It increases its body weight ten thousandfold within 20 days, from 0.5mg at hatching to 5g at fifth instar, by eating only 20g of mulberry leaf. Since the efficiency of silk worm protein synthesis is extremely high during the larval stage, several mgm of proteins could be synthesised within a day. Besides, silkworm larva could be viewed as an extremely sophisticated factory for the production of storage polypeptides which changes dramatically during metamorphosis. Haemolymph serves as an excellent medium for the secretion and transportation of proteins.

(iii) *Development of transgenic animals:* Gene transfer studies in silkworm egg have been carried out by scientists of CSR and TI, Mysore. The vectors, which include a stretch of homologous sequences and transposon tagged *Baculovirus* genome, are being used for this purpose. Initial results show that the introduced genes have transient expression in the extrachromosomal state and some nuclei also integrate into the genome.

(iv) *Analysis of genetic basis of defense mechanism in silkworm:* Since the silkworm is a host to many bacterial and viral diseases, studies were initiated to understand the molecular basis of immune response in silkworm. Injection of live bacteria, induced anti-bacterial activity in the haemolymph, which was detected by petridish zonal inhibition as well as bactriolytic assay. The purified substance showed anti-bacterial activity.

(v) *Development of immunodiagnostic kit:* The silkworm, *B. mori* is susceptibile to many diseases caused by pathogens resulting in heavy crop loss. This is largely due to non availability of specific detection methods in the early stages of these diseases. Specific monoclonal and polyclonal antibodies may be raised against various silkworm pathogens and purified antibodies can be employed to develop diagnostic methods such as agglutination test, ELISA in the form of 'easy to use' kits for the early detection of pathogens in the egg production and young silkworm rearing units.

(vi) *Biological control of uzifly:* The uzifly, is a well known endoparasitoid of silkworm, causing considerable loss. Recently, several parasitoids *i.e., Nesolynx thymus, Dirhinus,*

Trichopria spp. have been noticed to be highly destructive in the maggot and pupal stages of uzifly.

8.6.2 Improvement of Mulberry through Application of Biotechnology

Mulberry, the host plant of silkworm, *B. mori*, belongs to the genus *Morus*. Many new methods that can contribute to plant breeding are now available through emerging applications of tissue culture. Many novel plants have been produced through protoplast fusion and electroporation of desired genes in many plant species.

(i) *In vitro* **screening for tolerance to specific stress conditions of somaclonal variation** : These techniques could be efficient to detect the mulberry genotypes for tolerance to alkalinity, salinity and draught conditions.

(ii) **Somatic hybridization through protoplast fusion**: Somatic hybridization through isolation, *in vitro* culture and fusion of protoplast of two plants have made it possible to produce novel plants which are unattainable through conventional breeding. In mulberry, this technique has better scope as the high biomass production and regeneration ability of tropical genotypes can be combined with the superior nutritional quality of the temperate genotypes.

(iii) **Evolution of haploids through anther pollen culture:** Mulberry is a highly heterozygous dioecious tree veriety. It is a difficult task to raise pure lines, though pure lines are an invaluable prerequisite for an understanding of the genetic mechanism underlying the inheritance of characters for exploiting hybrid vigour and for obtaining novel recombinants.

(iv) **Development of suitable mycorrhizal system for mulberry production** : VAM (Vesicular arbuscular mycorrhizal) association and its beneficial effects on mulberry for phosphate assimilation and uptake are important areas of research.

8.7 PEST MANAGEMENT

A pest can be defined as "anything, such as an insect, animal or plant that causes injury, loss, or irritation to a crop, stored goods, an animal or people". From an ecological stand point, a pest is nothing more than a competitor, parasite or predator of humans or something that humans use.

Our ancestors found that environmental sanitation and personal hygiene were important pest control measures. They used a variety of biological, cultural and chemical pest control measures.

A pesticide is any chemical used to kill organism unwanted by humans. But, three very important problems can develop when pesticides are used as the primary means of pest management :

(i) Pests can eventually develop resistance to specific pesticides

(ii) Population of insects can rebound (resurgence) to even higher levels after pesticide applications

(iii) Other (secondary) pests can arise to take the place of the original pest

8.7.1 Chemical Control of Pests

During 1940s, a number of chemical insecticides were developed as a means of controlling the proliferation of noxious insect populations. One of these was the chlorinated hydrocarbon DDT. It proved to be exceptionally effective in killing and controlling many species of insects. Chlorinated hydrocarbons like DDT function by attacking the nervous system and muscle tissue of insects. Other chlorinated hydrocarbons like dieldrin, aldrin, chlordane, lindane and toxophene have since been synthesized.

Another class of chemical insecticides is called organophosphates and includes malathion, parathion and diazinon. They are used to control insect populations by inhibiting the enzyme acetylcholinesterase which hydrolyzes the neurotransmitter acetylcholine. These insecticides disrupt the functioning of motor neurons and neurons in the brain of the insect.

Drawbacks: As the targeted pest populations become increasingly resistant to treatment with many chemical pesticides, higher concenterations of the insecticides had to be appplied. In addition the chemical insecticides lacked specificity. Beneficial insects were killed along with pests.

8.7.2 Biological Control of Pests

Biological control of pests includes the use of one living plant or animal to control another unwanted plant or animal. Biological control can be carried out as follow.

A. Sex attractants or pheromones. Pheromones are hormones that produce sexual attraction between insects. They are the principal means of communication between insects and are useful in sending alarm signals, marking trails or attracting mates.

A control of the mating behaviour of the insect through the use of pheromones the population size of the concerned insect pest can be reduced. Numerous synthetic pheromone dispensers are placed within the crop so that the level of female sex pheromone in orchard or field becomes higher than background level. It is reduced in the subsequent generations.

In the past years, it has been the odour of the female insect that attracted the male. But in 1987, scientists synthetically duplicated a pheromone, odour of the male papaya fruit fly to lure female fruit flies to a sticky surface. As of 1987, only one other synthetic male pheromone was being used to attract female boll weevils into a trap. At that time it was discovered that the synthetic male pheromone attracted both female and male boll weevils. This pheromone is now used in cotton fields by insect scouts to trap the cotton boll weevil. Such hormones are used to control insect populations in two ways:

Trapping and confusion: Pheromones are secreted by the females and detected by the antennae of the males. They are species specific. Traps containing pheromones of the gypsy moth are placed in the infested fields. The males fly to the traps and are lured into the hollow cylinder coated inside with a sticky substance. They are, thus, not available for reproduction.

Confusion technique : Another of practical pheromone use is called mating disruption (previously sometimes called "**male confusion**"). It has been successfully applied in fields to a number of moth species such as pink bollworm (*Pectinophora gossypielle*) in cotton.

In this, large amounts of a hydrophobic paper containing the sex attractant is dropped over a cropped area. The males are no longer able to locate the females as the characteristic smell is spread all over.

B. Use of juvenile hormone analogues. Insect hormones, such as juvenile hormone and the molting hormone (ecdysone) are required for proper metamorphosis of the young ones to the adult stage. For example, the insect cuticle must be shed periodically in the process of growth. This is called **"molting"** and involves the hormone ecdysone. Juvenile hormones must be present in the early stages to prevent early maturation. But if juvenile hormone analogues are given, the transformation of the larvae to an adult insect is prevented, thereby, inhibiting its proper development.

Genetic control of Pest

Some plants can be selected and or bred to be resistant to certain insects, diseases and or insecticides. This is known as genetic control. It has now been confirmed that interplanting even different cultivars of the same plant species reduces insect damage. This may be called "genetic diversity". For example, in California it has been reported that mixing wheat varieties can reduce certain rusts.

The lacewing, larvae are relatively tolerant to insecticides such as chlorinated hydrocarbons, pyrethroids and many microbials; however, most individuals are easily killed by organophosphates and carbamates. If some individual lacewings can be found in nature to be resistant to sevin (a carbamate), these could be reproduced and distributed in areas where sevin was used. This research was reported for the first time in California in 1986.

Genetic Control of a Disease Carrying Tick

Bovine babesiosis is a tick-borne protozoa disease that causes cattle fever, anaemia, loss of apetite and a drop in milk production. It is transmitted by the bite of a tick, genus *Babesia*. The tick and its associated disease also affects humans. A new integrated pest management technique being developed to control bovine babesiosis consists of :

- crossing two species of *Babesia, B. annulatus* and *B. microphes* to produce sterile male hybrid ticks, however, female hybrids are fertile,
- mass producing large number of males and females of both species and crossing them,
- when the sterile male hybrids mate with normal females, the eggs are infertile (will not hatch),
- although hybrid females are fertile, their male offsprings are sterile.

Of all the extensively cultivated crops, cotton seems to require the most pesticides. Monsanto chemicals has developed a transgenic cotton cultivar, resistant to insect damage and it successfully passed its fruit field tests, in 1990, in Mississippi, Texas, Arizona and California.

Scientists spliced into a cotton plant a gene from a bacterium that is the natural enemy of the cotton bollworm and the cotton pink bollworm.

The 'new' cotton plant contained the same toxin as the bacterium and its sap killed the larvae of the two species of cotton bollworms. Unfortunately, it will take several more years of field testing to determine:

- if the "new" cotton will produce economic yields of cotton lint,
- if total yields are economical, will the quality of the cotton fibre be satisfactory.

Moreover, there is every reason to assume that the pests will simply develop resistance to the transgenic toxin, just as they would to synthetic peptides

8.7.3 Baculoviruses in Pest Control

Given the drawbacks of chemical insecticides, alternative means of controlling harmful insects have been sought using insecticides produced naturally, by either microorganisms or plants. In particular, the insecticidal activities of the bacteruim *Bacillus thuringiensis* and insect baculovirus system appear to be safe, specific and effective.

The principle target of a *Baculovirus* is the larval (caterpillar) stage of the host *i.e.*, the insect. Viruses are ingested together with food. The polyhedrin protein is removed by proteolytic digestion when the virus reaches the alkaline environment of the caterpillar midgut. Infectious virus particles are believed to initially infect the columnar cells in the epithelium of the caterpillar's midgut. Virus DNA enters the cell nucleus and replication ensues. In *Lepidoptera*, non-occluded virus particles are released from these cells to spread the infection via the haemocoel to cells in other tissues of the larva. Late in the infection course, inclusion bodies (polyhedra) are produced. The virulence and extent of viral infection induces death of the host species.

The first release (1986-1987) used a genetically marked *Autographa californica NPV*. The results demonstrated that an innocuous piece of DNA, appropriately positioned in the AcNPV genome, was an effective means to tag the virus without affecting its phenotype, thereby, allowing it to be identified in bioassays of plant and soil samples. The second and third releases (1987, 1988) involved a genetically marked, "self-destructive" virus from which the gene coding for the protective polyhedrin protein of the virus has been removed. The field data obtained with this virus showed that it did not persist in the environment, neither in the soil, nor on vegetation, nor in the corpses of caterpillars. The fourth release (1988) involved a polyhedrin - negative virus that contained a "junk" (β-galactosidase) gene as phenotype marker.

Baculovirus has been shown to be useful as insecticide for pest control. In some cases they have been employed as cost effective and environmentally acceptable alternatives to chemical insecticides. The objectives of the programme of genetic engineering of baculovirus insecticides is to improve their speed of action, while maintaining their host specificity and other attributes that make them desirable alternatives to chemical pesticides. This is desirable since during their normal infection process, a baculovirus undergoes several cycles of replication. These cycles take time-upto several days, or weeks, depending on the virus, the host and the environmental conditions. By contrast, most chemical insecticides act quickly, killing the target insect in a matter of hours. Using genetic engineering procedures, it should be possible to minimise the time taken by viral insecticides to act by incorporating other genes (*e.g.,* insect hormone genes, *etc.*) into the viral genome.

Risk Assessment

(i)   A major issue in terms of risk is whether the host range of an engineered virus differs from that of the present virus.

(ii) Another consideration of risk associated with an engineered organism is the question of its genetic stability *i.e.,* whether the introduced gene induces genetic instability.

(iii) A third consideration of risk concerns the possible spread of virus from the field site.

(iv) Another risk is the question of exchange of genetic information. DNA transfer involving the aquisition by a baculovirus of genetic information from cells of a host species could occur and produce an altered phenotype.

Factors affecting success of baculoviruses in pest management

Several factors will affect the degree of success of baculovirus in the various approaches to insect control (Bishop and Possee, 1990).

These include:

- Economics and specificity
- Speed of activity
- Reliability and efficacy
- The ecosystem and environment
- Delivery of pathogen to the insect
- Resistance to the virus
- Persistence and safety to the environment
- User acceptance

8.7.4 Insecticidal Toxin of *Bacillus thuringiensis*

The most studied and effective microbial insecticides are the toxin synthesized by *B. thuringiensis*. This bacterium comprises a number of different strains and subspecies, each of which produces a different toxin that can kill certain specific insects for example, *B. thuringiensis subsp. kurstaki* is toxic to lepidopteran larvae. The insecticidal activity (toxin) of *B. thuringiensis* contained within a very large structure called the parasporal crystal, which is synthesized during bacterial sporulation. The cyrstal is an aggregate of protein that can generally be dissociated by mild alkali treatment into subunits. The parasporal crystal does not usually contain the active form of the insecticide; rather, once the crystal has been solubilized, the protein that is released is a protoxin. When a parasporal crystal is ingested by a target insect, the protoxin is activated within its gut by the combination of alkaline pH (7.5 to 8.0) and specific digestive proteases, which convert the protoxin into an active toxin. The toxic protein inserts itself into the membrane of the gut epithelial cells of the insect and creates an ion channel through which there is excessive loss of cellular ATP. After this ion channel forms, cellular metabolism ceases, and the insect stops feeding, becomes dehydrated, and eventually, dies. Because the conversion of the protoxin to the active toxin requires both alkaline pH and the presence of specific proteases, it is unlikley that non target species such as human and farm animals will be affected.

The mode of action of *B. thuringiensis* toxin imposes certain constraints on its application. To kill an insect pest, *B. thuringiensis* must be ingested. A second limiting feature of the action of the *B. thuringiensis* toxin is that it can kill a susceptible insect only during a specific developmental stage.

Under normal conditions, most *B. thuringiensis* toxins are synthesized only during the sporulation phase of growth. When a DNA fragment containing a toxin gene that lacked its native promoter was cloned into a plasmid under the control of a continuously active, constitutive promoter from a tetracycline resistance gene that had been originally isolated from a *B. cereus* plasmid and reintroduced into *B. thuringiensis*, active toxic protein was continuously synthesized, both during vegetative and sporulation phases.

Because many crops are attacked by more than one insect species, it could be advantageous, if feasible, to create microbial insecticides that are each effective against a broad spectrum of target insects. Such a broad specificity compound could be obtained by:

- transferring the gene for a particular toxin into a *B. thuringiensis* strain that normally synthesizes a different species specific toxin,

or

- by fusing portions of two species specific toxin genes to one another so that a unique dual acting toxin is produced.

8.8 BIOFERTILIZERS IN PLANT GROWTH

The fertilizers help the crop plant to give higher yield. The industrial production of fertilizers is based on energy source, the cost of which day by day increases making chemical fertilizers out of reach of the cultivators. Because chemical fertilizers and pesticides have caused tremendous harm to the environment. Even the industries are sources of air, water and soil pollution, so an alternative source for fertilizers, becomes a necessity.

The remedy for this is the biofertilizers, the environment friendly fertilizers being used in many countries.

Biofertilizers are cultures of microorganisms used for inoculating seed or soil or both, under ideal conditions to increase the availability of plant nutrients such as nitrogen and phosphorus. The organisms enrich the nutrient quality of the soil. The major sources of biofertilizers are bacteria, fungi and cyanobacteria. Biofertilizers can help to solve problems of :

(*a*) Increased salinity of soil.

(*b*) Chemical run-off from agricultural fields.

(*c*) Plant nutrition, disease resistance and tolerance to adverse soil and climatic conditions.

The knowledge about association of leguminous plants with microbes which can fix nitrogen formed the basis of biofertilizers. The beneficial effects of microbes as fertilizers are manifold.

Biological fertilizers promote plant growth in a number of ways. They.

- aid in replenishing and maintaining long-term soil fertility by providing optimal conditions for soil biological activity.

- suppress pathogenic soil organisms and degrade toxic organic chemicals.

- stimulate microbial activity around the root system, significantly increasing the root mass and improving plant health.

- by stimulating the growth of natural soil microorganisms they increase the available nitrogen for plants far in excess of their own content. These soil microorganisms metabolize nitrogen from the air to multiply. When they die (some microorganisms have a life-span of less than 1 hour), the nitrogen is then released to the soil in a form that is readily available to the plants.

- supply essential nutrients such as nitrogen, phosphorous, calcium, copper, molybdenum, iron, zinc, magnesium and moisture to the plants by interaction with other soil organisms and biodegradable components in the soil.

- aid in solubilizing manganese. A significant role in both disease resistance and plant growth is played by manganese.

- they increase crop yields by both enhanced growth and by protection the enhanced plant growth is accompanied by reduced stress and improved disease resistance.

- initiate and accelerate the natural decomposition of crop residue turning it into humus.

- provide protection against disease associated with numerous fungi. They produce peptides which inhibit the growth of fungi in some environments. In others, through a process known as mycoparasitism, they grow toward the hyphae of fungi, coil around them and degrade the cell walls.

- significantly increase yield and reduce incidents of disease in fruit, vegetables, root crops, flowers, trees, shrubs, turf, grain, ornamental crops *etc.*

- improve soil porosity, drainage and aeration, reduce compaction and improve the water holding capacity of the soil, thereby, helping plants resist drought and produce better crops in reduced moisture conditions. It has been indicated that a 5% increase in organic matter quadruples the soils ability to hold and store water.

- stimulate and improve seed germination and root formation and growth.

- increase the protein and mineral content of most crops.

- produce thicker, greener and healthier crops with reduced inputs costs

- produce plants with increased sugar flavour and nutrient content.

- improve oxygen assimilation in plants.

- aid in balancing soil pH and also in reducing soil erosion.

Various micro-organisms having the potential application as biofertilizers (Table 8.3) include:

1. Bacteria : The different species used as biofertilizers are *Rhizobium*, *Azospirillum* and *Azobactor.*

2. Fungi : Mycorrhizae like *glomus*.

3. Blue Green Algae : The various species are *Anabaena, Nostoc*.

In addition a water fern *Azolla* containing symbiotic *Anabaena azolla* is also used.

Table 8.3. Some Important Microorganisms with Potential Application as Biofertilizers

Organism	Activity	Association, if any	Crops
Rhizobium (*leguminosarum, japonicum, phaseoli* etc.)	N_2-fixation	Symbiotic	Legumes (pulses, oilseeds, forage crops)
Azospirillum	N_2-fixation	Associative	Graminaceous crops like wheat, rice, sugarcane, jowar
Azotobacter	N_2-fixation	Asymbiotic	Wheat, rice, vegetables
Blue-green algae (*Anabaena, Nostoc, Plectonema* etc.)	N_2-fixation	Asymbiotic	Rice
Azolla-Anabaena complex	N_2-fixation	Symbiotic	Rice
Phosphate solublizing bacteria (*Thiobacillus, Bacillus* etc.)	Phosphate solublization	Asymbiotic	Many crops
Mycorrhiza (*Glomus*)	Phosphate solublization	Associative	Many crops including pulses

8.8.1 Technology used in the Production of Biofertilizers

Biofertilizers are generally carrier based and always in powder form. The commonly used carrier is lignite that has high organic content. It can hold more than 200% water which enhances growth of microorganisms. Slurry is made before use and is applied to the seed. The process of coating seeds with bacteria is called **bacterization.**

Another technique of dry complex fertilizer for direct soil application has been developed. It consists of granules of 1 to 2mm diameter made from tank bed clay (TBC). The granules are baked at 200°C in a muffle furnace that sterilizes them and also gives porosity. They are soaked in suspension of N_2 fixing bacteria grown in a suitable medium. These granules under aseptic conditions are air dried at room temperature. These granules are used for field applications with seeds.

8.8.2 Microorganisms used as Biofertilizers

8.8.2.1 *Rhizobium*

In leguminous plants, the nitrogen uptake is facilitated by gram negative soil bacteria *Rhizobium* which forms root nodules (Fig. 8.13). However, in some cases stem nodules are also formed. There are many species of this genus inhabiting various legume species which they nodulate. Some of them are given with their host plant:

Rhizobium leguminosarum	Pea
R. trifolii	Trifolium sp.
R. lupini	Lupins
R. phaseoli	*Phaseolus sps.*
R. melilotii	*Melilotus species*

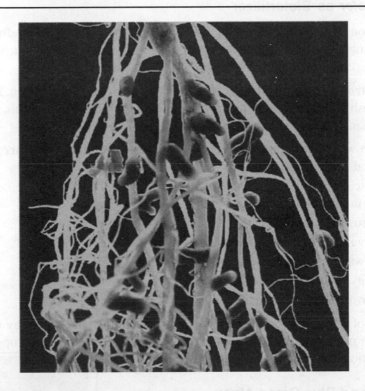

Fig. 8.13. Root nodules formed by *Rhizobium*; It is a nitrogen fixing bacteria.

The gene for nitrogen fixation (*nif* gene) is present on a megaplasmid of rhizobium cells. The bacteria invade the roots through root hair cells which are considered to secrete some chemical that attracts the bacteria. The interaction between root and bacteria ultimately leads to the formation of nodule. Within the nodule, the bacterial cells become non-dividing bacteroides and produce nitrogenase enzyme which converts free nitrogen into ammonia.

The ability of nitrogen fixation varies in different strains of bacteria. It is also subject to nature of genes of host legumes. For more efficient nitrogen fixation some mutant strains have also been developed. The host provides energy and carbon requirements of bacteria, while ammonia is made available to the host by the bacteria.

The use of *Rhizobium* as biofertilizers in fields has shown 10-15% increase in yield of inoculated pigeon pea and chick pea over uninoculated ones. The efficiency of selected strain for use as biofertilizers depends upon.

(*i*)   its tolerance to elevated temperatures.

(*ii*) its capability to out compete the native Rhizobia present in soil.

(*iii*) long shelf life while preparing inoculum.

8.8.2.2 *Azobacter* as Biofertilizer

Azobacter non-symbiotically fixes atmospheric nitrogen. All plants therefore get benefited. *Azobacter* also increases germination of seeds. It can fix upto 30kg of nitrogen per hectare/year. These bacteria are tolerant to high salts.

Culture of *Azobacter* is maintained on a solid medium free of nitrogen. It is periodically transferred to fresh solid medium to renew the growth. The production of this organism can be carried out on a small or large scale.

(*i*)   Pure growth of an organism on a small scale is called **mother culture**. It is prepared in a conical flask of 500/1000 ml capacity.

(*ii*) Large scale production of *Azobacter* is carried out by two ways *i.e.* fermenter and shaker, in sterilized conditions.

To support the growth of an organism the carrier must have

(*a*)   high organic matter

(*b*)   higher water holding capacity and

(*c*)   capability

Lignite is most frequently used carrier.

Inoculation of *Azobactor* can be done to seeds, seedlings and tubers by coating, dipping roots or tubers in mixture of biofertilizers in water respectively. For fruit crops, sugarcane and trees soil application is also done.

8.8.2.3 *Azolla* and Blue Green Algae

Azolla is an aquatic free floating, heterosporous pteridophyte with many branches and long roots. It is a nature's gift to mankind in harvesting atmospheric nitrogen. Its leaves have two lobes, the dorsal and the ventral. The dorsal lobes are chlorophyllous and house algal symbionts, the blue green algae (cyanobacteria) capable of harvesting nitrogen *e.g.*, *Anabaena azolla*. The symbiotic association with algae aids in creation of huge biomass on the surface of water which is harvested and used as biofertilizers (Fig. 8.14).

Other species of cyanobacteria used as biofertilizers are *Nostoc* and *Plectonema*. In addition to nitrogen fixation, cyanobacteria are useful in following ways:

Fig. 8.14 Azolla leaf.

(*i*)   They accumulate biomass that improves physical properties of soil.

(*ii*) They produce growth promoting substances.

(*iii*) They bring reanimation of alkaline soils.

Cyanobacteria are used for rice crops and inoculum is introduced in the field about 10 days after transplantation.

8.8.2.4 Fungi

Certain fungi form associations with plant roots called **mycorrhiza.** The fungus may be located at the root surface (ectomycorrhiza) or it may be present inside roots (endomycorrhiza). The fungal symbiont gets shelter and food from plant which acquires an array of benefits like better uptake of phosphorus, salinity and drought tolerance, maintenance of water balance and overall increase in plant growth and development. Mycorrhizal fungi absorb phosphorus from the soil and pass it on to the plant. These microorganisms are not yet used at commercial scale. They also produce growth promoting substances and protect against soil pathogens (Fig. 8.15).

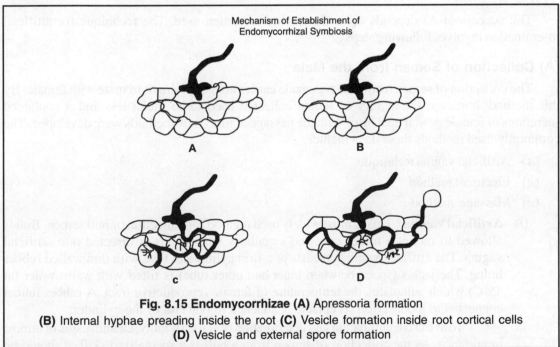

Fig. 8.15 Endomycorrhizae (A) Apressoria formation
(B) Internal hyphae preading inside the root **(C)** Vesicle formation inside root cortical cells
(D) Vesicle and external spore formation

8.8.2.5 Phosphate Solubilizing Bacteria

Non-available inorganic phosphorus present in soil is converted into an available and utilizable form by some bacteria *e.g., Thiobacillus, Bacillus* etc. In addition, these bacteria produce iron chelating substances *e.g.* pseudobactins called **siderophores.** They chelate iron present in the root zone and thus, iron becomes non-available to harmful organisms. The plants become protected from harmful bacteria.

8.8.3 Limitations of Biofertilizers

Although biofertilizers have manifold prospects, their acceptability has been low because :

(*i*) They do not produce quick and spectacular responses.

(*ii*) The amount of nutrient provided does not adequately meet the total needs of crops for high yields.

There is thus, need to develop a rational combination of biofertilizers and conventional fertilizers to get better crop yields.

8.9 ARTIFICIAL INSEMINATION IN LIVESTOCK AND OTHER ANIMALS

AI is used in animals to propagate desired characteristics of a male to many females particularly in horses, cattle, pigs, dogs *etc*. It is also used to overcome breeding problems in these animals. Artificial insemination of farm animals is quite common especially for breeding dairy cattle and swine in the developed world. It is an economical method to breed animals with desirable traits.

8.9.1 Technique

The success of AI depends upon the quality of semen used. The technique for artificial insemination involves following steps :

(A) Collection of Semen from the Male

The collection of semen from bull was made earlier after allowing it to mate with female. By this method, however, the quality of semen collected used to be much less and it contained secretions of female genital tract also. With the passage of time, new methods were developed. The commonly used methods these days include:

(*a*) Artificial vagina technique

(*c*) Electrical method

(*c*) Massage method

(*i*) **Artificial vagina :** This is most widely used method for collection of bull semen. Bull is allowed to mount a teaser cow and ejaculates when penis is directed into artificial vagina. The artificial vagina consists of a firm cylindrical tube with thin walled rubber lining. The jacket (space between inner and outer tube) is filled with warm water (at 45°C) which simulates the temperature of female reproductive tract. A rubber funnel connected to a collection receptacle is attached to one end of the cylinder.

To avoid contamination and deterioration of semen quality, cleanliness is of utmost importance. As the collection of semen from a bull is a specialized skill, it should be attempted by those with proper training and experience. Further, there should be adequate facilities for controlling the bull and teaser animal so that there is a minimum danger of injury to the person as well as animal.

(*ii*) **Electrical method :** In this method of semen collection, a multipolar electrode is introduced into the rectum of a bull. When electric current is passed, the ampulla and seminal vesicles of bull contract and semen comes out through penis that is collected in a container.

The method is used if the bull is old or incapable of mounting due to illness. However, frequent use of this method is not advisable.

(iii) **Massage method :** In this method, the semen is collected by massaging ampullae and seminal vesicles of bull through the wall of rectum. The penis protrudes and releases semen.

(B) Semen Quality

Before using semen for insemination, it must be subjected to detailed study. The good results are obtained only if composition and properties of semen and conditions of spermatozoa are optimum. Microscopic examination can reveal motility of spermatozoa, sperm count, percentage of abnormal and dead spermatozoa as well as infection, if any.

Healthy semen is milky white to cream, while yellowish green or light green colour indicates diseased condition.

(C) Extension and Storage of Semen

A normal ejaculation contains 5-10 billion sperms sufficient to insemination 300-1000 cows if extended (diluted). A healthy and normal semen is therefore extended (diluted) for the following purposes:

(i)   Inseminate a number of cows per ejaculation

(ii) Maintain viability of spermatozoa by providing nutrients

(iii) Nullify the effect of lactic acid produced due to anaerobic metabolism in spermatozoa

(iv) Protect it from bacterial infections

(v)   Preserve it for longer periods

There are several semen extenders. They are made from egg yolk or pasteurized, homogenized milk. The extender must contain nutrient buffer and antibiotics for nourishment and neutralization of acids respectively.

The ratio of semen and extender is generally 1: 50. The diluted semen is stored under frozen conditions. There has been revolution in the AI in cattle with the discovery that bull semen could be successfully frozen and stored for indefinite periods. In 1949, British scientists discovered that if glycerol is added to semen extender, the resistance of sperm to freezing is improved. It prevents formation of cellular ice crystals which would damage the sperm by replacing water prior to freezing. There are two methods of storing semen:

(i)   Dry ice and alcohol (–100°F)

(ii) Liquid nitrogen (–320°F)

Liquid nitrogen method is preferred as in dry ice alcohol, fertility gradually declines while in liquid nitrogen, there is no deterioration.

Semen is generally stored in glass ampules.

(D) Insemination technique

The inseminating technique requires adequate knowledge, experience and patience. The semen must be deposited within the tract of the cow at best location and time to obtain acceptable conception rates. All the efforts to obtain conception can be negated by the improper technique. Two methods of insemination are employed.

(E) Recto-Vaginal Method

In this method, a sterile, disposable catheter containing thawed semen is inserted into vagina and guided into cervix by means of a gloved hand in the rectum (Fig. 8.16). The catheter is passed through the spiral folds of the cow's cervix into uterus. The semen is partly deposited inside the uterus and remainder in cervix as catheter is withdrawn.

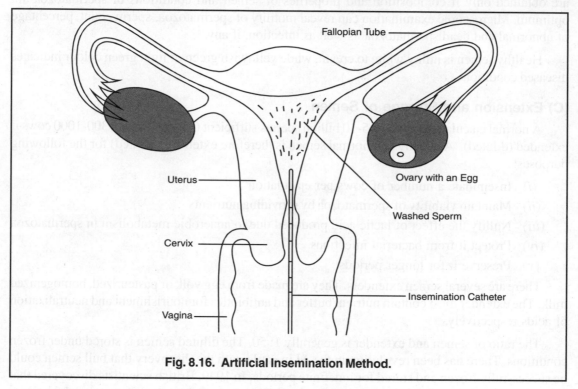

Fig. 8.16. Artificial Insemination Method.

8.9.2 Precautions

While performing this method the following precautions must be taken:

(a) To avoid sperm losses semen expulsion should by accomplished slowly.

(b) The body of uterus is short so catheter should not be penetrated too deeply as it may cause physical injury.

(c) The catheter should not be forced through cervix in previously inseminated animals.

Use of disposable catheter and proper sanitation measured should be followed.

As some variations in conception rates have been reported when semen is placed in cervix, uterine body or uterine horns, incomplete penetration is recommended by some so that semen is deposited in cervix.

(ii) Speculum Method

This is rather easier method to be learnt, but proper cleaning and sterilization of equipment is essential. In this method, a sterile speculum is introduced through vagina using a light source, cervix is located and semen is deposited using catheter and syringe.

8.9.3 Advantages of AI

Artificial insemination has the following advantages:

(i) It makes possible maximum use of superior sires. Natural service would limit the use of one bull to less than 100 matings per year. However, AI makes it possible to use one dairy sire to provide semen for about 60,000 services.

(ii) It makes possible the improvement of genetic features of cattle.

(iii) The old or crippled bulls otherwise unable to inseminate can be used for collecting semen.

(iv) The cost and trouble involved in maintaining a herd sire is saved by using artificial insemination method.

(v) It reduces the possibility of genital diseases, if followed under strict hygiene and good sanitary condition.

(vi) It can promote cross-breeding.

8.9.4 Disadvantage of AI

Following are the disadvantages of AI.

(i) Artificial insemination requires special training, skill and practice

(ii) Utilization of a few sires, as occurs with AI, reduces genetic base. The efforts should be made to sample as many young sires as possible.

(iii) Improper cleaning of instruments and insanitary conditions may lead to low fertility rates and may also cause infections.

8.9.5 Artificial Insemination in Humans

This process in humans is part of an infertility treatment. For this purpose sperms can be used from :

(i) woman's partner (artificial insemination by husband, AIH)

(ii) a donor (artificial insemination by donor AID)

Procedure

Using ovulation kits, ultrasound or blood tests the woman's menstrual cycle is observed. When ovum is released semen from the donor is inserted into the body. To ensure high sperm count, the donor is not allowed to ejaculate for a few days before the procedure. After donation, sperms are immediately washed in the laboratory. If the procedure is successful, the woman conceives and bears a term baby as normal.

8.10 LEGAL TRENDS IN AGRICULTURAL BIOTECHNOLOGY

We are using transgenic animals for the production of foods and drugs.

No matter how transgenic is applied, the Food and Drug Administration is to play a key role in regulating the products resulting from this transgenic animal. This means that any drug or biological material created through transgenic techniques will need to undergo the same FDA scrutiny as any other treatment that a company wants to market, including clinical trials that demonstrate safety and effectiveness. And while it's still too soon to tell how quickly foods derived from transgenic animals will move to the market, FDA has already begun to focus on how it will ensure that they meet the same safety standards as traditional foods.

The application of agriculture biotechnology is influenced by certain legal trends. In addition to traditional issues of intellectual property and environmental safety, the practical regulatory issues also restrict biotechnology research and application.

Thus while biotechnology research can be made compatible with sustainable development, public input is required to guarantee that the new products do not become the undoing of land, water, genetic resources and the individualized farming medicine systems *etc*. Once these precautions are taken, applied biotechnology is bound to revolutionize the society. Indeed, the revolution will likely be of global significance.

Questions

1. What do you mean by transgene, transgenesis and transgenics ? Explain with the help of examples.
2. Explain in detail how the Dolly was produced ? What are the limitations in producing transgenic animals?
3. What are transgenic plants ? Give their current and potential applications.
4. Describe the applications of biotechnologies in Agriculture.
5. Give the role of biotechnology in :
    (a) Nitrogen fixation
    (b) Herbicide Resistance
    (c) Salt Tolerance.
6. What is sericulture ? Give the role of biotechnology in sericulture.
7. Write what you know about artificial insemination?

Biotechnology and Environment

9.1 INTRODUCTION

We are in a state of dynamic interaction with the environment. We affect the environment and in turn environment affects us. One of the basic components of our living environment next to air, is water. Human culture and civilizations are virtually linked with water resources. Human indifferences, ignorance and greed combine globally to waste it. Due to continuous industrial revolution for creation of conveniences for human, a number of harmful chemicals are released in air and water (by industrial effluents) either without any treatment or with improper treatment.

The human activities such as spraying the toxic chemicals on crops and burning of fossil fuels are increasing the atmospheric burden of particulate matters and disturb the gaseous equilibrium (Table 9.1). With all these activities, the ratio of carbon dioxide to oxygen in the atmosphere continues to increase at alarming rate; the greenhouse effect is an inevitable result because of the change in this ratio. There is a need for long range environmentally sound approach to industrialization so as to greatly reduce and even reverse the rate at which the earth is surely becoming unfit habitat for life. Though this will be ideal, it would not probably be enough. The biological revolution is expected to provide the technological tools to prevent and cure the destructive effects of pollution.

Today biotechnological processes are playing a significant role in the protection and rehabilitation of environment and are advantageous alternative or adduct to physical and chemical treatment technologies.

A variety of enzymes and microbes are employed in a wide range of pollution control processes and also reclaim waste land. Even certain plants that remove and/or degrade toxic compounds are also being used.

In fact environment biotechnology is concerned both with its implications and applications in the wider context of the environment because increasing industrialization, urbanization and other developments are constantly threatening the environment and the natural resources are being depleted. Most essential and urgent environment problems which need attention include:

(i)   chemical wastes,
(ii) ground and surface waters contaminated with chemicals from the increasing use in agriculture,
(iii) effects on wild life and flora with decreasing diversity because of agricultural use of pesticides,
(iv) heavy metals such as mercury, cadmium and lead spread as fertilizers, through gasoline consumption,
(v)   increasing municipal organic wastes.

Table 9.1: Some of the Major Types of Wastes and Pollutants Generated by Various Human Activities (Adopted and modified from literature and Internet).

| Human activity | Wastes | Pollutants | | |
		Air	Surface water	Soil/ground water
A. Agriculture and Dairy				
(a) Agriculture	Crop residues, pesticides, plastics	Pesticides	Pesticides NO_3^-, PO_4^{3-}	Pesticides NO_3^-, PO_4^{3-}
(b) Dairy	Animal refuse,	NH_3^-, CH_4^-	NO_3^-, PO_4^{3-} (from manure)	NO_3^-, PO_4^{3-} (from manure), feed residues
Transport	Scrapped vehicles used up oil, greese etc.	CO_2, CO, SOx NOx, volatile hydrocarbons aerosols, lead	Oil products	Oil prodcuts
B. Energy production	Fly ash, heavy metals, radioactive waste, etc.	CO, CO_2, SOx NOx, aerosols, water vapours	Heat, coal residues, fly ash *etc.*	Fly ash, coal residues
C. Building and Domestic Waste				
(a) House-building	Building and demolition refuse wood, metals, asbestos, fly ash, etc.	CO_2, CO, SOx NOx, heavy metals, volatile organic compounds	NO_3^-, PO_4^{3-} O_2^- binding compounds	Metals, wood, fly ash etc.
(b) Domestic	Sewage, garbage, scrapped utensils appliances, etc.	CO_2, CO, SOx NOx, C.F.C.**, CH_4	Percolation water, NO_3^-, PO_4^3	Solid refuse, percolation water
D. Manufacturing				
(a) Chemical industry	Chemical refuse (solid/liquid)	Volatile organic compounds, SOx*, NOx*	Organic and inorganic compounds, heavy metals, cyanides	Organic and inorganic compounds, heavy metals, cyanides
(b) Oil refinery	Waste water	Volatile hydrocarbons, SOx, NOx, aerosols etc.	Oil, acids, heavy metals, phenolics etc.	Oil, acids, phenolics, etc.
(c) Fertilizer industry	Contaminated gypsum	SOx, NOx, NH_3	Inorganic chemicals	Inorganic chemicals

Research is being carried out to develop various processes and techniques to deal with these problems. The biggest impact of biotechnology could be in developing a new strategy for the utilization of natural resources, especially biomass. It could bring a shift away from resource strategies based on exhaustible fossils such as crude oil, coal and natural gas, towards biomass. An increased application of biotechnology within the framework of this new resource strategy would alter the pressure on the environment and could contribute to harmonizing resource use with the production, consumption and disposal of wastes.

The environmental biotechnology according to Prof. Moser includes:

(i) **Pollution clean up** *e.g.,* clean up of oil spills and detoxification of contaminated soil, treatment of domestic and industrial waste water supplies.

(ii) **Pollution control** *e.g.,* recovery of heavy toxic metals from mining water, use of enzymes rather than chlorine in pulp and paper manufacture.

(iii) **Pollution protection** *e.g.,* closed cycling practice at enzyme production plants implies that raw materials are renewable, waste material is a biodegradable sludge that can be used as a local fertilizer.

9.2 MANAGEMENT OF WASTE

For managing the wastes produced by different factories, the following strategies are needed to be applied:

(*a*) Minimize the amount of waste generated

(*b*) Reduce the toxicity of waste that is generated

(*c*) Find more satisfactory ways of disposing off and degrading wastes which have been generated

(*d*) Find new alternatives for the treatments causing pollution

Biotechnology is playing an important role in every phase of the above mentioned points by using the following technologies:

(*a*) Point source reduction/process plant and off site utility

(*b*) Utilization of by-product

(*c*) "End of pipe" treatment

9.2.1 Point Source Reduction

Introduction

A large area of the earth's surface and the oceans and other waterways have been contaminated with oil-derived compounds and toxic chemicals. The contamination of soil normally results from a range of activities related to our industrialized society. Contaminated land contains substances that when present in sufficient quantities or concentrations can probably cause harm to human beings, directly or indirectly, and to the environment in general. Many xenobiotic industrially derived compounds can show high levels of recalcitrance and while in many cases, only small concentrations reach the environment, they are biomagnified *i.e.,* an increase in the concentration as it passes through the food chain.

Bioprocess Technology

The bioprocess technologies, therefore, are a set of technologies that use biocatalysts that are living cells or their enzymes to :

- produce a molecular product
- change one molecule into another
- degrade a molecule

Advantages of bioprocess over chemical process

(i) Biocatalysts are the ultimate in sustainability. Microbes and cells reproduce, so they continually renew themselves and their molecular component.

(ii) Unlike many chemical processes, bioprocesses must occur under conditions that are compatible with life. They, however, also produce waste, less destructive to the environment than those of chemical processes.

(iii) Because enzyme-catalyzed reactions are very specific, they generate undesirable byproducts than chemical process.

Disadvantages of bioprocess technology

Many of the advantages that biocatalysts provide become limitations, if the process must be carried out under extreme conditions.

Now certain systems that involve biological catalysts to degrade, detoxify or accumulate contaminating chemicals, have been developed. The application of biological agents, mostly microorganisms, for the treatment of environmental chemicals has been directed mostly towards remedial activities.

9.2.1.1 Bioremediation

Biological methods of remediation variously termed **bioremediation**, **biorestoration**, **bioreclamation** or **biotreatment** are now replacing physical and chemical methods of separation and/or removal of the pollutants.

Bioremediation employs living organisms, such as microbes and plants, to extract, eliminate and/or bind toxins in forms that are not harmful to the environment.The organisms used include :

(i) **Microalgae:** These are used in ponds to eliminate nitrogen and phosphorus.

(ii) **Aquatic plants** (*e.g.*, water lentils): These are used to extract heavy metals in industrial effluents.

The modern biotechnology techniques have enhanced the performance of these natural processes in pollution control. For example, mercury is a highly toxic metal that accumulates in the food chain when released in water. In the Minamata accident, inhabitants of the Japanese island of Kyushu suffered the toxic effects of fish poisoned by mercury-rich industrial effluents. Since naturally thriving mercury-tolerant bacteria are rare and cannot be easily grown in culture, the metallothionein gene has been inserted into *Escherichia coli,* at Cornell University. Such genetically modified bacteria grow well in culture. The genetically engineered bacteria are placed inside bioreactors that efficiently remove mercury from water. The bacteria are later incinerated.

The accumulated mercury is then recovered (European Commission, 2002). Existing techniques of mercury removal are expensive and inefficient.

Each year more than 2 million tones of oil enters the sea. Half of these are derived from industrial effluents, sewage and river outflows and the rest from non-tanker shipping and natural seepage from below the seafloor. Only about 18% of the total comes from refineries, off-shore operations, and tanker activities. Generally most oils do have a relatively low toxicity to the environment. It can, however, have catastrophic and immediate effects on bird and animal life associated with water.

To deal with contaminated sites, following approaches are made:

(a) Assessment of the nature and degree of the hazards,

(b) The choice of remedial action.

Clean-up operation of contaminated soil can involve:

(i)   In-site processing,

(ii) *In situ* treatment,

(iii) Off-site processing.

As mentioned earlier, the different kinds of pollutants found in the environment are:

Inorganics

Metals : Cd, Hg, Ag, Co, Pb, Cu, Cr, Fe

Nitrates, nitrites, phosphates

Radionuclides

Cyanides

Asbestos

Organics

Biodegradable sewage : domestic, agricultural and process waste

Petrochemical wastes: oil, diesel, BTEX (benzene, toluene, ethylbenzene, xylenes)

Synthetic: pesticides, organohalogens, PAHs (polyaromatic hydrocarbons)

Biologicals

Pathogens : bacteria, viruses

Gaseous

Gases: SO_2, CO_2, NO_2, methane

Volatiles: chlorofluorocarbons, volatile organic compounds (VOCs),

They are present as a result of industrial effluents, outfalls, mine waters, landfill run-offs, waste tips and accidental spillages. These pollutants can be found in all regions of the environment: marine, estuaries, lakes and soil.

The simple principle of bioremediation is to optimize the environmental conditions so that:

(*i*)   microbial biodegradation can occur rapidly

(*ii*) as completely as possible.

For the biological transformation of xenobiotic compounds that are introduced into the ecosystem, naturally present microbes in soil and water environments are potential candidates. Microbial populations in natural environments exist in a dynamic equilibrium. This can, however, be altered by modifying environmental conditions such as availability of nutrient so that microbes work efficiently.

It has been observed that not the individual strains, but consortia of microorganisms act on pollutant molecules. The metabolic effects of microorganisms on pollutants can take forms and are not always to the environmental advantage of the ecosystems.

Bioremediation to environmental clean-up can be applied as follows :

9.2.1.1.1 *In situ* promotion of microbial growth

It can be achieved by the addition of nutrients. When the indigenous microbial populations are exposed to specific polluting compounds for longer periods, sub-populations develop a limited metabolic ability to utilize and, thus, degrade the offending pollutant. The growth of these particular microbes, however, will invariably be nutrient limited. When essential growth nutrients such as nitrogen and phosphorus are added, normally the growth stimulates and there is an increase in the breakdown of pollutant. Nutrient supplementation is basically done for the following purposes:

(a)   To enrich the useful microbes

(b)   To scale-up from the mixture by bioreactor cultivation

(c)   To reinoculate large quantities of the 'cocktail' of microbes into the contaminated site.

This method was successfully applied to beaches along Prince William Sound and the Gulf of Alaska in 1989-1990 to clean up the oil spillage from the oil tanker Exxon Valdez. The application of fertilizers (nutrients) in various formulations to the beaches stimulated the indigenous microorganisms to degrade the oil to less harmful products that subsequently became part of the food chain. This method is used to dechlorinate recalcitrant chemicals such as polychlorinated biphenyls (PCBs) (considered being highly toxic and indestructible industrial pollutants).

The extraction of the contaminants and their treatment on the surface in bioreactors, the stimulation of microbial growth *in situ* by providing extra nutrients, and the supplementation of the microbial population, are the other processes that can be applied.

The indigenous microbial population can be stimulated by the addition of nitrogen and phosphorus containing substrates. The addition of a second carbon source to stimulate co-metabolism can be another alternative.

Over the years, a number of *in situ* aerobic co-metabolism processes have been demonstrated. The major problem with this process is to ensure that the substrates added are distributed through the soil and not just close to the injection bore. To reduce recirculation, systems have been investigated for trichloroethylene removal where oxygen and methane, the co-metabolic substrate, were added to the subsoil.

Bioaugmentation. The addition of microorganisms is a process known as bioaugmentation. It has been suggested that the best approach is to stimulate the indigenous microbial populations, but this is not suitable for all occasions. Some degradative pathways can produce intermediates which trap the pathways or transform the pollutants into non-toxic compounds. Thus there may be situations which can be improved by bioaugmentation with selected or genetic manipulated organisms. A number of fungal inocula have been used to bioaugment soils contaminated with PCP and this procedure removed 80-90% of PCP within 4 weeks. A selected strain, in a field study *Methylosinus trichosporium* was used. Here the organism was selected for its high TCE transformation rate under low copper conditions. Once the organisms had been injected, 50% of the cells attached to the soil, formed a biofilter which was efficient for the transformation of TCE.

All the soil contaminants are not easy to be removed or readily available for degradation. Many are insoluble in the aqueous phase so that a number of methods have been used for their extraction.

(a) *Application of biosurfactants:* These are surface-active molecules that solubilize, emulsify, disperse and act as detergents. These surfactants represent a spectrum of structures and are generally more complex than chemical surfactants. The addition of biosurfactants to oil-contaminated soil has resulted in an increased rate of degradation. Various biosurfactants have been used to solubilize oils and xenobiotics such as PCBs and organophosphates. These biosurfactants can be produced *in situ* by added microorganisms.

(b) *Use of liquid carbon dioxide:* This has been used for the removal of compounds. The method has been successfully used for diesel removal.

(c) *Soil washing:* This has been used to remove pollutants such as pentachlorophenol. The soil is generally removed and washed by a series of scrubbing and physical separation techniques. In this technique the contaminants are partitioned into the liquid phase and the fine particles which were collected for biotreatment or disposal. The water can be treated in a number of bioreactors which can be run in sequence.

In fact many microbes are capable of generating ferric ions which act as oxidizing agents and transform solid metals (sulfide minerals) into water soluble metal ions and elemental sulfur. A microorganism capable of removing a portion of organic sulfur (thiophenic sulfur) has been developed and patented in USA. Many such organisms involving the removal of additional organic sulfur (aromatic sulfides) from coal are being developed.

No cost effective technology exists for the pre-combustion and desulfurizaton of coal, but biotechnology may provide a solution. Various strains of bacteria *Sulfolobus acidocaldarius, Pseudomonas* species (TG 232, CB1) are being developed which can remove organic sulfur and yet maintain the fuel value of coal (Kilbane, 1989).

Deutrofication

River and lake waters get eutroficated by phosphates coming from detergents in waste water. A number of microbial species which are also able to produce biosurfactants *e.g., Nocardia erythropolis* are being exploited to get rid of eutrofication.

Spilled Oil in Waters

Petroleum-based wastes. Crude oil is an extremely complex and variable mixture of organic compounds. The majority of the compounds in crude oil are hydrocarbons which can range in molecular weight from the gas methane to the high molecular weight tars and bitumens. These hydrocarbons can also come in a wide range of molecular structures: straight and branched chains, single or condensed rings and aromatic rings. The two major groups of aromatic hydrocarbons are:

(a) monocyclic such as benzene, toluene, ethylbenzene and xylene (BTEX) and

(b) the polycyclic hydrocarbons (PAHs) such as naphthalene, anthracence and phenanthrene.

Crude oil has accumulated underground as a result of the anaerobic degradation of organisms over a very long time. Under the conditions of high temperature and pressure, the organic material has been converted to natural gas, liquid crude oil, shale oil and tars. At the underground temperatures, shale oils and tars do not flow, but crude oil is liquid and unless contained, will escape to the surface, where the volatiles evaporate forming a tar bed. Apart from this, release of crude oil, the main source of crude oil and oil products, such as BTEX and PAHs released into the environment comes from the leaking of storage tanks, spillages and accidents during its transport. These BTEX compounds, although not miscible with water are mobile and can contaminate the ground water. The volatile components can be lost to the atmosphere if the leak is on the surface, but if the leak is below soil level, the mobile components can migrate down through the soil to the water table (Fig. 9.1).

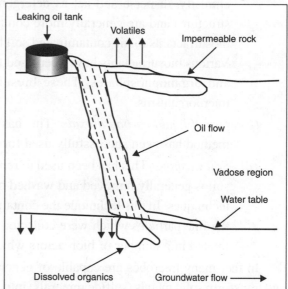

Fig. 9.1. Crude oil: Spilling of crude oil on soil leads some of it to penetrate till the water table, a part is evaporated and rest is spread on the soil. (from Internet)

As indicated above the crude oil when released at sea will not mix with sea water and will float on the surface. This allows the escape of the volatile components those of 12 carbons and below. The floating oil, if it does not reach the shore, will be dispersed due to the action of waves.

The dispersion will allow naturally occurring hydrocarbon-degrading organisms to break down the oil. Oil breakdown will occur at the interface between the oil and water and therefore, the better the oil dispersion, the greater the area, the faster the degradation. The more complex and less soluble oil components will be degraded much more slowly than the lighter oils. It is these high molecular weight components that will persist on the sand and rocks if the spill reaches shore.

(a) *The first stage in the recovery and clean-up of an oil spill.* It require to stop the release and contain the spill. The surface oil can be mechanically removed with skimmers and other machines. Chemical dispersants can be used on both floating oil and oil which has reached a rocky shore, but care has to be taken as the detergents can be as harmful to the environment as the oil. Although both physical and chemical methods are efficient, not all the oil can be removed and it is the remaining high molecular weight oil that may need to be removed by some form of bioremediation (Fig. 9.2).

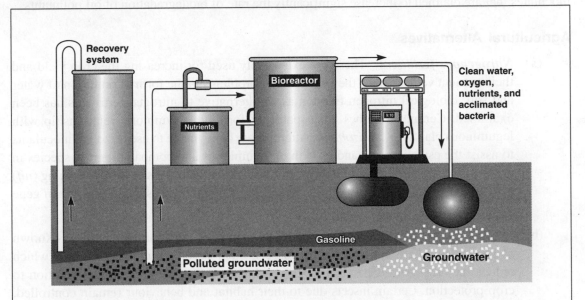

Fig. 9.2. Bioremediation of a gasoline spill. Gasoline from an underground storage tank seeps through the soil to the water table. After the leak is stopped, the free-floating gasoline is pumped out to a recovery tank. Polluted ground water is pumped into a bioreactor tank with oxygen, nutrients, and hungry microbes. After the microbes eat the gasoline the mixture of clean water, nutrients, and microbes is pumped back into the ground so that more of the pollutant can be degraded (Modified from Kreuzer and Massey, 2003).

(b) *Addition of microbial population.* Crude oil can be degraded by the indigenous microbial population in the sea and on the shore, as microorganisms capable of using hydrocarbons as a source of carbon compounds and energy are widely found. However, it is thought that the supply of utilizable phosphorus and nitrogen compounds is limited in most marine environments, so as to encourage degradation. Slow release fertilizers have been added to oil slicks at a nitrogen to oil ratio of 1:100.

Oil spills also destroy the habitat of the aquatic animals and fishes and cause long-term damage to the environment. Chemical dispersants were used to remove these, but they cause major pollution problem in shallow waters due to their toxic nature and persistence in the environment. In recent years, by use of genetic engineering, oil utilizing microorganisms have been successfully produced, which would rapidly grow on oil. Many oil utilizing microbes produce surface active compounds that can emulsify oil in water and, thus, facilitate the removal of oil.

A strain of *Pseudomonas* has been developed by Dr. Ananda Mohan Chakrabarty that produces a glycolipid emulsifier capable of reducing the surface tension of oil-water interface and, thus, helps in removal of oil from water. This microbial emulsifier is non-toxic and biodegradable.

This has been quite successful in some sites. Some companies are now marketing the microbial inoculates that are claimed to increase significantly the rate of biodegradation of oil pollutants.

Agricultural Alternatives

(a) *Nitrogenous fertilizers* : These are extensively used for increasing the crop yield and their nutrient values. Since they cause lot of problems in the form of pollution of water and soil, biological nitrogen fixation as an alternative to nitrogen fertilizers has been examined. Certain microbes are capable of forming a symbiotic relationship with leguminous plants *e.g., Rhizobium*. Attempts are being made to use rhizobial inoculants to assist the establishment and growth of leguminous and companion grass species in areas where soil nitrogen is quite low. It is now possible to transfer nitrogen fixing (*nif*) gene by conjugation in *Klebsiella pneumoniae*. It might be possible to transfer *nif* gene from *Rhizobium* to non leguminous plants.

(b) *Pesticides and insecticides* : Field pests are controlled by applying chemicals known as pesticides. These are used to kill insects, other pests and combat diseases which reduce the crop yield. Chemical pesticides have not provided a complete solution to crop protection. Certain insects due to their habitat and behaviour remain controlled. Some of the insects also develop resistance to pesticides.

The residues of these pesticides can for sometime be present on foliage, in the plants and in the soil. Pesticide residues in soil interact with constituents like clay and organic matter. Adsorption of a pesticide onto a clay surface and interactions with organic matter can result in partial or even complete inactivation. The nature of soil pesticide interactions can in turn determine the extent to which a pesticide is leached into soil. Certain persistent pesticides such as DDT and associated breakdown product 1.1-dichloro-2,2-diethane (DDE) can remain in soil for considerable periods of time and accumulate in soil fauna which form part of food chain for higher animals.

An alternative to chemical pest management can be developed by manipulation of bacteria and fungi for obtaining microbial products with fungicidal, herbicidal and insecticidal properties.

The following microbes are used in the management of pests :

(*i*) **Bacteria as pesticides:** These belong to the families, Pseudomoniaceae, Lactobacilliaceae, Micrococcaceae and Bacilliaceae. Majority of commercial strains belong to genus *Bacillus*. The other examples are :

   (*a*) *Metarhizium anisopliae*: It infects a number of different insects including spittle bugs in pasture and sugarcane for which it is commercially used in Brazil.

   (*b*) *Beaureria bassiane*: It attacks Colorado potato (produces cyclodepsipeptide toxin). It is used in Russia under the name Boverin.

   (*c*) *Hirsutella thompsoni*: It is used for the control of citrus mites.

      Bacillus thuringiensis is a gram positive bacterium that produces parasporal, crystalline inclusions during sporulation. These inclusions consist of proteins (insecticidal crystal proteins ICPs or delta-endotoxins) having molecular masses of 70 to 130 kDa. Most *B. thuringiensis* strains that have been characterized produce ICPs that exhibit highly specific activity against some lepidopteran (caterpillar), dipteran (mosquito and blackfly) or coleopteran (beetle) larvae. In addition, some recently discovered strains are reported to be toxic to protozoan pathogens, nematodes and mites. *B. thuringiensis* ICPs are environmentally sound alternatives to chemical insecticides. It is expected that their use in insect control will increase in the near future. They are applied either as a spray of natural organisms or in genetically modified organisms such as plant-associated bacteria, cyanobacteria or transgenic plants. In this context, experimental work has shifted towards developing crops transformed with *ICP* genes to become insect resistant. The use of these insecticidal transgenic plants will eliminate stability problems and should significantly reduce cost of insect control. More than 50 crops such as cotton, corn, soyabean and rice have now been successfully transformed.

(*ii*) **Viruses as pesticides:** Baculoviruses are commonly used as pesticides. These are DNA viruses and can be placed in 3 groups :

   (*a*) Nuclear Polyhedrosis viruses (NPVs)

   (*b*) Granulosis viruses (GVs)

   (*c*) Nonoccluded viruses (NOVs)

There are certain other groups of viruses that belong to this category. These are :

   (*a*) Cytoplasmic polyhedrosis viruses

   (*b*) Entomopox viruses

   (*c*) Iridoviruses

Viruses have undergone extensive evaluation and have been registered for use against insects. Due to their specificity and effectiveness, viruses have greatest potential for use in the control of integrated pest management.

Among several types of insect viruses are nuclear polyhedrovirus (NPU) and Granulosis virus (GV). Insect viruses belong to the faimly Baculoviridae. Among viral insecticides occluded viruses are beneficial as they are made up of proteinaceous and silicaceous covers, which are resistant against harsh environmental factors such as temperature, UV radiation *etc.* Approximately 280 insect viruses have been described.

Nuclear polyhedrosis viruses are rod-shaped viral particles embedded within distinct polyhedra inclusion bodies. These inclusion bodies are 0.5-3.0 micrometer and are visible under microscope. The nucleocapsid or virion in each polyhedron is made up of double-stranded DNA. Majority of the NPV viruses with narrow range, insecticidal activity. However, some isolates from alfa alfa caterpillar *Autographa californica* has a broad spectrum insecticidal activity and shows target upto 40 lepidopteran species.

Since virus requires living media to grow, large scale production is still in infant stages. The needs of larvae for NPV virus culture have been registered in USA. Several European countries like Denmark, Sweden, USA. Australia have already implemented practice of using viral insecticides. In India, some promising works have been carried out on the use of viruses on some species.

The recombinant DNA technology is being used to insert new sequences into the gene for polyhedral proteins which can be used for the synthesis of novel proteins. These new proteins could include the protein toxins of *Bacillus thuringiensis*, by this potential toxic effect of viruses is enhanced. Although most viruses grow in living insects, a few can be grown in insect cell culture providing more uniformity.

Bioremediation of soils

Soils contain a very large number of microorganisms. These include a number of hydrocarbon utilizing bacteria and fungi representing 1% of the total population of some 10^4–10^6 cells per gram of soil. In addition, cyanobacteria and algae have also been found to degrade hydrocarbons. Hydrocarbon-contaminated soils have been found to contain more microorganisms than uncontaminated soils, but the diversity of the microorganisms was reduced.

The fate of organic compounds in the environment is affected by a number of factors which can be grouped into those affecting:

(a) the growth and metabolism of the microorganisms
(b) the growth and metabolism of the microorganisms
(c) the compound itself

Factors affecting the growth of microorganisms are as follows:

• Presence of other biodegradable organic material
• Presence of nitrogen- and phosphorus containing inorganic compounds
• Oxygen levels
• Temperature

- pH
- Presence of water, soil moisture
- Number of types of microorganisms present
- Presence of heavy metals.

Factors affecting degradation of the compound are as follows:

- Bacterial growth and metabolism
- Chemical structure of the organic compound
- Availability and/or solubility
- Photochemistry

The biodegradation of the hydrocarbons is associated with microbial growth and metabolism and therefore, any of the factors affecting microbial growth will influence degradation. Some other growth substrate will be needed, if the microorganisms cannot use the hydrocarbons as their sole source of energy and carbon skeletons. In some cases, if another substrate is present, the microorganisms may use this in preference to the hydrocarbons. The microorganisms may also require supplementation with nitrogen and phosphorus containing compounds. Aerobic degradation of hydrocarbons is considerably faster than the anaerobic process so that a supply of oxygen will be needed to maintain aerobic conditions, if rapid degradation is required. Soil with an open structure will encourage the oxygen transfer. A waterlogged soil will have the reverse effect. The microbial growth is affected by the temperature. At low temperatures, the rate of degradation will be slow. Nutrient addition to soils at temperature of 4-10°C has been shown to have little effect as the low temperature has reduced growth of such a low level. The pH of the soil will affect both the growth and the solubility of the compound to be degraded. In the beginning, the presence of a large number of hydrocarbon-degrading microorganisms in the soil will be of advantage start. As, however, most soils contain these types of organisms, growth will soon increase the numbers, so that seeding with specific hydrocarbon-degrading organisms will probably not be needed. Hydrocarbon contamination may also be associated with high levels of heavy metals which may inhabit microbial growth. It depends on the concentration and type of metals.

Pathways of Degradation

Soil microorganisms degrade petrochemicals, PAHs and BTEX compounds. These use them as a source of both energy and carbon compounds for cell synthesis. Hydrocarbons are stable reduced compounds. Their degradation generally proceeds by oxidation under either aerobic or anaerobic conditions. Many attempts have been made to study the microbial degradation of monocyclic and polycyclic aromatic hydrocarbons. Aliphatic hydrocarbons are converted into alcohols and then sequentially oxidized to carboxylic acids which are β-oxidized.

Monocyclic aromatics, are first hydroxylated by a dioxygenase enzyme to cis-1, 2-dihydroxy-1, 2-dihydrobenzene, which is then converted to catechol.

One of the two pathways can be taken by the subsequent metabolism of catechol:

(a) Ortho cleavage yields, *cis*-muconate,
(h) Meta cleavage, however, yields 2-hydroxymuconic semialdehyde. Both pathways lead to compounds which can enter the Krebs cycle.

The first two stages of the degradation of benzene are common for the breakdown of many other monocyclic and polycyclic aromatic hydrocarbons.

In general aromatic ring hydroxylation is followed by ring cleavage and the oxygenases carry out both of these reactions. The incorporation of two oxygen molecules causes the introduction of two hydroxyl groups which can undergo either meta- or ortho-cleavage. Incorporation of a single oxygen molecule is catalyzed by monooxygenases and both enzymatic systems can be used to degrade polycyclic aromatic hydrocarbons (Fig. 9.3).

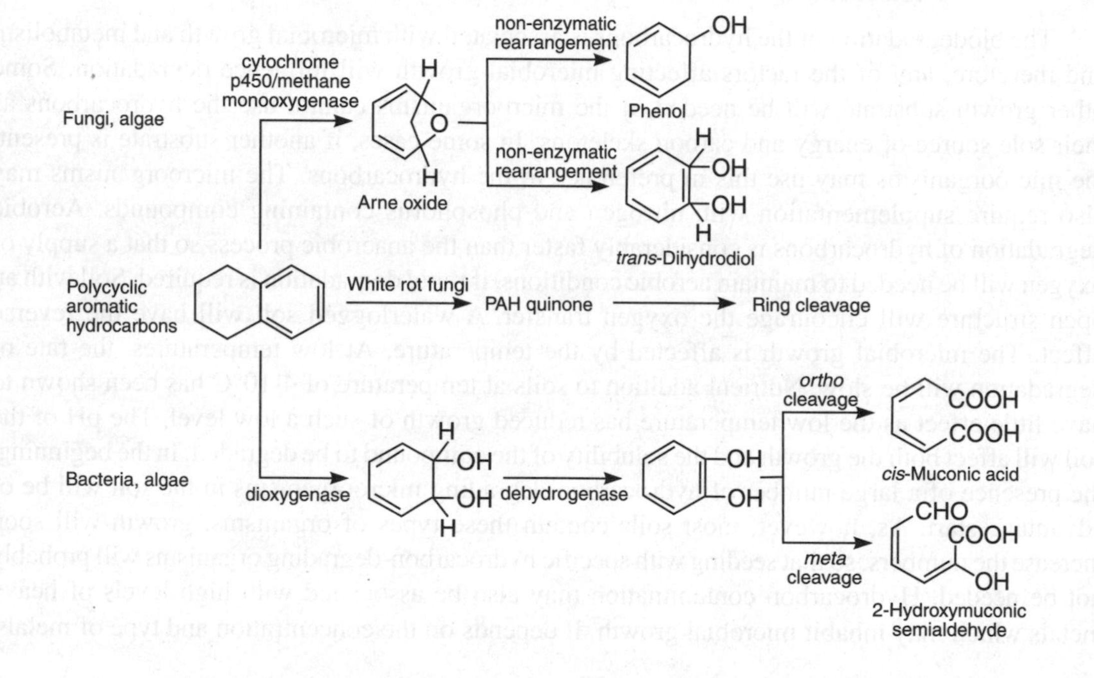

Fig. 9.3. Ring Cleavage Pathway

Many of the monocyclic aromatic hydrocarbon-degrading bacteria have chromosomal genes coding for the enzymes of the pathway for the degradation of hydrocarbons, but plasmids have been found that code for the degradation of compounds such as camphor, octane, toluene, naphthalene and some herbicides and pesticides.

Some plasmids code only for some part of the pathway. It is helpful to the soil to be able to transfer the degradative ability between bacteria, as this allows the rapid adaptation of the population to a particular compound. The best studied plasmid-based pathway has been the toluene degradation by the bacterium *Pseudomonas putida* mt-2 and the Plasmid TOL. The TOL plasmid has been shown to confer the ability to degrade toluene and other benzene derivatives. The *xyl* genes are arranged into two groups or operons:

(i)   the *xylCAB* codes for the degradation to benzoic acid

(ii) the other group *xylXYZLEGFJIH* code for the breakdown of benzoate into pyruvate and acetaldehyde (Table 9.2).

The bioremediation of hydrocarbon-contaminated soil cannot always be maintained under aerobic conditions. It is because of the following :

(a) Waterlogging

(b) The fine particle structure of the soil and blocking of the soil pores with the biomass itself.

However, aliphatic, monocyclic and polycyclic aromatic hydrocarbons can be anaerobically degraded provided oxygen can be obtained from water under methanogenic conditions and from sulfate under sulfur-reducing conditions. The hydrocarbons are converted to central metabolic intermediates by the following mechanisms :

(a) Hydration

(b) Dehydration

(c) Reductive dehydroxylation

(d) Nitroreduction

(e) Carboxylation

The central intermediates and benzoyl CoA and sometimes resorcinol are reduced and hydrolyzed and finally transformed to compounds which can enter the Krebs cycle. The only disadvantage with anaerobic degradation is that the process is much slower than the aerobic pathway.

Table 9.2: The Genes and their Enzymes Involved in Xylene Metabolism Pathway (From Internet)

Stages	Gene	Enzymes
Earlier stages	*xylA*	Initial part of pathway
	xylB	Xylene oxygenase
	xylC	Benzyl alcohol dehydrogenase
		Benzaldehyde dehydrogenase
Later part of pathway		
	xylX,Y,Z	Toluate dioxygenase
	xylE	Catechol 2,3-dioxygenase
	xylF	2-Hydroxymuconic semialdehyde
Hydrolase	*xylG*	2-hydroxymuconic semialdehyde
Dehydrogenase	*xylH*	4-Oxalocrotonate tautomerase
	xylI	4-Oxalocrotonate decarboxylase
	xylJ	2-Oxopent-4-enoate hydratase
	XylK	2-Oxo-4-hydroxypentenoate aldolase
	xylL	Dihydroxycyclohexadiene carboxylate
Dehydrogenase	*xylR*	Regulatory protein
	xylS	Regulatory protein

The degradation of both hydrocarbons and other organic molecules involve the ability of some enzymes to function with compounds other than their normal substrate. This condition often known as gratuitous metabolism, is probably a result of broad enzyme specificity. Co-metabolism describes the metabolism of a substrate not required for growth in which no apparent benefit is derived, but the co-metabolism in organisms is an undefined term describing the imprecise specificity of enzyme system and their induction system. For trichloroethylene the co-metabolic substrates of phenol and toluene have been described in *in situ* aerobic co-metabolism.

Microbial degradation of xenobiotics

Xenobiotics enter the environment through different paths, some as component of fertilizers, pesticides and herbicides are distributed by direct applications, others, such as the polycyclic aromatic hydrocarbons, dibenzo-p-dioxins and dibenzofelltrans, are released by combustion process. Many kinds of xenobiotics are found in the waste effluents produced by the manufacture and consumption of all the commonly used synthetic products. Xenobiotics released to the environment on a large scale include halogenated aliphatic and aromatic compounds, nitroaromatics, phthalate esters and polycyclic aromatic hydrocarbons. The local concentration of xenobiotics in environment depends on the following factors :

  (i)   The amount of compound released
  (ii) The rate at which it is released
  (iii) The extent of its dilution in the environment
  (iv) The mobility of the compound in a particular environment
  (v)   Its rate of degradation.

Many toxic xenobiotics present in the environment can become progressively more concentrated in each link of a food chain, a process called **biomagnification.** These include DDT, phthalate esters and polychlorobiphenyls (PCBs). Xenobiotics may be biodegradable, persistent or recalcitrants. Some microorganisms are metabolically able to carry out biodegradation of some of these compounds. Phenols in aqueous solution can be removed by treatment with mushroom tyrosinase. The reduction or order of substituted phenol is catechol > *p.* cresol > *p.* chlorophenol > *p.* methoxy phenol. In the treatment of tyrosinase alone, no precipitate was formed but colour changed from colourless to dark brown. The coloured compounds are removed by chitin and chitosan which are available in shellfish waste. Tyrosine immobilized on cation exchange resin can be repeatedly used. In just 2 hours, 100% of phenol can be removed by this treatment. In this enzymatic reaction phenol and other aromatic amines are polymerized to form less water soluble compounds and then precipitates are separated by filtration.

Biodegradation in a particular environment requires the presence of suitable microorganisms. This may involve a complex microbial community. The environment must also be suitable for the growth of these organisms and for any chemical transformation reaction to proceed at a significant rate. Important factors include:

  (i)   the concentration of toxic chemicals,
  (ii) the presence of other substrates and nutrients,

(iii) temperature,

(iv) pH,

  (v) oxygen concentration, *etc.*

A number of possible mechanisms lead to active biodegradation of xenobiotics. Gratuitous biodegradation occurs when an enzyme is able to transform a compound other than its natural substrate. The prerequisite for this is that the unnatural substrate should bind to the active site of an enzyme in such a manner that the enzyme can exert its catalytic activity. Co-metabolism is the process in which a substrate is modified often by stoichiometric conversion to a single product but is not utilized for growth by an organism that is grown on or metabolizing other substrate. The degradation of a range of compounds proceeds more readily with a mixed culture of organisms

The problem of toxic waste disposal is enormous. Worldwide production in 1985 of just one chemical that is released into the environment pentachlorophenol was more than 50,000 tons. Incineration and chemical treatment have been used to breakdown toxic chemicals, but these methods are costly and often create new environmental difficulties. With the discovery in the mid 1960s of a number of soil microorganisms that are capable of degrading xenobiotic ("unnatural", synthetic; from xenos, meaning "foreign") chemicals such as herbicides, pesticides, refrigerants, solvents and other organic compounds, the notion that microbial degradation might provide an economic and effective means of disposing of toxic chemical wastes, gained importance.

Members of the genus *Pseudomonas* are the most predominant group of soil microorganisms that degrade xenobiotic compounds. Biochemical assays have shown that various *Pseudomonas* strains can breakdown and as a consequence, detoxify more than 100 different organic compounds. In many cases, one strain can use several different related compounds as the sole carbon source.

The biodegradation of complex organic molecules generally requires the concerted efforts of several different enzymes. The genes that code for the enzymes of these biodegradative pathways are sometimes located in the chromosomal DNA although they are more often found on large (approximately 50 to 200 kilobase [kb]) plasmids. In some organisms, the genes that contribute to the degradative pathway are found on both chromosomal and plasmid DNA.

Degradative bacteria, in most cases, enzymatically convert xenobiotic, non-halogenated aromatic compounds to either catechol or protocatechuate. Then, through a series of oxidative cleavage reactions, catechol and protocatechaute are processed to yield either acetyl coenzyme A (acetyl-CoA) and succinate or pyruvate and acetaldehyde, compounds that are readily metabolized by almost all organisms (Fig. 9.4). Halogenated aromatic compounds which are the main components of most pesticides and herbicides, are converted to catechol, protocatechuate, hydroquinones or the corresponding halogenated derivatives by the same enzymes that degrade the non-halogenated compounds. However, for the halogenated compounds, the rate of degradation is inversely related to the number of halogen atoms that are initially present on the target compound. Dehalogenation, the removal of halogen substituent from an organic compound – is the critical requirement for detoxification and often occurs by a non-selective dioxygenase reaction that replaces the halogen on a benzene ring with a hydroxyl group. This step may occur either during or after the biodegradation of the original halogenated compound.

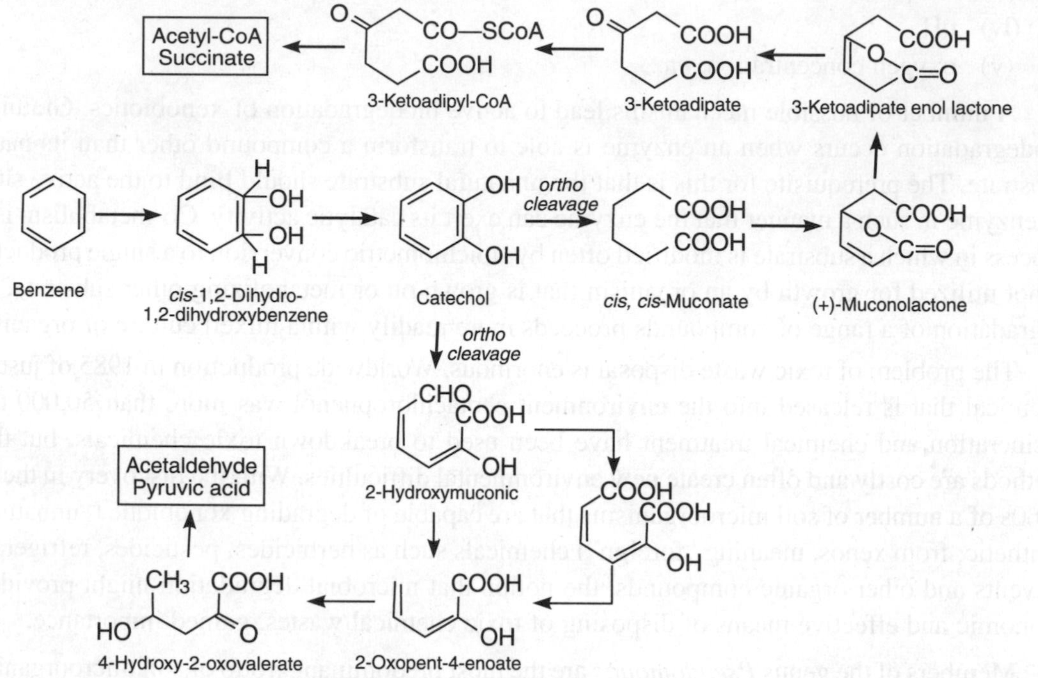

Fig. 9.4. Catechol metabolism pathways.

Genetic engineering of biodegradative pathways

Despite the ability of many naturally occurring microorganisms to degrade a number of different xenobiotic chemicals, there are limitations to the biological treatment of these waste materials. For example :

  (i)   no single microorganism can degrade all organic wastes,

  (ii) high concentrations of some organic compounds can inhibit the activity or growth of degradative microorganisms,

  (iii) most contaminated sites contain mixture of chemicals, and an organism that can degrade one or more of the components of the mixture may be inhibited by other components

  (iv) many non-polar compounds absorb onto particulate matter in soils or sediments and become less available to degradative microorganisms,

  (v)   microbial biodegradation of organic compounds is often quite slow.

Some of these problems can be addressed transferring by conjugation into a recipient strain plasmids that carry genes for different degradative pathways. If two resident plasmids contain homologous regions of DNA, recombination can occur and a single, large "fusion" plasmid with combined functions can be created. Alternatively, if two plasmids do not contain homologous region and in addition belong to different incompatibility groups, they can coexist within a single bacterium.

Synthetic organic compounds

Some organic compounds, many of which are on the priority pollutant lists, also contaminate the environment. Many of these compounds do not occur naturally. These essentially new organic compounds are often referred to as xenobiotics, from the Greek *xenos*, meaning new. Many thousands of compounds have been synthesized and some of these find their way into the environment. Most biocides and herbicides are released into the environment by direct use. Others such as polychlorobiphenyls (PCBs) which are used as hydraulic fluids, plasticizers, adhesives, lubricants, flame retardants and dielectric fluids in transformers are released into the environment during production, from spillages and disposal. The chlorinated solvents such as trichloroethene, carbon tetrachloride and tetrachloroethene constitute another group of compounds that frequently contaminate ground water. Trichloroethene is a priority pollutant and contamination is found at industrial, commercial and military sites as a result of disposal and spillage. During the combustion of PAHs, the contaminants such as dibenzo-p-dioxins and dibenzofurans can be formed. The release of some of these xenobiotics into the environment can occur on a very large scale.

Many of the xenobiotic compounds released into the environment can be microbially degraded. The others, however, are removed only slowly and in some cases, so slowly as to render them effectively permanent. There are number of unwanted consequences of the persistence of organic molecules in the environment. These include protracted exposure to the compound by organisms in the environment, which can increase its toxic effect. The very low rates of degradation mean that organisms tend to accumulate the compound, in a process known as **biomagnification.** In this case, the first organism may be the prey of another and as a consequence, the concentration of the compound will be increased in the second organism. Thus, a very low level of compound in the environment, under certain circumstances, can be magnified to a high and toxic levels. An example of this is the accumulation of DDT in grebes (fish-eating water birds) which were at the top of the food chain on a lake treated with 0.01-0.02 ppm DDT to control gnats. The levels found in the grebes were some 100,000 times the DDT level applied. Another problem with these persistent compounds is that in many cases there is little information on their toxicity and long-term effects.

The persistence of xenobiotics in the environment is influenced by the following:

(a) The structure and properties of the molecules

(b) Their toxicity to microorganisms

(c) The environmental conditions

Natural complex polymeric molecules, such as lignin, will be slow to degrade. Many of the xenobiotic compounds are also complex in structure and, therefore, slow to degrade. For many of the xenobiotics, their activity and persistence are also linked to the presence of halogens in the compound.

The number of halogen molecules, their position and the type of halogen, the toxicity and persistence of the organohalogen are influenced. Due to their hydrophobic properties organochlorines are effective biocides. It allows them to pass through or into membranes and disrupt cellular activity. They are also effective because of their ability to inhibit oxidative phosphorylation. As the chlorine content increases, the solubility of organochlorines decreases. It often increases their toxicity. Degradation of these organochlorines by microorganisms has to overcome their increased toxicity and their low solubility in the aqueous phase. The problem may also be increased as some organochlorine compounds dare a mixture of isomers, as found with PCBs which decreases the rate of degradation.

The rate of degradation of xenobiotics in soil and water is also dependent on the presence of microorganisms with the enzymatic capability to degrade the molecule. As xenobiotics are not normally found in nature, the level of the degrading microorganism may be very low. In many cases, before degradation occurs, a period of adaptation is required. The environmental conditions which affect the degradation of xenobiotics are essentially the same as those which affect the degradation of hydrocarbons.

Pathways of degradation

Organisms capable of degrading xenobiotics have been found in soil and sediment, particularly from contaminated sites. As mentioned earlier these include bacteria, fungi and algae. The degradation pathways for some of the chloroaromatics have been determined under both aerobic and anaerobic conditions. The chloroaromatics are normally cleaved by monooxygenases and dioxygenases similar to those found with the degradation of PAHs. Superimposed on the degradation is the process of dehalogenation which can be by four mechanisms :

- **Oxidative dehalogenation** – The halogen is removed and replaced by two hydroxyl ions.
- **Eliminative dehalogenation** – The simultaneous removal of the halogen and an adjacent hydrogen ion.
- **Hydrolytic (substitutive) dehalogenation** – The substitution of the halogen with a hydroxyl ion.
- **Reductive dehalogenation** – The halogen is replaced by a hydrogen ion.

The following are examples of the degradation of organochlorine. These illustrates individual pathways and general processes.

Pentachlorophenol (PCP). It is a herbicide and fungicide used for the preservation of wood and is a priority pollutant. A number of microorganisms which can degrade PCP under aerobic and anaerobic conditions, including *Flavobacterium, Arthrobacter, Rhodococcus* and the white rot fungus *Phaerochaete chryososporium* have been isolated. In most of the pathways, the breakdown of chlorophenols consists of the dechlorination and hydroxylation of the aromatic ring followed by ring cleavage. The oxygenases catalyze both hydroxylation and ring cleavage. In PCP degradation where aerobic degradation starts with reductive dehalogenation, the first step appears to be the rate-limiting step (Fig. 9.5).

The use of polychlorinated biphenyls (PCBs) has been banned for over 20 years due to their toxicity, but as their use was extensive, contaminated sites and sediments remain. PCBs can be aerobically degraded and the first step is similar to that for monocyclic aromatics involving a dioxygenase which is similar in nature to the dioxygenases for naphthalene and toluene.

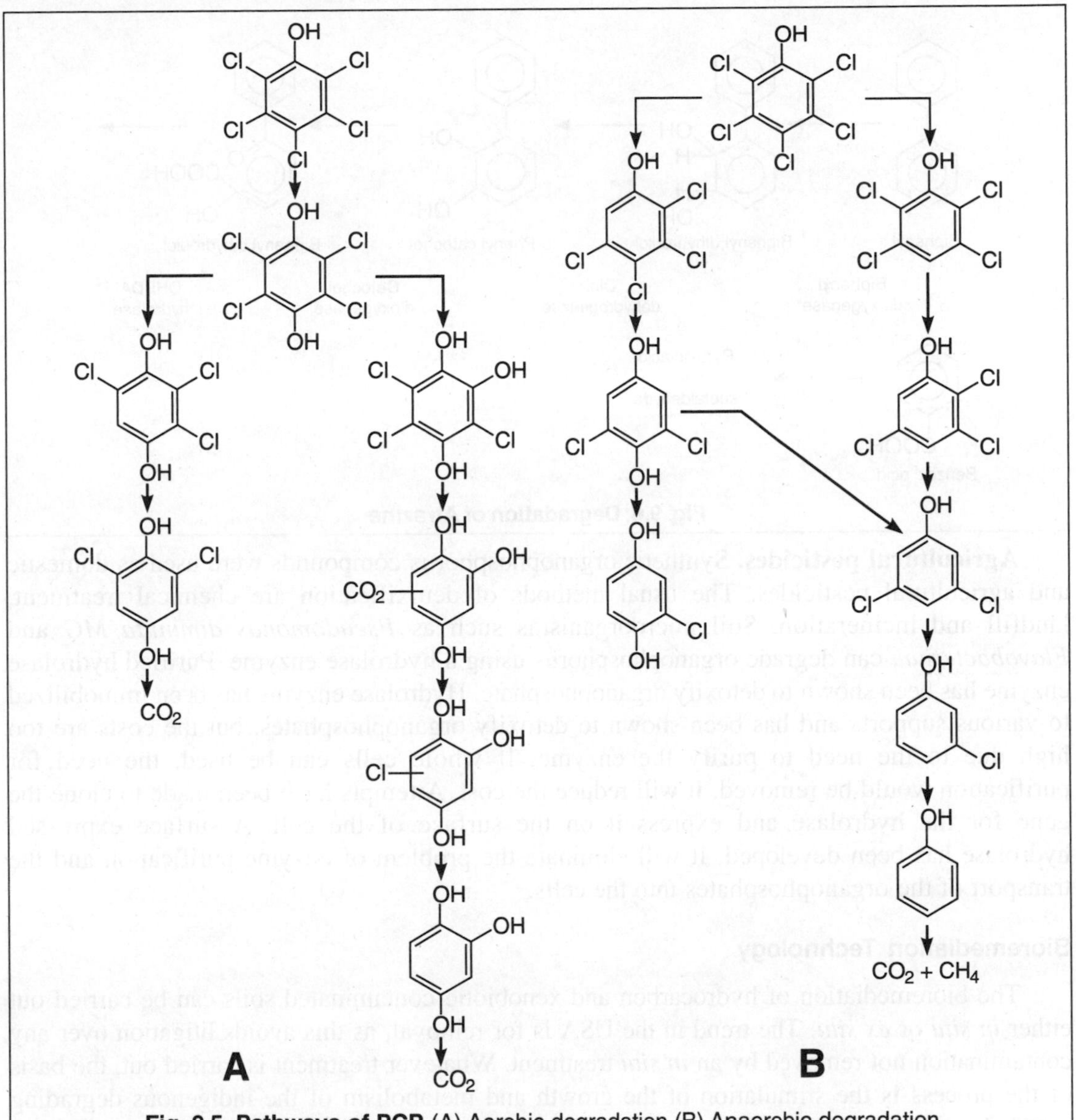

Fig. 9.5. Pathways of PCP (A) Aerobic degradation (B) Anaerobic degradation.

Atrazine. The most widely used triazine herbicide is atrazine. It is effective against broadleaved weeds. It had been employed for some 40 years and was considered to be recalcitrant. However, pure cultures of bacteria have been isolated which can degrade atrazine, although bacterial consortia have also been reported to degrade atrazine. Atrazine is converted to cyanurinc acid in three steps and the cyanurinc acid can be converted to CO_2 and NH_2. Cyanuric acid can also be metabolized by soil bacteria that cannot degrade it. In coding for the enzymes that convert atrazine to cyanuric acid, three genes are involved. In *Pseudomonas* spp. these genes are found on a plasmid. The isolation of these three enzymes has allowed the determination of the form of sharing of metabolism that occurs in a bacterial consortium which can degrade atrazine (Fig. 9.6).

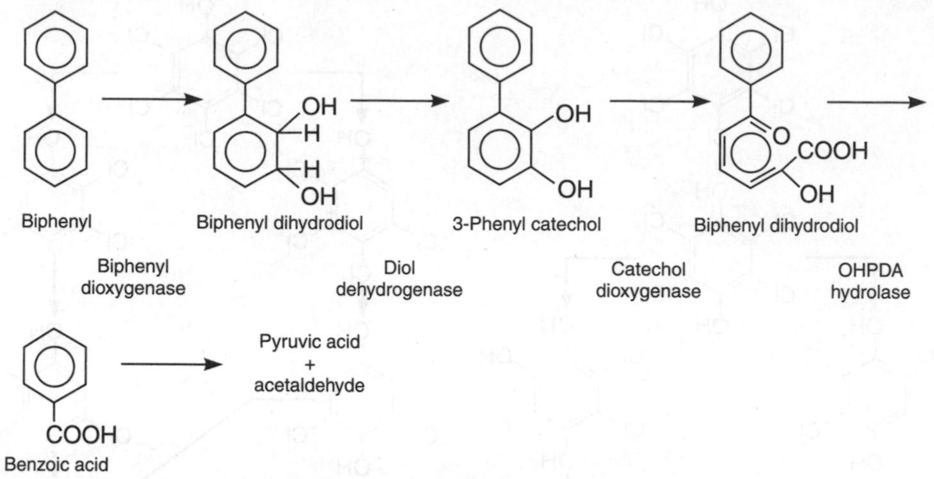

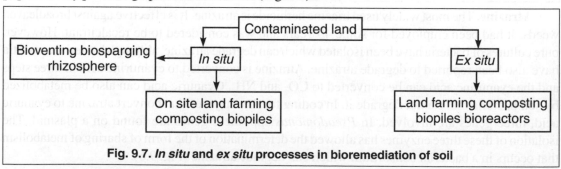

Biphenyl   Biphenyl dihydrodiol   3-Phenyl catechol   Biphenyl dihydrodiol

Biphenyl   Diol   Catechol   OHPDA
dioxygenase   dehydrogenase   dioxygenase   hydrolase

Pyruvic acid
+
acetaldehyde

Benzoic acid

Fig. 9.6. Degradation of Atrazine

Agricultural pesticides. Synthetic organophosphorus compounds were used as domestic and agricultural pesticides. The usual methods of detoxification are chemical treatment, landfill and incineration. Soil microorganisms such as *Pseudomonas diminuta MG* and *Flavobacterium* can degrade organophosphorus using a hydrolase enzyme. Purified hydrolase enzyme has been shown to detoxify organophosphate. Hydrolase enzyme has been immobilized to various supports and has been shown to detoxify organophosphates, but the costs are too high due to the need to purify the enzyme. If whole cells can be used, the need for purification would be removed. It will reduce the cost. Attempts have been made to clone the gene for the hydrolase and express it on the surface of the cell. A surface expressed hydrolase has been developed. It will eliminate the problem of enzyme purification and the transport of the organophosphates into the cells.

Bioremediation Technology

The bioremediation of hydrocarbon and xenobiotic contaminated soils can be carried out either *in situ* or *ex situ*. The trend in the USA is for removal, as this avoids litigation over any contamination not removed by an *in situ* treatment. Whatever treatment is carried out, the basis of the process is the stimulation of the growth and metabolism of the indigenous degrading population by providing optimum conditions (Fig. 9.7).

```
                                    Contaminated land
                                    /               \
Bioventing biosparging  ←  In situ                  Ex situ
    rhizosphere                |                        |
                               ↓                        ↓
                      On site land farming      Land farming composting
                        composting biopiles       biopiles bioreactors
```

Fig. 9.7. *In situ* and *ex situ* processes in bioremediation of soil

Ex situ treatment of contaminated land involves excavation of the soil for treatment or disposal elsewhere. Bioremediation process is land treatment or landfarming, where regular tilling of the soil increases aeration and supplementation with fertilizer. This encourages the microbial population. To retain any contaminants that leak out often, the treatment area is lined and dammed. The rate of degradation depends on the following facts :

(i)   The microbial population

(ii) The type and level of contamination

(iii) The soil type

With this type of system, the average half life for degradation of diesel fuel and heavy oils is in the order of 54 days.

Another solid-phase treatment carried out after extraction is **composting.**

(a) Composting material such as straw, bark and wood chips is mixed with the contaminated soil and piled into heaps. The process works in the same way as the normal compost system with a rise of temperature to 60°C and above caused by microbial activity. The growth of thermophilic bacteria is encouraged by the higher temperature. The increased costs of this type of system restrict it to highly contaminated materials, although the process is more rapid than land farming.

(b) Biopile process. In this the soil is heaped into piles within a lined area to prevent leaching. The piles are covered with polythene and liquid nutrients applied to the surface. By applying suction to the base of the pile as in a composting system, aeration can be improved. Any leachate formed is collected by pipes at the base and if necessary can be recycled. This type of system can be used when space is limited. When vapour emission need to be restricted, some form of biofilter can be added to the system (Fig. 9.8).

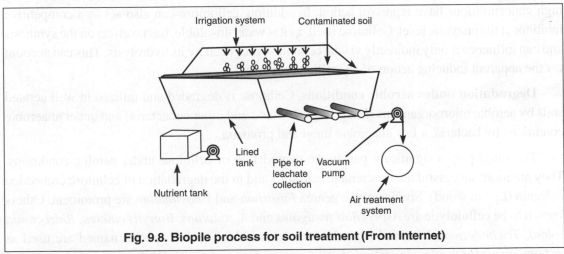

Fig. 9.8. Biopile process for soil treatment (From Internet)

(c) Bioreactors. Soil extracted from a contaminated site can also be treated as a solid waste (slurry) or as a liquid leachate in bioreactors of various designs. The use of bioreactors gives control of parameters such as temperature, pH mixing and oxygen supply, which can improve degradation rates. A wide range of bioreactor designs can be used and these include prepared bed reactors, slurry reactors, biofilters (fixed film) or anaerobic disasters

The processes of land farming, composting, biopiles and bioreactors can be carried out on site as well as on a soil removed from the contaminated land. These have been discussed in detail at the end of the chapter.

Degradation of Cellulose

Cellulose is the basic component of plant material and is produced in greater quantities than any other substance. Thus, about half of photosynthetically produced biomass consists of cellulose. Plant residues in soil consist of 45% (average) to 90% (cotton cellulose which is therefore of central importance – next to carbon dioxide – in the carbon cycle) cellulose.

The enzymatic cleavage of cellulose is catalyzed by cellulose. According to investigations on fungi, the cellulose system consists of at least three enzymes.

(1) Endo-β-1,4-glucanases attack the β-1,2-bonds in the centre of the macromolecule and produce long chain fragments with free ends.

(2) The exo-β-1,4-glucanases remove disaccharide cellobiose from the ends of the cellulose chains.

(3) The β-glucosidases hydrolyze cellobiose with formation of glucose.

The regulation of cellulose synthesis. This can be brought about either by catabolite repression or through substrate induction by cellobiose, whilst small amounts of cellulose are produced constitutively. The concentrations of cellobiose that have inducing or repressing effects vary in different organisms. In general, low concentrations of cellobiose act as inducer and relatively high concentrations have repressor action. In addition, cellobiose can also act as a competitive inhibitor at the enzymic level. Cellulose itself, as it is water-insoluble, has no effect on the synthesis and can influence it only indirectly via the cellobiose produced by its hydrolysis. This can account for the apparent inducing action of crystalline cellulose.

Degradation under aerobic conditions. Cellulose is degraded and utilized in well aerated soils by aerobic microorganisms (Fungi, myxobacteria and other eubacteria) and under anaerobic conditions by bacteria, a few anaerobic fungi and protozoa.

The fungi play a significant part in the degradation of cellulose under aerobic conditions. They are more successful than bacteria in acid soils and in the degradation of cellulose embedded in lignin (*i.e.,* in wood). Species of the genrea *Fusarium* and *Chaetomium* are prominent. Others known to be cellulolytic are *Aspergillus fumigatus* and *A. nidulans, Botrytis cinerea, Rhizoctonia solani, Trichoderma viride* and *Myrothecium verrucaria.* The last species named are used as test organisms for the demonstration of cellulolytic activity, respectively, for testing impregnation media used on cloths and coverings to prevent damage by cellulolytic microorganisms. The fungi excrete cellulases which can be isolated from the mycelium and the culture medium.

Cytophaga and *Sporocytophaga* are the most easily isolated of the aerobic cellulolytic bacteria. This is done by enrichment culture in liquid medium. Very little is known about the utilization and initial attack of cellulose by myxobacteria. No extracellular cellulose or initial cleavage products of cellulose have ever been demonstrated. The bacterial cells are attached to the celluose fibre with their longitudinal axis parallel to that of the fibre, they apparently hydrolyze cellulose only when they are in close contact with it and they immediately absorb the hydrolysis products. In addition to *Cytophaga* species, myxobacteria producing fruiting bodies of the genera *Polyangium, Sporangium* and *Archangium* are also able to grow on cellulose.

The ability to grow on cellulose as substrate is also quite common among many aerobic bacteria which could almost be considered 'omnivores'. Some of these attack cellulose only in the absence of other carbon sources. Some *Pseudomonas* like bacteria were formerly collected in a Cellvibrio group. *Psedomonas fluorescens* var *cellulose* has also been described. *Cellulomonas* should be mentioned among the coryneform bacteria.

A few cellulolytic species have also been described among the actinomycetes.: *Micromonospora chalcea, Streptomyces cellulose, S. sporangium.*

Degradation under anaerobic conditions. Under anaerobic conditions, cellulose is degraded by mesophilic and thermophilic eubacteria and a few fungi and protozoa. The termophile *Clostridium thermocellum* grows in a simple synthetic medium with cellulose or cellibiose as substrate and ammonium salts as the only nitrogen source; glucose and many other sugars cannot be metabolized. The degradation of cellulose is preceeded by the secretion of a yellowish, carotinoid-like substance, as has been observed in other cellulolytic bacteria. This increases the affinity of the cellulolytic enzyme for cellulose. The yellow colouration of the cellulose is a good indicator for the initiation of cellulose hydrolysis. In the case of *C. thermocellum,* the multienzyme complexes described above are organized in so called cellulosomes which can, have molecular weights of several millions.

The products of this cellulose fermentations are ethanol, acetate, lactate, formate, molecular hydrogen and carbon dioxide. Extracellularly, the cellulose is probably converted to glucose. The long rods known as *Bacillus cellulose dissolvens* appear closely apposed to cellulose fibres, similar to those of *Cytophaga* species, and do not excrete any cellulose into the medium.

The role of fermentation in the biodegradation of organic pollutants in ground water

Organic pollution from industry has increasingly affected such ecosystems as water, land, and the subsurface. Many theories and techniques have been developed for the treatment of waste water and surface water contaminated by organic pollutants. However, it has now been found difficult to successfully apply these theories and techniques to clean up the subsurface and ground water. Almost all the engineering techniques for the treatment of subsurface and ground water, including pump-and-treat, air sparging, chemical flooding, are high cost, technologically difficult and other failing to achieve desired clean up targets. Since American scientists developed a theory of natural attenuation or intrinsic remediation for the remediation of subsurface and ground water in the middle of 1990s, intrinsic and enhanced bioremediation have been increasingly applied to the treatment of ground water and subsurface sediments. However, the mechanism including some microbial processes in the natural attenuation has still remained unclear although many efforts have been made to study the respiratory processes from aerobic, nitrate through iron and sulphate reduction to methanogenesis. Fermentation could probably be one of the most important processes in natural attenuation which has not yet been understood clearly.

Ground water systems are mostly anaerobic, and particularly when the introduction of organic pollutants depletes dissolved oxygen quickly and develops the anaerobic conditions. Many microorganisms that inhabit anoxic environments can obtain their energy for growth through fermentation of organic carbon. In fermentation, organic compounds serve not only as electron donors but also as electron acceptors. In natural environments, fermentation almost always occurs coupled with such respiration processes as nitrate, Fe (III) and sulphate-reduction to drive the biotransformation of many natural organic compounds. Since organic compounds act as both electron donors and acceptors, strictly fermentative reactions should not include any other acceptors. However, in sediments and groundwater systems, fermentation usually exists together with respiration reactions, in which various inorganic electron acceptors normally participate. The conceptual basis is the methanogenic degradation of aromatic compounds since the methanogenic reactions occur without any other inorganic external electron acceptors participating. For example, benzoate acid is degraded by methanogenic microbes :

Fermentation step : $C_7H_6O_2 + 12\ H_2O \rightarrow 7\ CO_2 + 30\ [H]$

Methanogenesis : $30\ [H] + 3.75\ CO_2 \rightarrow 3.75\ CH_4 + 7.5\ H_2O$

Overall reaction : $C_7H_6O_2 + 4.5\ H_2O \rightarrow 3.25\ CO_2 + 3.75\ CH_4$

CO_2 is produced from fermentative degradation of benzoic acid although it acts as the electron acceptor and carbon source of the subsequential syntrophic methanogenic reaction. The fermentation reaction goes to completion due to hydrogen being used as a favourable energy source.

Anaerobic Digestions

It is used for the waste water treatment. This is very useful for highly polluted waste water with greater excess of carbon in relation to nitrogen and phosphorus. The microorganisms use carbon in organic molecules as electron acceptors. Sludge digestion is the most widely used anaerobic process. The anaerobic process is used because of the following reasons :

(a) Sludge product is reduced many folds *e.g.,* 0.5-1.5Kg of sludge solids for one kg of BOD removed in aerobic process as compared to 0.1-0.2 kg for each kg of BOD removed in anaerobic process

(b) There is a drastic reduction in the energy input

(c) There is a production of amount of biogas (methane and CO_2) with highly loaded waste water.

After a period of aerobic digestion, sewage sludge is digested under anaerobic conditions as well. Anaerobic digestion is also used to treat materials with a high content of insoluble organic matter, such as cellulose and to degrade concentrated industrial wastes such as those from the food processing industry. Anaerobic degradation is a complex multisubstrate multiorganism reaction. It involves following stages:

(*i*) **Stage 1:** Polymeric materials like carbohydrates, proteins and lipids are hydrolyzed to monomers like amino acids, sugars, fatty acids by biomass (X_{HY}) *i.e.,* hydrolyzing microbes.

(*ii*) **Stage II Acidogenesis:** The monomeric compounds are fermented to organic acids like propionic butyric acetate by XAG *i.e.,* acid generating biomass.

(*iii*) **Stage III Acetogenesis:** The organic acids with more than 3C per molecule are converted to acetic acid and H_2O.

(*iv*) **Stage IV:** Conversion of acetic acid to CH_4 and water along with organic acids with less than 3 carbons by methanogenic bacteria.

The major metabolic stages in the decomposition of organic wastes into methane and CO_2 have been summarized below:

Complex organic compounds
(polysaccharides, fats, proteins)
↓
Hydrolysis by extracellular
bacterial enzymes (Hydrolysis)
↓
Monomeric compounds
(Sugars, fatty acids, amino acids)
↓ Acidogenesis
Higher organic acids
↓ Acetogenesis
Acetic Acid, H_2, CO_2
↓ Methanogenesis
CH_4

The degradative and fermentative reactions in the anaerobic secondary treatment process can be divided into two stages:

(a) Acid forming

(b) Methane forming

Acid forming stage: In this, complex organic polymers including carbohydrates, fats and proteins are hydrolyzed by extracellular hydrolytic enzymes (polysaccharidases, lipases and proteases) and converted to volatile short chain fatty acids, alcohols and ketones. The fatty acids are fermented to acetate, carbon dioxide and hydrogen. Numerous different bacterial species participate in acid forming stage. These species include some that are known to be obligate aerobes, but that may grow by utilizing alternative electron acceptors, such as nitrate.

Methane forming stage: It restricts anaerobes of genera *Methanobacterium. Methanobacillus, Methanococcus* and *Methanosarcina* convert the acetate, hydrogen and carbon dioxide to methane. The success of anaerobic digestion depends on the cooperative interaction between microorganisms with different metabolic capabilities. An important feature of anaerobic digestion is that most of the free energy present in the substrate is conserved in the methane that is produced. Use of methane for energy generation offsets the cost of the sewage treatment process.

Retention time is significant in anerobic stage sludge digestion. It is because of relatively long mean generation time of methanogens and acetogens. Therefore, reactors in which biomass are retained, wash out of slower growing microbes is avoided. In anaerobic contact process, sludge is recirculated from a settling tank back to fully mixed digester.

By increasing temperature on the digester, reaction rates are improved. For this purpose warm influent is used in food industry or separated digester heating is carried out.

9.2.1.1.2 Limitations of Bioremediation

There are, however, certain constraints of this technology :

(*a*) Indigenous degradative microbes are fully adapted to the specific environment to be treated.

(*b*) Introduced 'foreign' microbes must be able to survive in the new environment and be able to compete with the established indigenous microbes.

(*c*) There is a need of close contact of added inocula with the pollutant and in aqueous environments dilution has to be avoided.

Attempts are now being made to genetically engineer microorganisms to be able to degrade organic pollutant molecules that they are unable to attack at present. This has posed certain problems such as:

(*i*)   genetic stability,

(*ii*) survival of the 'new' microbe in a hostile environment,

(*iii*) some legislative, ethical and perceptional problems concerned with their release into environments such as sewage systems, soils and oceans.

9.2.1.2 Phytoremediation: Smart Plants

It is the use of plants to remove pollutants from water and soils. There are about 1.4 million polluted sites in Western Europe alone. Current techniques are costly and destroy soil structure. The use of plants that can store 10 to 500 more pollutants in their leaves and stems is cheaper, it also stabilizes the soil structure.

The metals can be recovered from ashes and reused. There are many hyperaccumulating plants. These plants accumulate lead, zinc, nickel, copper and cobalt, among others, at levels toxic to other plants. For example, *Sebertia acuminata* can contain up to 20% of nickel in its sap (nickel is generally toxic to plants at a concentration of 0.005%). Similarly, the fern *Pteris vittata* accumulates arsenium, while conserving a very rapid growth and a high biomass. The commercialization rights of the fern (now called edenfern™) for use in phytoremediation have been obtained by the firm Edenspace. The potential market for phytoremediation in the United States alone was estimated at $100 million in 2002 (Tastemain, 2002).

Accumulation of metals by some plants

Plant	Metal
Thlaspi Caerulenscens	Cd
Lpomoea alphina	Cu
Haumanisatrum robertii	Co
T.rotundifolium	Pb
Macadamia neurophylla	Mn
Psychotria douarrei	Ni
Sebertia accuminata	Ni
T. caerulenscens	Zn

Genetic engineering is being used to enhance the ability of plants to accumulate heavy metals. Now the "smart plants" (plants that can detect and/or remove toxins and pollutants) have been developed by genetic engineering. They are expected to have role in :

(i)   improving environment

(ii)   enhancing health

(iii) promoting national security

"Smart plants" in combination with bioremediation can also be used to detect heavy metals, microorganisms and other environmental changes. These plants could be used in phytoremediation, monitoring the changes in quantities or presence of toxins or pollutants. For example, the introduction of two *Escherichia coli* genes into the genome of *Arabidopsis thaliana* which encodes an enzyme involved in transformation of arsenate into arsenite enables the plant to accumulate three to four times more arsenium than the normal plants (Tastemain, 2002). Aresao Biodetection (www.aresa.dk/), a biotechnology company in Denmark, has engineered plants for the detection of landmines and/or explosives. The plant when comes into contact with explosive materials changes colour from green to red. The company is also developing plants for the detection of heavy metals.

Similarly, tobacco plants with bacterial genes that control the synthesis of an enzyme-detoxifying trinitrotoluene, TNT, have been developed. Through the introduction of genes controlling specific degradation pathways, these initiatives could improve bioremediation processes.

9.2.1.3 Use of Enzymes

Several attempts have been made to minimize waste at source especially in food processing by applying enzymes. These include the following :

(*a*) **Hemicellulases and cellulases:** These are used to hydrolyze hemicelluloses in pulp in order to reduce consumption of bleaching chlorine in the pulping process. There is, thus, a reduction in the generation of chlorinated organic wastes in bleached pulp product.

(*b*) **Invertases:** These are used in the recycling of effluents from food and dairy industries where residual starch content is 5-10%. It allows the growth of various harmful microbes

Effluents → Biofilters → Starch → Microbial invertase → Glucose → Animal feed

(c) **Disulfurization of coal:** Coal during combustion as a fuel, releases pollutants into environment. SO_2 and nitrogen oxides affect air quality. Sulfurous emissions from the combustion of coal are cited as the leading contributors to acid rain. It is the largest problem in the current use of coal and biggest constraint on its increased use. Attempts are being made to bioprocess coal by its :

(i) beneficiation to remove undesirable contaminants (S, N and other trace metals).

(ii) conversion (by microbial liquification, microbial solubilization and gasification) into non-pollutant form.

Leather and textile are among the major environmental polluting industries. The industrial discharge can be reduced by the use of enzyme through recycling of water. It will cut down the electrical and water bills and improve the quality of the final products. For this purpose, only simple adjustments are required in the plants besides replacements of 140 harsh chemicals with biological systems. Enzymes and microbes could easily be produced locally with minor additions. The industry will be set to become competitive with a reduced clean-up bill, increased earnings and turnover.

9.2.2 Utilization of Byproduct

Various byproducts released during usual processes can cause pollution of air, water and soil as they act as potential source of growth of various harmful microorganisms. By using various biotechnological methods, these byproducts of chemical, food and dairy industries can be used for various purposes.

9.2.2.1 Whey

It is a byproduct of cheese making industry. It is left after the separation of curd (solidified casein and butter fat). Large units release 0.5-1.5 million litre of whey per day. In fact solid content of whey is very little and is a good substrate of various microbes which can cause problems. Whey can be converted to various products.

(a) **Whey molasses:** Viscous liquid produced by concentrating whey to 70% solid after using reverse osmosis and evaporation.

(b) **Whey permeate:** Liquor containing ~30-35% byproduct of whey protein.

(c) **Whey powder:** In addition to this, the microbes which are able to hydrolyze lactose, can grow and release sugar and used in generation of food yeast, SCP or ethyl alcohol.

9.2.2.2 SCPs (Single Cell Proteins)

These can be produced in single or two stages.

(a) **Single stage process:** *Kluyveromyces fragilis, Candida intermedia* are used for ethanol production.

(b) **Two stage process:** The lactic bacteria are used to transform the lactose which is then utilized by *Candida krusei* under aerobic conditions.

In addition to this, by using various types of microflora, various products can be obtained.

9.2.2.3 Molasses

300-360 kg of molasses is formed for each tonne of sugar produced. A number of products using microbes can be produced from molasses. These include :

(a) Alcohol: using *Saccharomyces cerevisiae*

(b) Baker's yeast : using cane/beet molasses. Under highly aerobic conditions, strains of *Saccharomyces cerevisiae* can be grown

(c) Citric acid : using fungus *Aspergillus niger*

(d) Glutamic acid : using *Corynebacterium glutamic*

(e) Acetone-Butanol : using *Clostridium acetobutylicum*

For the production of bulk products like SCPs, various gases like CO can be used with the help of carboxydotrophic bacteria which synthesize gases (CO and H_2) from coal, lignite *etc.*

$$C + H_2O \longrightarrow CO + H_2$$
$$C + \tfrac{1}{2}O_2 \longrightarrow CO$$

These carboxydotrophic bacteria can be used for the production of enzymes, insecticides and pharmaceutical products.

The anaerobic fermentation of organics, in aqueous waste to methane is a biotechnological application which can be applied to byproduct streaming of other chemicals.

9.2.2.4 Disposable Plastic Starch

The development of starch with improved physical and chemical properties, a substitute for disposable plastic, can potentially provide a source of inexpensive raw material for the production of ethanol for liquid fuel. The research is on in this area.

9.2.3 "End of Pipe" Treatment (*i.e.,* Waste Destruction Technique)

This treatment involves the removal of waste or pollutants at the stage of release in the environment. This depends on the type of wastes encountered whether gaseous, liquid or solid.

9.2.3.1 Gaseous Waste Treatment

Many processes cause air pollution in the form of odour nuisance. For this, very often organic compounds are responsible. These are perceptible and therefore, nuisance at very low concentration is caused. Biological treatment can control point sources of odour nuisances. In this, wastes subjected to purification are passed through biologically active filters having microbes immobilized on organic or inert material.

For this purpose, wet or dry type reactors are used :

(a) **Wet reactor or bioscrubber:** It operates as packed bed with a counter current of liquid (very often sewage) and gas (the contaminated air). Because of controlled loading, growth rate is limited. The odourous compounds in the gas are transferred to liquid phase and then oxidized by flora within biofilm. This has certain benefits. These include the following:

(i) The efficiency of scrubbing is high because the concentration of odouriferous molecules by biooxidation in the liquid phase reduces nearly to zero, thereby, mass transfer from gaseous phase is enhanced,

(ii) There is a great reduction of liquid volumes needed for scrubbing,

(iii) There is no ultimate problem of effluents.

(*b*) **Dry reactors:** In these the beds are packed with biologically active sorptive material (compost, peat *etc.*). The contaminated gases are blown upwards through beds.

Biofilters allow the purification (removal of toxic materials) from waste gases having phenol and its derivatives, toluene, esters like ethyl and butyl acetate. This process is used to purify alcohols like methanol, ethanol, butanol and ketones like acetones from various industrial activities such as chemicals (in polymeric resin production), petrochemicals (oil refining), pharmaceuticals, metallurgical and pesticide product *etc.* Microbes such as *Pseudomonas putida, Candida tropicalis, Fusarium flocciferium* are used as biofilters for phenol.

An important feed stock solvent is methanol. It cleans burning automotive and turbine fuel. During its manufacture, use and disposal, there is loss of methanol vapours to atmosphere. Immobilized microbial cells and stationary aqueous phase in a porous medium are used in a column for methanol vapour.

Methanol is utilized as growth substrate by methylotrophic microorganisms such as *Methylomonas, Aeromonas, Achromobacter, Pseudomonas, Flavobacterium etc.*

The combustion process produces various gases which are colourless, odourless and extremely toxic *e.g.,* CO (carbon monoxide). These include blast furnance (25-30%), automobile exhaust (0.5-12%), faulty domestic heating system and small amount from cigarette, cigar and pipe smoking. A number of bacterial species utilize CO as energy source. Such aerobic bacteria are collectively known as Carboxydotrophic *e.g., Pseudomonas carboxydovorans, P. carboxydohydrogens, P. compransoris, Bacillus schlegelli.*

9.2.3.2 Liquid Wastes

Various biotechnological waste water treatments are available. The amounts of waste materials discharged into bodies of water around the world are steadily rising. Increases in population, the dependence of agriculture on massive amounts of fertilizers and pesticides, expansion of the food industry and growth of other industrial processes, all contribute to the volume of sewage and wastewater and to the content of undesirable substances. The waste water treatment aims to remove :

(a) compounds with a high biochemical oxidation demand

(b) pathogenic organisms and viruses

(c) a multitude of human made chemicals

Recycling of human, animal and vegetable wastes has been practiced for a long time. In many cases, it provides fertilizers or fuel. It, however, has also been a source of diseases to humans and animals by residual pathogenicity of enteric (intestinal) bacteria. Concurrently efficient waste collection and specific treatment processes for deleterious wastes have been developed in urban communities, because it is not possible to discharge high volumes of waste into natural land and waters.

9.2.3.2.1 Waste Water Treatment

A variety of biological treatment systems has been developed. These range from cesspits, septic tanks and sewage farms to gravel beds, percolating filters and activated sludge process coupled with anaerobic digestion. These systems or bioreactors are used to alleviate health hazards and reduce the amount of biologically oxidizable organic compounds. They produce a final effluent or outflow that can be discharged into the natural environment without any adverse effects.

The efficiency of such bioreactor assemblies rely on the metabolic versatility of mixed microbial populations (microbial ecology). These bioreactors contain a range of microorganisms with the overall metabolic capacity to degrade most organic compounds entering the system.

In the industrialized nations, the largest biotechnological industry has been biological treatment of domestic waste-waters and sewerage. There is virtual elimination of such waterborne diseases as typhoid, cholera and dysentery by the controlled use of microorganisms in these communities. Major epidemics may quickly develop, if water and sewage treatments are seriously interrupted. Biotechnology, therefore, besides producing a whole new range of useful products also plays an important role in the reduction of infectious diseases of humans and animals through water and sewage treatment processes.

The biological disposal of organic wastes involves three stages *i.e.,*

   (i)   Primary processing
   (ii) Secondary processing
   (iii) Microbial digestion.

A stage involving chemical precipitation may also be included.

   (i)   **The primary processing:** It is performed to remove coarse particles and solids. This leaves the dissolved organic materials to be degraded or oxidized by microorganisms in a highly aerated, open bioreactor.

   (ii) **The secondary processing:** It needs considerable energy input to drive the mechanical aerators that actively mix the whole system. This ensures regular contact of the microorganisms with the substrates and air.

   (iii) **Microbial digestion:** The microorganisms multiply and form a biomass or sludge that can either be removed and dumped, or passed to an anaerobic digester (bioreactor) that will reduce the volume of solids, the odour and the number of pathogenic microorganisms. Methane or biogas, may be generated there. This can be used as a fuel. The value of biogas, however, is marginal because of its content of CO_2 and hydrogen sulfide.

Dilute organic liquid wastes can be degraded by the **percolating** or **trickling filter** bioreactor. The liquid in this system flows over a series of surfaces, which may be stones, gravel, plastic sheets *etc.* The microorganisms attached to them remove organic matter which is used by them for their essential growth. Excessive microbial growth can be a problem. It creates blockages and loss of biological activity. Such techniques are widely used in water purification systems.

Trickling filter tank is 3-10 feet deep and consists of a bed of crushed stone, gravel, slag or some other inert support material on which the microbial population forms a thin film. The liquid waste is evenly applied to the top of the filter and percolates downwards, depositing organic matter on the support and microbial film. Aerobic conditions are maintained by the upward flow of air through the spaces between the particles of the packed solid support.

Heterotrophic organisms in the upper part of the filter obtain energy and nutrients by oxidizing the organic matter. Their number, thus, multiplies and a new film is formed. The mixed microbial population in the upper layers of trickling filters mainly consists of aerobic chemoheterotrophic bacteria and fungi like *Zoogloea, Pseudomonas, Alcaligenes, Ascoidea, Fusarium, Sepedonium* and *Geotrichum.* Certain chemoheterotrophs degrade many organic nitrogen containing substrates with concomitant release of ammonia. In the lower layers of the filter, the autotrophic nitrifying bacteria oxidize ammonia to nitrite (*Nitrosomonas*) and then nitrate (*Nitrobactor*). BOD is reduced to about 95% in trickling filter tank.

In activated sludge process, the sewage that has received primary treatment is mixed with activated sludge (an inoculum of microorganisms) and continually aerated. The organisms that degrade organic matter are the same as those in the trickling filter tank, but slime forming bacteria, primarily *Zoogloea remigera,* play a particularly important role. As these organisms grow, they form clumps called **flocs** to which soluble organic matter, as well as protozoa and other organisms become attached. When the effluent from an activated sludge tank passes into a sedimentation tank, these flocs sediment out and are transferred to an anaerobic sludge digestor. Much of the reduction of BOD in this process occurs due to floc dependent removal of organic matter.

A major field of biotechnological interest in the future is microbiological effluent treatment. For treating complex wastes, integrated systems are being developed.

Combined algal/bacterial systems for waste and water treatments are developed in countries with high annual hours of sunlight. Such processes can lead to the formation of relatively pure water and algal/bacterial biomass. These may be used for animal feeding and biogas formation. Even the bulk organic chemical formation may be there.

A deep-shaft fermentation system has now been developed for waste-water treatment. The deep shaft is, in fact, a hole in the ground (upto 150m in depth), devised to allow the cycling and mixing of waste-water, air and microorganisms. It is most economical in land use and power and produces much less sludge than do conventional systems. Various biotechnological waste water treatments are now available.

9.2.3.2.2 Biological Pulping in Paper Manufacture

Making paper from wood requires separation of wood fibres from each other and then reforming them into a sheet. The objective of all pulp processing operations is to efficiently remove lignin which glues together the fibres, without damaging the valuable cellulose fibres. This is done by various mechanical and chemical processes called **luping,** but they suffer from disadvantages like damage to cellulose fibres, high costs, high energy use and corrosion.

A lignin degrading and modifying enzyme has been isolated from *Phanerochaete chrysosporium* and is found to be useful since it may reduce energy costs and corrosion thus increasing the life of the system. These also reduce environmental hazards associated with bleech plant effluents. Treatment of pulp with lignolytic fungi such as *Trametes versicolor*, *P. chrysosporium* and *Pleurotus ostreatus* also increases the tensile strength of the paper. Research is being carried out in this area to:

(i)   develop fermentation process, including optimizing conditions for the action of the fungus,

(ii)   design a reactor that provides the optimum conditions, and

(iii)   scale up the fermentation.

9.2.3.2.3 Paper Production Plants

In some developing countries, these have either been closed or are uncompetitive. Many advantages that were never available before, have been provided by the use of microbes and enzymes that could replace chemicals. This will result in saving of water and heat and improve the quality of paper.

Designer wood: It can be produced by genetic engineering. This will grow faster and when processed, requires few steps. It will result in extra savings and improved quality of paper. Many of the paper manufacturing plants that are currently uncompetitive could soon become exporters of paper.

Addition of biotechnology in this sector will provide more value to their raw materials.

9.2.3.2.4 Heavy Metal Treatment

A major problem is pollution of aquatic environment by toxic metals and radionuclides. Heavy metals even at or below current permissible levels, depress soil and aquatic microbial biomass, ATP content and N_2 fixation in soil.

To abate the toxic metal pollution by selectively using and enhancing the natural processes, diverse biotechnological approaches are being utilized by which microorganisms interact with toxic metals.

Detoxification of metal pollution can be achieved by :

(i)   biosorption

(ii)   extracellular precipitation

(iii)   uptake by purified biopolymers and other specialized molecules derived from microbial cells.

Microorganisms and microbial products can be efficient accumulators of soluble and particulate forms of metals and radionuclides particularly from dilute external concentrations. A variety of mechanisms, ranging from physicochemical interactions (biosorption) to process dependent on metabolism (*e.g.*, transport or protein synthesis) may be involved.

It has been reported that a resistant phosphatase in *Citrobacter* species cleaves HPO_4^{2+} from phosphate containing organic molecules and serves to precipitate the metals as cell bound metal phosphates ($MHPO_4$). Metal phosphate loads can reach upto 9 g per kg biomass. This system can be applied to the treatment of recalcitrant uranium/tributyl phosphate wastes as well. Microbes can accumulate gold and silver and sequester them inside their cell membrane. Gallium can be complexed with bacterial and fungal siderophores normal for iron chelation.

(i) Biosorption

Metal uptake takes place by whole biomass (living/dead) via physico-chemical mechanisms such as adsorption or ion exchanges. The metal toxicity may leads to inactivation of living cells, therefore, most of the living organisms are exploited to decontaminate effluents containing metals at subtoxic level *e.g.,* various algal and cyanobacterial blooms encouraged by the addition of sewage effluents, reduce level of Cd, Cu, Zn, Hg and Fe in mining effluents. Metals are removed from water column with 99% efficiency.

The nuclear industry currently relies on physiochemical techniques to treat radioactive effluents like ion exchange, ultrafilteration and flocculation. Some high activity effluents may be stored without processing. Most appropriate technologies for development and application to radionuclide removal should be based on biosorption by biomass in the short term, by modified or unmodified biopolymers in the medium term. In the longer-term, biopolymers like lignin, chitin, chitosan and cellulosics have some potential (but this may depend on the application of genetic and protein engineering *e.g.,* for the production of appropriate metallothioneins and siderophores). The major opportunity in the short-term lies with dead biomass, particularly since waste effluents from the nuclear industry are toxic.

Other mechanisms such as precipitation, particulate entrapment *etc.* can also be used in such complex systems. By these, metals in the sediments are concentrated in a form that greatly reduces environment mobility and biological availability.

For biosorption, fungi including yeasts are also used. It is because of the fact that waste fungal biomass arises as a byproduct from several industrial fermentations. Fungal biomass can remove thorium from some very acidic process streams. The biosorptive role of components within fungal cell walls such as chitin and chitosan, a potentially useful biosorptive agent, is based on chitin from crustacean shells. Metals containing particles *e.g.,* Zn dust, magnetite and other metal sulfides can be removed by fungal mass such as *Aspergillus niger,* citric acid fermentation waste by combination of absorption and precipitation.

Biosorption by immobilization of biomass : The microbial biomass is immobilized in particles of specified size and mechanical strength with favourable mass transport properties and very low weight percentage of inactive additives. The examples of this category are :

- granulated *Bacillus* preparation for removing Cd, Cu, Cr, Hg, Zn at a range of concentration.
- for small volumes, a fixed bed reactor containing 20 kg of granules is used. A large fluidized pulsed bed system containing 80-90kg of biomass is used for larger flows.

After loading, metals are stripped from biomass.

- immobilized particles of diameter 0.7-1.3 mm containing *Rhizopus arrbhizus* biomass have 12-23% added polymers. There was an improved removal of uranium at low polymer contents and at low particle diameter.

- for the biosorption of Cu^{++}, Pb^{++}, Zn^{++}, Au^{++} from mixtures, fluidized bed reactor of algenated and polyacrylamide immobilized algae *e.g., Chlorella vulgaris* and *Spirulina platensis* have been used.

Biosorption using inactive (dead microbial) biomass is useful because

(a) solution toxicity does not affect the biomass biosorptive uptake

(b) there is no biomass requirement

(c) there is no problem of maintenance of culture purity

(ii) Transformations of Microbial Metal

A number of metals and metalloid species can be transformed by microbes. This involves oxidation, reduction, methylation and dealkylation *e.g.,* Hg^{++} to Hg(0) with mercuric reductase.

Au^{+++} can be reduced to elemental Au and Ag$^+$ to elemental Ag, which are deposited on culture vessels by many bacterial, algal, fungal species.

Bacteria, *Zoogloea ramigera* in activated sludge has metal binding capacity. Other bacterial polymers *e.g.,* polysaccharides from *Klebsiella aerogenes, Arthrobacter viscous, Pseudomonas* species, *Aureobasidium pullulans* also have the metal binding capacity.

Detoxification of Metal Ions

Demethylation and reduction of mercury: Several hundred million tonnes of toxic mercury-containing wastes are discharged into our environment. The ionic forms of mercury such as Hg$^+$ and Hg^{2+} and organic mercury (R-Hg$^+$) such as methyl mercury and phenylmercury are extremely toxic and even mutagenic to living organisms. Many microorganisms are resistant to the toxic ionic forms of mercury. Mercury resistance found in bacteria can be categorized into two classes, narrow-spectrum resistance (resistant to inorganic mercury only) and broad-spectrum resistance (resistance to inorganic and organic mercury). Among these microorganisms, the mercury resistance of Gram-positive strains such as *Staphylococus* spp. *Bacillus* Spp. and Gram-negative bacteria such as *Pseudomonas* spp. and *Shigella* spp. are extensively studied. Organic mercury has been reported to be more toxic than its inorganic forms since the penetration of organic mercury into microbial cells is comparatively easier and more rapid than inorganic mercury. However, most of the discharged inorganic mercury can be rapidly and easily transformed biologically into organic forms in the environment. Therefore, the broad spectrum mercury resistant microorganisms are important to detoxify mercury-contaminated sites in our environment.

What is the difference between mercury narrow-spectrum resistance and broad-spectrum resistance in microorganisms? Many studies on the biochemistry and mechanisms of the mercury resistance of Gram-negative and Gram-positive bacteria reveal that the narrow-spectrum resistance bacterium has only an enzyme called **mercury reductase** that is used to catalyze the reduction of Hg^{2+} to Hg^0. Further studies on the mercury resistance genes of *Pseudomonas stutzeri* indicate that a plasmid pPB confers broad-spectrum mercury resistance. The plasmid is self-transmissible. Two pPB regions are separated by 25-30 Kb and share homology with the transposon, Tn501 *mer* (Hg detoxification) genes. Two regions were cloned separately and each was shown to carry a cluster of functional and independently regulated *mer* genes. One cluster confered resistance only to inorganic mercury (with atleast *mer*A and other mercury transporting genes such as *mer*T for narrow-spectrum mercury resistance) and other gene cluster, upstream from *mer*A has a novel *mer*B gene encoding for organomercurial lyase.

The microorganisms with broad-spectrum mercury resistance have, in addition to the mercury reductase, an organomercurial lyase which removes the organic residue(s) of organic mercury. The function of organomercurial lyase is to remove the methyl group of methyl mercury and to produce free methyl group (CH_3) and inorganic mercury (Hg^{2+}). The inorganic mercury can be further transformed into Hg^0 by the reduction mediated by the mercury reductase. Thus the broad spectrum mercury resistant microorganisms can grow in the presence of organic and/or inorganic mercury.

In addition, the elementary mercury (Hg^0) is volatile and most of it leaves the aquatic environment and enters the aquatic and atmospheric environment. Although the total mercury content of the aquatic and atmospheric environments remains unchanged, the availability of mercury of organisms in aquatic environment, mainly in ionic form, will be greatly reduced. Thus the reduction of Hg^{2+} to Hg^0 (also called mercury volatilization) is one of the most effective methods to detoxify mercury in aquatic environment.

Although many microorganisms contain mercury reductases and organomercurial lyases, only few mercury reductases and quite a few organomercurial lyases have been purified from bacteria. Most of the properties of these enzymes are based on the purified enzymes from Gram-negative bacteria (*e.g.*, *Pseudomonas* spp.).

Mercury reductase

Mercury reductases purified from hte cellular membrane of Gram-negative bacteria are either dimeric and trimeric with monomers of 58,700 daltons. The native enzymes require FAD (flavin) for reduction function and preferentially use NAD(P)H as electron donor. A reducible active-site disulfide (cys-135, cys-14) and a C-terminal pair of cysteines (cys-558, cys-559) were identified. All 4 cysteines are required for efficient Hg^{2+} reduction. These cysteine residues at its active site and the electrons donated by NAD(P)H are transferred via FAD to reduce the disulfur bond (-S-S-) of two adjacent cysteine residues at the active site; and converted them into two cysteine residues (with -SH groups). The electrons captured by the enzyme are then used to reduce Hg^{2+} to Hg^0.

Organomercurial lyase

The enzyme is inducible by the presence of organic mercury. It has been reported to be located in the cytoplasm of the bacterial cell. The enzyme is 19,000-20,000 daltons (19-20KD). Its function is the cleavage of the C-Hg bond of organic mercury by protonolysis, *e.g.*, methylmercury is detoxified by demethylation to mercuric ion by bacterial enzyme.

Besides the mercury reductase and organomercurial lyase, the enzymes involved in the transportation of Hg^{2+} such as *mer*T and *mer*P are also important to the mercury resistance of bacteria. *Mer*T encodes a transmembrane protein. The molecular biology of *mer*T and *mer*P proteins of Tn501 isolated from *Escherichia coli* carrying the pPB plasmid has been studied. For transport of Hg^{2+} through the cytoplasmic membrane, an active site with cys-24 and cys-25 in the first transmembrane region of *mer*T protein was essential. A cys-33-ser mutation in *mer*P appears to block transport to Hg^{2+} by *mer*T protein. Deletion of whole *mer*P only slightly reduced Hg^{2+} resistance of the bacterium. It suggests that a functional *mer*T protein is sufficient for Hg^{2+} transport across the cytoplasmic membrane.

Reduction of Chromium

Hexavalent chromium ion (Cr^{6+}) is a common and toxic pollutant in soils and waters. Trivalent chromium ion (Cr^{3+}) is less toxic as compared with its oxidized form (Cr^{6+}). Chemical reduction of Cr^{6+} is feasible and some operations have been practiced. However, the addition of reduced compounds such as NAD(P)H that are potential toxicants for living organisms render the large-scale application of chemical method to detoxify Cr^{6+}. Biological detoxification of Cr^{6+} becomes a promising approach to solve the pollution problem of toxic chromium ion. In some cases, chromate (*i.e.*, Cr^{6+}) is used as terminal electron acceptor during anaerobic respiration by microorganisms. Once Cr^{6+} in chromate is reduced to Cr^{3+} Equation (1), the toxicity of the chromium is greatly reduced.

$$Cr^{6+} 2O_7^{2-} \rightleftharpoons Cr^{3+} O_4^{2-} \qquad \text{(Equation 1)}$$

In addition, since Cr^{3+} compounds are insoluble and tend to precipitate out from the aqueous phase, the reduced chromium ion (Cr^{2+}) will not be available for uptake by microorganisms as well as other organisms in the contaminated environments.

There are many cases using bioreactors with bacteria to detoxify Cr^{6+} in aqueous solution. Many bacteria such as *Bacillus* spp., *Pseudomonas* spp. and *Thiobacillus ferrooxidans* are able to reduce Cr^{6+}. Both monoculture of Cr^{6+}-reducing bacteria or bacterial consortium have been used in detoxification of Cr^{6+} in aqueous solution. Bacterial cells immobilized by appropriate matrices (*e.g.*, DuPont Bio-Sep beads) in either packed-bed or in fixed-film (biofilm) reactors efficiently reduced Cr^{6+} in the presence of reducing compounds such as sulfite or thiosulfate. Under selected conditions, up to 200 mg/L of Cr^{6+} can be completely reduced to Cr^{6+} within 24 hours. Further study indicates that a chromate reductase is associated with the Cr^{6+}-reducing bacteria such as *Pseudomonas putida* PRS2000. Both cell suspension and cell free extract can reduce Cr^{6+}. The enzyme is a soluble protein and the crude enzyme activity is heat labile and with a Km of 40 μM of chromate and is NAD(P)H dependent. Neither sulfate nor nitrate affects chromate (Cr^{6+}) reduction either *in vitro* or with intact cells. In the reduction of Cr^{6+} by *Pseudomonas ambigua* G-1, an intermediate, Cr^{5+} was identified by electron spin resonance during the enzymatic reduction of Cr^{6+}. This suggested that the chromate reductase in this bacterium reduced Cr^{6+} to Cr^{3+} with at least two reaction steps via Cr^{5+} as an intermediate.

9.2.3.2.5 Precipitation of Phosphate

Immobilized cells of *Citrobacter sp.* have been used for this purpose. *Citrobacter* has surface located acid type phosphatase enzyme that releases HPO_4^{2-} from a supplied substrate *e.g.,* Glycerol 2 phosphate precipitates divalent cations (M^{2+}) as $MHPO_4$ at cell surface. This process is used where phosphates containing organic substrates are present in metal and radionuclide effluents.

Removal of phosphorus

One of the plant nutrients that contribute to the eutrophication of lakes is phosphorus. Raw sewage from household detergents as well as from sanitary wastes contains about 10 mg/L of phosphorus. The phosphorus in wastewater is primarily in the form of organic phosphorus and as phosphate (PO_4^{3-}) compounds. In a conventional secondary sewage treatment plant, only about 30 per cent of this phosphorus is removed by the bacteria leaving about 7 mg/L of phosphorus in the effluent.

A tertiary treatment process is added to the treatment plant when stream or effluent standards require lower phosphorus concentrations. In this process chemical precipitation of the phosphate ions and coagulation is brought about. The organic phosphorus compounds are entrapped in the coagulant flocs that are formed and settle out in a clarifier.

The chemical frequently used in this process is aluminum sulfate (Al_2SO_4) or alum which is also used to purify drinking water. The aluminum ions in the alum react with the phosphate ions in the sewage, to form the insoluble precipitate called **aluminum phosphate.** Other coagulating chemicals that may be used to precipitate the phosphorus include ferric chloride ($FeCl_3$) and lime (CaO).

The overall reliability for phosphorus reduction increases significantly by adding the coagulant downstream of the secondary processes. It not only removes about 90 per cent of the phosphorus, but it removes additional TSS and serves to polish the effluent as well. However, when applied, as a third or tertiary treatment step, additional flocculation and settling tanks are required. In some cases, even filters may have to be added to remove the nonsettleable floc.

By adding coagulant for phosphorus removal at some point to wastewater in the conventional process itself the need of additional tanks and filters can be avoided. For example, alum may be added just before the primary settling tanks. The resulting combination of primary and chemical sludge would be removed from the primary clarifiers. Or, in activated sludge plants, the coagulant may be added directly into the aeration tanks. So that the precipitation and flocculation reactions occur along with the biochemical reactions. Sometimes, the coagulant may be added to the wastewater just before the secondary or final clarifiers. It has been observed that the total volume and weight of sludge requiring disposal increases significantly regardless of the point in the process at which coagulant is added (Fig. 9.9).

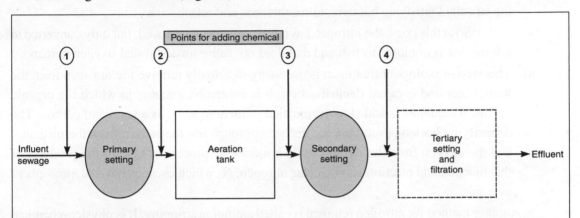

Fig. 9.9. Removal of phosphorus. Phosphorus can be removed from sewage by chemical precipitation. The chemical usually alum, can be added at one of four different points in the process. Point 2 is the most common point of application.

9.2.3.2.6 Control of N$_2$ in Environment

Due to the excessive chemical synthesis of nitrogen fertilizers, use of internal combustion engines and of xenobiotic compounds, there is an increased N$_2$ in waste water. As a result, there is an increase in concentration of ammonia (NH$_3$) which has offensive smell and also causes eutrophication of water resources.

Nitrogen can exist in wastewater in the form of organic nitrogen, ammonia or nitrate compounds. The effluent from a conventional activated sludge plant contains mostly the ammonia form of nitrogen NH$_4^+$. More of the nitrate form, NO$_3^-$ may be present in effluents from a trickling filter or rotating biodisc. This is because the nitrifying bacteria, those microbes that convert ammonia to nitrate, have a chance to grow and multiply on some of the surfaces in the trickling filter or biodisc units. They do not survive in a mixed-growth aeration tank, where they are crowded out by the faster growing bacteria that consume carbonaceous organics.

Nitrogen in the form of ammonia can be toxic to fish, and it exerts an oxygen demand in receiving waters as it is converted to nitrate. Nitrate nitrogen is one of the major nutrients that cause algal blooms and eutrophication. For these reactions, it is sometimes necessary to remove the nitrogen from the sewage effluent before discharge. This is particularly important if it is discharged directly into a lake.

The removal of nitrogen is brought about mainly by a method called biological nitrification-denitrification. It involves two basic steps.

(i)   The first step is called nitrification. In this ammonia nitrogen is converted to nitrate nitrogen, producing a nitrified effluent. For this the secondary effluent is introduced into another aeration tank, trickling filter, or biodisc. With most of the carbonaceous BOD already been removed, the microorganisms that will now thrive in the tertiary step are the so called nitrifying bacteria, nitrosomoneas and nitrobacter.

      So at this point, the nitrogen has not actually been removed, but only converted to a form that is not toxic to fish and that does not cause an additional oxygen demand.

(ii) This step of biological treatment is necessary to actually remove the nitrogen from the wastewater and is called denitrification. It is an aerobic process in which the organic chemical methanol is added to the nitrified effluent to serve as a source of carbon. The denitrifying bacteria pseudomonas and other groups use the carbon from the methanol and the oxygen from the nitrates in their metabolic processes. One of the products of this biochemical reaction is molecular nitrogen, N_2 which escapes into the atmosphere as a gas.

      Another method for nitrogen removal is called ammonia stripping. It is physico-chemical rather than a biological process, consisting of two basic steps. First, the pH of the wastewater is raised in order to convert the ammonium ions, (NH_4^+) to ammonia gas, (NH_3). Second, the wastewater is cascaded down through a large tower, this causes turbulence and contact with air, allowing the ammonia to escape as a gas. Large volumes of air are circulated through the tower to carry the gas out of the system. The combination of ammonia stripping with phosphorus removal using lime as a coagulant is advantageous, since the lime can also serve to raise the pH of the wastewater. Ammonia stripping is less expensive than biological nitrification-denitrification, but it does not work very efficiently under cold weather conditions (Fig. 9.10).

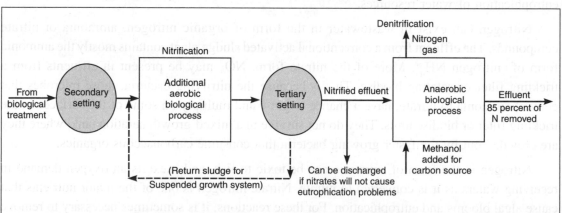

Fig. 9.10. Nitrification-Denitrification. The biological process of nitrification and denitrification must be carried out after the basic activated sludge process is complete.

      The bacteria with appropriate enzyme complement use ammonia as electron donor for denitrification in the absence of O_2.

$$5NH_4^- + 3NO_3^- \longrightarrow 4N_2 + 9H_2O + 2H^+$$
$$\Delta G^0 = -1483.5 \text{ KJ}$$

Here chemolithotrophic bacteria are involved. They are used in Anamox TM process. In this O_2 is not essential for nitrification, therefore, it is a total anaerobic waste water treatment process that is capable of achieving nitrification and denitrification simultaneously.

A number of various bacteria have been found to degrade nitrogen containing heterocyclic pesticides like dixuron, metamitron *etc.* Metamitron is degraded by organisms like *Rhodococcus* species.

9.2.3.2.7 Removal of Cyanide

Cyanide rich wastes are yielded during the production of electrophoretic objects, plastic and steel. They often get mixed to various degree with phenol, thiocyanate, heavy metals and organic nutrients. Different types of cyanogenic glycosides are produced by >200 plant species *e.g.,* *Falk, Sorghum, Cassava etc.*

Certain microbes have been found to be able to degrade cyanide to non-toxic products. Three enzymes able to degrade cyanide have been identified in *Chromobactrium violacerum.* Most important is cytoplasmic B cyanoalanine synthetase.

Other cyanide degrading enzymes are rhodanase, cyanide hydratase, nitrogenases and cyanide oxidase. Cyanide hydratase is secreted by *Glieocercopora sorghii* (leaf spot of *Sorghum*).

9.2.4 Solid Wastes

As a result of protection of aquatic system many solid wastes are generated. Thus, pollution problem is transferred from water to land. Besides, a great amount of waste is produced by industries, hospitals, domestic sources *etc.* The following methods are used to meet the problem. While part of this solid waste material is made up of glass, plastics, *etc.*, a considerable proportion is decomposable solid organic material such as paper, food wastes, wastes from large scale poultry and pig farms *etc.*

9.2.4.1 Landfill Technologies

In large urbanized communities, the essential disposal of such wastes is problematic and one well-used system is by low cost anaerobic **landfill technology**. The solid wastes are deposited in low lying, low value sites and each day's waste deposit is compressed and covered by a layer of soil. Depending on the size of site and flow of wastes, the complete filling of such sites can take months or years. If improperly managed, they can be badly smelling and unhygienic. Toxic wastes can also create severe problems both to the microbiological process occurring in the site and with toxic run-off.

Improperly prepared and operated landfill sites may result in toxic heavy metals, hazardous pollutants and products of anaerobic decomposition that may leach into urban water supplies.

Properly constructed and sealed landfill sites can be used to generate methane gas for commercial use. Much effort is now made to use strong, impermeable liners to avoid leachates damaging surrounding land and water courses.

The landfill sites are required to be air and water-tight, to protect the environment. Regular monitoring is necessary to detect contamination of groundwater, surface water and surrounding air.

Landfill sites in the past were essentially seen as 'dumping' sites or storage vessels, where the waste was essentially sealed from the surrounding environment. Now the landfilling sites are managed as bioreactor vessels, where correct stabilization enhancement systems are operated during the working life of the landfill site. Attempts are being made to reduce the amount of waste to be landfilled and to increase the safety of the operation.

9.2.4.2 Soil Treatment

New area of application of biotechnology is soil treatment. At present following biological treatment methods are available :

   (i)   Land farming

  (ii) *In situ* biorestoration

 (iii) Bioreactors

 (iv) Composting

   (*i*) **Land farming:** It is used for clean up of contaminated soil. The contaminated soil is spread over the sand lay with a drainage system to a depth of 40 cm. The land farming is also carried out under the so called **controlled circumstances** *i.e.,* control of parameters like oxygen, water contents, temperature *etc.* Soil contaminated with oil and aromatic hydrocarbons can be treated with this.

  (*ii*) ***In situ* biorestoration:** For highly permeable sandy soils, this seems to be satisfactory. In Netherlands, ground water is usually used as a medium for the addition of oxygen, nutrients and eventually microbes to stimulate biodegradation of contaminants in the soil. The pumped up water contains the dissolved degradation products and a part of contaminant is treated above the ground and usually recycled. The additional oxygen is found to be important for the stimulation of biodegradation.

      The *in situ* treatments of contaminated soil are

    (a) bioventing,

    (b) biosparging,

    (c) extraction, and

    (d) phytoremediation.

Bioventing in an *in situ* process which combines an increased oxygen supply with vapour extraction. A vacuum is applied at some depth in the contaminated soil which draws air down into the soil from holes drilled around the site and sweeps out any volatile organic compounds. Nutrient supplementation can be provided by running nutrients into trenches dug across the site. The rate of natural degradation by the aerobic microorganisms can be increased by the increased supply of air. This is effective only for reasonably volatile compounds and where the soil is permeable. However, the vapour extracted may need some form of treatment and one biological solution is the use of biofilters.

Biosparging. It is a process to increase the biological activity by increasing the supply of oxygen to the soil by sparging air or oxygen into the soil. Air injection was tried at first, but was replaced by pure oxygen in order to increase the degradation rates. The expense of this type of treatment has limited its application to highly contaminated sites, but the cost can be reduced by on site generation of oxygen. Hydrogen peroxide has been used on a number of sites, but even at low concentration, it can be toxic to microorganisms. This process is similar to soil vapour extraction which can be used for volatile contaminants.

Both in bioventing and biosparging, the structure of the soil can be the predominant factor in its success. The major engineering consideration with *in situ* bioremediation is the delivery of the additions and supply to the contaminants which is affected in the soil. The fate of contaminants in soil is affected by channels and pores in the soil structure, diffusion into closed pores and soil organic matter. Other features are adsorption onto mineral surfaces and partitioning into organic matter.

(*iii*) **Bioreactors:** Two types of bioreactors are used:

- *Dry reactors*: These are used for treatment of sandy soil. The optimum soil moisture content is 10-20%.
- *Slurry reactors*: In these, all types of soil may be treated. Slurry has 50-60% solids. Intensive mixing of soil (slurry) and sufficient oxygen is required for rapid degradation of contaminants. Depending on chemical nature of contaminants, aerobic/anaerobic mode can be chosen. More than 90% of explosive is degraded within 18 days in the thermophillic pile maintained at 55°C.

(*iv*) **Composting:** Solid wastes from sewerage sludge domestic refuse/agricultural waste can be converted to compost by well established techniques. It has been used by farmers, mushroom growers and sub-urban gardeners for years. Vermiculture is one of the examples.

Composting can be defined as 'the biological stabilization of wastes of biological origin under controlled aerobic conditions and is a process that most gardeners are familiar with. In composting, the organic component of municipal waste is decomposed by a mixture of microorganisms under warm, moist and aerobic conditions. This is in contrast to the anaerobic landfill conditions where decomposition is slow and methane is formed. In composting, the matrix or conditions should be open to allow an adequate supply of oxygen. In the presence of oxygen, the rapid metabolic processes produce heat. Not all the heat can be dissipated and the compost will rise in temperature to 50°C and above. The high temperatures are responsible for the inactivation of pathogens producing a very acceptable final product

Once the compost system has been assembled, aerobic mesophilic microorganisms start the degradation, but as temperature rises in the compost, thermophilic organisms take over. The pattern of growth can be seen where growth has been followed by the formation of carbon dioxide. Most pathogens and viruses can be inactivated by keeping them at 55°C for three days. The system with sufficient oxygen will be rapid (11-16 weeks) with low energy consumption giving a hygienic standard end product. There are a number of composting systems or processes which can be divided into two groups, **open** and **closed systems**.

The simplest open system is the Windrow where the waste to be composted is piled in long heaps, often covered with straw to conserve heat and aeration is achieved by periodic turning of the heaps. In other static systems aeration is either by blowing air in at the base of the heap or by suction from the base.

The closed systems are normally situated in buildings and up to seven different systems have been described in which some form of mechanical means gives aeration and mixing in either continuous or discontinuous mode. An example is the Dano biostabilizer which is a sloping plug-flow bioreactor where mixing and aeration is carried out by rotating the cylinder. This system combines the removal and recycling of ferrous metals with composting. The residence time in the cylinder is about 1-5 days which is not enough to give complete composting so that the partially composted material is passed into a Windrow system for 1-6 weeks depending on conditions and starting material.

The Silo system consists of a vertical plug flow vessel where air is introduced at the base and the compostable material pumped in at the top, which gives both mixing and aeration. There is a facility for the recycling of the compost and the residence time in the first vessel is about 14 days. After this time, the compost is moved to the curing vessel where the high temperature developed in the first vessel inactivates any pathogens present. About 25% of the curing vessel is removed after 14-20 days incubation to be replaced by compost from the first vessel. The compost formed is used for horticulture.

Factor affecting composting

The following factors influence the process of composting :

  (i) Elevated temperature increases the process (enzyme kinetics)
  (ii) Opportunity for coordination may be enhanced due to the range of alternate substitutions present.
  (iii) Modification in the physical/chemical microenvironment within the composting mass can increase the density of the microflora to which the contaminant is exposed.

There is an increasing interest in the feasibility of composting as process for detoxifying, degrading or inactivating hazardous wastes.

It is thought to provide a controlled system for containment of toxic constituents which may be subject to vitalization or leaching, drug degradation, composting of urban wastes and the use of finished compost as organic fertilizer. This has been generally successful in many countries.

The process of composting has many advantages. These are :

(i)   Composting is the biochemical degradation of organic material to a sanitary nuisance free humus like material.

(ii) Plants produce a usable end product that may be sold thus reducing cost. Normally composting offers favourable conditions for salvage of rags, glass, cardboard, paper cans and metals.

(iii) Whether does not affect on enclosed composting plant although heavy rains adversely affect most kinds of outdoor composting.

(iv) It can be used to dispose off such industrial wastes as those from meat packing plant, paper mills, saw mills, tanneries, stock yards, canneries, etc.

The material is usually used as potting mixture. Sometimes waste chemicals contain compounds which are highly toxic, their direct application to soil is undesirable or their reverse physical, chemical or microbiological properties would prevent or limit the extent to which they could be composted alone. It may be feasible to mix them with sewerage sludge, refuse or even manure or crop residue to provide a more compostable mixture.

The process of composting has certain disadvantages. These are :

(i)   Capital and operation costs apparently are relatively high

(ii) Trained personnel are required

(iii) Refuse that damage grinder such as tins, heavy stones, mattresses must be removed and decomposed off separately

(iv) Not as economical as chemical fertilizers in enhancing plant growth

(v)   Not as effective as animal manure in improving the physical conditions and fertility of soils

(vi) It contains particles of glass and plastic that are objectionable especially in home garden, where hand manipulation of soil is common.

Compost system can be classified on three general bases, namely oxygen usage, temperature and technological approach (aerobic and anaerobic, mesophilic and thermophilic, open or window respectively).

Aerobic composting requiring activity of aerobic microbes is characterized by high temperature absence of foul odours and it is more rapid than anaerobic. Under upon or window process, entire process is carried out in the open. In mechanical systems, greater part of the initial composting takes place in an enclosed unit, the digester.

Since, it is a biological operations, factors and requirements peculiar to the maintenance of biological activity such as a suitable microbial population must be present. The rate of efficiency of the process are functions of rate and efficiency of microbial activity. The substrate subject to composting generally must be organic.

Role of microorganisms in composting

Utility of isolation : Isolating and identifying organisms is useful because if one knows the important microorganisms in a compost pile, it is easy to determine the optimum growth condition for them as well as their growth kinetics by examining pure cultures and can design and plan the plant and the operation accordingly.

However, isolating an organism in higher number does not rule out presence of others. It all depends on time and type of material. Isolation procedure also creates some artifacts which hinder in identifying the type of microorganisms.

Inoculums : The number of bacteria is rarely a limiting factor in composting, because bacteria are always present in great abundance on all exposed objects, especially on municipal refuse. In the composting process inoculation would be of value only if the bacterial population is in any emerging environment were unable to develop rapidly enough to take full advantage of the capacity of environment to support it. Commercial inocula the components of which are known only to its manufacturers, are available. These are mixtures of several pure strains of laboratory cultured organisms especially prominent in decomposition or organic matter and nitrogen fixation.

Other commercially available inocula contain enzyme systems, hormones, preserved living organisms, activated factors, biocatalysts, *etc.*

Types of microorganisms : It has been found that certain groups of microorganisms are associated with different stages of the process, facultative and obligate aerobic representatives of bacteria, fungi and actinomycetin etc. are very much associated with composting. At the start of the process, bacteria are predominant, in 7 to 10 days, fungi appear. In the final stages actinomycetes become conspicuous. Bacteria are found in all parts of pile, whereas actinomycetes and fungi are encountered or confined to sharply defined outer zone to 2 to 5 inch in thickness. The limitation of these two groups in the outer zone could be a function of temperature or of aeration.

Substrate

Nature : The nature of the substrate in composting is more complex than that of the wastes being processed. With rare exceptions, it must be organic (paper, wood, manures, food preparation wastes, crop wastes *etc.*) as we decompose the materials to simpler and yet more simple forms.

1. Proteins → peptides → amino acids → ammonium compounds → bacterial protoplasm → atmospheric nitrogen or ammonia.
2. Carbohydrates → simple sugars → organic acids → CO_2 and bacterial protoplasm.

An important feature of a substrate as far as composting is concerned, is the availability to the microorganisms of the nutrients tied up in the substrate. Therefore, if first succession of microbes is missing nothing would happen.

Nutrient balance : To keep reproducing and, thus, bringing about decomposition, microorganisms, indeed all organisms, must have minimum supply of all the elements, of which, their cellular matter is composed. In addition, they also require elements that enline their metabolic activities as an energy source or enzyme constituents.

One of the important balance with respect to composting is the Carbon-Nitrogen balance (or C/N ratio). Optimum C/N ratio with most waste falls within 20 or 25 to 1. Wastes such as paper, wood, skin fibre have C/N ratio higher than 25/1.

Nitrogen concentration can be determined by the Kzeldahl method. Carbon content can be roughly determined by formula.

$$\% \ \text{Carbon} = \frac{100\% \ \text{ash}}{1.8}$$

Modern municipal wastes generally have nitrogen content too low for best, or even practical composting, because very few food products each the "garbage", can packaging of food has reduced food related waste generated at homes.

Too low C/N ratio affects loss of nitrogen through the production of ammonia and its subsequent volatilization.

Controlling Factors : Since the compost process is essentially biological, many environmental factors are those which influence biological activity. These are moisture, temperature, pH level, nutrient concentration and availability and concentration of oxygen.

Theoretically optimum moisture content is 100%, however it is a function of aeration, capacity of equipment and of the composted material nature.

Temperature has been a key factor affecting biological activity. Each group of organisms has optimal temperature and any deviation from the optimum is manifested by a decline in growth and activity of organism.

Optimum pH range for bacteria is between 6 and 5.75, whereas fungi can tolerate vary wide range of pH from 5.5 to 8.0. Changing pH requires addition of reagents which in turn involve two fold cost; that of reagent and that of applying it. Composting is almost universally aerobic, oxygen is essentially an important factor.

Genetic trait, though not an environmental factor constitute the ultimate rate limiting factor, since the capability of microorganisms depends on their genetic make up.

Nature and value of finished compost

Soil conditioners : Finished compost may be designated by the general term "humus". It has many characteristics beneficial both to the soil itself and to growing vegetation.

Organic acids resulting from metabolic break down of organic materials form complex with the inorganic phosphate. In this form, phosphorus is more readily available to higher plants, both phosphorus and nitrogen are involved in a stormy effect peculiar to humus. The physical effects of humus on the soil are perhaps more important than nutrient effects, such as acidity, alkalinity and neutralization of toxic substances.

Land reclamation : Application of compost is the obvious solution in the areas where the top soil has been lost due to strip mining, soil erosion etc. For crop production, compost may be used as mulch and as a soil conditioner, it releases absorbed water more easily than peat, moss and bark. The C/N ratio of the compost is an important determinant of the immediate utility in crop production.

Thermal processing : Incineration and pyrolysis

Incineration of solids wastes : It provides a means of disposing of refuse by high temperature oxidation. Its primary function is to reduce the volume of wastes. It can also serve as the secondary function of energy production or destruction of hazardous or putrescible wastes.

The combustion of solid wastes is associated with certain questions. These are:

(a) At what volume reduction is attainable ?

(b) How much air must be supplied for efficient combustion ?

(c) What are the emissions that leave the furnace enclosure which must be removed by air pollution control devices ?

(d) How much energy can be recovered from the combustion of gases ?

(e) How much cooling must be provided ?

Reduction of weight and volume on incineration : Incineration of solid wastes will leave a residue approximately equal in mass to the inert content of the wastes, decrease in the mass of residue may occur due to the vapourization. A small increase also results as a consequence of either oxidation of some of the metals or incomplete combustion of the organic content of the waste.

Residue from the combustion will vary widely depending on the composition of the wastes. Paper and card board will yield ash as little as 1% of the mass of the waste.

The residue does not undergo anaerobic digestion when placed in landfill; consequently settlement and gas generation do not occur. The incineration of low-hazard solid wastes with high organic content in large purpose built facilities is useful. There is considerable reduction in the volume of solid material that is achieved by the process. In the case of domestic refuse this is generally about 75%.

Effects of incineration

Air pollution : Smoke stacks from incinerators may emit oxides of nitrogen and sulphur that lead to acid rain, heavy metals such as lead, cadmium, mercury and carbon-dioxide. In modern incineration facilities, smoke stacks are fitted with special devices to trap pollutants, but the process of pollutant abatement is expensive.

Many times the solid wastes sent for incineration contain chlorione containing organic substance such as polyvinyl chloride (PVC). The burning of these leads to the formation of trace amounts of polychlorinated dibenz-p-dioxides and polychorinated dibenzofurnas (PCDDs and

PCDFs) some of which are highly toxic. The presence of these materials in the emissions and ashes produced by incinerators of all kinds have lead to opposition of their use.

The major drawback of incineration is the generation of fuel gases and particulates. This can be minimized the application of technologies that are essentially the same as used to clean the stack gases of static fossil fuel burning facilities.

A critical requirement of modern incinerators is that they meet the most stringent air pollution codes. Primary concern with emission from incinerators has been and continues to be the particulate matter emitted which is although relatively small, can account for as much as 20% of the total loading. The emission of pollutants, such as carbon monoxide, hydrocarbons and soot, may be controlled in well designed units by improvement of the combustion process within the furnace itself.

The particulate emission is related to the fine ash content of the refuse to the degree of incinerator *etc*. Partial control of emission may be achieved by reducing the under ground air velocity and/or bed agitation, thus reducing the emission from the combustion chamber.

Water pollution : Nearly all incinerator require water, the residue is quenched to facilitate handling except for small unit and some batch type unit, where the refuse is allowed to burn down and the residue is removed after it has cooled. Therefore, extensive use of water is a normal feature of the present incineration and is potentially a water pollution problem.

Burning destroys about 2% of solid waste. It includes any kind of backyard open fire or a huge furnace operated by municipality for the successful incineration system. It is necessary that for carrying out certain essential functions, the following provisions be included :

(i) Measuring the quality of refuse received
(ii) Receiving loads of mixed refuse at varying rates of supply
(iii) Storing in a sanitary, accessible, nuisance free manner
(iv) Feeding the refuse to the system at the controlled rate
(v) Drying the refuse sufficiently to permit ignition of combustibles
(vi) Dissipating the heat of combustion
(vii) Burning the refuse to produce essentially inert solid residue and clean gases
(viii) Cooling, collecting and recovering non-combustibles
(ix) Controlling the process for safety, efficiency, economy and community or plant acceptance
(x) Protecting personnel equipment from the elements, dangerous refuse and careless operation.

The process of incineration has many advantages. These are :

(i) Relatively small plot of land as compared to sanitary landfills is required.
(ii) It can be located in industrial area near the centre of the service area and collection routes.
(iii) Inclement weather does not interrupt its operation.

   (iv) To accommodate the variation in refuse generation, operation time can range upto 24 hrs. a day.

   (v) Residue generated is stable and nearly inorganic

   (vi) They can be less expensive then sanitary landfills.

   (vii) Practically any kind of refuse can be burnt by incineration.

The disadvantage of the process include :

   (i) High initial cost on establishment is required.

   (ii) Incinerators emit pollutants into the atmosphere.

   (iii) Operating costs are relatively high, although manpower required is lower than it is for the other methods of disposal.

   (iv) Maintenance and repair costs may be high because of the high temperature required for the burning of the dirty and damaging nature of the refuse and residue.

Pyrolysis

Pyrolysis is destructive distillation often with carbonization in the absence of air. By the application of indirect heat, it can convert trash, waste after drying and contaminated paper into usable fuels or it may be employed as modified incineration to produce ash slag and ferrous metals from solid or in chemical plants, moist and liquid residue.

Pyrolysis of the refuse yields a solid residue and gas, each with significant thermal value. Water in the feed stock and the produced in pyrolysis were recovered as an aqueous fraction containing pyrolygenous products – aldehydes, alcohols, acids, ketones and so on.

9.3 ROLE OF MICROBES IN GEOLOGICAL PROCESS

In certain geological processes microbes act as catalytic agents. These include mineral formation, mineral degradation, sedimentation, weathering and geochemical cycling.

When bituminous coal seams are exposed to air and moisture during mining, microbial pyrite oxidation takes place. Huge volumes of sulfuric acid produced in this way have created pollution on an unprecedented scale. Microbial weathering of building stone such as limestone, leading to defacement or structural changes are other detrimental acts of microbes.

9.3.1 Bioleaching

Microbes are increasingly used to extract commercially important elements by solubilization (bioleaching). Using microbial agents, metals such as cobalt, copper, zinc, lead or uranium can be more easily separated from low-grade ores.

In extractive metal leaching, the biological reactions are usually concerned with the oxidation of mineral sulfides. These specific reactions can be carried out by bacteria, fungi, yeasts, algae and even protozoa. Many minerals exist in close association with other substances such as sulfur *e.g.,* iron sulfide that must be oxidized to free the valuable metal. A bacterium *Thiobaccillus ferro oxidans* can oxidize both sulfur and iron; the sulfur in the ore wastes is converted by the bacteria to sulfuric acid. Simultaneously, this enhances the oxidation of iron sulfide to iron sulfate.

The crushed ore (normally in large heaps) is repeatedly washed with a bioleaching solution containing live microorganisms and some essential nutrients (phosphate/ammonia) to the metal that can easily be separated (downstream processing) from the sulfuric acid into which it has been extracted.

Canada, India, the USA and the countries of the former Soviet Union are using bacterial leaching to recover uranium from low-grade ore (0.01 to 0.5% U_3O_8). 4000 tonnes of uranium is extracted per year by USA alone in this manner. It is primarily used as fuel in nuclear power generation. The microbial recovery of uranium from otherwise useless low-grade ores is an important contribution to energy production. Bioleaching of uranium ores is an important contribution to the economics of nuclear power stations. Bioleaching also helps in the recovery of uranium from low grade nuclear waste.

Bioleaching technology is, thus, of significant importance for extracting the increasingly scarce metals necessary for modern industry. However, this process is relatively slow.

As much as 40% of organic sulfur has been removed in bench process. The process is bacterial leaching *i.e.,* extraction or solubilization of metals from minerals contained either in rock matrix or in the form of concentrated mineral products from processing plants. Since most of the metals occur mainly as sulfides, acidophilic thiobacilli from family Thiobacteriaceae *e.g., Thiobacillus ferrooxidans, T. thiooxidans, T. acidophilus. Thiobacillus* substrates are used for desulfurization. For example:- oxidation of pyrite (FeS_2 minerals form of ferrous sulfide) to ferro sulfide.

Pyrites are oxidized as follow:

(*a*) Autooxidation of ferrous sulfide in the presence of O_2, H_2O and H_2SO_4.

$$2FeS_2 + 7O_2 + 2H_2 \longrightarrow 2FeSO_4 + 2H_2SO_4 \qquad \text{(i)}$$

$$2FeS_2 + 2H_2SO_4 + O_2 \longrightarrow 2FeSO_4 + 2H_2O + 4S \qquad \text{(ii)}$$

(*b*) Oxidation of pyrite in the presence of bacteria

$$4FeSO_4 + 2H_2SO_4 + O_2 \longrightarrow 2Fe_2(SO_4)_3 + 2H_2O \qquad \text{(iii)}$$

$$2S + 3O_2 + 2H_2O \longrightarrow 2H_2O + 2H_2SO_2 \qquad \text{(iv)}$$

$$FeS_2 + Fe_2(SO_4)_3 \longrightarrow 3FeSO + 2S \qquad \text{(v)}$$

Overall Reaction

$$4FeS_2 + 5O_2 + 2H_2O \longrightarrow 2Fe_2(SO_4)_3 + 2H_2SO_4$$

As in these reactions, S is not produced as an end product, thus, production of toxic gases like SO_2 can be prevented.

Ferric sulfate is a strong oxidizing agent and is able to dissolve several economically important copper sulfide minerals by the following reactions

$$CuFeS_2 + 2Fe_2(SO_4)_3 \longrightarrow CuSO_4 + 5FeSO_4 + 2S^0$$
Chalcopyrite

$$Cu_2S + 2Fe_2(SO_4)_3 \longrightarrow 2CuSO_4 + 4FeSO_4 + S^0$$
Chalcocite

$$Cu_5FeS_4 + 6Fe_2(SO_4)_3 \longrightarrow 5CuSO_4 + 13FeSO_4 + 4S^0$$
Bornite

Leaching by $Fe_2(SO_4)_3$ is called **indirect** because it is independent of the presence of oxygen or microbial action. The rate of metal extraction from sulfide ores depends on the concentration of ferric ion (Fe^{3+}). In the absence of catalysis, the rate of oxidation of Fe^{2+} is very slow and leaching would likewise be very slow.

T. ferrooxidans increases the rate of Fe^{2+} oxidation by ten folds. Indirect leaching depends on the ability of microbes to supply the necessary $Fe_2(SO_4)_3$ by oxidizing Fe^{2+} to Fe^{3+}.

T. ferrooxidans also derives energy from the oxidation of elemental sulfur (S^0) to sulfuric acid.

$$2S^0 + 3O_2 + 2H_2O \longrightarrow 2H_2SO_4$$

The sulfuric acid maintains the low pH that is optimal for the acidophilic microbe and suppresses the loss of ferric sulfate by hydrolysis.

$$Fe_2(SO_4)_3 + H_2O_2 \longrightarrow 2Fe(OH)SO_4 + H_2SO_4$$

The sulfuric acid also leaches various copper oxide minerals.

$$Cu_3(OH)_2(CO_3)_2 + 3H_2SO_4 \longrightarrow 3CuSO_4 + 2CO_2 + 4H_2O$$
Azurite

$$CuSiO_3\,2H_2O + H_2SO_4 \longrightarrow CuSO_4 + SiO_2 + 3H_2O$$
Chrysocolla

In the direct leaching process *T. ferrooxidans* becomes attached to the mineral particles, and then enzymes associated with the cell membrane catalyze oxidative attack on the crystal lattice of the metal sulfide. Oxidation of minerals proceeds in two steps :

$$CuS + 0.5O_2 + 2H^+ \longrightarrow Cu^{2+} S^0 + H_2O$$
$$S^0 + 1.5O_2 + H_2O \longrightarrow H_2SO_4$$

Other acidophilic chemolithotrophs believed to be important to the leaching process include *T. thiooxidans, T. acidophilus, T. organoparus, Leptospirilum ferroxidans* and species belonging to genus *Sulfolobus*.

Metal can also be concentrated from dilute solution using microorganisms (bioaccumulators). The microorganisms, bacteria, yeasts and molds can actively take up the metals by various ways and such processes have a potential use in extracting rare metals from dilute solution. In a similar way, microorganisms are being used to extract toxic metals from industrial effluents and reduce subsequent environmental poisoning. As mentioned earlier, certain plants called smart plants are also helping us in this direction.

Bacterial bioleaching also helps in the removal of sulfur-containing pyrite from high sulfur coal. As high sulfur coal produces sulfur dioxide pollution after burning, it is not in use. But it is not feasible to ignore the high sulfur coals because of its increasing availability. Here, the bacterial removal of pyrite (which contains most of the sulfur) from high sulfur coal is bound to help and the process is expected to be economically viable.

In order to prospect for petroleum deposits, aliphatic hydrocarbon-utilizing bacteria are also being used. Microbes are used to release petroleum products from oil shelf and tar sands. In all these systems, the natural geological site acts as a bioreactor. It allows water and microorganisms to flow over the ore and collected after natural seepage and outflow. Recycling by mechanical pumping can also be used.

In all these activities, multidisciplinary approaches are required and new biotechnological techniques such as designing an organism for a specific function could yield further benefits. The overall picture of this area of biotechnology is one of the rapid and exciting developments.

Biofeedstock

Most chemical feedstocks are also derived from petroleum. Many petrochemicals, such as the molecular feedstocks for polyethylene and plastic are small organic (carbon-containing) molecules that are hooked together to create polymers.

The lactic acid molecules are hooked together to create the polymer, polylactic acid, which is marked under the name Natureworks. Natureworks is a biodegradable polymer that can be used as a starting material for manufacturing packaging materials and fibres for clothing, pillows and comforters. The bedding company Pacificcourt feather plans to replace 80% of its polyester products with polylactic acid over the next 5 years.

9.4 ENERGY

Non-renewable energy sources like fossil fuels are depleting at a rate 100,000 times faster than they are being formed. A threat to global ecosystem is in terms of climatic changes and health hazards, the depleting energy resources coupled with increasing demand for energy. It is because of this non-conventional sources of energy, such as solar and geothermal energy are being developed.

Solar energy can be harvested in the form of :

(*i*) biomass through the use of waste material and

(*ii*) by growing energy crops

It can also be harvested through the use of microorganisms, which can convert solar energy into chemical energy in the form of molecules like glycerol.

These energy sources are renewable and less polluting. A great deal of research is being carried out for increasing the production and utilization of biomass as a source of renewable energy.

The common examples of biomass are wood, grass, herbage, grains or bagasse. The main sources of biomass are :

 (i) waste materials derived from agriculture, forestry and municipal wastes
 (ii) growing energy crops involving short rotation forestry plantation.

9.4.1 Biogas Production

Biogas represents a mixture of different gases in varied composition, produced as a result of action of anaerobic microorganisms on domestic and agricultural wastes. It contains methane in bulk (50-68%) and other gases like CO_2 (25-35%), H_2 (1-5%), N_2 (2-7%) and O_2 (0-0.1%). A large number of biogas plants is being used by villagers in India as a cheap source of energy and for improvement of health and sanitation conditions. Biogas plants also provide highly enriched organic fertilizer. Production of biogas involves three steps. These are :

 (*i*) **Hydrolysis :** It converts organic polymers into monomers with the help of hydrolytic bacteria.
 (*ii*) **Acid formation** : It involves conversion of monomers into simple compounds such as CO_2, NH_3 and H_2 by acid forming bacteria.
 (*iii*) **Methane formation :** It involves conversion of simple compounds into methane and CO_2, utilizing anaerobic methanogenic bacteria.

Biomass refers to all the materials produced during the production and processing of agricultural and food materials that is usually discarded as waste. This often takes the form of indigestible portions of plants. This includes woody portions of plants which are composed of lignin and cellulose. Tremendous quantities of these wastes are produced. They are generally not toxic or hazardous, they are simply wastes because they cannot be used in the products.

Besides waste from the timber industry and paper products, indigestible material left from the processing of foods can be a source of biogas. They all contain large amounts of cellulose. As we know now cellulose is a biopolymer of connected glucose molecules attempts are being made to degrade cellulose and use it as valuable source of glucose, a sugar that can be readily used in a variety of ways.

Many organisms can degrade cellulose through the use of enzymes known as **cellulases**. The identification, isolation, and purification of cellulases make up an important avenue of research.

It is being tried to treat waste paper (office paper, newspaper, cardboard) with cellulases and then use the released glucose in yeast fermentations. One major end product of such fermentation is alcohol, an important industrial raw material and also a potential fuel source.

The attempts are being made to reduce the pressure on fossil fuel which is fast decreasing. Certain microbes such as fungi are capable of producing cellulases. These microbes are being manipulated to increase their potential so that more and better cellulases are produced.

For commercial biogas production in India, cattle waste is the prime source.

9.4.2 Large Scale Growing of Energy Crops

Sugar crops like sugar cane and sugar beet and starch crops like corn are valuable as solar energy converters. An effective use of these renewable sources needs their conversion into several products that can be used as fuels, chemicals and other useful products. These crops give sugar which may be directly used to convert to ethanol, valuable byproducts such as bagasse which can be used as solid fuel.

Other energy crops include a large number of *Euphorbia* and *Asclepeus* species which bear latex and 35% of their dry weight can be extracted as organic extracts.

9.4.3 Short Rotation Forestry

It involves growth of forest species like *Eucalyptus*, *Populus*, *Pinus etc*. which give rapid juvenile growth and can be regrown from harvested stumps. Micropropagation using tissue culture techniques is very useful in propagation of these trees.

9.4.4 Hydrogen Gas Production by Microorganisms

Hydrogen gas production is an important area of biotechnological research with respect to environment friendly non-conventional and renewable source.

9.5 ENVIRONMENTAL SUSTAINABILITY

Attention has been drawn to various biotechnologies that are being used to reduce the impact of wastes of societies on the environment. As it is doubtful that this can be achieved we must:

  (i) either accept an increasing erosion of environmental values and standards world-wide or

  (ii) make processes more efficient by minimizing mass and energy flows.

Some decisions have to be taken now.

All environmental biotechnology processes and products can have negative implications and such risks must be balanced against ensuring benefits. While such processes must always be put through risk assessment, it is clear that, in the light of legislative awareness and technological realism, a great amount of existing and forthcoming environmental biotechnologies should be capable of achieving maximum environment safety.

There is a need of protection of ecosystem from the adverse environmental effects associated with increased urbanization and industrialization. This will involve the following:

(i)   Creative management of effluents and emissions

(ii) Reduction of waste generation

(iii) Production of reliable and clean technologies

Thus, there is now an increased awareness that rather than attempts to remediate the process, the problem needs to be tackled at source. Biotechnology will increasingly be seen as a means to improve many types of existing biological and chemical engineering processes that presently generate environmental damaging byproducts. There will be an increased integrated bioprocess design that is expected to develop high quality bioprocesses that are efficient, controllable and clean. Biotechnology will play a central role in 'Eco-Tech'–a new technology concept *"embedding technology into the ecosphere and human culture by using the whole range of biodiversity in a holistic and low-invasive way in order to achieve benefits for humankind obeying ecological principles"* (International Organization for Biotechnology and Bioengineering, 1994).

For any technology to be considered sustainable, it must not degrade the environment through either the overuse of resources or the creation of unbearable ecological burdens.

It is becoming increasingly evident that man's activities within the environment are far exceeding the sustainable capacity of the earth. In essence, the environmental load equals the size of the world's population X the prosperity or welfare per head of population X the environmental use per unit of prosperity (welfare). It is now apparent that there will be an unavoidable requirement to reduce the environmental load 20-50 times in 50 year's time.

Biotechnology in environment in fact invites its specific use in management of environmental problems including waste treatment and pollution control and their integration with non-biological technologies. However, the safety of the use of the products of biotechnology is being questioned. The release of genetically engineered microorganisms to the environment is under discussion. The application of these techniques may solve some, but at the same time cause some serious environmental problems. If the methodological framework is adequately developed, technology assessment can ensure the former and help avoid the latter.

Much attention from the press and the public has been on the contribution that biotechnology has made to health care and agriculture, the society has now started to see the benefits of biotechnology's "third wave"-industrial and environmental biotechnology. This third wave of biotechnology is already successfully competing with traditional manufacturing processes and has shown promise for achieving industrial sustainability.

Biodegradable Plastics

We use materials that are synthesized from petrochemicals such as polyethylene, polyvinylchloride and polystyrene as plastic wares in our daily life. The disposal of these solid wastes becomes a major problem in waste management in many modern cities because of the synthetic and recalcitrant properties of petroleum-derived plastics. The production of photodegradable plastics becomes meaningless since many cities use landfill as the means for disposal of solid wastes. Once photodegradable plastics are disposed of in a landfill and covered with soil, the photodegradable plastics cannot be exposed to sunluight and thus remain intact in landfill sites.

Attempts are being made to produce biodegradable plastics. The first biodegradable plastic was plastic polymer blended with a biologically degradable polymeric compound such as starch. Once the biodegradable plastic was disposed of, microorganisms in the natural environment can degrade the starch molecules in the copolymer and result in the disintegration of the plastic polymer. However, the physical and chemical properties of the starch-substituted biodegradable plastics are not suitable for practical usage. In addition, a large part of non-degradable plastic polymers remains residues that can accumulate in our enviornment and cause pollution problems.

The production of another types of biodegradable plastics that uses PHAs, which has the physical and chemical properties very similar to those of polypropylene, has been studied extensively. PHAs are microbial storage polyesters. In the presence of excessive carbon sources, many microorganisms form PHA granules intracellularly. Many companies are engaged in the large-scale production of PHA biodegradable plastics. One of the examples is "Biopol™", which is a copolymer of P(3HB-Co-HV) [poly(3-hydroxoybutyrate-co-hydroxyvalerate)], produced by Zeneca Bio Products (formerely Imperial Chemicals Industries) in the UK. Later, Monsanto, a leading agricultural chemical manufacturer in the US, applied genetically engineered oilseed rape plant to produce PHA biodegradable plastic polymer. However, the PHAs produced by the plants have physical and chemical properties different from those produced by microorganisms. Another large chemical manufacturer in the US, Union Carbide has also produced microbial biodegradable PHA plastic called "Tone Polymer", while a chemical company in Japan, Showa Denko has produced "Bionolle" which is a chemically synthesized biodegradable PHA plastic. Among these biodegradable plastics, PHB (poly-β-hydroxybutyrate) and its copolymer, P(3HB-Co-HV) [poly (3-hydroxybutyrate-co-hydroxyvalerate)] are the most widespread and thoroughly characterized PHAs. A number of bacteria, including *Alcaligenes* spp., *Pseudomonas* spp. and a number of filamentous genera such as *Nocardia* spp. produce these PHAs when they encounter unfavourable growing conditions.

In general, a specific microorganism, which occurs widely in nature, is utilized in a novel transgenic fermentation process to convert carbohydrates into the PHA polymers. After the fermentation, the microorganisms have accumulated 80% of their dry weight as PHA polymer. The polymer is purified by breaking open the cells and then harvesting the PHA using an aqueous based extraction method. Owing to their amorphous character, these biopolymers can be ideally handled as lattices. The biopolymer can be extracted with organic solvents and then purified to the usage form. A nonsolvent based process for recovery of PHAs from bacteria was reported. The process is to solubilize the bacterial biomass by sequential treatments of heating protease digestion and detergent and leave the bacterial peptidoglycan in cell wall intact for the ease of separation. Then the PHA granules will be purified by filtration and the final product comes in a latex form. The estimated cost of production and extraction of PHA by this route will be $5/kg, which is still higher than that of synthetic plastics.

PHA polymer can be compounded with safe, biodegradable additives to tailor procesing and mechnical properties before being converted into plastic articles using standard techniques. Although stable in normal use, the PHAs can be biologicaly degraded when deposited in an active microbial environment such as compost. Certain microorganisms can metabolize and consume PHA polymers as their nutrients. Biodegradation of the PHAs under aerobic conditions yields carbon dioxide and water in the same amounts originally used during photosynthesis, completing the PHA biocycle. In order to test the biodegradability of the PHA-plastics, a series of tests under laboratory and field are being conducted. Standard methods have been published by the American Society for Testing and Materials (ASTM) for the degradation of synthetic and biodegradable plastics under laboratory conditions. A series of microorganisms, including bacteria and fungi, has been characterized as being able to degrade PHA-plastics. Most of these microorganisms have been isolated from aerobic environment. Over 10 different extracellular PHA depolymerases have been purified and characterized.

The critical factor limiting the use of PHA plastics is their high production cost. Alternatively, one produce novel PHAs with improved physical and chemical properties that suit the specific needs of certain usages. In Hong Kong, there are several studies on the production of biodegradable PHA-plastics by microorganisms. Studies have been conducted to investigate the use of synthetic media with various amounts of organic acid (*e.g.,* butyric acid) as the sole carbon source to produce PHA with improved physical and chemical properties by selected microorganisms such as *Alcalcigenes eutrophus* and *Klebsiella pneumoniae.* By varying the concentrations of nutrient supplied, the bacteria produced PHA-plastics with HB:HV (hydroxbutyrate : Hydroxvalerate) the ratios of 97:3 and 79:21. These PHA-plastics have different thermal stabilities. The amount and ratio of various organic acids added to control the polymeric composition of PHAs, change the physical and chemical properties of the PHA-plastics.

Another approach is to produce PHA-plastics with a more economical substrate such as wastewaters from food and chemical industries with xenobiotic organics. The soya wastes from a soya milk diary and malt waste from a beer brewery have been successfully used as feedstock to produce PHAs from microorganisms in activated sludge. Microorganisms, mainly bacteria, in the activated sludge fed with various amounts of wastewater from food and chemical industries produced copolymer (p(3HB-Co-HV)) of PHA-plastics. Various types of PHAs were produced when the C:N ratio of the wastewater varies, with the amount of PHAs increased as the C:N ratio increased. The result was due to the wide variety of bacteria in the activated sludge that produce various types of PHAs. Different bacteria have specific PHA synthetases with different substrate specificities and produce different types of PHAs. Sugars such as maltose, sucrose and fructose present in the industrial wastewaters are used as carbon sources for the bacteria to produce PHAs. The use of activated sludge to convert carbon substrates into PHAs not only can produce biodegradable plastics, but also serves as a wastewater treatment in which the problem of disposal of activated sludge is solved.

Recent advances in understanding the metabolism, molecular biology and genetics of the PHA-producing bacteria and cloning of more than 20 different PHA biosynthesis genes allow the construction of various recombinant strains that are able to produce a much larger amount of biodegradable plastics with different types of PHA copolymers. These advances lead to the production of biodegradable plastics with costs that are comparable to or lower than that of synthetic plastics and with greatly improved physico-chemcial properties.

Summary

The techniques of biotechnology are providing us with novel methods for diagnosing environmental problems and assessing normal environmental conditions so that we can be better-informed environmental stewards. Methods for detecting harmful organic pollutants in the soil using monoclonal antibodies and the polymerase chain reaction have been developed. The scientists in government labs have produced antibody based biosensors that detect explosives at old ammunition sites. These methods are not only cheaper and faster than laboratory methods that require large and expensive instruments they are portable too.

Industries that Benefit

The following industries are to gain from the contributions of biotechnology:

- *The chemical industry:* These can use biocatalysts to produce novel compounds. This will reduce waste byproducts and improve purity of chemicals.

- *The plastics industry:* There will be a decrease in the use of petroleum for plastic production by making "green plastics" from renewable crops such as corn or soybeans.

- *The paper industry:* There will be an improvement in manufacturing processes, including the use of enzymes to lower toxic byproducts from pulp processes.

- *The textiles industry:* There will be less toxic byproducts of fabric dying and finishing processes. With the addition of enzymes to their active ingredients, fabric detergents are becoming more effective.

- *The food industry:* Baking processes will be improved. There will also be an improvement in fermentation-derived preservatives and analysis techniques for food safety.

- *The livestock industry:* To increase nutrient uptake and decrease phosphate byproducts enzymes can be used.

- *The energy industry:* Enzymes can be used to manufacture cleaner biofuels from agricultural wastes.

Questions

1. What is the role of biotechnology in the management of environment? Explain.

2. What is end of pipe treatment ? Explain in detail.

3. How the solid wastes can be degraded using biotechnology? Explain.

4. In what way biotechnology helps in energy production ? Describe

5. What are Bioplastics ? Write in detail.

Chapter 10

Biotechnology and Aquaculture

10.1 INTRODUCTION

Aquaculture is the farming of aquatic plants and animals. It is broadly defined as artificial propagation and culture of aquatic organisms. It mainly comprises commercial production of food for human consumption. It also includes the culture of species for the enhancement or conservation of natural aquatic resources. It, therefore, pertains to production of useful aquatic plants and animals such as fishes, prawns, crayfish, lobsters, crabs, shrimps, mussels, oysters and seaweeds by proper utilization of available waters in the country. Aquaculture contributes more than 16 million tones of fish and shell fish annually to the world. Aquaculture is still a growing young science with a lot of potential for innovation. It has thrown a great challenge and an opportunity to the aquaculture scientists to unravel the mystery of this fascinating field of science.

The oceanic organisms are of enormous scientific interest for two major reasons :

(i) They constitute a major share of the earth's biological resources.

(ii) Marine organisms often possess unique structures, metabolic pathways, reproductive systems and sensory and defense mechanisms, because they have adapted to extreme environments ranging from the cold polar seas at $-2°C$ to the great pressures of the ocean floor, where hydrothermal fluids spew froth.

Overfishing and unsustainable fish-catching practices pose a serious threat to the natural aquatic ecosystem and future of the aquatic resources. Global efforts to promote sustainable fishing have been difficult to fulfill because of the lack of monitoring facilities and implementation of sanctions.

Fish farming is the world's fastest growing sector of the agricultural business. There is an increase in the demand by the consumer for fish products. At the same time, as a result of overfishing, environmental degradation and disease there is an alarming decline in commercial fisheries. According to the Food and Agriculture Organization (FAO) of United Nations, nearly 70% of world's commercial marine fish species are now fully exploited, overexploited or depleted.

While the yields of fish-catching have reached a ceiling at around 100 million tons per year in the last decade, demands from aquaculture are still growing rapidly. Worldwide, it is estimated that in order to reduce the increasing gap between demand and supplies of fisheries products, there is a need to augment the production of aquaculture three to four-fold during the next two decades. To meet that challenge, the industry has to be more efficient, reliable and cost effective. For this purpose, aquaculture must overcome some major obstacles related to the biology and life cycle of the farm organisms as well as become fully compatible with the marine and coastal environments. The decline in world fish stocks, by overfishing and climate changes, can only be substituted by aquaculture.

A sustainable aquaculture is only possible if several challenges are addressed. These include development of technologies based on sound scientific knowledge in areas such as handling of organisms, both in experimental and production aquaria, nutrition, health and genetics, increase production efficiency, a disease resistance and expanded ecological range. India has great potential for aquaculture because of lengthy coastline and innumerable inland water spreads.

As with other forms of agriculture, this endeavour has also got the impact of biotechnology. Production of diagnostic kits to identify pathogens, genomic vaccines and probiotics will be other achievements. Genetic engineering has yielded transgenic organisms with improved capabilities such as fast growing and cold-resistance varieties.

On the other hand, there is a considerable concern among the players in the fish industry about consumer acceptance of transgenic fish and among environmentalists about potential environmental hazards resulting from the possibility of transgenic fish escaping into the aquatic environment. Moreover, there are socio-economic risks for developing countries that must be taken into account.

Despite all these, interests in the use of biotechnology in aquacultures have been growing in the recent years.

The five research priorities are :

(i) To develop a fundamental understanding of the genetic, nutritional and environmental factors that make the production of primary and secondary metabolites in aquatic particularly marine organisms, a basis for developing new and improved products.

(ii) Identify bioactive compounds and determine their mechanisms of action and natural functions, to provide models for new lines of selectively active materials for application in medicine and the chemical industry.

(iii) Develop bio-remediation strategies for application in the world's coastal oceans, where multiple uses including waste water disposal, recreation, fishing, and aquaculture-demand prevention and remediation of pollution and develop bioprocessing strategies for improving sustainable industrial process.

(iv) Use the tools of modern biotechnology to improve health, reproduction, development, growth and overall well-being of cultivated aquatic organisms and promote interdisciplinary development of environmentally sensitive, sustainable system that will enable significant commercialization of aquaculture.

(v) Improve understanding of microbial physiology, genetics, biochemistry and ecology in order to provide model system for research and production systems for commerce and to contribute towards understanding and conservation of the seas.

Aquaculture utilizes three main groups of multicellular organisms :

(i) Fish (vertebrates),

(ii) Shellfish (invertebrates),

(iii) Seaweeds (non-vascular plants)

Besides microorganisms are being exploited for various products and applications. Cells of the these multicellular organisms can be cultured *in vitro*. Two broad types of applications of cell culture of these organisms are :

(1) to gain knowledge and products that can be used to manipulate the whole organism to achieve the farming goals,

(2) to be a better source of biochemical products than the whole organism.

10.2 ROLE OF BIOTECHNOLOGY IN AQUACULTURE

As mentioned above biotechnology is used in several different ways in aquaculture. Transgenics are used to introduce desirable genetic traits into the fish, thereby, creating hardier organisms. These are used:

- for a stock development
- as a bio-reactor for manufacturing drugs
- as model for studying various diseases such as cancer *etc.*
- testing gene therapy of inherited diseases in humans
- to look for new feed sources to improve the composition of the feed
- for growth rate improvement and control of reproductive cycles through hormone therapy
- for production of new vaccines
- for development of disease resistance in fish

10.2.1 Cell Culture

10.2.1.1 Fish

Fish or piscine cell culture can be initiated in a number of ways to yield cultures and cell lines. They have originated from a variety of tissues, but only from relatively few of the estimated 20,000 species of bony (teleost) fish. The simplest method for fish culture is referred to as 'explant outgrowth'. The following steps are followed for this purpose:

a) Finely cut pieces of tissue are placed in a growth medium and cells are allowed to migrate out onto a growth surface.

(b) Tissue pieces are dissociated into suspension of single cells with the help of enzymes like trypsin and collagenase.

(c) These like mammalian cells require a solid substrate to attach, spread and grow.

Plastic tissue cultureware that has been developed for mammalian cells appears to be adequate for piscine cells. Contamination is prevented by antibiotics like penicillin and streptomycin. Maintenance and growth occurs over a wide temperature range. The initial cultures are referred to as **primary cultures** and can be subcultured with trypsin into new culture flasks. This can result in either a finite or continuous cell line that can respectively be grown for either a definite or indefinite period. Although rarely studied, most fish cell lines appear to be capable of being grown indefinitely. Very few examples of fish cell lines expressing differentiated functions have been reported.

The complete medium for fish cell growth has two parts :

(i) A basal medium

(ii) Undefined supplement

(i) **A basal medium:** It is an aqueous solution of nutrients and buffering agents that satisfy the requirements for growth in the presence of appropriate hormones and growth factors. The components of a basal medium are bulk ions, trace elements, amino acids, vitamins and a carbohydrate. A variety of basal media have been developed for mammalian cells and most of these are adequate without modification for teleost cells.

(ii) **Undefined supplements :** The appropriate hormones and growth factors are supplied by an undefined supplement which for fish cells usually has been some type of bovine serum. Other successful supplements include trout embryo extract and a complex mixture of lactalbumin hydrolysate, trypticase, soy broth, bactopeptone and yeastolate. Very little work has been carried out on the development of completely defined media for piscine cells.

10.2.1.2 Shellfish Cell Cultures

Shellfish cell cultures are difficult to prepare. Primary cultures have been developed, but not continuous cell lines. Explant outgrowth from haematopoietic tissues and ovaries have yielded primary cell cultures from a variety of shrimps. They have been maintained in the basal medium, Leibovitz's-15, supplemented with foetal bovine serum (FBS) and shrimp haemolymph. Primary cell cultures have also been obtained from bivalve molluscs.

The main impetus for the development of these cultures has been the identification and isolation of viruses that cause diseases in commercially important shellfish. Oyster cell cultures have been used to show that American oyster is unable to synthesize sterols.

Mollusc cell culture has also been carried out to make it a source of adhesive proteins. Mussel and barnacle adhesive proteins account for the moisture resistant mechanism by which these organisms naturally adhere strongly to under water surfaces. These proteins are being explored for use in manufacturing and in medicine. Already a mussel adhesive protein has been tested in ophthalmology for treatment of corneal perforation. Mussel adhesive protein is also commercially available as Cell Tac which is sold as attachment factor for culturing mammalian cells.

Shrimp aquaculture in India : Marine Product Export Development Authority, Cochin (MPEDA) started promoting shrimp farming in 1979 and it could catch momentum in the late eighties. Nutritional requirements of shrimps have been well studied and it is a question of least cost fed formulation from the indigenous raw materials, without reducing the nutritional properties so as to achieve a growth of minimum 30 g size in the period of 4 months.

Another area of research is shrimp disease and very little work has been carried out in India on it. Sometimes shrimps show stunted growth. Shrimps are often affected by protozoan, bacterial, fungal and viral diseases, especially in intensive culture system.

10.2.1.3 Sea Weed Cell Cultures

The routine culturing of cells of multicellulare marine algae has been achieved only recently. Seaweeds can be dissociated into cell and protoplasts by digesting algal cell walls with crude enzyme preparations from abalones, limpets and seahorse. With adequate light, *in vitro* cultures can be maintained in enriched seawater without an organic carbon source.

It is attempted to have multiplication and production of improved strains of seaweeds. Single cell and protoplast techniques are being used to determine nutritional requirements to select for disease resistant strains and to genetically engineer and propagate seaweeds.

Cell culture offers advantages as a source of commercial red algal products, polysaccharides. Currently, the agars and carrageens are extracted from red algae for use in food and other industries.

10.2.2 Transgenesis

With diminishing natural resources due to the result of overexploitation, transgenic fish research will play a key role in aquaculture industry in the near future and transgenic fish will be extensively used as model for basic research.

Following are the essential steps in the production of transgenic fish:

(i)   Construction of artificial or fusion genes (These genes are purpose built for a specific trait such as fast growth rate).

(ii) Once gene is made, it is put into fish eggs and eggs are allowed to develop.

(iii) Embroys, fry and adult fish can be sampled to determine the presence of foreign genes in fish. In general, the fish carrying the foreign gene are identified by analyzing a small piece of fin tissue.

(iv) The expression of gene may be detected by biochemical methods.

(v)   Once identified, the transgenic fish can be used for genetic crosses to determine the transmission of foreign gene to its progeny and the genetic trait expressed.

Gene Transfer Methods in Fish

Gene transfers have been successful in several fishes *e.g.,* common carp, rainbow trout, Atlantic salmon, catfish, goldfish, loach, medaka, tilapia, zebra fish *etc.*

There are various reports of different gene transfers by different scientists. Some of these are as follows:

(i) Stage-dependent expression of the chicken delta-crystalline gene in transgenic fish embryos,

(ii) Regulation of antifreeze gene expression in winter flounder and in transgenic fish cells. Attempts have been made to produce a breed of more freeze resistance by transferring antifreeze protein (AFP) genes to the genome of Atlantic salmon. The gene coding for the major AFP found in winter flounder (*Pseudopleuronectes americanus*) was microinjected into fertilized Atlantic salmon eggs. Experiments are in progress to determine the inheritability of the transferred AFP gene and to increase the level of AFP expression in transgenic salmon to attain physiologically significant amount,

(iii) Integration and germ-line transmission of human growth hormone gene in medaka,

(iv) Expression and transmission of melanogenic phenotype in the orange coloured mutant of medaka fish introduced with mouse tyrosinase gene,

(v) Integration, expression and inheritance of rainbow trout growth hormone gene in carp and catfish,

(vi) Characterization of three cDNAs of insulin-like genes from the liver of rainbow trout,

(vii) Highly sensitive detection system for luciferase gene in transgenic fish. Luciferase assay offers very sensitive, rapid and inexpensive determination system for transcription and translation activities. Firefly luciferase catalyzes light production with energy transfer from ATP. A very sensitive light detection system made up of photomultiplier tube, power supply and amplifier are linked to microcomputer. Maximum light emission in luciferase cloned *E. coli* is obtained within 0.3 sec.

The following methods are used to achieve transfection in fish :

I. Microinjection

There are some difficulties associated with the application of gene transfer technologies to eggs of fish.

(i) It is difficult to visualize pronuclei or the nucleus within fertilized fish eggs due to the dense cytoplasm and large yolk mass.

(ii) The large size of the egg further aggravates this problem necessitating the use of low resolution stereomicroscopy when performing the microinjection.

As a result of these limitations, it is tried to transfer the gene in fish by microinjecting DNA into the cytoplasm of the developing egg near the centre of the germinal disc. This has proven to be very successful in a number of fish species.

In this method, genetic material is injected into newly fertilized fish eggs one at a time (typically 20 µl DNA solution per ovum; it would contain 10^6–10^8 copies of linearized gene construct). About 35-80% of the embryo survives microinjection, of which 10-70% may be transgenic. Transgenes show typical Mendelian inheritance.

Disadvantages

This method has following disadvantages:

- This method is time-consuming and tedious one *i.e.,* very labour intensive.
- Tansfection efficiency is not so high.
- The presence of acellular external protective coating (the chorion). This structure which is analogous to the *zona pellucida* of mammalian embryos, undergoes a process called **water hardening** at the time of fertilization. Water hardened eggs of some fish such as salmonids are extremely rigid and difficult to puncture with a microinjection needle. However, some people have tried gene transfer experiments by transferring DNA into very early stage zygotes.

II. Chromosome-mediated gene transfer

Chromosome mediated gene transfer is generally considered as *in vitro* technique, whereby genes are transferred between cell lines on isolated metaphase chromosomes. The technique used involve fertilizing albino rainbow trout eggs with gamma-irradiated brook trout sperm and then heat shocking to induce second polar body retention. Electrophoretic analysis of over 500 newly hatched transgenic offspring confirmed the expression of paternal Ldh-4, Mdh-3, Aal-1-2, 6-pgd. The transfer of these loci varied from 2-17%. Cytogenetic analyses of embryos from these crosses revealed variable number and types of chromosome fragments within individuals. Survival and pigment transfer in the resultant gynogenetic offspring considerably varied with the individual female used in a particular cross.

III. Electroporation

The procedure is based on the incubation of fish sperm cells in DNA solution followed by application of high field strength electric pulses (electroporation) to increase DNA binding and uptake. Gene expression assay and slot blot hybridization DNA analysis prove the presence of the transgene.

A recombinant plasmid pMV, GH containing rainbow trout growth hormone cDNA fused to mouse metallothionein I promoter was introduced into medaka by electroporation of 3109 fertilized eggs treated with electric pulses (750 V/cm, 5 times), 783 (25%) hatched out. 4% of hatchlings were transgenic. To obtain transgenic lines, 180 hatchlings were maintained and 35 of them grew into adult fish.

Interest in the use of gene transfer in fish is for the increasing importance of fish proteins in human diets. In recent years, per capita consumption of fish has been increasing on world-wide basis. Aquaculture and mariculture, have emerged in recent years as major agrobusinesses targeted at meeting this increased demand for fish protein. Interest in the use of gene transfer in this is to enhance production of fish reared under intensive culture conditions.

Transgenic fish have been created, but the technology for their creation could be improved. Fish lend themselves to experimental introduction of novel genes. The following are advantageous points in fish transgenesis :

(i) Fertilization of egg is external and is easily carried out by artificial stripping of cock and henfish and mixing of eggs and milt immediately or after some delay

(ii) There are numerous eggs and in many species quite large. This renders injection of material by micromanipulation relatively easy

(iii) After fertilization eggs are easily maintained and in many warm water species. It is followed by rapid development

(iv) It is possible to predict the timing of fertilization in many species with accuracy. This permits the gene transfer to be performed at a prescribed time

(v) The culture conditions required to support development of eggs of many species before, during and after experimental treatment are simple and well documented

(vi) Typically, the eggs are incubated in a circulating bath of filtered, oxygenated water at a desired salinity and temperature.

10.3   APPLICATIONS OF FISH CELL CULTURE AND TRANSGENIC FISH IN BASIC RESEARCH

A diversity of questions can be addressed to piscine cell cultures. These include applications in biomedical research, toxicology in which cultured fish cells are used for ecotoxicity testing of aquatic pollutants and basic research. The fundamental research areas whose findings might have applications in aquaculture are growth, reproduction and health. Some uses of cell cultures in these areas are given below:

Fish cell cultures can contribute in very different ways to methods of manipulating growth.

(a) They can be used to determine cellular nutrition requirements to supplement research on whole fish nutrition.

(b) Cell cultures are being used to understand the regulation of fish growth better.

(c) At the level of pituitary, primary cell cultures are being used to study the growth hormone (GH) secretion in response to factors such as growth hormone releasing factor and dopamine.

(d) Other contributions are to methods for permanently altering the genetic make up of fish and to understand growth factors and the regulation of growth.

(e) Cell cultures can be developed as complement recombinant DNA technologies for the transfer of genes to fish, such as those for growth hormone (GH).

Growth

Attempts have been made to have improved growth in channel cat fish, common carp, gold fish, medaka, rainbow trout, salmon and tilapia by transferring salmon or rainbow trout growth hormone (GH) gene. Transgenic families of a fish differ in the level of expression of GH as well as their body growth; this is most likely the consequence of positional effect due to random integration of transgene. This effect is presumed to arise due to the following reasons.

(i) The regulatory elements present in the region flanking the transgene.

(ii) General availability of the region for transcription.

(iii) Restriction of the expression of transgene to specific tissues or cell types due to the presence of *cis* and *trans* acting regulatory-elements conferring tissue-specific expression.

Some transgenic families show upto 60% increase in body growth. A combination of family selection with mass selection among the transgenic individuals has yielded the fastest growing fish lines. In order to have rapid growth, monosex fish cultures are developed.

A major obstacle for the production of monosex population in fish is the usual absence of both sex-linked markers and differentiated sex chromosomes. This can be resolved by the isolation of sex-specific DNA probes. Sex differences in the repetitive genomic DNA bands of the rainbow trout, the common carp *Cyprinus carpio*, *Oreochromis miloticus* and *Leporinus elongates* (Anostomidae) were not observed in agarose gels stained with ethidium bromide. Southern hybridization with heterologous DNA probes (Bkm, hHPRT, SRY) did not reveal any sex specific band either.

T.J. Pandian and K. Varadaraj employed the induction of ploidy and administration of steroids to produce 100% male *Tilapia*.

Reproduction

Cell cultures are being used to understand the hormonal regulation of fish reproductive cycles. Both pituitary and gonadal cell cultures have been prepared. Primary pituitary cell cultures are being used to study the regulation of gonadotropin hormone (GTH) secretion in response to gonadotropin releasing hormone (GnRH). A carp pituitary cell that releases GTH has also been established. Primary cultures of trout steroidogenic testicular cells are being used to study the regulation of steroidogenesis and to obtain the information about the local control of spermatogenesis with primary cultures of trout sertoli and germ cells. Ovarian follicle cultures have been used to study steroidogenesis and steroid secretion.

Health

A number of health issues can be addressed with fish cell lines. The most important and traditional application is in the diagnosis of viral diseases. Viruses require cells in order to replicate and reveal themselves by causing characteristic deterioration of cell cultures. Other uses can be broadly classified as diagnostic. Piscine cell lines have been used to detect the cytotoxicity and adhesion of pathogenic bacteria and to monitor antiviral activity. Attempts are being made to develop rapid screening methods that use fish cell lines for monitoring aquatic pollutants.

Primary cell cultures are an important tool in understanding the fundamental aspects of fish immunology. Certain continuous blood cell lines have been established as well primary cultures have been used for studies on the mitogenic response of fish lymphocytes, the *in vitro* generation and detection of antibody secreting cells and factors influencing the migration of white blood cells. Macrophage cultures have been used for the following purposes:

(a) Phagocytosis

(b) The respiratory burst

(c) The formation of multinucleate giant cells

(d) Bactericidal activity associated with fish pathogens

(e) Identification of fish cytokines and antibacterials

The products obtained through fish cell culture are being used or being considered for use in mammalian medicine, but are poorly characterized in fish. An interleukin-1 that elicits a response from catfish peripheral blood lymphocytes is produced by a carp epidermal cell line. An interferon gamma like molecule has been detected in the medium of rainbow trout leucocytes stimulated by concanavalin-A.

Stress affects the immune response and health of fish in aquacultural system. Aquacultural practices such as handling, crowding *etc.* evoke the clinical signs of stress that are mediated through endocrine system. It has been found that brief exposure of fishes to stressors results in a transitory physiological response of rather short duration. However, continuous cortisol suppression of immune system may have detrimental effect in terms of disease resistance perhaps weeks after the removal of hormones. Testosterone, a hormone that becomes elevated during maturation of both male and female salmonids, appears to be the immunosuppressive agent.

Fish Pathology

Methods have been developed for rapid identification of pathogens by using DNA probe hybridization. A specific DNA fragment was randomly cloned from chromosomal DNA of *Pasteurella piscicida. Streptococcus species* and *Vibrio anguillarum.* These probes only hybridized to themselves and did not hybridize to any other fish pathogens. The method is most useful for identification of each pathogen. Three haemolysin genes were cloned from *Aeromonas hydrophila* and an equal number from *A. salmonicida.*

Various genes conferring resistance to different diseases have been introduced in fish :

• Genetic resistance to microbial diseases in fish.

• Lysozyme from rainbow trout *Salmo gairdneri* as an antibacterial agent against fish pathogens.

• Specific DNA probes for the diagnosis of bacterial kidney disease caused by gram positive diplobacillus *Renibacterium salmoninarum* have been developed.

• Bacterially expressed nucleoprotein of infectious haematopoietic necrosis virus augments protective immunity induced by the glycoprotein vaccine in fish.

Fish Features

With worldwide growth in aquaculture, there is considerable demand for aquaculture produce to meet the world sea food requirement. New plant based protein sources and feed enzymes will help to make aquaculture an environmentally sound and sustainable farming operation.

By transferring a growth hormone gene, transgenic fish with faster growth rates than the non-transgenic siblings have been produced. Getting the fusion gene working in the fish eggs is a step forward towards the ultimate goal of producing a superior broodstock for the aquaculture industry. Using molecular techniques, finfish and shellfish with accelerated growth, disease resistance, efficient food conversion *etc.* are produced. In addition to transgenic fish, attempts are being made enhance salmon stock by artificial sperm selection using electric shock. Another thrust area of research is to develop microsatellite markers for lobster stock.

Products from Fish Cell Cultures

Piscine cell cultures are used to purify viruses for use as vaccines. Other useful products from fish cell cultures are proteins for use in fish reproduction, growth and health care. These are tried from three polypeptide groups : **gonadotropins, growth factors** and **cytokines.** Generally, these polypeptides are too large to be synthesized chemically. Therefore, they are purified from tissue and cell culture system. With the development of efficient methods for transferring and expressing genes, fish cell culture systems are used for this purpose. Fish genes for some potential products have been expressed in bacteria and yeast.

The application of growth hormones for increased aquaculture production has become possible due to cloning and sequencing of a number of fish growth hormone genes and development of transgenic techniques for fish. Growth hormone is an important regulatory molecule which controls a wide variety of physiological and developmental processes as well as growth. It is difficult to control regulation of expression of growth hormone gene, once it has been inserted into the fish chromosome. An alternative approach is to use recombinant fish growth hormone produced in bacteria as a fish food supplement. This allows the amount of hormone administrated to be carefully regulated.

Oral administration has been shown to be effective in promoting fish growth since hormone is taken into the blood stream through the digestive tract. Growth hormone is needed only in small amounts to promote growth. As a result bacteria containing active recombinant growth hormone may be grown in sea waters and directly used as food. This approach avoids need for overexpression and purification of growth hormone or the need to develop and maintain transgenic fish. A marine bacterium (photosynthetic) *Rhodobacter* has been used as an expression vector for the expression of yellow tail growth hormone (YI-GH).

The cytokines are large diverse group of polypeptides that are being considered for use in human and veterinary medicine because they help in the maintenance of defense mechanisms. Fish cell cultures are being exploited for this purpose.

Fish can be a source of proteins for applications in human health care and in the food industry. Atleast two fish proteins are employed in medicine :

(i) Protamine is used for reverse anti-coagulation during procedures such as vascular surgery and cardiac catheterization.

(ii) Salmon calcitonin is used to treat Paget's disease of bone and certain forms of osteoporosis and is more effective than mammalian calcitonin.

Fish anti-freeze proteins can be used to cryopreserve human organs and in food industry to preserve frozen foods.

As a source of biologically active polypeptide hormones and cytokines, fish cells have some advantages and disadvantages.

Advantages : These include the synthesis of polypeptides with all the appropriate post-translational modifications necessary for biological activity and secretion of these completed proteins into the medium. For example, glycosylation which is post-translational modification, is essential for biological activity of many eukaryotic proteins including GTH (Gonadotrophin). It is species specific and in some cases, tissue and cell-type specific within a given species.

Disadvantages : These include low level of production or expression and expensive growth media.

Fish as a Model System

Fish represent the largest and most diverse group of vertebrates. Their evolutionary position relative to other vertebrates and their ability to adapt to a wide variety of environments make them ideal for studying both organismic and molecular evolution. A number of other characteristics make them excellent experimental model for studies in embryology, neurology, environmental biology and other areas.

Research techniques that enable scientists to make isogenic lines in a single generation, create and maintain mutants, culture cells and transfer cloned genes into embryo, signal an increasing role for fish as experimental models.

Better Fish Feed

Biotechnology is also helping to answer some of the technical and environmental concerns of fish farming. The most common protein source for many fish diet is fish meal. Fish meal, a byproduct of fish processing, is used because of its high quality and high protein content. It, however, has some disadvantages :

(i) **It is expensive.** It sells for about $800-$1200 a tonne. So any cheaper alternative protein source would be preferred.

(ii) **Unstability of supply.** Fish meal comes from the byproducts of wild fish, but world fish stocks are declining. At the same time, fish farming is on the rise, and demand for fish meal is increasing. It is, therefore, unlikely that there will be enough wild fish to meet the increasing demand for fish meal.

(iii) **Environmental concerns.** The use of fish meal in aquaculture causes certain environmental concerns. It contains level of phosphorus far beyond the requirement for optimal growth in fish. The excess phosphorus goes into the water causing problems such as eutrophication or excess algae growth.

As a result of these concerns with fish meal, biotechnology is being used to produce alternative plant-based protein sources. Plant protein has the potential to address the problem of phosphorus pollution since plants do not contain such high phosphorus levels. The use of plant protein in aquaculture would help take away the pressure on wild fish stocks. Research is focusing on investigation of crops as new source for fish feed protein. Some of the potential fish meal replacements include distiller's byproducts, pulse crops and canola. Wheat and canola oil are already being used to some extent in feed for aquaculture.

Fish are very efficient in their growth requiring much less energy than other animals. In the laboratory, researchers are able to give fingerling fish 700 g of feed and obtain 1,000g of growth. It is because most of the growth is in muscle which is mainly water. The fish have such low energy needs that they can use high levels of dietary protein often upto 40-50% of the diet.

For prairie crops to be used as the main protein source for fish, they must be processed into a concentrate. Biotechnology is often used in this processing. Plant proteins also require processing because plants contain certain compounds called as **anti-nutritional** compounds as a defense mechanism. These compounds must be destroyed during processing or they could harm the fish. Research is also going on to deal with these anti-nutritional factors by producing feed enzymes to counteract them *e.g.,* phytase. This enzyme would help fish make the best use of the phosphorus available in a plant-protein based feed.

10.4 USES OF OTHER MARINE ORGANISMS

Recent research has discovered unicellular to multicellualr microorganisms that are unique to the marine world. In fact marine bacteria are emerging as a significant chemical resource.

It is important to cultivate marine microorganisms that produce novel products. Attempts are being made to transfer genes of interest into non-marine microorganisms *e.g.,* the capability to produce a marine polysaccharide, a complex molecule that could be useful as food additive or a water resistant adhesive-could be transferred to an easily grown bacteria (*e.g.,* E. coli or *Bacillus subtilis*). The following materials are being explored from marine organisms.

- Pharmaceuticals
- Enzymes
- Biomolecular materials
- Biomonitors
- Biopesticides
- Biomass for energy production

Pharmaceuticals

Therapeutic natural products found in terrestrial plants and microorganisms were the basis of early drug development (Antiviral from phototrophic green and purple sulfur bacteria).

Many bioactive substances from the marine environment have already been isolated and characterized, several with great promise for the treatment of human diseases. The compound manoalide from a pacific sponge has spawned more than 300 chemical analogs, with a significant number of these have been subject to clinical trials as anti-inflammatory agents.

Enzymes

Enzymes produced by marine bacteria are important in biotechnology due to their range of unusual proteins. Some are salt-resistant, a characteristic that is quite useful in industrial processes. The extracellular proteases are of particular importance. These can be used in detergents and industrial cleaning applications, such as in cleaning reverse-osmosis membranes. *Vibrio* species have been found to produce a variety of extracellular proteases.

The presence of unique haloperoxidases (enzymes catalyzing the incorporation of halogen into metabolites) has been indicated in algae. These enzymes are valuable products because halogenation is an important process in the chemical industry.

Methods to induce a marine algae to produce large amount of enzyme superoxide dismutase, which is used in enormous quantities for a range of medical, cosmetic and food applications have been developed in Japan.

Thermostable enzymes offer distinct advantages, many still to be discovered, in research and industrial processes. Thermostable DNA-modifying enzymes, such as polymerase, ligase and restriction endonucleases have important research and industrial applications. Hot springs in Yellow Stone National Park provided the first archaeon (*Thermus aquaticus*) from which thermostable DNA polymerase was isolated. These novel enzymes (the taq polymerases) became the basis for the polymerase chain reaction (PCR). Attempts are on to get more enzymes with differing characteristics for use in industry.

Biomolecular Materials

It has been demonstrated that marine biochemical process can be exploited to produce new biomaterials *e.g.,* a corporation in Chicago is commercializing a new class of biodegradable polymers modeled on natural substances that form the organic matrices of mollusc shells. The understanding of mechanisms used by marine diatoms, coccolithophorids, molluscs and other marine invertebrates to generate elaborate mineralized structures on a nanometer scale is of great importance.

Biomaterials also hold promise for counteracting biofouling, which has long been recognized as an extensive and costly problem. Bacterial biofilms form slime layers that increase drag on moving ships, interfere with transfer on heat exchangers, block pipelines and contribute to corrosion on metal surfaces. Bacterial and microalgae colonization of surfaces is accompanied by settlement of invertebrate larvae and algal spores eventually leading to "hard fouling" and the need for costly cleaning.

It is now possible to determine the genes and pathways involved in regulation and synthesis of bacterial adhesive polymer. Considerable progress has been made in understanding the nature and expression of surface polymer produced by microorganisms such as the nitrogen-fixing *Rhizobium* species and the opportunistic pathogen *Pseudomonas aeroginosa.*

Biomonitors

Marine organisms can provide the basis for development of biosensors, bio-indicators and diagnostic devices for medicine, aquaculture and environmental monitoring. One type of biosensor employs the enzymes responsible for bioluminescence. The *lux* genes, which encode these enzymes have been cloned from marine bacteria such as *Vibrio fischeri* and transferred successfully to a variety of plants and other bacteria. The *lux* genes typically are inserted into a gene sequence or operon, that is functional only when stimulated by defined environmental features. The enzymes responsible for toluene degradation are synthesized only in the presence of toluene. When *lux* genes are inserted into a toluene operon, the engineered bacteria glow yellow-green in the presence of toluene.

This genetically engineered system reports that biodegradation of a specific chemical, in this case toluene, is proceeding.

Biopesticides

Natural marine products have the potential to replace chemical pesticides and other agents used to maximize crop yields and growth. We may soon have natural pesticides that would provide greater specificity and fewer harmful side effects than to conventional synthetic agents of carbon into marine biomass annually.

Bioremediation

Bioremediation shows great promise for addressing problems in marine environments and in aquaculture. These problems include the following:

(a) Catastrophic skills of oil in harbors and shipping lanes and around oil platforms
(b) Movement of toxic chemicals from land, through estuaries into the coastal oceans
(c) Disposal of sewage sludge
(d) Bilge waste and chemical process wastes
(e) Reclamation of minerals, such as manganese and management of aquaculture and sea food processing waste

The full potential for marine organisms and process to contribute new waste treatment and site remediation technologies cannot be realized without enhanced understanding of the unique conditions in marine environments *e.g.,* oxidation-reduction (redox).

An example of a marine biopesticide in use today is Padan™, which was developed from a bait worm's toxin known to ancient Japanese fisherman. This natural pesticide has demonstrated activity against larvae of the rice stem borer, the rice plant skipper and the citrus leaf miner, among other pests.

Novel compounds in marine algae and sponges containing symbiotic microorganisms have now been discovered in Montana. These compounds promoted growth and stimulated germination and increased root and coleoptile lengths in test plants.

Several sponge and nudibranch species produce terpenes, a broad class of aromatic compounds used in solvents and perfumes and known to deter feeding by fish. Extracts derived from these sponge and nudibranch species show powerful insecticidal activity against two species, grasshoppers and the tobacco hornworm.

Biomass for Energy Production

Approximately 40 per cent of all primary energy production or photosynthesis occurs in the sea. In this process, oceanic plants (phytoplankton, seaweeds, seagrasses) take up CO_2 and with light energy from the sun, convert it into organic carbon (primarily sugars) and oxygen.

The oceans contain 50 times as much CO_2 as that in the atmosphere and it is estimated that primary production incorporates 35 gigatons (1 gigaton = 1×10^{15} gm). It can fluctuate in coastal aid estuarine sediments. The impact of changing redox conditions on biodegradation of environmental contaminants must be understood before waste management and remediation strategies and predictive models can be developed for contaminated sediments.

The chemical composition of biomass can be altered to make it more suitable for particular application *e.g.,* marine microalgae are being genetically engineered to boost their lipid content, with the aim of providing a source of alternative fuels that is more economical than are conventional sources. Biotechnology is being used to convert biomass to ethanol and other alternative forms of energy and chemical feedstocks.

Bioprocessing

The bioprocess engineering involves the application of biological science in manufacturing the biopharmaceuticals and natural bioactive agents. Bioprocess engineering requires an understanding of the biological system employed (such as marine organisms), isolation and purification of a product and its translation into a stable, efficacious, and convenient form.

An emerging area of interest is the potential of marine bacteria/fungi to produce unusual chemical structures with no parallels in terrestrial organisms.

In conclusion, it has been realized that biotechnology offers substantial opportunities to improve the health and well being of cultivated aquatic organisms. More than 50 diseases affect fish and shell fish cultured in the United States. This causes losses of tens of millions of dollars annually. Biotechnology can improve the survival, growth, vigour and well-being of cultivated and wild stocks. Potential products resulting from this research include gene therapy technique, brood stock free of pathogens, safe, effective prophylactic agents, including immune modulators, antigen and vaccines, safe effective therapeutic agents, and improved systems for administering prophylactic and therapeutic agents.

Moreover, the aquatic organisms are increasingly gaining importance because of their proving to be a source of innumerable molecules useful for life. It is because of this, lot much attention is required for improving and refining technology. Rapid developing assay technology can facilitate exploration of the bioactivity of newly discovered compounds. These assay methods which employ specific receptors for known physiological agents, require only minute amounts of a test substance and can be automated.

Todate exploitation of natural agents from the sea has been hindered by problems with limited or sporadic distribution and production. Much more research must be conducted to determine what seasonal factors and life cycle or reproduction states are linked with natural production of an agent.

With technological improvements more products will become feasible and new products will be revealed.

Questions

1. What is aquaculture ? Describe the role of biotechnology in aquaculture.
2. (a) What are the requirements of culturing fish cells in laboratory ? Describe.
   (b) How can we introduce a gene of choice in fish ? Explain.
3. Write an assay on the role of biotechnology in agriculture.
4. Write short notes on
   (a) Biomass of marine organisms for energy production
   (b) Bioremediation and marine life
   (c) Bioprocessing

Chapter 11

Biotechnology and Food

11.1 INTRODUCTION

Food and health have always been closely linked. Both foodstuffs and medicines have the power to heal bodily dysfunctions. An imbalance, however, can disrupt our well-being. The ecosystem of intestinal microorganisms lies at the very foundation of human health and well-being. This complex microflora, comprising a wide range of different bacterial species plays following roles :

(i) It supplies the human host with additional value from foodstuffs
(ii) It protects against intestinal infections
(iii) It contributes to the development of immune system.

We are quite aware of many health-improving properties of certain foodstuffs (European Commission 2002).

(i) The dairy products, fruits and vegetables contain vitamins that protect humans against infections by strengthening the human immune system.
(ii) Proteins important for the growth and development of the young body are provided by meat and flesh.
(iii) Fibre-rich foodstuffs are important for the intestinal transport of digested food.
(iv) Plant hormones have a long-term protective function against cardiac diseases and probably cancer.

Food biotechnology uses what is known about plant and animal sciences and genetics to improve food and how it is produced. The food biotechnology is a burgeoning field that transcends many specific disciplines. Using modern biotechnology, it is possible to move genes for valuable traits from one plant or animal into another. This way, crops and animals that are protected from diseases and capable of producing more food can be developed. The food may also be improved for better taste or nutrition. Microbes and plant cells can be programmed to overproduce natural products. Antibodies targeted to minute portion of a protein molecule, can be "custom made" within a matter of weeks.

Food processes for example, cheese making, brewing, fermentations and other bio-conversions have been used for centuries in the production of wines, beers, fermented milk and vegetables. Such bio-conversions have long been an inherent part of food manufacturing and biotechnology is intimately related to it. With a few exceptions, most short-term results of modern biotechnology applied to food production will be invisible to the consumer's eye. However, indirect effects on existing products, such as cost savings and product improvements, will be far reaching.

For centuries, farmers have moved and changed genes to produce better food more efficiently. The process began when our ancestors settled in one place to grow food. Early farmers learned to combine plants and provide new varieties of corn, soybeans, sunflowers, tomatoes, and other crops. This process is called **crossbreeding.** Although crossbreeding works, it takes a lot of time and efforts. For example, if farmers want tomato plant A to have the colour and size of tomato plant B, they crossbreed them to produce a new variety of plant. However, to get the two traits, from plant B, farmers must also take the rest of plant B's 100,000 genes. To get rid of the unwanted genes, farmers use a process called **back-crossing,** which can take 10 to 12 years. With modern biotechnology scientists can choose a specific trait and move only the genes for that trait into another plant. The process is precise and fast.

Genetic engineering techniques, can theoretically produce strains of food plants specifically designed to possess almost any combination of characters (Table 11.1).

Table : 11.1 Goals for Biotechnology in Improvements of Crop Plants

S.No.	Goals	Remarks
1.	Resistance to disease, insect pests, microbial pathogens, competing weeds	Several companies are already trading plants with these characteristics
2.	Resistance to drought and soil salinity agriculture	Will offer improved prospects for third world
3.	Nitrogen fixing ability	Long-term goal
4.	Cheaper to grow, requiring less herbicide, pesticide	Less environmental damage
5.	Higher yielding-more tons per acre (the traditional goal)	Traditional methods of plant breeding have given higher yield of wheat in Europe and India, Rice, palm oil
6.	Easier handling, rapid and synchronous ripening	Cloned sesame seeds, tomatoes
7.	Better nutritional properties	Higher protein content, increased amount of essential amino acids
8.	Better storage properties	Reduced post harvest losses
9.	Absence of allergens	Gluten free grains
10.	Novel products	Unique tasty processed vegetable snack 'Vagisnax' is available in USA

The first stage of development of genetically engineered strains does not rely on sexual processes, but on the insertion of specific genes derived from donor sources. The second stage requires the subsequent growth of whole plants from those modified undifferentiated cells.

11.2 DEVELOPMENT OF NOVEL FOODS AND FOOD INGREDIENTS

11.2.1 Single Cell Proteins (SCPs)

The term single cell protein (SCP) was coined in 1966 to describe the dried cells of microorganisms put to human and animal use. This term is now universally accepted. However, it refers to a whole microbial biomass *i.e.,* to a complex mixture of proteins, nucleic acids, carbohydrates, lipids and other cell constituents. The utilization of microbial biomass for food and feed purposes dates back to the origin of modern microbiology. Various groups of microorganisms, including bacteria, yeast, algae, molds and higher fungi have been considered for use as source of proteins, the dried cells of these microorganisms being referred to as single cell proteins (SCPs). Since the early 1960s, the idea of SCP as a non-agricultural means of producing foods or feed has become prominent. Although SCP was one of the most fashionable areas of applied biotechnology during the early mid 1970s, when several major oil companies were developing process to produce microbial biomass from petroleum fractions, interest waned in the latter part of decade as oil prices soared. However, SCP has enjoyed a limited 'renaissance' in recent years with the realization that these can be derived from feed stock of very low or even negative commercial value (*i.e.,* wastes) and, thus, represents a means of adding value to wastes and by-product streams. Additionally, SCP has the advantage of converting substances which are not normally food materials, into high quality protein-something which no other food processing operation is able to do.

The potential and products manufactured specifically for the food industry are likely to increase rapidly (Table 11.2).

Table 11.2: Targets for SCP and Food Ingredients

1.	Animal feed additives
2.	Textured proteins for human food
3.	Food ingredients. These include :
	products with special functional properties, texturization through fibre formation, dispensability, emulsification, gelation and thickening *e.g.,* polysaccharides, new low calorie sweeteners, inherent flavour modification, nutritional improvement, food supplements, improvement of colour, water or fat binding capacity, improvement of dough elasticity, heat coagulation and whippability

Following are some of the organisms used in SCP production.

11.2.1.1 Bacteria and Actinomycetes

(i) Raw materials : Various species of bacteria can utilize a wide range of carbon and energy sources, including sugars, cellulose either in pure form as agricultural or forest products (hydrocarbons and petrochemicals).

Carbon to nitrogen ratio in the growth medium is maintained at around 10 : 1. Carbon concentration is lower in continuous than in batch cultures. Anhydrous ammonia or ammonium salts are suitable nitrogen sources. Mineral salts in water supplies are usually adequate for growth, although sometimes, iron, magnesium and manganese as the sulphates or hydroxides are required. The pH is controlled in the range of 5-7 by the addition of ammonia and phosphoric acid.

For SCP production under aerobic conditions, especially from hydrocarbon methanol, ethanol and hydrogen oxygen transfer in growing cells is an important factor.

(ii) Process outline : In a continuous production, a suspension containing about 3 per cent dry weight of bacterial cells is removed continuously. The cells are then concentrated by flocculation and floatation process and the product dried. The product, a Pruteen contains 72% protein and 8.6% total lipids and has an amino acid profile high in lysine and methionine, being comparable to the fish meal.

(iii) Product recovery : In most SCP processes, the dry cell concentrations are in the range of 10-20 g/L. Therefore, large volumes of water must be handled.

Plate and frame filter presses are non-amenable to continuous processing. The vacuum filters also pose difficulty. Filter aid flocculants cannot be used as these would contaminate the product.

(i)   The two zone froth floatation process (*Acinetobacter; Micrococci*) is used.

(ii) The biomass is subsequently concentrated in decanter type centrifuges and then dried (*Pseudomonas methylotrophus*).

(iii) Electrochemical coagulations and centrifugation followed by spray drying are also used (*Methylomonas clara*).

11.2.1.2 Algae

(i) Raw materials: Algae can be grown either photosynthetically or heterotropically. Photosynthetic and autotrophic growths require either artificial illumination or sunlight and carbon dioxide, whereas heterotrophic growth occurs in the dark with organic carbon and energy sources.

In photosensitive algal growth, illumination is the timing factor and hence outdoor cultivation is restricted to the use of shallow ponds of 20-30 lagoons sited latitudes 35°. The efficiency of light energy conversion is low, resulting in a requirement of approximately 35 bwH for the production of 1 kg of algae.

The carbon dioxide content of air is low, being 0.3 per cent. Natural alkaline waters which contain high concentration of bicarbonates may be used to enhance algal growth, alternatively additional CO_2 may be supplied using combustion gases *etc.* For this purpose, pH of the medium needs to be alkaline in order that bicarbonates and the dissolved CO_2 tension are independent of its partial pressure in the gas phase.

(ii) Process outline: Generally Texaco plant is used for the cultivation and harvesting the products of *Spirulina* species. The cultivation units are 60-90 cm deep, 10 ha pond. The pond is fertilized using nitrate and iron.

(iii) Product recovery: Separation of algal biomass from the suspending liquid presents a major problem owing to the low dry weight biomass concentration of 1-2 g/l and the low settling rates of these cultures.

Cells are recovered by concentrating, dewatering and drying flocculants such as aluminium sulphate, calcium hydroxide and separated from the biomass. Algae may flocculate in shallow ponds at pH 9.5 or above without the addition of flocculants.

11.2.1.3 Yeast

(i) Raw materials : The substrates for yeast growth are n-alkanes, methanol, ethanol, diesel oil, gas oil, brewery wastes, sulphites, waste liquor, starch, anaerobic digester supernatant, molasses, cheese, whey and domestic sewage. The most commonly used species of yeasts as single cell proteins include *Candida, Hanesnula, Rhodotorula* and *Torulopsis*. However, for food related fermentation, strains of *Saccharomyces cerevisiae* are usually grown on sugar containing media, such as molasses, wood sugars or spent sulphite liquor.

For SCP production, the carbon to nitrogen ratio of the medium is in the range of 1-10. Concentration of carbohydrates in batch cultures is about 1-5%. To keep pH in the range of 3.5-4.5, anhydrous ammonia together with phosphoric acid is used. For aerobic growth on hydrocarbons, the oxygen requirement is 1g/gm dry weight of biomass and for growth on n-alkanes, it is about 2g/gm dry weight of cells.

(ii) Process outline : Various processes are performed for different products and yeast species

(a) Process for the production of *Candida* utilizes paper pulp sulphate

(b) n-alkanes are separated from gas oil by molecular sieves.

(c) Lovera process is used for Baker's yeast and food yeast

(iii) Product recovery : Yeasts have a size in the range of 5-8 mm and a density of 1.04-1.09 g/cm³. The cells can be readily separated from the growth medium by continuous centrifugation, the final creams contains 15-20% solids.

For separation of *Saccharomyces fragilis* after growth, the growth medium is passed through a 3 stage evaporator to concentrate the solids from 0 to 20% resulting in a feed-grade product. Multistage centrifugation is also used. The separated cells are then dried or spray dried.

SCPs can be industrially produced from a large variety of raw materials including agricultural and industrial wastes. However, petroleum methanol and natural gas are the only substrates from which proteins can be obtained in quantities corresponding to the present and future needs of the world.

Existing processes, particularly those based on solid state fermentation can be operated in integrated agrosystem. It combines the production of raw material with protein enrichment and utilization of the product for human or animal feeding.

In India, little attention has been paid to the production of SCPs, though mushroom cultivation started early in 1950. National Botanical Research Institute (NRBI), Lucknow and Central Food Technological Research Institute (CFTRI), Mysore have established Centres for mass production of SCPs from *Cyanobacteria*. At NBRI, SCP is produced from sewage which is further utilized as animal feed.

Therefore, in the light of protein shortage, microorganisms offer many possibilities for protein production. They can be used to replace totally or partially valuable amount of conventional vegetable and animal protein food. For this, development of technologies to utilize the waste product would play a major role for the production of SCP.

A process based on the fermentation of methanol by the bacterium *Methylophilus methylotrophus* has been developed by ICI Agricultural division at Burmingham, U.K. Years of development of the process, together with toxicological and nutritional studies, led to the commissioning and running of a plant capable of producing 50,000 tonnes/annum of the SCP product Pruteen (its trade name) which is sold either as a powder or as granules to animal feed manufacturers and has amino acid profile similar to that of fish protein.

There have been numerous attempts to develop processes that yield SCP from waste materials, many of which have reached pilot scale. Wastes that are evaluated include, sulfite waste liquor from the wood/pulp processing industry, meat packaging wastes, confectionary wastes, cheese, whey, cannery wastes and even domestic sewage. The vast majority of these processes were conceived to supply low-grade protein for use as animal fodder.

A process has been developed to make high quality SCP from surplus starch of bread making. This SCP has been approved as fit for human use. This product 'Mycoprotein' is SCP derived from *Fusarium germinarium,* a mold and has advantages over other SCPs of having functional properties. Due to mycelial nature, the mold is capable of being texturized and can be spun into various shapes and flavours to simulate meat or vegetable products.

Mycoprotein underwent customer acceptability trials during 1986. It has now reached the open market in U.K. as 'Quoran'. It is in fact being seen as 'premium' food product, rather than low-grade protein supplement.

11.2.2 Gene Transfer and Food Biotechnology

Genetic engineering of plants and animals and even their cells has been used to alter the quantity and quality of food. The original gene insertion work used undifferentiated naked plant cells, the protoplasts into which the plasmid containing the desired gene was inserted and those cells were encouraged to multiply and differentiate into plants. However, with this method, many plant variations appeared which subsequently had to be bred out.

An alternative technique, which has been successfully applied is the use of tissue discs (explants) of the leaves of dicotyledonous plants (*e.g.,* tobacco and tomato) for genetic engineering and plant regeneration. However, limitation is the poor expression of transferred genes which requires the transfer of genetic switch as well as the gene itself.

Mutagenesis breeding creates new genes through mutations. Certain desirable traits cannot be found in the gene pools available to plant breeders even when they implement the unnatural crossbreeding tricks. In the 1940s, crop developers began to create new genes in crops and related plants by using mutagenic agents, such as X-rays and mutagenic chemicals.

11.2.2.1 Improvement of Nutritive Value in Staple Crops

The main direct value of amino acids is as nutritional additive

(i)   to cereal and other foods low in essential amino acids, such as L-lysine, L-threonine and L-methionine,

(ii) to food requiring special supplements such as baby food, beer, special diets for old and the convalescent.

Other applications are the use of the reducing power of L-cysteine in dough to improve quality of bread, and L-lysine and L-threonine may be added to animal feeds to improve the protein utilization. Some amino acid productions like that of phenylalanine and aspartic acid will be directed towards those needed for the production of aspartame, a dipeptide of both these amino acids.

Most amino acids are produced by fermentation, their excretion from the cells (principally of genus *Corynebacterium*) and their accumulation in the supernatant fluid requires separation of cells from the supernatant. The Japanese have used mutagenesis to enhance yields. Some amino acids can be clearly produced by the use of immobilized enzymes of bacterial cells where the conversion requires a single step from substrate to amino acids. Genetically engineered organisms may be similarly used to affect such conversions to produce phenylalanine, L-tryptophan, L-tyrosine and 5-hydroxy L-tryptophan.

The essential amino acid concentration can be increased by providing a crop plant with many copies of the genes encoding enzymes that synthesize the amino acid it lacks. A staple crop may be provided with genes encoding biosynthesis enzymes from the pathway shared by a number of essential amino acids. But in practical sense this may not be possible because amino acid synthesis is a reasonable strategy for increasing essential amino acid concentration. It might be possible to provide a crop plant with many copies of the genes encoding enzymes that synthesize the amino acids it lacks. Why not load up a staple crop with genes encoding biosynthesis enzymes from the pathway shared by a number of essential amino acids. In theory, this might work, but not in practice, because amino acid synthesis is tightly regulated through end product feedback inhibition. To increase the essential amino acid end product, researchers must release the pathways from regulatory control. Using genetic modification to shift existing enzymatic pathways so that certain products are favoured over others, is known as **metabolic engineering.**

Metabolic engineering means to shift the existing enzymatic pathway to produce desirable products with the help of genetic engineering. This phenomenon is true theoretically, but not practically as in body every enzyme is under feed back inhibition. Hence, genetic modifications are done to get desired product.

It is now possible to improve the amino acid content of certain foods. Potato is the most common non-cereal staple crop used all over the world. It is due to its easy availability and least maintenance. The gene encoding a protein which is rich in essential amino acids lysine, methionine and tyrosine has been isolated from amaranth and added into potatoes to increase the yield up to three times with concentration of amino acids increasing to 2.5-4.0 times.

A similar attempt has also been made in Brazil to improve methionine content in poultry feed. The gene was isolated from Brazil nut and was incorporated into soybean. However, it was found that methionine rich protein triggered allergic reactions in some people. So, its large scale production has been terminated. But this has helped in finding out the toxic allergenic protein present in Brazil nut through gene silencing. The information can be utilized to improve the protein by removing the toxic amino acid.

11.2.2.2 Healthy Food Habits

The human diet includes three important biological molecules *i.e.,* carbohydrates, fats and proteins. Of these, fats were earlier considered as highest calorific diet with more calories to burn than to consume. It has, however, been reported that decreased consumption of fats in USA in diet made them 100% obese. Besides this, a number of scientific studies have shown that populations with the lowest rate of heart diseases have the highest fat intake. This indicates that some fats decrease heart disease risks, while other increase them. Hence all fats do not affect in same manner. The same can be said of cholesterol, a lipid. People associate cholesterol with health problems because high cholesterol levels lead to increased risk of heart disease. The most useful health measure is not the absolute level in blood, but the ratio of good (HDL) to bad (LDL), cholesterol. HDL actively protects against heart disease by removing excess cholesterol from the blood. The probabilities of heart disease in two patients with cholesterol levels of 200 mg/dl vastly differ.

A healthy diet must include some lipids, because they are key components of cell membranes, enclose nerve cells in protective sheath and carry messages within and between cells. In addition, just like dietary amino acids, the body cannot synthesize all lipids it needs, so diets must include these "essential lipids" by making smart nutritional choices. In general, this translates into avoiding saturated fats and partially hydrogenated oils and using monounsaturated and polyunsaturated vegetable oils. It is being tried to help people to eat healthier food without having to change ingrained dietary preferences through metabolic engineering of fatty acid synthesis pathways in crop plants that provide oils.

11.2.2.3 Metabolic engineering of fats

Most of the fatty acids in oil seed crops are chains of 16 and 18 carbons, but the relative proportions of those that are saturated, monounsaturated and polyunsaturated vary from one crop to the other. Using molecular genetics, the fatty acid profiles of oilseed crops can be changed to minimize the intake of unhealthy fatty acids.

Fatty acid synthesis to synthesize fatty acid, a 2C molecules was used. The two carbon molecules get attached with the help of fatty acid synthesis enzymes (FAsyns) to produce eight carbon fatty acid. For that 3 different FAsyn enzymes (FAsyn1, FAsyn2 and FAsyn3) are used.

$$C-C + C-C \xrightarrow{\text{FAsyn1}} C-C-C-C +C-C \xrightarrow{\text{FAsyn2}} C-C-C-C-C-C + C-C$$

$$\xrightarrow{\text{FAsyn3}}$$

$$C-C-C-C-C-C-C$$

To maximize the efficiency of fatty acid synthesis, cells organize FAsyn enzymes into a cluster, so that each FAsyn enzyme can pass on its product to the next one. To produce 16 C fully hydrogenated saturated fatty acid *i.e.,* palmitic acid, following changes may occur

First step. The palmitic acid is the end product from the FAsyn enzyme complex called as palmitic-releasing enzyme.

Second step. The enzyme palmitic desaturase removes two hydrogens from palmitic acid converting it into a 16-carbon monounsaturated fatty acid.

Third step. More carbon units can be added to palmitic acid resulting in 18-carbon saturated fatty acid *i.e.* stearic acid which can be further converted to unsaturated forms by desaturase enzymes.

To change the ratio of saturated to unsaturated fats in salad oils, gene silencing technology can be used in which certain enzymes are blocked, while others get activated. In the above examples it is used to block the palmitic – releasing and stearic – releasing enzymes. This blockage diverts 18-carbon saturated fatty acid, stearic acid, to take desaturase route. Stearic acid desaturase removes two hydrogens, creating the 18-carbon monounsaturated acid oleic acid which provides more health benefits. Blocking both these releasing enzymes, decreases the percentage of saturate fats in soybean oil by 50 to 60%. Other scientists add palmitic desaturase genes from yeast and bacteria which have higher activity levels. These genes convert palmitic acid into a monounsaturated 16-carbon fatty acid.

Eliminating trans-fatty acids. It has been observed that there is a link between saturated fatty acids and heart diseases. It is because of this the food processing units have started using vegetable oils instead of saturated animal fats. But the high proportion of polyunsaturated acids makes vegetable oils unstable at the high temperatures because they become rancid when exposed to oxygen. In addition, liquid nature of vegetable oil is also of concern to food processors. To solve this, partial hydrogenation of polyunsaturated fatty acid is done at room temperature by bubbling hydrogen through them. The partially hydrogenated oils caused health problems due to trans fatty acid formation. Now, the polyunsaturated oils are changed to monounsaturated fats genetically rather than through chemical hydrogenation. These monounsaturated oils have same desirable properties as partially hydrogenated oils-they are solid at room temperature and stable when exposed to oxygen or heated during frying and baking and they have a long shelf life. Plant breeders at the US Department of Agriculture (USDA) altered the sunflower's fatty acid profile through random mutagenesis and selective breeding. They have decreased the polyunsaturated fatty acid amount to 20%, increasing the monounsaturated fatty acid concentration to 80%. But these techniques do not provide any information about changes in fatty acid synthesis pathway. Breeders only respond for the higher concentration of monounsaturated fatty acid.

Recombinant DNA technology has also helped in increasing the concentration of monounsaturated oils in soybean. The process includes gene silencing by which desaturase enzyme which converts monounsaturated to polyunsaturated oil, is blocked. The concentration of saturated oil is decreased and that of monounsaturated fatty acids increased by to 75-80%.

11.2.2.4 Biofortification

Micronutrients include vitamins and minerals. These are needed in small amount and they play key roles in specific cellular processes such as oxygen transport (iron), hormone function (iodine), enzyme catalysis (zinc), or vision (vitamin A). Researchers from around the world are using recombinant DNA technology and plant breeding to increase the amounts of micronutrients in staple grains. This is known as **biofortification.** This can be explained with following examples:

(i) Enhancement of the amount and availability of iron. Iron deficiency is not at all dependent on poverty, but also on the poor food choices. About 2 billion people worldwide are suffering from its deficiency. It leads to impair in the mental and physical health of persons and make them more susceptible to infections. Iron is not produced in plants. It is absorbed from soil and is stored in specialized form after binding with organic molecules.

Rice contains almost no usable iron. The amount and availability of iron in rice has been improved by using recombinant DNA technology. A gene for an iron storage protein, ferritin was introduced to increase the amount of iron in each grain. Each ferretin molecule can store 4,500 iron atoms. Two additional genes have also been introduced to improve iron's bioavailability. It is a measure of the amount of iron that is digested and absorbed from the small intestine in a form cells can utilize. In rice plants, as much as 95% of their iron is attached to phytate, an organic molecule humans cannot digest, as phytate-bound iron passes through the digestive tract and is excreted. A gene for a fungal enzyme that degrades phytate to increase the iron avialable for absorption hase been introduced in rice. The protein metallothionein helps in iron absorption and transport in the bloodstream. Such a rice has been named as **Golden rice.**

Vitamins can be produced from a variety of microbial sources. These include bacteria, yeast and fungi. Algae, however, have the potential to offer the widest range of vitamins as food additives. The microscopic alga *Dunaliella salina* produces very high quantities of vitamin A precursor β-carotene.

Attempts are being made at the International Rice Research Institute (IRRI) in Philippines to increase provitamin A amounts further by incorporating the golden rice genes into local varieties and conducting field tests. The IRRI plans to distribute this golden rice variety to farmers for free. Since developing countries have poverty and poor infrastructure, problems like many NGOs oppose the production of golden rice and encourage the distribution of vitamin A pills and promote to grow palm oil, a crop rich in vitamin A in developing countries.

11.2.2.5 Functional Foods

Functional food stuffs are capable of modifying one or more organic functions in addition to their nutritional effects in the body. These are also known as **neutraceuticals.**

These are functional foodstuffs capable of modifying one or more organic functions favourably, in addition to their nutritional effect. In the 1990s, research on functional foods led to products that were found to be "anti-cholesterol" (oil derived from maize) and "anti-oxidant" (a grapevine synthesizing more resveratrol) among others.

As there are more and more demands for healthier products, neutraceutics are extensively being explored. Some of the neutraceutics are :

(i) *Resveratrol*

The life of a yeast (*Saccharomyces*) cell is lengthened by resveratrol, 80% activities of enzymes that prevent cancer, stave off cell death and boost cellular repair systems. This naturally occurring molecule builds up in undernourished animals and plants attacked by fungi.

(ii) *Omega 3 and 6 containing milk*

Nestle has launched a bottle of milk containing fatty acids "omega 3 and 6" for the prevention of coronary diseases.

Unilever has commercialized a "hypocholesterol" margarine which helps prevent the accumulation of "bad" cholesterol.

A healthy life depends on a healthy diet, but the constituents of a healthy diet vary from one person to the next. The variations seen among people regarding dietary requirements are due to genetic variations. Genetic differences lead not only to differences in other protein-mediated cell processes that affect nutrition, such as absorption from the intestine, assimilation of nutrients into tissues and nutrient storage capacity. The study of the interaction between someone's genetic make up and nutrition is known as **nutrigenomics** Nutrigenomics recognizes that genetic variation can affect the safety and efficacy of the nutrients people consume. Very few of the gene variants associated with the variable health effects of diet have been identified.

In spite of the very small number of diet and health associated genes that have been identified, a number of companies have begun making individualized nutritional profiles directed to consumers. Campbell, Kellogg and Quaker have developed soups, beverages and cereals, these can help digestion, prevent cardiovascular diseases and hypertension. The world market for nutraceutics was estimated to be about $14 billion in 1997 and with a growth rate of about 20% per annum, according to Arthur D. Little. Larger firms such as Novartis AG, Danone, Unilever, Nestlé and Campbell, Monsanto Co., Johnson and Johnson and Dupont Co., all are focusing on this growing market with various products.

For example, Danone launched Actimel in 1995 as a small bottle corresponding to an individual dose. By 1999, more than 600 million bottles had been sold worldwide. In France, Actimel was dubbed the "morning health gesture".

11.2.2.6 Nanotechology

Today, industrial biotech companies are exploiting the physio-chemical activities of cells to accomplish tasks at nano (10^{-9}) scale. Some are taking genomics and proteomies one step further and exploring how to apply this knowledge gained in the organic world to the inorganic world of carbon and silicon e.g., Genencor International and Low-corning have partnered to combine their respective expertise in protein engineering systems and silicon. Such convergence of biotech and nanotech promises to yield many exciting and diverse materials and products.

Nano-biotech applications use knowledge about protein engineering to pre-engineered nanostructures for specific tasks. For instance, we know that certain genes in aquatic microorganisms code for proteins that govern the construction of inorganic exoskeleton.

Food firms are, thus, using biotechnology techniques to help them capitalize on emerging demands by consumers concerned about the quality of food and its impact on health. This is especially important in rich countries where many consumers take a lot of vitamin pills and other food additives to correct the deficiencies of an unbalanced diet. Genetic engineering could be useful for producing food ingredients deprived of some undesirable elements or enriched with healthy substances to qualify as nutraceutics.

11.2.2.7 Probiotics and Prebiotics

These microbial food additives or ingredients restore a good balance of beneficial microbial flora in the gut. They mainly consist of lactic acid bacteria, bifidobacteria and yeasts. They have health-promoting impacts in the oral cavity, stomach, small and large intestine and the vagina.

Lactobacilli (bacteria) are known to affect immunomodulation and were reported to reduce the risk of STD infections, including HIV in women. The selection of probiotic strains is based on their enhanced protective or therapeutic effect and increased understanding of the molecular factors of these organisms to contribute towards health effects.

Prebiotics : Unlike probiotics, these are non-digestible carbohydrates such as fructo- and galacto-oligosaccharides. They exert health-promoting effects by improving the characteristics of intestinal flora. Prebiotics, like dietary fibres, act as anti-constipation, faecal bulking and pH reducing ingredients. However, the major mechanism lies in their support of probiotics. The gastrointestinal tract functionality and human health are being studied by European Commission's Cluster. The aims of the cluster are to provide :

- a clearer understanding of the relationship between food and intestinal bacteria, and human health and disease,
- new molecular research tools for studying the composition and activity of intestinal microbiota,
- new therapeutic and prophylactic treatments for intestinal infections, chronic intestinal diseases and for healthy ageing,
- a molecular understanding of immune modulation by probiotic bacteria and examination of probiotics as vaccine-delivery vehicles,
- process formulation technologies for enhanced probiotic stability and functionality,
- commercial opportunities for the food and pharmaceutical industries.

Probiotics and prebiotics can provide benefit to consumers, but these depend essentially on their successful processing, viability, stability and functionality, as well as on storability. Attempts are being made to explore optimal processes and formulation technologies for use in processing probiotics (European Commission, 2002).

11.2.2.8 Modification of Taste of Foods

Many foods have poor taste. This is due to marked accentuating or blocking certain elements in the food (Day, 2003). This is useful since large amounts of salt is used to mask the bitter tastes. For this purpose, sometimes salt and sugar/sweeteners are used. For example, grapefruit juice is not sweet without added sugar. Processed foods, such as canned soups, sauces and snacks, and soft drinks are sweetened to tone down the bitter taste of caffeine. Addition of salt is required to make potato chips flavourful. In fact in many cases, the search is on for molecules capable of tricking the taste buds on the tongue. For example, an acidic taste is blocked by the compound adenosine 5'-monophosphate (AMP), which occurs naturally (*e.g.* in human breast milk). When AMP is added to certain foodstuffs, such as coffee and citrus juice, it prevents some of the acidic tastes from being felt by the tongue (Day, 2003). Such activities could have applications in manufacturing of medicine as well. Several food firms, such as Coca-Cola Co., Kraft Foods and Solae, are interested in food flavour and have major deals with research firms involved in tricking the receptors on the tongue. The current interest is in molecules that could trick the taste buds (Day, 2003). About 20 compounds that block bitter tastes have been discovered by Linguagen Corp. They have been granted patents to use four of the compounds as bitter blockers. Since humans have more than 30 different bitter taste receptors, it is quite difficult to find a universal bitter blocker. Linguagen have also discovered a natural sweetener to replace artificial ones such as aspartame or saccharine, which often leave a bitter aftertaste. Senomyx, blocks bitterness and unpleasant smells and increase the salty taste.

11.2.2.9 Protection of the Environment

Scientists have made some foods, like papayas and potatoes more resistant to disease. These crops need less chemical spray to protect them from harmful insects or viruses, which is better for water and wildlife. Other crops are protected from herbicides that are used to control weeds. Better weed control allows farmers to conserve soil by tilling the ground less often. Farmers also use biotechnology to help plants survive. For example, new varieties of corn and cotton ward off harmful insects and improved soybeans can tolerate herbicides. Farmers can be expected to harvest more crops from these hardier plants.

11.2.2.10 Fresher Foods

Sweeter peppers and tomatoes that ripen more slowly are just two examples of how biotechnology can produce fresher and better-tasting food.

11.2.2.11 Food Borne Illnesses

These occur due to pathogenic microbes which may be fatal in certain cases. About 75 million cases of diarrhoea and 5000 deaths occur annually in US due to microbial contamination. The main cause of this illness is the consumption of undercooked meat as well as raw fruits and vegetables. In recent years, food borne illnesses are increasing due to increased consumption of raw foods like sea food and globalization of food supply. More meals are eaten outside especially at unhygienic conditions which aggravate the food borne illnesses. More use of control measures makes the microorganism more resistant and hence number of new food borne pathogens has increased in recent years which has emerged as a new problem to control this illness.

11.2.2.12 Improvement in Animal and Plant Health

The incidence of microbial contamination of foods is to be controlled at every step of food production as well as distribution. The improvement in animal health care is as important as that of the human. A number of biotechnology based diagnostic kits are helping the farmers to identify the sick animal as early as possible, to prevent spreading of the pathogen. Besides this, new animal vaccines boost the animal immune system and fight the infectious diseases. A number of recombinant techniques are also used to improve plant varieties which are susceptible to infections. Although plant infections do not directly affect the human the mycotoxins produced by them, kill the animals hence affect the human.

Different analytical and microbiological testing techniques are used for all the products to prevent the food contamination. A number of diagnostic techniques have increased the detection sensitivity and decreased the detection time from days to hours. These kits do give rapid and accurate results and are portable and cheaper and easy to handle. Recently due to development of microarray technology, companies are creating gene chips to test many food borne pathogens simultaneously.

11.2.2.13 Food Allergies

Food allergy is an abrupt and abnormal response of immune system due to naturally occurring and wildly consumed foods. Any food is capable of triggering an allergic response, but only eight foods are considered as major allergens and cause 90% food allergies. These are peanut, soybean, egg, milk, wheat, fish, shellfish and brazil nuts. The remaining 10% are due to other 170 minor allergens which affect the immune response from a few people to single response. The allergenicity of food vary among races with different diets *e.g.,* rice allergies are common in Japan and not in North America. Approximately 200 people with food allergies are allergic to a few specific proteins within one or two foods. These proteins trigger the production of allergen specific antibodies like the immune response to other antigens. The initial exposure does not cause reaction, but subsequent exposures trigger the released histamine within minutes. Food intolerance like that of lactose and food sensitivities are not food allergies as they do not involve any antigen antibody reaction.

11.2.2.14 Polysaccharides

Many processed foods require the addition of thickeners and stabilizers for which various gums, starches and other polysaccharides are used. They have large market. Alginates are widely used as finings in beer production, as thickeners in ice-cream and in other foods, particularly those where severe thermal processing is used *e.g.,* pet foods. These are now used in reassembled edible foods. Alginate can be produced from micro-algal culture, but the process is costly as compared to that from macroalgae (seaweed kelp). Many of the uses to which alginates are put, can be performed by xanthan gums. They are products of fermentation of certain bacteria and fungi, of which *Xathomonas campestris* is the most widely used.

Gellan : It is produced by *Pseudomonas elodea*. It forms thermoreversible gels and has the potential for use in a wide range of food systems having a highly gelled structure, or which requires a specific texture. But gellan, like other microbial polysaccharides has not yet received approval for use as food in U.K.

11.2.2.15 New Low-Calorie Sweetners

High intensity sweetners have been widely used in diabetic foods. These have a big market in Europe, Canada and USA, because the overfed, but health-conscious public is always seeking ways to reduce caloric intake, without loosing sweetness. Zero-sweetness compounds (sugars) have the bulking properties of sucrose. These include low calorie sweetener such as aspartame, polydextrose, and sucralose. The West African fruit *Thaumatococcus denielli* Benth contains a sweet-tasting compound called **thaumatin.** It has a sweetness intensity 100,000 times that of sucrose on a molar basis. The gene for thaumatin has been cloned and expressed and its three-dimensional structure has been elucidated.

11.2.2.16 Naturally-Produced Flavour Modifiers

Many food ingredients particularly flavour compounds such as lactones, esters, acetoin, pyrazines, diacetyl terpenes and volatile fatty acids, are derived from plant and microbial cells.

For example, a lipase, that accelerates cheddar flavour development was discovered by incubating culture supernatants of many bacteria with cream and screening for aroma generation.

The method to produce flavour modifier sodium glutamate from wheat or soybean protein was developed in early twentieth century. In the late 1950s, this was superceded by the microbial production of L glutamic acid and later of 5-inosine monophosphate (5-IMP) and 5-guanosine monophosphate (5-GPM). These nucleotides are produced when microbial strains secrete the nucleoside guanosine and inosine into the growth medium. They are then separated from culture broth, concentrated and phosphorylated to give nucleotides directly into medium. To find such strains, the selection of mutants or engineering of strains whose cytoplasmic membranes are permeable to the nucleotides are required. Since bacteria in nature frequently produce low levels of desirable products, the levels are increased by genetic manipulation. If genetics of the organism is poorly understood, strains with higher activity may be generated by mutagenesis and selection.

11.2.2.17 Food Supplements

Another biotechnologically derived health product is eicopentanenoic acid (EPA). It is a polyunsaturated acid and is used to make food healthier. It has potential to reduce the cholesterol levels in the blood and so hopefully will help reduce the incidence of heart disease. It is present in fish oil, but actually it is derived from the algae which the fish eat. EPA can be directly produced from the culture of micro-algae or from direct enzymic extraction from fish oil, and new products such as oil, salad dressing are appearing in the market which incorporate it and sell to the health food market.

The active ingredient of evening primrose oil α-linoleic acid (GLA) is useful in curing rheumatoid arthritis, some forms of heart diseases, pre-menstrual stress and other conditions. GLA is normally extracted from seeds of evening primrose, but Sturge Chemicals (UK) developed a fermentation process using an undisclosed mould to produce it.

Food colouring : Attempts are being made to produce food colourings extracted from natural sources such as flowers, leaves, stems and roots or are chemically synthesized. There is a controversy over the role of some food colours in food allergy, hypersensitivity and food intolerance.

Instead of direct extraction processes, many food colours could be produced with greater efficiency by fermentation of plant cell cultures or extracted from micro-algae. Targets include carotenoids (yellow/orange), xanthophylls (orange), canthaxanthin (red), phycocyanin (blue) and chlorophyllin (green). The genes controlling production and excretion of indigo (the dye for dying Jeans blue) have been transferred into bacteria in order to produce the dye by fermentation, rather than by extraction.

Water binding agents : In many foods, propylene glycol (propane- 1,2 diol) is used as a humectant and emulsifier. Propylene glycol is exclusively made from petrochemical sources. It is however, possible to produce it by fermentation of the readily available cheap substrates of D-glucose, or D-xylose by strains of *Clostridium thermosaccharolyticum.*

Raw material conversion : Acids, alcohols and other microbial inhibitors provide a preservative action on food that will otherwise spoil. These also create textures, flavours. Sometimes alcohols have a unique appeal. The organisms which are responsible for the fermentations may be present in the raw materials, but increasingly they are added as inocula (microbial starters) which rapidly and efficiently convert the raw ingredients into the required end products through growth. With the desire for more reproducible processes, it is being attempted to define the role that microorganisms play and subsequently through genetic engineering, organisms are tailored to those requirements.

11.2.2.18 Fermented Foods and Microbial Starters

Microbial fermentations have played important roles in food processing for thousands of years. These processes cover the whole spectrum of technological sophistications.

Dairy products : The dairy industry has been plagued by the loss of lactose-fermenting ability and viability in culture bacteria for more than 60 years. An increased understanding of genetics has yielded solutions to this problem. For example, it was found that the gene coding for lactose metabolism in *Streptococcus lactis* is carried on the plasmid. Bacteria that lose this plasmid also lose the ability to ferment lactose. Lactose utilization was stabilized by moving gene from plasmid to *Streptococcus lactis* chromosome.

Lactobacilli are also important to many fermentations and have benefitted from genetic scrutiny. Many lactobacilli carry multiple (20-50) small cryptic (unknown function) plasmids some of these have been cloned from *Lactobacillus casie* into *Escherichia coli* and *Streptococcus sanguis*.

In dairy industry the problem is attack by bacteriophages that kill bacterial starter cultures. Increased understanding of phage molecular biology has provided many ways of combating phage attacks. The realization that calcium is required for phage adsorption led to the development of phage inhibitory media containing calcium chelators. Multiple strain starter cultures composed of strains having different phage susceptibilities, also decreases susceptibility to attack from a given phage.

Meat products : There have been dramatic benefits of the application of most elementary principles of modern fermentation to meat fermentation. At the time of *Staphylococcal* food poisoning outbreaks in fermented meats in 1977, only 52% of meat processors used any form of controlled acidulation. Because acidification was the sole mechanism for preventing *Staphylococcal* growth, the National Academy of Sciences (NAS) recommended that the fermentation be started with known microbial cultures rather than relying on indigenous microflora or inoculating with undefined strains generated from previous batch. There were many unanticipated benefits. The starter cultures not only stopped the *Staphylococcal* poisonings, but there was improvement in quality and consistency of product. The processing time was cut from an average of 3.5 days to 6.8 hours.

Cheese flavour : The flavour of diary products goes far beyond the production of lactic acid from lactose. Important flavour compounds are produced by bacterial esterases, lipases and proteases. They are active long after the actual "fermentation" is over. Since aging takes time and money, there is a considerable interest in using enzymes to modify cheese flavour more rapidly.

Vegetable products : The adaptation of new technologies has improved the vegetable fermentations. Low-cost energy efficient means of preserving vegetables, cereals and legumes are provided by lactic acid bacteria. Preservation is achieved through pH reduction, substrate removal and the production of hydrogen peroxide, bacteriocins and other antimicrobial compounds. It is possible to alter microbial end products so as to use fermentation as a preservation method. The complete substrate utilization by homofermentative bacteria is more suitable for this task. *Lactobacillus cellobiosis,* therefore, can remove all fermentable sugars form green beans, cucumbers, peppers and green tomatoes and produce a palatable shelf stable product.

Attempts are being made to develop novel vegetable fermentations. During the fermentation, corn of *Lactobacillus plantarum* excretes lysine. It improves protein quantity and quality.

11.2.3 Production and Improvement of Enzymes

Enzymes can be used in many processes in the food industry because of their specificity of reaction and low working temperatures. These include viscosity reduction, improved extractions, (Table 11.3) bioconversions and synthesis, change in functionality and flavour modification.

Table 11.3: Major Enzymes Used by the Food Industry

1. Proteases	
(a) Papain, bromelian,	Meat tenderization, haze removal and chill proofing
(b) Rennin	cheese making
2. Glycosidases	
(a) Amylases (alpha-beta)	Baking, brewing, sweetener gluco and debranching production
(b) Cellulases/Xylanases	Biomass conversion, Juice clarification
(c) Glucanases	Brewing
(d) Glucose isomerase	High fructose corn syrup production
(e) Glucose oxidase/catalyase	Desugaring of egg whites, oxygen removal
(f) Glycolytic enzymes	Fermentation-carbon dioxide and ethanol production
(g) Invertase	Candy making
(h) Lactase (B-galactosidase)	Low lactose dairy products
3. Lipases	
Acylglyceride hydrolases and phospholipases	Texture modification, flavour generation

Advantages : These include the following :

(a)   Specificity

(b)   Low optimal working temperature (20-40°C) implying low energy requirements.

(c)   Can carry out reactions unattainable by other techniques.

Disadvantages : These include the following :

(a) Satisfactory conversion requires hours or days or for shorter periods, high enzyme concentration

(b) Enzymes are soon liable to be classified as food additives. Added enzymes contaminate the system in which they operate

(c) Enzymes currently exploited are usually extracellular, derived from deep tank and/ or surface fermentations of microorganisms and are sold as powders with little or no purification. Genetic engineering of organisms is leading to new and cheaper sources of enzymes, enzymes of greater heat stability, enzyme modification to change functionality and to use enzyme in new and more efficient techniques.

11.2.3.1 Immobilization of Enzymes and Cells

The addition of enzymes to the material undergoing processing allow to execute their reaction and then they get inactivated (usually by heating).

The advent of immobilization techniques, has opened up additional opportunities for the application of enzyme technology to food processing. These techniques, however, allow irreversible adhesion (immobilization) of enzymes or indeed whole cells, to inert support materials which can then be packed into bioreactors, composed of beds, columns or stirred tanks. Raw materials can then be passed through such reactors and undergo conversion, yet remain uncontaminated by enzymes/cells themselves. The ability to reuse the immobilized preparation repeatedly or use it continuously, is economically feasible.

The commercial potential of immobilized enzyme technology is wide, particularly for liquid feed stocks which can be continuously pumped through the bioreactors. A number of laboratory/ pilot scale conversions of food materials have been tried and some of them prove to be commercially viable. The convertion of bulk liquid waste (*e.g.,* starch waste, whey) into useful products is economically beneficial. Immobilization also offers a new means of making established products (such as beer and cheese). It provides a method to upgrade feedstock by the removal of unwanted components (*e.g.,* removal of lactose form milk using β-galactosidase, prevention of oxidative deterioration by the removal of oxygen, using glucose oxidase) and novel products may be made from proteins whose functionality has been altered.

A large amount of starch waste is produced in food processing. Most of it is converted into high-fructose corn syrups, in some cases using immobilized glucoamylase and glucose isomerase. The starch waste can be converted into ethanol. The hydrolysis of cheap fats to higher grade end products such as savoury and cheese flavours, unsaturated fats, emulsifier and cocoa substituents can be done by using immobilized microbial lipases. The isomerization of D-glucose to D-fructose which is important in the production of high fructose syrups (HFS) from glucose syrups can be achieved by the use of glycose isomerase. An enormous boost to HFS manufactures has been provided by the decision of Pepsi Cola and CocaCola companies to use HFSs in their Cola products. The demand for glucose isomerase preparations, which are intracellular enzymes derived from *Streptomyces* which is important in the production of HFS from glucose syrups that themselves have been enzymatically produced from starch is increasing.

11.2.4 Joint Food Quality

It has been estimated that food industry spends 1.5% of the value of its total sales on quality appraisal and loses a further 3.7% of sales on products quality failure. There is an increasing demand for on-line monitoring and for rapid, preferably immediate, result. Radical advances in chemical and microbiological monitoring of foods are being made through the development of biosensors.

Biosensors : These are electronic devices which contain a biological sensing element (enzymes, antibodies, DNA organelles, whole cells, tissues) linked to a suitable transducer, and electrochemical device. The principle of sensors is that analyte molecule interacts directly with receptor component to give output signal to the sensor.

In 1985, over ninety substances including amino acids, gases, cofactors, amides, heterocyclic compounds, carboxylic acids, carbohydrates alcohols, phenols and inorganic ions had become detectable by prototype biosensors. Some complex variables such as antibiotics, 'freshness' of meat, and biological oxygen demand (BOD) were also quantifiable. The range and sensitivity of biosensors is of great value in the detection and quantification of analytes whose presence, even at very low concentrations is very significant. Aflatoxins for example fall into this category.

Spoilage of fish has been determined by the measurement of hypoxanthine, using xanthine oxidase peroxidase electrode, spillage determinants using aldehydes for oils (lactate for canned foods) are being developed using lactate oxidase membranes, lactate in whey from several dairy and other foods has also been quantified. Glucose contents of products such as honey and jam can be measured and recorded in analyzers equipped with specific biosensors.

11.2.5 Testing of Food Safety

The most significant food safety issue that food producers face is microbial contamination. It can occur at any point from farm to table. Any biotechnology product that decreases microbes found on animal products and crop plants will significantly improve the safety of raw materials entring the food supply. Improved food safety through decreased microbial contamination begins by farming transgenic disease resistant and insect resistant crops.

Biotechnology is improving the safety of raw materials by helping food scientists discover the exact identity of the allergenic protein in foods such as peanuts, soybeans *etc.*, so they can then remove them with biotechnological techniques. It has been possible by biotechnology to block or remove allergenicity genes in peanuts, soybeans and shrimp.

Finally, biotechnology is helping us to improve the safety of raw agricultural products by decreasing the amount of natural plant toxins found in foods such as potato and cassava.

Biotechnology is also helping us to enhance the safety of the food supply. It is providing us with many tools to detect microorganisms and the toxins they produce. Monoclonal antibody tests, biosensors, PCR, methods and DNA probes are being developed that will be used to determine the presence of harmful bacteria that cause food poisoning and food spoilage such as *Listeria* and *Clostridium botulinum.*

We can now distinguish *E. coli 0157 : H7,* the strain of *E. coli* responsible for several deaths in recent years, from the many other harmless *E. coli* strains. These tests are portable, quicker and more sensitive to low levels of microbe contamination than previous tests because of the increased specificity of molecular techniques *e.g.,* the new diagnostic tests for *Salmonella* yield results in 36 hours compared to with the three-four days, the oldest detection method required. We can detect toxins such as alfa-alfa toxins produced by fungi and molds that gorw on crops.

Immunoassays : Food industry is suppose to ensure a wholesome food supply that is free of pathogens and toxins. Immunoassays may be developed to detect any component for which a specific antibody can be produced. The two most common immunoassays are the Enzyme-Linked Immuno-sorbent assay (ELISA) and Radio Immuno Assay (RIA). Immunoassays have been developed to measure a broad spectrum of components such as microbial pathogens, mycotoxins, molds, hormones and enzymes. Many of these assays are using highly specific monoclonal antibodies.

With regards to foods, the immunoassay application that has been given the most publicity is the monitoring of food borne infectious agents such as *Salmonella.* Immunoassays have the advantage of giving results after only 1-2 days, compared to conventional culture techniques which may take four or more days.

Regulatory considerations : A number of laws potentially apply to applications of food biotechnology, such as Federal Meat and Poultry Products and Food and Drug Administration (FDA) authority to ensure the safety of all human and animal foods (including food ingredients). There are several criteria for adulteration. A food is adulterated, if it contains any poisonous or deleterious added substance which may render it injurious to health. "An added substance" includes one that is not inherently a constituent of food or is present in food at higher levels than normal because of some technological modification. Finally, since modern biotechnology involves the transfer of a wide variety of nucleic acid sequences, not all of which necessarily present any health or environmental concerns, a distinction can be made as to the types of sequences that may be transferred without triggering new regulatory requirements.

11.2.6 Benefits of Food Biotechnology

About six billion people live on earth today. By the year 2020, that number will increase to nine billion.Using biotechnology, farmers may be able to produce more crops on the land they already have. If we can get the food we need from these crops, we won't have to devote more land to farming. Developing countries will benefit most from this modern technology, since they will have the largest population growth.

It will be possible to more accurately detect unwanted viruses and bacteria that may be present in food. We will, therefore, have an even lower risk for food-borne illnesses.

Enhancing altering the component of some foods using biotechnology may help lower our risk for chronic diseases like cancer and heart disease and other health conditions. For example, some fruits and vegetables will contain more antioxidants, vitamin A, and vitamin B.

- Cooking oils will be made from plants that contain fewer saturated fats.
- Peanuts may contain less of the proteins that cause allergies.

11.3 CONCLUSIONS

Biotechnology is an emerging field. Although numerous food applications have been envisioned, it is clear that many technical obstacles remain to be overcome before they are ready for commercialization. Food manufacturing represents a market for biotechnological innovation. Attempts are on to improve manufacturing of food. Food manufactures are looking towards the following aspects :

(a) Reduction in costs of raw material

(b) Increased reliability of supply

(c) Increased varieties of suppliers

(d) Increased consistency of quality

(e) Availability of novel materials leading to new products and many other putative benefits

Just as the food manufacturers are a market to the biotechnology companies, so is the consumer to food manufacturer. The consumer is influenced by advertising-sales show, but when it comes to health and safety, the public is persuaded by facts issued by 'neutral bodies' such as World Health Organization than by information put out by the experts of food industry.

Questions

1. Give the historical account food biotechnology.

2. Write what you know about the following:

(a) Neutraceuticals

(b) Pre and probiotics

(c) Food safety testing

3. What are the benefits of food biotechnology? Explain.

Chapter 12

Biotechnology in Developing Countries (Including Biotechnology in India)

12.1 INTRODUCTION

Initially, the biotechnology was a domain of the developed countries. The developing countries, however, were skeptical of adopting it. It was primarily due to non-availability of technology and its high cost. Biotechnological developments need high inputs of finance and skilled work forces – both of which were in short supply in most developing nations. Many countries have limited human resources, institutional capacity, and legal and regulatory regimes to actively pursue research in biotechnology.

In fact formidable, but not impossible task to become real players in the new technology-driven economy is faced by developing countries. The greatest challenge lies in the following:

(i) Massing sufficient human capital capable of sustaining scientific enterprise

(ii) Having the ability to seek unique, but efficient research, development, production and marketing strategies

(iii) Developing the legal and regulatory regimes that would promote technology development and safeguard the public interest and above all the political will and management foresight required to harness the developing technologies

A few developing countries are able to carry out the whole spectrum of research and development activities leading to the commercialization of genetically engineered organisms. But it must be clear that biotechnological developments not only will depend on scientific and technological advances, but will also be subject to considerable political, economic and above all, public acceptance. Many developing countries do not have enough human, financial and institutional resources to compete or meet the investments required to participate in biotechnology research. This situation is compounded by the following factors :

Most of the information given in chapter has been obtained from various sites of internet, the notable being on Promises of Biotechnology, Biotechnology in India etc. The exact sites have been mentioned at appropriate places.

(i)   Multiplicity of useful biotechnology techniques

(ii) Multiplicity of protocols to achieve the same objective

(iii) Large number of competing and pressing health problems

However, with the understanding of global needs, particularly with respect to decrease in the gap between rich and poor nations, the perception has changed.

Today, there is hardly any country which has not understood the importance and urgency of adopting biotechnology. In fact, many developing nations had been successfully collaborating with Western biotechnology companies. Between 1986 and 1991, the percentage of arrangements implemented by US biotechnology companies with developing countries dropped by 10%. There has been realization that the ability of developing nations to avail themselves of the many promises of new biotechnology to a large extent, depend on their capacity to integrate modern developments of biotechnology within their own research and innovation systems, in accordance with their own needs and priorities.

It is important to select the tools that could be used to meet the research priorities.

The tools may be selected on the basis of their ability to make a significant difference in improving health, address the most important issues and meet objectives within a realistic time frame (Daar *et al.*, 2002). Technologies may also be selected on the basis of their ability to create new knowledge, economic implications and social acceptability. In a recent study, scientists ranked different biotechnology techniques on the basis of their ability to meet the needs of the poor in developing countries.

Biotechnology activities in a number of developing countries have yet not reached the advanced technology end (such as genetic engineering and genomics). Countries such as Brazil and China, however, have contributed to genome sequencing efforts. The main hindrance is the lack of funding for institutional development (infrastructure and personnel development). To overcome this hurdle, some Governments have formed biotechnology venture capital firms (*e.g.,* Chrysalis Biotechnology and Bioventure in South Africa) or provided direct finances to the institutions (*e.g.,* Republics of Korea and India). The need to reduce cost for biotechnology development concerns many policy makers. It is difficult and expensive to build one state-of-the-art facility to meet all the biotechnology needs. Countries with limited biotechnology capabilities could use universities and other such centres for research purposes and industrial partners for development, production and marketing requirements as long as regulations clearly stipulate the relationships, benefits and privileges of the various players. Just for reference Table 12.1 highlights the advances in biotechnology and progress made by some selected developing countries.

**Table 12.1 : Advances in Biotechnology and Progress made by
Selected Developing Countries**

Regional leaders	Genetic engineering	Genomics	Cell technology
Africa	Egypt		
	South Africa		
	Zimbabwe		
Asia	India	India	India
	China	China	China
	Thailand		
	Philippines		
	Republic of Korea		Republic of Korea
Latin America	Argentina	Argentina	
	Brazil	Brazil	
	Mexico		

Source: Modified from Country reports (as available in the internet : Promises of Biotechnology).

The FAO document has recently compiled biotechnology profiles of developing countries (http://www.fao.org/biotech/inventory_admin/dep/default.asp). FAO member countries recognize that biotechnology, when coupled with other technologies, offers considerable potential and opportunity for new solutions to some of the old problems hindering sustainable rural development and achievement of food security, but it should be noted that it is an area where there is a growing gap between developing and developed countries. There is a need to promote information exchange in the areas of biotechnology policy and regulation as well as on the extent of biotechnology research, development and application between member countries. Such information is readily available from developed countries, but scarce and dispersed for most developing countries. It is because of this they have compiled biotechnology profiles of these countries. The objective of the profiles was to provide a platform on which biotechnology-related policies, regulations of developing countries and activities can be readily accessed, directing the user to have updated sources of information.

For this purpose a number of databases have been developed. One such database FAO-BioDeC is meant to gather, store, organize and disseminate, updated baseline information on the state-of-the-art of crop biotechnology products and techniques, which are in use, or in the pipeline in developing countries. The data base includes about 2000 entries from 70 developing countries, including countries with economies in transition.

12.2 VARIOUS AREAS OF CONCERN IN DEVELOPING WORLD

Biotechnology has entered into almost all the areas concerning man say agriculture, medicine, environment *etc*.

12.2.1 Agri-food Production

The economies of many developing countries still largely depend on agriculture (for food supply, export and employment) and do not have or have very small subsidies from government.

The following challenges are faced by the farmer :

    (i)   To increase their productivity and competitiveness at national, regional and international levels (within the framework of fair trade regulations)

   (ii)   To protect environment and biological diversity, while reducing agricultural inputs (water, fertilizers and biocides) so as to:

       (a) improve soil fertility and conservation (*e.g.,* biological nitrogen fixation),

       (b) increase nitrogen and phosphorus absorption by crops.

It is diversified to meet the changing needs of the consumers and food industry.

It was realized that in the developed world, agricultural sciences are well developed, producing an abundance of high quality products. Agricultural biotechnology will further improve the quality, variety and yield. New plant species, improved by genetic engineering have found their way to the developing countries. This ensures the following:

    (a) Higher productivity,

    (b) Greater resistance to disease,

    (c) More market value.

Current genetic modifications (engineering) have provided protection against herbicides (for easy management of weeds) and some pests. Work is going on to protect plants against devastating bacterial and fungal diseases as well as drought, cold and salt levels.

In addition, food crops with improved nutritional or health values are being designed. Biotechnology has to help, especially resource poor farmers, to produce more food in a sustainable way. The use of all technologies available to enhance food production may have a greater impact than one technology that may not suit different societies. If any of the biotechnology-related products is successfully adopted, it may lead to acceptance of other advanced biotechnology products. Unless producers/farmers see the benefits of the new technologies in terms of increased yields, the productivity of crops and animals, they remain skeptical in their adoption.

Farmers use a particular crop strain that performs well. It limits the number of varieties of that particular crop. There are 50,000 edible plant species on earth, but more than 90 per cent of caloric value of foods produced is due to only 29 of them. The success of these particular crops means less usage of other potential crops.

It is, however, questioned that what will happen when the affluent nations will become increasingly well endowed with an abundance of food? There will be enough food for all world-wide, but will it always continue to be disproportionately distributed ?

12.2.2 Food Products

A variety of foods are produced by fermentation. Fermentation is a process mediated by or involves microorganisms in which a product of economic value is obtained. Each food is produced by a highly specific process using the concerned microorganism. Some of the processes are now used worldwide. It was initially, noted that natural growth of certain microbes on grains, mostly cereals, improved their flavour, texture and nutritional value. Most fermented foods are produced by solid substrate fermentation and several of them are processed on an industrial scale. These processes are called Koji in Japan. Some of the Koji processes have been developed in the west as well *e.g.,* Tempeh. Miso, is another rich, salty condiment and used as flavouring agent. Natto is made of fermented, cooked soybeans and has a sticky viscous coating with chessy texture. The fermented foods offer the following advantages :

(i) Improved flavour

(ii) Elimination of undesirable flavours

(iii) Improvement in texture of food

(iv) Enhanced nutritional value

(v) Increased digestibility

(vi) Reduced cooking time

12.2.3 Medicine and Public Health

Most developments in biotechnology had originated for their potential applications in health care of both humans and animals, but primarily the former.

The potential of biotechnology-related tools for improving the health of mankind is high. However, meeting the health needs of the millions of people living in poor nations is a challenge. Three technologies appear to be of utmost importance in human health. These are : **diagnostics, recombinant vaccines** and **drug/vaccine delivery**. A list of the top 10 biotechnology tools on the basis of their usefulness and likelihood of achieving significant health improvement within 5 to 10 years has been given in Table 12.2.

Table 12.2. Ranking of Biotechnologies expected to Improve Health in Developing Countries

Rank	Biotechnology technologies
1	Modified molecular diagnostic techniques for infectious diseases
2	Technologies for recombinant vaccine development for infectious diseases
3	Technologies for drug and vaccine delivery
4	Bioremediation to improve environmental quality
5	Sequencing genomes of pathogens to improve diagnosis/vaccine/drug development
6	Controlled systems against sexually transmitted diseases
7	Bioinformatics for drug target identification
8	Nutrient-enriched transgenic plants to counter deficiencies
9	Recombinant technology for therapeutic product development
10	Combinatory chemistry for drug discovery

(*Source*: http://www.utoronto.ca/jcb/).

Drugs like insulin and interferon synthesized by bacteria have already been released for sale. A large number of vaccines for immunization against deadly diseases, DNA probes and monoclonal antibodies (including ELISA test) for diagnosis of various diseases, and human growth hormone and other pharmaceutical drugs for treatment of diseases are being released or are in the process of their release.

In 1988, in an experiment, introduction of lymphocytes containing a bacterial gene in humans had been approved for patients who were in the terminal stages of cancer and had no chance to survive.

In 1990-92, patients suffering from some lethal diseases were subjected to gene therapy. Since then many successful attempts have been made in this direction. DNA fingerprinting and autoantibody fingerprinting are used for the identification of criminals like murderers and rapists through the study of DNA or antibodies from blood and semen stains, urine, tears, saliva, perspiration or hair roots *etc.* These techniques are also proving a great boon in forensic medicine in developing countries.

Many improvements regarding medicine are made in the developed world. The developing nations, however, have potential to bring about improvements in medical care by using advances in biotechnology. In most of the cases, the premier research institutes working on particular diseases are located where they are highly prevalent. It might be simple for developing country to import the necessary expertise and make the required technological investments in well established research institutes. The raw materials which are close to new biotechnological production facilities within a developing country provide both jobs and scientific expertise. This helps to improve the economic situation through biotechnology.

The number of medical disorders or illnesses that need attention is very large. However, research priority setting gives both a focus and benchmark to be attained. Countries have used this approach to create institutions whose research area is limited to one crop (*e.g.,* three Cassava Research Centres in Zambia), one animal diseases (*e.g.,* Trypanosomiasis research facility in Kenya) or one objective (*e.g.,* Vaccine production institute in Cuba). This approach increases specialization, but it suffers from the inability to benefit from expertise/facilities in other areas. The centre can be made irrelevant by a change in the research agenda.

Impact of genomics on the health scenario: After the coming in of the results of human genome project, the priorities of many developing countries have changed. In fact in March 2002, the subject of the Africa Genome Policy Forum (AGPF) held in Nairobi was the health-related biotechnology research priorities to be addressed by new developments in genomics. The forum consisted of various representatives of Southern, Western, Eastern and Northern African countries. The three major research priority diseases to be addressed by genomics are malaria, HIV/AIDS and tuberculosis. AGPF also identified challenges such as capacity-building in research and development, policy development, technology foresight and financial investment. The aim in setting research priority was to help achieve some depth and avoid rediscovery of what is readily available. The recommendations are being pursued through the New African Partnership for Development (NEPAD), the Joint Centre for Bioethics (Toronto University) and African Centre for Technology Studies Acts (ACTS, 2002).

Many developing countries, such as India have developed their own health related programmes. India for example has made all round strides to combat the diseases by using biotechnological techniques at various public and private institution sectors.

12.2.4 Bioconversion and Recycling of Materials

A great deal of research is going on for increasing the production and utilization of biomass as a source of renewable energy. The common examples of biomass are wood, grass, herbage, grains or bagasse.

12.2.4.1 Conversion of Waste into Useful Products

Waste is produced in a number of ways :

The plant wastes include wood, green plant matter, straw, paddy husk, rice bran, saw dust *etc.*

Waste wood can be converted into methanol liquid fuel.

Green plant matter and straw are used as animal feed or converted into an efficient energy source through anaerobic digestion.

Paddy husk can be converted into smokeless solid fuel briquettes. These are suitable for use in domestic cooking, hotels, kilns and boilers.

Saw dust is also converted into low calorific value producer gas for thermal power generation.

Animal waste is also a potential energy source, although the high water content in the animal waste makes it unsuitable for many conversion processes, except anaerobic digestion used for biogas production.

Domestic and municipal wastes are also used as alternative energy sources.

This may not prove to make any significant contribution in western countries, but will be of much significance in Asian countries where it is available in huge quantity due to population explosion. Such improvements are inexpensive ways to effectively use waste materials.

Food waste may be broken down into amino acids, fuels and fertilizers. These would benefit the rural and urban poor. Many developing countries can easily enter this area. Microbes and enzymes play a major role in this process. The developing countries need to seize the opportunities.

12.2.4.2 Biofertilizers and Biopesticides

Nitrogen supply is a key limiting ingredient in crop production. It is often not available and/or beyond the reach of many poor farmers, especially those in rural areas. However, biological nitrogen fixation (BNF), the fixing of atmospheric nitrogen by microbes and making it available to plants, could be harnessed to improve the soil fertility and productivity of crops (Mekonnen *et al.*, 2002). These microorganisms are often referred to as biofertilizers.

Biofertilizers also include microorganisms that solubilize phosphorus to make it available for plants. Many microorganisms have the ability to fix nitrogen. These include *Azospirillum, Azotobacter, Rhizobium, Sesbania*, algae and *Mycorrhizae*, while *P. striata*, and *B. megaterium* and *Aspergillus* are among other microorganisms that solubilize phosphorus. These organisms are provided with a favourable habitat and a carbon source by the plant in a symbiotic relationship. This relationship is critical in seeking to broaden the use of biofertilizers in association with many food crops.

Biofertilizers have been used in Kenya, the United Republic of Tanzania, Zambia and Zimbabwe (Juma and Konde, 2002). They are easily produced locally and the technology needed to produce them is not complex. In some countries, the demand has often exceeded production of the pilot plants. Expansion of these pilot plants could help improve food productivity in Africa.

The use of biopesticides in the control of pests is well established. For example, sterile *Tse tse* flies (the vector of sleeping sickness) have been used to control and eliminate their population on the island of Zanzibar. Similarly, the cassava mealy bug, *Phenacoccus manihoti*, has been effectively controlled by the use of a wasp, *Apoanagyrus lopezi*, from Latin America, and this work has been awarded the World Food Prize.

The bacterium *Bacillus thuringiensis* (Bt) has been used by farmers to control worms and insects for many years. Nematodes, bacteria, fungi and viruses may be used to control industrial, home and farm pests.

The use of biopesticides has not been so great. It represents only a small fraction of the global $8 billion pesticide market. *Bacillus thuringiensis (Bt)* alone accounts for 90% of the $160 million biopesticide market (Jarvis, 2000). The biopesticide market is driven by consumer, retail and government demands for reduction in use of chemical fertilizer use. The following factors limit their use :

   (i) Lack of spectrum (few targets),

  (ii) Slow killing rate,

 (iii) Batch variations,

 (iv) High sensitivity (to soil types, chemicals, temperature and moisture content)

  (v) Low stability (short shelflife and high storage needs).

Biofertilizers and biopesticides have provided an excellent opportunity to the developing countries to enhance their crop yields. Countries such as Bangladesh, Brazil, Kenya, the United Republic of Tanzania, Zimbabwe and Zambia have had successful pilot plants for the production of biofertilizers, and demand has often exceeded production.

Biopesticides too, can help increase crop yield, reduce import bills and increase export earnings. In fact, they could provide an affordable source of agricultural inputs, especially in rural areas where chemical inputs are unavailable or resource-poor farmers cannot afford them.

12.2.4.3 Biofuels

The first car of Henry Ford was empowered by ethanol. However, very few cars today use ethanol (alcohol). Brazil uses fuel blends with up to 20% of ethanol, while in the United States, nearly a tenth of all motor vehicle fuel sold is blended with up to 10% ethanol. In Brazil, ethanol is produced from cane sugar and in the United States from maize. The US ethanol production is expected to reach 75 billion litres a year by 2020 from the current 9 billion. Iogen, a Canadian firm in January 2003, opened a pilot plant that converts straw into ethanol using cellulase. Shell, Petro-Canada and the Government of Canada are Iogen's main partners and investors in the Eco-Ethanol project. But need of this hour is that developing countries must adopt this technology.

12.3 PRIORITIES IN BIOTECHNOLOGY OF VARIOUS DEVELOPING COUNTRIES

Biotechnology is a diverse field in its application and multiplicity of procedures in each given area. Therefore, research priority setting is very important. Because of limited resources and manpower, and overwhelming conflicts of interests, even when a country chooses to focus on one field, however, is difficult .

The process of research priority setting should be seen to be legitimate and fair (Daniels and Sabin, 1997). Following are certain prerequisites of the research priorities :

(i) Recognition of authority under which the research priorities are set; unless it is done, even good intentions will not be realized.

(ii) The research priorities may have to be acceptable, if they have to gain support.

(iii) They should have reasonable and justifiable goals.

These values are not necessarily the values of good science and do not ensure success. Scientists and their institutions quite often conduct research based on assumptions built on current knowledge. The new developments may, however, change the objectives. For example, if vaccine development is the main aim and we overlook the promising leads for new drug developments, it is quite possible that the benefits of an effective drug may be lost, if a vaccine is not developed in time. While setting priority, there should always be a scope for flexibility to pursue promising leads. New development should not, however, overshadow the original goal.

Transgenic crops were developed by small biotechnology firms in the United States. It was done to generate resources to finance development of pharmaceutical products that take a long time to bring to market due to stringent regulatory regimes (Schimmelpfennig *et al.,* 2000). A similar strategy could be adopted by developing countries. They may focus on biotechnology development niches to have significant short-term returns to the national economy.

The experiences gained therein may help to develop more sophisticated tools, products and services. Technological niches help those lagging behind to catch up with the leaders, employing a unique strategy. In genomics *e.g.,* technological niches have been exploited. It can be done by Brazil that has been able to develop its genomic programme in a very short period of time. They combined the genomic and information technology to develop virtual Genomic institutions. More than a million human expressed sequence tags (ESTs) have been contributed by them in a very short time. Besides, genomes of three whole organisms have been sequenced. In this way, it has become one of the most productive genome sequencing and analysis centres in the world.

The biotechnology research priorities of many developing countries do not seem to differ very much in ranking because of the breadth of their goals. The top three priorities include **agricultural**, **medical** and **industrial** biotechnology. Only those that have developed a biotechnology base could be said to have clear priorities. Some of the examples have been given below :

(i) Cuba's biotechnology has a great emphasis on medical applications rather than agriculture or industrial applications. This is because of country's emphasis on good health standards for its people. Other areas such as agriculture and fisheries have benefited from Cuba's health biotechnology development. The country has produced transgenic fish and plants.

(ii) Africa and Latin America biotechnology development is focusing on agriculture even when health is ranked very high. In these countries, most of the programmes on capacity building and policy aspirations seem to target agricultural biotechnology. The research institutions have acquired significant capacity in agricultural sectors.

Various countries have set biotechnology research priorities relevant in meeting national aspirations using different biotechnology tools. This can be illustrated by the country reports available in the literature. The priorities of some of the countries have been discussed below.

12.3.1 Biotechnology Profiles in Brazil

The details of biotechnology programs as compiled by FAO have been reproduced below :

Biotechnology Research Policy	Responsible for biotechnology research in agriculture is the Federal Ministry for Agriculture and Food Supply (MAPA), Ministry for Health (MS) and the Ministry of Science and Technology (MCT); responsible for environmental aspects of biotechnology is the Ministry for Environment		
Research Capacity	Key Institutions	An overview on agricultural research institutes can be obtained from the AROW and ASTI database Brazilian Agricultural Research Corporation Agricultural Research and Rural Extension Institute (EPAGRI) Instituto de Tecnologia de Alimentos (ITAL) Instituto Rio Grandense do Arroz Universidade de Brasília; Faculdade de Agronomia e Medicina Veterinária University of São Paulo; Faculty of Veterinary Medicine and Zootechnology Escola Superior de Agricultura "Luiz de Queiroz"	
	Summary of Major Research Programmes	There are research projects on biotechnology in some of the institutes above; for an overview on agricultural research projects in general see WISARD directory	
Biotechnology Regulatory Framework	Biosafety	Act No 8974 and Decree No 1752 (1995) and Medida Provisória 2-191-9 (2001) creating CTNBio (MCT, MAPA, MS, MMA members), in charge with biotechnology activities Bio safety policy, institutions coordination, regulation, and risk assessment; an overview on biosafety legislation can be obtained from CTNBio, BINAS and the Biosafety Clearinghouse; party of the the Convention on Biological Diversity	
	Food Safety	Act No 8974 and Decree No 1752 (1995) re-gulate biotechnology activities; an overview on other regulations related to food safety can be obtained from FAOlex and ECOLEX	
	IPRs	Patents	Patenting is regulated by Law No 9279 (1996) and Presidential Decrees Nos 2553 and 3201 (1999); more information on IP legislation and on PVP can be obtained from the WIPO Guide to Intellectual Property
		PVP	PVP is regulated by Law No 9456 (1997), Decree No 2366 (1997) and Ministerial Ordinance No 503; member of UPOV
	Genetic Resource	Plant Genetic Resources	A summary on PGR activities can be obtained from WIEWS portal
		Animal genetic resources	
Biotechnology Application	There is a number of application available from the FAO BioDeC database		
Publications and links	Links on PGR related to Brazil can be obtained from IPGRI Some useful information can be obtained from IICA Brazil. Publications on agricultural research policy can be obtained from ISNAR. A Document on National Biotechnology Strategy - Bio-Industry Development Policy (2006) is available from the Biotechnology in Food and Agriculture Web Page.		

(Source : http://www.fao.org/biotech/inventory_admin/dep/default.asp)

The main earlier thrust had been ethane production. Ethanol for human consumption has been manufactured as a component of alcoholic fermentation since prehistoric times. In recent years, it has also found its use as an important chemical feedstock and as a fuel supplement. In Brazil, in the years 1980s, 20% of the petroleum import was replaced by ethanol produced from sugarcane.

Currently, hydrous ethanol production is decreasing. On the other hand, the production of anhydrous ethanol has increased. Ethanol unlike fossil fuels such as gasoline is used as being renewable and cleaner burning material. It produces no green house gases on combustion.

Approximately 80% of the world supply of alcohol is produced by fermentation, although in countries with advanced technology, such as United States, almost all of the industrial ethanol is made from ethylene derived from petroleum sources. 1.9 kg of ethylene is used to make 3.8 litre of ethanol. With the increased cost of imported petroleum, prices of industrial alcohol have sharply risen.

The interest in US is focused on a mixture called **'gasohol'.** It is a blend of 90% unleaded gasoline and 10% ethanol. For this purpose, anhydrous ethanol is used as a gasoline additive. In Brazil, 22% anhydrous ethanol is mixed with gasoline and the mixture can be used in any type of automobile, truck or bus engine without modification. On the other hand, hydrous alcohol can be used in its pure stage (without mixing), but only in specially designed engines.

Extensive efforts are required for a country wide transformation. Following are the requirements for this purpose :

(i)   The appropriate amount of sugar cane is required to be planted, cultivated and harvested for ethanol production

(ii) Vehicles are modified in such a way so as to use ethanol more efficiently as fuel

(iii) The population is being educated about the advantages of the progress and use of ethanol gasoline as a viable alternative fuel.

Extensive efforts have been made by Brazil to produce ethanol from cane sugar molasses, cane juice and cassava starch. It has been done to lower its dependence on imported oil. Attempts are being made to replace gasoline with an 80/20 blend of gasoline and ethanol in motor cars. This progress with extensive government supports is well on its way.

12.3.2 Biotechnology in Malaysia

Malaysian government and people of Malaysia by their tremendous efforts have increased their standard of living by becoming technologically more capable. Many industries are dependent on Malaysian economy. Palm oil and rubber industries will be greatly affected by the present and future developments in biotechnology. The biotechnology profiles in Malaysia as compiled by FAO have been reproduced below :

Biotechnology Research Policy	Responsible for biotechnology research in agriculture is the Ministry of Agriculture; responsible for environmental aspects of biotechnology is the Ministry of Science, Technology and the Environment (MOSTE)		
Research Capacity	Key Institutions	An overview on agricultural research institutes can be obtained from the AROW and ASTI database and from NARS Academy of Sciences Malaysian Agricultural Research and Development Institute (MARDI) Forest Research Institute of Malaysia (FRIM) University Kebangsaan Malaysia (UKM); Faculty of Science and Technology, Centre for Gene Analysis and Technology (CGAT) Malaysian Rubber Board (MRB) Palm Oil Research Institute of Malaysia (PORIM)	
	Summary of Major Research Programmes	Research Programme on biotechnology within the 8th Malaysian Research Plan by the National Biotech Directorate; additional research programmes on biotechnology within the institutes above; for an overview on agricultural research in general see also WISARD directory	
Biotechnology Regulatory Framework	Biosafety	Biosafety is regulated by Guideline for the Release of GMOs; an overview on biosafety regulations can be obtained from BINAS; party of the Convention on Biological Diversity and the Cartagena Protocol. MoSTE takes advises from the GMAC and the NBC. Informations on biotechnology are also available on the BIC site.	
	Food Safety	Food safety is regulated by Food Act (1983), its Amendment (2001) and Food Regulations (1985) which do not contain regulations on biotechnology or GMO; an overview on food safety regulations can be obtained from ECOLEX and FAOlex	
	IPRs	Patents	Patenting is regulated by Patents Act No 291 (1983) latest amendment by Patents Amendment Act (2000); more information on IP legislation can be obtained from the WIPO Guide to Intellectual Property
		PVP	PVP is regulated by the Protection of New Plant Varieties Act (2004); consultations with UPOV are in progress
	Genetic Resource	Plant Genetic Resources	Malaysia has accession to ITPGR. A summary on PGR activities can be obtained from WIEWS portal.
		Animal genetic resources	
Biotechnology Application	There are a number of records found in the FAO BioDeC database		
Publications and links	Background information on PGR can be obtained from IPGRI Malaysian Biotechnology Information Centre (MABIC). A Document on National Biotechnology Policy (2005) is available from the Biotechnology in Food and Agriculture Web Page.		

(Source : http://www.fao.org/biotech/inventory_admin/dep/default.asp)

One of the world's largest suppliers of palm oil is Malaysia. This oil is used in the production of cooking oils, margarines, soaps and detergents. The oil is obtained from the Karnel of palm oil seed. One of the more valuable oils derived from palm kernel is the lauric acid. This acid is used in soaps and detergents. Oils isolated from plants contain different fatty acids, some with short and others with long chains. Short chain fatty acids are liquid at room temperature, whereas the long chain ones are solids or semisolids.

Palm oil is obtained by harvesting seeds from palm tree plantations. The Palm Oil Research Institute of Malaysia (PORIM), is the government research institute concerned with the development of palm oil industry.

Biotechnology is likely to affect Malaysia's ability to remain competitive in the world market. The genetically engineered rapeseed plant developed by Calgene, a U.S. plant biotechnology company produces oil, which contains a large amount of lauric acid. This plant was developed by insertion of a gene that cuts off the synthesis of any fatty acid longer than lauric acid. The resulting plants are healthy and produce nearly as much seed per acre as non-engineered plants.

The production of acid from this genetically engineered rapeseed plant is cheaper than producing it from palm kernel oil. Calgene, which is also working with rapeseed plant, produces large amount of myristate oil and is expected to make raw materials for soaps and personal care products.

As a result of extensive negotiations between Calgene and Malaysian palm oil industry to determine the feasibility and economic desirability, it has been revealed that Malaysia, a small country with a few resources, might not be able to effectively diversify its commitment to oil production as compared to United States because of its limited resources and acreage. Even Malaysia may not be able to respond to market change because palm trees may take long period for growing up and production.

12.3.3 Industrial and Environmental Biotechnology in Islamic Republic of Iran

The priority of Iran is also agricultural, industrial and environmental biotechnology. It is because of the structure of its economy. The economy heavily relies on the oil industry, which accounts for more than 85 per cent of its exports. The Islamic Republic of Iran, therefore, is likely to get greater benefits from industrial and environmental biotechnology. The Biotechnology Centre of the Iranian Research Organization for Science and Technology (IROST) has five research areas, two of which focus on application of biotechnology in processing and engineering, and on environmental remediation.

The Environmental Biotechnology Centre has been experimenting with 52 microorganism isolates for desulfurization using the Gibb's assay. The FMF strain of *Rhodococcus*, has been found to be useful. It has also been investigating the ability of microorganisms from the Persian Sea to clean up oil spill (biodegradation of oils). The microorganisms (*e.g., Aspergillus niger* and *Bacillus coreas*) are also employed to decolourize textile effluents.

The Bioprocess and Bioengineering Unit has installed a distiller that is currently being used at a pilot plant. It has also produced bioreactors with 1-20 litres capacity, while a 3000-litres stirred fermentation system has also been developed. The National Research Centre for Genetic Engineering and Biotechnology has introduced programmes in plant and industrial biotechnology in 1998. Later in 1999 animal and marine biotechnology programmes were introduced. Medical programmes have also now been introduced.

The textile industry has a long history and is probably the second largest contributor to foreign currency (it earned about 2 per cent in 2000). This along with other industries is environmentally unfriendly. Therefore, attempts are being made to make investment on cleaner processing and reclamation of polluted lands.

The research priorities match national aspirations. Projects run by the Biotechnology Centre of IROST Technologies developed in these areas, have been diffused to other sectors as well.

12.3.4 Eastern and Central Africa: Agricultural Biotechnology

In Africa, agriculture stands very high in development agenda. Almost 50 per cent of the population for their survival depends on agriculture. Therefore, African countries have made agricultural biotechnology their top priority. The Eastern and Central African region is composed of Burundi, Democratic Republic of the Congo, Eritrea, Ethiopia, Kenya, Madagascar, Rwanda, Sudan, the United Republic of Tanzania and Uganda. These countries have formed an Association for Strengthening Agricultural Research in Eastern and Central Africa (ASARECA). ASARECA, in conjunction with the Agricultural Biotechnology Support Project (ABSP) at Michigan State University, commissioned a study that developed a list of agricultural research priorities for the region (Johanson and Ives, 2001). The cereal yields in Africa are lowest and there is limitation of aerable land. The population of Africa in the last four decades has increased threefold, while there is only 2 fold increase in cereal production. As a result, cereal production per capita dropped from 183 kg in 1962 to 143 kg in 2000. The research priorities of african countries have been summarized by ABSP as follow (Table 12.3).

Table 12.3. Research Priorities in African Crops

(African crops, current production and research constraints being targeted)

Crop	Production (Metric tonnes)	Research priority targets
Maize	18 402 504	Yield, disease, pest, storage, weed
Beans	1 820 271	Disease, pests, N & P deficiency, drought
Sorghum	4 89 409	Weed, genetic base, pest, disease, acid tolerance
Bananas	5 660 575	Pest, diseases, processing, genetic base
Wheat	2 297 345	Yield, disease, soil fertility, weed, drought, tillage
Potatoes	12 080 990	Disease, soil fertility, storage
Coffee, Green	787 378	Soil fertility, disease, pest
Seed Cotton	1 168 853	Disease, pests
Rice, Paddy	3 832 051	Weed, soil fertility, pests, disease
Cassava	45 495 641	Disease, pests

Source: ABSP and FAOSTAT (2002). (Adopted from Promises of Biotechnology obtained from internet)

Some of the country reports which have been discussed above reveal two factors. :

  (i) Most of the biotechnology research is concentrated in public institutions and Governments are setting their research agenda

 (ii) Most of the biotechnology research efforts have been added to existing institutions except where countries have significant investment resources.

The research priorities of many countries within given regions appear to be quite similar and thus, there should not be any problem in developing regional alliances. These have been successfully used in some agricultural, veterinary and medical projects. The regional alliances can help countries with limited financial and human resources by sharing information, human resources and facilities.

From the above discussion, it is evident that biotechnology has impregnated the R & D programs of a large number of developing countries including India. But it must be kept in mind that the scientific foundation upon which biotechnology industry development flourishes, has to be solid. Therefore, assessing and building strong scientific and entrepreneural bases may be important. Above all, technology is a product of humans. Many factors such as passion, profits and excellence drive the innovations.

The lessons from developing countries and countries with economies in transition that have built significant capacity in biotechnology, could help guide other countries to model their industrial development accordingly to meet their unique needs and status.

Developing countries in fact constitute a group of countries at different levels of development. However, it is possible that the basic factors that may determine quick growth of biotechnology in countries at different levels of developments may be similar.

12.4 SETTING RESEARCH PRIORITIES

National Biotechnology Capabilities

The assessment of national capabilities in biotechnology is an important step for developing countries. It has also been performed to fulfil the requirement of the Cartagena Protocol on Biosafety. The protocol requires countries to "take necessary and appropriate legal, administrative and other measures to implement its obligations under [the] Protocol" and "ensure that the development, handling, transport, use, transfer and release of any living modified organisms are undertaken in a manner that prevents or reduces the risks to biological diversity" (see www.biodiv.org/biosafety). The assessment of national capabilities should also help policy makers, funding and investors, in mapping out biotechnology development plans. A number of countries have used public institutions, private companies, regional and international centres and combinations of the four to foster development of biotechnology. The aim has been to accelerate delivery of products and services to the market place.

12.5 MANAGMENT OF CAPACITY DEVELOPMENT

Biotechnology also involves development of entrepreneurs. In many developed countries, technology transfer offices are now available in most research facilities operated by universities, non-profit and government-funded institutions. They :

(i) Identify the inventions

(ii) Determine the value

(iii) Define protection of inventions and suggest alternatives to commercialization

These efforts help to elevate the profiles of the institutions, open up new sources of funding and leverage the institution's bargaining power. Some developing countries have developed similar mechanisms in their research centres.

Most of the poor nations do not have well-established technology transfer, management and marketing systems. Biotechnology has, most often, been treated as research tool rather than an industry. Therefore, political leaders in certain countries do not see biotechnology as another tool in their efforts to industrialize.

12.6 DEVELOPMENT OF REGULATORY CAPACITY

There is a need to develop strong, flexible and effective regulatory regimes. Regulatory policies are still emerging in all countries. This is irrespective of their levels of economic and scientific development. However, at national, regional and international levels, basic regulatory procedures are emerging. These encompass the following:

  (i)   Biosafety
  (ii) Intellectual property rights
  (iii) Trade in various biotechnology products

(i) The Biosafety and Bioethics Regulatory Capacity

It is still in its infancy in developing countries. To cut the cost of biosafety review processes and development, and concentrate limited human resource and facilities, it is possible for countries to establish a regional biosafety regime. This will encourage trade in regional biotechnology products and services of those imported into the region. The safety (or risk) of products and services derived using biotechnology-related techniques has great international interest. Biosafety is largely expressed in terms of risk assessment and risk management. The biosafety framework is a set of regulatory instrument(s) designed to promote the safe use, distribution and application of biotechnology methods, products and services aimed at minimizing the likelihood of causing harm to the health of humans, plants, animals and even that of environmental health. At the international level, the biosafety issues are dealt with through the Sanitary and Phytosanitary (SPS) and Technical Barriers to Trade (TBT). Agreements of the World Trade Organization (WTO) and the Cartagena Protocol on Biosafety to the Convention on Biological Diversity (CBD). However, Cartagena Protocol on Biosafety have provided general comprehensive guidelines for the trade in and use of living transgenic products. The member States are required to establish a biosafety framework or administration. These institutions are responsibile to monitor use, generation, movement and release of living transgenic organisms. The review processing body that allows the movement and commercialization of transgenic products needs to be informed, competent, transparent and trustworthy.

The Convention on Biological Diversity (CBD) emphasizes the need to balance the risk and benefits of modern biotechnology products and services. The United Nations Environmental Programme (UNEP) and the Global Environmental Facility (GEF) have developed a biosafety project: the UNEP-GEF Project on Development of National Biosafety Frameworks. It will benefit over a hundred developing countries.

The countries must build national biosafety regulations into their national legislations. This will also help bring the Biosafety Protocol into force.

(ii) Intellectual Property Regimes

Its development has remained contentious. Many countries are in the process of incorporating or extending patent protection to include living forms. It is a fact that there is a need to have balance between protection to encourage innovations, public access to advanced technology and protection to conserve traditional knowledge. It is difficult to define protection of traditional knowledge.

(iii) Trade in Various Biotechnology Products

To meet the minimum international norms to enhance trade and development of biotechnology products and services, there may be a need to harmonize local regulations. Strong and trusted regulatory regimes should be transparent enough to dispel suspicions, especially in the wake of bioterrorism and abuse of intellectual property, be it traditional or modern. Weak regulatory regimes may lead to indiscriminate distribution of biotechnology products, while a strict regulatory regime may hinder technology transfer, adoption and development.

While the IPR regimes may exist both at national and regional levels, biosafety and bioethics have remained at the national level even in developed nations.

12.7 INDIAN SCENARIO

Traditional biotechnology had been in use in India from time immorial. It includes the use of fermentation products, such as alcohol, yoghurt, fermented foods (Dossa *etc.*). India has also adopted modern biotechnology in a bigger way. After becoming an IT doyen, India is now shifting its focus to the next promising industry, biotechnology. Numerous companies have sprung up to take a piece of the exponentially growing biotech market. The ever-decreasing physical boundaries enable biotech companies to tap large markets around the world. India to this extent holds a good advantage over other countries.

India has many advantages over many other countries of the world. These include the following (source : Various sites on internet) :

1. **Large population :** India has a large population of over a billion people. They provide huge market for products and services. The demography of India's population is very interesting. It has created almost a perfect environment for biotech companies to start their ventures in India.

2. **Large and diverse biodiversity :** Indian sub-continent occupies only 2.4% of the total global surface area. It, however, has the most varied species of flora and fauna. India has about 7.6% of total mammalian species, 12.6% of bird species, 11.7% of fishes and roughly 6.0% of total flowering plants that are present in the world. The biotech companies can take the benefit of this extensive biodiversity. They can easily find samples and also conduct field research much more efficiently.

3. **Largest agriculture sector :** India has one of the largest agriculture sectors in the world with varied climatic zones. Both these can help in research and development of different agrobiotech products applicable world wide. India no doubt today holds a small share of the global biotech market, but it is bound to have a leading role. In the next decade, the consumption of biotech products in India is expected to quadruple. The human and animal segments of the industry alone is growing by at least 20%.

4. **Rich human capital :** India has a rich human capital. There has been all round acceptance of biotechnology in different sections of society. India according to Confederation of Indian Industry estimates, produces roughly 2.5 million graduates in IT, engineering and life sciences, about 650,000 postgraduates and nearly 1500 PhDs in biosciences and engineering each year.

5. **Readily acceptable to changes :** India has proved its competency in selected areas of biotechnology such as :

   (i)   **capacity** in bioprocess engineering

   (ii)   **skills in** gene manipulation of microbes and animal cells

   (iii) **capacity** in downstream processing and isolation methods

   (iv)   **its competence in** recombinant DNA technology of plants and animals

Assisted stem cells are the prospects of the biotech industry in India.

The French Embassy in India prepared a document Biotechnology in India a couple of years ago. The following description is based on some of the information cited in that paper (Source: Internet).

India has been one of the first few countries, among the developing countries, to have recognized the importance of biotechnology as a tool to advance growth of agricultural and health sectors as early as in 1980s. India's Sixth Five Year Plan (1980-85) was the first policy document to cover biotechnology development in the country. The plan document proposed to strengthen and develop capabilities in areas such as immunology, genetics, communicable diseases, *etc.* In this context, referring to the Council of Scientific and Industrial Research (CSIR), the document suggested to ensure coordination on inter-institutional, inter-agency and on multi-disciplinary basis, full utilization of existing facilities and infrastructures in major areas including biotechnology. Programmes in the area of biotechnology included the following :

   a) Tissue culture application for medicinal and economic plants,

   b) Fermentation technology and enzyme engineering for chemicals, antibiotics and other medical product development,

   c) Agricultural and forest residues and slaughterhouse wastes utilization,

   d) Emerging areas like genetic engineering and molecular biology.

The existing national laboratories under the S & T agencies, such as Indian Council of Medical Research (ICMR) and Council for Scientific and Industrial Research (CSIR) had initiated several research programmes to fulfill the above plan objectives.

The Biotechnology profile of India as compiled by FAO has been reproduced below.

Biotechnology Research Policy	Responsible for biotechnology research in agriculture is the Ministry of Agriculture, Department of Agriculture Research and Education, Indian Council of Agricultural Research and the Ministry of Science and Technology, Department of Biotechnology and Council of Scientific and Industrial Research Indian Council on Medical Research, the Department of Atomic Energy, the Department of Biotechnology; responsible for environmental aspects of biotechnology is the Ministry of Environment and Forests.		
Research Capacity	Key Institutions	An overview on the many agricultural research institutes can be obtained from the AROW and ASTI database and from NARS National Research Centre on Plant Biotechnology, University Grants Commission Indian Agricultural Research Institute Central Agricultural Research Institute Central Tuber Crops Research Institute Indian Institute of Science (IIS) Assam Agricultural University (AAU) Punjab Agricultural University (PA U) Kerala Agricultural University (KAU) Indian Council on Agriculture Research (ICAR); Council of Scientific and Industrial Research (CSIR); Genetic Engineering Approval Committee (GEAC), under Conservation and Survey Division of the MoEF issues decisions concerning commercial use production and release of GMOs or derivates.	
	Summary of Major Research Programmes	There is a number on research programmes on biotechnology on the national and institute level; for an overview on agricultural research projects see WISARD directory. Worldbank projects database.	
Biotechnology Regulatory Framework	Biosafety	Biosafety is regulated by Rule on (...) GMOs (1989) and Recombinant DNA Safety Regulations; an overview on biosafety regulations can be obtained from BINAS; party of the Convention on Biological Diversity and the Cartagena Protocol	
	Food Safety	Food safety is regulated by Guidelines on the Evaluation of Toxicity and Allergenicity of Transgenic Plants; the Atomic Energy Control of Irradiation of Food Act (1996); an overview on foodsafety regulations can be obtained from ECOLEX and FAOlex	
	IPRs	Patents	Patenting is regulated by Patents Act No 39 (1970) and Patents Amendment Act (1994), Patents rules 2005 and Patents ordinance 2004; more information on IP legislation can be obtained from the WIPO Guide to Intellectual Property and the Patent Office of India.
		PVP	PVP is regulated by Protection of Plant Varieties and Framers Rights Act (2001); consultations with UPOV are in process
	Genetic Resource	Plant Genetic Resources	Plant Protection Act (2001), the Biological Diversity Act (2002); the Draft Biological Diversity Rules are available on the Ministry of Environment and Forests website under the Conservation and Survey Division. A summary on PGR activities can be obtained from WIEWS portal
		Animal genetic resources	
Biotechnology Application	There is a number of records on the FAO BioDeC database		
Publications and links	Biotech policy papers by regional governments are available from the FAO biotechnology site for the regions Andhra Pradesh, Karnataka, Maharashtra and Tamil Nadu Publications on agricultural research system and policy can be obtained from ISNAR Background information on PGR can be obtained from IPGRI; law Portals for India.		

(Source : http://www.fao.org/biotech/inventory_admin/dep/default.asp)

Some of the initiatives taken by the Government of India are as follows:

- Outlays have been increased to provide financial support to this area. A venture capital fund, to support small and medium enterprises has been set up.
- Good regulatory framework has been set up for approval of GM crops and rDNA products.

- Changes in the Drugs and Cosmetics Act have been made to make it globally more compatible.
- Indian Patents Bill recently passed by the Parliament allowing 20-year patent term, is in line with provisions made by WTO and TRIPS.
- Sound and widely acknowledged framework of bio-safety guidelines has been developed to deal with evaluation, monitoring and release of genetically engineered organisms and there are more than 106 institutional bio-safety committees.

In the 1980s, the Government of India considered the need for creating a separate institutional framework to strengthen biology and biotechnology research in the country. Scientific agencies supporting research in modern biology include:

a) Council of Scientific and Industrial Research (CSIR)

b) Indian Council of Agricultural Research (ICAR)

c) Indian Council of Medical Research (ICMR)

d) Department of Science and Technology

e) University Grants Commission

Biotechnology was given an important boost in 1982 with the establishment of the National Biotechnology Board. Its priorities were human resource development, creation of infrastructure facilities, and supporting research and development (R&D) in specific areas. The success and impact of the National Biotechnology Board prompted the Government to establish a separate Department of Biotechnology (DBT) in February 1986. This is now a central agency, responsible for the following :

(i) Policy

(ii) Promotion of R & D

(iii) International cooperation and manufacturing activities. There have been major accomplishments in areas of basic research in agriculture, health, environment, human resource development, industry, safety, and ethical issues.

Dr. Manju Sharma former Secretary, Department of Biotechnology has compiled an exhaustive review about the achievements in this area. According to her the Department of Biotechnology also acts as the government's agent for biotechnology imports. The department supports two autonomous laboratories, the National Institute of Immunology in New Delhi and the National Facility for Animal Tissue and Cell Culture in Pune, Maharashtra. Research and development takes place in the areas of burn, heart, and cornea treatment; germ plasm banks for plants, animals, algae and microbes; viral vaccines; animal embryo technology; animal and human fertility control; communicable and genetic disease prevention and biofertilizers, biocontrol and biomass agents. The department also controls state-owned enterprises involved in vaccine production, the Bharat Immunologicals and Biological Corporation and the Indian Vaccines Corporation, both in New Delhi.

The DBT has about 50 approved MS, Postdoctoral, and MD training programs in biotechnology in progress in different institutions and universities covering most Indian States.

Short-term training programs, technician training courses, fellowships for students to go abroad, training courses in Indian institutions, popular lecture series, awards, and incentives form an integral part of the human resource development activities in India. National Bioscience Career Development Awards have been instituted. Special awards for women scientists and scholarships to the best students in biology help promote biotechnology in India and give recognition and reward to the scientists.

A National Facility for Plant Tissue Culture Repository has been organized at the National Bureau of Plant Genetic Resources (NBPGR), New Delhi with the objective of germplasm conservation of clonally propagated crops.

Three National Gene Banks for medicinal and aromatic plants have been created at NBPGR, New Delhi, Central Institute of Medicinal and Aromatic Plants (CIMAP), Lucknow and Tropical Botanical Garden and Research Institute (TBGRI), Trivandrum. These gene banks have facilities for tissue culture repository and cryobank.

The National Centre for Plant Molecular Biotechnology at New Delhi and National Brain Centre, New Delhi have also been set up by DBT.

CSIR oversees numerous subnational biotechnology institutions. These include the Central Drug Research Institute in Lucknow, Uttar Pradesh; the Centre for Cellular and Molecular Biology in Hyderabad, Andhra Pradesh; the Indian Institute of Chemical Biology in Calcutta; and the Institute of Microbial Technology in Chandigarh among others.

India has been engaged in strengthing basic scientific research which makes the basics of the modern biology. It is important in modern biology including development of the tools to identify, isolate, and manipulate the individual genes that govern the specific characters in plants, animals, and microorganisms. India led through the work of G.N. Ramachandran, in which he elucidated the triple helical structure of collagen. This discovery of Ramachandran plot has helped to solve the protein structure. The priority areas include biosystematics using molecular approaches, mathematical modeling, and genetics including genome sequencing for human beings, animals, and plants. Impact of genome sequencing is increasingly evident in many fields. As an increasing number of new genes are discovered, short, unique, expressed sequenced tag segments are used as signatures for gene identification. New avenues in biosciences are being opened by the power of high throughput sequencing, together with rapidly accumulating sequenced data.

The sequencing of *Arabidopsis* and rice genome has been completed. Most advanced research work in biosciences is being done in the institutions under the CSIR, ICMR, ICAR, DST, and DBT have established a large number of facilities. Much success has been achieved in the identification of new genes, development of new drug delivery systems, diagnostics, recombinant vaccines, computational biology, and many other related areas. According to Dr. Sharma, the break through includes studies on the three-dimensional structure of a novel amino acid, a long protein of mosquito (University of Poona), and demonstration of the potential of the reconstituted Sendai viral envelops containing only the F protein of the virus, as an efficient and site-specific vehicle for the delivery of reporter genes into hepatocytes (Delhi University)

- India is the second largest food producer. Thus, there is a huge market for agribiotech products.
- India has an excellent scientific infrastructure in agriculture, rich bio-diversity and skilled and low cost human-power.
- According to Ernst and Young (2001), at present the market for Nutraceuticals is roughly US $ 532 - 638 million and is growing.
- India has a rich aquaculture with its 8000 kilometer of coastline including Andaman and Nicobar and Lakshwadeep islands. There is a great potential in marine resource development.

In the area of agriculture, new hybrids including apomixis, genes for abiotic and biotic resistance, have been developed. Planting materials with desirable traits have been developed. Genetic enhancement of all important crops has been done. From the view-point of sustainable agriculture, soil fertility, and a clean environment, attempts are being made to have integrated nutrient management and develop new biofertilizers. Studies are being carried out on stress biology, marker-assisted breeding programs, and analysis of important genes. Attempts are on to switch to organic farming practices, with greater use of biological software on a large scale. The cloning and sequencing of at least six genes, development of regeneration protocols for citrus, coffee, mangrove species, and new types of biofertilizer and biopesticide formulations, including mycorrhizal fertilizers, are some of the steps forward. Genetically improved (transgenic) plants for brassicas, mung bean, cotton, and potato are being developed. Tissue culture pilot plants in the country, one at TERI in New Delhi and the other at NCL in Pune now serve as Micropropagation Technology Parks. These act as a platform for effective transfer of technology to entrepreneurs, including training and the demonstration of technology for mass multiplication of horticulture and trees. Cardamom and vanilla have been produced by tissue culture with increased yield.

According to Dr (Mrs) Sharma in just eight countries, the area covered by new genetically improved transgenic plants has increased from 16.8 to 27.8 million hectares in between 1996 and 1998 (James, 1998). Some of the main crops grown are soybean, corn, canola, cotton, and potato. The new plants exhibited herbicide, insect, and viral resistance, and overall improvement in product quality.

Molecular fingerprinting and areas of genomics and proteomics are entering the barriers of fertilization to allow transfer of important characters from one plant to another. By identifying appropriate determinants of male sterility, the benefits of hybrid seeds can be extended to more crops. The farmer must be helped by ensuring hybrid vigour generation after generation. Such possibilities can be opened up by additional research on apomixis. A National Plant Genome Research Centre has been set up at Jawaharlal Nehru University. Besides a number of centres for plant molecular biology in different parts of the country had been responsible for training significant number of people in crop biotechnology. There are large number of possibilities of a producing more proteins, vitamins, pharmaceuticals, colouring materials, bioreactors, edible vaccines, therapeutic antibodies and drugs. A number of genetically improved crops are ready for field trials of transgenic plants. Work is being carried out on developing transgenic cotton, brassica, mung bean, and potato. The Green Revolution gave us self-reliance in food as the livestock population has provided a "White Revolution," with 80 per cent of the milk in India coming from small and marginal farms. Much success has been achieved in the application of embryo transfer technology. PRATHAM the world's first IVF buffalo calf was born through embryo transfer technology at the National Dairy Research Institute, Karnal. Multiple ovulation and embryo transfer, *in vitro* embryo production, embryo sexing, vaccines and diagnostic kits for animal health have also been developed. Waste recycling technologies that are cost effective and environmentally safe have been developed. Attempts are being made to exploit more than 8,000 kilometers, and two island territories of Andaman and Nicobar and Lakshadweep for marine resource development and aquaculture. There is an annual target production of 10 million metric tons of fish. There are attempts to improve production of seeds, feed and health products.

Biotechnology is being exploited to develop healthier and more nutritious food. With the advent of gene transfer technology and its use in crops, it is hoped to achieve higher productivity and better quality, including improved nutrition and storage properties. It is also hoped to ensure adaptation of plants to specific environmental conditions:

    (i)   To increase plant tolerance to stress conditions

   (ii)   To increase pest and disease resistance

  (iii)   To achieve higher prices in the marketplace

All these will help the malnourished people by providing them good nourishing food as about 2.7 million of these children die in India. More than half of these deaths result from inadequate nutrition.

Genetically improved foods are under adequate regulatory processes. There is a need to develop the genetically improved foods, but their safety and proper labeling must be ensured, so consumers will have a choice. Biotechnology is providing new tools to breeders to enhance plant capacity because due to extensive urbanization, the per capita availability may be reduced from 2.06 hectares to 0.15 hectare by 2050.

Medical Biotechnology

Medical biotechnology mainly emphasizes on thrust areas like molecular diagnostics, genetic vaccines, molecular characterization of pathogens, novel drug delivery systems, cancer markers, stem cell biology *etc.* Criteria for health related biotechnology programmes include disease burden, cost effectiveness, emerging and reemerging diseases, disorders because of prevalent lifestyle. These programmes have been classified into two categories :

    (i)   Core

   (ii)   Non-core

Core programme : These include:

    (a)   biotechnological approach to emerging and reemerging diseases

    (b)   development of affordable vaccines

    (c)   application of stem cell biology for treatment of diseases (*e.g.,* thalassemia)

    (d)   development of diagnostic tools

(e) development of drugs

(f) sequencing and molecular typing of infectious agents

(g) phage display technology and protein engineering

(h) neurodegenerative diseases

(i) research on reproductive human health, edible vaccines.

Non-core programmes : These include:

(a) applications of biotechnology tools for identification of susceptible and resistant vectors

(b) expression systems for better yields of biomolecules

(c) prevention and cure of diseases

(d) establishment of transgenic animals for strengthening biomedical research

(e) use alternative approaches to minimize use of animals in biomedical research

A major responsibility of biotechnologists in the 21st century will be to develop low-cost, affordable, efficient, and easily accessed health care systems. Advances in molecular biology, immunology, reproductive medicine, genetics, and genetic engineering have revolutionized our understanding of health and diseases and may lead to an era of predictive medicine. Genetic engineering promises to treat a number of monogenetic disorders, and unravel the mystery of polygenetic disorders with the help of research on genetically improved animals. Globally, there are about half a century biotechnology-derived therapeutics and vaccines in use and more than 500 drugs and vaccines are at different stages of clinical trials. Every year about 12 million people die of infectious diseases. The main killers according to WHO are acute respiratory infections, diarrhoeal diseases, tuberculosis, malaria, hepatitis, and HIV-AIDS. There are attempts to develop vaccines for many diseases, and diagnostic kits for HIV, pregnancy detection, and hepatitis. The technologies have been transferred to industry. The Department of Biotechnology has developed guidelines for clinical trials for recombinant products, which have now been accepted by the Health Ministry and circulated widely to industry. Promising leads now exist to develop vaccines for rabies, *Mycobacterium tuberculosis,* cholera, JEV, and other diseases. Recombinant hepatitis B vaccine and LEPROVAC are already in the market. There is a Jai Vigyan technology mission on the development of vaccines and diagnostics. A National Brain Research Centre has been established to improve knowledge of the human brain and the brain diseases. The discovery of new drugs and the development of the drug delivery system are increasingly gaining importance. Bioprospecting for important molecules and genes for new drugs has begun as a multi-institutional effort. The age-old system of Ayurveda practiced in India is being popularized and made an integral part of health care. The global market for herbal products may be around US$5 trillion by 2050.

Vaccines

Immuvac : An immunomodulator based on killed *Mycobacterium* has been developed by NII and is used as an adjunct to MDT, for prophylaxis in leprosy patients. It is upscaled by M/S Candela Pharmaceuticals as IMMUVAC[R] and has shown profound immunomodulatory effects in cancer, HIV and TB infections. The main issue of concern that leads to the development of IMMUVAC included increase in the number of leprosy patients in India to put Mw as an adjunct to MDT and leprosy. Current development involves investigation of Mw for breast cancer, HIV, therapeutic vaccine *etc.*

Rabies vaccine : Mainly developed by National Jai Vigyan Mission Programme, the main objectives of this programme are preclinical toxicity and protective efficacy of a rabies DNA vaccine in mice, monkey and dogs undertaken by Indian Institute of Science (IISC), Bangalore. It is involved in the production of plasmid DNA from shake cultures and commercial production of rabies DNA vaccine in fermenter; large scale downstream processing technology. Indian immunologicals, Hyderabad are involved in the development of combined DNA rabies vaccine. Rabies 'G' protein expression vector as per WHO/USFDA guidelines has been prepared for use in clinical trails. This vaccine has obtained RCGM clearance and experiments indicate that CRV is stable at room temperature.

HIV/AIDS subtype 'C' vaccines : The development of this vaccine is undertaken by R & D, JVM. The VAP. R & D project involves multicentric programmes on HIV-I genotypes in India and revealed that 80% of HIV is of subtype C and belongs to quasi species C3. National Jai Vigyan Science and Technology Mission includes a project at AIIMS towards development of recombinant DNA vaccines for HIV-1 subtype 'C'. Indo US VAP a project to develop infectious clone of SHIV-C (Simian – Human Immunodeficiency virus clade 'C') and cloning of 3 kb gene fragment encoding lat-rev-env gene of HIV – 1 subtype C from Indian patients.

Japanese encephalitis : Development of its vaccine involves a multicentric program which focuses on collection of Indian strains, improvement of existing vaccine and development of newer vaccines. The achievements of this programme include a prototype list system which is transfered to industry. DNA and synthetic peptide vaccine strategies are also being tried. Immunogenicity of the Vero-cell grown virus is being studied

Cholera : Its main thrust of activity involves the development of indigenous vaccine. An indigenous recombinant oral vaccine based on VA 1.3 strain on *V. cholerae* is being jointly developed by NICED, Kolkata, IMTECH, Chandigarh, SAR, Kolkata, SGPGIMS, Lucknow and PGIMER, Chandigarh. Phase I trials have been conducted and vaccine has been found to be safe. Vaccine under trials has an ampicillin gene marker, though IMTECH Chandigarh is also working on strain minus amplicillin markers.

Tuberculosis : The development of TB vaccine focuses on three main components of research *i.e.,* repository of well characterized *Mycobacterium* isolates, diagnostics involving serological and PCR techniques, vaccine development approaches. *Mycobacterial* respository is established at Central JALMA institute of leprosy, Agra and currently 300 strains of mycobacteria are well characterized. The usage of atypical mycobacteria such as *M. habara* gave encouraging results. A multisubunit vaccine based on two immunodominant mycobacterial *i.e,* 71 kDa cell wall and 70kDa secretory proteins are effective against BCG.. Workers at Delhi University are developiong recombinant BCG vaccine by expressing 6 antigens 85A, 85 B, 19kDa, 58 kDa, ESAT of *M. tuberculosis.* They are also working on DNA vaccine.

Malaria : Three projects are in progress for the development of malaria vaccine. The target of these projects comprises the production of recombinant PfMSP 1_{19} and PfF2 under GLP/

CGMP conditions and conducting pre-clinical toxicity testings. The project on development of candidate antimalarials based on new drug targets implemented at IISc, Bangalore has made considerable progress in the synthesis of PCR primers to detect K76T mutation in PfCrT gene and N86Y mutation in Pfmdr gene.

Edible vaccines: Attempts are being made to develop cholera vaccine in tomato plants at Delhi University. A rabies glycoprotein gene of ERA strain rabies virus alongwith a *rptII* marker gene is transformed into *Cucumis melovar.* Drug development project has been undertaken by the Indian Institute of Chemical Biology (IICB) Kolkata. A novel anti-leishmanial compound designated as S-4 has been synthesized. It is significantly active in inhibiting promastigotes and amastigotes of *L.donovani.*The demand for healthcare services in India has grown from $ 4.8 billion in 1991 to $ 22.8 billion in 2001-02, indicating a compounded annual growth rate of 16 per cent. The healthcare industry accounted for 5.2 per cent of India's GDP in 2002, and this figure could reach $ 47 billion or 6.2–7.5 per cent of GDP by 2012. On the one hand, the Indian middle class, with its increasing purchasing power, is more willing than ever before to pay more for quality healthcare. On the other, the supply of healthcare services has steadily grown, as the private sector becomes more involved in owning and running hospitals (Healthcare Industry in India PSi, Inc., Oct 2004). The upcoming field of medical biotechnology is being applied to various other areas like oncology, histocompatibility and transplantation and goat hepatocyte transplantation.

Medical biotechnology has also played a significant role in the development of various diagnostic assays.

A. *Western Blot :* This technique is used for the diagnosis of HIV 1 and 2. It is an *in vitro* qualitative immunoassay for detection of antibodies of HIV 1 and 2 in human serum or plasma. It shows 100% sensitivity and specificity and is being successfully used in private hospitals, nursing homes and clinics.

B. *NEVA (Naked Eye Visible Agglutination) Test for HIV 1 and 2.* It comprises whole blood assay for detection of antibodies against HIV-1 and HIV-2 in a drop of blood using recombinant proteins. It is a rapid, cost effective and most suited serving of the blood banker and is being marketed by M/S Candila Pharmaceuticals, Ahmedabad.

C. *PCR based diagnostic method for detection of STD infections.* It is used to detect *Neisseria gonorrhoeae* and *Chlamydia trachomati.*

India being on the 'threshold of biotech revolution' has about 300 biotech and 200 biosuppliers contributing to the total biotech market more than US $100 billion. The country has a global market more than $91 billion and there is a scope for cheap R&D through bio-partnering and co-developing technologies mainly with Chinese and American companies. Already the pharma companies of world are seeking India to set their research and development centres here. Moreover, to facilitate foreign investment, capital and government policies are being revised.

Many multinational companies have penetrated into India with an aim to market drugs and conduct clinical trials. Thus, impetus has been received by Indian pharmaceutical research, manufacturing and outsourcing. This has created an image of a potential healthcare market and a number of opportunities in pharmaceuticals. The pace of development of the pharmaceutical industry has hightened by the economic liberalization policies which came to force in the 1990s and the strong emergence of private sector in the Indian economy has heightened and will continue to do so.

The Indian pharmaceutical market has rapidly grown in the recent past. The market of approved recombinant therapeutics in India was estimated to be about 109 millions in 2001 which represented 3.2% of the total Indian pharmaceutical market and 1.6% of the world market for recombinant therapeutics. According to McKinney (2002), Indian Pharma industry is expected to grow to an innovation-led US $25 billion industry by 2010 with a market capitalization of almost US $ 150 billion from the current US $ 5 billion generic based drug industry. According to him, there will be roughly 20% growth in the vaccine market (Promises in Biotechnology, Internet).

It has been found that Indian biotech firms are making a strong position, developing business models and improving product commercialization capabilities. With this, India could make about $5 billion by 2010.

A market research report, "Indian Biotech Industry (2005)" has provided insights into the Indian biotech industry including product estimates, market segmentation, and regulatory mechanisms, pricing policies, patents and opportunities in international ventureship. The report says how in 2004-05, 20 privately helped drug discovery groups had started to operate in India with their business primarily to create innovations and IPR. The biotech industry is likely to draw huge FDI to the Union Government.

The market research report also details the success of the biotech firms in terms of their business profiles, export outlets, financial performances and future plans. It also has the contact details of the major industry associations.

The recent regulatory and much awaited patent law changes have lead the Indian Pharmaceutical Industry towards exploring newer avenues of drug development, thus, promising higher capital investment in the pharmaceutical industry in the near future. The Indian pharmaceutical research is strongly supported by Government and availability of surplus skilled technical workers at lower costs.

At a growth rate of 9 per cent plus per year, the pharmaceutical industry in India is well set for rapid expansion. As a result of this, the Indian pharmaceutical and healthcare market is undergoing a spurt of growth in its coverage, services and spending in the public and private sectors. The healthcare market has opened a window of opportunities in the medical device field and has boosted clinical trials in India.

"Indian Pharmaceutical and Healthcare Market—Annual Review (2005)" presented information on the Indian pharmaceutical research. It highlighted the important factors that draw the foreign investors towards the Indian pharmaceutical market to establish themselves. This report provides an in-depth analysis of the business prospects. It interprets the key issues that influence the success of a pharmaceutical company involved in research and development of drugs (Indian Pharmaceutical and Healthcare Market Annual Review, 2005). The key players in the Indian Pharmaceutical Industry and Companies are :

- Aarti Drugs
- Abbott India
- Ajanta Pharma
- Alembic
- Astrazeneca Pharma
- Aurobindo Pharma
- Aventis Pharma
- Bangalore Genei,
- Bharat
- Biological E.
- Biocon
- Cadila Health
- Cipla
- Dr. Reddy
- Elder Pharma
- German Remedies
- Genoptypic
- Glaxo Smithkline
- Nicholas Piramal
- Ind Swift Lab
- Ingenvois
- Ipca Laboratories
- J B Chemical
- Jagson Pharma
- K D L Biotech
- Kopran
- Krebs Biochem
- Lupin
- Lyka Labs
- Medicorp Tech

- Merck
- Monsanto
- Natco Pharma
- Nicholas Piramal
- Novartis
- Orchid Chemicals
- Organon
- Panacea Biotech
- Pfizer
- Pharmacia
- Rallis
- Ranbaxy
- R P G Life Sciences
- SP Biotech Park
- Shanta
- Shasun Chemicals
- Siris Limited
- Sterling Biotech
- Strand Genomics
- Strides Arcolab
- Sun Pharma
- Suven Life Sciences
- Themis
- Torrent Pharma
- Unichem Lab
- Wockhardt
- Wyeth Ltd
- Xcyton
- Zandu Pharma

A large number of industries have come forward to have R & D in Biotechnology.

The companies belonging to these groups have developed valuable assets in fermentation and downstream processes. With the entry of several firms from different origins in the coming years, the Indian market for biogenerics will become more and more competitive. Besides these, several small sized companies in India have been created with business models based on partnership in fields related to manipulation of genetic information. All these companies do not only focus the sector of drug developments, some of them like Avesthagen and Metahelix focus on plant genome knowledge application.

At the public sector also, the following institutions have been identified as National Centres of Excellence.

1. Centre of Cellular and Molecular Biology (CCMB) (CSIR), Hyderabad
2. Centre for Biochemical Technology (CBT) (CSIR), New Delhi
3. Indian Institute of Science (IISc) (UGC), Bangalore
4. International Centre for Genetic Engineering and Biotechnology (ICGEB) New Delhi
5. All India Institute of Medical Sciences (AIIMS) (ICMR), New Delhi

Beside these five nodal centres, the following institutions are carrying out biotechnology programmes in a big way :

1. Anna University, Tamil Nadu
2. Centre for DNA Fingerprinting, Hyderabad
3. Central Drug Research Institute (CDRI) Lucknow
4. Central Institute of Brackish Water Aquaculture (CIBA), Chennai
5. International Crops Research Institute for the Semi-Arid Tropics (ICRISAT)
6. Indian Institute of Chemical Biology (IICB), Calcutta
7. Indian Institute of Chemical Technology (IICT), Hyderabad
8. Indian Institute of Technology (IIT) Themis
9. Institute of Microbial Technology (IMT) Chandigarh
10. Jawaharlal Nehru University (JNU) Delhi
11. Madurai Kamaraj University (MKU) Madurai
12. National Centre for Biological Sciences (NCBS), Bangalore
13. National Dairy Research Institute (NDRI), Karnal
14. National Institute of Immunology (NII), Delhi
15. National Institute of Mental Health and Neuro Sciences (NIMHANS) Bangalore
16. National Research Centre for Plant Biotechnology (NRCPB), Delhi
17. Osmania University, Hyderabad
18. Tata Memorial Centre (TMC), Mumbai
19. Tata Energy Research Institute (TERI), Delhi
20. University of Agricultural Science, Bangalore
21. Panjab University, Chandigarh
22. Bhabha Atomic Reserach Centre (BARC) Mumbai

There is in fact active interaction between private sector and the institutes mentioned above so as to have product oriented R & D programs. The market opportunity for recombinant therapeutics manufactured in India is now well known and there is a large demand from all the local companies with connected competencies for the technology enabling them to produce such products.

Shanta has collaborative programs with the IISc Bangalore, Jawaharlal Nehru University (JNU) Delhi, Bhabha Atomic Research Centre (BARC) Mumbai, Centre for Cellular and Molecular Biology, Hyderabad, All India Institute of Medical Sciences (AIIMS) New Delhi, National Institute of Immunology (NII) New Delhi, Indian Institute of Chemical Biology (IICB) Calcutta, Anna University Chennai, Tata Memorial Hospital Mumbai, National Dairy Research Institute (NDRI), Karnal, International Vaccine Institute, Korea.

A concrete example of the role such a company can play in local as well as international public-private collaborations has been provided by the study of the ongoing collaborative projects of Bharat. Bharat is involved in two joint programs for the development of a rotavirus and a malarial vaccine and include both industrial and academic partners. Both projects are funded by the Bill Gates Foundation (the first one through the Children Vaccination Programme and the other one through the Malarial Vaccine Program). The joint program on rotavirus involves the DBT, AIIMS, NIH, CDC Atlanta, and the University of Stanford. Bharat which is the commercial partner of the project provides the Good Medical Practices (GMP) samples during the development studies. The project on malarial vaccine is carried out in collaboration with the ICGEB, New Delhi. The institute is incharge of cloning the gene and inserting it in bacteria in order to express the protein. Bharat is producing GMP samples and clinical trials are carried out by AIIMS, New Delhi. The Intellectual Property Right (IPR) is to be transferred to Bharat.

Bharat is also involved in bilateral collaborations. A world wide patent was filed by Bharat in October 1999 on a vaccine against Lysostaphine. This project was initiated 4 years ago as collaboration with the Centre for Biochemical Technology (CBT), New Delhi.

The pharmaceutical giant Wockhardt has, however, entered into a partnership with a public research institution, the Trieste branch of the ICGEB in 1993. The financial contribution from Wockhardt was supposed to be Rs. 50 M for a partnership. As the desirable results were not obtained, it was stopped after 3-4 years, but the Wockhardt had already spent about Rs. 20 M by then. Wockhardt then reoriented its strategy towards private-private technology transfer and signed an agreement with the German Biotech Company, Rhein Biotek.

Many other big players in pharmaceutical industry have entered into agreement with various institutes.

Environmental biotechnology : India is becoming more conscious about the pollution control. The priorities are to use biotechnology for cleaning up the large river systems and ensuring the destruction of pesticide residues in large slums in the city. Phytoremediation is being carried out to remove the high levels of explosives found in the soil. Although it was known that some microbes can denitrify the nitrate explosives in the laboratory, they could not thrive on site. This degradative ability had been transferred from microbe to tobacco plants by French and others (1999), who have produced a microbial enzyme capable of removing the nitrates.

Biodiversity : There is a threat to biodiversity. About 2000 species of vascular plants are under threat. There is a need to understand the scale of this destruction and extinction. Attempts are being made to carry out research on forests, marine resources, bioremediation methods, restoration ecology, and large-scale tree plantations. Many goods and benefits including bioactive materials, drugs, and food items are provided by marine resources. They must be characterized and conserved.

Industrial biotechnology : Extensive efforts have been made in the diagnostic area. It is expected that sales will rise from about US$235 million to US$470 million in the next century. The consumption of biotechnology products is expected to increase from US$6.4 billion to about US$13 billion by 2000. Industrial enzymes have emerged as a major vehicle for improving product quality. A number of workers are involved in producing industrial enzymes such as alpha-amylase, proteases, and lipases. As many as 13 antibiotics are produced by India by fermentation. There is a capacity to produce important vaccines such as DPT, BCG, JEV, cholera, and typhoid. Cell culture vaccines such as MMR and rabies, and hepatitis-B, have also been introduced. AIDS vaccines produced in India are under clinical trails.

Bioinformatics

A National Bioinformatics Network with ten Distributed Information Centres (DICs) and 35 sub-DICs have been established by DBT. A Jai Vigyan Mission on establishment of genomic databases has been started. It has a number of graphic facilities throughout the country.

Ethical and Biosafety Issues

A three-tier mechanism of Institutional Biosafety Committees has been instituted in India

(a) The Review Committee on Genetic Manipulation

(b) The Genetic Engineering Approval Committee

(c) The State Level Coordination Committees

To allay their fears, the people must be provided with a clear explanation of the new biotechnologies. New models of cooperation and partnership are being established to ensure close linkages among research scientists, extension workers, industry, the farming community, and consumers. Though the aims and objectives of new technology are quite good and the tools are available, it, however, needs a cautious approach following appropriate biosafety guidelines. Worldwide more than 25,000 field trials of genetically modified crops have been conducted. The anticipated benefits are better planting material, savings on inputs, and genes of different varieties can be introduced in the gene pool of crop species for their improvement. The following issues must be addressed by coming together of researchers, policymakers, NGOs, progressive farmers, industrialists, government representatives, and all concerned players.

• Environmental safety

• Food and nutrition security

• Social and economic benefits

• Ethical and moral issues

• Regulatory issues

Some Special Programs

To benefit the poor and weaker sections, programs for women biotechnology-based activities have been launched. A Biotechnology Golden Jubilee Park for Women, which will encourage a number of women entrepreneurs to take up biotechnology enterprises that benefit women in particular has been established. Biotechnology-based activities are being developed by states. The States of Uttar Pradesh, Arunachal Pradesh, Madhya Pradesh, Kerala, West Bengal, Jammu and Kashmir, Haryana, Mizoram, Punjab, Gujarat, Meghalaya, Sikkim and Bihar have already started large-scale demonstration activities and training programs.

Almost all the state governments in India have also realized the benefits/importance of this industry. They have created the state level Departments of Biotechnology to promote the R & D in this area. Such state level departments by their own resources and with Central Government funding, are doing well in the field of research for the welfare of human society.

Currently, the concept of setting up of Biotech parks has been adopted by many states.

Services Segment

- Pharmaceutical companies are finding it difficult to conduct entire drug discovery process-in-house. They are, therefore, trying to find out ways to minimize costs. India has become a very attractive base because the cost of infrastructure is relatively lower as compared to other nations.
- Cheaper qualified workforce available in India is another attraction for foreign companies.
- Each year India produces enough qualified graduates. The companies looking to expand their operations, can do so without facing a shortage in labour.

12.8 CONCLUSIONS

According an article on Promises of Biotechnology (2003), it is expected that the quality of life is bound to be improved by future biotechnology products and services. There will, however, be no barrier of any product and services developed for markets in the North to find applications in the South. The cost of biological research to develop their own products may become affordable for developing countries. The increase in information exchange and access will empower innovators and the public to seek better products and services. There is a need of rethinking on formal and informal educational systems to meet the current and future challenges. Developing countries will lose most of their current market to the developed countries and countries with economies in transition. This is based on the fact that the technology market is imperfect. It is poorly characterized and highly regulated in favour of developers. It is difficult and time consuming to identify the best, cost-effective and useful technology, while the fast turnover of technology makes even new technology platforms obsolete, sometimes before products and services are generated. The role of technology in industrial competitiveness is now well established and is a distinguishing agent between the poor and the rich. The replacement of older technology products and services with newer ones, is part of the free market. These forces are not restricted to biotechnology alone nor are they new. For example, the silk market was taken by synthetic fibres, while artificial sweeteners and syrups have displaced some of the sugar market. Similarly, technology that enables crops and animals to perform well in environments previously thought to be hostile will be both useful and harmful for different societies that depend on them. Biotechnology, therefore, has to impress upon the countries to be at gain.

Many developing countries aspire industrialization. The driving force is provided by the desire to shift from raw material exports to processed products and increase their share of international trade. The ability to export finished products is partly dependent on technologies used, which may determine the quality of the products. A basis for facilitating the transition towards industrial processing is provided through the trade incentives such as the US Africa Growth and Opportunity Act of April 2000. Besides improving the quality of products, biotechnology promises to cut the costs of investments. It also promises to provide flexible processing and manufacturing platforms that could easily be modified and adapted. In addition, it reduces industrial stockfeed quality demands, waste production, energy, water and use of hazardous materials in production among others. These countries are forced by the demands to meet market needs some with emerging and others with already growing markets.

Questions

1. Why biotechnology had remained domain of the developed countries? Explain with the help of examples.
2. Give various areas of concern in developing countries that need attention of biotechnology. Explain in detail.
3. Give the priorities of biotechnology in various developing countries.
4. Describe the status of biotechnology in India.

Chapter 13

The Ethical and Social Implications of Genetic Engineering

13.1 INTRODUCTION

Right from the earliest days of genetic engineering, there has been concern about the potential hazards and ethical issues associated with this new branch of biology. Original concerns in 1971 focused on plans to clone cancer genes from viruses into *E. coli*. It was argued that if the genetically altered *E. coli* escaped from the laboratory, they might spread the gene into *E. coli* that live in the human gut by transferring the plasmids, for example by conjugation. It was argued that human DNA, and the DNA of other mammals like mice, could contain cancer-causing genes (oncogenes) and that these might inadvertently get transmitted with neighbouring piece of DNA being used for genetic engineering.

In February 1975, a group of more than 100 internationally well-known molecular biologists met in California and decided that until the risks could be more precisely estimated, certain restrictions should be placed on genetic engineering research. This was in fact a self-imposed brake on scientific progress, not imposed by the scientists themselves and by the Government. Work on cancer viruses was stopped. Non-scientists who had been part of the debate were invited to join the Genetic Manipulation Advisory Group (GMAG) in UK. There was a halt of about two-years during which safe procedures were established. Debate continued about how strict the rules and regulations should be. There has been more resistance to developments in Europe than in the USA.

In the 1980s, there was an explosion of activity. Manufacturing companies quickly began investing billions of dollars into not just genetic engineering, but into all the 'biotechnologies' which were emerging from molecular biology. Biotechnology industry catering to agriculture, medicine, industry and waste treatment was born.

13.2 ETHICS OF BIOTECHNOLOGY

Although biotechnology provides the benefits, there are certain dilemmas too. Certain groups in society would greatly restrict the uses of biotechnology and who might prefer that the techniques of recombinant DNA had never been developed. But to others, the benefits are much more than the liabilities and that by fairly simple legal means, any potential for harm can rather easily be overcome. Many others though have strong opinions about specific applications, take no stand on biotechnology in general.

According to MacDonald and Dhanda (2006) the first – and perhaps most difficult – element of ethical decision-making is recognizing an issue as an ethical issue. It pays to think broadly and imaginatively, here. Some biotech executives will want to assume that their company's work is unlikely to involve ethical issues or to attract public scrutiny. Such an attitude risks being both naïve and dangerous because the people affected by corporate practices are the very people who decide its propriety and its social / market context – from the regulatory and financial agencies that oversee business practices to the consumers who decide which competitor creates the most desirable products (MacDonald and Dhanda, 2006).

According to the authors signs of the potential for ethical problems can be spotted by asking the following questions to ourself:

- Is anyone to be harmed or helped by this decision?
- Is there a question of trust?
- Are there fiduciary obligations at stake? Other kinds of obligations?
- Is someone's autonomy – their right to choose – at risk?
- Is there a question of fairness?
- How will the costs and benefits of this research and/or product be distributed?
- Are important relationships in jeopardy?
- Do the products meet a social need?
- Are there popular arguments in society (the press, social activists, *etc.*) that relate to the core research?

Once an ethical issue has been recognized as such, there is a need to begin to think in terms of time frame. Is the issue here (a) an ethical crisis, (b) an ongoing ethical debate, or (c) a distant possibility of ethical controversy?

(a) **Ethical crises.** It the Tylenol poisoning case, in which Johnson and Johnson quickly increased its market share, arguably because their response was immediate, honest, and put public safety first. It has been shown by experience that when scandals or other crises arise, openness and honesty are the key.

(b) **Ongoing ethical debate** – The debate over stem cell research or somatic cell nuclear transfer – requires that companies understand that the ethical context of their research is sensitive in the here-and-now, despite the lag time before products are developed. A wary investment community has been created by debates over cloning and stem cells. The government organizations like the National Institutes of Health (NIH) are unlikely to provide funding to firms or research labs developing these technologies. Those companies that have engaged in ethical reflection as a long-term project, one that requires immediate and ongoing consideration rather than post-marketing public relations, are better positioned vis-à-vis the investment community and other sources of funding.

(c) A different kind of planning is required for issues perceived as involving the possibility of eventual ethical controversy. If any lesson has been learned from the genetically modified food sector, it is that regulatory agencies are not the sole decision makers regarding the marketability of new bioscience-related products. When regulatory policies lag behind technological trends, it is possible that corporations will have to form their own responsible policies to oversee the introduction and continued evaluation of their technologies; for instance, reflective information management procedures provide a firm grounding for strategic planning and ethical deliberation. MacDonald and Dhanda (2006) have rightly mentioned that as the technologies grow, expand and revolutionize, corporations will need to continue to involve themselves in that process, as well as to interact with the consumers and intermediaries affected by and affecting the technology.

From an ethical point of view, there is no one right way to evaluate the options. Some advocate a principle-based approach, evaluating ethical issues in terms of social benefit, avoidance of harm, respect for autonomy, and justice. Others favour a "stakeholder" approach, advising companies to think imaginatively about the range of groups and individuals who might have a "stake" – a legitimate ethical interest – in the decisions they take.

13.2.1 Some Uses of Biotechnology have Social Impacts

The people who do not want biotechnology, fear that it may be socially undesirable.

In the late 1980s, a biotech firm began to mass-produce bovine growth hormone by recombinant DNA techniques. Growth hormone, when injected into cows, enhanced milk production. Some people were vigorously opposed to this practise because of following reasons :

(i) The recombinant hormone may change the composition of milk.

(ii) Increased milk production may lead to an increased incidence of udder infections in cows. This effect might not only be harmful to the cows, but might also lead to increased antibiotic resistance among microbes.

(iii) There was already a milk glut in both America and Europe. Many dairy farmers opposed the use of recombinant growth hormone because they felt that greater milk production per cow might bring down prices even further and put some small farmers out of business.

Experts from the National Institute of Health and the Food and Drug Administration had, however, concluded that the changes were minor and posed no threat to human health. In fact the concerns like these do not appear to have any direct bearing on biotechnology. Millions of cattle and other mammals are slaughtered for meat every year, suppose that a cheap way was found to extract growth hormone from their pituitary glands. The same arguments would arise without biotechnology being part of the issue at all. In instances such as this, society may wish to intervene in a specific use of biotechnology for reasons of public health, economic issues, or social justice. The outcome, not the methodology, is the hearth of the matter.

Bioengineered organisms are considered to be problematic. Many crop plants such as oranges and tomatoes are quite sensitive to frost. Small particles of dust or other matter act as "seeds" for ice-crystal formation in the air, in water or on plant surfaces. In the absence of "seed particles," ice crystals do not form even at temperatures somewhat below freezing. One of the primary sources of ice seeding on plants is a common bacterium, *Pseudomonas syringae*. Ice formation is promoted by specific protein on the surface of the bacterium. Genetic engineering is trying to remove the gene encoding this protein from the bacterium and replace the "wild-type" bacterium on crops with the new "ice-minus" bacterium. Though the genetic engineering was accomplished in the mid-1983s, there were several years of controversy and litigation before field trials were permitted.

13.2.2 Human Safety

The first food containing genetically engineered DNA to be approved for marketing was the 'Flavr Savr' tomato. This is a useful case study of concerns about the safety of foods.

One of the main concerns relates to the vectors used for transforming plant cells. These contain genes for antibiotic resistance, most frequently kanamycin resistance. These genes enter the transformed plants with the desired gene. Flavr Savr tomatoes contain one of these antibiotic resistance genes. The concern is that when the tomato is eaten, the gene may pass from tomato to the *E. coli* bacteria in the gut making them resistant to kanamycin and related antibiotics. Since bacteria leave the gut in the faeces, the gene may spread to other potentially harmful bacteria in the environment which if infect humans, would be antibiotic resistant. In practice, likelihood of its being digested once eaten is less and even if the chances of the gene passing through a series of organisms is extremely remote. Also, the kanamycin-resistance gene is already common in the environment. Nevertheless, scientists are trying to find ways to remove the marker genes after transformation.

In 1996, the European Union allowed genetically modified maize to be imported from the USA. The maize has a bacterial gene which increases its resistance to pests and diseases. It also has a gene for resistance to the antibiotic ampicillin. Greenpeace opposed this introduction and even threatened the legal action.

The public are extremely wary of genetically engineered products because of the publicity which has surrounded such issues as the use of growth hormones. Companies that have invested millions of pounds in research and development, cannot afford to make mistakes, they have a strong vested interest in making sure that their products are safe *e.g.*, PPL manufacturers of the anti-emphysema drug AAT imported all the sheep from New Zealand to ensure that they were scrapie-free (Scrapie causes a disease in sheep similar to mad cow disease). Transgenic goats being used by Genzyme Transgenics for the production of monoclonal antibodies in milk are fed the food free of pesticides and herbicides. No protein or animal fat additives to their food are allowed so that any possibility of transfer of disease to humans from other animals is prevented. The products are probably safer than many traditional ones which we are happy to accept.

13.2.3 About Virus Resistant Plant

There is a fear with regard to the production of virus resistant crop. It is felt that different virus might infects the crop and have its genetic code (RNA or DNA) wrapped in the protein coat of TMV instead of its own normal protein coat. It may then invade all the crops that are invaded by TMV. It is because of this, they are being tried in the natural conditions so as to be sure that no problem arises later on.

There are very strict regulations in North America and Europe regarding the release of genetically engineered organisms (GEOs) into the environment.

In the European Union, each member country has its own authority which oversees all releases of GEOs. In the UK, the authority is jointly held by the Department of Environment and the Ministry of Agriculture, Fisheries and Food (MAFF).

13.2.4 Animals and Ethics

Humans often think of themselves as being superior to other animals (not to mention of plants, fungi, bacteria and so on) and, therefore, have the 'right' to exploit other organisms for their own benefit. However, in recent years, there has been a growing trend to challenge the human-centred (anthropocentric) view of our relationship with other species. There is particularly a concern about the way we exploit animals for food and for the development of medical products. One aim of genetic engineering is to increase the growth rate and yield of animals like cattle, pigs and poultry. The harmful effects of unregulated production of growth hormone on the health of pigs and sheep have been described. Use of BST in dairy cattle in the United States carries increased risk of mastitis. There appears to be not much concern about whether the animals are biologically 'designed' to withstand the additional stress of increased production of milk, meat, eggs and other products. The study on Hermann, a transgenic bull born in Holland in 1999 provides an intersting example. Hermann contains a gene which, if passed on to his female offspring, will enable them to produce a human milk protein (lactoferrin) in their milk. Environmental groups threatened to boycott companies that sponsored the work. All these forced a Dutch producer of baby foods to withdraw from the project.

An important motive for producing modified animals for food is commercial profit. Where there is an additional motive, such as preventing or treating disease, the issues get even more complex because the well-being of the animal has to be balanced against the well-being of humans. There is no doubt that medical experiments involve a certain amount of animal suffering *e.g.*, oncomouse. It was the first animal to be patented. The oncomouse is a transgenic mouse to which an oncogene, a gene that causes cancer has been added. The mice develop tumours much more frequently than normal and are used in cancer research. Some people argue that patenting animals itself is unethical, because it reduces them to the level of objects. Others believe that experiments such as those with oncomice cause suffering and should, therefore, be banned. In January 1993, two UK Animal Rights groups, the British Union for the Abolition of Vivisection (BUAV) and Compassion in World Farming (CIWF) joined with other European groups, launched an appeal against the European patent for the oncomouse, which was granted in 1992. The European Patent Office held public hearings from November 1995, but ran out of time before a judgement could be made, and the issue could not be resolved. Patenting animals, it is argued, makes their production more profitable. If patenting is prevented, it will reduce exploitation of animals. However, there is no guarantee that this would happen. In fact, some patents that have run out have been allowed to lapse. The Cancer Research Campaign (which has reduced its animal experimentation enormously in recent years) says its policy now is 'not to patent transgenic animals after consideration of the moral, scientific and utility issues'. The utility issues may include a feeling that it is not commercially worth patenting the animals. Public opinion may be a part of the reason.

It was believed that the insertion of genes from other species into laboratory animals is an ethical minefield. The directions in which that approach to life can lead open up a Pandora's box. In response to BUAV's claims, it can be argued that the research is contributing to our understanding of diseases like cystic fibrosis, heart disease, AIDS, multiple sclerosis and cancer for which cures will only come about if their genetic basis is understood. Transgenic animals are protected by the same laws used for all laboratory animals. It may be possible that there will be a rise in the number of genetically engineered mice, because of their help to us in understanding genetic diseases.

At the other extreme of animal welfare are the animals that are used for pharmaceutical products such as AAT and factor IX that are probably the best cared-for farm animals in the world. It is because of the fact that they are most valuable.

The concerns have now increased with a feeling that genetic engineering now offers the potential to change the human genome. It is not yet possible to eliminate, say, insulin-dependent diabetes by cutting the defective gene out of fertilized eggs and inserting a functional one. But if it does become possible, we probably could agree on diabetes, cystic fibrosis and muscular dystrophy. There is, however, a concern with respect to suitable changes involving the genes that might alter personality, length of life, predisposition to take risks? It depends on the way we go ahead for the betterment of life? or creation of problem (Prospects in Biotechnology, Internet).

13.2.5 Release of GEOs

As mentioned earlier, many crop plants such as oranges and tomatoes are quite sensitive to frost. Genetic engineering is trying to remove the gene encoding this protein from the bacterium and replace the "wild-type" bacterium on crops with the new "ice-minus" bacterium. Though this was achieved by genetic engineering in the mid-1983s, there were several years of controversy and litigation before field trials were permitted. The controversy was with respect to the following:

(i) What will happen if the ice-minus strain, once released into the environment, drifted beyond the fields onto which it was sprayed ?

(ii) What will happen if the bacterium established in the wild become widespread? Across the planet susceptibility to frost is a major determining factor in the distribution of plants. The ice-minus bacteria might trigger massive changes in vegetation. Like the ice on leaves, raindrops also usually form around "seeds" of dust or debris in the atmosphere.

(iii) If wild-type bacteria significantly contribute to normal rain seeding–droughts might become frequent, if ice-minus bacteria became the norm?

Genetically engineered organisms, such as the ice-minus bacteria, are likely to be genetically inferior to the wild-type counterparts and, therefore, will get competed out. But still some risk remains there. Although generally supportive of properly controlled releases of genetically engineered organisms, in 1989 a committee of prominent ecologists wrote the following warning in the journal Ecology: "The direct effects of introduced self-replicating organisms may not necessarily decrease with time or with distance from the point of introduction. The absence of an immediate negative effect does not ensure that no effect will ever occur."

Other crops that are resistant to disease, drought or other types of environmental stress, might overrun agricultural areas very rapidly. There have been many hundreds of releases of transgenic plants elsewhere in the world. In China, virus-resistant tobacco is being grown commercially. None of these releases has caused any known harmful environmental effects.

Genetic engineers are of the view that we have been practicing crude genetic engineering for millennia, breeding plants and animals with desired properties. Modern biotechnology is merely a faster, more precise version of standard agriculture practice. Various forms of genetic recombination occur all the time in nature, there is no need to single out modern genetic engineering be as a special threat ?

13.2.6 Use of Herbicide Resistant Plants

The use of herbicide-resistant plants is also under fire. It is felt that extensive herbicide use is hazardous to human health and the environment and herbicide-resistant crops would almost certainly lead to use of greatly increased herbicide. It is also felt that herbicide-resistance genes may be transferred from crops to weeds, for example by a plant virus. Such "lateral transfers" to plant genes may be possible. If such a transfer occurs, herbicides might ultimately become completely useless.

In 1994, the first approval for unrestricted release of a GEO in Britain was given by the Department of the Environment Advisory Committee. The organism was produced by the Belgian Company 'Plant Genetic Systems' (PGS). It was a new type of oilseed rape which contains genes for resistance to the herbicide Basta. By that time, over 60 small-scale field trials of GEOs had taken place in Britain and over 1000 in Europe and North America. It was thought that developing herbicide-resistant plants may encourage the use of greater amounts of herbicides, particularly Basta, although the companies concerned argued that it may lead to less herbicide spraying because it will be more effective and that older, more harmful herbicides will be phased out. The release of the rapeseed was, however, opposed by greenpeace an environmental organization. There are far more potential dangers once unrestricted release is granted. Rapeseed, for example, can become a weed in hedgerows and would be impossible to control by Basta. It was later on cleared that the environmental risks with rapeseed were negligible

These examples *i.e.,* ice-minus bacteria and herbicide-resistant plants have put the release of genetically engineered organisms into the environment at test. Genetically engineered fish *e.g.,* the giant salmon poses a serious threat. The fish are contained so that in theory they should not escape. However, young, small fish have been known to be carried away by birds and dropped in local waters, and larger fish have been known to escape. There are many examples from the past of newly introduced animals causing great ecological damage, such as the rabbit in the UK and in Australia. If the Scottish salmons escape into the sea, where they migrate as adults, there are fears that they may affect the balance of already endangered wild salmon populations. They might also affect food chains in unpredictable ways. Already more than 90% of salmon in some Scottish streams are descended from salmon that had escaped from fish farms in Norway. We need to ensure that genetically released organisms, or the modified genes they carry, will not cause ecological harm?

According to ecologists, ecological disasters of varying magnitude do occur *e.g.,*

(a) the invasion of pastureland in the western United States by cheatgrass and knapweed,

(b) the displacement of bluebirds by sparrows,

(c) the evolution of antibiotic-resistant bacteria,

(d) the mutation of a monkey virus into the AIDS virus.

Bioengineering on the other hand offer the possibility of correcting the ecological disasters of major proportions. It is, therefore, needed to devise the ways that how biotechnology can help in preventing and even curing the ecological disaster if at all it occur.

13.2.7 Human Genome Alterations by Biotechnology

The greatest concern of many observers is that genetic engineering offers the potential to change the human genome. It is not yet possible to eliminate, say, insulin-dependent diabetes by cutting the defective gene out of fertilized egs and inserting a functional one. But suppose that it does become possible (few molecular geneticists would be willing to bet against it). Then what? Could humankind agree on what constitutes a "bad" gene? Even if we could, would it be right? Probably most people could agree on diabetes, cystic fibrosis and muscular dystrophy. But what about more subtle changes? What about genes that might alter personality? or length of life? or predisposition to take risks? Are we wise enough to direct our evolution – or, if we develop the capability, are we wise enough to refrain. All these questions need brain storming session to find answers to them.

13.3 SOCIAL ACCEPTANCE OF BIOTECHNOLOGY

13.3.1 Transgenic Crops

Governments, farmers, consumers and, to a lesser extent, scientists fundamentally disagree on the risks and benefits of transgenic crops. The reasons for opposition include:

(a) safety,

(b) ownership (patents) of life, the influence and role of multinational firms and economic muscle or control (transgenic crops being another way of controlling food supply, and a threat to agricultural diversity),

(c) the neglect of interest of small-scale and poor farmers.

Therefore, social acceptance is not simply based on the strength of scientific evidence and perceived benefits and risks, but also issue of self-empowerment.

All sections of society have their own way of thinking.

(a) Consumers wish to choose what they eat

(b) Farmers think about what they grow

(c) Governments what they regulate

(d) Citizens what science they support

As a result, all these stakeholders (or at least a majority of them) should agree on the dissemination of transgenic crops.

World Trade Organization (WTO) regime recognizes threat to human health or the environment as the basis on which a country can refuse to admit a product. Countries cannot openly express the full range of their concerns about transgenic crops, because such fears have no legal standing in WTO. This could force nations to inflate concerns about human health and the environment. Without an open debate, democratic decision-making would not easily occur. Although it is different to find solutions, it may not be useful to avoid the cultural issues that lie behind the public arguments (Raynes, 2003). The biotechnology applications may help meet issues of food security affecting a number of developing countries, especially Africa. However, the current transgenic crops on the market were not developed to help feed the developing world, although they may help the commodity sector in a number of ways. There are only a few major public sector programmes that target crops and livestock of interest to developing countries, because about 70% of investment in agricultural biotechnology R&D comes from the private sector. From the early 1990s, major biotechnology firms targeted products for lucrative markets of Western Europe, the United States and Canada were made infinite promises about profit were made. A similar dedication to develop genetically improved lines of African staple crops such as sorghum, cassava, yams, pearl millet, pigeon pea, chickpea, groundnut and cowpea would make a big impact and possibly increase yields by 10 to 15%, if properly adopted. For example, transgenic rice varieties, developed through technologies patented by private firms, have the potential to improve yields by as much as 20% by resisting disease, and yet no field testing was done till 2003 (Piore, 2003). The acceptance of transgenic crops widely varies between and across nations. For example, in a study carried out in 1999 by Environics International, 79% of Chinese held a favourable view of agricultural biotechnology to create pest-resistant crops. This percentage was even higher than the 78% registered in the United States, and 63% of Japanese and 36% of British (holding similar views). Another survey of Beijing residents found that a large majority of shoppers were quite willing to buy transgenic foodstuffs, with many even willing to pay a premium for such products, if there were noticeable benefits. Such attitudes facilitate the Government's plans for expanded use of transgenic crops. Another reason for resistance is the lack of consumer benefits.

The World Bank launched a three-year review of all agricultural technologies used around the globe from transgenic crops to organic farming. "The key question about any review of transgenic organism was did it have the full ownership of the scientific community and those who make decisions about biotechnology", Bank's chief scientist, Bob Watson had expected that review would involve various biotechnology stakeholders, such as scientists, firm executives, farmers, consumers and NGOs (Mason, 2003).

13.3.2 Social Acceptance of Medical Biotechnology

In the case of medical applications, many people agree that health care is a top priority and anything that may improve it, is more than welcome, especially if it affects individuals directly and independently. In addition, access to drugs, diagnostics and vaccines remains limited for many life-threatening illnesses. It is expected that the medical biotechnology will play an important role to improve the accuracy and speed of diagnosis of diseases and identification of pathogens, and also the prevention (through efficient, cheap and safe vaccines) and management of illnesses (new drugs and genetic profiles). For these and other reasons, medical applications find significant support even when they are controversial. A firm involved in medical research, Geron has worked out how to lead embryonic stem cells to turn into seven different types of normal cell lines that may be used to repair damaged tissue (heart, muscle, pancreas, bone, brain, spinal injuries and liver). This could solve rejection of organs or tissue derived from stem cells of another organism.

The case of Molly Nash and Fanconi's (anaemia) disorder

This example has been cited in the paper on Promises of Biotechnology available on internent. Molly Nash (8 years in 2003) was born with a rare disorder Fanconi's anaemia, which causes bone-marrow cells to fail. Molly needed new cells from a donor who is an almost exact genetic map. Pre-implantation genetic diagnosis (PGD), a test performed using embryonic cells was used to save Molly. With the help of PGD, Molly's parents conceived their son Adam, who successfully donated umbilical-cord blood to save his 2.5 years old sister's life. However, PGD is transforming reproductive medicine by giving parents unprecedented control over what genes their offspring will have. The fear is that as other aspects of reproductive technology improve, PGD may be misused. The process starts with a single human stem cell, plucked from a three-day-old embryo or less. Although many clinics in London, Chicago, Tel Aviv and Brussels offer the PGD service, it is controversial since the cells come from fertility clinics, where would-be parents have their eggs harvested, fertilized and grown in the laboratory. By day three, the egg cell may divide, on an average, into only six stem cells. To find out if it carries the genes for Tay-Sachs or cystic fibrosis or sickle cell anaemia, the laboratory's researchers and technicians analyze the DNA. The number of people on lists for organ transplant remains high, but few viable organs are available to save these lives. Work is going on to make stem cells to grow into organs, overcome rejections of organs from other animals (*e.g.,* pigs) and target ailing tissues/the organ's ability to regenerate. Despite impressive results, there is no consensus on the use of embryonic cells and the limits to which such use is permitted. In 2002, the EU adopted a moratorium on the Commission's funding of research on embryonic stem cells until clear-cut and strict ethical rules have been set up. The United States has also cut down Federal funding for embryonic research. In addition, the UN postponed a decision on human cloning in 2003. However, the United Kingdom, Singapore and China are among countries having more liberal regulation on embryonic research.

The development of research on adult stem cells, spurred by therapeutic aims, may raise more formidable problems than those aimed at forbidding destruction or creation of embryos. There is not a clear definition of an embryo in science. Ethical reflection and debates seem to be based on the representation of life which the continuous discoveries put in question. At the same time, people want therapies without ethical dilemmas, the absence of risk without questioning our representations of life (Renard and Bonniot de Ruisselet, 2003). Throughout history, to meet new challenges the ethical codes have been continuously revised by mankind. The continuous creation of knowledge, innovation and technologies require the design of new procedures and the participation of academic, professional associations and legislatures. According to the accumulation of biological knowledge and its wide application will place a share of responsibility among the stakeholders Renard and Bonniot de Ruisselet (2003).

Despite the risks it possesses, gene therapy is highly supported. For instance, treatment of Parkinson's disease via gene therapy was tested for the first time on humans on 18th August, 2003. It consists of injection of a virus carrying the gene for the synthesis of dopamine (whose absence causes the disease) into the brain . The scientific community is divided about this approach. Some neurologists consider it to be highly risky. The results were expected within two months after the initial injection. Some recent gene therapy experiments dealing with the repair of major deficiencies of the immune system were interrupted after the death of the patients. The causes of fatality are being sought and experiments might be resumed if safety is ensured. On the other hand, because of its eugenic approach, extending gene therapy to germ cells to stop the disease being passed on, is controversial.

13.3.3 Acceptance of GM Crops for Food and Pharmaceutical Production

Biopharming, which critics call Pharmageddon, worry consumer advocates. It is feared that crop products carrying drugs, vaccines and industrial chemicals will end up on their dinner tables. This fear is increased by the discovery of transgenic maize variety containing a vaccine against pig diarrhoea in a soya field that had previously been used as testing site by ProdiGene Inc. USDA Animal and Plant Health Inspection Service (APHIS) instructed ProdiGene, Inc. to remove the maize plants from the field. Despite the fact the plants had no viable seeds, it constituted a failure by the firms to destroy all the crops as demanded in field trials. However, the soybeans were harvested before all of the transgenic maize was removed to a storage facility. The soybeans had to be stopped from entering the human or animal food chains. Another breach of the US regulations was discovered at a ProdiGene, Inc. test site in Iowa in September 2002 and the maize plants were removed from the field earlier in the season. The contaminated soya batches did not enter the human or animal food-supply chain. ProdiGene, Inc. was fined more than $3 million for breaching the Plant Protection Act and paid a civil penalty of $250,000 as well as all costs for collecting and destroying the contaminated soybeans, and cleaning the storage facility and all equipment were reimbursing to USDA. ProdiGene, Inc. also agreed to a $1 million bond and higher compliance standards, including additional approvals before field testing and harvesting transgenic material. It was expected that the company develops a written compliance programme with the USDA to ensure that its employees, agents, cooperators and managers are aware of, and comply with the Plant Protection Act, Federal regulations and permit conditions. Such lapses reinforced fears that managing the segregation of transgenics for industrial chemical crops meant for food may not be feasible. The concerns in biopharming are similar to those of GM food plants and animals. Issues of gene flow, resistance to drugs, contamination of non-GM plants and animal and biodiversity concerns plague both GMOs for food and pharmaceuticals or industrial chemicals. Some key players, including Monsanto Co. and Dow Agrosciences, have chosen to grow their pharma-maize in isolated areas, such as Arizona, California and Washington State, instead of the Corn Belt. In response to these breaches, the USDA created a new Biotechnology Regulatory Services (BRS) Unit for regulating and facilitating biotechnology within the Animal and Plant Health Inspection Service (APHIS). Draft guidance to industry on drugs, biologicals and medical devices derived from bioengineered plants for use in humans and animals was published in September 2002. The USDA also set up a new unit in the Foreign Agricultural Service to deal with biotechnology trade issues. Gene escape from a biopharmed crop towards a conventional one would occur only if a certain gene from the crop confers a selective advantage on the recipient or has the ability to reproduce. However, many plant varieties for biopharming are less fit and less able to proliferate than conventional ones. Various countries are putting in place legislation and guidelines for biopharming with a high degree of responsibility of being placed on the firms.

13.3.4 Social Acceptance of Industrial Biotechnology

Industrial biotechnology enjoys a positive social acceptance because it is environmentally friendly and contributes to sustainable development.

Biotechnology may become acceptable to all by:

(a) providing new materials and fuels that are not derived from petrochemical processes,

(b) improving and enhancing the bioremediation of water, soils and ecosystems at large, and

(c) trying to use less fossil-fuel energy.

Industrial biotechnology has benefits from the facts, that it may simplify by processing, improve efficiency, reduce waste production and increase productivity. It meets both the demands of shareholders by being cost-effective and profitable, and those of environmental advocates by being environment friendly. Therefore, firms may be compelled to adopt industrial biotechnology applications because of

i) their simplicity,

ii) reduced initial investment capital, and

iii) flexibility or legal requirements rather than because they are environmentally friendly (OECD, 2001).

However, the release of genetically-modified organisms used in industrial biotechnology applications into the environment raise fears about their potential impact on biological diversity. That is why the industrialists are rather using them in confined environments, such as their factories and greenhouses under strict biosafety regulations.

In conclusions, despite all the dilemmas, controversies for and against, biotechnology is gaining acceptance in all the fields, be it agriculture, medicine or environment. It is increasingly becoming part of our everyday life. However, cautious implications are the need of the time.

Questions

1. What ethics are involved in biotechnology ? Explain.

2. What is making biotechnology being increasingly accepted at various levels of the society? Describe

3. Write short notes on:

   (a) ProdiGene controversy

   (b) Animals and Ethics

   (c) What is case of Molly Nash and Fancani's anaemia ?

Intellectual Property Rights in Biotechnology

14.1 INTRODUCTION

Apart from the ethical aspects of genetic engineering (Biotechnology), there are other related issues *e.g.,* patenting. Although, history of patenting is more than 600 years old, yet the grant of patents for life form or biotechnological inventions is of recent origin. In 1992, an American company attempted to patent genetically engineered cotton and soya plants. Farmer would then have to pay royalties to sow the crop. Although the patents were granted, they got challenged by the international community. The US National Institutes of Health tried to patent the human genome in 1991, but after more international protests, they withdrew their application. However, the human breast cancer gene (*BRCA1*) was patented in the US once its base sequence had been determined. This was followed by the patenting of other genes in the subsequent years.

14.1.1 Intellectual Property Rights

Intellectual property rights are the rights through which creators can be given the right to exclude others from using their registered inventions, designs or other creations and to use that right to negotiate payment in return for allowing others to use them. These rights are statutory rights granted by the Governments to creators as an incentive for patenting their inventions that will benefit society as a whole. However, the extent of protection and enforcement of these rights varied widely around the world. Governments encourage filing of patents so that there is an increase in the interest for making innovation and investment in the research and development activities. This eventually leads to economic, industrial and technological development of the country.

Intellectual property is like other tangible and corporeal property generated from man's thinking power and covers industrial property and copy rights.

These are assets like other properties and cannot be defined or identified by its own physical parameters, but becomes valuable in tangible form as products. To protect it, intellectual property has to be expressed in some discernible way.

"Intellectual Property refers to all categories of Intellectual property that are subject of section 1 through 7 of Part-II or TRIPS agreement" *viz.,* copyright and related rights (Copy rights act, 1999); trademarks (Trademarks act, 2000); geographical indications (Geographical indication act, 2000); industrial designs (Designs act, 2000, 2003, 2005); patents (patents amendment act, 2002); layout designs of integrated circuits; and undisclosed information (for which no legislation exists in India).

14.1.2 Protection of Intellectual Property

The intellectual property is protected and governed by appropriate national legislations. The national legislation specifically describes the inventions which are the subject matter of protection and those which are excluded from protection. For example, methods of treatment of the humans or animals by surgery or therapy, and inventions whose use would be contrary to law, or inventions which are injurious to public health are excluded from patentability in the Indian legislation.

Forms of Protections available in India :

- Patents
- Copyright and related rights
- Trademarks
- Designs
- Geographical indications
- Layout design for ICs

14.1.2.1 Patents

These are most important form of protection available for inventions. Such inventions must fullful the important criteria of patentability, *i.e.,* novelty, non-obviousness and industrial applicability. The subject matter of the invention should also be permitted under the Indian patent act. As per the extent provisions, patents are granted for a period of 20 years on all products in all areas for human endeavour except exclusions provided under the patent law. For computer software if it is an embedded software then patent is granted. For plain software india grants only copyright protection.

14.1.2.2 Copyrights

Copyrights are concerned with the right of preventing copying of physical material existing in the field of literature and arts. Copyright subsists only in original work *i.e.,* the work must be an expression of inventive or original thought. These protect creative works that are musical, literary, artistic, lectures, plays, art reproductions, models, photographs, computer software *etc.* The term of protection available in India to 60 years from the end of calendar year in which the author died. For related rights this term is restricted to 20 years.

14.1.2.3 Trademarks

A trademark is a visual symbol in the form of a word, a device (slogan) or a label applied to articles of commerce with a view to indicate to the purchaser that they are the goods manufactured or dealt in by a particular person as distinguished from similar goods manufactured or dealt in by other persons. Registerable Marks (words/signs/or combinations) distinguish the goods or services in connection with which they are used in course of trade, material mode of manufacture of goods or performance of services. In some countries, distinctive sound or smell characterizing the product can also be registered as a trademark *e.g.,* "type that smells like a rose".

14.1.2.4 Geographical Indication/ Origin of Appellation

The term refers to those goods which owe their particular quality, reputation or other characteristics to a particular region, territory or place. GI if not protected in the country of origin, cannot be protected outside and if not protected, it becomes a generic name. It gives collective rights to a number of people. India now has a GI Act. Basmati is GI, not a trade name.

14.2 PATENT AND PATENTING

A patent is a right granted to an inventor to exclude others from using the invention by way of making, selling, improving *etc.* for a limited time period (20 years). The owner of the patent does not necessarily 'own' the objects that make up the invention itself, nor does the granting of the patent indicates that the invention has any commercial value or merit. It, however, grants the patentee the right to take legal action to prevent others from exploiting the invention without his or her consent. The inventor in return for the protection and exclusive rights must disclose details of the invention. This widens scientific and technological knowledge and expands the 'state of the art'. The knowledge can be used by others for research or after expiry of the patent, for commercial purposes. Patents, therefore, encourage openness and freedom of information.

14.2.1 Rationale of Patents

The rationale of patents is to stimulate investment in invention and innovation through a temporary monopoly, while ensuring disclosure of technical information. No inventor would be willing to invest any significant sum of money on making and developing an invention without the monopoly given by a patent. The inventor's idea can be taken by anyone after the disclosure of invention. Third parties could easily undercut the innovator without any expenditure on research. Patents, therefore, prevent others from commercially benefiting from the innovator's research. So without patents, there would be very little invention and certainly no new drugs. There would also be more secrecy in research.

14.2.2 Incentives for Innovation

Patent system represents a fair balance. During the period of the patent, temporary protection against direct competition is provided to the inventor. He, therefore, gets time to market and commercialize the invention to recover the costs of research and development. By this, society at large gains from new products. It also benefits from the disclosure of information which can support technological development. Therefore, patents are incentives for innovation, research and investment. They encourage high financial risk in long term research and development. A patent, however, lasts for a time period of 20 years.

14.2.3 Novelty, Inventive Step and Applicability

Patentable subject matter must meet three requirements *i.e.,* **utility, novelty** and **non-obviousness**. Since patents are awarded only for inventions, an invention to be patented must be novel. An invention, therefore, cannot be patented if

(i)   it is already known to the public or

(ii) lacks an "inventive step" or is obvious to a person of ordinary skill experienced in the particular technology.

(iii) does not have any assignable use or industrial applicability.

An invention must also be patentable subject matter. For example, methods of treatment of the human or animals by surgery or therapy or diagnostic procedures cannot be patented. It is because of the fact that there have been problems in proving novelty and in defining the limits of statutory subject matter as applied to biotechnological products.

Who may apply for patent ?

An application for the ordinary patent for an invention may be made by any person, whether a citizen of India, or not, claiming to be the true or first inventor of the invention or his assignee or by the legal representative of any deceased person who immediately before his death was entitled to make an application for patent. The application for patent may be made by any of the said persons either alone or jointly with any other person. According to section 2 (1) (Y) of the Act, the first importer of an invention into India or the first communicate of an invention from outside India cannot be considered as the 'true and first inventor'. A company or a firm cannot be named as the true and first inventor.

An application for a patent of addition may be made only by the applicant for the original patent to which it is an addition, if the application for the original patent is pending or by the registered proprietor of such original patent, if it has been granted.

A convention application may be made by any person who has made on application for a patent in respect of that invention in a convention country or by his assignee or his legal representative.

There are certain provisions regarding persons employed in Government Service.

What may be patented ?

Any invention which satisfies the definition of 'invention' given in the Act, *i.e.,* art, process, method or manner of manufacture, machine, apparatus or other article, substance produced by manufacture and includes any new and useful improvement of any of them and an alleged invention, may be patented.

Certain inventions are not patentable: As provided in Section 4 of the Act, no patent shall be granted in respect of an invention relating to atomic energy falling within sub-section (1) of section 20 of the atomic energy Act, 1962.

Patents do not result in the ownership of life and they do not monopolize life itself. Patents do not confer any right of ownership in the things being patented, any more than copyright of ownership of a book or CD embodying the copyright work. For example, if a patent is awarded on a revolutionary new washing machine which can wash clothes more effectively, using less water and detergents, then when a customer buys the new washing machine, he or she owns it, not the patent holder. The patent only provides the patent holder the right to prevent other manufacturers from exploiting commercially his or her invention for a limited period of time.

Patent law is concerned with inventions which are the outcome of research carried out on or with the source material. It is not possible to patent the source material itself, in its natural state. The isolation of an active principle or a genetic component of the original material leading to a new and useful application, however, is patentable.

Term of the Patent

Subject to the payment of prescribed renewal fee within the prescribed period, the term of every patent granted under the Act shall be 20 years from the date of filing or date of priority, whichever is earlier. US patent law has brought about some changes is the terms of the patent in 2007.

Fees : Fees as prescribed by the Rules, are payable for filing an application for a patent and in respect of subsequent proceeding thereon.

A proceeding in respect of which a fee is payable is of no effect unless the fee has been paid.

General procedure for obtaining a patent :

The following are the successive stages of the procedure for obtaining a patent.

(i) Filing an application for a patent accompanied by either a provisional specification or a complete specification.

(ii) Filing the complete specification, if a provisional specification accompanied the application.

(iii) Examination of the application.

(iv) Acceptance of the application and advertisement of such acceptance in the official Gazette.

(v) Overcoming opposition, if any, to the grant of a patent.

(vi) Sealing of the patent

14.2.4 Applications

Patenting of Drugs

New medicines need to be patented. It is because of the fact that their research and development involves enormous investment. About 8-15 years and between US $100 to $ 300 million are needed to bring one medicine through to the market. This long time scale means that the remaining life of a patent is considerably reduced. Infact there is a very short period to recover the research and development costs. It would be impossible for companies to recoup their initial investment without patents. The companies, therefore, would be reluctant to invest in the research and development needed to find new cures for the many diseases which afflict millions of people world-wide. This is especially the case in promising new fields of research, such as biotechnology. Biotechnology has already led to the development of a range of new medicines and also to better production methods of pre-existing medicines.

Biotechnology Patents: Contrary to belief, patent protection of biotechnology inventions is not new. Many patents have been granted in many countries world-wide for biotechnology products and processes such as yeast culture, anti-toxic sera, viral vaccines and the production or preparation of bread, beer and vinegar. The modern biotechnology products and processes which have been patented include industrial proteins, laundry enzymes, pharmaceuticals, new diagnostic tests and nucleotide sequences derived or isolated from the human body. These include new and safer methods for producing existing products in large quantities, and for the development of new medicines for diseases which until now, have been untreatable.

Today, patents have been granted for a range of different biotechnological processes and products. These include :

   (i)   Therapeutically active proteins

   (ii) The processes used to manufacture them

  (iii) Monoclonal antibodies

  (iv) DNA sequences isolated from organisms

   (v) Cell lines

  (vi) Microorganisms

 (vii) Transgenic organisms, including both plants and animals

However, patent law does incorporate limits to what is patentable. A recent directive on the legal protection of biotechnological inventions harmonized the laws of the member states of the European Union. Human beings and elements of the human body such as body organs, limbs and body fluids in their natural state cannot be patented. Accordingly, nucleotide sequences elucidated by human genetic research and other molecules identified by such research in the human body are not patentable in their natural state, but if these are isolated or otherwise produced by means of a technical process, they can be patented.

14.2.4.1 Patenting of Human Genes and DNA

For over ten years, genes, nucleotide sequence and DNA derived from human genetic research have been patented in all the major territories of the world, including Europe. There is, however, opposition to this in some quarters. Certain people feel that they should not be patented as they are part of "nature". They oppose their use in the treatment and curing of disease.

There are about 30,000 genes in the human genome and by preserving, copying and passing on information, they allow the body to grow and reproduce. In talking about "human genes", it is important to understand that structure and chemical properties of the DNA encoding human genetic information are indistinguishable from DNA from other species.

Human DNA and nucleotide sequences may be patented when they satisfy the basic criteria of patentability. They, however, have to be **novel, inventive** and **industrially applicable** and this distinction marks the difference between discoveries and inventions.

(i) **Novelty :** Any attempt to patent the existing human gene in its natural surroundings falls at this hurdle. A gene in a body like arm or a leg can not be patented. Patentability of human genes is, thus, limited to artificial structures outside the human body containing information about the human genome. This rather dull reality contrasts with the imaginative, but misplaced accusation that "people are being patented" if patent protection is being granted on a human gene.

(ii) **Inventive step :** It is required to prevent patents being granted for trivial or obvious developments of something that has been done before. This means that there has to be a solution to a technical problem.

(iii) **Industrial applicability :** It requires that the way in which the invention is to be applied industrially must be known (*e.g.,* therapeutic use).

There are differences between a discovery and an invention, particularly in relation to isolated human gene sequences. A new enzyme and its sequence and that of the gene that codes for it, are all discoveries. The inventions, however, can be based on them. Following are some of the examples :

(i) Knowledge of the enzyme's existence and structure may allow it to be produced in purified form on an industrial scale and used to cure a disease. The isolated and purified enzyme is a new object, useful in industry.

(ii) The nucleotide sequence (genes) determination allows it to be isolated or synthesized and transferred to other organisms. The isolated gene, therefore, is a new entity and it may be claimed as an invention.

(iii) A discovery would be the isolation of a nucleotide sequence and description of its structure without determining a utility.

On the other hand, an invention would require additional information, such as the specification of the DNA sequence as coding for, say, interferon and that interferon was found to be indicated to be useful for example against cancer. Anything lacking a utility would fail the requirement of specific industrial application.

There have been discussions on the patentability of partial gene sequence of unknown origin. It, however, is clear that gene without indication of a function, whether human or not, are not patentable.

14.2.4.2 The Patentability of Transgenic Organisms

In several countries including USA, Japan, Europe *etc.* transgenic animals can be protected through patent claims. Animals and plants have been modified through breeding programmes for many centuries. In conventional breeding, organisms are modified through the random mixing of two or more parent genomes (which corresponds to a mixing of a multitude of genes), carefully chosen by the breeder and taking into account the specific properties of a parental line.

In transgenic animals and plants, one or several specific, well-determined DNA coding sequences are introduced by direct gene transfer. These nucleotide sequences code for a specific property which either could not be introduced by conventional breeding or only done so less rapidly or less predictably.

Transgenic animals can now be bred with specific genetic characteristics. These help in the following way:

(i) To investigate new medicines – for treatment of various diseases like multiple sclerosis, cancer or diabetes,

(ii) Can produce therapeutic proteins – goats and sheep can be used as production host for life-saving medicines, thereby, improving the quality of life,

(iii) Other techniques of genetic engineering of an animal make possible tests of promising new cancer treatments and at the same time reduce the number of animals needed for specific safety and screening tests.

Research on transgenic plants generally is directed to

(i) improve the food supply as well as the quality of plants and their ingredients,

(ii) provide valuable industrial feedstocks as well as new properties *e.g.*, disease or insect resistance which helps plants to protect themselves.

Although animal and plant varieties are excluded from patentability as are essentially biological methods (breeding methods) both transgenic animals and plants are patentable for their production characteristic. Patenting of transgenic animals and plants is justified to prevent copying. It encourages investment in developing new products and processes.

Patenting of transgenic animals does not mean, however, that it is owned through the patent. It has long been accepted that outright ownership of animals and plants had been acceptable in the society. Farmers had been owning crops and livestock and same is the case of owning pests and plants. Patent, however, does not give right of ownership to the subject matter that they protect (animal or plant).

It is not possible to patent life. The word "life" denotes an abstraction from concrete realities– life is only ever encountered as embodied in living things. No one has ever sought to patent such an abstraction, nor could they, even if they wanted to, since "life" is not new. It follows that "patenting life" is a misleading slogan or concept since patent law does not aim to cover abstractions.

The patenting of transgenic animals has been safeguarded from the ethical point of view also.

(i) Animal welfare and protection is already regulated in Europe.

(ii) Before any animal experimentation can be undertaken, licenses are required in EU member States and in Switzerland. Prior to any patenting activity, such licenses were procedural. Even in India every life sciences or biotechnological research institution which use animals for research purposes has to have Animal Ethics and Biosafety Committees which examine each and every procedure and protocol related touse of animal in R and D activities.

(iii) Processes for modifying the genetic identity of animals which are likely to cause them suffering without any substantial medical benefits (in terms of research, prevention, diagnosis) are excluded from patentability.

14.2.5 Patenting and Biodiversity

Throughout the world "genetic resources" are the totality of the germ plasm of existing plants, animals and other organisms. The issue of rights deriving from the ownership of source material for example human cells from an individual person or plant tissue from a particular geographical location is a separate issue from patentability. It is often handled under national legislation.

Many areas have generated discussion in this respect.

(i) **Bio-prospecting :** It refers to the surveying and analysis of the wealth of species available in areas such as the tropics. The fear is that scientists take what they want from such areas without payment of any kind. The issues of soveregnity over natural resources have been addressed by Convention on Biological Diversity. It provides that access to genetic resources be subject to prior informed consent of the party providing the resources (usually a country) and the interested party (often a company) on mutually agreed terms.

(ii) **Reduction in genetic diversity :** By patenting, there could be a reduction in

(a) diversity of species (plants, animals and microorganisms)

(b) diversity within species

(c) diversity of ecosystems (for example, salt marshes and coastal systems)

Since the beginning of the last century, genetic diversity has been decreasing. It has been due to deforestation, agricultural practices, over-population and expanding deserts. The generation of intellectual property in agricultural biotechnology has only been a recent phenomenon and so cannot be a cause in the reduction of biodiversity nor are there credible reasons why patents would increase this trend.

As patents can only be awarded for inventions, it means that the natural resource or genetic material is not destroyed through patenting. Biotechnology practice is expected to lead to improved practices of conservation and provide improved means of preserving global biodiversity.

Provisions of CBD (Earth Summit, 1992)

The submit visualized that the member states have sovereign rights over their own biological resources. Biological diversity is being significantly reduced by certain human activities. There is fundamental requirement for the conservation of biological diversity

(a) in the *in situ* conservation of ecosystems and natural habitats,

(b) maintenance and recovery of viable population of species in their natural surroundings, and

(c) taking up *ex situ* measures for protection of biodiversity preferably in the country of origin.

This act seeks to recognize the following advantages for India:

- It is one of the 12 mega biodiversity countries of the world
- It has only 2.5% of the land area and accounts for 7-8% of the recorded species of the world
- It is a signatory of the convention on Biological diversity which interalia, joins the member states to facilitate access to genetic resources by other parties for environmentally sound purpose, subject to national legislation on mutually agreed terms.

Indian Biological Diversity Act 2002

As a founder member of the Convention on Biological diversity, India has enacted its own biological Diversity Act, which was :

- Passed by Lok Sabha on December 2, 2002
- Passed by Rajya Sabha on December 22, 2002
- Presidential assent on February 5, 2003

Salient Features of the Act

To comply with CBD requirement, Indian biodiversity act, has the following salient features:

- To regulate access to biological resources of the country with the purpose of securing equitable share in benefits arising out of the use of biological resources, and associated knowledge relating to biological resources.
- To conserve and sustainably use elements of biological diversity
- To respect and protect knowledge of local communities related to biodiversity
- To secure sharing of benefits with local people as conservers of biological resources and holders of knowledge and information relating to the use of biological resources.
- Conservation and development of areas important from the standpoint of biological diversity by declaring them as biological diversity heritage sites.
- Protection and rehabilitation of threatened species
- Involvement of institutions of self-Government in the board scheme of the implementation of the act.

Requirement Envisaged in the Indian Biodiversity Act 2002

It provides for the establishment of national biodiversity authority, State biodiversity boards, Biodiversity management committees, Biodiversity funds and National heritage sites.

The national biodiversity authority : It will deal with matters relating to request for access by foreign individuals, institutions or companies and all matters relating to transfer or result of research to any foreigner, imposition of terms and conditions to secure equitable sharing of benefits and approval for seeking any form of collaboration outside India for an invention based on research or information pertaining to a biological resources obtained from India.

State biodiversity boards : They will deal with matters relating to access by Indians for commercial purposes and restrict any activity which violates the objectives of conservation, sustainable use and equitable sharing of benefits.

Biodiversity management committees : These are set up by institutions of state Governments in their respective areas. Their functions will relate to the conservation, sustainable use, documentation of biodiversity and chronicling of knowledge relating to biodiversity. These committees are to be consulted by the National Biodiversity Boards on matters related to use of biological resources and associated knowledge within their jurisdiction.

Biodiversity funds : These shall be set up at central, state and local levels. The monetary benefits, fees and royalties received as a result of approvals by national biodiversity authority will be deposited in national biodiversity fund. The fund will be used for conservation and development of areas from where resources have been accessed.

National heritage sites : These are important from the standpoint of biodiversity. These shall be notified by state government in consultation with institutions of local self Governments.

Protection of Plant Varieties and Farmers' Right Act, 2002

The plant varieties protection and farmers' right act, 2002 specifies rights of farmers, breeders, researchers through a *sui generis system*.

Farmers' Rights

- The farmer ….." shall be deemed to be entitled to save, use, sow, resow, exchange, share or sell his farm produce including seed of a variety protected under this act in the same manner as was entitled before the coming into force of this act
- Provided that the farmer shall not be entitled to sell branded seed of a variety protected under this act"
- Branded seed means any seed put in a package or any other container and labeled in a manner indicating that such seed is of a variety protected under this act"

Farmers' Rights Include the following:

- Breeders wanting to use farmers varieties for creating essentially derived varieties shall have to seek permission of the farmers

- Data about the parentage of the new variety shall have to be disclosed while applying for breeders certificate
- Breeders can not use sterile seed technologies
- No fee to be paid by farmers for examining the documents
- Protection against innocent infringement
- Use of farmers' varieties to breed new varieties shall have to be paid for protection against supply of bad/stale seed.

Breeders' Rights

- Right to produce, sell, market, distribute, import or export a variety
- Deceptive, similar packaging shall invite action
- Burden of proof on the violater
- Breeder to pay 'Royalty' for using farmers' parental stock for breeding of new varieties.

Right of Researchers'

Scientists and breeders shall have free access to registered varieties for research, provided the use is occasional and not repetitive, in which case authorization of the breeder whose variety is being used, is to be sought (Repetitive use means when the registered variety needs to be used repeatedly as a parental line for commercial production of another variety).

The IPR protection in medicinal plants, therefore, has to be achieved through protection of our vast biodiversity, new varieties and technologies generated.

14.2.6 Benefits of Patents

A patent rewards the investment of time, money and efforts associated with research and stimulates further research. It encourages

(i) innovation,

(ii) investments in patented inventions,

(iii) quick commercialization and increase in the general pool of public knowledge.

14.2.7 How to Get a Patent in India

(i) The grant of patents for inventions in India is governed by the Patent Acts, 1970 (hereinafter referred to as the Act) and the Patents Rules, 1972 (hereinafter referred to as the Rules) and its subsequent amendments. Patent Amendment Act (1994), Patent rules 2005 and Patent Ordinacne 2004.

   The patents granted under the Act are operative in the whole of India.

(ii) **The Patent Office :** The patent office has been established by the Government of India for granting patents for inventions under the Act, for the registration of industrial designs under the Design Act, and for other purposes specified in these two Acts.

The Patent Office is not concered with the registration of Trademarks under the Trade and Merchandise Marks Act or with the artistic or literacy copyright under the Indian Copyright Act, or with the registration of patent or proprietary medicines under the Drugs Act. The Head office of the Patent office is loated at Kolkota.

At present, the Patent office has Branch offices at Mumbai, Chennai and Delhi.

(iii) **Types of Patents:** The following kinds of patents are granted under the provisions of the act namely :

- Ordinary applications (which could be a complete or provisional specification)
- A patent of addition for improvement or modification of an invention for which invention a patent has already been applied for or granted. A patent remains in force only as long as the patent for the original invention remains in force and no renewal fee are payable is respect thereof.
- A patent granted in respect of a conventional application filed under section 135 of the Act which is based on an application made in the convention country as notified under section 133 of the Act in respect of the same invention.
- PCT International application
- PCT National Phase application
- Divisional application

The convention application has to be made within one year from the date of the first application made in a convention country in respect of that invention.

Structure of Patent Application

- **Bibliographic :** It contains the title of the invention, inventor's name, and indication of its technical field, date of filing, country of filing *etc.*
- **Description of the invention :** In this the inventor describes in clear language and provides enough details of his invention duly supported by a series of workable examples alongwith diagrams/charts, if needed. The invention has to be described in complete details, so that any person with an average understanding of the field could use or reproduce the invention.
- **Background of the invention or state of the art :** In this the inventor lists the state of the art available on the date of filing application for his invention.
- **Discrepancies in the available state of the art :** Here the inventor lists the shortcomings/drawbacks found in the state of the art and define his problem.
- **Claims :** The inventor has to bring out a series of claims establishing his rights over the state of the art. It is this portion, upon which the protection is granted and not on the description of the invention. This has to be carefully drafted.

Appropriate office for filing an application in India

- **Patent Office Branch, Mumbai** for the States of Gujarat, Goa, Maharashtra, Madhya Pradesh, Chhatisgarh and the Union Territories of Daman and Diu and Dadra and Nagar Haveli.

- **Patent Office Branch, New Delhi** for the States of Haryana, Himachal Pradesh, Jammu & Kashmir, Punjab, Rajasthan, Uttar Pradesh, Uttarakhand and Delhi, and the Union Territory of Chandigarh.
- **Patent Office Branch, Chennai** for the States of Andhra Pradesh, Karnataka, Kerala, Tamilnadu and Pondicherry and the Union Territories of Laccadive, Minicoy and Aminidivi Islands.
- **Patent Office** (Head Office), Kolkota for rest of India.

14.2.8 Patents and Ethics

Since the evolution of patent legislation, there has been a greater input on the role of ethics in the patenting process. In Europe, an invention is excluded from being patented if its publication or exploitation would generally be considered immoral or otherwise contrary to public order. Inventions which offend human dignity or which encompass the human body, at any stage in its formation and development are unpatentable. More specifically patent legislation prohibits protection for inventions in the following fields :

- Processes for the reproductive cloning of human beings
- Processes for modifying the germ line genetic identity of human beings (in other words, processes which could enable genetic modifications to be passed on by parents to their children)
- Use of human embryos for industrial or commercial purposes
- Processes for modifying the genetic identity of animals which are likely to cause them suffering without any substantial medical benefit (in terms of research, prevention, diagnosis)

CONCLUSION

Patenting system is designed to cover newly emerging technologies including biotechnology. Many of the concerns and misunderstandings that people may have about biotechnology can be answered through a careful and sensible application of current patent law. Specific legislations have been introduced in areas where more specific concerns are needed to be addressed.

Questions

1. What do you mean by IPR ? Explain in detail.
2. In what way intellectual property can be protected ? Explain
3. Write short notes on :
    (a) Patent
    (b) Biotechnology patent
4. Write an essay on biodiversity and role of patent law.

References

Athel Cornish Bowden, *Fundamentals of Enzyme Kinetics*, Portland Press Ltd. London.

Bernard R. Glick and Jack Pasternak, 1998, *Molecular Biotechnolgoy*, ASM Press.

Colin Ratledge, Bgorn Kristianses, 2001, *Basic Biotechnology*, Cambridge University Press.

Gera, V.K., 2006, *Genetic Engineers and Biotechnology*, Dhruv Publications.

Hanson Peter, Schmander, 1997, *Methods in Biotechnology*, Taylor and Francis.

Helon Kreuzer and Adrianne Massey, 2005, *Biology and Biotechnology Science Applications & Issues*. ASM Press.

Higgins & Taylor, 4th Impression 2006, *Bioinformatics*, Oxford University Press, New Delhi.

Higgins and Taylor, 3rd Impression, 2004, *Bioinformatics*, Oxford University Press.

Ian Freshney, 2005, *Culture of Animal Cells*, John Wiley.

Jean L. Mark, 1989, *A Revolution in Biotechnology*, Cambridge university Press, Cambridge.

John Endecle, Susen Blanchacd, 2005, *Introduction of Biomedical Engineering*, Accademic Press, New York.

Keith Walsa and John Walker, 2005, *Biochemistry and Molecular Biology*, Cambridge University Press, Cambridge.

Keith Wilson & John Walker, *Biochemistry and Molecular Biology*, Replica Press.

Kuby, Jains, 6th Edition 2007, *Immunology*, WH Freeman and Company, New York.

Kumar, V., Abas A. K. and Fauston, N. 2007, *Pathologic Basics of Disease*, Elsevier.

Lee Yuar Kun, 2006, *Microbial Biotechnology : Principles and Applications (2nd Edition)*, World Scientific.

Lodish, Berk, 2003, *Molecular Cell Biology*, WH Freeman and Company.

Maarten J., Chrispeels & David E. Sadava, *Plants Genes and Crop Biotechnology*, Jones and Bartlett Publishers.

McGarvey and Yusibov, *Plant Biotechnology*, Springer.

Monroe W. Strickberger, 3rd Edition 2006, *Genetics*, Prentice Hall of India, New Delhi.

Normann. Potter Joseph H. Hotchkiss, Reprint 2006, *Food Science*, CBS, New Delhi.

Primrose, S. B. Twyman 7th Edition 2006, *Principles of Gene Manipulation and Genomics*, Blackwell, USA.

S. Agarwal and Sita Naik, 2007, *Fundamentals of Immunogenetics : Principles and Practices*, International Book Distributors.

S. B Primrose, 2nd Indian Reprint 2001, *Molecular Biotechnology*, Panima, Prakash, New Delhi.

Satya Dass, 2004, *Essentials of Biotechnology for Students*, Peepee Publishers.

Setty and Veena, *Biotechnology-I*, New Age International Publishers, New Delhi.

Srivastava, Narula, 2003, *Plant Biotechnology and Mol. Markers*, Manish Sejwal for Anamaya Pulishers, New Delhi.

Stephen L. Wolfe, 1993, *Molecular and Cellular Biology*, Wadsworth, California.

Strachan T. and Read A.P., 2004, *Human Molecular Genetics (3rd Edition)*, Taylor and Fransic Group, London.

Index

B

I

M

T

It is worth noting for the special case of a circular tube ($a = b = r$), the axisymmetric nature of the problem dictates that $\partial(\,\cdot\,)/\partial\eta = 0$ and so Equations 34 and 35 lead to

$$\omega = \frac{trf_1}{12L^2\sqrt{1-v^2}} + \frac{L^2\sqrt{1-v^2}\,f_5}{rt} + \frac{krL^2\sqrt{1-v^2}\,f_6}{Et^2} \quad (38)$$

which is the same as that derived by Bradford et al. (2006).

For a given bucking problem, the buckling coefficient ω is a function of the vector $\mathbf{r} = (\beta, L)$; the relationship defining a surface in $\Re^3$. A convenient strategy for determining the critical value ω_{cr} (defined by $\omega_{cr} = \omega(\mathbf{r}_{cr})$ where $(\partial\omega/\partial\mathbf{r}) = 0$ is evaluated at the value of $\mathbf{r} = \mathbf{r}_{cr}$) is to assume an 'equivalent diameter' $d = 2a^2/b$ (Zhu & Wilkinson 2006) and $\beta = 1$ as a starting point. Equations 34 to 37 may then be solved numerically with β kept constant for small increments of L until a localised minimum is found and recorded. This loop can be nested inside of another with β being incremented in the neighbourhood of $\mathbf{r}$ chosen on the basis of a circular tube.

4 ILLUSTRATION

Two cases have been considered for verification; one for an ellipse without infill ($\alpha = 1/2$) and one for an ellipse with a rigid infill ($k \to \infty$, $\alpha = 0$). The two ellipses used had the dimensions $a = 150$ mm, $b = 100$ mm and $a = 200$ mm, $b = 100$ mm (this second ellipse was analysed using ABAQUS by Zhu and Wilkinson (2006)), and the results are shown in Tables 1 and 2 for different values of the tube thickness. The tables also show the approximate buckling stress σ_{app} that can be determined from the closed form solution (Bradford et al. 2006) using an equivalent diameter of $d = 2a^2/b$. The notation $195 \div 5$ (and similar) in the ABAQUS column of Table 1 indicates that 5 wavelengths occurred along a length of 195 mm, producing $L = 39$ mm.

It can be seen from the tables that the results of the present method are consistent with the ABAQUS results, and those of Zhu and Wilkinson (2006), for a hollow elliptical tube. Disparities would be expected because the Rayleigh-Ritz solution technique uses only one harmonic function for the buckled shape lengthwise, and a single cubic function in for the projection of the meridional buckle onto the z-axis. For the case of an ellipse with a rigid concrete infill, using a buckling stress of (Bradford et al. 2002, 2006)

$$\sigma_{0\ell,\text{approx}} \approx \frac{E}{\sqrt{1-v^2}}\left(\frac{t}{r}\right) \quad (39)$$

that uses $\omega \approx 1$ (Table 1) and $r = a^2/b$ produces acceptable buckling stresses for preliminary engineering design.

Table 1. Results for elastic buckling with $a = 1.5b = 150$ mm.

t	Result	$\alpha = 0$	$\alpha = 1/2$	ABAQUS
0.5	ω_{cr}	1.029	0.599	
	β	0.45	0.51	
	σ_{0l}(N/mm^2)	479.4	279.12	279.38
	σ_{app}/σ_{0l}	0.972	0.964	
	$2a/t$	600	600	600
	L (mm)	27.65	36.3	$195 \div 5$
1.0	ω_{cr}	0.999	0.603	
	β	0.482	0/543	
	σ_{0l}(N/mm^2)	930.7	562	561.41
	σ_{app}/σ_{0l}	0.961	0.958	
	$2a/t$	300	300	
	L (mm)	50.3	51.30	$275 \div 5$
5.0	ω_{cr}	1.041	0.636	
	β	0.695	0.777	
	σ_{0l}(N/mm^2)	4850	2963	2879.2
	σ_{app}/σ_{0l}	0.922	0.908	
	$2a/t$	60	60	60
	L (mm)	111.3	113.2	$600 \div 5$
10.0	ω_{cr}	1.148672	0.686	
	β	0.93	0.99	
	σ_{0l}(N/mm^2)	10,703	6396	6156.8
	σ_{app}/σ_{0l}	0.871	0.841	
	$2a/t$	30	30	30
	L (mm)	119.4	156.3	$320 \div 2$

Table 2. Results for elastic buckling with $a = 2.0b = 200$ mm.

t	Result	$\alpha = 1/2$	(Zhu)	ABAQUS
0.5	ω_{cr}	0.603		0.608
	β	0.47		
	σ_{0l}(N/mm^2)	158.0	–	159.26
	σ_{app}/σ_{0l}	0.958		0.950
	$2a/t$	800		800
	L (mm)	48.35		
1.0	ω_{cr}	0.607	0.620	0.621
	β	0.503		
	σ_{0l}(N/mm^2)	318	325	325.8
	σ_{app}/σ_{0l}	0.951	0.931	0.929
	$2a/t$	400	400	400
	L (mm)	68.3		
5.0	ω_{cr}	0.667	0.636	0.638
	β	0.80		
	σ_{0l}(N/mm^2)	1746.8	1668	1671
	σ_{app}/σ_{0l}	0.8662	0.907	0.905
	$2a/t$	80	80	80
	L (mm)	148.3		
10.0	ω_{cr}	0.712	0.657	0.654
	β	0.92		
	σ_{0l}(N/mm^2)	3729.1	3445	3430
	σ_{app}/σ_{0l}	0.812	0.878	0.882
	$2a/t$	40	40	40
	L (mm)	206.3		

947

5 CONCLUDING REMARKS

This paper has considered an analytical modelling of the elastic local buckling of thin-walled elliptical tubes containing an elastic infill. Despite considerable research data being discoverable for buckling of hollow elliptic tubes, little seems to have been reported hitherto on elliptic tubes with a rigid infill and none on elliptic tubes with an elastic infill, despite the potential benefits that can result from this application.

The analysis used here formulated the strain energies and work done associated with bifurcation buckling ignoring imperfection sensitivity, and made recourse to the Rayleigh-Ritz technique with an assumed displacement function that incorporated a localisation of the buckle at the vertices of the ellipse where the radius of curvature is greatest. It was shown that the solution reduces to a dimensionless representation for which the relevant geometrical and material properties that govern the local buckling coefficient can be identified. A numerical example was undertaken and the results compared, where possible, with solutions derived by ABAQUS and with solutions reported elsewhere. The agreement was shown to be satisfactory.

ACKNOWLEDGEMENT

This paper was supported in part by The Australian Research Council through its Federation Fellowship and Discovery Project Schemes.

REFERENCES

ABAQUS. 2006. Users Manual – Version 6.1-1. Pawtucket: Abaqus Inc.

Bortolotti, E., Jaspart, J.E., Pietrapertose, C., Nicaud, G., Petitjean, P.D. & Grimault, J.P. 2003. Testing and modelling of welded joints between elliptical hollow sections. *Proceedings of 10th International Conference on Tubular Structures*, Madrid.

Bradford, M.A. & Roufegarinejad, A. 2006. Elastic buckling of thin circular tubes with an elastic infill under uniform compression. *Proceedings of 11th International Conference on Tubular Structures*, Quebec, 359–364.

Bradford, M.A. & Vrcelj, Z. 2004. Elatic local buckling of thin-walled square tubes containing an elastic infill. Proceedings *4th International Conference on Thin-Walled Structures*, Loughborough, UK, 893–900.

Bradford, M.A., Loh, H.Y. & Uy, B. 2002. Slenderness limits for circular steel tubes. *Journal of Constructional Steel Research* 58(2): 243–252.

Bradford, M.A., Roufegarinejad, A. & Vrcelj, Z. 2006. Elastic buckling of thin-walled circular tubes containing an elastic infill. *International Journal of Structural Stability and Dynamics*, 6(4): 457–474.

Chan, T.M. & Gardener, L. 2006. Experimental and numerical studies of elliptical hollow sections under axial compression and bending. *Proceedings of 11th International Conference on Tubular Structures*, Quebec, 163–170.

Corus 2004. *Celsius 355® Ovals – Sizes and Resistances Eurocode Version*. Corus Tubes: Structures and Conveyance Business.

Gardner, L. & Chan, T.M. 2006. Cross-section classification of elliptical hollow sections. *Proceedings of 11th International Conference on Tubular Structures*, Quebec, 171–177.

Hutchinson, J.W. 1968. Buckling and initial post-buckling behavior of oval cylindrical shells under axial compression, Transactions, ASME.

Hyer, M.W. & Vogl, G.A. Response of elliptical composite cylinders to a spatially uniform temperature change. *Composite Structures* 51(2): 169–179.

Kempner, J. 1962. Some results on buckling and post buckling of cylindrical shells. *NASA* TN D-1510.

Maguerre, K. 1951. Stability of cylindrical shells of variable curvature. *NACA* TM 1302.

Meyers, C.A. & Hyer, M.W. 1999. Response of elliptical composite cylinders to axial compression loading. *Mechanics of Composite Materials and Structures* 6(2): 169–194.

Teng, J.G. 1996. Buckling of thin shells: recent advances and trends. *Applied Mechanics Reviews, ASME* 123: 1622–1630.

Steel and Composite Structures – Wang & Choi (eds)
© *2007 Taylor & Francis Group, London, ISBN 978-0-415-45141-3*

Lateral buckling behaviour and design of the new LiteSteel beam

C.W. Kurniawan & M. Mahendran

Queensland University of Technology, Brisbane, Queensland, Australia

ABSTRACT: The flexural capacity of the new hollow flange steel section known as LiteSteel beam (LSB) is limited by lateral distortional buckling, which is characterised by simultaneous lateral deflection, twist and web distortion. Recent research into this unique buckling behaviour has developed appropriate design rules for the member capacity of LiteSteel beams. However, these rules are limited to a uniform bending moment distribution and do not take into account the effects of non-uniform moment distributions and load heights. Many steel design codes have adopted equivalent uniform moment distribution and load height factors to accommodate different loading conditions. However, past investigations have shown the inadequacy of these factors to different cross sections and materials. This paper describes the effects of non-uniform moment distribution and load height on the lateral buckling strength of LSBs based on finite element analyses of LiteSteel beams. It presents the finite element analysis results and a discussion of suitable design methods for LSBs subject to non-uniform distributions and varying loading positions.

1 INTRODUCTION

LiteSteel beam is a new cold-formed high strength steel section developed by an Australian company, Smorgon Steel, using its patented dual resistance welding techniques. This unique section has a mono-symmetric channel shape comprising two rectangular hollow flanges and a slender web (see Figure 1).

Recent research has identified that the structural performance of LSBs for intermediate spans is governed by their lateral distortional buckling (LDB). Under flexural action, the presence of two stiff hollow flanges and a slender web leads to this buckling mode for which a web distortion occurs in addition to the lateral deflection and twist that occur in the common lateral torsional buckling (LTB) mode. The major implication for LDB is a significant lateral buckling strength reduction at intermediate spans than that of LTB. For long spans, the critical buckling mode is LTB as for other open steel sections.

Since there were inadequate design rules that include the effects of LDB, Mahaarachchi and Mahendran (2005b) developed new design rules that allow the effects of LDB in the member capacity of LSBs. These rules have been adopted in the Design Capacity Tables for LiteSteel Beams and in the new AS/NZS 4600 (SA, 2005). However, they are limited to a uniform bending moment. Uniform bending moment conditions rarely exist in practice despite being the worst case due to uniform yielding across the whole length. Further, loading position or load height also significantly affects the

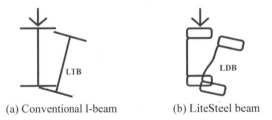

| (a) Conventional I-beam | (b) LiteSteel beam |

Figure 1. Lateral buckling modes.

beam strength. Loading above the shear centre would result in unfavourable destabilising effects while loading below the shear centre results in the opposite. Appropriately addressing loading conditions including non-uniform moment distributions and load heights will lead to accurate assessment of the member capacities of LSBs and thus an economical and safe design.

Equivalent uniform moment or moment modification factors have been a popular approach in many steel design codes to allow for different loading conditions (non-uniform moment) while load height factors are used to allow for the effect of loading positions. However, they are adopted mostly based on simply supported hot-rolled doubly symmetric I-beams and past research has suggested that they are dependent on the type of cross-section and materials used. LSB made of high strength steel with unique stress-strain curves has a unique cross section with specific residual stresses and geometrical imperfections along with

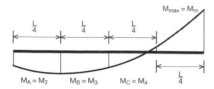

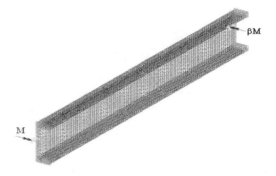

Figure 2. Moment diagram for Equations (1) to (3).

a unique buckling mode. This has therefore initiated the following investigation of the effects of loading condition on the lateral buckling and flexural strength behaviour of LSBs.

Figure 3. Overall view of finite element model.

2 CURRENT DESIGN CODES

2.1 Australian design code

Tables 5.6.1 and 5.6.2 of AS4100 (SA, 1998) provide a list of moment modification factors (α_m) for various common loading conditions. It also gives a simple approximation using Eq. (1) to accommodate any bending moment distribution.

$$\alpha_m = \frac{1.7M_m^*}{\sqrt{\left(M_2^*\right)^2 + \left(M_3^*\right)^2 + \left(M_4^*\right)^2}} \le 2.5 \tag{1}$$

AS4100 Clause 5.6.3 allows the effect of load height by increasing the effective length with a factor of 1.4 for top flange loading and 1.0 for the bottom flange loading.

2.2 American design code

ANSI/AISC 360 (AISC, 2005) provides a moment modification (C_b) equation, originally developed by Kirby and Nethercot (1979) to allow for various shapes of moment diagrams and also the mono-symmetric effect (R_m).

$$C_b = \frac{12.5M_{max}}{12.5M_{max} + 3M_A + 4M_B + 3M_C} \le 2.27 \tag{2}$$

ANSI/AISC 360 does not account for the destabilising and stabilising effects of load height.

2.3 British design code

BS5950-1 (BSI, 2000) provides a moment modification (m_{LT}) equation analogous to AISC equation that applies to various shapes of moment diagrams.

$$m_{LT} = 0.2 + \frac{0.15M_2 + 0.5M_3 + 0.15M_4}{M_{max}} \ge 0.44 \tag{3}$$

The destabilising effect of load height is taken where a load is applied at the top flange by increasing

the effective lengths by a factor of 1.2 in calculating the elastic buckling resistance. The stabilizing effect is conservatively treated using a factor of 1.0.

2.4 Cold-formed steel design codes

The three design codes discussed above are hot-rolled steel structural design codes. The current cold-formed steel design codes generally adopt the equivalent uniform moment and load height factors used in the hot-rolled steel design codes although there is limited research in this area. However, in general they use the equivalent uniform moment factors differently. Hot-rolled steel codes apply them to the member capacity (M_b) whereas cold-formed steel codes apply them to the elastic buckling moment (M_{cr}). The latter method is more conservative and does not take into account the important effect of localized yielding in inelastic buckling (Trahair, 1993). While the main reason is unknown, it could possibly be as a safety measure as these factors were originally developed for hot-rolled steel sections.

3 FINITE ELEMENT MODELLING

The finite element analysis (FEA) program ABAQUS was used to develop the nonlinear inelastic finite element (FE) model of LSB (Fig 3). This model is a modification of the earlier model developed by Mahaarachchi and Mahendran (2005a), which has been validated by experiments.

ABAQUS S4R5 shell element was selected to allow for buckling and yielding effects in LSB model. This element is a thin, shear flexible, isoparametric quadrilateral shell with four nodes and five degrees of freedom per node, utilizing reduced integration and bilinear interpolation schemes. The ABAQUS classical metal plasticity model was used to model the material non-linearity. An elastic perfect plasticity model based on a simplified bilinear stress-strain curve without strain hardening was used.

This study only included an "Idealised" simply supported boundary condition, that is both ends fixed against vertical deflection, out-of plane deflection and twist rotation, but unrestrained against in-plane rotation, minor axis rotation, warping displacement, and only one end fixed against longitudinal horizontal displacement. A system of Multiple Point Constraints (MPC) was used to achieve this boundary condition. This end support system and all the loads are applied at the cross section shear centre to eliminate any torsional loading effects.

Using the imperfection measurements of LSB's, Mahaarachchi and Mahendran (2005a) stated that the local plate imperfections of LSB are within the manufacturer's fabrication tolerance limits while the overall member imperfections are less than L/1000 (AS4100 fabrication tolerance). This initial imperfection of L/1000 was adopted in the modelling (but not less than 3 mm). The critical imperfection shape was introduced via ABAQUS *IMPERFECTION option with the lateral buckling eigenvector obtained from an elastic buckling analysis. The residual stresses were not considered in all the cases.

The effect of loading conditions on LSB was studied using two methods of analysis, elastic buckling and non-linear static analysis. The first is to determine the critical buckling moment and shape. This method is generally used for the development of currently used factors. The latter is to investigate the effects up to the ultimate moment. The solution was achieved using the Newton-Raphson method, in conjunction with modified RIKS method.

To include the effect of section geometry into the investigation, three LSB sections were chosen, LSB $125 \times 45 \times 2.0$, $250 \times 60 \times 2.0$ and $300 \times 75 \times 3.0$. Based on AS4100 rules, they are classified as compact, non-compact and slender sections, respectively. The beam lengths were also varied from intermediate to long spans to observe the relationship of lateral buckling modes (LDB vs. LTB) to the loading conditions. Three loading conditions were studied in this paper, beams subjected to moment gradient, mid-span point load and uniformly distributed load.

4 MOMENT DISTRIBUTION EFFECTS

The elastic buckling results of moment distribution factors (α_m) under moment gradient are summarised in Figure 4. It shows that the moment gradient increases the buckling strength with increasing end moment ratio (β).

It appears that α_m factors are influenced by the lateral buckling mode, ie. it varies with the length. For higher β (>-0.2), LSBs subjected to LDB mode have lower α_m factors compared to LTB mode. The reduction rate of α_m factor depends on the level of web

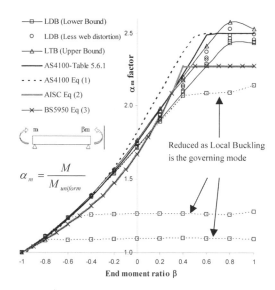

Figure 4. α_m factors for moment gradient cases based on elastic buckling analyses.

distortion in LDB, and thus a variation exists in the results where LSB with severe LDB mode gives the lower bound results. In contrast, this effect is almost negligible at lower β (<-0.2).

However, there is a concern that the moment gradient action may cause additional local buckling (LB) near the support in the case of intermediate beam slenderness (Fig. 4). Interaction of local buckling near the support and lateral buckling mode has a small effect on the α_m factor if the latter mode is governing. But with increasing β, the local buckling (particularly the web) may practically govern and reduces the α_m factor (Fig.4). The reduced α_m factors due to the governing LB mode are not considered here as the α_m factors are for lateral buckling.

A comparison of α_m results with the current design equations 1 to 3 appears to indicate that they do not provide accurate predictions, ie. conservative or unconservative. However, the AS4100 method based on Table 5.6.1 predicted closer results with the upper bound factors, which implies that it may only be applicable for LSBs subjected to LTB.

The α_m factors for transverse loading are shown in Table 1 and they are also unfavourably influenced by the level of web distortion of LDB. The reduction was found significant especially for slender and non-compact LSBs due to their high section slenderness. At lower spans, the web distortion may lead to lower α_m factors, but as the web distortion diminishes with increasing span, the α_m factor increases to that predicted by Table 5.6.1 of AS4100. This further implies its suitability for beams subjected to LTB. All the current factors are constant and independent of section

Table 1. α_m factors for transverse loading based on elastic buckling analyses.

Mid-span point load (PL)

Section	L m	Buckling Mode	α_m Factors				
			FEA	(a)	(b)	(c)	(d)
$125 \times 25 \times 2.0$	1.5	LDB	1.331	1.35	1.39	1.32	1.18
	2.5	LDB	1.336	1.35	1.39	1.32	1.18
	6	LTB	1.342	1.35	1.39	1.32	1.18
	10	LTB	1.345	1.35	1.39	1.32	1.18
$250 \times 60 \times 2.0$	2.5	LDB*	1.311	1.35	1.39	1.32	1.18
	4	LDB	1.339	1.35	1.39	1.32	1.18
	6	LDB	1.336	1.35	1.39	1.32	1.18
	10	LDB[1]	1.339	1.35	1.39	1.32	1.18
$300 \times 75 \times 3.0$	2.5	LDB*	1.268	1.35	1.39	1.32	1.18
	4	LDB	1.335	1.35	1.39	1.32	1.18
	6	LDB	1.338	1.35	1.39	1.32	1.18
	10	LDB[1]	1.339	1.35	1.39	1.32	1.18

Uniformly distributed load (UDL)

Section	L m	Buckling Mode	FEA	(a)	(b)	(c)	(d)
$125 \times 25 \times 2.0$	1.5	LDB	1.115	1.13	1.17	1.14	1.08
	2.5	LDB	1.123	1.13	1.17	1.14	1.08
	6	LTB	1.125	1.13	1.17	1.14	1.08
	10	LTB	1.125	1.13	1.17	1.14	1.08
$250 \times 60 \times 2.0$	2.5	LDB	1.087	1.13	1.17	1.14	1.08
	4	LDB	1.117	1.13	1.17	1.14	1.08
	6	LDB	1.121	1.13	1.17	1.14	1.08
	10	LDB[1]	1.122	1.13	1.17	1.14	1.08
$300 \times 75 \times 3.0$	2.5	LDB	1.054	1.13	1.17	1.14	1.08
	4	LDB	1.115	1.13	1.17	1.14	1.08
	6	LDB	1.121	1.13	1.17	1.14	1.08
	10	LDB[1]	1.123	1.13	1.17	1.14	1.08

(a) AS4100 Table 5.6.1 (b) Equation 1 - AS 4100
(c) Equation 2 - AISC (d) Equation 3 - BS5950
* Associated with a localised web buckling at midspan
[1] LDB with negligible web distortion (close to LTB mode)

slenderness and member length, showing its inadequacy for LSBs. However, BS5950 factors may be suitable as the lower bound.

The results from FEA have confirmed the limitation of the current α_m factors, and therefore they may not be safe or economical for LSBs. New factors may be developed using the results above, but the "real beam strength" may be different and thus it should be investigated. The nonlinear finite element analysis results were used to determine the strength ratio (M_{bnon}/M_b), where M_{bnon} and M_b are member moment capacities under non-uniform and uniform moment distributions, respectively. Tables 2 and 3 compares these moment capacity ratios with α_m factors from buckling analyses.

Table 2 demonstrates that applying the α_m factors directly to LDB's design curve (member capacity) may over-predict the strength ratios. Moment gradient is usually favourable because it ensures yielding only

at a short region closer to the support where the rest remains elastic. LSBs subjected to LDB mode under uniform moment is usually governed by uniform yielding on the compressive web (due to torsionally stiff compressive flange) and tension flange.

But moment gradient causes web yielding on the compression and tension sides (cross section yielding) near the support as opposed to the conventional I-beam with only flange yielding (Figure 5). This appears to limit the usual benefits of moment gradient for LSBs, particularly for lower beam slenderness. When a moment is applied at the end support, the monosymmetric shape causes asymmetric rotation of the two stiff flanges near the support. This is augmented by moment gradient (higher moment at one end), and while the flanges are very stiff, local distortion and yielding occur in the web near the support, causing premature failure. This local web distortion is identical to

Table 2. Comparison of M_{bnon}/M_b with α_m factors from elastic buckling analyses for moment gradient cases.

Section	L m	Failure mode	FE M_{bnon}/M_b β			FEα_m factors β		
			−0.4	0.0	0.6	−0.4	0.0	0.6
125 × 25 × 2.0	2.5	LDB	*1.02**	*1.04**	*1.04**	1.38	1.72	2.27
	6	LTB	1.24[+]	*1.30**	*1.32**	1.39	1.75	2.37
	10	LTB	1.29[+]	1.51[+]	1.74[+]	1.39	1.76	2.41
250 × 60 × 2.0	6	LDB	1.30[+]	1.42[+]	N/A	1.38	1.71	N/A
	10	LDB[1]	1.33[+]	1.59[+]	1.89[+]	1.38	1.74	2.32
300 × 75 × 3.0	4	LDB	*1.15**	*1.18**	N/A	1.38	1.73	N/A
	6	LDB	1.25[+]	*1.32*[+]	*1.35**	1.38	1.72	2.28
	10	LDB[1]	1.34[+]	1.58[+]	1.77[+]	1.38	1.74	2.33

* Severe web yielding near one of the supports [+] Less severe web yielding near one of the supports N/A Neglected due to interaction of LDB + LB modes [1] LDB with negligible web distortion (close to LTB mode)

Table 3. Comparison of M_{bnon}/M_b from FEA and LSB's member capacity equation for moment gradient cases.

Section	Length m	FE M_{bnon}/M_b β			M_{bnon}/M_b of LSB Eq. β		
		−0.4	0.0	0.6	−0.4	0.0	0.6
125 × 25 × 2.0	2.5	1.02	1.04	1.04	*1.18**	*1.31**	*1.51**
	6	1.24	1.30	1.32	1.18	*1.32**	*1.44**
	10	1.29	1.51	1.74	1.24	1.40	1.64
250 × 60 × 2.0	6	1.30	1.42	N/A	1.18	1.31	N/A
	10	1.33	1.59	1.89	1.37	1.54	1.78
300 × 75 × 3.0	4	1.15	1.18	N/A	*1.18**	*1.32**	N/A
	6	1.25	1.32	1.35	1.18	1.31	*1.51**
	10	1.34	1.58	1.77	1.21	1.36	1.58

* Overestimate the actual strength ratios

the small web local buckling (interaction buckling) in the elastic buckling analysis. As discussed previously it has negligible effect, but it seems that yielding in the inelastic region exaggerates the effect. The effect is more severe for less slender section and shorter beam as the resulting rigidity may attract higher stress to the flanges near the support. One such example is 2.5 m LSB125x 25x2.0 which has severe effects in both uniform moment ($\beta = -1$) and moment gradient cases, and hence only a small benefit is gained (Fig.5).

Table 3 compares the moment capacity ratios from the FEA results with LSB's member capacity equation (M_b) which uses a modified elastic buckling moment based on the α_m factors from FEA buckling analyses. It shows that the use of cold-formed steel code's design approach is inadequate for some cases. It appears that LSBs require a more complex method to allow for the moment gradient benefits.

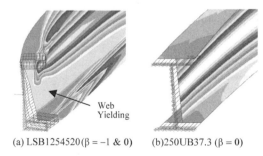

(a) LSB1254520 ($\beta = -1$ & 0)   (b)250UB37.3 ($\beta = 0$)

Figure 5. Yielding distribution (non-linear analysis) at failure.

These complex and contradicting results may also place the modelling accuracy in question, indicating that experimental evidence is important to confirm the findings of this paper. Suitable design rules for

Table 4. Comparison of M_{bnon}/M_b with α_m factors from elastic buckling analyses for transverse loading cases.

Section	L (m)	Failure mode	FE M_{bnon}/M_b		FE α_m factors	
			PL	UDL	PL	UDL
$125 \times 25 \times 2.0$	2.5	LDB	1.62	1.58	1.336	1.123
	6	LTB	1.36	1.17	1.342	1.125
	10	LTB	1.30	1.12	1.345	1.125
$250 \times 60 \times 2.0$	2.5	LDB	*N/A*	1.46	1.311	1.087
	4	LDB	1.48	1.37	1.339	1.117
	6	LDB	1.37	1.19	1.336	1.121
	10	LDB[1]	1.29	1.12	1.339	1.122
$300 \times 75 \times 3.0$	2.5	LDB	*N/A*	1.31	1.268	1.054
	4	LDB	1.52	1.36	1.335	1.115
	6	LDB	1.43	1.25	1.388	1.121
	10	LDB[1]	1.35	1.15	1.339	1.123

[1] LDB with negligible web distortion (close to LTB mode)

moment gradient effects in LSBs will be developed once experimental results are available.

While the level of web distortion of LDB unfavourably affects the buckling strength of LSBs under transverse loading, non-linear analyses indicate the opposite trend for ultimate strengths as shown in Table 4. LDB mode appears to be more beneficial with transverse loading because of the favourable yielding effect. In the uniform moment case, a combination of compressive web and tension flange yielding reduces the strength substantially below elastic buckling. But the transverse loading provides twofold benefits of confining those yielding on both compressive web and tension flange to shorter region (naturally less for UDL case), thus results in less strength reduction below the elastic buckling as shown in Figure 6 (note that high slenderness beam may have post buckling strength). Hence the yielding effect in the inelastic region is also important for transverse loading cases.

The non-linear analysis results indicate that for lower beam slenderness, higher α_m factors than that reported from elastic buckling analysis can be safely used. Therefore, new factors are proposed (slightly less than the current factors) for use with the LSB's member capacity (M_b) equation (adopting the hot-rolled steel code approach).

$$M_{bnon} = \alpha_m M_b \leq M_y$$
Where; α_m = 1.30 for mid-span point load
$\quad\quad\quad \alpha_m$ = 1.12 for uniformly distributed load $\quad$ (4)
$\quad\quad\quad M_y$ = Section moment capacity of LSB

This proposed method deliberately neglects the actual variation due to buckling mode (Fig. 6) as it is convenient for practical design and also conservative for lower beam slenderness. Based on the moment

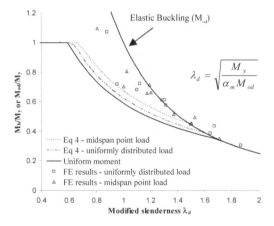

Figure 6. Application of Equation 4 with LSB's design curve.

gradient findings, experimental evidence is required to justify this design recommendation.

5 LOAD HEIGHT EFFECTS

The FEA elastic buckling results for loading above (TF) and below (BF) the shear centre are summarised in Figure 7. As expected, the destabilising effect of loading above the shear centre increases the twisting and reduces the buckling resistance, while for the the loading below the shear centre is always beneficial. The effect is more important for LSBs with higher modified torsion parameter ($K_e = \sqrt{(\pi^2 EI_y h^2 / 4GJ_e L^2)}$), indicating that it is dependent on the level of web distortion (a function of beam slenderness) as for moment distribution

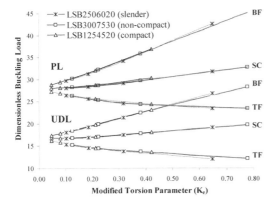

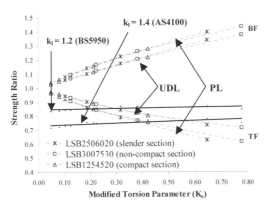

Figure 7. Effects of loading above and below the shear centre based on elastic buckling analyses. The dimensionless buckling load is $QL^2/\sqrt{(EI_y GJ_e)}$ for PL and $QL^3/\sqrt{(EI_y GJ_e)}$ for UDL.

effects. The load height effect for the PL case was found to be slightly greater than that for the UDL case which indicates that other transverse load types may be worthwhile considering in this study.

A comparison with the current methods is presented as a buckling moment ratio for loading at the shear centre, but with loads at the middle and top and bottom flange levels (Fig.7). Using the higher load height factor (k_l) of 1.4, AS4100 prediction for loading above the shear centre (top flange loading) is more conservative than BS5950. But, they are not in good prediction to represent the actual load height effect variation and often too conservative for lower torsion parameter and unconservative for the opposite case. Loading below the shear centre effect is always ignored in the current design methods, demonstrating that it can be used safely for LSB with some conservatism.

Previous discussion has shown the importance of inelastic behaviour on the moment distribution effects. This also highlights the need for non-linear analysis for load height effects. The non-linear analysis is currently

being undertaken and its results including suitable design methods will be presented at the conference.

6 CONCLUSIONS AND RECOMMENDATIONS

This paper has described a series of finite element analyses undertaken to investigate the moment distribution and load height effects for a new cold-formed section known as LiteSteel Beam. Three loading types were studied, moment gradient, mid-span point load and uniformly distributed load. The results show that the effects of non-uniform moment distribution and load height are dependent on the level of web distortion of LDB mode. This effect is more significant than for LSBs subjected to LTB mode but it diminishes with increasing beam slenderness (less web distortion).

The currently used α_m factors in AS4100, ANSI/AISC 360 and BS950 were found to be of limited use to LSBs subject to LDB mode. For moment gradient case, new factors have not been developed yet because the unfavourable yielding effect in the inelastic region appears to limit the benefits of moment gradient. Experimental works are recommended to confirm the findings before developing new factors. However, for the transverse load case, new factors are proposed here including the recommendation for its application to design.

The current methods (AS4100 and BS950) for load height effects are inadequate for LSBs with loading above the shear centre. The current study on the load height effect only presents the elastic buckling analysis results. A new design method will be attempted once the non-linear analysis results are available. As in the current design methods, it is recommended to ignore the favourable effect of loading below the shear centre for LSB design. This paper has reported the work completed to date. More results will be presented at the conference.

REFERENCES

ABAQUS. 2004. *ABAQUS Online Documentation Version 6.5*. ABAQUS, Inc.
American Standard. ANSI/AISC 360–05. 2005. *Specification for Structural Buildings*. Chicago: American Institute of Steel Construction.
Australian Standard. AS4100. 1998. *Steel Structures*. Sydney: Standards Australia.
Australian/New Zealand Standard. AS/NZS4600. 2005. *Cold Formed Steel Structures*. Sydney: Standards Australia.
British Standard. BS5950-1. 2000. *Structural Use of Steelwork in Buildings – Part1*. British Standards Institution.
Nethercot, D.A. 1975. Inelastic buckling of steel beams under non-uniform moment. *The structural Engineer* 53(2): 73–78.

Kirby, P. A., and Nethercot, D. A. 1985. *Design for Structural Stability*. London: William Collins Sons & Co Ltd.

Kitipornchai, S., and Trahair, N. S. 1975. Buckling of inelastic I-beams under moment gradient. *ASCE Journal of Structural Division* 101(5): 991–1004.

Mahaarachchi, D., and Mahendran, M. 2005a. *Finite element Analysis of LiteSteel Beam Sections, Research Report No.3*. Brisbane: Physical Infrastructure Centre, Queensland University of Technology.

Mahaarachchi, D., and Mahendran, M. 2005b. *Moment Capacity and Design of LiteSteel Beam Sections, Research Report No.4*. Brisbane: Physical Infrastructure Centre, Queensland University of Technology.

Pi, Y.-L., and Trahair, N. S. 1997. Lateral-distortional buckling of Hollow Flange Beams. *Journal of Structural Engineering* 123(6): 695–702.

Publication. 2005. *Design Capacity Tables for LiteSteel Beam*. Australia: Smorgon Steel Tube Mills Pty Ltd.

Trahair, N. S. 1993. Flexural-Torsional Buckling of Structures, 1st Edition. London: Chapman & Hall.

Steel and Composite Structures – Wang & Choi (eds)
© 2007 Taylor & Francis Group, London, ISBN 978-0-415-45141-3

Compressive behaviour of cold-formed thin-walled steel studs with perforated web at ambient temperature and in fire

B. Salhab & Y.C. Wang

School of MACE, The University of Manchester, Manchester, UK

ABSTRACT: Perforated channels are made by slotting the web of solid channels and they are used in the cold environment to reduce heat loss and the risk of thermal bridging. Introducing perforations not only weakens the structure, but also affects its fire performance under fire conditions, these perforations may prevent the fast spreading of temperature, thus prolong the period till failure. On the other hand, perforations in a steel section will reduce its capacity to resist loads. This paper presents the results of an experimental and numerical study to investigate the thermal and structural performance of small size specimen and the structural performance of a full scale panel under compression in ambient temperature and in fire as well as introducing a new method for calculating a reduced thickness for the perforated web where a solid web with an equivalent thickness can substitute the original perforated web. Thus, all design method can be applicable to the new equivalent solid section.

1 INTRODUCTION

Cold-formed thin-walled steel sections are increasingly used in residential, industrial and commercial buildings in many parts of the world as primary members as well as secondary members of building structures, e.g. as purlins to support the roof and cladding of a building.

Whilst slotting studs will improve thermal insulation and reducing the effect of thermal bridging, it will also reduce the structural resistance of the studs.

This paper presents a summary of the results of an experimental and numerical investigation (Salhab 2007) of structural and fire behaviour of thermal studs. The experimental work includes fire tests on unloaded small panels (300 × 300 mm) to obtain temperature distributions in the thermal studs, one load test of a large panel (2 × 2 m), and one fire test on a loaded large panel (2 × 2 m). The numerical simulations were carried out using the general finite element package ABAQUS.

The main emphasis of this research is to develop a new method in which the equivalent thickness of a solid plate would give the same elastic local buckling strength under compression as a perforated plate. ABAQUS simulations have been performed to check the applicability to the entire cross-section. For thermal studs under compression, the equivalent thickness equation has been found to be applicable over a wide range of values for different design parameters under ambient temperature and fire conditions.

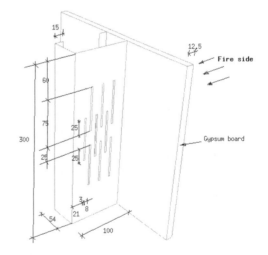

Figure 1. Dimensions of test specimens.

2 RESULTS & ANALYSES OF STRUCTURAL AND FIRE TESTS ON PANELS MADE OF PERFORATED CHANNELS

2.1 *Fire tests on unloaded small panels*

Four fire tests were carried out on small panels (300 × 300 mm) in the fire-testing laboratory at The University of Manchester. The test specimens were unloaded and the objective of these tests was to obtain temperature distribution in the panels. The

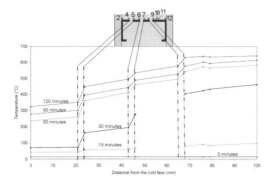

Figure 2. Temperature distributions in 1.2 mm section with 1-layer of gypsum board.

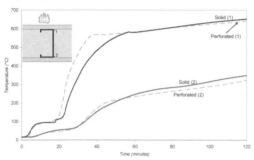

Figure 3. Temperature vs. Time at point 1(the hot web), point 2 (the cold web).

specimens were made of perforated cold-formed lipped channel sections. Two tests used channel section $100 \times 54 \times 15 \times 1.2$ mm and the other two used $100 \times 54 \times 15 \times 2$ mm. Figure 1 gives dimensions of the channel and shows locations and sizes of the perforations.

Isowool Spacesaver 1000 mineral wool manufactured by British Gypsum Limited was used as internal insulation in all tests. The gypsum boards were 300×300 mm FireLine Gyproc with a thickness of 12.5 mm (also manufactured by British Gypsum Limited) and were attached to both sides of the test sections by bolts in the middle.

Fire exposure was from one side and the fire temperature – time curve followed the standard one (ISO 834). Figure 2 shows typical temperature distributions throughout the perforated steel channel at different times. It can be seen that perforations can substantially reduce steel temperatures. Nevertheless, due to quick heat conduction of the solid steel, figure 3 shows that the flanges of the perforated section attained very similar temperatures as those of the solid section.

2.2 Ambient temperature and fire structural tests

The panel was 2 m high and 2.2 m width and used three $100 \times 54 \times 15 \times 1.2$ mm lipped cold-formed

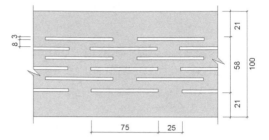

Figure 4. Perforation dimensions (mm).

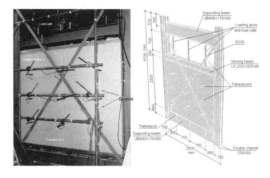

Figure 5. Test panel and displacement transducers.

steel channels. FireLine gypsum boards with a thickness of 12.5 mm and manufactured by British Gypsum Limited were attached to both sides of the perforated channels by screws at 300 mm spacing. The specimen consisted of 3 cold-formed lipped channels with perforations along the entire web length.

Figure 4 shows one perforated channel with dimensions and figure 5 shows one test specimen in the reaction frame before testing.

In a previous project by Feng (2004), the same set up was used but solid sections were used instead of perforated sections. The ambient temperature failure load of the author's specimen with perforated sections is about 5.5% lower than that of Feng's solid section specimen, the failure load in one channel section was 54.32 kN and 57.5 kN respectively. The failure mode of the three channels in the test specimen was varied, being distortional and global flexural buckling. This was attributed to different initial imperfections. The authors' test channel behaviour was modelled using ABAQUS (HKS 2003), the results of which indicate that by using different initial imperfections, it was possible to obtain the various observed failure modes. However, there was very little difference in the simulated failure loads.

Figure 6 shows a comparison between the ambient temperature test results and simulation for axial displacement using nominal maximum global imperfection of $L/1000$ and distortional imperfection of

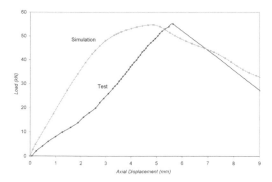

Figure 6.   Load-axial deformation relationship.

$L/1000$. It can be seen that ABAQUS can be used to accurately simulate the test column behaviour.

The structural fire test used exactly the same setup as the ambient temperature structural test. For the fire test, the reaction frame containing the test specimen was placed in front of the large fire test furnace in the Fire Testing Laboratory of the University of Manchester for fire exposure on the test specimen from one side. Again, the standard fire exposure condition was followed. After the applied load in each jack had reached 23 kN, the furnace was ignited. After 21.17 minutes of the fire test, the applied load on the left channel could not be maintained and the panel may be considered to have failed. The corresponding failure time for the solid panel of Feng was 25.5 minutes. Thus perforating the web reduced the panel fire resistance by 17%.

After the fire test, it was found that the gypsum plasterboard on the fire exposed side suffered severe damage as shown in figure 7. However, it was difficult to determine the precise time and extent of damage during the fire exposure. Therefore, four sets of ABAQUS simulations were conducted to represent detachment of gypsum plasterboard from the steel sections at different times:

(1) detachment of gypsum plasterboard from the beginning of the fire test;
(2) detachment of gypsum plasterboard at 10 minutes of fire exposure;
(3) detachment of gypsum plasterboard at 20 minutes of fire exposure;
(4) gypsum plasterboard remained effective throughout the fire test.

Detachment of gypsum plasterboard was simulated by removing the lateral restraints presented by the gypsum plasterboard to the steel section at the interconnections.

Figure 8 compares the four simulation results with the fire test results. The vertical axis represents the axial displacement of the column during fire exposure and the horizontal axis represents fire exposure time.

Figure 7.   Gypsum board behaviour.

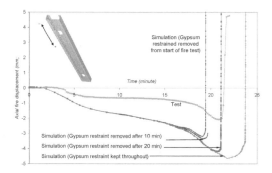

Figure 8.   Comparison between test and simulation results for perforated section, gypsum plasterboard restraining effect removed at different times of simulation, axial deformation.

As expected, the test failure time is between the two extreme cases (cases 1 and 4), indicating it is not appropriate to either assume the gypsum plasterboard being removed at the beginning of the fire test nor being kept throughout the fire test. The fire test failure time is closest to the simulation failure time of case (3) in which the gypsum plasterboard was assumed to have detached from the steel section at 20 minutes. Under this circumstance, since the assumed time of gypsum plasterboard detachment is very close to the actual fire failure time, it may be argued that because the structure was approaching failure, the deformation in the structure was accelerating, which caused the gypsum plasterboard to detach. Therefore, in order to accurately determine the time of gypsum detachment, one approach could be to include the gypsum plasterboard in the structural analysis. However, this could make the analysis extremely difficult to perform. At present, it is not possible to specify a simple and reliable approach to deal with gypsum plasterboard detachment and a research project has recently been started to resolve this issue.

Perforations have an important influence on the structural behaviour, failure modes and failure time of the panel under exposure to fire. The thermal and structural behaviour of a panel with perforated channels in fire are complex and exact treatment of the two aspects (thermal and structural behaviour) will be difficult.

Section 4 will investigate how to best approximate this type of behaviour for practical engineering applications.

3 EQUIVALENT THICKNESS

3.1 Assumptions and main parameters

Using a solid web with reduced thickness to represent a perforated web is extremely attractive in dealing with the structural behaviour of a perforated thermal stud.

In order to eliminate or reduce complications in the solution process it is necessary to make some simplifying assumptions.

Different buckling modes play different important roles in the behaviour of thin-walled structures. Strictly speaking, it would be necessary to use different equivalent thicknesses to account for the effects of the following possible different buckling modes of a thin-walled web plate as part of an open channel section: flexural bending about the major axis, overall torsional buckling, distorsional buckling, and local buckling. Global buckling about the minor axis can be ignored because this is unlikely to occur in the construction system that is concerned in this study owing to restraints of the plasterboards. Since local buckling, global torsional and distorsional buckling modes of the cross-section depend largely on the transverse bending stiffness of the perforated web, the equivalent thickness for these three modes of buckling should be similar. In addition, global flexural buckling is controlled by the second moment of area of the cross-section about the major axis, to which the web (whether perforated or not) will contribute very little. In other words, the global flexural buckling resistance will be insensitive to the equivalent thickness of the web. Therefore, the simulations in this paper will only deal with one mode of plate behaviour being the local plate buckling under compression. Consequently, to enable a large number of simulations to be performed, it is possible to analyse only the web plate of the thermal stud. Thus, the equivalent thickness is defined as that of a solid plate which would give the same elastic local buckling load as the perforated plate under compression.

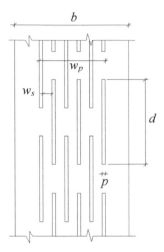

Figure 9. Parameters.

Figure 9 shows the parameters used in this study where: w_p is the width of the perforation region; b is the plate width; t is the plate thickness; p is the width of a single perforation; d is the length of a single perforation; pb is the sum of single perforation widths in a section containing all perforations lines $p_b = \sum_{i=1}^{n} p_i = p \cdot n$; and w_s is the solid width between every two adjacent perforation lines.

To obtain a valid equivalent thickness formulation for general use, extensive numerical simulations are performed to cover a range of values for the following different parameters: plate width, length, thickness, boundary conditions, stress distribution, and patterns of perforation. The simulation studies were carried out by varying one variable at a time and keeping others constant at the basic values. The sensitivity of the equivalent thickness to these different parameters is then assessed to identify the most important variables, which are: t/b, w_p, and p_b.

Regression analysis is then performed to obtain general expressions to calculate the equivalent thickness of perforated plates with different width, thickness and patterns of perforation.

By using a reduced equivalent thickness, a perforated plate is replaced by a solid plate with the reduced equivalent thickness.

3.2 Results of regression analysis

After an extensive regression analysis, it has been found that the equation defining the equivalent thickness ratio can be given in two ranges:

$$Y = (\frac{P_b}{b})^2 (\frac{w_s}{w_p})^{1/2} \tag{1}$$

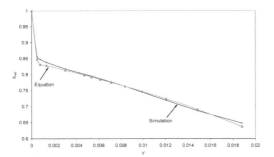

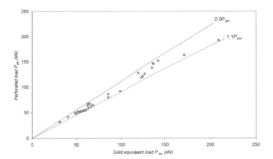

Figure 10. Comparison between regression equation and numerical results for $t/b = 0.01$.

Figure 11. Comparison of all results between using perforated section and equivalent solid section.

For $Y \geq 0.00133$

$$k_{red} = \frac{t_{eq}}{t} = \sqrt[3]{c_1 + c_2 Y} \qquad (2a)$$

where $c_1 = 0.608 - 1.7\frac{t}{b}$ and $c_2 = 70.256\frac{t}{b} - 18.33$

For Y<0.00133

$$k_{red} = \frac{t_{eq}}{t} = \sqrt[3]{1 + c_1 Y + c_2 \sqrt{Y}} \qquad (2b)$$

where

$c_1 = 247.13 - 3057.75\frac{t}{b} + 2025.11(\frac{t}{b})^{1/2}$ and

$c_2 = -19.48 + 94.5\frac{t}{b} - 84.375(\frac{t}{b})^{1/2}$

Figure 10 compares the results of the regression equation and finite element simulations, indicating close agreement between the two sets of results.

4 APPLICATIONS OF THE NEW METHOD ON COMPLETE SECTIONS

This section presents the results of numerical simulations to confirm that the equivalent thickness method can also be applied to analyze the ultimate strength of thermal studs at ambient temperature. To apply the equivalent thickness method to thermal studs, the perforated web of a thermal stud is replaced by a solid web with the reduced equivalent thickness.

4.1 Ambient temperature

In order to ensure that the newly developed equivalent thickness method presented in the previous section has the widest range of applicability, the validation study is performed for thermal studs with a range of values of flange width, web depth, thickness, column length, axial load eccentricity, and perforation pattern.

Figure 11 compares all the simulation results using perforated sections with those using the corresponding equivalent solid sections. It can be seen that the maximum difference is less than 10%, indicating very good accuracy of the equivalent thickness method.

In summary, the results of this numerical study confirm that the equivalent thickness method based on local buckling of a perforated plate under compression can be used to predict the ultimate strengths of compression members using channel cross-sections with perforated webs with good accuracy for a very wide range of design parameters.

4.2 Elevated temperatures

For simplicity of ABAQUS simulation, instead of calculating the column fire resistance at fixed applied load, the column ultimate strength will be calculated at elevated temperatures corresponding to standard fire exposure times of 15, 30, 45 and 60 minutes, which are the realistic design fire resistance times for the type of structure under consideration.

When a perforated section is exposed to fire from one side, the temperature distribution in the structure is non-uniform in 3 dimensions. This makes the problem extremely difficult to deal with in design calculations. Therefore, it is necessary to simplify the temperature profile. For a perforated section exposed to fire, there are a number of possible simplification procedures:

(1) To represent the 3-D temperature profile by a 2-D temperature profile based on a representative section of the structure. After calculating the 3-D temperature distributions, the temperature of the representative cross-section will be applied to the entire length of the structure.

(2) To simplify the 2-D temperature profile according to Feng (2004). In a previous study by Feng on fire resistance of solid channel sections, it has been established that the 2-D temperature profile of a solid channel section may be represented by a simple temperature profile as shown in figure 12,

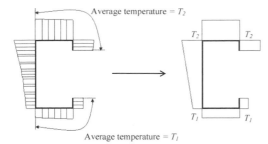

Average temperature = T_2

Average temperature = T_1

Figure 12. Simplified temperature profile.

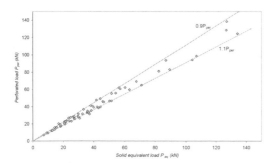

Figure 13. Comparison of all results between using perforated section and equivalent solid section.

where the temperature is linear on the web and constant on the lips and flanges.

In figure 12, $T2$ is the hot face temperature, obtained as the average of temperatures of the hot side flange and lip and $T1$ is the cold face temperature which is obtained by averaging the flange and lip temperatures on the cold side.

(3) In the previous two methods, the temperature analysis should still be carried out for the perforated section. In a further simplification, the 2-D temperature profile may be calculated using the solid section with the original thickness or the solid section with the equivalent thickness in the web.

The simplest simplification for predicting fire resistance of perforated steel studs in compression is to use the equivalent solid section in both thermal and structural simulations, and to further simplify the temperature distribution in the cross-section according to figure 12.

Figure 13 compares the calculated column strengths using perforated section and using the equivalent solid web thickness for all the results changing various parameters, including different cross-section depth, thickness, column height, perforation pattern, and flange width.

Considering all different parameters at elevated temperatures, the difference between the two sets of results is less than 10%.

5 CONCLUSIONS

The following conclusions may be drawn:

– For both the small scale and large scale fire tests, perforating the webs can reduce steel temperatures on the unexposed surface of the panels, thereby increasing temperature gradients in the steel cross-section.
– The effect of gypsum plasterboard failure on the structural behaviour of the perforated channel structure may be simulated by removing its restraints to the structure.
– The equivalent thickness method, based on elastic local buckling analysis of perforated and solid web plates, is applicable to entire sections for compression members at ambient temperature.
– The equivalent thickness method can also be applied for elevated temperature analysis under fire exposure. Once a perforated section is converted into an equivalent solid section with reduced web thickness, the equivalent section can be used for both thermal analysis and structural analysis. In addition, it is acceptable to use the temperature simplification method proposed by Feng (2004) for solid sections.
– By using the equivalent thickness method, the perforated web can be transformed to a solid web with reduced thickness afterwards existing design methods can be used on the equivalent solid section.

REFERENCES

British Standrad 1999. Fire resistance tests - Part 1: General requirements. EN 1363-1:1999. In London, U.K.
Davies, J. M., Leach, P. & Taylor, A. 1997. The design of perforated cold-formed steel sections subjected to axial load and bending. Thin-Walled Structures, Vol. 29. Issues 1–4, pp.141–157.
Eurocode 3. 2001. Design of steel structures, Part 1.2: General rules – structural fire design, ENV 1993-1-2:2001
Eurocode 3. 2001. Design of steel structures, Part 1.3: General rules – Supplementary rules for cold-formed thin gauge members and sheeting, ENV 1993-1-3:2001
Feng, M. 2004, Numerical and experimental studeies of cold-formed thin-walled steel studs in fire, PhD thesis, the University of Manchester, U.K.
HKS 2003. ABAQUS Standard User's Manual. Volumes I-II-III Version 6.4, Habbit, Karlsson & Sorenson Inc.
ISO 834. 1975. Fire resistance tests – Elements of building constructions. International Standard ISO 834, Geneva.
Kesti, J. 2000. Local and distortional buckling of perforated steel wall studs. PhD Thesis, Helsinki University of Technology, Finland.
Salhab, B. 2007, Behaviour of Cold-Formed Thin-Walled Steel Studs with Perforated Web at Ambient Temperature and in Fire, PhD thesis, the University of Manchester, U.K.

Steel and Composite Structures – Wang & Choi (eds)
© 2007 Taylor & Francis Group, London, ISBN 978-0-415-45141-3

Thermal performance and axial compressive behavior of slotted C shape cold-formed steel stud walls

H. Yang & S.M. Zhang
Harbin Institute of Technology, Harbin, China

B. Han
Shanghai Municipal Engineering Design General Institute, Shanghai, China

ABSTRACT: Slots in the studs' web improve the thermal performance of cold-formed steel stud walls, while they weaken the cross-sectional integrality, distortional resistance and ultimate bearing capacity of the studs. Therefore, the thermal and mechanical behaviour of the cold-formed steel studs should be studied together. The slotted C shape cold-formed steel stud walls are studied in this paper, not only on the thermal performance but also their failure modes, ultimate bearing capacities of the studs under axial compressive forces. The finite element analysis models are developed using ANSYS, the sheathing effect by gypsum wallboards is taken into account in the model. Then parametrical analysis is carried out on the U-value of the walls and axial strength of the studs. It is found that for the thermal performance, row number and length of slots are the important parameters; while for the axial compressive strength, row number of slots is the important one in the given value ranges for the slot parameters.

1 INTRODUCTION

Light gauge steel frame building systems are becoming more popular in commercial, industrial and residential constructions. In the light steel framed buildings, the thin-walled steel structural elements are incorporated into wall, floor and roof panels in which plasterboard sheathing is fastened to cold-formed thin-walled steel channels (Fig. 1). To reduce the thermal bridge effect, the slotted steel stud is developed in some North European countries (Hoglund,

and Burstrand, 1998). By slotting the web the thermal bridges are significantly reduced, however, its mechanical behaviour is affected by the slot setting. Therefore it is necessary to study both the thermal performance including thermal bridge, temperature distribution and U-value of the walls, and the mechanical behaviour including buckling modes, ultimate bearing capacities of the studs.

2 THERMAL PERFORMANCE

2.1 Finite element analysis model

The finite element analysis package ANSYS is used to predict the thermal performance of the slotted cold-formed steel stud walls. A model is developed to calculate the two-dimensional steady stage temperature distribution and heat transfer in the walls with the following hypotheses: indoor and outdoor temperatures remain constant; heat transferring manner is simplified to be convection; contract thermal resistances between different materials are neglected; thermal contribution of the polyethylene film is neglected.

A unit of 600 mm width is selected from the wall, shown in figure 2. Heat insulating boundaries are applied on the two ends of the unit. According to the dimensions and calculation demands, Shell57 and Solid 70 are respectively employed for steel stud and

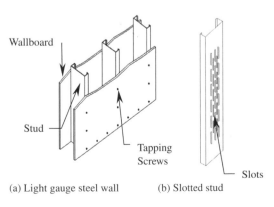

Wallboard

Stud

Tapping Screws

Slots

(a) Light gauge steel wall   (b) Slotted stud

Figure 1. Sketch of light gauge steel wall.

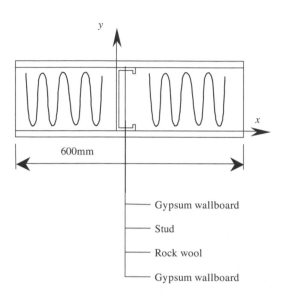

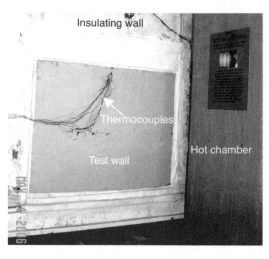

Figure 2. Analysis unit of the slotted panel.

— Gypsum wallboard

— Stud

— Rock wool

— Gypsum wallboard

Figure 3. Experiment setup before heat transfer test.

Table 1. Thermal conductivities and heat transfer coefficients.

Material	Thermal conductivity W/(m·K)
Steel stud	58.2
Gypsum wallboard	0.33
Rock wool	0.05
EPS board	0.042

Location	Heat-transfer coefficient W/(m²·K)
Cold side	23.0
Room side	8.7

the other components including gypsum wallboard, rock wool and expanded polystyrene (EPS) board if existing. The material thermal conductivities and heat transfer coefficients, according to the Thermal Design Code Civil Building (GB50176-93) are shown in table 1.

To test the developed model and study the thermal improvement of the walls due to slots, an experimental investigation was carried out by Yin (2006). One type of C-shaped lipped studs, $205 \times 40 \times 15 \times 1.5$ mm, was used and its average measured dimension was $205.1 \times 39.5 \times 15.0 \times 1.5$ mm. Each stud has five rows of narrow slots which is die cut by the numeric control machine. 19 thermocouples and 21 thermocouples were respectively embedded in wall I and wall II. To measure the thermal blocking of slots, about one third thermocouples were arranged along the middle steel stud. And some thermocouples were placed

away from the stud to compare the heat transfer of the quartered cross-section.

2.2 Verification of numerical analysis

An evaluation of the experimental and numerical thermal analysis is done by comparing the measured and predicted temperature and overall heat transfer coefficients of the walls, given in figure 4. Further evidence for the comparison can also be found in it, and the test was carried out by C. Barbour & J. Goodrow (1995) on a wall without slots. From these comparisons, it can be judged that a satisfactory finite element analysis model is developed in this paper and capable to predict the thermal performance of both slotted and non-slotted walls. However, it should be point out that there is a general temperature difference especially in the boundaries between the stud flanges and gypsum wallboards. It is mainly caused by ignoring the contract thermal resistance between the different materials.

2.3 Parametrical analysis

To study the thermal performance of the slotted walls, a parametrical analysis is carried out. The concerned parameters are row number of slots n, slot length lu, slot width lv, longitudinal distance of slots du, vertical distance of slots dv. If there are no special declarations, the values of parameters are shown in table 2, where h, b, a, t is the web height, flange width, lip height and stud thickness respectively, and t_p is the thickness of gypsum wallboard.

Figure 5 shows the U-value of the cold-formed steel stud walls with different slot parameters. It is found that the wall with more slot rows in the stud web get less U-value. For example, the U-value of the walls

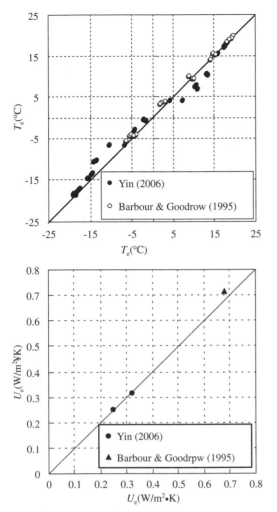

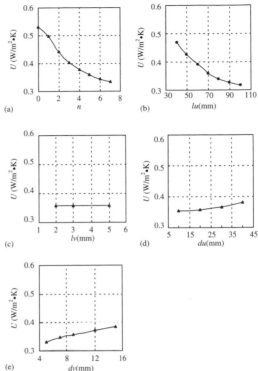

Figure 5. U-value of the overall walls with different slot parameters: (a) slot width, (b) slot length, (c) slot width, (d) longitudinal distance of slots, (e) vertical distance of slots.

distance increasing respectively. Among the five slot parameters, the slot row number and length affect the thermal performance of the overall walls significantly, while the other parameters play little role on it within their set values.

Figure 4. Measured and calculated results: (a) temperatures and (b) heat transfer coefficients of the overall panels.

Table 2. Physical dimensions of the parameters.

Parameter	Dimension mm	Parameter	Dimension mm
h	205	lu	70
b	40	lv	3
a	15	du	20
t	1.5	dv	9
t_p	12	n	5 (No unit)

with 7-row slots is only about 60% of that without slots. But this degressive trend by slot row number becomes gentle when n is large enough. The U-value decreases if slot length increases, and it increases gently as slot width, longitudinal distance and vertical

3 AXIAL COMPRESSIVE BEHAVIOUR

3.1 Mechanical analysis model

ANSYS is used to predict the axial compressive behaviour of C shape cold-formed steel studs in this paper. Due to the symmetry, only half longitudinal length of stud is analyzed. Element shell181, being capable of being used in the plasticity and large deflection analysis, is used to for the C-shape steel studs. A dual-linear stress-strain relationship model is employed for the steel studs. 1/1000 of deformation of first order buckling model is applied to create the initial geometric imperfections in the studs. As the two flanges of studs are usually braced laterally by sheets of plasterboard which restrains the studs from overall buckling greatly, the studs are supposed to be

965

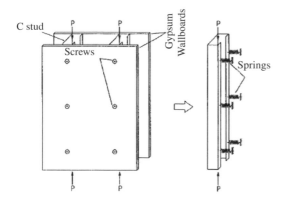

Figure 6. Simplified mechanical analysis model.

Table 3. Axial strength of studs with different spring rigidities.

No.	$h \times b \times a \times t$ mm	L mm	d mm	k N/mm	N_u kN
S3	$205 \times 40 \times 15 \times 1.5$	3000	300	80	52.7
				120	52.8
				160	52.9
				200	52.9

L is the effective length of stud. d is the fastener distance. N_u is the ultimate bearing capacity of stud.

fixed with a series of lateral springs at the right site of fasteners, shown in figure 6.

3.2 Restraining effect

Miller and Pekoz (1994) carried out a series of wallboard-fastener connection tests to evaluate the restraining effect from gypsum wallboards. It is found that spring rigidity k is varied from 80 N/mm to 200 N/mm which is depended on the quality, thickness of the sheathing materials and even depended on the ambient humidity. To discuss the restraining effect on the studs, the ultimate bearing capacities with different values of k are predicted in Table 3. It is found that the capacities are not sensitive to k if its value is varied from 80 N/mm to 200 N/mm. Therefore its lower limit value, 80 N/mm, is taken as the spring rigidities in the following studies.

Figure 7 shows the deformation and stress distribution of the C-shape studs marked as S3. It is found that, for the former stud without restraint by the wallboards, overall buckling develops clearly and therefore the stress level is comparative low. While overall buckling in the latter stud is restrained due to the wallboards, and stress in many points is close to or even larger than yielding stress ($f_y = 235$ MPa). Stress redistribution

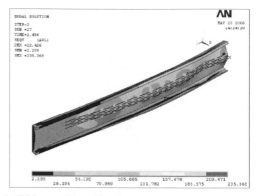

(a) Without wallboard bracing

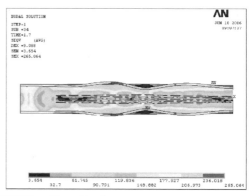

(b) With wallboard bracing

Figure 7. Deformation and Mises stress of S3.

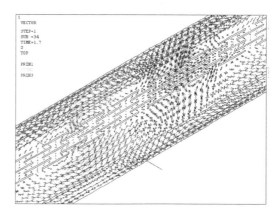

Figure 8. Principal stress of S3 with wallboard bracing.

occurs in the web with local buckling and post-buckling strength is fully developed, shown as figure 8. The ultimate bearing capacities are increased significantly compared with those neglecting the restraining effect, shown in Table 4.

Table 4. Ultimate bearing capacity comparison.

No.	$h \times b \times a \times t$ mm	n	N_u' kN	N_u kN	N_u/N_u'
S1	$140 \times 40 \times 15 \times 1.5$	5	11.13	53.22	4.8
S2	$205 \times 40 \times 15 \times 1.5$	3	14.92	53.47	3.6
S3	$205 \times 40 \times 15 \times 1.5$	5	13.50	52.73	3.9
S4	$205 \times 40 \times 15 \times 1.5$	7	12.45	51.85	4.2
S5	$255 \times 40 \times 15 \times 1.5$	5	9.89	55.64	5.6

N_u' and N_u are the ultimate bearing capacities of studs without restraint and with restraint by wallboards, respectively.

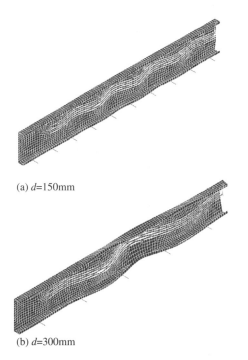

(a) d=150mm

(b) d=300mm

Figure 9. Buckling modes of S3 member with different fastener distance.

3.3 Distance of fasteners

Fastener distance affects severely the failure modes and ultimate bearing capacities of the slotted studs. For S3 with five-row slots in the web, distortional buckling is restrained if the fastener distance $d = 150$ mm, shown in figure 9. Its ultimate bearing capacity $N_u = 59.47$ kN which is close to that of non-slotted studs; as $d = 300$ mm, distortional buckling occurs in the same stud and the corresponding bearing capacity decreased by 11%, shown in figure 10.

3.4 Slots analysis

For the slotted C-shape studs, the slot parameters, such as slot row n, length lu, width lv, longitudinal distance

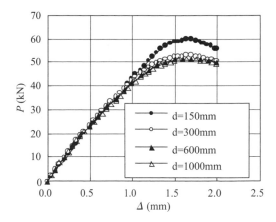

Figure 10. Influence of different distance between fasteners on the axial strength of S3 member.

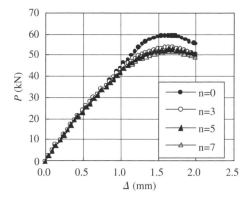

Figure 11. Axial force versus compressive deformation of studs with different slot row number.

du and vertical distance dv, reduce their strength and even change the failure modes. At the same time, slots reduce heat loss through the studs by prolonging their heat transfer paths. So it is expected that the walls can meet thermal requirement by arrange enough slots with an acceptable strength reduction.

Figure 11 shows the axial forces versus compressive deformation relation curves of the studs with different slot row number. It is found that slots reduce the axial strength of studs by about 10% with 3-row slots. The reason for that is the restraint between the weakened web and flanges is reduced, distortional buckling instead of local buckling occurs in the studs. But the strength of slotted studs, with the row number increasing from 3 to 7, changes little due to their same failure mode.

Slot length, width, longitudinal distance and vertical distance have little effect on the ultimate bearing capacity of the studs if n is constant. It means these four

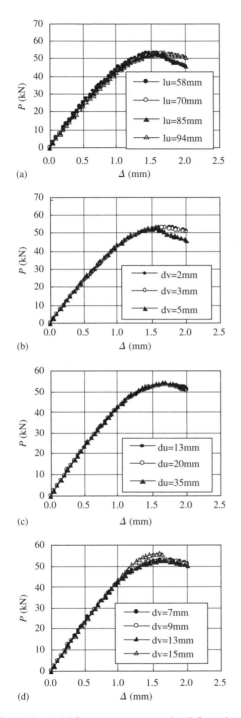

(a)

(b)

(c)

(d)

Figure 12. Axial force versus compressive deformation of studs with different slot parameters: (a) slot length (b) slot width (c) longitudinal distance of slots (d) vertical distance of slots.

parameters could be set mainly considering the thermal requirements or the else possible demands within their discussed value ranges.

4 CONCLUSION

The finite element models for predicting the thermal performance and axial compressive behaviour of the C shape cold-formed steel stud walls are developed using ANSYS. Parametrical analysis is then carried out to discuss the influences of the slots on the U-values and ultimate bearing capacities of the walls. The following conclusions can be drawn.

Slots reduce heat bridge effect from the steel studs and save great heat energy through the cold-formed steel stud walls.

Slotting the stud web could change their buckling modes, and weaken the axial strength. But if a series of reasonable slot parameters is set, this reduction could be controlled within about 10%.

Restraining action from the sheathing materials enhances the axial strength of the studs greatly, and it is suggested to be taken into account in the relative designs.

The row number and length of slots are the important parameters on the U-values of overall walls, and the row number of slots also affects the axial strength of the studs.

REFERENCES

Hoglund, T. & Burstrand, H. 1998. Slotted steel studs to reduce thermal bridges in insulated walls. *Thin-Walled Structures* 32: 81–109.

Yin, D. W. 2006. Thermal and flexural property of light-gauge steel stud wall. *Master Dissertation of Harbin Institute of Technology*, China (In Chinese).

Barbour, E. & Goodrow, J. 1995. The Thermal Performance of Steel-Framed Walls. *ASHRAE Transactions Symposia* 5(2): 766–775.

Miller, T & Pekoz T. 1993. Behavior of Cold-Formed Steel Wall Stud Assemblies. *Journal of Structural Engineering, ASCE* 119(2): 641–651.

Ministry of Construction, China. 1993. Thermal Design Code Civil Building, GB50176-93, China (In Chinese).

Steel and Composite Structures – Wang & Choi (eds)
© *2007 Taylor & Francis Group, London, ISBN 978-0-415-45141-3*

Numerical analysis of the non-linear behaviour of CFRP-strengthened cold-formed steel columns

N. Silvestre & D. Camotim
Department of Civil Engineering and Architecture, IST-ICIST, Technical University of Lisbon, Portugal

B. Young
Department of Civil Engineering, The University of Hong Kong, Hong Kong

ABSTRACT: This paper reports the results of a numerical investigation on the non-linear behaviour and load-carrying capacity of CFRP-strengthened cold-formed steel lipped channel columns – in particular, one studies a total of 16 short and long fixed-ended columns. The fully non-linear numerical analyses are based on shell finite element models, carried out in the code ABAQUS, and adopt an elastic-plastic constitutive law to describe the steel material behaviour. The columns are strengthened with carbon fibre sheets (CFS) glued at different outer surface locations (web, flanges and/or lips) and having fibres oriented either longitudinally or transversally – since the aim of the study is to assess the influence of the CFS on the column structural response, bare steel specimens were also analyzed. The numerical results, which consist of non-linear equilibrium paths (applied load vs. axial shortening) and ultimate strength values (most of them associated with local-plate or distortional failure mechanisms), are subsequently compared with experimental values obtained earlier. Finally, on the basis of both the numerical and experimental results, some relevant conclusions are drawn concerning the CFS location and fibre orientation that are most effective to strengthen lipped channel steel columns exhibiting either local-plate or distortional collapses.

1 INTRODUCTION

Given the considerable stiffness of cold-formed steel members, their strengthening by means of FRP composite requires the use of expensive high-strength fibres, a fact strongly affecting the economical viability of this procedure. However, the increasingly competitive cost of carbon fibres, together with their high stiffness and strength properties, has progressively altered this situation – indeed, carbon fibre sheets have been shown to be particularly well suited to reinforce steel plates. As far as the strength is concerned, the failure of CFRP-reinforced cold-formed steel members may stem from (i) local (local-plate or distortional) or global buckling of the steel-CFRP member, (ii) rupture or debonding of the CFRP sheet or (iii) a combination of both – thus, an efficient (safe and economical) design of such members must be based on an in-depth knowledge concerning all these potential failure modes. The objective of this work is to report the results of a numerical investigation on the non-linear behaviour and load-carrying capacity of CFRP-strengthened cold-formed steel lipped channel columns, a task accomplished by performing fully non-linear analyses based on shell finite element

models, carried out in the code ABAQUS. Finally, one should mention that this study is a part of a wider (experimental and numerical) investigation aimed at assessing how CFRP-strengthening enhances the non-linear behaviour and load-carrying capacity of cold-formed steel lipped channel columns. Since this paper concerns mainly numerical analysis and results, the details about the experimental results can be found in other papers by the authors (Young *et al.* 2006, Silvestre *et al.* 2007).

2 CHARACTERISATION OF THE COLUMNS

The numerical investigation involves a total of 16 lipped channel columns (8 short and 8 long) with fully fixed end supports. These columns were strengthened with CFS glued in different locations of their outer surfaces (web, flanges, lips) and having the fibres oriented either longitudinally ($\alpha = 0°$) or transversally ($\alpha = 90°$) – for reference purposes, a few bare steel columns were also analysed. The labelling provides information about (i) the column length (short or long) and (ii) the location and orientation of the CFS. The first letter identifies a short (**S**) or long (**L**) column and the following ones indicate if the specimen has no strengthening (**NIL**), CFS in the web (**W**), web and flanges (**WF**) or web, flanges and lips (**WFL**). The numbers specify the CFS orientation, which may be longitudinal (**0**) or transversal (**90**).

All columns have nominal web, flange and lip widths equal to $b_w = 125\,mm$, $b_f = 102\,mm$, $b_l = 14\,mm$, and inside corner radius r = 3.0 mm. The short columns are made of G550 steel sheets ($E = 207\,GPa$, $f_y = 610\,MPa$) with nominal thickness $t = 1.0$ mm and length $L = 600$ mm. As for long columns, they are made of G450 steel sheets ($E = 218\,GPa$, $f_y = 521\,MPa$) and have nominal thickness $t = 1.5\,mm$ and length $L = 2200\,mm$. Finally, the CFS material properties, taken from information provided by the fabricators, are: longitudinal modulus 235 GPa, tensile strength 4200 MPa and thickness 0.11 mm. In the tests, the carbon fibre sheets were attached to the zinc coated cold-formed steel columns by means of an epoxy resin with tensile strength *30 MPa* and modulus 3.5 GPa – a single CFS was attached to each specimen outer surface.

3 FINITE ELEMENT MODELLING

In order to analyse the post-buckling behaviour of the CFRP-strengthened columns, one uses the finite element code ABAQUS (HKS 2002). Concerning the performance of the analyses, the following issues deserve a few words:

(1) *FE discretisation*. To analyse both the local and global behaviours of a given thin-walled member, one must adopt a two-dimensional model to discretise its mid-surface, a task adequately performed by means of S4 elements (4-node isoparametric shell element with full integration). For the lipped channel columns dealt with in this work, it was found that it suffices to consider a cross-section discretisation into 36 FEs (12 in the web, 10 per flange and 1 per lip) – this corresponds roughly to adopting 10 *mm* wide FEs. In the short columns, a *8 mm × 10 mm* (length-width) mesh size was used, corresponding to a total of 1462 elements, 1505 nodes and 8622 degrees of freedom (d.o.f.). In the long columns, the adopted *30 mm × 10 mm* mesh size led to 2250 elements, 2625 nodes and 15342 d.o.f.. It is worth noting that the round corners were not modelled.

(2) *End support conditions*. In order to be able to make meaningful comparisons between numerical and experimental results, it is essential to ensure that an adequate modelling of the member end support conditions. Therefore, all the global and local displacements/rotations are prevented at the column end sections (obviously, the axial translation

of the loaded end section is free) – rigid plates attached to the column end sections were modelled by means of rigid three-node finite elements R3D3, thus preventing the warping and in-plane cross-section deformation at the supports. These end plates made it possible to prevent (i) all displacements at one end section cen-troid and (ii) all but one (axial displacement) at the other end section centroid.

(3) *Loading*. The compressive load was applied at the centroid of the axially free end section. To obtain the load vs. axial shortening equilibrium path, one assessed the free end section centroid axial displacement using the ABAQUS instruction "MONITOR".

(4) *Steel modelling*. The column (steel) material behaviour was always assumed to be homogeneous and isotropic and two constitutive laws were considered to model it, namely (i) a linear elastic law (bifurcation analysis) and (ii) a linear-elastic/perfectly-plastic law with no strain hardening (post-buckling analysis). The linear elastic behaviour is fully characterised by the Young's modulus (E) and Poisson's ratio (ν) values. As for the elastic-plastic behaviour, it is described by the well-known Prandtl-Reuss model (J_2-flow theory), combining Von Mises's yield criterion with an associated flow rule. These models are available in the ABAQUS material behaviour library and their implementation just involves providing the steel elastic constants and nominal yield stresses ($f_y = 550$ MPa and $f_y = 450$ MPa, respectively for the short and long columns).

(5) *CFS modelling*. The cross-section walls having CFS attached were modelled as double-ply plates with one ply of steel and the other of CFS. In order to have an indication about a possible CFS failure, and since the ABAQUS library does not include (defined *a priori*) constitutive laws for composite laminates accounting for material degradation, one has to resort to the *Maximum Stress* and *Tsai-Hill* failure criteria – this is achieved by using the commands "FAIL STRESS", "MSTRS" and "TSAIH". The *Maximum Stress* and *Tsai-Hill*[1] failure criteria involve failure indexes I_F, given respectively by

$$I_{F.MS} = max\left\{ \left|\frac{\sigma_{xx}}{\sigma_{xx}^{adm}}\right| ; \left|\frac{\sigma_{ss}}{\sigma_{ss}^{adm}}\right| ; \left|\frac{\tau_{xs}}{\tau_{xs}^{adm}}\right| \right\} \quad (1)$$

$$I_{F.TH} = \left(\frac{\sigma_{xx}}{\sigma_{xx}^{adm}}\right)^2 - \frac{\sigma_{xx}\sigma_{ss}}{(\sigma_{xx}^{adm})^2} + \left(\frac{\sigma_{ss}}{\sigma_{ss}^{adm}}\right)^2 + \left(\frac{\tau_{xs}}{\tau_{xs}^{adm}}\right)^2 \quad (2)$$

---

[1] The Tsai-Wu failure criterion is also available in the ABAQUS material behaviour library. However, for the problem under consideration, its application yields results very similar to the ones obtained with the Tsai-Hill criterion – thus, it has not been used here.

where σ_{xx}^{adm}, σ_{ss}^{adm} and τ_{xs}^{adm} are maximum admissible values for the CFS normal longitudinal stress, normal transversal stress and shear stress (the compressive and tensile values are deemed equal). It should be noted that (i) the *Maximum Stress* criterion assumes no interaction between the three failure modes and (ii) the *Tsai-Hill* criterion is an extension of *Von Mises* yield criterion to orthotropic materials. As long as the failure index remains below unity ($I_F < 1$), no CFS collapse occurs – *i.e.*, it only takes place for $I_F \geq 1$.

(6) ***Initial imperfections***. Initial geometrical imperfections can be incorporated in an ABAQUS member post-buckling analysis either (i) manually, by directly inputting an arbitrary initial deformed configuration, or (ii) automatically, through the *a priori* definition of a linear combination of normalised buckling mode shapes (yielded by a preliminary buckling analysis, based on a finite element mesh *identical* to the one adopted in the post-buckling analysis). In this work, the column initial geometrical imperfections were included *automatically*, as different linear combinations of the most relevant (critical) local-plate and distortional buckling mode shapes – these combinations are incorporated into the column initial geometry through a specific ABAQUS command. Since no column initial geometric imperfections were measured, the approach employed involved performing several preliminary *elastic* non-linear analyses, each incorporating initial imperfections that (i) equally combine the normalised local-plate and distortional buckling mode shapes and (ii) have different overall amplitudes. The criterion adopted to select the appropriate imperfection to include in the test simulation (geometrically and physically non-linear analysis) was the "close vicinity" between the initial portions of the experimental and numerical equilibrium paths P vs. u – the imperfection amplitudes yielded by this approach were found to agree fairly well with the ones obtained by adopting the methodology proposed by Schafer and Peköz (1998). For long columns, one considers local and distortional imperfection amplitudes similar to $d_1 = 0.66t$ (type 1) and $d_2 = 1.55t$ (type 2), values corresponding to a cumulative distribution function $P(\Delta < d) = 0.75$. For short columns, one uses mainly local imperfection amplitudes similar to $d_1 = 1.98t$ (type 1), value associated with a cumulative distribution function $P(\Delta < d) = 0.96$.

(7) ***Solution techniques***. Performing a buckling analysis requires solving an eigenvalue problem, defined by the column (discretised) elastic and geometric stiffness matrices – ABAQUS employs the "subspace iteration method" to obtain this solution. As for the non-linear (post-buckling) equilibrium paths, relating the load parameter (applied

compressive load or imposed axial shortening) with a suitably chosen column displacement, they are determined by means of an incremental-iterative technique employing Newton-Raphson's method. Since the column elastic-plastic local-plate and distortional post-buckling behaviours exhibit limit points (ultimate strengths), the elastic-plastic analyses are performed with Riks's arc-length control strategy – this is automatically done in ABAQUS, on the basis of a pre-defined "calibration" increment and tolerance parameters.

4 NUMERICAL RESULTS

The performance of ABAQUS buckling analyses led to the buckling load values given in tables 1 and 2 (P_L and P_D – local and distortional buckling). The observation of these values prompts the following comments:

(1) for the short columns, the local buckling load values ($26.8 \leq P_L \leq 29.5 \, kN$) are much smaller than their distortional counterparts ($P_D \approx 130 \, kN$). Comparing the P_L values of all CFRP-strengthened and bare steel columns, one readily observes that the maximum benefit (*10.1%*) is achieved for the column **S-WFL-0**.

(2) for the long columns, the local buckling load values ($84.5 \leq P_L \leq 89.6 \, kN$) are very similar to their distortional counterparts ($84.4 \leq P_D \leq 90.3 \, kN$). Comparing the P_L and P_D values of all CFRP-strengthened and bare steel columns, one readily observes that the maximum benefits (6.0% and 7.0%, respectively) are achieved for the column **L-WF-90** – this assertion is in perfect accordance with the conclusions of an earlier parametric study by the authors (Silvestre *et al.* 2004), which showed that transverse CFS strengthening is more beneficial for distortional buckling.

(3) Figure 1(a)–(b) shows the buckling mode shapes obtained for long and short columns – unlike the buckling load values, these shapes do not vary with the type of strengthening. One notices that the short column buckling mode is basically local (5 half-waves), while its long column counterpart combines local (17 half-waves) and distortional (3 half-waves) modes – this inevitably leads to local/distortional mode interaction in the post-buckling range, a phenomenon that currently interests the cold-formed steel technical and scientific communities (*e.g.*, Yang & Hancock 2004, Dinis *et al.* 2007).

After performing the ABAQUS non-linear (post-buckling) analyses, one obtained the load vs. axial shortening curves $P(u)$ shown in figures 2(a)–(b) – short and long columns, respectively. Furthermore, the

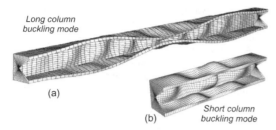

Long column buckling mode

(a)

Short column buckling mode

(b)

Figure 1. Buckling mode shapes of (a) long and (b) short columns.

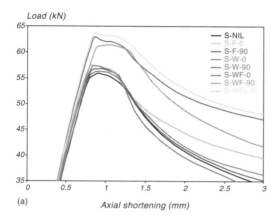

Load (kN)

— S-NIL
— S-F-0
— S-F-90
— S-W-0
— S-W-90
— S-WF-0
— S-WF-90
— S-WFL-0

(a)

Axial shortening (mm)

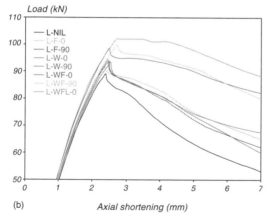

Load (kN)

— L-NIL
— L-F-0
— L-F-90
— L-W-0
— L-W-90
— L-WF-0
— L-WF-90
— L-WFL-0

(b)

Axial shortening (mm)

Figure 2. (a) Short and (b) long column $P(u)$ curves.

Table 1. Buckling and ultimate loads for short columns.

Column	Experimental		Numerical			P_{Num}/P_{Exp}
	P_{Exp}	Gain	P_L	P_{Num}	Gain	
S-Nil	54.9	–	26.8	55.9	–	1.018
S-F-0	56.0	2.0%	28.5	57.4	2.7%	1.025
S-F-90	56.4	2.7%	27.9	56.8	1.6%	1.007
S-W-0	55.9	1.8%	27.6	57.3	2.5%	1.025
S-W-90	55.6	1.3%	28.0	56.3	0.7%	1.013
S-WF-0	60.2	9.7%	29.2	62.8	12.3%	1.043
S-WF-90	63.2	15.1%	29.3	61.4	9.8%	0.972
S-WFL-0	61.4	11.8%	29.5	63.4	13.4	1.026

Table 2. Buckling and ultimate loads for long columns

Column	Experimental		Numerical				P_{Num}/P_{Exp}
	P_{Exp}	Gain	P_D	P_L	P_{Num}	Gain	
L-Nil	85.2	–	84.4	84.5	89.0	–	1.045
L-F-0	90.1	5.8%	84.7	87.8	91.8	3.2%	1.019
L-F-90	92.6	8.7%	85.5	86.9	93.7	5.3%	1.012
L-F-0	95.5	12.1%	86.5	85.6	94.3	6.0%	0.987
L-W-90	88.1	3.4%	89.0	87.2	93.5	5.1%	1.061
L-WF-0	98.6	15.7%	86.7	88.9	98.4	10.6%	0.998
L-WF-90	100.9	18.4%	90.3	89.6z	99.8	12.1%	0.989
L-WFL-0	102.1	19.8%	88.3	89.3	102.5	15.2%	1.004

column central zone. Moreover, notice also the remarkable similarity between the numerical (figure 3(a_1)) and experimental (figure 3(a_2)) deformed configurations of column **S-F-0** at the onset of collapse.

(2) As mentioned before, long columns buckle in mixed local-distortional modes. Then, it is not surprising to observe that the corresponding failure modes involve a distortional-type mechanism, which is very localised at the column mid-span zone – see figure 3(b), which concerns column **L-F-0**. Once more, one stresses the quite amazing similarity between the numerical (figure 3(b_1)) and experimental (figure 3(b_2)) deformed configurations of column **L-F-0** at the brink of collapse. Notice also the (a bit hard to distinguish) local buckles that appear in the web of the above column (both in the numerical and experimental configurations) and provide clear evidence concerning the occurrence of local/distortional mode interaction at failure (Yang & Hancock 2004, Dinis *et al.* 2007).

(3) All numerical equilibrium paths practically follow the linear relation $P/u = EA/L$ until a certain load value is reached ("proportional limit load"). In the short columns, this load value is close to the column ultimate strength, which means that there is very little axial strength degradation (due to local

corresponding ultimate load values P_{Num} are also given in tables 1 and 2. After observing the results presented in these figures and tables, the following remarks are appropriate:

(1) Despite buckling in local modes, the short columns exhibit a distortional failure mechanism – figure 3(a) shows the collapse mode of column S-F-0, and one readily observes the formation of a yield-line mechanism involving the plastification of the

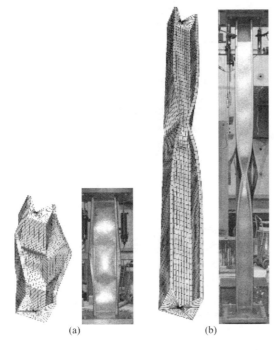

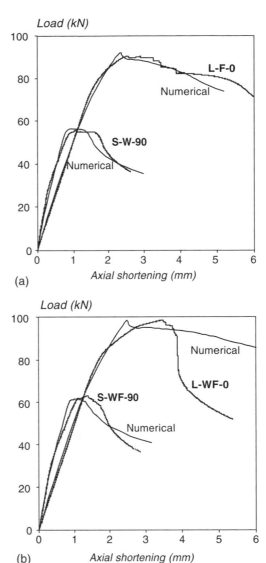

(a)   (b)

Figure 3.   Columns (a) S-F-0 and (b) L-F-0: numerical and experimental failure mechanisms.

buckling effects) prior to failure. Conversely, the long column proportional limit loads (≈ 60 kN) is far from the ultimate loads, thus leading to a rather strong axial strength degradation due to local-distortional interaction before collapse occurs – this behavioural aspect was also observed in the experimental results. By comparing the shape of the $P(u)$ curves concerning short and long columns (figures 2(a)–(b)), one realizes that the long column equilibrium path descending (post-ultimate) branches (i) drop more abruptly immediately after failure, but (ii) exhibit lower slopes at higher axial shortening values – this difference, which was not visible in the experimental results (Young *et al.* 2006, Silvestre *et al.* 2007), stems from the (inevitable) inadequate modelling of the column post-ultimate behaviour.

(4) In both the short and long columns, the effects (benefits) of the CFRP-strengthening detected by the numerical analyses are similar to those found in the experimental investigation (see tables 1 and 2). In particular, one should note that attaching the CFS to the short columns leads to similar buckling and ultimate load increases (approximately 10%). On the other hand, the addition of the CFS to the long columns is more beneficial for the ultimate loads than for their buckling counterparts (15% and 7% increases, respectively – one thus concludes that, at least for the particular cases investigated,

Figure 4.   Comparison between numerical and experimental $P(u)$ curves for columns (a) S-W-90, L-F-0 and (b) S-WF-90, L-WF-0.

the CFRP-strengthening of long columns is more effective for (axial) strength than for stiffness.

(5) The numerical ultimate load values (P_{Num}) compare fairly well with the experimental ones (P_{Exp}) – the maximum difference is *6.1%*. Figures 4(a)–(b) show comparisons between the experimental and numerical curves concerning the (representative) columns (i) **S-W-90** and **L-F-0**, and (ii) **S-WF-90** and **L-WF-0**. One notices that the short columns are axially stiffer in the initial loading stages, which is due to their higher steel Young's

973

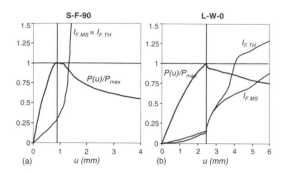

(a)

(b)

Figure 5. Variation of the maximum stress and Tsai-Hill failure indexes with u (axial shortening) for columns (a) S-F-90 and (b) L-W-0.

moduli and much lower lengths (the cross-section dimensions are similar). While the numerical and experimental descending branches of columns **L-F-0**, **S-W-90** and **S-WF-90** exhibit a fairly good correlation, the ones concerning column **L-WF-90** are qualitatively different – this is most likely due to the occurrence of *debonding* between the steel and CFS, a phenomenon that cannot be captured by the numerical analysis and provides a logical explanation for the steeper descending branch exhibited by the experimental curve.

Finally, one pays some attention to the CFS mechanical behaviour. Since ABAQUS provides information about its failure indexes I_F, associated with the *Maximum Stress* and *Tsai-Hill* failure criteria, figures 5(a)–(b) show the variation of the maximum mid-span values of $I_{F.MS}$ and $I_{F.TH}$ with the axial shortening u, for the two representative columns **S-F-90** and **L-W-0** – the numerical curves $P(u)/P_{Num}$ (dotted lines) are also represented (note that are always below 1.0). The observation of figure 5 leads to the following remarks:

(1) Regardless of the failure criterion adopted, the index I_F always increases with the axial shortening u. However, notice that this increase is much more pronounced after reaching the ultimate load – in fact, the failure index never exceeds *30%* before the column collapse, which means that the CFS remain fully effective at that stage. Concerning the two failure criteria considered, one obvious has $I_{F.MS} \leq I_{F.TH}$, since $I_{F.MS}$ does not account for the stress component interaction (as does $I_{F.TH}$).

(2) In column **S-F-90**, one has $I_{F.MS} = I_{F.TH}$ due to the (i) strong predominance and (ii) low admissible value of the longitudinal normal stress σ_{xx} (recall that $\alpha = 90°$) – then, the ratio $\sigma_{xx}/\sigma_{xx}^{adm}$ is very high for the two criteria (stress component coupling is marginal). As σ_{xx}^{adm} is small, both $I_{F.MS}$ and $I_{F.TH}$ increase steeply immediately after failure – e.g., one has $I_{F.MS} = I_{F.TH} = 1$

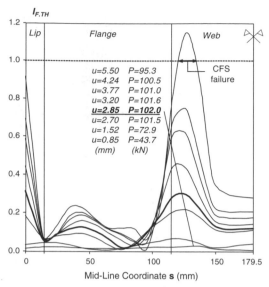

Figure 6. Distribution of the Tsai-Hill failure index along the mid-span cross-section contour of column L-WFL-0.

for $u = 1.33 \, mm$ and $P = 51.3 < P_{Num} = 56.8 \, kN$. Having $I_{F.MS} = I_{F.TH} > 1$ indicates failure due to an excessive σ_{xx} value (rupture of the CFS in the transverse direction).

(3) In column **L-W-0**, the $I_{F.MS}$ and $I_{F.TH}$ curves are different, due to stress component coupling – mainly σ_{xx} and σ_{ss}. In this case, cross-section deformation leads to the appearance of non-negligible values of the transverse normal stress ratio $\sigma_{ss}/\sigma_{ss}^{adm}$ (recall that $\alpha = 0°$), which "compete" with their $\sigma_{xx}/\sigma_{xx}^{adm}$ counterparts. Thus, the more conservative Tsai-Hill criterion indicates that, for this column, the CFS collapse ($I_F > 1$) only occurs for a quite high axial shortening (much larger than the same column **S-F-90** value).

(4) for illustrative purposes, figure 6 displays the variation of $I_{F.TH}$ along the column **L-WFL-0** mid-span cross-section mid-line (coordinate s), for different applied load levels – only half cross-section is shown and the curve corresponding to the column ultimate load is represented by a thicker solid line. Because this column failure is governed by local-distortional interaction, its $I_{F.TH}$ values increase significantly with P, *both* in the (i) web (near the web-flange corners) and (ii) lips. In the former case, this is mostly due to the predominance of the transverse normal stress ratio $\sigma_{ss}/\sigma_{ss}^{adm}$ associated with transverse web bending. In the latter second case, the key role is played by the significant longitudinal normal stresses σ_{xx}, arising from the warping displacements associated

with distortional buckling (Silvestre & Camotim 2006).

(5) Looking at items (2), (3) and (4), and recalling that the CFS failure was only experimentally detected in the equilibrium path descending branches, one may conclude that the two-layer (steel-CFS) numerical model adopted in this work provides very satisfactory column ultimate strength estimates. However, note also that the column load-carrying capacity stems mostly from the steel, since only one thin CFS was glued to the steel member – however, using more CFS layers to strengthen the column might lead to debonding at the adhesive interfaces prior to collapse (due to high shear and peeling stresses).

5 CONCLUDING REMARKS

This paper presented a numerical investigation to assess the benefits of CFRP-strengthening cold-formed steel columns – single CFS were glued at different locations of the column outer surface (web, flanges and/or lips) with the fibres oriented either longitudinally or transversally. The numerical results presented and discussed consisted of non-linear equilibrium paths and ultimate strength values and were compared with experimental values reported earlier. The following conclusions of this study deserve to be specially mentioned:

(1) The comparison between the ultimate loads of the otherwise identical columns with and without CFRP-strengthening showed that the carbon fibres increase the load-carrying capacity by up to (i) 15%, for short columns, and (ii) 20%, for long columns. Moreover, it is expected that adding more than one CFS to the columns will lead to more substantial strength and stiffness increases (however, debonding at the adhesive interfaces may become a problem).

(2) The numerical approach adopted to analyse the non-linear behaviour of CFRP-strengthened steel columns (two-layer steel-CFS model) yielded fairly good results concerning the column post-buckling behaviour (pre and post-collapse), ultimate load and failure mechanism. Moreover, the Maximum Stress and Tsai-Hill criteria ensured the validity of the numerical results up to the column collapse.

REFERENCES

Dinis, P.B., Camotim, D. & Silvestre, N. 2007. FEM-based analysis of the local-plate/distortional mode interaction in cold-formed steel lipped channel columns. *Computers & Structures*, accepted for publication.

HKS (Hibbit, Karlsson and Sorensen Inc.) 2002. *ABAQUS Standard* (version 6.3-1).

Schafer, B.W. & Peköz, T., 1998. Computational modeling of cold-formed steel: characterizing geometric imperfections and residual stresses. *Journal of Constructional Steel Research*, 47(3), 193–210.

Silvestre, N., Camotim, D., 2006. Local-plate and distortional post-buckling behavior of cold-formed steel lipped channel columns with intermediate stiffeners. *Journal of Structural Engineering* (ASCE), 132(4), 529–540.

Silvestre, N., Camotim, D. & Young, B. 2004. Buckling behaviour of cold-formed steel members strengthened with carbon fibre sheets. *Proceedings of 2nd International Conference on Steel & Composite Structures* (ICSCS'04 – Seoul, 2–4/9), C.K. Choi, H.W. Lee, H.G. Kwak (eds.), 148. (full paper: CD-ROM proceedings, 412–27)

Silvestre, N., Young, B. & Camotim, D. 2007. Non-linear behaviour and load-carrying capacity of CFRP-strengthened lipped channel steel columns, *submitted for publication*.

Yang, D.M. & Hancock, G.J., 2004. Compression tests of high strength steel channel columns with interaction between local and distortional buckling. *Journal of Structural Engineering* (ASCE), 130(12), 1954–1963.

Young, B., Silvestre, N. & Camotim, D. 2006. Experimental and numerical analysis of the structural response of FRP-strengthened cold-formed steel columns. *Composites in Civil Engineering* (CICE 2006 Miami, 13–15/12), A. Mirmiran, A. Nanni (eds.), International Institute for FRP in Construction, 725–728.

Steel and Composite Structures – Wang & Choi (eds)
© 2007 Taylor & Francis Group, London, ISBN 978-0-415-45141-3

Behaviour of Profiled Steel Sheet Dry Board (PSSDB) composite floor system with foamed concrete infill material

W.H. Wan Badaruzzaman, A. Aszuan & W.H.M. Wan Mohtar

Department of Civil & Structural Engineering, Faculty of Engineering, Universiti Kebangsaan Malaysia, Malaysia

ABSTRACT: The Profiled Steel Sheet Dry Board (PSSDB) system is a type of lightweight composite structural system constructed from profiled steel sheeting and dry board, connected by self-drilling and self-tapping screws. This paper focuses on the enhancement of the structural performance of PSDDB floor system with the introduction of foamed concrete as an infill material in the trough of the profiled steel sheet. Experimental results showed that the introduction of foamed concrete increased the stiffness value by almost 40% compared to the no-infill PSSDB sample. In addition, the ultimate load at failure was observed to be higher by about 35% for the samples with foamed concrete compared to the no-infill sample. Experimental results were also compared with results from simple beam theory based on full interaction behaviour for simplicity reason, even though quite significant differences in results were observed as already expected. It can be concluded that the use of foamed concrete, as an infill material will enhance the rigidity and load bearing capacity of the PSSDB system.

Keywords: PSSDB floor system, EI value, foamed concrete infill.

1 INTRODUCTION

Studies on the behaviour of a typical PSSDB system (without infill) as floor panels have been reported in earlier publications (Ahmed et al. 1996,a,b, Wan Badaruzzaman et al. 1996). The studies have successfully established that the system is structurally capable of being utilised as a load bearing floor system in buildings. Various non-structural performances of the PSSDB floor panels, such as fire and thermal resistance properties, acoustic, vibration, and waterproofing have also been looked into. The structural behaviour of the PSSDB system depends on the properties of the basic components forming the system, and the degree of interaction between them. The degree of interaction can either be full or partial, depending on the connectors' modulus and spacing (Ahmed et al. 1996).

Many previous publications have dealt with the theoretical modeling that catered for the partial interaction and non-linear behaviour of the PSSDB system (Ahmed et al. 2005, 2003, Wan Badaruzzaman et al. 2003, Akhand et al. 2004). However, this paper, for simplicity reason has adopted full interaction beam analysis to give predicted theoretical results.

The structural and non-structural performances of the PSSDB system could also be improved by introducing infill material in the troughs of the profiled

steel sheet. As a guideline, the floor in a building shall be rigid and strong to support the dead and imposed load. The maximum allowable deflection of floor slabs is taken as span/250. This paper focuses on the effect of foamed concrete as an infill material in the PSSDB floor system and compare with other type of infill that has been previously studied (Khadijah 2002, Harsoyo 2003, and Normelia 2002). The increase in stiffness and strength of the system is expected from the bonding between the profiled steel sheet and the foamed concrete infill. Prediction of the PSSDB floor system's performance will be verified using experimental results.

1.1 *PSSDB floor panel system*

The basic system consists of profiled steel sheet and dry board, connected mechanically by self-drilling, self-tapping screws. Figure 1 shows a typical PSSDB floor panel.

The profiled steel sheet used was PEVA 45, whilst Cemboard, a type of cement bonded rubber wood board was used as the dry board component. The foamed concrete used was of 1500 kg/m^3 density produced by normal process using foaming agent. Based on the laboratory test done, the compression strength for this foamed concrete was 24.5 MPa.

2 EXPERIMENTAL INVESTIGATION

An experimental study was conducted to study the effect of using foamed concrete as infill material on the structural behaviour of the PSSDB floor system. All samples were 840 mm wide with a span of 1000 mm. Table 1 gives the specifications of the samples. The cross-sectional illustrations for the two types of samples are shown in Figures 2(a) and (b). The first sample

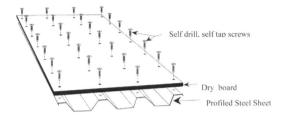

Figure 1. Typical PSSDB floor panel.

Table 1. Specifications of the samples.

Sample	Profiled Steel Sheet	Dryboard	Screw Spacing	Infill
A	PEVA 45 1 mm thick	Cemboard 16 mm thick	100 mm c/c	–
B	PEVA 45 1 mm thick	Cemboard 16 mm thick	100 mm c/c	Foamed Concrete
C	PEVA 45 1 mm thick	Cemboard 16 mm thick	100 mm c/c	Foamed Concrete

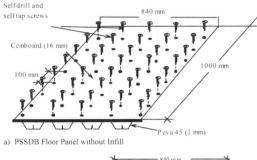

a) PSSDB Floor Panel without Infill

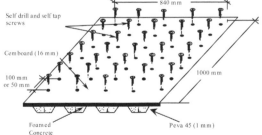

b) PSSDB Floor Panel with Foamed Concrete Infill

Figure 2. Cross section of PSSDB floor test samples.

is a standard PSSDB which consists of 1 mm thick Peva 45, a type of profiled steel sheet manufactured by Asia Roofing Sdn. Bhd., 16 mm Cemboard, a type of cement bonded rubber wood board manufactured by Hume Cemboard Berhad, and self-tapping and self-drilling screws with 100 mm spacing. Samples B and C are the same as the previous sample except that foamed concrete is used as an infill material and the connectors spacing has been reduced to 50 mm for Sample C.

The loading system and load increment are as given in Table 2. A concentrated point loading at mid-span was incremented to investigate the flexural behaviour of the floor system.

3 INSTRUMENTATION AND TEST PROCEDURES

The test was based on BS 8110 Part 2: Section 9 (1985). A deflection transducer was placed at the expected position of maximum deflection, i.e. at mid-span, mid-width of the panel. Additional deflection transducers were positioned at either side of the centre point of the test panels along both the longitudinal and transverse directions. These symmetrically positioned transducers were used to check the expected symmetrical behaviour of the panels. The loading system and the transducers were assembled on the support and specimen frames. Loads were then applied on the sample until failure. The loading value including the ultimate load at failure and the corresponding deflections were recorded. These values were then used to conclude on the behaviour of the panels.

4 PREDICTION OF RESULTS

In predicting its behaviour, the system was assumed to be in full interaction behaviour, i.e. no slip was assumed at all the interfaces between the various layers of components. The floor system was assumed to be behaving as a beam, thus the simple beam theory was employed in predicting its behaviour. The neutral axis and bending stiffness in the strong direction of each of the floor panel were calculated from the assumed fully interacting sections with the introduction of the modular ratio wherever necessary. However, it will be

Table 2. Loading system.

Type of loading	Setup & load procedure
Concentrated	Mid span point load was generated manually using a hand jack/pump and load cell connected to computerised data logger to measure the load. The increment load was 2 kN until specimen failure.

shown later that this assumption is not really true in the real situation. More accurate methods of predicting the behaviour including classical partial interaction analysis (Newmark et al. 1951, folded plate method of analysis (Wan Badaruzzaman 1994), and finite element modelling (Ahmed 1999) of such a system are readily available and have been presented in other publications.

5 RESULTS AND DISCUSSION

From the results obtained experimentally, all tests conducted were seen to be exhibiting similar load-deflection characteristic. Load-deflection curves (see Figure 3) plotted from all tests show that the initial load-deflection response was linear and elastic, and this elastic response continued until just before failure. The final failure occurred when the upper flanges of the steel sheeting buckled. In obtaining the stiffness value from experimental work, the slope of elastic phase was taken into consideration. A summary of the stiffness values for all tests from the experiment together with the full composite (calculated based on all equivalent steel section) values is shown in Table 3. If the connection between the boarding and steel sheeting had been infinitely stiff, duc to either a very stiff

connector modulus, or a very closely spaced connectors, a full connection would occur between the two layers. Table 3 shows the comparison between the theoretical full interaction stiffness value and those found experimentally from all tests. It can be concluded that firstly, full interaction behaviour had not been achieved in all tests with the percentage differences in results from the full interaction value ranging between 40.6 to 61.9%. The screw spacing played a very important role in determining the degree of interaction between the various components forming the system. What is important at this stage is the high observed stiffness value achieved by the PSDDB system, which is 814.28 kN m^2/m (taken from Sample C). This is very much higher than 200 kN m^2/m obtained by the original PSSDB system (Wan Badaruzzaman et. al. 1996).

Samples B and C exhibited an increment of 32.5% and 36.4%, respectively of the stiffness value compared to the control sample, Sample A. These results are obtained due to the presence of foamed concrete as the PSSDB infill material. Sample C showed 5% increment in its stiffness value compared to Sample B. As connector spacing for Sample C is 50 mm c/c compared to 100 mm for Sample B, it is confirmed that closer connector spacing will give higher stiffness value for the PSSDB system.

Table 4 showed the maximum load and the maximum deflection value for all tested samples. The maximum deflection recorded for all samples was at mid-span of the panels. The highest ultimate load was 31 kN observed for Sample C. This is 35% higher than the ultimate load observed for Sample A (the sample without infill). Results for Sample C showed an increase in ultimate load of 7% comparcd to Sample B.

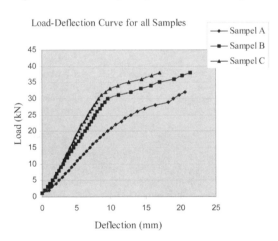

Figure 3. Load-deflection curves for all samples.

Table 4. Ultimate load and maximum deflection value for all samples.

Sample	Ultimate Load (kN)	Maximum Deflection (mm)
A	23	11.23
B	29	9.61
C	31	11.31

Table 3. Comparison of EI value for all samples.

Sample	Experimental Stiffness, EI (kNm^2/m)	Fully Composite Stiffness, EI (kNm^2/m)	% Difference between exp. and theory	% Difference with Sample A
A	521.85	1369.2	61.89	–
B	774.05	1371.3	43.55	32.5
C	814.28	1371.3	40.62	36.4

Table 5. Maximum allowable span for each sample.

Sample	Load (kN/m^2)	Composite Stiffness, EI$_{experiment}$ (kNm^2/m)	Max. Allowable Span (m)
A	2.5	521.9	3.38
B	2.5	774.1	3.85
C	2.5	814.3	3.93

Table 6. Comparison of EI values with a previous study.

Research Source	System adopted	Max. EI value (kNm^2/m)
This study (2005)	'Single' PSSDB system with foamed concrete	814.3
Normelia (2002)	'Double' PSSDB system with grade 30 concrete	995.6

All the above indicated that with foamed concrete as an infill material, and with closer connector spacing, the stiffness and load bearing capacity of the PSSDB system would be further enhanced.

By assuming a deflection limit of span/250, the maximum allowable span can be calculated using simple beam deflection formula. Table 5 shows the maximum predicted span that can be achieved by each sample for typical uniform live loads of 2.5 kN/m^2.

Sample C showed the maximum allowable span of 3.93 m, which is only 2% increment from Sample B. Therefore, in real design, a connector spacing of 100 mm c/c would be sufficient for practical purposes.

6 COMPARISON FROM PREVIOUS STUDIES

Table 6 shows the comparison of results with a previous research work conducted using concrete grade 30 concrete as the infill material (Normelia 2002). In the previous case, one major difference was in the use of 'double' profiled steel sheeting instead of 'single' in the case reported in this paper. This comparison is for the sake of benchmarking.

The above comparison has given a very good indication of the accuracy of the results obtained in the tests reported in this paper. The higher stiffness value when using the 'double' PSSDB system is as expected. What is to be highlighted here is the values obtained by both researchers are within the same range.

7 CONCLUSION

Sample C has the highest stiffness value (EI) of 814.3 kN m^2/m. Samples B and A possessed the EI

values of 774.1 kN m^2/m and 521.9 kN m^2/m respectively. Comparison of results with simple beam theory (assuming full-interaction behavior) gave a range of discrepancies between 40.6% (Sample C) – 61.9% (Sample A). It is observed that, with closer spacing of screws and the presence of foamed concrete as an infill material (Sample C), the discrepancy in results became lower (40.6%). This actually indicated that the system is coming closer to a full-interaction (zero-slip) system; hence results were more comparable as compared to the control sample (Sample A).

Two kinds of PSSDB floor panels have been tested and analysed successfully. The results indicates that some kind of infill materials, such as foamed concrete used in the described tests can have a very significant contribution in increasing the stiffness and load carrying capacity of the PSSDB floor panels. The deflection of the panel with infill, at the same load has been shown to be less than that of the panel without infill by about 30% for the test. This indication sheds some light to any further studies along the same line.

It can be concluded that the use of foamed concrete, as an infill material will enhance the rigidity and load bearing capacity of the PSSDB system. The potential of this new system in replacing the traditional systems looks promising. In addition to the system being structurally sound, it has other added advantages such as being lightweight, easily transportable and easily assembled by semi-skilled labour.

REFERENCES

Akhand A.M., Wan Badaruzzaman W.H., and Wright H.D., 2004: Combined Flexure and Web Crippling of a Low-Ductility High Strength Steel Decking: Experiment and a Finite Element Model, *Thin-Walled Structures*, 42(7), pp. 1067–1082.

Ahmed, E. 1996. *Behaviour of Profiled Steel Sheet Dry Board Panel*. MSc Thesis. Universiti Kebangsaan Malaysia, Bangi, Selangor Darul Ehsan, Malaysia.

Ahmed, E., Wan Badaruzzaman, W.H. and Rashid, A.K. 1996a. A Simplified Elastic Composite Floor Section Analysis with Incomplete Interaction. *Engineering Journal UKM* **8**, 67–78.

Ahmed, E., Wan Badaruzzaman, W.H. and Rashid, A.K. 1996b. Composite Partial Interaction of Profiled Steel Sheeting Dry Board Floor Subject to Transverse Loading. *Proceeding of the CIB International Conference on Construction Modernization and Education,* Beijing, China.

Ahmed E. 1999, *Behaviour of PSSDB folded plate structures*, PhD thesis, Universiti Kebangsaan Malaysia.

Ahmed E. and Wan Badaruzzaman W.H., 2003: Equivalent Elastic Analysis of Profiled Metal Decking using Finite Element Method, *International Journal of Steel Structures* 3(1), pp. 9–17.

Ahmed E. and Wan Badaruzzaman W.H., 2005: Finite Element Prediction on the Applicability of Profiled Steel Sheet Dry Board Structural Composite System as a Disaster Relief Shelter, *Journal of Construction and Building Materials*, 19(4), pp. 285–295.

Harsoyo M.S. 2003. *Peningkatan Prestasi Sistem Lantai Keluli Berprofil/ Papan Kering dan Isian Konkrit*. PhD Thesis. Universiti Kebangsaan Malaysia.

Newmark, N.M., Siess, C.P. and Viest, I.M. 1951. Tests and Analysis of Composite Beams with Incomplete Interaction, *Proc. Society for Experimental Stress Analysis*, 9(1):75–95.

Normelia Abdul Hamid. 2002. *Kajian Kelakuan Sistem Lantai Komposit Dengan Dek Berkembar*. BEng (Hons) Thesis. Universiti Kebangsaan Malaysia.

Siti Khadijah Mohd Noor. 2002. *Kajian Kelakuan Sistem Lantai Komposit Kepingan Keluli Berprofil/ Papan Kering Secara Ujikaji*. BEng (Hons) Thesis. Universiti Kebangsaan Malaysia.

Wan Badaruzzaman, W.H. 1994. *The Behaviour of Profiled Steel Sheet/Dryboard System,* PhD Thesis, University of Wales Cardiff, U.K.

Wan Badaruzzaman, W.H., Ahmed, E. and Rashid, A.K. 1996. Out-of Plane Bending Stiffness along the Major Axis of Profiled Steel Sheet Dryboard Composite Floor Panels. *Jurnal Kejuruteraan UKM*, No.8, 79–95.

Wan Badaruzzaman W.H., Zain M.F.M., Akhand A.M., and E. Ahmed, 2003: Dry Board as Load Bearing Element in the Profiled Steel Sheet Dry Board Floor Panel System – Structural Performance and Applications, *Journal of Construction and Building Materials*, 17(4), pp. 289–297.

Steel and Composite Structures – Wang & Choi (eds)
© *2007 Taylor & Francis Group, London, ISBN 978-0-415-45141-3*

Buckling behavior of cold-formed sigma-purlins partially restrained by steel cladding

Zhan-Jie Li & Xiao-Xiong Zha
School of Civil Engineering, Harbin Institute of Technology, Shenzhen Graduate School, Shenzhen, P.R. China

ABSTRACT: This paper presents an analysis of the influences of the cladding on the buckling behavior of cold-formed sigma purlins subjected to downward loading or/and uplift loading. The restraint of the cladding to the purlin is simplified by using two springs representing the translational and rotational restraints. The analysis is performed using finite strip method. The results highlight the differences of the cladding restraints and illustrate how they affect the local, distortional, and lateral-torsional buckling of the section for the two loading cases. Moreover, this paper also examines the difference between the pure bending and uniformly distributed load in a purlin-sheeting system in terms of critical loads of local, distortional, and lateral-torsional buckling.

Keywords: cold-formed steel; Purlin; Steel; Thin-walled; Restrained; Cladding; Sigma.

1 INTRODUCTION

Cold-formed steel members commonly used as both secondary and primary structural members today may exhibit local, distortional and lateral-torsional buckling. It is well known that thin-walled members must carefully consider the role of cross-section instability in their design. Most design methods used in current design standards and specifications that account for local and distortional buckling are based on the effective width concept for stiffened and unstiffened elements. The key part to calculate the effective width is the selection of the technique to determine the critical loads related to local and distortional buckling. Currently, analytical methods (Rhodes & Lawson 1993, Hancock 2003, Schafer, Pekoz 1999) based on various simplified models and the finite strip methods (Cheung 1998, Hancock 1997, Schafer 2003) are widely used to calculate these critical loads.

Cold-formed sigma sections are usually used as purlins or rails, the intermediate members between the main structural frame and the corrugated roof or wall sheeting in buildings for farming and industrial use. Practically, the performance of purlins may be influenced by the cladding system. Early research on the cold-formed steel section was focused on the buckling behaviors of various individual sections, in which the restraint of the cladding on the purlins was assumed to be either ignored or fully provided (Hancock 1997). Recently, a general finite strip buckling analysis package was developed by Schafer (Schafer 2003), in which nodal spring restraints can be added, and can be

used to carry out local, distortional and lateral-torsion buckling analyses of purlin-sheeting systems for any prescribed pre-buckling stresses.

Historically, the load-bearing capacity of the purlin-sheeting system has been determined by full-scale testing (Davies 2000). However, with the increasing use of various new cladding systems, and the ever increasing cost of testing, there is now a trend to analytical design procedure. Most recently, Li developed an analytical model for predicting the lateral-torsional buckling of cold-formed zed-purlin partial-laterally restrained by metal sheeting (Li 2004). Chu et al. examined the partial restrained channel-section beams (Chu et al. 2004). Ye et al. analyzed buckling behavior of clod-formed zed-purlin partially restrained by steel sheeting (Ye et al. 2002).

In this paper, the influence of the cladding on the buckling behavior of the sigma section is analyzed. The analyses are conducted in two cases. One is for downward loading where the up flange is in compression; the other is for uplift loading where the up flange is in tension. Moreover, for the purlin-sheeting system, it is very common that the cold-formed member is subjected to a uniformly distributed transverse load. Also, sigma sections are more effective in terms of the web depth because of the stiffeners used. When the sections are used together with anti-sag bars the dominant buckling is likely to be the distortional buckling, which has relatively longer half-wave length than that of local buckling, so the varying stresses may have significant influence on it. Therefore, the buckling problem of purlin-sheeting system subjected to

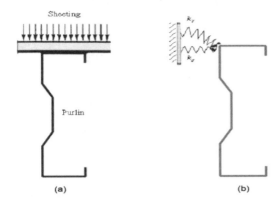

Figure 1. (a) Purlin-sheeting system (b) Analysis model for purlin-sheeting system

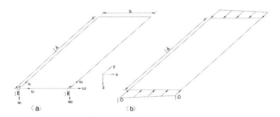

Figure 2. (a) Displacement field for a typical simply supported finite strip and (b) strip with edge traction.

uniformly distributed transverse load are also studied by using finite strip methods.

2 ANALYTICAL MODEL AND ELASTIC BUCKLING ANALYSIS

The buckling analysis performed here is based on the classical theory of linear buckling analysis (Cheung 1998). The spring model used for the buckling analysis is shown in Fig. 1. The restraints provided by the sheeting to the sigma purlin are assumed to be two forms, a translational spring, k_s, and a rotational spring, k_r, as shown in Fig. 1. Consistent with the notations used in the finite strip method, the local coordinate system is shown in Fig 2.

The three displacement of the strip at a point (x, y) can be expressed in terms of the nodal displacements as follows:

$$\left\{\begin{matrix} u(x,y) \\ v(x,y) \end{matrix}\right\} = \sum_{m=1} \begin{bmatrix} \sin\frac{m\pi y}{a} & 0 \\ 0 & \cos\frac{m\pi y}{a} \end{bmatrix} \begin{bmatrix} 1-\frac{x}{b} & 0 & \frac{x}{b} & 0 \\ 0 & 1-\frac{x}{b} & 0 & \frac{x}{b} \end{bmatrix} \begin{Bmatrix} u_{1m} \\ v_{1m} \\ u_{2m} \\ v_{2m} \end{Bmatrix} \quad (1)$$

$$w(x,y) = \sum_{m=1} \sin\frac{m\pi y}{a} \begin{bmatrix} 1-\frac{3x^2}{b^2}+\frac{2x^3}{b^3} & x-\frac{2x^2}{b}+\frac{x^3}{b^2} & \frac{3x^2}{b^2}-\frac{2x^3}{b^3} & \frac{x^3}{b^2}-\frac{x^2}{b} \end{bmatrix} \begin{Bmatrix} w_{1m} \\ \theta_{1m} \\ w_{2m} \\ \theta_{2m} \end{Bmatrix} \quad (2)$$

The assumed displacement functions satisfy the simply supported boundary conditions. Thus the element stiffness matrix can be obtained by substituting the displacement functions (1) and (2) into the strain energy equation and the element geometric stiffness matrix can be derived from the work done by membrane stresses through the nonlinear strains of the buckling displacements. When deriving the geometric stiffness matrix, the function defining the variation of the longitudinal stress along the y-axis is f(y) = 1 for either pure compression or pure bending and f(y) = 4(ay−y²)/a² for a uniformly distributed transverse load. Note that when f(y) is not equal to 1 the element geometric stiffness matrices of different wave numbers are coupled with each other.

After assembly of element matrices the following global matrix equation can be obtained:

$$[K_M]\{d\} = \lambda[K_G]\{d\} \quad (3)$$

where $[K_M]$ is the stiffness matrix, $[K_G]$ is the geometrical stiffness matrix, λ is the loading proportional factor (the actual load is $q = \lambda q_0$, where q_0 is the referenced loading density), and $\{d\}$ represents the vector of displacements describing the modes of buckling (the eigenvector). The critical load can be obtained by solving the eigenvalue matrix equation (3).

For the analytical model of the purlin-sheeting system, the finite strip equation for the linear elastic buckling analysis is shown as the following algebraic eigenvalue problem:

$$([K_M]+[K_S])\{d\} = \lambda[K_G]\{d\} \quad (4)$$

where $[K_S]$ is the stiffness matrix corresponding to the springs. Full derivation of this semi-analytical finite strip method and the expressions of $[K_G]$ can be found in references (Cheung 1998, Schafer 2003, Chu et al. 2004). The expression of $[K_M]$ can be found in references (Cheung 1998, Hancock 1997, Schafer 2003).

3 NUMERICAL EXAMPLES

For the purpose to illustrate the buckling behaviors of cold-formed sigma beams under pure bending and uniformly distributed loads, and also to demonstrate the application and to provide a better understanding of capabilities of the approach presented in the previous section, a numerical example is studied in this section. The numerical example refers to a sigma section

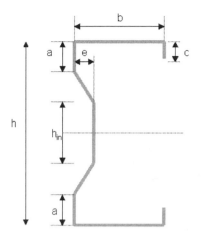

Figure 3. The shape of the sigma section and the notations used.

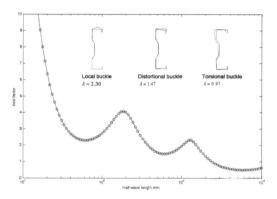

Figure 4. Relationship between critical stress and the half-wave length ($k_s = 20.0$ N/m, $k_r = 400$ kN-mm).

shown in Fig 3. The geometrical parameters are given as: h = 260 mm, b = 70 mm, a = 65 mm, e = 16 mm, c = 20 mm, t = 2.0 mm and the angle between the vertical and inclined webs is 150°. For the chosen section, two different loading cases, downward loading and uplift loading (applied at shear centre), are discussed.

For any given spring constants k_s and k_r, a typical buckling curve for the purlin with spring constants k_s is 20.0 N/m² and k_r is 400 kN is shown in Fig. 4. It can be seen from the figure that there are two minimum points that represent a local buckle, occurring at a critical load $\lambda = 2.27$ over a half-wavelength of 61 mm and a distortional buckle, occurring at a critical load $\lambda = 1.47$ over a half-wavelength of about 610 mm. Lateral–torsional buckling is found to occur at relatively long half-wavelengths (>1500 mm) with lower critical loads ($\lambda < 1.47$).

In the analytical model of the purlin-sheeting system subjected the downward load the influence of the two spring constants on the buckling behaviors of the

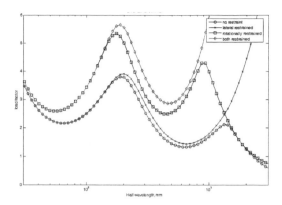

Figure 5. Buckling behavior of the purlin-sheeting system under the downward load.

purlin is shown in Fig. 5. The results show that, for local buckling only the rotational spring k_r has significant influence, the translational spring k_s has no influence at all. This may be explained that the rotational spring k_r restrains the rotation of the flange and web at the corner so the elastic critical load is increased. For the distortional buckling k_r and k_s have a mixed influence, however, the rotational spring k_r is more influential than the translational spring k_s. The effect of the rotational spring k_r to the distortional buckling is similar to that of the local buckling except that the translational spring k_s also provides restraint to restrict the distortional buckling. However, the effect of the translational spring k_s is not significant as that of rotational spring k_r. It is also worthy to be pointed that when the rotational spring k_r exists, the translational spring seems to have increased influence on the critical load. For the lateral-torsional buckling the translational spring k_s has the dominant influence while the influence of the rotational spring k_r seems not significant. This is due to the great restraint effect provided by the translational spring k_s when the sigma purlin tends to lateral buckling. The results imply that the worst situation is no restraints to the purlin.

Figure 6 shows the buckling behavior of the sigma section under the uplift load in the analytical model of the purlin-sheeting system. The results illustrate the influence of the two springs on the buckling behavior. It can be seen that when the purlin-sheeting system subjected to uplift load, the two springs have no influence on the local and distortional buckling at all due to the fact that the buckling region for local and distortional bucking is in the lower part of the web and the lower flange so the restraints of the two springs at the upper corner between the web and flange have no effect at all. However, for the lateral-torsional buckling, the rotational spring k_r surely has influence but not so much significant. Moreover, the translational

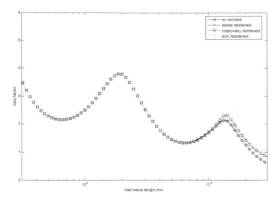

Figure 6. Buckling behavior of the purlin-sheeting system under the uplift load.

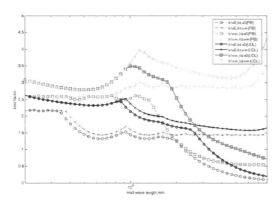

Figure 7. Critical load curves of different restraints under uniformly distributed downward loads.

spring k_s has no influence at all unlike the case of the downward load.

To demonstrate the buckling behavior of purlin-sheeting system subjected to uniformly distributed transverse load (downward load) and also illustrate the influence of the two springs on this loading case, the same sigma section is studied by using the theory explained in the above section. The critical load curves are shown in Fig. 7. It can be seen that the influences of the springs are similar with that of the buckling curves of the pure bending case shown in Fig. 5. Clearly, when the translational spring k_s and the rotational spring k_r are fully effective, the lateral-torsional buckling does not occur. Also to illustrate the difference between the pure bending and uniformly distributed load, the critical load curves of the same sigma section under pure bending are also superimposed in the figure. It can be seen that when the span length exceeds 1 m and does not exceed 3 m which means the buckling mode of the purlin is distortional buckling for most cases,

the critical load of uniformly distributed loading is significantly higher than that of pure bending.

For the uplift loading case, the influences of the two springs are not significant. The difference between pure bending and uniformly distributed loading is similar with the case of downward loading with no restraint.

4 CONCLUSIONS

The buckling behaviors of cold-formed steel sigma section beams restrained by the steel cladding have been investigated using a semi-analytical finite strip method. As the results show, when the purlin-sheeting system is subjected to downward load, the two springs have significant influence on the buckling behavior of the purlin. For local buckling only the rotational spring k_r has significant influence, the translational spring k_s has no influence at all. For the distortional buckling k_r and k_s have a mixed influence, however, the rotational spring k_r is more influential than the translational spring k_s. For the lateral-torsional buckling the translational spring k_s has the dominant influence while the influence of the rotational spring k_r seems not to be significant. When the purlin-sheeting system is subjected to uplift loading, the two springs have no influence on the local and distortional buckling at all. However, for the lateral-torsional buckling, the rotational spring k_r surely has influence but not so much significant in this loading case. Moreover, the critical load for uniformly distributed load is general higher than that for pure bending, especially when the beam is dominated by distortional buckling. This illustrates that the moment gradient along the longitudinal axis have significant influence on the critical load for distortional buckling for the longer half-wave length distortional buckling possesses. Finally, it should be pointed out that the approximation method presented in this paper is only suitable for the beam when the load position is at the shear center. If the load position is not at the shear center, the warping stress effect should be taken into consideration when analyzing the influence of the two springs.

REFERENCES

Rhodes, J Lawson RM 1993. Design of structures using cold formed steel sections. SCI Publication, 089; The Steel Construction Institute.

Hancock GJ 2003. Cold-formed steel structures. Journal of Constructional Steel Research; 59(4):473–87.

Schafer BW, Peköz, T 1999. Local and distortional buckling of cold-formed steel members with edge stiffeners. ICSAS'99, fourth international conference on lightweight steel aluminum structures, Espoo, Finland; 20–23:89–97.

Cheung, Y.K., Tam, L.G. 1998. Finite Strip Method, Chemical Rubber, Boca Raton, Fla.

Hancock G 1997. The behavior and design of cold-formed purlins. Steel Construction; 15(3):2–16.

Schafer BW 2003. Elastic buckling analysis of thin-walled members using the classical finite strip method. CUFSM Version 2.6.: Johns Hopkins University.

Davies JM 2000. Recent research advances in cold-formed steel structures. Journal of Constructional Steel Research;55:267–88.

L.Y Li 2004. Lateral-torsional buckling of cold-formed zed-purlins partial-laterally restrained by metal sheeting. Thin-Walled Structures; 42(7):995–1011.

Xiaoting Chu, Roger Kettle, Longyuan Li 2004. Lateral-torsion buckling analysis of partial-laterally restrained thin-walled channel-section beams. Journal of Constructional Steel Research; 60; 1195-1175.

Zhiming Ye, Roger J. Kettle, Longyuan Li, Benjamin W. Schafer 2002. Buckling behavior of cold-formed zed-purlins partially restrained by steel sheeting. Thin-Walled Structures; 40; 853–864.

Xiao-ting Chu, Zhi-ming Ye, Roger Kettle, Long-yuan Li 2005. Buckling behavior of cold-formed channel sections under uniformly distributed. Thin-walled Structures; 43; 531-542.

Structural analysis

Steel and Composite Structures – Wang & Choi (eds)
© *2007 Taylor & Francis Group, London, ISBN 978-0-415-45141-3*

A co-rotational formulation for curved triangular shell element

Z.X. Li

Department of Civil Engineering, Zhejiang University, Hangzhou, China

ABSTRACT: A six-node curved triangular shell element formulation based on a co-rotational framework is proposed to solve large-displacement and large-rotation problems, where the two smallest of the three components of the mid-surface normal vector at each node are defined as vectorial rotational variables, rendering all nodal variables additive in an incremental solution procedure. Different from most existing co-rotational element formulations, all nodal variables in the present triangular shell element are commutative in calculating the second derivative of the strain energy with respect to nodal variables, as a result, the element tangent stiffness matrix is symmetric. To alleviate membrane and shear locking phenomena, assumed strains are introduced to replace the conforming membrane strains and out-of-plane shear strains in calculating the element strain energy. Finally, several examples of elastic shell problems with large displacements and large rotations are analyzed to demonstrate the reliability, efficiency, and convergence of the present formulation.

1 INTRODUCTION

Shell structures have been widely used in engineering practice. So developing a reliable and efficient solution procedure is important in shell structural analysis, especially for thin shell problems. Lee and Bathe (2004) divided thin shell problems into three categories: membrane-dominated, bending-dominated, and mixed shell problems. Shear and membrane locking phenomena are serious in displacement-based thin shell finite elements for bending-dominated shell structures. These locking phenomena are due to overestimating the contribution of the membrane strain and shear strain in element strain energy, thus overestimating the element membrane and shear stiffness (Huang, 1987).

Reliable and computationally efficient curved triangular shell elements have important applications, as these elements offer significant advantages in modeling arbitrary complex shell geometries. Furthermore, in many cases, triangular elements are always used in conjunction with quadrilateral elements in modeling complex engineering shell structures. Many solution procedures have been proposed to improve the performance of curved triangular shell elements. Up to now, there does not exist, however, a satisfying locking-free triangular shell element. Developing an accurate and computationally efficient triangular element continues to be a challenging undertaking.

In this paper, the Reissner-Mindlin theory is introduced to a six-node co-rotational curved triangular shell element formulation. To alleviate the membrane and shear locking phenomena, the membrane strains and the out-of-plane shear strains are replaced with assumed strains in calculating the element strain energy, and the strategy employed in the MITC approach (Lee and Bathe, 2004) is adopted. In other existing co-rotational element formulations, most researchers adopted non-vectorial rotational variables, and enforced the semi-tangential behaviour of nodal moments through a correction matrix to the conventional geometric stiffness matrix (Yang et al, 2000; Crisfield, 1991,1996), however, due to the non-commutativity of finite rotations about fixed axes, this always leads to an asymmetric tangent stiffness matrix (Simo, 1992; Teh and Clarke, 1999; Izzuddin, 2001). Compared with other existing co-rotational element formulations, the present curved triangular shell element formulation has several features: 1) The vectorial rotational variables are defined, and all nodal variables are additive; 2) all nodal variables are commutative in calculating the second derivatives of the element strain energy with respect to nodal variables, resulting in symmetric element tangent stiffness matrices in the local and global coordinate systems; 3) the element tangent stiffness matrix is updated using the total values of the nodal variables in an incremental solution procedure, making it advantageous for solving dynamic problems.

2 DESCRIPTION OF THE CO-ROTATIONAL FRAMEWORK

In developing a curved triangular shell element formulation, the Reissner-Mindlin theory is adopted. The

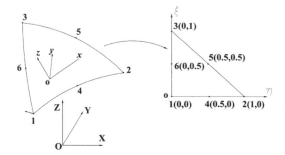

Figure 1.　Definition of local and global coordinate systems.

local and the global Cartesian coordinate systems, and the natural coordinate system are defined respectively as Figure 1, where, the direction of the local x-axis is coincident with Vector $\mathbf{v}_{120}$ at the undeformed configuration, the z-axis is orthogonal to the plane defined by Vector $\mathbf{v}_{120}$ and $\mathbf{v}_{130}$, and the y-axis is orthogonal to the $x-o-z$ plane. Two axes of the natural coordinate system run along two edges of the curved triangular shell element. Vectors $\mathbf{v}_{120}$ and $\mathbf{v}_{130}$ can be calculated from

$$\mathbf{v}_{120} = \mathbf{X}_{20} - \mathbf{X}_{10}, \qquad \mathbf{v}_{130} = \mathbf{X}_{30} - \mathbf{X}_{10} \qquad (1)$$

where, $\mathbf{X}_{i0}$ ($i = 1, 2, 3$) is the coordinates of Node i in the global coordinate system. In the undeformed configuration, the orientation vectors of the local axes are defined by

$$\mathbf{e}_{z0} = \frac{\mathbf{v}_{120} \times \mathbf{v}_{130}}{\left|\mathbf{v}_{120} \times \mathbf{v}_{130}\right|}, \quad \mathbf{e}_{x0} = \frac{\mathbf{v}_{120}}{\left|\mathbf{v}_{120}\right|}, \quad \mathbf{e}_{y0} = \mathbf{e}_{z0} \times \mathbf{e}_{x0} \qquad (2)$$

the local coordinate system will rotate with the element rigid-body motion.

In a deformed configuration, $\mathbf{v}_{12}$ and $\mathbf{v}_{13}$ are calculated as below,

$$\mathbf{v}_{13} = \mathbf{X}_{30} - \mathbf{X}_{10} + \mathbf{d}_3 - \mathbf{d}_1 \qquad (3)$$

$$\mathbf{v}_{12} = \mathbf{X}_{20} - \mathbf{X}_{10} + \mathbf{d}_2 - \mathbf{d}_1 \qquad (4)$$

where, $\mathbf{d}_i$ ($i = 1, 2, 3$) is the displacement vector at Node i. The local coordinate system is defined by

$$\mathbf{e}_z = \frac{\mathbf{v}_{12} \times \mathbf{v}_{13}}{\left|\mathbf{v}_{12} \times \mathbf{v}_{13}\right|}, \quad \mathbf{e}_x = \frac{\mathbf{v}_{12}}{\left|\mathbf{v}_{12}\right|}, \quad \mathbf{e}_y = \mathbf{e}_z \times \mathbf{e}_x \qquad (5)$$

There are 30 degrees of freedom for each element in the local coordinate system. The vector of local nodal variables is given as,

$$\mathbf{u}_L^{\mathrm{T}} = \left\langle u_1 v_1 w_1 r_{1,x} r_{1,y} \cdots u_6 v_6 w_6 r_{6,x} r_{6,y} \right\rangle \qquad (6)$$

where, (u_i, v_i, w_i) are three translational displacements at Node i, and ($r_{i,x}, r_{i,y}$) are two components of the

mid-surface normal vector $\mathbf{p}_i$ in the local coordinate system, they are two vectorial rotational variables.

There are 30 degrees of freedom in the global coordinate system. The nodal variable vector is given as,

$$\mathbf{u}_G^{\mathrm{T}} = \left\langle U_1 V_1 W_1 p_{1,n_1} p_{1,m_1} \cdots U_6 V_6 W_6 p_{6,n_6} p_{6,m_6} \right\rangle \qquad (7)$$

where, (U_i, V_i, W_i) are three translational displacements at Node i; ($p_{i,ni}, p_{i,mi}$) are two vectorial rotational variables, they are the two smallest components of the mid-surface normal vector $\mathbf{p}_i$ at Node i in X, Y, Z directions. This definition can avoid ill-conditioning in the resulting equations.

The relationships between the local and global nodal variables are given as,

$$\mathbf{t}_i = \mathbf{R}(\mathbf{d}_i + \mathbf{v}_{i0}) - \mathbf{R}_0 \mathbf{v}_{i0} \qquad (8a)$$

$$\mathbf{\theta}_{i0} = \mathbf{R}_{h0} \mathbf{p}_{i0} \qquad (8b)$$

$$\mathbf{\theta}_i = \mathbf{R}_h \mathbf{p}_i \qquad (8c)$$

where,

$$\mathbf{t}_i^{\mathrm{T}} = \left\langle u_i \, v_i \, w_i \right\rangle, \qquad \mathbf{d}_i^{\mathrm{T}} = \left\langle U_i \, V_i \, W_i \right\rangle \qquad (9a)$$

$$\mathbf{R}^{\mathrm{T}} = \left\langle \mathbf{e}_x \, \mathbf{e}_y \, \mathbf{e}_z \right\rangle, \qquad \mathbf{R}_0^{\mathrm{T}} = \left\langle \mathbf{e}_{x0} \, \mathbf{e}_{y0} \, \mathbf{e}_{z0} \right\rangle \qquad (9b)$$

$$\mathbf{R}_h^{\mathrm{T}} = \left\langle \mathbf{e}_x \, \mathbf{e}_y \right\rangle, \qquad \mathbf{R}_{h0}^{\mathrm{T}} = \left\langle \mathbf{e}_{x0} \, \mathbf{e}_{y0} \right\rangle \qquad (9c)$$

$$\mathbf{v}_{i0} = \mathbf{X}_{i0} - \mathbf{X}_{10} \qquad (9d)$$

$\mathbf{\theta}_i$ and $\mathbf{\theta}_{i0}$ are respectively the sub-vectors of the deformed and the initial mid-surface normal vectors $\mathbf{r}_i$ and $\mathbf{r}_{i0}$,

$$\mathbf{\theta}_i^{\mathrm{T}} = \left\langle r_{i,x} \, r_{i,y} \right\rangle, \quad \mathbf{\theta}_{i0}^{\mathrm{T}} = \left\langle r_{i0,x} \, r_{i0,y} \right\rangle \qquad (10)$$

$\mathbf{p}_{i0}$ are the initial mid-surface normal vectors at Node i in the global coordinate system. In calculating the internal force vector and element tangent stiffness matrix in the local coordinate system, rigid-body rotations $\mathbf{R}_{h0} \mathbf{p}_{i0}$ are excluded in advance.

In the local coordinate system, the coordinates and displacements at any point in the triangular shell element are interpolated by

$$\mathbf{x} = \sum_{i=1}^{6} h_i \mathbf{x}_{i0} + \frac{1}{2} \zeta \, \mathrm{a} \sum_{i=1}^{6} h_i \mathbf{r}_{i0} \qquad (11)$$

$$\mathbf{t} = \sum_{i=1}^{6} h_i \mathbf{t}_i + \frac{1}{2} \zeta \, \mathrm{a} \sum_{i=1}^{6} h_i (\mathbf{r}_i - \mathbf{r}_{i0}) \qquad (12)$$

where, h_i is the interpolation function at Node i; $\mathbf{x}_{i0}$ is the coordinate vector; a is the thickness of the element; ζ the natural coordinate in the direction of the element thickness.

3 ELEMENT FORMULATION

The Green-Lagrange strains specialized for the shallow curved shell (Crisfield, 1991) are adopted. For convenience, the strains are split into three parts of membrane strains $\boldsymbol{\varepsilon}_m$, bending strains $z_l \boldsymbol{\chi}$, out-of-plane shear strains γ_{xz} and γ_{yz},

$$\boldsymbol{\varepsilon} = \boldsymbol{\varepsilon}_m + z_l \boldsymbol{\chi} \tag{13a}$$

$$\gamma_{xz} = \bar{r}_x - \bar{r}_{x0} + \frac{\partial w}{\partial x} \tag{13b}$$

$$\gamma_{yz} = \bar{r}_y - \bar{r}_{y0} + \frac{\partial w}{\partial y} \tag{13c}$$

where,

$$\bar{r}_x = \sum_{i=1}^{6} h_i r_{i,x}, \quad \bar{r}_{x0} = \sum_{i=1}^{6} h_i r_{i0,x} \tag{14a}$$

$$\bar{r}_y = \sum_{i=1}^{6} h_i r_{i,y}, \quad \bar{r}_{y0} = \sum_{i=1}^{6} h_i r_{i0,y} \tag{14b}$$

$$w = \sum_{i=1}^{6} h_i w_i \tag{14c}$$

The total potential energy Π of the 6-node curved triangular shell element is calculated from,

$$\Pi = \frac{1}{2} \int_V (\boldsymbol{\varepsilon}_m + z_l \boldsymbol{\chi})^T \mathbf{D}_1 (\boldsymbol{\varepsilon}_m + z_l \boldsymbol{\chi}) dV + \frac{1}{2} \int_V \boldsymbol{\gamma}^T \mathbf{D}_2 \boldsymbol{\gamma} \, dV - W_e \tag{15}$$

where, V the volume of the element; W_e the work done by external forces; $\mathbf{D}_1$ and $\mathbf{D}_2$ the elastic constant matrices.

By enforcing the variation of the potential energy Π with respect to $\mathbf{u}_L$, the internal force vector in the local coordinate system is achieved,

$$\mathbf{f} = \mathbf{f}_{ext} = \int_V (\mathbf{B}_m + z_l \mathbf{B}_b)^T \mathbf{D}_1 (\boldsymbol{\varepsilon}_m + z_l \boldsymbol{\chi}) dV + \int_V \mathbf{B}_\gamma^T \mathbf{D}_2 \boldsymbol{\gamma} \, dV$$

$$= \int_V \left(\mathbf{B}_m^T \mathbf{D}_1 \boldsymbol{\varepsilon}_m + (z_l)^2 \mathbf{B}_b^T \mathbf{D}_1 \boldsymbol{\chi} \right) dV + \int_V \mathbf{B}_\gamma^T \mathbf{D}_2 \boldsymbol{\gamma} \, dV \tag{16}$$

By differentiating the internal force vector $\mathbf{f}$ with respect to $\mathbf{u}_L$, the element tangent stiffness matrix is achieved,

$$\mathbf{k}_T = \int_V \left(\mathbf{B}_m^T \mathbf{D}_1 \mathbf{B}_m + (z_l)^2 \mathbf{B}_b^T \mathbf{D}_1 \mathbf{B}_b + \frac{\partial \mathbf{B}_m^T}{\partial \mathbf{u}_m^T} \mathbf{D}_1 \boldsymbol{\varepsilon}_m + \mathbf{B}_\gamma^T \mathbf{D}_2 \mathbf{B}_\gamma \right) dV \tag{17}$$

(16) and (17) are the conforming element formulations for the 6-node curved triangular shell element in the local coordinate system. Due to the commutativity of the local nodal variables in calculating the

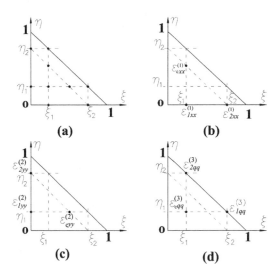

Figure 2. (a) membrane strain tying positions (b) 3 tying points for $\tilde{\varepsilon}_{xx}$ (c) 3 tying points for $\tilde{\varepsilon}_{yy}$ (d) 3 tying points for $\tilde{\varepsilon}_{qq}$.

second derivatives of Π, the resulting element tangent stiffness matrix $\mathbf{k}_T$ is symmetric.

In solving thin shell problems, membrane and shear locking phenomena will deteriorate the convergence and computational efficiency of the element. To improve the performance of the present triangular shell element, the membrane strains and out-of-plane shear strains, and their first and second derivatives with respect to $\mathbf{u}_L$ are replaced with assumed strains and their derivatives with respect to $\mathbf{u}_L$. The strategy used in the Mixed Interpolation of Tensorial Components (MITC) approach (Lee and Bathe, 2004) is employed in calculating assumed strains.

The assumed membrane strains are linearly interpolated by using the corresponding conforming strains at the well-chosen tying points, and 9 tying points (see Figure 2) on the shell mid-surface are defined. ξ_i and η_i ($i = 1, 2$) are the natural coordinates of the tying points,

$$\xi_1 = \eta_1 = \frac{1}{2}\left(1 - \frac{1}{\sqrt{3}}\right), \quad \xi_2 = \eta_2 = \frac{1}{2}\left(1 + \frac{1}{\sqrt{3}}\right) \tag{18}$$

ε_{qq} is defined as,

$$\varepsilon_{qq} = \frac{\varepsilon_{xx} + \varepsilon_{yy}}{2} - \gamma_{xy} \tag{19}$$

Three assumed membrane strains are defined by

$$\tilde{\varepsilon}_{xx} = a_1 + b_1 \xi + c_1 \eta \tag{20a}$$

$$\tilde{\varepsilon}_{yy} = a_2 + b_2 \xi + c_2 \eta \tag{20b}$$

$$\tilde{\varepsilon}_{qq} = a_3 + b_3 \xi + c_3 (1 - \xi - \eta) \tag{20c}$$

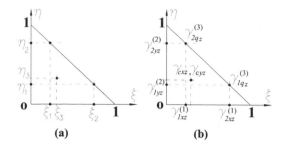

(a)   **(b)**

Figure 3. (a) Transverse shear strain tying positions (b) Transverse shear strains at 7 tying points.

and the in-plane shear strain can be calculated from them,

$$\widetilde{\gamma}_{xy} = \frac{1}{2}\left(\widetilde{\varepsilon}_{xx} + \widetilde{\varepsilon}_{yy}\right) - \widetilde{\varepsilon}_{qq} \tag{21}$$

Let the assumed membrane strains be equal to the conforming strains at 9 tying points, the unknown coefficients a_i, b_i, c_i ($i = 1, 2, 3$) can be achieved.

7 tying points (see Figure 3) are adopted in calculating assumed transverse strains. The tying positions are defined by the well-chosen parameters

$$\xi_1 = \eta_1 = \frac{1}{2} - \frac{1}{2\sqrt{3}} \tag{22a}$$

$$\xi_2 = \eta_2 = \frac{1}{2} + \frac{1}{2\sqrt{3}} \tag{22b}$$

$$\xi_3 = \eta_3 = \frac{1}{3} \tag{22c}$$

The assumed transverse strains are defined by

$$\widetilde{\gamma}_{xz} = a_4 + b_4\xi + c_4\eta + d_4\xi\eta + e_4\xi^2 + f_4\eta^2 \tag{23a}$$

$$\widetilde{\gamma}_{yz} = a_5 + b_5\xi + c_5\eta + d_5\xi\eta + e_5\xi^2 + f_5\eta^2 \tag{23b}$$

To calculate the coefficients in (23a, b), linear transverse shear strains along edges are assumed, and two tying points are chosen at each edge. In addition, a tying point inside the element is defined to express the quadratic variation of the transverse shear strains inside the element. The conforming shear strain and the assumed shear stain at the hypotenuse of the right-angled triangle are calculated respectively from those at two right-angled edges,

$$\gamma_{qz} = \frac{1}{\sqrt{2}}\left(\gamma_{yz} - \gamma_{xz}\right) \tag{24a}$$

$$\widetilde{\gamma}_{qz} = \frac{1}{\sqrt{2}}\left(\widetilde{\gamma}_{yz} - \widetilde{\gamma}_{xz}\right) \tag{24b}$$

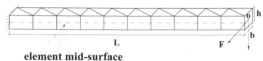

element mid-surface

Figure 4. A cantilever subject to in-plane/out-of-plane shear bending at free end.

By enforcing all above conditions, the coefficients in (23a, b) are achieved.

The improved element formulations are given as,

$$\mathbf{f} = \int_V \left(\widetilde{\mathbf{B}}_m^{\mathrm{T}} \mathbf{D}_1 \widetilde{\boldsymbol{\varepsilon}}_m + (z_l)^2 \mathbf{B}_b^{\mathrm{T}} \mathbf{D}_1 \boldsymbol{\chi}\right) \mathrm{d}V + \int_V \widetilde{\mathbf{B}}_\gamma^{\mathrm{T}} \mathbf{D}_2 \widetilde{\boldsymbol{\gamma}} \, \mathrm{d}V \tag{25}$$

$$\mathbf{k}_{\mathrm{T}} = \int_V \left(\widetilde{\mathbf{B}}_m^{\mathrm{T}} \mathbf{D}_1 \widetilde{\mathbf{B}}_m + \mathbf{B}_b^{\mathrm{T}} \mathbf{D}_1 \mathbf{B}_b + \frac{\partial \widetilde{\mathbf{B}}_m^{\mathrm{T}}}{\partial \mathbf{u}_L^{\mathrm{T}}} \mathbf{D}_1 \widetilde{\boldsymbol{\varepsilon}}_m + \widetilde{\mathbf{B}}_\gamma^{\mathrm{T}} \mathbf{D}_2 \widetilde{\mathbf{B}}_\gamma\right) \mathrm{d}V \tag{26}$$

After introducing the assumed strains, the resulting element tangent stiffness matrix is still symmetric.

The internal force vector $\mathbf{f}_G$ of the element in the global coordinate system can be calculated from that in the local coordinate system,

$$\mathbf{f}_G = \mathbf{T}^{\mathrm{T}}\mathbf{f} \tag{27}$$

where, $\mathbf{T}$ is the transformation matrix. It can be calculated from the relationships of the local and global nodal variables (see (8a–c)).

The element tangent stiffness matrix $\mathbf{k}_{\mathrm{TG}}$ in the global coordinate system can be calculated by differentiating $\mathbf{f}_G$ with respect to $\mathbf{u}_G$,

$$\mathbf{k}_{\mathrm{TG}} = \frac{\partial \mathbf{f}_G}{\partial \mathbf{u}_G^{\mathrm{T}}} = \mathbf{T}^{\mathrm{T}}\mathbf{k}_{\mathrm{T}}\mathbf{T} + \frac{\partial \mathbf{T}^{\mathrm{T}}}{\partial \mathbf{u}_G^{\mathrm{T}}}\mathbf{f} \tag{28}$$

In the right side of (28), the first term is symmetric. The second term includes the second derivative of $\mathbf{u}_L$ with respect to $\mathbf{u}_G$. Considering the commutativity of $\mathbf{u}_G$ in calculating the differentiation, the second term is also symmetric, thus $\mathbf{k}_{\mathrm{TG}}$ is symmetric.

4 ANALYSES OF ELASTIC PLATE AND SHELL PROBLEMS

4.1 *In-plane and out-of-plane shear bending of a cantilever beam*

A cantilever beam with a squared cross-section is subjected to a large transverse shear bending at its free end. Its material properties are $E = 10^7$ and $\mu = 0.3$, and the geometrical parameters are respectively $L = 10$ and $b = h = 0.1$.

10×2 mesh of the present triangular shell element (it is abbreviated to "ASTR" in the following examples) is employed (Figure 4), and two different

994

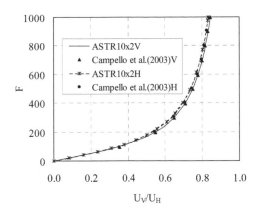

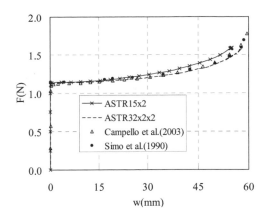

Figure 5. Load-deflection curves of a cantilever plate subject to transverse shear loading.

Figure 7. Load-deflection curves at the in-plane loading point of the L-shaped plate strip.

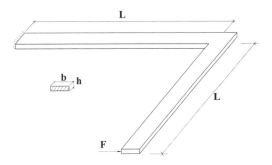

Figure 6. A flat L-shaped plate strip subject to an in-plane load.

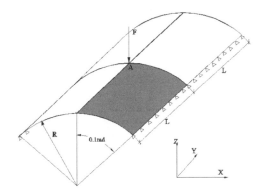

Figure 8. Geometry of a cylindrical shell subject to a point load.

situations are considered: (i) in-plane bending with the load applied in the same plane of the element mid-surface and (ii) out-of-plane bending with the load applied in the out-of-plane direction of the element mid-surface. This scheme aims to verify the capability of the present element in solving large in-plane and out-of plane rotation problems. The load-deflection curves at the free end of the beam are depicted in Figure 5, where the results for both cases are nearly identical. Meanwhile, the present results fit well with those from Campello *et al.* (2003) by using 10×2 T6-3i elements.

4.2 *Lateral buckling of an L-shaped plate strip*

A flat L-shaped plate strip is fully clamped in one edge and subjected to an in-plane point load at the free end (Figure 6). Its material properties are $E = 71240 \, \text{N/mm}^2$ and $\mu = 0.3$.

To investigate the lateral stability of the plate, a very small perturbation load is imposed on the free edge in the out-of-plane direction to initialize the post-critical lateral deflection. The in-plane load against out-of-plane deflection curves calculated by using 15×2 and

$32 \times 2 \times 2$ "ASTR" element meshes fit very well with the results from Campello *et al.* (2003) and Simo *et al.* (1990) (see Figure 7).

4.3 *Hinged cylindrical shell subject to a point load*

A cylindrical shell is hinged along the two straight edges, and is free along the other two edges (see Figure 8). The radius of the cylinder shell is $R = 2540 \, \text{mm}$, its length is $L = 254 \, \text{mm}$, the angle is $\theta = 0.1 \, \text{rad}$, and the shell thickness $t = 12.7 \, \text{mm}$. The Young's modulus and Poisson's ratio are $E = 3105 \, \text{N/mm}^2$ and $\mu = 0.3$, respectively. A concentrated load F is applied at Point A in the middle of the shell. Considering the geometrical symmetry and loading condition, a quarter of the shell (i.e. the colored part) is studied.

The load-deflection curves at the midpoint of the shallow shell calculated by using different "ASTR" element meshes and the results from Moita & Crisfield (1996) and Oliver & Oñate are presented in Figure 9. A nearly converged solution can be

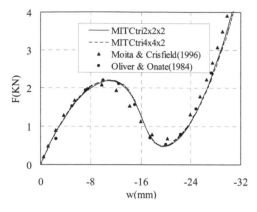

Figure 9. Load-deflection curves at Point A of the hinged cylindrical shell.

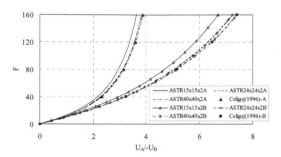

Figure 11. Load-deflection curves of a pinched hemispherical shell.

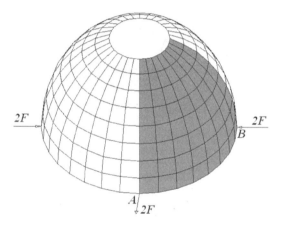

Figure 10. Pinched hemispherical shell with an 18°-hole.

achieved by using $2 \times 2 \times 2$ "ASTR" elements, and the curves from present study agree well with the results from Moita & Crisfield (1996) and Oliver & Oñate.

4.4 *Pinched hemispherical shell with an 18⁰-hole*

A hemispherical shell has an 18°-hole at its top (Figure 10). The shell radius $r = 10$, and its thickness $h = 0.04$. The elastic modulus and Poisson's ratio are $E = 6.825 \times 10^7$ and $\mu = 0.3$, respectively. It is subjected to four concentrated loads symmetrically on the bottom edge. Considering the symmetry of the hemispherical shell and the loads, only a quarter of the shell (i.e. the colored part) is studied.

The load-deflection curves at the pinched points of the hemispherical shell calculated by using different "ASTR" element meshes are presented in Figure 11. For comparison, the results from Celigoj (1996) are

also depicted in this figure, the solutions employing fine "ASTR" element mesh can fit well with them.

5 CONCLUSIONS

A 6-node co-rotational curved triangular shell element formulation employing assumed membrane strains and assumed out-of-plane shear strains is presented. This formulation is element-independent, and has several features: 1) The vectorial rotational variables are defined, so all nodal variables are additive; 2) all nodal variables are commutative in computing the second derivative of the element strain energy with respect to the local nodal variables, and thus a symmetric element tangent stiffness matrix is obtained both in the local coordinate system and the global coordinate system; 3) the total values of the nodal variables are used in updating the element tangent stiffness matrix, thus making it advantageous in solving dynamic problems.

ACKNOWLEDGEMENTS

This work is supported by National Natural Science Foundation of China (50408022) and the visiting scholarship from the Future Academic Star Project of Zhejiang University. In addition, this research also benefits from the financial supports of the Scientific Research Foundation for the Returned Overseas Chinese Scholars, provided respectively by State Education Ministry and Zhejiang Province.

REFERENCES

Campello, E.M.B., Pimenta, P.M., Wriggers P. 2003. A triangular finite shell element based on a fully nonlinear shell formulation. *Computational Mechanics* 31: 505–518.
Celigoj, C.C. 1996. Strain-and-displacement-based variational method applied to geometrically non-linear shells. *International Journal for Numerical Methods in Engineering* 39: 2231–2248.

Crisfield, M.A. 1991. *Nonlinear Finite Element Analysis of Solid and Structures, Vol.1: Essentials*. Chichester: Wiley.

Crisfield, M.A. 1996. *Nonlinear Finite Element Analysis of Solid and Structures, Vol.2: Advanced topics*. Chichester: Wiley.

Huang, H.C. 1987. Membrane locking and assumed strain shell elements. *Computers and Structures* 27: 671–677.

Izzuddin, B.A. 2001. Conceptual Issues in Geometrically Nonlinear Analysis of 3D Framed Structures. *Computer Methods in Applied Mechanics and Engineering* 191: 1029–1053.

Moita, G.F. & Crisfeld, M.A. 1996. A finite element formulation for 3-D continua using the co-rotational technique. *International Journal for Numerical Methods in Engineering* 39(22): 3775–3792.

Oliver, J. & Oñate, E. 1984. A total Lagrangian formulation for the geometrically nonlinear analysis of structures using finite elements. Part 1. Two-dimensional problems: shell and plate structures. *International Journal for Numerical Methods in Engineering* 20: 2253–2281.

Simo, J.C. 1992. (Symmetric) Hessian for geometrically nonlinear models in solid mechanics. Intrinsic definition and geometric interpretation. *Computer Methods in Applied Mechanics and Engineering* 96: 189–200.

Simo, J.C., Fox, D.D. Rifai, M.S. 1990. On a stress resultant geometrically exact shell model. Part III: computational aspects of the nonlinear theory. *Computer Methods in Applied Mechanics and Engineering* 79: 21–70.

Teh, L.H. & Clarke, M.J. 1999. Symmetry of tangent stiffness matrices of 3D elastic frame. *Journal of Engineering Mechanics* -ASCE 125: 248–251.

Yang, H.T.Y., Saigal, S., Masud, A. & Kapania, R.K. 2000. Survey of recent shell finite elements. *International Journal for Numerical Methods in Engineering* 47: 101–127.

Steel and Composite Structures – Wang & Choi (eds)
© 2007 Taylor & Francis Group, London, ISBN 978-0-415-45141-3

Pre- and post-buckling analysis of structures by geometrically Nonlinear Force Method

J.Y. Lu & Y.Z. Luo*

Space Structures Research Center, Zhejiang University, Hangzhou, Zhejiang Province, China
* Correspondence Author, Luoyz@zju.edu.cn

ABSTRACT: The paper is mainly concerned with nonlinear behavior of elastic systems using geometrically nonlinear force method (NFM) instead of geometrically nonlinear finite element method (NFEM). The singular value decomposition (SVD) of the equilibrium matrix is carried out throughout. The arc-length incremental strategies are introduced in the procedure. The method can not only track the whole equilibrium path of structures (including infinitesimal mechanisms), but also reach the exactly singular points. Method separates system into two parts: topology relationship and constitutive relationship of structures. Method has its inherited advantages in bucking analysis. Two classical numerical examples are illustrated in the end.

1 INTRODUCTION

The nonlinear behavior of elastic systems can be summarized as the computations of nonlinear equilibrium paths through limit points and bifurcation points. NFEM that based on finite element method (FEM) is a common approach dealing with these nonlinear responses. Method consists of linearization technique and incremental iteration scheme such as Newton-Raphson's method. In order to analyze post-buckling response, Riks (1979) and Wempner (1971) first proposed arc-length method and Crisfield (1980, 1983) proposed several different versions of arc-length method by updating the constraint equations. However, the algorithms will be invalid when system reaches the exactly singular points although arc-length incremental strategies have been introduced (Papadrakakis 1981). NFEM is most efficient for statically determinate or redundant assemblies. When mechanism mobility exists, the algorithms will be invalid as well. The main reason is that the tangent stiffness matrix may not have full rank when structures reach limit points or mechanism mobility exists.

With the development of new spatial structures and the application of deployable structures and new constructional methods, structure consists of more and more infinitesimal mechanism, even finite mechanism. Kinematic analysis of system is very important in tracing equilibrium paths of these non-traditional structures. Traditional NFEM is always invalid and cannot get more internal information of system.

As Force Method (FM) is less convenient in terms of the matrix operation, it has been more neglected. However, in comparison with the displacement based method, the FM has its inherited advantages. The essential structural and kinematic properties can be interpreted clearly from the orthogonal subspace of the equilibrium matrix. These advantages have made the FM an important part of the curriculum in the subject of the structural mechanics. Avoiding calculating the inversion of stiffness matrix, the SVD of equilibrium matrix has been used by Pellegrino and Calladine (1986). Pellegrino (1993) and Kuznetsov (1988) then extended the method and first order infinitesimal mechanism displacements are analyzed. However, only linearity was considered in their analysis. Therefore, a more precise geometrically nonlinear analysis called geometrically Nonlinear Force Method (NFM) is proposed by Luo and Lu (2006).

In this paper, we will introduce FM's basic equations by analyzing the first variation of potential energy function. The relationship between equilibrium matrix in NFM and tangent stiffness matrix in NFEM is discussed. When arc-length incremental strategies are introduced, the NFM is able to produce accurate solutions in tracing the post-buckling behavior of structures. As it does not require the computation or factorization of any tangent stiffness matrix, the method can not only allow the limit points to be passed, but also, reach the exactly singular points. Two classical structural systems are used as illustrative examples in the end.

2 POTENTIAL ENERGY EQUATION AND MATRIX ANALYSIS OF NFM

2.1 Potential energy equation and equilibrium matrix

Supposing load P_i acts on generalized coordinates Q_i, the potential energy of the mechanism under kinematic constraint $F_k = 0$ is (Pellegrino 2001)

$$\Pi_R = -\sum_{i=1}^{n} P_i(Q_i - Q_i^0) + \sum_{k=1}^{c} \Lambda_k F_k, \qquad (1)$$

where Λ_k are Lagrange multipliers, n is the number of generalized coordinates, and c is the number of generalized constraints.

The equilibrium state of this system can be obtained by taking the first variation of Π_R as zero:

$$\delta\Pi_R = 0. \qquad (2)$$

So we can obtain the equilibrium equation

$$\left(\frac{\partial \Pi_R}{\partial Q_i}\right) = -P_i + \sum_{k=1}^{c} \Lambda_k \frac{\partial F_k}{\partial Q_i} = 0, \quad i = 1,\ldots,n, \qquad (3)$$

and the compatibility equation

$$\left(\frac{\partial \Pi_R}{\partial \Lambda_k}\right) = F_k = 0, \quad k = 1,\ldots,c, \qquad (4)$$

Equilibrium equation in matrix form is shown as

$$\mathbf{J}^T\mathbf{\Lambda} = \mathbf{P}, \qquad (5)$$

where $\mathbf{J}$ is Jacobian matrix.

2.2 Bar element

For structures consist of bars and pin-joints shown in Figure 1, F_k in Equation 1 can be expressed as

$$F_k = \sqrt{(x_i - x_j)^2 + (y_i - y_j)^2 + (z_i - z_j)^2} - L_k = 0. \qquad (6)$$

Substituting into Equation 5, we have

$$\mathbf{At} = \mathbf{P}, \qquad (7)$$

where $\mathbf{A}$ is equilibrium matrix (corresponding to $\mathbf{J}^T$), $\mathbf{t}$ and $\mathbf{P}$ are internal force of bar and nodal force, respectively. $\mathbf{A}$ can be expressed as

$$\mathbf{A} = \left(-c\alpha \quad -c\beta \quad -c\gamma \quad c\alpha \quad c\beta \quad c\gamma\right)^T, \qquad (8)$$

here

$$c\alpha = \frac{x_j - x_i}{l_k}, \quad c\beta = \frac{y_j - y_i}{l_k}, \quad c\gamma = \frac{z_j - z_i}{l_k} \qquad (9)$$

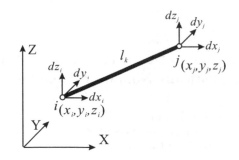

Figure 1. Bar element in real configuration.

Assume that the material of bar is linear elastic, we have

$$\mathbf{e} = \mathbf{e}^0 + \mathbf{Ft}, \qquad (10)$$

where

$$\mathbf{F} = diag(l_1^0/EA_1, l_2^0/EA_2, \cdots, l_i^0/EA_i, \cdots).$$

2.3 Force method and its geometrically nonlinear extension

FM's two basic equations are equilibrium equation shown in Equation 7 and compatible equation shown as follow

$$\mathbf{Bd} = \mathbf{e}, \qquad (11)$$

where $\mathbf{B} = \mathbf{A}^T$. They include the topological information of system and Equation 10 is constitutive relationship of system.

According to the Linear Algebra Theory, the SVD expression of $\mathbf{A}$ (with dimension $n_r \times n_c$ and rank r) is

$$\exists \quad \mathbf{U} \in C^{n_r \times n_r} \quad \mathbf{V} \in C^{n_c \times n_c} \quad \mathbf{S} \in C^{r \times r}$$

subject to $\mathbf{A} = \mathbf{U}\begin{bmatrix} \mathbf{S} & \mathbf{0} \\ \mathbf{0} & \mathbf{0} \end{bmatrix}\mathbf{V}^T, \qquad (12)$

$\mathbf{S} = diag(s_{11}, s_{22}, \ldots, s_{rr})$, $s_{11} \geq s_{22} \geq \ldots \geq s_{rr} \geq 0$. They are nonzero singular value. Then the solution of Equation 7 and Equation 11 give

$$\mathbf{t} = \mathbf{t}' + \mathbf{V}_s\alpha, \qquad (13)$$

$$\mathbf{d} = \mathbf{d}' + \mathbf{U}_m\beta. \qquad (14)$$

where

$$\mathbf{t}' = \sum_{i=1}^{r} \frac{\mathbf{u}_i^T\mathbf{P}}{s_{ii}}\mathbf{v}_i \quad \text{and} \quad \mathbf{d}' = \sum_{i=1}^{r} \frac{\mathbf{v}_i^T\mathbf{e}}{s_{ii}}\mathbf{u}_i.$$

With the consideration of large deformation of structures, geometrical non-linearity should be introduced with the change of topology, so nodal coordinates and internal forces are updated throughout.

The incremental forms of Equation 7 and Equation 11 give

$$\delta \mathbf{A} \cdot \mathbf{t} + \mathbf{A} \delta \mathbf{t} = \delta \mathbf{P},$$ (15)

$$\mathbf{A}^T \delta \mathbf{d} = \delta \mathbf{e},$$ (16)

respectively. And the increments of internal force and axial elongation have

$$\delta \mathbf{e} = \mathbf{F} \delta \mathbf{t}.$$ (17)

Either residual displacements or residual forces are used as convergence criterion for numerical iteration process (Luo & Lu 2006).

$$\frac{\left\| \delta \mathbf{d}^k \right\|_2}{\left\| \mathbf{d}^k \right\|_2} \le \Omega_d, \quad \frac{\left\| \delta \mathbf{P}^k \right\|_2}{\left\| \mathbf{P}^k \right\|_2} \le \Omega_P.$$ (18)

3 RELATIONSHIPS BETWEEN EQILIBRIUM MATRIX AND TANGENT STIFFNESS MATRIX

3.1 Tangent stiffness matrix of bar element

Total differential form of equilibrium equation can be expressed as

$$\delta \mathbf{A} \cdot \mathbf{t} + \mathbf{A} \mathbf{F}^{-1} \mathbf{A}^T \delta \mathbf{d} = \delta \mathbf{P},$$ (19)

$$\frac{\partial \mathbf{A}}{\partial \mathbf{d}} \delta \mathbf{d} \cdot \mathbf{t} + \mathbf{A} \mathbf{F}^{-1} \mathbf{A}^T \delta \mathbf{d} = \delta \mathbf{P}.$$ (20)

We have

$$\frac{\partial \mathbf{A}}{\partial \mathbf{d}} \mathbf{t} = \left(\mathbf{t} \overline{\mathbf{L}} \right) \mathbf{\Delta} - \mathbf{A} \left(\mathbf{t} \overline{\mathbf{L}} \right) \mathbf{A}^T,$$ (21)

where

$$\mathbf{\Delta} = \begin{pmatrix} \mathbf{I}_{3\times3} & -\mathbf{I}_{3\times3} \\ -\mathbf{I}_{3\times3} & \mathbf{I}_{3\times3} \end{pmatrix}, \quad \overline{\mathbf{L}} = diag(1/l_1, 1/l_2, \cdots)_{c \times c}.$$

Substituting into Equation 20 gives

$$\left(\left(\mathbf{t} \overline{\mathbf{L}} \right) \mathbf{\Delta} + \mathbf{A} \mathbf{F}^{-1} \mathbf{A}^T - \mathbf{A} \left(\mathbf{t} \overline{\mathbf{L}} \right) \mathbf{A}^T \right) \delta \mathbf{d} = \delta \mathbf{P},$$ (22)

$$\mathbf{A} \mathbf{F}^{-1} \mathbf{A}^T - \mathbf{A} \left(\mathbf{t} \overline{\mathbf{L}} \right) \mathbf{A}^T - \mathbf{A} \left(\frac{EA_k}{l_k^0} \quad \frac{t_k}{l_k} \right) \mathbf{A}^T.$$ (23)

Here $\frac{EA_k}{l_k^0} - \frac{t_k}{l_k} = \frac{EA_k}{l_k}$ is modified axial stiffness of bar element.

With the definition of tangent stiffness matrix, $\mathbf{K}_T$ in Total Lagrange configuration shows as

$$\mathbf{K}_T = \left(\mathbf{t} \overline{\mathbf{L}} \right) \mathbf{\Delta} + \mathbf{A} \mathbf{F}^{-1} \mathbf{A}^T - \mathbf{A} \left(\mathbf{t} \overline{\mathbf{L}} \right) \mathbf{A}^T$$ (24)

Expanding Equation 24 gives

$$\mathbf{K}_T = \frac{t_k}{l_k} \begin{pmatrix} \mathbf{I}_{3\times3} & -\mathbf{I}_{3\times3} \\ -\mathbf{I}_{3\times3} & \mathbf{I}_{3\times3} \end{pmatrix} + \frac{EA_k}{l_k} \begin{pmatrix} \overline{\mathbf{k}} & -\overline{\mathbf{k}} \\ -\overline{\mathbf{k}} & \overline{\mathbf{k}} \end{pmatrix},$$ (25)

where

$$\overline{\mathbf{k}} = \begin{pmatrix} (c\alpha)^2 & c\alpha \cdot c\beta & c\alpha \cdot c\gamma \\ c\alpha \cdot c\beta & (c\beta)^2 & c\beta \cdot c\gamma \\ c\alpha \cdot c\gamma & c\beta \cdot c\gamma & (c\gamma)^2 \end{pmatrix}.$$ (26)

Equation 25 is exact tangent stiffness which meets result in corresponding references.

3.2 Advantageous of equilibrium matrix method

NFM separates stiffness into two parts, which are derived from topology and material, and solving them respectively by SVD, see Equations 15–17. As it do not require the computation or factorization of any tangent stiffness matrix, the method can not only allow the limit points to be passed, but also, be valid when reach the exact singular points.

4 PROCEDURE OF BUCKLING ANALYSIS BASED ON NFM

4.1 Pre-buckling analysis

NFM is used as iterative algorithm in each load steps until solution converges (satisfy Equation 18). In order to save computer time and space, modified Newton-Raphson method is introduced. The equilibrium matrix $^t\mathbf{A}$ isn't reassembled in each iterative steps at t load step. So the SVD operation of $^t\mathbf{A}$ requires just one time.

4.2 Post-buckling analysis

In order to pass the singular point, arc-length incremental strategies are introduced. The incremental equation of system in k iterative step at t load step shows:

$$^t\mathbf{A} \cdot ^t\delta \mathbf{t}^k = ^t\lambda^k \cdot ^0\mathbf{P} - ^t\mathbf{F}^{k-1}$$ (27.a)

$$^t\delta \mathbf{e}^k = \mathbf{F} \cdot ^t\delta \mathbf{t}^k$$ (27.b)

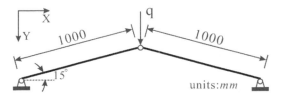

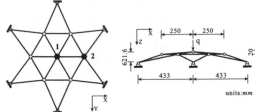

Figure 2. A two bar structure.

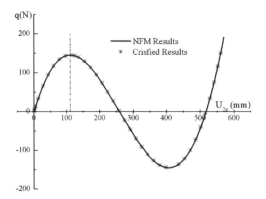

Figure 4. 24-bars dome.

Figure 3. Nodal displacement with load parameter.

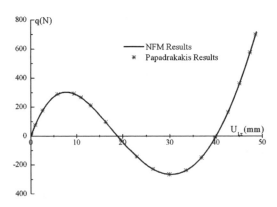

$$\left({}^t\mathbf{A}\right)^{\mathrm{T}}\cdot{}^t\delta\mathbf{d}^k={}^t\delta\mathbf{e}^k \qquad (27.c)$$

where ${}^t\lambda^k$ is load parameter. The nodal displacement and load parameter in k iterative step at t load step is

$${}^t\mathbf{d}^k={}^t\mathbf{d}^{k-1}+{}^t\delta\mathbf{d}^k \qquad (28)$$

$${}^t\lambda^k={}^t\lambda^{k-1}+{}^t\delta\lambda^k \qquad (29)$$

And ${}^t s_{rr}$ represents the geometrical stiffness of structure in total equilibrium path. Once ${}^t s_{rr}$ equals to zero, the structural system changes.

Figure 5. Y-DOF displacement of node 1 with load parameter.

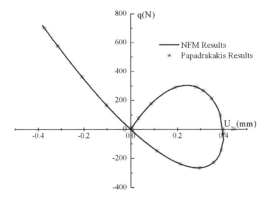

5 EXAMPLES

5.1 Planar two bar structure

Figure 2 shows a planar structure consisting of two bars under a vertically load. Let $EA = 2.1\mathrm{e}4\mathrm{N}$, initial length of bar is 1000 mm. By analyzing the equilibrium, we get $m = 0$, $s = 0$. It is kinematically determinate and statically determinate. We trace pre- and post-buckling path by NFM and the results shown in Figure 3 match the results by Crisfield's (1980) method.

5.2 24-bars dome

A dome with 24 bars is under a concentrate load acting at node 1, see Figure 4. The elastic mode of links

Figure 6. X-DOF displacement of node 2 with load parameter.

is $E = 3.03\mathrm{e}3$ MPa, and the area $A = 317\,\mathrm{mm}^2$. We get $m = 0$, $s = 3$. It is kinematically determinate and statically indeterminate. The buckling path results from NFM and NFEM (Papadrakakis 1981) are given in Figures 5–7. The minimum nonzero singular value of structure is tracked in Figure 8. It represents the geometrical stiffness while buckling.

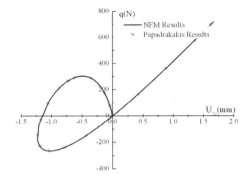

Figure 7. Z-DOF displacement of node 2 with load parameter.

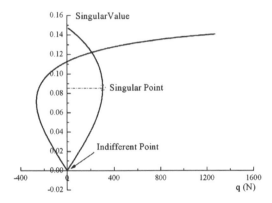

Figure 8. Minimum nonzero singular value with load parameter.

6 CONCLUSION

1. In comparison with the NFEM, the NFM has its inherited advantages: the essential structural and kinematic properties, e.g., the rigid body motions or infinitesimal mobilities, states of the self-stress, etc., can be interpreted clearly from the orthogonal subspace of the equilibrium matrix in the total equilibrium path.

2. Avoiding the computation or factorization of any tangent stiffness matrix, the NFM can not only allow the limit points to be passed, but also, reach the exactly singular points.

3. Like the NFEM, Newton-Raphson method is used in NFM's iteration procedure. When arc-length incremental strategies are introduced, the NFM is able to produce accurate solutions in tracing the post-buckling behavior of structures, even when the infinitesimal mechanisms exist in the assembly.

4. The NFM is less convenient in terms of the matrix operation. SVD operation requires more computer space and time. So in the field of methods for tracking nonlinear equilibrium path of structures, NFM is not a replacement of the FEM package, instead, it is more suitable for the non-traditional structures and structures where internal infinitesimal mechanism exists.

REFERENCES

Batoz, J.L. & Dhatt, G. 1979. Incremental Displacement Algorithms for Nonlinear Problems. *Int. J. Numerical Methods In Engineering* 14: 1262–1267.

Crisfield, M.A. 1980. A Fast Incremental/Iterative Solution Procedure That Handles "Snap-Through". *Computers & Structures* 13: 55–62.

Crisfield, M.A. 1983. An Arc-Length Method Including Line Searches and Accelerations, *Int. J. Numerical Methods In Engineering* 19: 1269–1289.

Kuznetsov, E.N. 1988. Underconstrained structural systems. *Int. J. Solids Structures* 24(2): 153–163.

Luo Y.Z. & Lu J.Y. 2006. Geometrically non-linear force method for assemblies with infinitesimal mechanisms. *Computers & Structures* 84(31–32): 2194–2199.

Riks, E. 1979. An Increment Approach to the Solution of Snapping and Buckling Problems. *Int. J. Solids Structures* 15: 529–551.

Papadrakakis, M. 1981. Post-buckling Analysis of Spatial Structures by Vector Iteration Methods. *Computers & Structures* 14(5–6): 393–402.

Pellegrino, S. & Calladine, C.R. 1986. Matrix Analysis of Statically and Kinematically Indeterminate Frameworks. *Int. J. Solids Structures* 22(4): 409–428.

Pellegrino, S. 1993. Structural Computation with the Singular Value Decomposition of the Equilibrium. *Int. J. Solids Structures* 30(21): 3025–3035.

Pellegrino, S. (ed.) 2001. Deployable Structures. NewYork: *CISM Course and Lectures No. 412. Springer, Wien.*

Wempner, G.A. 1971. Discrete approximations related to nonlinear theories of solids. *Int. J. Solids Structures* 7: 1581–1589.

Steel and Composite Structures – Wang & Choi (eds)
© 2007 Taylor & Francis Group, London, ISBN 978-0-415-45141-3

Modeling method for bionic latticed shell

N. Li & Y.Z. Luo*
Space Structures Research Center, Zhejiang University, Hangzhou, Zhejiang Province, China

ABSTRACT: The purpose of this paper is to describe a new modeling method for bionic latticed shell. This modeling method utilizes the theory of bionics and mathematics to do the modeling of complex free-form shells. Basic geometric features are extracted from the natural spatial shell patterns to reconstruct a free-form surface, and re-interpolation based on reflection meshing method is used to mesh the surface, and several meshing types and the grid quality criteria are adopted. In order to satisfy the engineering requirement, the model can be optimized by strain energy method. Then a modeling program orient to bionic latticed shell is developed and makes the rapid modeling possible.

1 INTRODUCTION

Spacial structures are landmarks and testimonials to the achievements of the structural engineering profession (Bradshaw et al. 2002). The forms of latticed shells increasingly tend to be diversified in the development, and particularly, the bionic shells, which imitate the natural patterns, develop quite rapidly since they vividly exhibit the power of life and the magic of nature. Balz & Böhm (2004) had practiced the realization of organic concrete shells, and in their opinion, as long as there are no structural limits from span or from thickness of the shell, there is an endless field of artistic design possibilities (Balz & Böhm 2004). However, the natural shapes are often free-form, and their modeling work is quite difficult because we do not have the precise analytic expression of the free-form shape. In engineering fields, accomplishing an objective with a minimum of effort, either in terms of material, time or other expense, is a basic activity. For this reason it is easy to understand the interest designers have in different techniques like bionics and mathematics theory (Stach 2004). Raup (1962) studied the seashell geometry and found that the surface of seashells could be expressed by four basic parameters. Stach (2004) described three familiar seashells with those basic parameters and obtained the corresponding analytical curves. Harald (2006) reported that digital workflows that link architectural form-finding, structural design and manufacturing processes are a basic condition to realize free-form architectures.

In this paper, the modeling method combined with the theory of bionics and mathematics imitates the natural spatial pattern and models the practical structure. In the process, the re-interpolation based on reflection meshing method is used to advance the general reflection meshing method, and the output of several meshing types is satisfied and can be evaluated by the grid quality criteria. Furthermore strain energy optimized method is adopted to guarantee the structural behavior. This particular process makes the modeling of bionic latticed shell interesting.

2 IMITATE THE NATURAL SPATIAL PATTERN

It is a really complicated task to describe the natural spatial shell pattern, whose surface is usually free-form. The aim of the bionic study is to extract geometric features rationally from free-form shape. Through observing the *scallop, pilosa* and other spatial patterns, it can be found that the geometric features can be described by base-lines and ridge-lines, as shown in Figure 1. The base-lines can determine the basic shape of the pattern as well as the ridge-lines can determine the curvature. To build base-lines is easy and after they are determined, the curvature of the ridge-lines could be adjusted arbitrarily to get certain patterns.

If the base-lines and ridge-lines have been established, data-points can be drawn. Figure 2 shows that the data-points are of the topological relationship and comprise points from base-lines and ridge-lines.

To get the ideal curvature, the data-points which are placed on ridge-lines can be adjusted in two ways, which are separately named as z-coordinate offset and expansion coefficient set, as shown in Figure 3.

---

* Correspondence Author, Luoyz@zju.edu.cn

by the corresponding data-points placed in the same z-coordinate.

3 SURFACE RECONSTRUCTION AND MESHING METHOD

After data-points acquisition, using B-spline interpolation and reflection meshing method approached by finite element theory can reconstruct the surface and model the practical structures.

3.1 B-spline interpolation surface

The B-spline interpolation surface can be expressed by the following equation:

$$p(u,v) = \sum_{i=0}^{m} \sum_{j=0}^{n} d_{i,j} N_{i,3}(u) N_{j,3}(v) \quad 0 \le u,v \le 1 \quad (1)$$

where $u, v =$ third order parameters; $d_{i,j}(i = 0, 1, \ldots, m; j = 0, 1, \ldots, n) =$ control nodes; and $N_{i,3}(u)$, $N_{j,3}(v) =$ normative B-spline functions. Totally, there are $(m + 1) \times (n + 1)$ control points needed to build a control grid. $N_{i,3}(u)$ and $N_{j,3}(v)$ are third order polynomial related to node vectors as $U = [u_0, u_1, \ldots, u_{m+4}]$ and $V = [v_0, v_1, \ldots, v_{n+4}]$.

Apparently, node vectors and control nodes are not pre-determined, and to generate the required surface, the unknown variables such as node vectors and control nodes should be solved by prepared data-points first, and then B-spline interpolation utilizes the recursive algorithm to compute the free-form surface (Shi 1994).

3.2 Re-interpolation based on reflection meshing method

To get the regular grids, using reflection meshing method is a effective way, which is developed from the finite element theory. As it is known, this method requests a regular parameters' field. The B-spline interpolation surface needs a rectangular field, so reflection meshing method can be used to calculate the target grid from parameters' field to the actual surface, and the basic procedure (Zheng et al. 1998) is as follows:

1 Mesh the rectangular parameters' field and get nodes (u_i, v_j).
2 Substituting each nodes into Equation 1, gives the corresponding $p(u_i, v_j)$.

Reflection meshing method is very simple, and if the parameters can describe the surface very well, it will produce a good division. However, the inherent flaw in the places whose curvature changes hugely

a. *Scallop.*

b. *Pilosa.*

Figure 1. Natural shell structures. The dashed lines denote the base-lines, and the solid lines denote the ridge-lines.

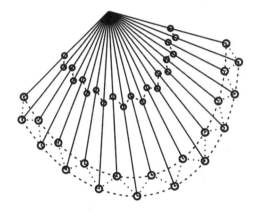

Figure 2. Data-points. The circles express the places of data-points.

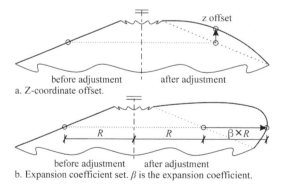

a. Z-coordinate offset.

b. Expansion coefficient set. β is the expansion coefficient.

Figure 3. The ways of adjustment.

In Cartesian coordinates, z-coordinate offset modifies z coordinate of the data-points to the target place, and the expansion coefficient set enlarges or reduces the size of the adjustable ring, which is composed will show uneven grids (Fig. 4a), with the following reasons:

1 In the pre-processing of B-Spline interpolation, the data-points are parameterized by two-direction average normative cumulative chord length parameter algorithm, which is an approximate method to express the precise arc length parameter algorithm. If the surface curvature changes hugely, the parameters will become anamorphic.
2 When the number of required grids is much more than the number of data-points, it is not realistic for every grid to have good parameter in computation.

Therefore, the re-interpolation reflection meshing method is proposed to advance the pre-processing of data-points and can mesh the surface better. The procedure is as follows:

1 Using B-spline interpolation and reflection meshing method to process the data-points, and get the output of enough re-data-points, which are rectangular topology and clearly show the features of the surface.
2 Re-data-points are processed by the same procedure showed above, and the output will be the grid-points.

This approach combines with the B-spline interpolation and the reflection meshing method, so it is similar to the precise arc length parameter algorithm. Since re-data-points have good parameters, the final grid-points can be distributed along the arc length approximately. As shown in Figure 4b, in great curvature changing area, grids tend to be more homogeneous.

3.3 Meshing types

To facilitate modeling, meshing types can be designed as the following solutions (Shen & Chen 1997):

1 Ribbed domes:
   – Set the number of the round and radial grids. This solution is suitable when the grid number has been determined.
   – Set the expectation of the element length, then it can be automatically meshed. It is suitable when the length of the element has been restricted.
   – In addition, the latticed grids can be reduced rationally to avoid too many grids centralizing on the top of ribbed dome.
2 Kiewitt domes:
   – Set the number of central round and radial grids.

Take the latticed shell of circular base-lines for example. The domes which are meshed by the above solutions are shown in Figure 5.

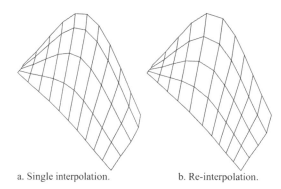

a. Single interpolation.   b. Re-interpolation.

Figure 4.   Comparison of meshing methods.

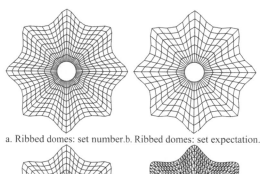

a. Ribbed domes: set number. b. Ribbed domes: set expectation.

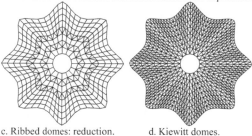

c. Ribbed domes: reduction.   d. Kiewitt domes.

Figure 5.   Meshing types.

3.4 Grid quality criteria

For the latticed shells, it is usually to expect the grids are regular, and the types of elements are few. Therefore, the grid quality criteria are composed of grid shape and element length as follows:

1 The sample of grid quality criterion is grid shape quality coefficient $(G_1, G_2, \ldots, G_m)$, and the mean value is:

$$\overline{G} = \frac{1}{m} \sum_{i=0}^{m} G_i \qquad (2)$$

The variance is:

$$S_G^2 = \frac{1}{m-1} \sum_{i=0}^{m} (G_i - \overline{G})^2 \qquad (3)$$

1007

2 The sample of the other criterion is element length $(L_1, L_2, \ldots, L_n)$, and the mean value is:

$$\bar{L} = \frac{1}{n} \sum_{i=0}^{n} L_i \qquad (4)$$

The variance is:

$$S_L^2 = \frac{1}{n-1} \sum_{i=0}^{n} (L_i - \bar{L})^2 \qquad (5)$$

According to statistical theory, the bigger the mean value of grid quality coefficient, the better the grid quality, and the smaller variance means less grid quality difference and the fewer types of elements.

4 STRAIN ENERGY METHOD

When the surface has been reconstructed, with considering structural behavior, the latticed shell needs to be optimized, which is named as shape optimization. In this phase, the aim is to find the best materials distribution on the structure under a given set of loads function and boundary conditions.

The strain energy of the whole structure evaluates the energy state. The lowest strain energy means the stress distribution of the structure is in even state, and the stiffness is the largest. So the optimization based on the strain energy is to seek a state of which the strain energy E is lowest under certain conditions.

In the geometric parameters, expansion coefficient β can adjust the radial curvature easily (Cen 2006). So β is chosen as the design variable, and its range is given as $\{\beta_1, \beta_2, \ldots, \beta_n\}$. At the same time, the other parameters of the surface are fixed. The aim function is strain energy E, then we can get the following optimized mathematical model (Ossenbruggen 1984):

$$\left. \begin{array}{ll} seek & \beta \\ min. & E \\ subject\ to & \beta \in \{\beta_1, \beta_2, \cdots, \beta_n\} \end{array} \right\} \qquad (6)$$

5 PROGRAM AND EXAMPLES

According to the above modeling method for bionic latticed shell, the computer program has been developed. The program makes the modeling work rapid and efficient. There are two examples as follows:

1 *Scallop*-shaped latticed shell (Fig. 6)
  – Ribbed domes: set grid number.
  – Through the strain energy method optimizing, the expansion coefficient β is 1.44.
  – Rational reduction.

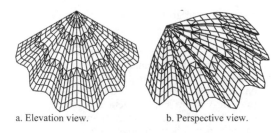

a. Elevation view. b. Perspective view.

Figure 6. *Scallop*-shaped latticed shell.

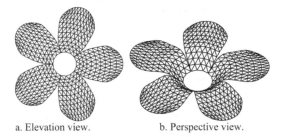

a. Elevation view. b. Perspective view.

Figure 7. *Pilosa*-shaped latticed shell.

  – The variance of element length: 0.250.

2 *Pilosa*-shaped latticed shell (Fig. 7)
  – Kiewitt domes: set the number of central round and radial grids.
  – The mean value of grid quality coefficient: 0.784.
  – The variance of grid quality coefficient: 0.022.
  – The variance of element length: 0.239.

6 CONCLUSION

When utilizing the modeling methods for bionic latticed shell, there are some advises that may be noticed:

1 Bionic latticed shell shows both vogue and natural beauty. However, natural shell cannot be imitated effectively until the basic geometric features have been extracted rationally. Especially for complex shell surface, features are often scattered. In order to get sufficient data-points, 3D scanner can be used to solve the problem.
2 The re-interpolation reflection meshing method overcomes the inherent flaw of reflection meshing method. Though the surface obtained from the re-interpolation is not exactly the same as the initial interpolation, with sufficient re-data-points the output of re-interpolation can express the original surface in enough precision.
3 The meshing types and grid quality criteria adopted in this paper are necessary in the practical application.

With above modeling method for latticed shell structures, the modeling design will become rapid and facilitative, and it will provide a new way to construct the bionic structures.

REFERENCES

Balz, M. & Böhm, J. 2004. Generating shell models and their realization by photogrammetric measurement. In R. Motro (ed.), *Shell and spatial structures from models to realization; Proc. Intern. Symp., Montpellier, 20–24 September 2004*.

Bradshaw, R.R. Campbell, D. Gargari, M. Mirmiran, A. & Tripeny, P. 2002. Special structures: past, present, and future. *Journal of Structural Engineering* 128(6): 691–709.

Cen Pei-chao. 2006. *Parametric Description and Meshing Algorithm for the Surface of Spatial Structure. Ms. D. Dissertation*. College of Civil Engineering and Architecture, Zhejiang University. Hangzhou, Zhejiang Province, China. (in Chinese)

Harald, K. 2006. Structural design of contemporary free-form-architecture. *New Olympics new shell and spatial structures; Proc. Intern. Symp., Beijing, 16–19 October 2006*.

Ossenbruggen, P. J. 1984. *Systems analysis for civil engineering*. New York: John Wiley & Sons Ltd.

Raup, D.M. 1962. Computer as aid in describing form in gastropod shells. *Science* 138: 150–152.

Shen Zu-yan & Chen Yang-ji. 1997. *Space Truss and Latticed Shell*. Shanghai: Tongji University Press. (in Chinese)

Shi Fa-zhong. 1994. *CAGD&NURBS*. Beijing: Higher Education Press. (in Chinese)

Stach, E. 2004. Form-optimizing in biological structures: the morphology of seashells. In R. Motro (ed.), *Shell and spatial structures from models to realization; Proc. Intern. Symp., Montpellier, 20–24 September 2004*.

Zheng Zhi-zhen Li Shang-jian & Li Zhi-gang. 1998. Classification and comparison of algorithm for surface mesh generation. *Computer Aided Engineering* 1: 53–58. (in Chinese)

Steel and Composite Structures – Wang & Choi (eds)
© *2007 Taylor & Francis Group, London, ISBN 978-0-415-45141-3*

Dynamic post-buckling analysis of the fullerene symmetric structures based on the block-diagonalization method

I. Ario, A. Watson & M. Nakazawa

ABSTRACT: To overcome the problems of space in urban areas the subterranean use of dome structures gives an opportunity for exploiting new spaces. In particular the form of the Buckminsterfullerene geodesic dome and other fullerene molecular structures has become popular among architects. The icosahedral symmetry of the structure, which is rarely found in nature, is difficult to analyse. The analysis is further complicated by the subterranean use of such structures. Post-buckling analysis is needed in order to exploit this new frontier of space. The analysis of these highly symmetric systems gives rise to a non-linear problem with a singularity. The authors therefore put forward a method to analyse such structures. In this paper, the authors analyse the dynamic and static large finite displacement post-buckling of the elastic perfect geodesic system using the block diagonalization method of group theory. By applying the analysis to the sphere-fullerene dome, the authors show the mechanism of "symmetry breaking" of axi-symmetric structures.

1 INTRODUCTION

As cities evolve and become ever more densely populated there is a need for new structural forms to overcome the problems associated with high density living. The form of *the Buckminsterfullerene* (5) geodesic dome and other fullerene molecular structures (6) have become popular among architects and physicists. To overcome these problems of space the authors give consideration to the subterranean use of *the Buckminsterfullerene* geodesic dome and examine the behaviour of the dome in this situation where the structural members are highly stressed and subjected to large soil and/or water pressures. This subterranean use of dome structures deep underground represents a new frontier for the use of space (1). These structural forms require more sophisticated structural analysis and includes nonlinear dynamic behaviour. It is well-known that these axi-symmetric systems have critical structural instabilities which can precipitate collapse of the structure. The authors extend this unstable bifurcation problem to the geometrical nonlinear post-buckling behaviour of truss domes.

The authors present the dynamic and static large finite displacement post-buckling analysis of the elastic perfect geodesic system using *the block diagonalization method (BDM)* of group theory. By applying the analysis to the sphere-fullerene dome, the authors show the mechanism of "symmetry breaking" of axi-symmetric structures. The authors show two dynamic post-buckling results for the perfect sphere-fullerene structure subjected to a symmetric distributed load.

Although it is difficult to establish the complex equilibrium path for this axi-symmetric structure the use of *the Block-Diagonalization Method* of group theory makes it possible to decompose several invariant nonlinear equilibrium equations. The method is used to solve for the multiple singularity 0 eigenvalues of nonlinear FEM (2; 3; 4) for axi-symmetric systems. The authors show that it is possible to create highly-symmetric and highly stressed structures for use deep underground using dynamic and static structural analysis.

2 CHARACTERISTIC OF THE SPHERE FULLERENE STRUCTURE

In this section, the characteristics of sphere fullerene structure are described.

2.1 *The sphere fullerene structure*

In order to maintain the mechanical equilibrium of a sphere and its many symmetries, it is necessary to discretise the sphere into uniform elements/shapes. The use of uniform elements ensures that the stress can be uniformly distributed through the structure. The discretisation of the sphere into elements will reduce the number of symmetries of the structure. In addition the fundamental behaviour of the discretised model is not identical to the fundamental behaviour of the fullerene structure and it may be possible to miss critical failure modes. In this paper, the sphere and is represented by

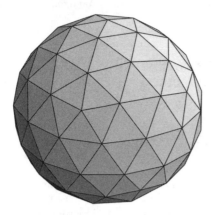

Figure 1. Sphere fullerene structure.

the polyhderon (a sphere fullerene structure) shown in **Fig. 1**. The nodes of fullerene structure would contact the sphere from the inside[1].

It can be seen that pentagonal and hexagonal cells can be used to model the fullerene structure. There are different stresses between these patterns, however since these cells have uniform symmetry the stress distribution is uniform.

3 NONLINEAR BUCKLING ANALYSIS

3.1 *The search of the singular points*

Let the total potential energy function[2] of a discretized system be denoted by $U(f, u)$. f is the load pattern vector and u is the displacement vector. The (N-dimensional) equilibrium equation of the perfect system becomes:

$$\left(\frac{\partial U}{\partial u}\right)^{\mathrm{T}} \equiv F(u, f) = 0 \qquad (1)$$

where, F is a vector function, u, the displacement vector, $\in R^N$, f is the load vector parameter. The solutions (u, f) that satisfy this equilibrium equation form a curve.

To examine the nature of a particular equilibrium point (u, f), it is necessary to differentiate Eq. (1) with respect to u and f to give,

$$\frac{\partial F}{\partial u}\tilde{u} + \frac{\partial F}{\partial f}\tilde{f} + \text{h.o.t.} = 0 \qquad (2)$$

---

[1] The large-scale sphere fullerene has the same symmetry, but with more orbits.

[2] Although the discussion presented is applicable for non-potential systems, use of the total potential is made to make the discussion more comprehensive.

Here, $\tilde{u}$, and $\tilde{f}$ denote the increasing displacement and the increasing load, respectively. The Jacobian matrix ($R^{N \times N}$) is obtained by partially differentiating the equilibrium equation by the displacement u, i.e.

$$J = J(u, f) \equiv \frac{\partial F}{\partial u}.$$

Substituting in equation (2), and rearranging we obtain

$$\tilde{u} = -J^{-1}\frac{\partial F}{\partial f}\tilde{f}$$

Here if J is a regular matrix the solution can be obtained, i.e. if the matrix is not regular the solution cannot be obtained. The stability of the system is assessed by examining the matrix J and whether it is regular or not. From the determinant of the matrix J we see the following:

$$\det J \begin{cases} \neq 0 \bullet \quad \bullet \text{for equilibrium point} \\ = 0 \bullet \quad \bullet \text{for searching singular point} \end{cases}$$

The stability of the system is found from the solution to the eigenproblem. The first singular point is important for the bifurcation analysis.

If this problem is applied directly using dynamic analysis then the following differential equation needs to be expressed as an ODE and/or PDE which depends on time,

$$M\ddot{u}(t) + C\dot{u}(t) + F(u(t), f) = 0$$

where, M and $C \in \mathbf{R}^{N \times N}$ denote mass matrix and damping matrix respectively. The solution to the dynamic analysis is obained by a numerical method. However the nature of the equilibirun path would not be obtained. To overcome this we can use the index of structural stability from the condition of Jacobian matrix.

4 BLOCK DIAGONALIZATION

This section describes the geometrical symmetry of structures with reference to published work (2; 3; 4).

4.1 *Equivariance of equilibrium equation*

In describing the symmetry of the equilibrium equation consider a group G composed of geometric transformations g (such as reflections and rotations). It is assumed that when an element g of G acts on an N-dimensional vector u or F then u is transformed into $g(u)$ and F into $g(F)$. An N-dimensional ($N \times N$)

representation matrix $T(g)$ that describes the action of g on the corresponding vector space is defined by

$$T(g)u = g(u), \quad T(g)F = g(F), \quad g \in G \quad (3)$$

The representation matrix of the load vector[3] f is defined by

$$\widetilde{T}(g)f = g(f), \quad g \in G \quad (4)$$

The representation matrices $T(g)$ and $\widetilde{T}(g)$ are assumed to be orthogonal.

The symmetry of this system is expressed in terms of the invariance of the total potential energy U with respect to the transformation by all elements g of group G. U is called invariant with respect to G, when

$$U(\widetilde{T}(g)f, T(g)u) = U(f, u), \quad g \in G \quad (5)$$

Such invariance is inherited to F such that

$$T(g)F(f, u) = F(\widetilde{T}(g)f, T(g)u), \quad g \in G \quad (6)$$

Eq. (6) is a general symmetry condition applicable for non-potential systems, and is called the equivariance of F to G. **Eq. (6)** means that the transforming of independent variables f and u by $\widetilde{T}(g)$ and $T(g)$ respectively is the same as the transforming of the whole equation F by $T(g)$.

By virtue of this condition, the linear stiffness matrix J in (1) satisfies the symmetry condition

$$T(g)J = JT(g), \quad g \in G \quad (7)$$

and hence can be block-diagonalized by means of a suitable geometric transformation.

The equivariant of the equilibrium equation of a group can be transformed into a set of independent equations corresponding to the irreducible representations of the group. The forms of the transformation matrix and block-diagonal matrix varies with individual groups.

Hence the irreducible representation matrices of group G are defined by

$$T^{\mu}(g) = T_i^{\mu}(g), \ i = 1, \cdots, a^{\mu}, \ g \in G, \ \mu \in R(G) \quad (8)$$

---

[3] In general, the representation matrix of f is different from that of u and F because the dimension of f in general is different from that of u and F.

4.2 Symmetry of sphere fullerene structure

The sphere fullerene structure has the same symmetry as the carbon C60 molecule which was discovered by chemical scientists H.W.Kroto et al(6). This structure is made by truncating regular pentagonal corns of an icosahedron. The symmetry of this fullerene is described by group I_h, and includes groups D_5, D_3 and D_2 [4] This D_n is called the dihedral group, and it is defined as

$$D_n \equiv \langle r, s | r^n = 1, s^2 = 1, rs = sr^{-1} \rangle. \quad (9)$$

This group has $2n$ orders of symmetry and includes rotational group C_n where

$$C_n \equiv \langle r | r^n = 1 \rangle, \quad (10)$$

where r^k is the rotational transformation $r(2\pi k/n)$ $(k = 0, \ldots, n-1)$, and s is the mirror transformation $s^2 = 1$. The icosahedron group with its symmetry has several dihedral groups as subgroups

$$\{D_n \in I_h \mid n = 2, 3, 5\} \quad (11)$$

The whole of the irreducible representation of the dihedral group is defined as

$$R(D_n) = \{\mu \equiv (d, j) \mid j = 1, \cdots, m_d; d = 1, 2\} \quad (12)$$

For example, D_5-symmetry has subgroups D_5, C_5, D_1, C_1, D_1^2, C_1. The irreducible representations corresponding to this symmetry are

$$\mu = (1, 1)_5, (1, 2)_5, (2, 1)_5, (2, 2)_5.$$

and the number m_d that depend on the order d become

$$m_1 = 1, \quad m_2 = 2$$

The fullerene structure has symmetry group I_h given by the systematic irreducible representation of the dihedral group. Group I_h contains twelve subgroups D_5, twenty subgroups D_3 and fifteen subgroups D_2. These relationships are shown in **Fig. 2**.

4.3 Block diagonalization for C_{60}

The geometric transformation matrix which decomposes equilibrium equation (1) into the components of irreducible representations is defined by

$$\begin{aligned} H &\equiv [\cdots, H^{\mu}, \cdots] \\ &= [\, H^{(1,1)_5}, \cdots, H^{(2,2)_5} \,] \end{aligned} \quad (13)$$

---

[4] D_n is the number n of mirror symmetries these symmetries follow Schöenflies notation.

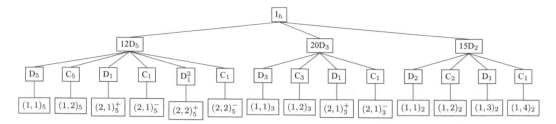

Figure 2. Relationship of I_h – symmetric structure and the irreducible representation.

Here $H^{(1,j)_5}$ indicates blocks for the one-dimensional irreducible representations $(1,j)_5$. Further $H^{(2,j)_5^+}$ and $H^{(2,j)_5^-}$ indicate blocks for the two-dimensional irreducible representations $(2, j)_5$. By means of this transformation matrix H, the stiffness matrix can be transformed into a block-diagonal form:

$$\tilde{J} = H^{\mathrm{T}}JH = \mathrm{diag}[\cdots, \tilde{J}^\mu, \cdots]$$
$$= \mathrm{diag}[\tilde{J}^{(1,1)_5}, \cdots, \tilde{J}^{(2,2)_5}] \quad (14)$$

where $\mathrm{diag}[\cdots]$ denotes a block-diagonal matrix with the diagonal blocks in the parentheses. It is to be noted that two identical diagonal blocks correspond to a two-dimensional irreducible representation. It is computationally efficient to compute each diagonal block using the formula

$$\tilde{J}^\mu = (H^\mu)^{\mathrm{T}}JH^\mu, \quad \mu \in R(\mathrm{D}_5) \quad (15)$$

which exploits the orthogonality of H^μ of the irreducible representations.

The coordinate system associated with the irreducible representation is defined by

$$\boldsymbol{u} = H\boldsymbol{w} = \sum_{\mu \in R(G)} H^\mu \boldsymbol{w}^\mu \quad (16)$$

The equilibrium equation (1) can be decomposed into a set of independent equations

$$(H^{(d,j)_5})^{\mathrm{T}}\boldsymbol{f} = \tilde{J}^{(d,j)_5}\boldsymbol{w}^{(d,j)_5} \quad (17)$$

through the irreducible representation of **Eq.** (16). Thus we obtain the differential equation for each irreducable representation of BDM,

$$M^\mu\ddot{\boldsymbol{w}}^\mu(t) + C^\mu\dot{\boldsymbol{w}}^\mu(t) + \boldsymbol{F}^\mu(\boldsymbol{w}^\mu(t), f^\mu) = \boldsymbol{0}^{\mu}.$$

4.4 Coordinate system and orbits

The geometric orbits of the fullerene structure are shown in **Fig. 3**. Figure (a) shows pentagonal symmetry (D$_5$) as viewed from the top, Figure (b) shows

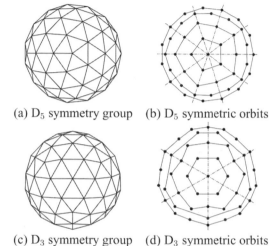

(a) D$_5$ symmetry group (b) D$_5$ symmetric orbits

(c) D$_3$ symmetry group (d) D$_3$ symmetric orbits

Figure 3. Symmetric orbit of the fullerene structure.

nodal network (orbit) keeping the D$_5$ symmetry group with 4 orbits. The center to the outside orbits are defined as the first, second, third orbit etc., respectively. This sphere fullerene has also D$_3$ symmetry shown in figure (c), and the orbital structure shown in figure (d) with 4 orbits. In the next section we obtain the mechanical behaviour of the orbits.

5 POST-BUCKLING ANALYSIS FOR FULLERENE STRUCTURE

We consider the post-buckling analysis of the FEM models for both a part of fullerene-structures and the full fullerene structure. This gives an estimate of the ultimate force and deformation of two different orbits.

5.1 Analysis of the part of fullerene

There were two different types of orbits in this fullerene defined in the previous section namely orbits D$_3$ and D$_5$ shown in the **Fig. 4** (a) and (b) as models 'D5' and 'D3' respectively. Although a range of load cases are considered here we show only the vertical vector force for each node. The point load is defined as 'A' (**Fig. 4** (a) and (b)) and the equivalent distributed load case is shown as 'B' (**Fig. 4** (c) and (d)).

All the models considered in this paper assumed the same boundary and loading conditions.

The equilibrium paths shown in **Fig. 5** are obtained from the numerical results for the models with the D5 and D3 orbits for each case of the load patterns A and B. These figures 5(a) and 5(b) show the relationship between load parameter and the vertical displacement at the top node of the model. Figure 5(a) is for the load pattern A and figure 5(b) is for the load pattern B. We can see the snap-through at the first step on the equilibrium curves for 5(a). It appears at the limit points (LP) LP1 and LP2. The model D3 is stronger

with approximately 1.2 times the stiffness of model D5. The first local snap-through also is longer than D5. On the other hand, there is simple nonlinear behaviour for D5B shown in figure 5(b). Model D3B has complex behaviour with bifurcation point BP and its LP1 is sharply pointed. Hence model D3B has a nonlinear instability around the local buckling BP that branches the bifurcation path from primary equilibrium path for this model.

5.2 Analysis condition for full fullerne structure

There are various collapse mechanisms for the buckling of a sphere shell structure, indeed because of its geometric symmetry this structure has an infinite number of bifurcation paths. Of course, this means it is not possible to analyze all the solutions to obtain all the equilibrium paths. In this paper the analysis of the sphere shell is carried out on the discrete truss element model. This discretised approach enables the analysis of a sphere structure to find post-buckling mechanism and buckling modes.

We assume that the stiffness of all members is unitary stiffness EA, and the radius of a sphere is normalized to $R = 1$ for the discretised model. The fullerene includes two kinds of D_5 and D_3 symmetric surfaces which are analyzed under the unitary vertical loads for all nodes by nonlinear finite displacement analysis. For the boundary conditions it is assumed that the bottom nodes are translationally fixed and the other nodes are free to move in all 3 translational dimensions.

5.3 BDM for Fullerene C_{60}

The stiffness equation (1) of the fullerene truss structure transforms to the block-diagonalization by means

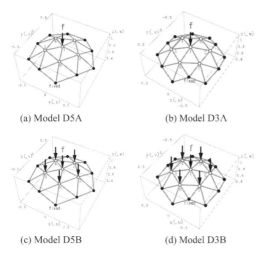

(a) Model D5A   (b) Model D3A

(c) Model D5B   (d) Model D3B

Figure 4. A part of structural models for D_5 and D_3.

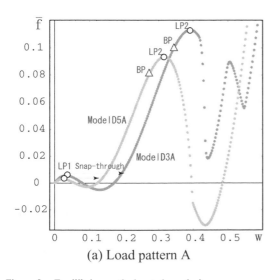

(a) Load pattern A

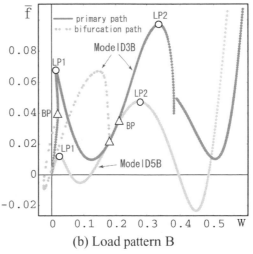

(b) Load pattern B

Figure 5. Equilibrium paths by static analysis.

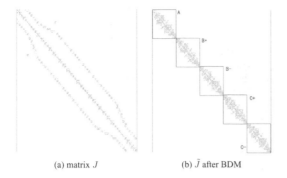

(a) matrix J (b) $\tilde{J}$ after BDM

Figure 6. Before and after the transformation of BDM.

of a suitable geometric transformation for each irreducible representation. The image of the stiffness matrix for the fullerene structure is shown in **Fig. 6**(a). On the other hand, in this case, it is possible to decompose the stiffness matrix to the irreducible representations $(1, 1)_5$, $(2, 1)_5$, $(2, 2)_5$ of a subgroup D_5, When a local transformation matrix corresponding to the non-zero elements of each node is used and converted efficiently then the resultant transformation in the stiffness matrix shown in **Fig. 6**(b). This figure shows a set blocks. The block A corresponds to the irreducible representaion $(1, 1)_5$, the block B corresponds to $(2, 1)_5$ and the block C corresponds to $(2, 2)_5$. Also shown are the orthogonalization blocks $+$ and $-$ for B and C. The dihedral symmetry is given by each irreducible representation; $D_5 \oplus C_5, D_1, C_1, D_1^2, C_1^2$.

5.4 Dynamic analysis of Sphere Fullerene under water pressure

This model has the I_h symmetry group for the loading conditions. The relationship of its symmetric structure and the irreducible representation is shown in **Fig.2**. The high-symmetric model of the sphere is shown in **Fig. 1**. The nodes are subject to same load. The equilibrium path using both the dynamic loading control method and the displacement control method are obtained and shown in **Fig.7**. The figure shows two equilibirium paths: the solid line is for dynamic loading control method; and the other line is for the dynamic displacement control method. Local snap-through behaviour occurs at L.P. where the two different solutions separate. The second L.P., after the local snap-through, is a sharp singular point and appears as an unstable state. The motion associated with this state leads to collapse. The equator of fullerene crushes in this analysis and only the vertical axis remains. Thus, after D_5 symmetry is lost the structure becomes unstable. Therefore attention is needed to examine the sudden loss of strength in this sysem where the singular points become very unstable bifurcation buckling modes.

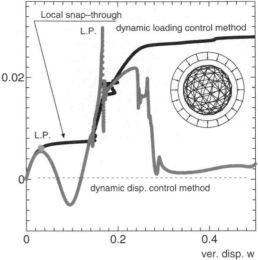

Figure 7. dynamic equilibrium paths.

6 REMARKS

The solution to finding the bifurcation paths of highly symmetric structures is a complex problem. The solution shows that there are many singular points including for the fullerene. In this paper dynamic analysis has been used with the BDM to find the nonlinear equilibrium paths of the sphere fullerene. It is recognized that this analysis is well suited for the singular problem when obtaining the solutions for stable state although we include instability.

REFERENCES

[1] Ario, I. : Fundamental research of the parallel FEM analysis by the irreducible representation decomposition of the fullerene structures, Journal of Structural Engineering, Architectural Institute of Japan (in Japanese), Vol. 45B, pp. 229–235, 1999.

[2] Murota, K., and Ikeda, K. Computational use of group theory in bifurcation analysis of symmetric structures, SIAM, *Journal on Scientific and Statistical Computing*, 12(2), pp. 273–297, 1991.

[3] Ikeda, K., and Murota, K. Bifurcation analysis of symmetric structures using block-diagonalization, *Computer Methods in Applied Mechanics and Engineering*, 86(2), pp. 215–243, 1991.

[4] Ikeda, K., Ario, I., and Torii, K. Block-diagonalization analysis of symmetric plates, *International Journal of Solids and Structures*, 29(22), pp. 2779–2793, 1992.

[5] Fuller, R.B. Synergetics, Macmillan, New York, 1975.

[6] Kroto,H.W., Heath,J.R., O'Brien, S.C., Curl, R.F., and Smally, R.E. C_{60} : Buckminsterfullerene, Nature, **318**, 162–3, 1985.

Steel and Composite Structures – Wang & Choi (eds)

Structure optimization with discrete variables based on heuristic particle swarm optimal algorithm

F. Liu, Z.B. Huang & L.J. Li

Faculty of Construction Engineering, Guangdong University of Technology, Guangzhou, China

ABSTRACT: Pin-connected structures with discrete variables are optimized in this paper based on the heuristic particle swarm optimization (HPSO) algorithm. The HPSO algorithm is based on the standard particle swarm optimizer and the harmony search (HS) scheme. The simulation results show that the HPSO algorithm is able to accelerate the convergence rate effectively and has the fastest convergence rate among the three algorithms used. The research results show the HPSO algorithm can be effectively used to solve optimal problems for steel structures with discrete variables.

1 INTRODUCTION

Many algorithms have been developed to solve the structural engineering optimization problems with continuous variables. However, in reality, the design variables of optimization problems such as the cross-section areas are discretely valued. Traditionally, the discrete optimization problems are solved by mathematical methods by employing round-off techniques based on the continuous solutions. However, the solutions obtained by this method may be infeasible or far from the optimum solutions.

Recently, some papers, published on the subject of the structural engineering optimization, are about the evolutionary algorithms such as the genetic algorithm (GA) (Wu et al. 1995) and the particle swarm optimizer algorithm (PSO) (Li 2006, Kennedy 2001). Most of the applications of the PSO algorithm to structural optimization problems are based on the assumption that the variables are continuous (Li 2007). Based on the HPSO algorithm for continuous variables. This paper improves the HPSO algorithm to deal with discrete valued structural optimization problems. The research results show the HPSO algorithm can handle optimal problems with discrete variables, and has faster convergence rate than the PSO and the PSOPC algorithms, especially in the early iterations (Li 2007).

2 THE OPTIMIZATION FORMULAS

A pin-connected structure optimization problem with discrete variables can be formulated as a nonlinear programming problem and can be expressed as follows: $\min f(x^1, x^2, \ldots, x^d)$, $d = 1, 2, \ldots, D$ subjected to: $g_1(x^1, x^2, \ldots, x^d) \leq 0$, $d = 1, 2, \ldots, D$, $q = 1, 2, \ldots, M$
$x^d \in S_d = \{X_1, X_2, \ldots, X_p\}$
where $f(x^1, x^2, \ldots, x^d)$ is the structure's weight function, which is a scalar function. And $x^1, x^2, \ldots, x^d$ represent a set of design variables. The design variable x^d belongs to a scalar S_d, which includes all permissive discrete variables $\{X_1, X_2, \ldots X_p\}$. The inequality $g_q(x^1, x^2, \ldots, x^d) \leq 0$ represents the constraint functions. The letter D and M are the number of the design variables and inequality functions respectively. The letter p is the number of available variables.

3 THE PARTICLE SWARM OPTIMIZER (PSO) ALGORITHM FOR DISCRETE VARIABLES

The optimization problem with discrete variables is a combination optimization problem which obtains its best solution from all possible variable combinations. The scalar S includes all permissive discrete variables arranged in ascending sequence. Each element of the scalar S is given a sequence number to represent the value of the discrete variable correspondingly. It can be expressed as follows:

$$S_d = \{X_1, X_2, \cdots, X_j, \cdots X_p\}, \quad 1 \leq j \leq p$$

A mapped function $h(j)$ is selected to index the sequence numbers of the elements in set S and represents the value X_j of the discrete variables correspondingly.

$$h(j) = X_j$$

Thus, the sequence numbers of the elements will substitute for the discrete values in the scalar S. This method is used to search the optimum solution, and makes the variables to be searched in a continuous space.

The PSO algorithm includes a number of particles, which are initialized randomly in the search space. The position of the ith particle in the space can be described by a vector x_i,

$$x_i = \left(x_i^1, x_i^2, \cdots, x_i^d, \cdots, x_i^D \right), 1 \le d \le D, \quad i = 1, \cdots, n$$

where D is the dimension of the particles, and n is the sum of all particles. The scalar $x_i^d \in \{1, 2, \ldots, j, \ldots, p\}$ corresponds to the discrete variable set $\{X_1, X_2, \ldots, X_j, \ldots X_p\}$ by the mapped function $h(j)$. Therefore, the particle flies through the continuous space, but only stays at the integer space. In other words, all the components of the vector x_i are integer numbers. The positions of the particles are updated based on each particle's personal best position as well as the best position found by the swarm at each iterations. The objective function is evaluated for each particle and the fitness value is used to determine which position in the search space is the best of the others. The swarm is updated by the following equations:

$$V_i^{(k+1)} = \omega V_i^{(k)} + c_1 r_1 \left(P_i^{(k)} - x_i^{(k)} \right) + c_2 r_2 \left(P_g^{(k)} - x_i^{(k)} \right) \quad (1)$$

$$x_i^{(k+1)} = INT \left(x_i^{(k)} + V_i^{(K+1)} \right) \quad (2)$$

$1 \le i \le n$

where $x_i^{(k)}$ and $V_i^{(k)}$ represent the current position and the velocity of each particle at the kth iteration respectively, $P_i^{(k)}$ the best previous position of the ith particle (called $pbest$) and $P_g^{(k)}$ the best global position among all the particles in the swarm (called $gbest$), r_1 and r_2 are two uniform random sequences generated from U(0, 1), and ω the inertia weight used to discount the previous velocity of the particle persevered (Shi et al. 1997). The object function and the constraint functions can be expressed by the scalar x_i^d as follows:

$$f \left(h(x_i^1), h(x_i^2), \cdots, h(x_i^d), \cdots h(x_i^D) \right)$$
$$g_q \left(h(x_i^1), h(x_i^2), \cdots, h(x_i^d), \cdots h(x_i^D) \right)$$

4 THE PSOPC ALGORITHM FOR THE DISCRETE VALUED VARIABLES

It is known that the PSO may outperform other evolutionary algorithms in the early iterations, but its performance may not be competitive when the number of the iterations increases (Angeline 1998). An improved particle swarm optimizer with passive congregation (PSOPC) is first used in optimization problems with discrete variables (He et al. 2004). The formulations of the PSOPC algorithm for discrete variables can be expressed as follows:

$$V_i^{(k+1)} = \omega V_i^{(k)} + c_1 r_1 \left(P_i^{(k)} - x_i^{(k)} \right) + $$
$$c_2 r_2 \left(P_g^{(k)} - x_i^{(k)} \right) + c_3 r_3 \left(R_i^{(k)} - x_i^{(k)} \right) \quad (3)$$

$$x_i^{(k+1)} = INT \left(x_i^{(k)} + V_i^{(k+1)} \right) \quad (4)$$

$1 \le i \le n$

where R_i is a particle selected randomly form the swarm, c_3 the passive congregation coefficient, and r_3 a uniform random sequence in the range (0, 1): $r_3 \sim U(0, 1)$.

5 CONSTRAINT HANDLING METHOD: FLY-BACK MECHANISM

The most common method for the evolutionary algorithms to handle the constraints is to use penalty functions. However, some experimental results indicate that such technique will reduce the efficiency of the PSO, because it resets the infeasible particles to their previous best positions $pbest$, which sometimes prevents the search from reaching a global minimum (He et al. 2005). He and Wu introduced a new method that is called 'fly-back mechanism' to handle the constraints (He et al. 2005). The particles are initialized in the feasible region. When the optimization process starts, the particles fly in the feasible space to search the solution. If any one of the particles flies into the infeasible region, it will be forced to fly back to the previous position to guarantee a feasible solution. This makes the particles to fly to the global minimum in a great probability. Therefore, such a 'fly-back mechanism' technique is suitable for handling the optimization problem containing the constraints.

6 THE HEURISTIC PARTICLE SWARM OPTIMIZER (HPSO) FOR THE DISCRETE VARIABLES

The heuristic particle swarm optimizer (HPSO) algorithm is introduced by Li (2007) and was first used in continuous variable optimization problems. The HPSO is based on PSO algorithm and the harmony search (HS) algorithm, the latter is based on natural musical performance processes that occur when a musician searches for a better state of harmony, such as during jazz improvisation (Geem 2001). The HPSO algorithm for the discrete valued variables can be expressed as follows:

$$V_i^{(k+1)} = \omega V_i^{(k)} + c_1 r_1 \left(P_i^{(k)} - x_i^{(k)} \right) + \\ c_2 r_2 \left(P_g^{(k)} - x_i^{(k)} \right) + c_3 r_3 \left(R_i^{(k)} - x_i^{(k)} \right) \quad (5)$$

$$x_i^{(k+1)} = INT\left(x_i^{(k)} + V_i^{(k+1)} \right) \quad (6)$$

$1 \le i \le n$

After the (k + 1)th iterations, if $x_i^d < x^d (LowerBound)$ or $x_i^d > x^d (UpperBound)$.

Where x_i is the vector of a particle's position, and x_i^d one component of this vector.

The scalar x_i^d is regenerated by selecting the corresponding component of the vector from *pbest* swarm randomly, which can be described as follows:

$$x_i^d = \left(P_b \right)_t^d, \quad t = INT\left(rand(1, n) \right)$$

where $(P_b)_t^d$ denotes the dth dimension scalar of *pbest* swarm of the tth particle, and t a random integer number.

7 NUMERICAL EXAMPLES

In this paper two pin-connected structures with discrete variables are optimized by the HPSO algorithm. The algorithm proposed is coded in Fortran language and executed on a Pentium 4, 2.93 GHz machine. The PSO and the PSOPC algorithms are applied to the examples to compare the performance of the HPSO algorithm. For all these algorithms, a population of 50 individuals is used. The inertia weight ω, which starts at 0.9 and ends at 0.4, decreases linearly, and the value of acceleration constants c_1 and c_2 are set to 0.5. The passive congregation coefficient c_3 is set to 0.6 for the PSOPC and the HPSO algorithms.

7.1 A 15-bar planar truss structure

A 15-bar planar truss structure, shown in Figure 1, has previously been analyzed by Zhang (2003).

The material density is 7800 kg/m³ and the modulus of elasticity is 200 GPa. The members are subjected to stress limitations of ±120 MPa. All nodes in both directions are subjected to displacement limitations of ±10 mm. There are 15 design variables in this

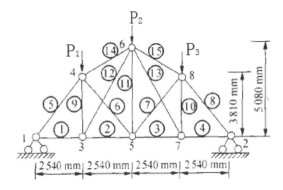

Figure 1. A 15-bar planar truss structure.

Table 1. Comparison of optimal designs for the 15-bar planar truss structure.

Variables (mm²)	Zhang (2003)	PSO	PSOPC	HPSO
A_1	308.6	185.9	113.2	113.2
A_2	174.9	113.2	113.2	113.2
A_3	338.2	143.2	113.2	113.2
A_4	143.2	113.2	113.2	113.2
A_5	736.7	736.7	736.7	736.7
A_6	185.9	143.2	113.2	113.2
A_7	265.9	113.2	113.2	113.2
A_8	507.6	736.7	736.7	736.7
A_9	143.2	113.2	113.2	113.2
A_{10}	507.6	113.2	113.2	113.2
A_{11}	279.1	113.2	113.2	113.2
A_{12}	174.9	113.2	113.2	113.2
A_{13}	297.1	113.2	185.9	113.2
A_{14}	235.9	334.3	334.3	334.3
A_{15}	265.9	334.3	334.3	334.3
Weight (kg)	142.117	108.84	108.96	105.735

example. The discrete variables are selected from the set D = {113.2, 143.2, 145.9, 174.9, 185.9, 235.9, 265.9, 297.1, 308.6, 334.3, 338.2, 497.8, 507.6, 736.7, 791.2, 1063.7} (mm²). Three load cases are considered: Case 1: $P_1 = 35$ kN, $P_2 = 35$ kN, $P_3 = 35$ kN; Case 2: $P_1 = 35$ kN, $P_2 = 0$ kN, $P_3 = 35$ kN; Case 3: $P_1 = 35$ kN, $P_2 = 35$ kN, $P_3 = 0$ kN. A maximum iteration step of 500 is imposed. Optimal results and convergence rates by various methods are shown in Table 1. and Figure 2 respectively.

After 500 iterations, three algorithms have obtained good results, which are better than the Zhang's (Zhang 2003). Figure 2 shows that the HPSO algorithm has the fastest convergence rate, especially in the early iterations. Table 1 shows the HPSO algorithm has the best optimal result.

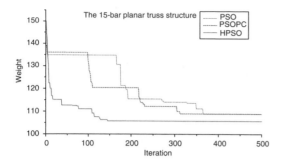

Figure 2. Comparison of convergence rates for the 15-bar planar truss structure.

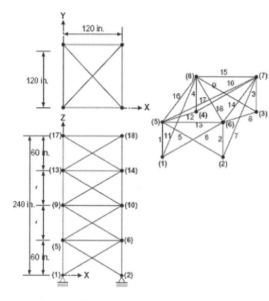

Figure 3. The 72-bar spatial truss structure.

7.2 A 72-bar spatial truss structure

A 72-bar spatial truss structure, shown in Figure 3, has been studied by Wu (1995) and Kang (2005). The material density is 0.1 lb/in.^{3} and the modulus of elasticity is 10,000 ksi. The members are subjected to stress limitations of ±25 ksi. The uppermost nodes are subjected to displacement limitations of ±0.25 in. both in x and y directions. Two load cases are listed in Table 2. There are 72 members, which are divided into 16 groups, as follows: (1) $A_1 \sim A_4$, (2) $A_5 \sim A_{12}$, (3) $A_{13} \sim A_{16}$, (4) $A_{17} \sim A_{18}$, (5) $A_{19} \sim A_{22}$, (6) $A_{23} \sim A_{30}$ (7) $A_{31} \sim A_{34}$, (8) $A_{35} \sim A_{36}$, (9) $A_{37} \sim A_{40}$, (10) $A_{41} \sim A_{48}$, (11) $A_{49} \sim A_{52}$, (12) $A_{53} \sim A_{54}$, (13) $A_{55} \sim A_{58}$, (14) $A_{59} \sim A_{66}$ (15) $A_{67} \sim A_{70}$, (16) $A_{71} \sim A_{72}$. There are two optimization cases to be implemented. Case 1: The discrete variables are

Table 2. The load cases for the 72-bar spatial truss structure.

| | Load Case 1 | | | Load Case 2 | | |
| | P_X (kips) | P_Y (kips) | P_Z (kips) | P_X (kips) | P_Y (kips) | P_Z (kips) |
Nodes						
17	5.0	5.0	−5.0	0.0	0.0	−5.0
18	0.0	0.0	0.0	0.0	0.0	−5.0
19	0.0	0.0	0.0	0.0	0.0	−5.0
20	0.0	0.0	0.0	0.0	0.0	−5.0

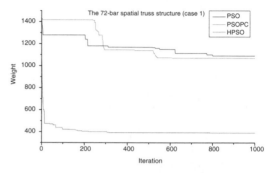

Figure 4. Comparison of convergence rates for the 72-bar spatial truss structure (case 1).

Table 3. Comparison of optimal designs for the 72-bar spatial truss structure (case 1).

Variables (in^2)	Wu (1995)	Kang (2005)	PSO	PSOPC	HPSO
$A_1 \sim A_4$	1.5	1.9	2.6	3.0	2.1
$A_5 \sim A_{12}$	0.7	0.5	1.5	1.4	0.6
$A_{13} \sim A_{16}$	0.1	0.1	0.3	0.2	0.1
$A_{17} \sim A_{18}$	0.1	0.1	0.1	0.1	0.1
$A_{19} \sim A_{22}$	1.3	1.4	2.1	2.7	1.4
$A_{23} \sim A_{30}$	0.5	0.6	1.5	1.9	0.5
$A_{31} \sim A_{34}$	0.2	0.1	0.6	0.7	0.1
$A_{35} \sim A_{36}$	0.1	0.1	0.3	0.8	0.1
$A_{37} \sim A_{40}$	0.5	0.6	2.2	1.4	0.5
$A_{41} \sim A_{48}$	0.5	0.5	1.9	1.2	0.5
$A_{49} \sim A_{52}$	0.1	0.1	0.2	0.8	0.1
$A_{53} \sim A_{54}$	0.2	0.1	0.9	0.1	0.1
$A_{55} \sim A_{58}$	0.2	0.2	0.4	0.4	0.2
$A_{59} \sim A_{66}$	0.5	0.5	1.9	1.9	0.5
$A_{67} \sim A_{70}$	0.5	0.4	0.7	0.9	0.3
$A_{71} \sim A_{72}$	0.7	0.6	1.6	1.3	0.7
Weight (lb)	400.7	387.9	1089.9	1069.8	388.9

selected from the set D = {0.1, 0.2, 0.3, 0.4, 0.5, 0.6, 0.7, 0.8, 0.9, 1.0, 1.1, 1.2, 1.3, 1.4, 1.5, 1.6, 1.7, 1.8, 1.9, 2.0, 2.1, 2.2, 2.3, 2.4, 2.5, 2.6, 2.7, 2.8, 2.9, 3.0, 3.1, 3.2} (in^2); Case 2: The discrete variables are selected from the Table 4. A maximum iteration of 1000 is imposed.

1020

Table 4. The available cross-section areas of the ASIC code.

No.	in²	mm²	No.	in²	mm²
1	0.111	71.613	33	3.840	2477.414
2	0.141	90.968	34	3.870	2496.769
3	0.196	126.451	35	3.880	2503.221
4	0.250	161.290	36	4.180	2696.769
5	0.307	198.064	37	4.220	2722.575
6	0.391	252.258	38	4.490	2896.768
7	0.442	285.161	39	4.590	2961.284
8	0.563	363.225	40	4.800	3096.768
9	0.602	388.386	41	4.970	3206.445
10	0.766	494.193	42	5.120	3303.219
11	0.785	506.451	43	5.740	3703.218
12	0.994	641.289	44	7.220	4658.055
13	1.000	645.160	45	7.970	5141.925
14	1.228	792.256	46	8.530	5503.215
15	1.266	816.773	47	9.300	5999.988
16	1.457	939.998	48	10.850	6999.986
17	1.563	1008.385	49	11.500	7419.340
18	1.620	1045.159	50	13.500	8709.660
19	1.800	1161.288	51	13.900	8967.724
20	1.990	1283.868	52	14.200	9161.272
21	2.130	1374.191	53	15.500	9999.980
22	2.380	1535.481	54	16.000	10322.560
23	2.620	1690.319	55	16.900	10903.204
24	2.630	1696.771	56	18.800	12129.008
25	2.880	1858.061	57	19.900	12838.684
26	2.930	1890.319	58	22.000	14193.520
27	3.090	1993.544	59	22.900	14774.164
28	1.130	729.031	60	24.500	15806.420
29	3.380	2180.641	61	26.500	17096.740
30	3.470	2238.705	62	28.000	18064.480
31	3.550	2290.318	63	30.000	19354.800
32	3.630	2341.931	64	33.500	21612.860

Table 5. Comparison of optimal designs for the 72-bar spatial truss structure (case 2).

Variables (in²)	Wu (1995)	PSO	PSOPC	HPSO
$A_1 \sim A_4$	0.196	7.22	4.49	4.97
$A_5 \sim A_{12}$	0.602	1.80	1.457	1.228
$A_{13} \sim A_{16}$	0.307	1.13	0.111	0.111
$A_{17} \sim A_{18}$	0.766	0.196	0.111	0.111
$A_{19} \sim A_{22}$	0.391	3.09	2.620	2.88
$A_{23} \sim A_{30}$	0.391	0.785	1.130	1.457
$A_{31} \sim A_{34}$	0.141	0.563	0.196	0.141
$A_{35} \sim A_{36}$	0.111	0.785	0.111	0.111
$A_{37} \sim A_{40}$	1.800	3.09	1.266	1.563
$A_{41} \sim A_{48}$	0.602	1.228	1.457	1.228
$A_{49} \sim A_{52}$	0.141	0.111	0.111	0.111
$A_{53} \sim A_{54}$	0.307	0.563	0.111	0.196
$A_{55} \sim A_{58}$	1.563	1.990	0.442	0.391
$A_{59} \sim A_{66}$	0.766	1.620	1.457	1.457
$A_{67} \sim A_{70}$	0.141	1.563	1.228	0.766
$A_{71} \sim A_{72}$	0.111	1.266	1.457	1.563
Weight (lb)	427.2	1209.5	941.8	933.1

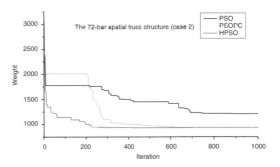

Figure 5. Comparison of convergence rates for the 72-bar spatial truss structure (case 2).

Figure 4 shows the convergence rates and Table 3 shows the optimal results.

For both of the cases, it seems that Wu's results (1995) achieve smaller weight. However, we discovered that both of these results do not satisfy the constraints, the results are unacceptable.

In case 1, the HPSO algorithm gets the optimal solution after 1000 iterations and shows a fast convergence rate, especially during the early iterations. For the PSO and PSOPC algorithms, they do not get optimal results when the maximum number of iterations is reached. In case 2, the HPSO algorithm gets best optimization result comparatively among three methods and shows a fast convergence rate.

8 CONCLUSIONS

In this paper, a heuristic particle swarm optimizer (HPSO) handling discrete variables is presented. The HPSO algorithm has all the advantages that belong to the PSO and the PSOPC algorithms, and has faster convergence rate than the other two methods. Two pin connected structures are optimized with the HPSO algorithm. In particular, the HPSO algorithm has a much faster convergence rate which means that it is practical for complicated engineering problems. Structures with bending members are not considered yet now.

ACKNOWLEDGEMENTS

We would like to thank Guangdong Natural Science Foundation and Guangzhou Bureau of Science and Technology, Peoples' Republic of China, for partially supporting this project (Project number is 032489, 06104655, 03Z3-D0221)

REFERENCES

Angeline, P. 1998. Evolutionary optimization versus particle swarm optimization: philosophy and performance difference. *Proceedings of the Evolutionary Programming Conference*, San Diago, USA.

Geem, Z.W., Kim, J.H. & Loganathan, G.V. 2001. A new heuristic optimization algorithm: harmony search. *Simulation,* 76, 60–8.

He, S., Wu, Q.H., Wen, J.Y., Saunders, J.R. & Paton, R.C. 2004. A particle swarm optimizer with passive congregation. *BioSystem,* 78,135–147.

He, S., Prempain & Wu, Q.H. 2005. An improved particle swarm optimizer for mechanical design optimization problems. *Engineering Optimization*, 5(36), 585–605.

Kang Seok Lee, ZongWoo Geem, Sang-Ho Lee and Kyu-Woong Bae, 2005. The harmony search heuristic algorithm for discrete structural optimization. *Engineering Optimization.* (37), 663–684.

Kennedy, J. & Eberhart, R.C. 2001. *Swarm Intelligence.* Morgan Kaufmann Publishers.

Li, L.J., Ren, F.M., Liu, F. & Wu, Q.H. 2006. An Improved Particle Swarm Optimization Method and Its Application in Civil Engineering, *Proceedings of the Fifth International Conference on Engineering Computational Technology.* 12–15 Sep. 2006. Spain. paper 42. ISBN 1-905088-11-6.

Li, L.J., Huang, Z.B. & Liu, F. 2007. A heuristic particle swarm optimizer (HPSO) for optimization of pin connected structures, *Computers and Structures*, in press.

Shi, Y. & Eberhart, R.C. 1997. A modified particle swarm optimizer. *Proc. IEEE Inc. Conf. On Evolutionary Computation,* 303–308.

Wu, S.J. & Chow, P.T. 1995. Steady-state genetic algorithms for discrete optimization of trusses. *Comput. Struct.,* 56(6), 979–991.

Zhang, Y.N., Liu, J.P., Liu, B., Zhu C.Y. & Li, Y. 2003. Application of improved hybrid genetic algorithm to optimize. *Journal of South China University of Technology,* 33(3), 69–72.

Steel and Composite Structures – Wang & Choi (eds)
© *2007 Taylor & Francis Group, London, ISBN 978-0-415-45141-3*

A new method for classifying the simple critical points of imperfect structures

Chang-gen Deng & Tian-ying Wang
Department of Building Engineering, Tongji University, Shanghai, China

ABSTRACT: The postbuckling behavior of imperfect structures with small initial geometric imperfections under conservative loading is studied with a perturbation method. Based on the higher order equilibrium equations and criticality equations, the slopes and curvatures of both the primary equilibrium paths and the secondary branch paths are solved. Discriminants to classify simple critical points and basic catastrophes are derived. Furthermore, the relationships between simple critical points and basic catastrophes are established. A simple model example is presented to illustrate the efficiency of the proposed classification criteria.

1 INTRODUCTION

In structural nonlinear stability analysis and optimization, locating and classifying the critical points are quite important. According to Koiter's (1945) classical postbuckling theory, the type of a critical point has a profound influence on the initial postbuckling behavior of an elastic structure, which determines the structural imperfection sensitivity. Based on the initial postbuckling characteristics, the strategy for tracing the postbuckling paths can be readily determined (Xia & Qian 1992). Moreover, Mang, Schranz, *et al* (2006) have shown that it is possible to convert imperfection-sensitive structures into imperfection-insensitive structures by means of modifications of the original design. Nowadays, people are seeking ways to improve and overcome the sever imperfection sensitivity of some structures, which is one the topics of passive buckling control.

There are a lot of methods for classifying critical points in literatures in general stability theory, bifurcation theory, catastrophe theory, etc. Among others, Koiter (1945) studied the stability of general continuous elastic systems using perturbation method in his doctoral thesis, classifying simple critical points into four types, namely limit point, asymmetric bifurcation point, stable-symmetric bifurcation point, and unstable-symmetric bifurcation point. Thompson (1984) studied the instabilities of discrete systems based on structural total potential energy function in terms of generalized coordinates, and established the relationships between critical points and basic catastrophes. Ikeda (1990) simplified a set of equilibrium equations to a single bifurcation equation with bifurcation theory and Liapunov-Schmidt method, and

constituted the bifurcation equation corresponding to the four simple critical points on the basis of investigating the simplified bifurcation equation in the asymptotic sense. Deng (1992) contributed a finer critical point classification via studying the coefficients of the various order equilibrium equations of perfect structures.

Deng's perturbation method is extended in this paper to study and classify the simple critical points of imperfect structures.

2 BASIC EQUATIONS AND CLASSIFICATION OF CRITICAL POINTS

2.1 *Basic equations and notations*

Practical structures inevitably contain a variety of initial imperfections, such as eccentric loads, residual stresses, material defects and geometric deviations. This paper will study and classify the simple (or distinct) critical points of geometrically nonlinear stability problems of imperfect structures with initial geometric imperfections.

Suppose that a set of n nonlinear algebraic equilibrium equations of imperfect structures under conservative loading are derived by the Raleigh-Ritz method, the Galerkin method, the least squares method (Deng, 1992) or a finite element method, and expressed symbolically as follows:

$$D_i(u_j, \lambda, \varepsilon_p) = 0, \quad (i = 1, 2, \cdots, n) \tag{1}$$

where $u_j (j = 1, 2, \cdots, n)$ are the state variables, λ is the loading parameter, and $\varepsilon_p (p = 1, 2, \ldots, n_p)$ are initial imperfection parameters representing the amplitudes of various geometric imperfection patterns and are assumed to be small and constant.

Suppose the tangent stiffness matrix $[D_{ij}]$ is diagonalized at the critical point $C(u_j^c, \lambda^c, \varepsilon_p)$, i.e., $D_{ij}^c = 0 (i \neq j)$. If the critical point is distinct, $D_{11}^c = 0$ and $D_{\alpha\alpha}^c \neq 0 (\alpha = 2, 3, \cdots, n)$.

The state variables are divided into two groups: u_1 is the active state variable that is the amplitude of the first buckling mode, and $u_\alpha (\alpha = 2, 3, \cdots, n)$ are the passive state variables. Greek letters $\alpha, \beta, \gamma, \sigma$ are used as subscripts of passive state variables ranging from 2 to n.

The equilibrium paths of an imperfect structure with constant imperfections ε_p are defined by the solutions of eq. (1), i.e., u_j in terms of λ. In order to analyze the imperfection sensitivity and trace the postbuckling equilibrium paths near the critical point $C(u_j^c, \lambda^c, \varepsilon_p)$, a proper path parameter s is introduced, e.g., $s = u_1 - u_1^c$ or $s = \lambda - \lambda^c$, and u_j and λ are expressed as the functions of s: $u_j = u_j(s), \lambda = \lambda(s)$.

For convenience, derivatives are denoted according to the following rules. Subscripts $i, j, k, l, \alpha, \beta, \gamma, \sigma$ of D denote partial derivatives of D with respect to the state variables with corresponding subscripts, and subscript λ of D denotes partial derivatives of D with respect to λ. Derivatives of J are denoted in the same way. Dot symbol $\ddot{y}$ over a variable represents derivative of the variable with respect to s. For example, $D_{ij} = \frac{\partial^2 D}{\partial u_i \partial u_j}, D_{i\alpha} = \frac{\partial^2 D}{\partial u_i \partial u_\alpha}, D_{i\lambda} = \frac{\partial^2 D}{\partial u_i \partial \lambda}, J_j = \frac{\partial J}{\partial u_j}, J_\lambda = \frac{\partial J}{\partial \lambda}$, $\dot{u}_\alpha = \frac{du_\alpha}{ds}, \dot{\lambda} = \frac{d\lambda}{ds}$.

After coordinate translation, the critical point $C(u_j^c, \lambda^c, \varepsilon_p)$, which is used as the reference point of potential function expansion on the equilibrium path, is located at $s = 0$. Superscript C or c denotes evaluation at the critical point C.

The dummy-suffix summation convention is employed except for explicit statement of exclusion.

2.2 The m-th order equilibrium equations and criticality equations

Upon introducing the path parameter s, the equilibrium equations become

$$D_i(u_j(s), \lambda(s), \varepsilon_p) = 0, \quad (i = 1, 2, \cdots, n) \tag{2}$$

and the criticality equation becomes

$$J(u_j(s), \lambda(s), \varepsilon_p)\big|^C = \big|D_{ij}(u_j(s), \lambda(s), \varepsilon_p)\big|^C = 0 \tag{3}$$

Calculating the m-th order derivatives of eq. (2) and (3) with respect to the path parameter s at the critical point $C(u_j^c, \lambda^c, \varepsilon_p)$, the m-th order equilibrium equations and the m-th order the criticality equations $(m = 1, 2, \cdots)$ are obtained:

$$\frac{d^m}{ds^m} D_i(u_j(s), \lambda(s), \varepsilon_p)\big|^C = 0, \quad (i = 1, 2, \cdots, n) \tag{4}$$

$$\frac{d^m}{ds^m} J(u_j(s), \lambda(s), \varepsilon_p)\big|^C = 0 \tag{5}$$

The m-th order $(m = 1, 2, 3, 4)$ equilibrium equations and the m-th order $(m = 1, 2)$ criticality equations are as follows.

$$\frac{dD_i}{ds}\big|^C = D_{ij}\dot{u}_j + D_{i\lambda}\dot{\lambda} = 0 \tag{6}$$

$$\frac{d^2 D_i}{ds^2}\big|^C = D_{ijk}\dot{u}_j\dot{u}_k + 2D_{ij\lambda}\dot{u}_j\dot{\lambda} + D_{i\lambda\lambda}\dot{\lambda}^2 + D_{ij}\ddot{u}_j + D_{i\lambda}\ddot{\lambda} = 0 \tag{7}$$

$$\frac{d^3 D_i}{ds^3}\big|^C = D_{ijkl}\dot{u}_j\dot{u}_k\dot{u}_l + 3D_{ijk\lambda}\dot{u}_j\dot{u}_k\dot{\lambda} + 3D_{ij\lambda\lambda}\dot{u}_j\dot{\lambda}^2$$
$$+ D_{i\lambda\lambda\lambda}\dot{\lambda}^3 + 3D_{ijk}\dot{u}_k\ddot{u}_j + 3D_{ij\lambda}\ddot{u}_j\dot{\lambda} + 3D_{ij\lambda}\dot{u}_j\ddot{\lambda}$$
$$+ 3D_{i\lambda\lambda}\dot{\lambda}\ddot{\lambda} + D_{ij}\dddot{u}_j + D_{i\lambda}\dddot{\lambda} = 0 \tag{8}$$

$$\frac{d^4 D_i}{ds^4}\big|^C = D_{ijklm}\dot{u}_j\dot{u}_k\dot{u}_l\dot{u}_m + 4D_{ijkl\lambda}\dot{u}_j\dot{u}_k\dot{u}_l\dot{\lambda} + D_{i\lambda\lambda\lambda\lambda}\dot{\lambda}^4$$
$$+ 6D_{ijkl}\ddot{u}_j\dot{u}_k\dot{u}_l + 12D_{ijk\lambda}\ddot{u}_j\dot{u}_k\dot{\lambda} + 6D_{ijk\lambda}\dot{u}_j\dot{u}_k\ddot{\lambda}$$
$$+ 6D_{ij\lambda\lambda}\ddot{u}_j\dot{\lambda}^2 + 12D_{ij\lambda\lambda}\dot{u}_j\dot{\lambda}\ddot{\lambda} + 6D_{i\lambda\lambda\lambda}\dot{\lambda}^2\ddot{\lambda}$$
$$+ 4D_{ijk}\dddot{u}_j\dot{u}_k + 3D_{ijk}\ddot{u}_j\ddot{u}_k + 4D_{ij\lambda}\dddot{u}_j\dot{\lambda}$$
$$+ 6D_{ij\lambda}\ddot{u}_j\ddot{\lambda} + 4D_{ij\lambda}\dot{u}_j\dddot{\lambda} + 4D_{i\lambda\lambda}\dot{\lambda}\dddot{\lambda}$$
$$+ 3D_{i\lambda\lambda}\ddot{\lambda}^2 + D_{ij}\ddddot{u}_j + D_{i\lambda}\ddddot{\lambda} = 0 \tag{9}$$

$$\frac{dJ}{ds}\big|^C = J_j\dot{u}_j + J_\lambda\dot{\lambda} = 0 \tag{10}$$

$$\frac{d^2 J}{ds^2}\big|^C = J_{jk}\dot{u}_j\dot{u}_k + 2J_{j\lambda}\dot{u}_j\dot{\lambda} + J_{\lambda\lambda}\dot{\lambda}^2 + J_j\ddot{u}_j + J_\lambda\ddot{\lambda} = 0 \tag{11}$$

All coefficients in the equations above are evaluated at the critical point C, e.g., $D_{ij} = D_{ij}^c = D_{ij}(u_j(0), \lambda(0), \varepsilon_p)$.

2.3 The classification criteria of critical points

The equilibrium equations are $D_i(u_j^c, \lambda^c, \varepsilon_p) = 0$ and the criticality equation is $J(u_j^c, \lambda^c, \varepsilon_p) = 0$. Equations (6) and (10) are the corresponding parametric forms of the first-order.

From eq. (6) for $(i = 1)$ and eq. (10), the first order equilibrium equation and criticality equation are simplified as linear equations in terms of $\dot{\lambda}, \dot{u}_1$:

$$D_{1\lambda}\dot{\lambda} = 0 \tag{12}$$

$$\bar{D}_{11\lambda}\dot{\lambda} + D_{111}\dot{u}_1 = 0 \tag{13}$$

where

$$\bar{D}_{11\lambda} = D_{11\lambda} - D_{11\alpha}D_{\alpha\lambda}/D_{\alpha\alpha} \tag{14}$$

The determinant of eqs. (12) and (13) is evaluated as:

$$\mathbb{D}_1 = \begin{vmatrix} D_{1\lambda} & 0 \\ \bar{D}_{11\lambda} & D_{111} \end{vmatrix} = D_{1\lambda}D_{111} \tag{15}$$

which is an important discriminant to classify simple critical points.

When $D_1 = 0$, the critical point C is a bifurcation point (BP). There are two cases: (a) if $D_{1\lambda} = 0$, $D_{111} \neq 0$, the critical point is an asymmetric bifurcation point; (b) if $D_{111} = 0, D_{1\lambda} \neq 0$, the critical point is a symmetric bifurcation point.

On the other hand, when $D_1 \neq 0$, namely, $D_{1\lambda} \neq 0$ and $D_{111} \neq 0$, the critical point C is a stationary point, i.e., limit point (LP) or inflexion point.

3 TYPES OF BIFURCATION POINTS

3.1 Bifurcation points when $D_{1\lambda} = 0$

In this case, eq. (12) is automatically satisfied. Higher order equilibrium equations have to be solved. From eq. (6) for ($i = \alpha$), $\dot{u}_\alpha$ are solved $\dot{u}_\alpha = -D_{\alpha\lambda}/D_{\alpha\alpha}\dot{\lambda}$.

From eq. (7) for ($i = 1$), a quadratic equation in $\dot{\lambda}$ and $\dot{u}_1$ is derived:

$$\bar{D}_{1\lambda\lambda}\dot{\lambda}^2 + 2\bar{D}_{11\lambda}\dot{u}_1\dot{\lambda} + D_{111}\dot{u}_1^2 = 0 \tag{16}$$

The discriminant of eq. (16) is:

$$\bar{\Delta}_2 = \left(\bar{D}_{11\lambda}\right)^2 - D_{111}\bar{D}_{1\lambda\lambda} \tag{17}$$

According to different values of the discriminant (17) and different coefficients of eq. (16), the solutions of eq. (16) fall into the following three cases pertaining to three different types of bifurcation points:

(1) Two distinct solutions indicate that two equilibrium paths intersect each other at the critical point, which is named as the intersectant bifurcation point.
(2) Two identical solutions indicate that two equilibrium paths are tangent at the critical point, which is named as the tangent bifurcation point.
(3) Undeterminable solutions indicate that higher order equilibrium equations have to be solved to determine the bifurcation paths. In this case, further analysis shows that there are three equilibrium paths that are intersectant or tangent at the critical point.

3.1.1 Intersectant bifurcation points when $\bar{\Delta}_2 > 0$
When $\bar{\Delta}_2 > 0$, eq. (16) has two distinct solutions, which can be discussed in four cases as follows.

(1) When $D_{111} \neq 0, \bar{D}_{1\lambda\lambda} \neq 0$,

$$\left(\dot{\lambda}/\dot{u}_1\right)_1 = \left[-\bar{D}_{11\lambda} + \sqrt{\bar{\Delta}_2}\right]/\bar{D}_{1\lambda\lambda}, \quad \left(\dot{\lambda}/\dot{u}_1\right)_2 = \left[-\bar{D}_{11\lambda} - \sqrt{\bar{\Delta}_2}\right]/\bar{D}_{1\lambda\lambda} \tag{18}$$

On both paths $(\dot{\lambda}/\dot{u}_1)_{1,2} \neq 0$, so the bifurcation is asymmetric. This critical point is called the first order linear intersectant bifurcation point.

(2) When $D_{111} \neq 0, \bar{D}_{1\lambda\lambda} = 0$,

$$\left(\dot{u}_1/\dot{\lambda}\right)_1 = 0, \quad \left(\dot{u}_1/\dot{\lambda}\right)_2 = -2\bar{D}_{11\lambda}/D_{111} \tag{19}$$

On the branch path $(\dot{\lambda}\dot{u}_1)_2 \neq 0$, so the bifurcation is asymmetric.

(3) When $D_{111} = 0, \bar{D}_{1\lambda\lambda} \neq 0$,

$$\left(\dot{\lambda}/\dot{u}_1\right)_1 = 0, \quad \left(\dot{\lambda}/\dot{u}_1\right)_2 = -2\bar{D}_{11\lambda}/\bar{D}_{1\lambda\lambda} \tag{20}$$

(4) When $D_{111} = 0, \bar{D}_{1\lambda\lambda} = 0$,

$$\left(\dot{\lambda}/\dot{u}_1\right)_1 = 0, \quad \left(\dot{u}_1/\dot{\lambda}\right)_2 = 0 \tag{21}$$

In cases (3) and (4), when $D_{111} = 0$, there is a branch path with $\dot{\lambda} = 0$, so the bifurcation is symmetric and the loading parameter reaches the limit value on the branch path at the bifurcation point.

It is obvious from the definition, $\Delta_1 = D_1 = 0, \Delta_2 = D_{11} = 0$. More discriminants are defined below:

$$\Delta_3 = D_{111} \tag{22}$$

Substituting $\Delta_3 = 0, \dot{\lambda} = 0, \dot{u}_\alpha = 0$ into eq. (7) for ($i = \alpha$), then $\ddot{u}_\alpha = -\left(D_{\alpha\lambda}\ddot{\lambda} + D_{11\alpha}\dot{u}_1^2\right)/D_{\alpha\alpha}$, and substituting it into eq. (8) for ($i = 1$), the curvature of the branch path is solved as $\ddot{\lambda} = -\Delta_4/(3\bar{D}_{11\lambda})\dot{u}_1^2$, where

$$\Delta_4 = D_{1111} - 3D_{11\alpha}^2/D_{\alpha\alpha} \tag{23}$$

If $\Delta_4 \neq 0$, then $\ddot{\lambda} \neq 0$. In this case the critical point C is second order limit intersectant bifurcation point.

Otherwise, if $\Delta_4 = 0, \ddot{\lambda} = 0$, eq. (8) for ($i = \alpha$) are solved. The solution process continues until certain order derivative is nonzero, namely $d^{m-1}\lambda/ds^{m-1} = 0$, $d^m\lambda/ds^m \neq 0$:

$$\begin{cases} \dfrac{d^k\lambda}{ds^k} = 0, \quad (k = 1, 2, \cdots, m-1), \\ \dfrac{d^m\lambda}{ds^m} = -\dfrac{\Delta_{m+2}}{(m+1)\bar{D}_{11\lambda}}\dot{u}_1^m \neq 0, \end{cases} \quad (m = 2, 3, \cdots) \tag{24}$$

If m is even, C is the m-th order limit intersectant bifurcation point; otherwise, C is the m-th order inflexion intersectant bifurcation point.

3.1.2 Tangent bifurcation points when $\bar{\Delta}_2 = 0$
When $\bar{\Delta}_2 = 0$, eq. (16) has two identical solutions, which can be discussed in three cases as follows.

(1) When $D_{111} \neq 0, \bar{D}_{11\lambda} \neq 0, \bar{D}_{1\lambda\lambda} \neq 0$:

$$\left(\dot{\lambda}/\dot{u}_1\right)_{1,2} = -\bar{D}_{11\lambda}/\bar{D}_{1\lambda\lambda} = -D_{111}/\bar{D}_{11\lambda} \tag{25}$$

(2) When $D_{111} \neq 0, \bar{D}_{11\lambda} = 0, \bar{D}_{1\lambda\lambda} = 0$:

$$\left(\dot{u}_1/\dot{\lambda}\right)_{1,2} = 0 \tag{26}$$

1025

(3) When $D_{111} = 0, \bar{D}_{11}\lambda = 0, \bar{D}_1\lambda\lambda \neq 0$:

$$\left(\lambda/\dot{u}_1\right)_{1,2} = 0 \tag{27}$$

In case (3), $\dot{\lambda} = 0, \dot{u}_1$ is arbitrary. Substituting $\dot{\lambda} = 0$ and $\dot{u}_\alpha = 0$ into eqs. (7) for $(i = \alpha)$ and (8) for $(i = 1)$, a pair of linear equations in $\ddot{\lambda}$ and $\ddot{u}_\alpha$ are derived

$$\begin{cases} D_{11\alpha}\dot{u}_1^2 + D_{\alpha\alpha}\ddot{u}_\alpha + D_{\alpha\lambda}\ddot{\lambda} = 0 \\ D_{111}\dot{u}_1^2 + 3D_{11\alpha}\ddot{u}_\alpha + 3D_{11\lambda}\ddot{\lambda} = 0 \end{cases} \tag{28}$$

Solving the above system of equation,

$$\begin{cases} \ddot{\lambda} = -\tilde{\Delta}_4/\left(3\bar{D}_{11\lambda}\right)\dot{u}_1^2 \\ \ddot{u}_\alpha = -\left(D_{11\alpha} - D_{\alpha\lambda}\tilde{\Delta}_4/\left(3\bar{D}_{11\lambda}\right)\right)\dot{u}_1^2/D_{\alpha\alpha} \end{cases} \tag{29}$$

When $\Delta_4 = 0$, $\ddot{\lambda}$ can't be determined. From eq. (8) for $(i = \alpha)$ and (9) for $(i = 1)$, a quadratic equation in $\ddot{\lambda}$ and $\dot{u}_1^2$ is derived

$$3\bar{D}_{1\lambda\lambda}\ddot{\lambda}^2 + 6\bar{D}_{111\lambda}\dot{u}_1^2\ddot{\lambda} + \Delta_5\dot{u}_1^4 = 0 \tag{30}$$

where

$$\bar{D}_{111\lambda} = D_{111\lambda} - \frac{D_{111\alpha}D_{\alpha\lambda} + 3D_{11\alpha}D_{1\alpha\lambda}}{D_{\alpha\alpha}} + \frac{3D_{1\alpha\beta}D_{11\alpha}D_{\beta\lambda}}{D_{\alpha\alpha}D_{\beta\beta}} \tag{31}$$

$$\Delta_5 = D_{11111} - 10D_{111\alpha}D_{11\alpha}/D_{\alpha\alpha} + 15D_{1\alpha\beta}D_{11\alpha}D_{11\beta}/\left(D_{\alpha\alpha}D_{\beta\beta}\right) \tag{32}$$

If $\Delta_5 \neq 0$, $\ddot{\lambda}$ has two nonzero distinct solutions, indicating that the two tangent branches have limit value at the bifurcation point. If otherwise $\Delta_5 = 0$, the two solutions of $\ddot{\lambda}$ are $\ddot{\lambda} = -2\bar{D}_{111\lambda}\bar{D}_{1\lambda\lambda}\dot{u}_1^2$ and $\ddot{\lambda} = 0$, respectively.

When $\bar{D}_{111\lambda} = 0$ and $\ddot{\lambda} = 0$, solving continually eq. (9) for $(i = \alpha)$ and the fifth order equilibrium equations for $(i = 1)$, λ is solved $\lambda = -\Delta_6/10\bar{D}_{111\lambda})\dot{u}_1^3$, where

$$\Delta_6 = D_{111111} - \left(15D_{1111\alpha}D_{11\alpha} + 10D_{111\alpha}^2\right)/D_{\alpha\alpha}$$
$$+ \left(45D_{11\alpha\beta}D_{11\alpha}D_{11\beta} + 60D_{111\alpha}D_{1\alpha\beta}D_{11\beta}\right)/\left(D_{\alpha\alpha}D_{\beta\beta}\right)$$
$$- \frac{15D_{\alpha\beta\gamma}D_{11\alpha}D_{11\beta}D_{11\gamma} + 90D_{1\alpha\beta}D_{11\alpha}D_{11\gamma}D_{11\beta\gamma}}{D_{\alpha\alpha}D_{\beta\beta}D_{\gamma\gamma}} \tag{33}$$

The general nonzero solutions are

$$\begin{cases} \dfrac{d^k\lambda}{ds^k} = 0, \ (k = 1,2,\cdots m-1), \\ \dfrac{d^m\lambda}{ds^m} = -\dfrac{\Delta_{m+3}}{C_{m+2}^2\bar{D}_{111\lambda}}\dot{u}_1^m \neq 0, \end{cases} (m = 3,4,\cdots) \tag{34}$$

where $C_{m+2}^2 = (m+2)!/(m+2)!$

If m is even, C is the m-th order limit intersectant bifurcation point; otherwise, the bifurcation point is the m-th order inflexion intersectant bifurcation point.

3.1.3 Multiple-branch bifurcation points when $D_{111} = \bar{D}_{11}\lambda = \bar{D}_1\lambda\lambda = 0$

When $D_{111} = \bar{D}_{11}\lambda = D_1\lambda\lambda = 0$, the solutions of eq. (7) can't be determined. Simplifying eqs. (8) for $(i = \alpha)$ and (8) for $(i = 1)$ yields

$$\bar{D}_{1\lambda\lambda}\dot{\lambda}^3 + 3\bar{D}_{11\lambda\lambda}\dot{u}_1\dot{\lambda}^2 + 3\bar{D}_{111\lambda}\dot{u}_1^2\dot{\lambda} + \Delta_4\dot{u}_1^3 = 0 \tag{35}$$

where

$$\bar{D}_{1\lambda\lambda} = D_{1\lambda\lambda} - \frac{3\left(D_{1\alpha\lambda\lambda}D_{\alpha\lambda} + D_{1\alpha\lambda}D_{\alpha\lambda\lambda}\right)}{D_{\alpha\alpha}}$$
$$+ \frac{3\left(D_{1\alpha\beta\lambda}D_{\alpha\lambda}D_{\beta\lambda} + 2D_{1\alpha\lambda}D_{\alpha\beta\lambda}D_{\beta\lambda} + D_{1\alpha\beta}D_{\beta\lambda}D_{\alpha\lambda\lambda}\right)}{D_{\alpha\alpha}D_{\beta\beta}}$$
$$- \frac{D_{1\alpha\beta\gamma}D_{\alpha\lambda}D_{\beta\lambda}D_{\gamma\lambda} + 3D_{\alpha\beta\gamma}D_{1\alpha\lambda}D_{\beta\lambda}D_{\gamma\lambda}}{D_{\alpha\alpha}D_{\beta\beta}D_{\gamma\gamma}}$$
$$- \frac{6D_{1\alpha\beta}D_{\beta\lambda}D_{\alpha\gamma\lambda}D_{\gamma\lambda}}{D_{\alpha\alpha}D_{\beta\beta}D_{\gamma\gamma}} + \frac{3D_{1\alpha\beta}D_{\beta\lambda}D_{\alpha\gamma\sigma}D_{\gamma\lambda}D_{\sigma\lambda}}{D_{\alpha\alpha}D_{\beta\beta}D_{\gamma\gamma}D_{\sigma\sigma}} \tag{36}$$

$$\bar{D}_{11\lambda\lambda} = D_{11\lambda\lambda} - \frac{2D_{11\alpha\lambda}D_{\alpha\lambda} + D_{11\alpha}D_{\alpha\lambda\lambda} + 2\left(D_{1\alpha\lambda}\right)^2}{D_{\alpha\alpha}}$$
$$+ \frac{D_{11\alpha\beta}D_{\alpha\lambda}D_{\beta\lambda} + 2D_{11\alpha}D_{\alpha\beta\lambda}D_{\beta\lambda} + 4D_{1\alpha\beta}D_{1\alpha\lambda}D_{\beta\lambda}}{D_{\alpha\alpha}D_{\beta\beta}}$$
$$- \frac{D_{11\alpha}D_{\alpha\beta\gamma}D_{\beta\lambda}D_{\gamma\lambda} + 2D_{1\alpha\beta}D_{1\alpha\gamma}D_{\beta\lambda}D_{\gamma\lambda}}{D_{\alpha\alpha}D_{\beta\beta}D_{\gamma\gamma}} \tag{37}$$

The discriminant of the solution is derived:

$$\bar{\Delta}_3 = \frac{\bar{q}_3^2}{4} + \frac{\bar{p}_3^3}{27} = \bar{D}_{1\lambda\lambda}^2\Delta_4^2 - 6\bar{D}_{1\lambda\lambda}\bar{D}_{11\lambda\lambda}\bar{D}_{111\lambda}\Delta_4$$
$$+ 4\bar{D}_{1\lambda\lambda}^3\Delta_4 - 3\bar{D}_{11\lambda\lambda}^2\bar{D}_{111\lambda}^2 + 4\bar{D}_{1\lambda\lambda}\bar{D}_{111\lambda}^3 \tag{38}$$

where

$$\bar{p}_3 = 3\left(\bar{D}_{1\lambda\lambda}\bar{D}_{111\lambda} - \bar{D}_{11\lambda\lambda}^2\right) \tag{39}$$

$$\bar{q}_3 = \bar{D}_{1\lambda\lambda}\left(\bar{D}_{1\lambda\lambda}\Delta_4 - \bar{D}_{11\lambda\lambda}\bar{D}_{111\lambda}\right)$$
$$- 2\bar{D}_{11\lambda\lambda}\left(\bar{D}_{1\lambda\lambda}\bar{D}_{111\lambda} - \bar{D}_{11\lambda\lambda}^2\right) \tag{40}$$

Similarly, according to different values of $\bar{\Delta}_3, \bar{p}_3, \bar{q}_3$, the solutions of eq. (35) have the following four cases representing different types of bifurcation points:

(1) When $\bar{\Delta}_3 < 0$, eq. (35) has three distinct sets of real solutions, indicating that there are three branches that intersect at the critical point.

(2) When $\bar{\Delta}_3 = 0$, eq. (35) has three sets of real solutions, two of which are identical, indicating there are two branches tangent at the critical point, while the other one intersects at the critical point.

(3) When $\bar{\Delta}_3 = 0$, and $\bar{p}_3 = 0, \bar{q}_3 = 0$, eq. (35) has three identical sets of real solutions, indicating three branches are tangent at the critical point.

Table 1. The relationship between simple critical points and basic catastrophes.

Discriminant Δ_m	Simple critical points				Basic catastrophes
	Limit points $D_{1\lambda} \neq 0, D_{111} \neq 0$	BP $D_{1\lambda} = 0$ or $D_{1\lambda} \neq 0, D_{111} = 0$			
		Intersectant BP $P_2 \neq 0$	Tangent BP $P_3 \neq 0$		
$\Delta_1 = \Delta_2 = 0, \Delta_3 \neq 0$	2nd order LP	1st order linear intersectant BP	1st order linear tangent BP		fold
$\Delta_1 = \Delta_2 = \Delta_3 = 0, \Delta_4 \neq 0$		2nd order limit intersectant BP	3/2 order cusp tangent BP		cusp
$\Delta_1 = \Delta_2 = \Delta_3 = \Delta_4 = 0, \Delta_5 \neq 0$		3rd inflexion intersectant BP	2nd order limit tangent BP		swallowtail
$\Delta_1 = \Delta_2 = \Delta_3 = \Delta_4 = \Delta_5 = 0, \Delta_6 \neq 0$		4th order limit intersectant BP	2nd order limit and third inflexion tangent BP		butterfly

(4) When $\bar{\Delta}_3 > 0$, eq. (35) has only one set of real solution.

When $\Delta_4 = 0$, one of the solutions of eq. (35) is $\ddot{\lambda} = 0$. Simplifying eqs. (8) for $(i = \alpha)$ and (9) for $(i = 1)$, $\dddot{\lambda}$ is solved as:

$$\dddot{\lambda} = -\frac{\Delta_5}{6\bar{D}_{111\lambda}} \dot{u}_1^2 \tag{41}$$

3.2 Bifurcation points when $D_{1\lambda} \neq 0, D_{111} = 0$

From eq. (12), $\dot{\lambda} = 0$ is obtained. From eq. (13), it is clear that $\dot{u}_1$ is arbitrary.

Making use of the conditions and the solutions above, $\ddot{\lambda} = 0$, $\dot{u}_\alpha = 0$ are obtained. Substituting the known solutions into eq. (7) for $(i = \alpha)$, $\ddot{u}_\alpha$ is solved

$$\ddot{u}_\alpha = -D_{11\alpha} \dot{u}_1^2 / D_{\alpha\alpha} \tag{42}$$

Substituting the solutions into eq. (11):

$$\left(J_{11} - J_\alpha D_{11\alpha}/D_{\alpha\alpha} \right) \dot{u}_1^2 = 0 \tag{43}$$

Since $J_{11} - J_\alpha D_{11\alpha} D_{\alpha\alpha} \neq 0 D_{\alpha\alpha} \neq 0$, there are two identical solutions $\dot{u}_1 = 0$, indicating two equilibrium paths that are tangent at the critical point. In this case, the critical point C is a tangent bifurcation point.

4 TYPES OF LIMIT POINTS

In this case, from eq. (12), the single solution $\dot{\lambda} = 0$ is obtained, and $\dot{u}_1$ is arbitrary. From eq. (6) for $(i = \alpha)$, $\dot{u}_\alpha = 0$. The above solutions are incorporated into eq. (7) for $(i = 1)$, $\ddot{\lambda} = -D_{111}/D_{1\lambda} \dot{u}_1^2$ is obtained. Since $D_{111} \neq 0$, then $\ddot{\lambda} \neq 0$, the critical point C is the second order limit point (LP). Substituting the above solutions into eq. (7) for $(i = \alpha)$, $\ddot{u}_\alpha = -(D_{\alpha\lambda}\lambda + D_{11\alpha}\dot{u}_1^2)D_{\alpha\alpha}$.

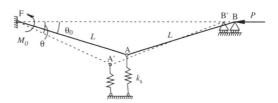

Figure 1. Deformed configuration of an imperfect 1-DOF model.

5 THE RELATIONSHIPS BETWEEN SIMPLE CRITICAL POINTS AND BASIC CATASTROPHES

Further analysis can be conducted to obtain other simple critical points. The most common simple critical points are summarized in Table 1, where the discriminant $\Delta_m(\Delta_1 = \Delta_2 = \cdots = \Delta_{m-1} = 0, \Delta_m \neq 0)$, which is the lowest level nonzero discriminant, together with $P_n(P_1 = D_{1\lambda}, P_2 = \bar{D}_{11\lambda}, P_3 = \bar{D}_{111\lambda})$ can be used to identify the types of critical points. The relationships between simple critical points and basic catastrophes are also listed in Table 1. It is seen that the discriminant Δ_m can readily be used to classify basic catastrophes.

6 AN EXAMPLE

The model shown in Figure 1 consists of two rigid bars of equal length L interconnected to each other at the middle point A by a frictionless hinge supported on a vertical movable linear elastic spring with stiffness k_s. The end B is a movable hinge, while end F is a fixed hinge. The model is subjected to a horizontal static force P at its end B and a small external moment M_0 which remains constant as P increases from zero. θ_0 and θ denote the initial angle imperfection (when

1027

the spring is unstressed) and the total angle (associated with the deformed position) respectively.

The total potential energy of the model is $\Pi = \frac{k_s L^2}{2}(\sin\theta - \sin\theta_0)^2 - 2PL(\cos\theta_0 - \cos\theta) + M_0(\theta - \theta_0)$ and the dimensionless total potential energy function of the model is

$$D = \frac{1}{2}(\sin\theta - \varepsilon_1)^2 + \lambda\cos\theta + \varepsilon_2\theta + \text{constant} \qquad (44)$$

where $\lambda = 2P/k_s L$, $\varepsilon_1 = \sin\theta_0$, $\varepsilon_2 = M_0/kL^2$.

The equilibrium equation and the criticality equation are easily derived:

$$D_1 = (\sin\theta - \varepsilon_1)\cos\theta - \lambda\sin\theta + \varepsilon_2 = 0 \qquad (45)$$

$$D_{11} = \cos^2\theta - (\sin\theta - \varepsilon_1)\sin\theta - \lambda\cos\theta = 0 \qquad (46)$$

From eqs. (45) and (46), the values of θ^c and λ^c can be obtained at the critical point C:

$$\theta^c = \sqrt{2(1 - \varepsilon_1/\varepsilon_2)} \qquad (47)$$

$$\lambda^c = \frac{(\sin\theta^c - \varepsilon_1)\cos\theta^c + \varepsilon_2}{\sin\theta^c} \qquad (48)$$

The discriminants are evaluated at the critical point C:

$$D_{1\lambda}^c = \left.\frac{\partial^2 D}{\partial\theta\partial\lambda}\right|^C = -\sin\theta^c \qquad (49)$$

$$D_{111}^c = \left.\frac{\partial^3 D}{\partial\theta^3}\right|^C = (-4\sin\theta^c + \varepsilon_1)\cos\theta^c + \lambda^c\sin\theta^c \qquad (50)$$

$$D_{1111}^c = \left.\frac{\partial^4 D}{\partial\theta^4}\right|^C = 4 - 8\cos^2\theta^c - \varepsilon_1\sin\theta^c + \lambda^c\cos\theta^c \qquad (51)$$

When $\varepsilon_2 = \varepsilon_1$, $\theta^c = 0$, then $D_{1\lambda}^c = 0$, so the critical point is a bifurcation point:

(1) If $\varepsilon_1 = \varepsilon_2 = 0$, $\Delta_3 = D_{111}^c = 0, \Delta_4 = D_{1111}^c \neq 0$, then the critical point is a symmetric bifurcation point.
(2) If $\varepsilon_1 = \varepsilon_2 \neq 0$, $\Delta_3 = D_{111}^c \neq 0$, then the critical point is an asymmetric bifurcation point.

When $\varepsilon_2 \neq \varepsilon_1$, $D_{1\lambda}^c \neq 0$, $D_{111}^c \neq 0$, the critical point is the second order limit point.

7 CONCLUSIONS

Deng's (1992) perturbation method and discriminants for the classification of the simple critical points of perfect structures are extended to study and classify the simple critical points of imperfect structures with small initial geometric imperfections under conservative loading.

The nonlinear equilibrium equations and the criticality equation are converted into the higher order equilibrium equations and criticality equations in terms of derivatives of the state variables and the loading parameter with respect to a path parameter, and the slopes and curvatures of both the primary equilibrium paths and the secondary branch paths are solved. Discriminants to classify simple critical points and basic catastrophes are derived. Furthermore, the relationships between simple critical points and basic catastrophes are established. Finally, a simple model example is presented to illustrate the efficiency of the proposed classification criteria.

This study could avail the studies on postbuckling analysis and optimization, and passive buckling control.

REFERENCES

Deng, C.G. 1992. The combination of weighted residual methods with catastrophe theory. Proc. of 4th National Conference on Methods of Weighted Residuals: 23–31. (in Chinese).

Thompson, J.M.T. & Hunt, G.W. 1984. Elastic instability phenomena. Chrichester: Wiley.

Ikeda, K. 1990. Critical initial imperfection of structures. Int. J. Structures 26: 865–886.

Koiter, A.N. 1945. On the stability of elastic equilibrium. Ph. D. Thesis. T. U. Delft, the Netherlands.

Mang, H.A., Schranz, C. & Mackenzie-Helnwein, P. 2006. Conversion from imperfection-sensitive into imperfection-insensitive elastic structures I: theory, Comput. Methods Appl. Mech. Engrg. 195: 1422–1457.

Xia, S.H. & Qian, R.J. 1992. The estimation of buckling type in nonlinear stability analysis of reticulated shells, Symposiums on new space structures. (in Chinese).

Steel and Composite Structures – Wang & Choi (eds)
© *2007 Taylor & Francis Group, London, ISBN 978-0-415-45141-3*

Thermal buckling of imperfect composite plates

B. Samsam Shariat

Electrofan Co., Tehran, Iran

ABSTRACT: Thermal buckling analysis of rectangular laminated composite plates with initial geometrical imperfections is presented in this paper. The equilibrium, stability, and compatibility equations of an imperfect composite plate are derived using the first order shear deformation plate theory. The plate is assumed to be under uniform temperature rise. Resulting equations are employed to obtain the closed-form solutions for the critical buckling temperature change of imperfect laminated composite plate.

1 INTRODUCTION

A comprehensive work on the buckling of structures is presented by Brush and Almroth [1]. They have examined the effect of initial imperfections on the critical loads. A review of research on thermal buckling of plates and shells is presented by Thornton [2]. He has described the elastic thermal buckling of metallic as well as composite plates and shells. Turvey and Marshall studied buckling and post-buckling of composite plates due to mechanical and thermal loads [3].

Initial geometric imperfections are inherent in many real structures. Therefore, many investigations are conducted on the stability analysis of imperfect structures. Elastic, plastic, and creep buckling of an imperfect cylindrical shell under mechanical and thermal loads is studied by Eslami and Shariyat [4]. Mossavarali et al. studied the thermoelastic buckling of isotropic and orthotropic plates with imperfections [5,6]. Murphy and Ferreira investigated the thermal buckling analysis of clamped rectangular plates based on the energy consideration [7]. They determined the ratio of the critical temperature for a perfect flat plate to the one for an imperfect plate as a function of initial imperfection size. The study includes experimental results. Eslami and Shahsiah reported thermal buckling of imperfect circular cylindrical shells based on the Wan-Donnell and Koiter imperfection models [8]. Recently, the present author and Eslami have presented the buckling analysis of imperfect functionally graded plates under mechanical and thermal loads [9,10].

In the present article, the influence of geometrical imperfections on thermal instability of laminated composite plates is investigated. The first order shear deformation plate theory is used with a double-sine function for geometrical imperfections along the x and y-directions. The boundary conditions along the four edges of the plate are assumed to be fixed-simply supported. The buckling load of the plate under uniform temperature rise is obtained in closed-form solution.

2 FUNDAMENTAL EQUATIONS

Consider a rectangular plate of length a, width b and thickness h, composed of laminated composite materials which is mid-plane symmetric. The plate is referred to the rectangular Cartesian coordinates (x, y, z) and is subjected to the thermal loading. The first order shear deformation theory, used in the present study, is based on the following displacement fields [11]

$$u(x, y, z) = u_0(x, y) + z u_1(x, y)$$

$$v(x, y, z) = v_0(x, y) + z v_1(x, y)$$

$$w = w(x, y) \tag{1}$$

Here, u, v, and w are the total displacement components along the x, y, and z-directions, respectively, u_0 and v_0 are the middle plane displacements, and u_1 and v_1 are the rotations about the y and x-axes, respectively. The constitutive relations of such plates are

$$N_x = A_{11}\epsilon_x + A_{12}\epsilon_y - N_x^T$$

$$N_y = A_{12}\epsilon_x + A_{22}\epsilon_y - N_y^T$$

$$N_{xy} = A_{66}\gamma_{xy} - N_{xy}^T$$

$$M_x = D_{11}k_x + D_{12}k_y - M_x^T$$

$$M_y = D_{12}k_x + D_{22}k_y - M_y^T$$

$$M_{xy} = D_{66}k_{xy} - M_{xy}^T$$

$$Q_x = A_{44}\epsilon_{xz} - Q_x^T$$

$$Q_y = A_{55}\epsilon_{yz} - Q_y^T \tag{2}$$

where A_{ij} and D_{ij} are the extensional and flexural stiffness matrix components respectively, notified as

$$(A_{ij}, D_{ij}) = \sum_{k=1}^{N} \int_{z_{k-1}}^{z_k} \left(Q'_{ij}\right)_k (1, z^2) dz \tag{3}$$

$$i, j = 1, 2, 4, 5, 6$$

Also, the stress resultants N_l^T, M_l^T, and Q_l^T are defined as

$$\left(N_l^T, M_l^T\right) = \sum_{k=1}^{N} \int_{z_{k-1}}^{z_k} -\left(Q'_{ij}\right)_k (1, z) \Delta T \left(\alpha_l\right)_k dz$$

$$i, j = 1, 2, 6 \quad l = x, y, xy \tag{4}$$

$$Q_l^T = \sum_{k=1}^{N} \int_{z_{k-1}}^{z_k} -\left(Q'_{ij}\right)_k (1, z) \Delta T \left(\alpha_l\right)_k dz$$

$$i, j = 4, 5 \quad l = x, y$$

And Q'_{ij} are the components of the general stiffness matrix, which relate the stresses to the strains. Also, the normal strains ϵ_i, the shear strains γ_i, and the curvatures k_i are specified as [1]

$$\epsilon_x = u_{0,x} + \frac{1}{2}w_{,x}^2$$

$$\epsilon_y = v_{0,y} + \frac{1}{2}w_{,y}^2$$

$$\gamma_{xy} = u_{0,y} + v_{0,x} + w_{,x}w_{,y} \tag{5}$$

$$\gamma_{xz} = u_{1,z} + w_{,x}$$

$$\gamma_{yz} = v_{1,z} + w_{,y}$$

$$k_x = u_{1,x}$$

$$k_y = v_{1,y}$$

$$k_{xy} = u_{1,y} + v_{1,x}$$

The equilibrium equations of an imperfect plate based on the first order theory are [11]

$$N_{x,x} + N_{xy,y} = 0$$

$$N_{xy,x} + N_{y,y} = 0$$

$$Q_{x,x} + Q_{y,y} + N_x(w_{,xx} + w_{,xx}^*) +$$

$$2N_{xy}(w_{,xy} + w_{,xy}^*) + N_y(w_{,yy} + w_{,yy}^*) = 0 \tag{6}$$

$$M_{x,x} + M_{xy,y} - Q_x = 0$$

$$M_{xy,x} + M_{y,y} - Q_y = 0$$

where $w^*(x, y)$ denotes a known small imperfection. This parameter represents a small deviation of the plate middle plane from a flat shape [1].

The stability equation of the plate may be derived by the adjacent equilibrium criterion [1]. Assume that the equilibrium state of the plate under thermal load is defined in terms of the displacement components u^0, v^0, and w^0. The displacement components of a neighboring stable state differ by u^1, v^1, and w^1 with respect to the equilibrium position. Thus, the total displacements of a neighboring state are

$$u = u^0 + u^1$$

$$v = v^0 + v^1 \tag{7}$$

$$w = w^0 + w^1$$

Similarly, the force resultants of a neighboring state may be related to the state of equilibrium as

$$N_i = N_i^0 + N_i^1 \qquad i = x, y, xy$$

$$M_i = M_i^0 + M_i^1 \qquad i = x, y, xy \tag{8}$$

$$Q_i = Q_i^0 + Q_i^1 \qquad i = x, y$$

where N_i^1, M_i^1, and Q_i^1 represent the linear parts of the stress resultants with respect to u^1, v^1, and w^1. The stability equation may be obtained by substituting Eqs. (7) and (8) in Eq. (6). Upon substitution, the terms in the resulting equation with superscript 0 satisfy the equilibrium condition and therefore drop out of the equations. Also, the nonlinear terms with superscript 1 are ignored because they are small compared to the linear terms. The remaining terms form the stability equations of an imperfect composite plate as

$$N^1_{x,x} + N^1_{xy,y} = 0$$

$$N^1_{xy,x} + N^1_{y,y} = 0$$

$$Q^1_{x,x} + Q^1_{y,y} + N^1_x(w^0_{,xx} + w^*_{,xx}) +$$

$$2N^1_{xy}(w^0_{,xy} + w^*_{,xy}) + N^1_y(w^0_{,yy} + w^*_{,yy}) + \quad (9)$$

$$N^0_x w^1_{,xx} + 2N^0_{xy} w^1_{,xy} + N^0_y w^1_{,yy} = 0$$

$$M^1_{x,x} + M^1_{xy,y} - Q^1_x = 0$$

$$M^1_{xy,x} + M^1_{y,y} - Q^1_y = 0$$

The superscript 1 refers to the state of stability and the superscript 0 refers to the state of equilibrium conditions. A stress function Φ may be defined as [6]

$$N^1_x = \Phi_{,yy}$$

$$N^1_y = \Phi_{,xx} \quad (10)$$

$$N^1_{xy} = -\Phi_{,xy}$$

Assume $\epsilon^1_x, \epsilon^1_y,$ and γ^1_{xy} as parts of the strain components which are linear in $u^1_0, v^1_0,$ and w^1. These strains may be written in terms of the displacement components, using Eqs. (2), (5), (7), and (8) with consideration of the imperfection term w^*, as

$$\epsilon^1_x = u^1_{0,x} + (w^0_{,x} + w^*_{,x})w^1_{,x}$$

$$\epsilon^1_y = v^1_{0,y} + (w^0_{,y} + w^*_{,y})w^1_{,y}$$

$$\gamma^1_{xy} = u^1_{0,y} + v^1_{0,x} + (w^0_{,x} + w^*_{,x})w^1_{,y} + \quad (11)$$

$$(w^0_{,y} + w^*_{,y})w^1_{,x}$$

Using Eqs. (11), the geometrical compatibility equation is written as

$$\epsilon^1_{x,yy} + \epsilon^1_{y,xx} - \gamma^1_{xy,xy} = 2(w^0_{,xy} + w^*_{,xy})w^1_{,xy}$$

$$-(w^0_{,xx} + w^*_{,xx})w^1_{,yy} \quad (w^0_{,yy} + w^*_{,yy})w^1_{,xx} \quad (12)$$

Using Eqs. (2), (10), and (12), the compatibility equation of an imperfect composite plate, including stress function Φ, is obtained

$$(A_{22}\Phi_{,yyyy} - 2A_{12}\Phi_{,xxyy} + A_{11}\Phi_{,xxxx}) \times$$

$$\frac{1}{A_{11}A_{22} - A^2_{12}} + \frac{1}{A_{66}}\Phi_{,xxyy} \quad (13)$$

$$-2w^1_{,xy}(w^0_{,xy} + w^*_{,xy}) + w^1_{,xx}(w^0_{,yy} + w^*_{,yy})$$

$$+w^1_{,yy}(w^0_{,xx} + w^*_{,xx}) = 0$$

3   PRE-BUCKLING ANALYSIS

The plate is considered to be fixed simply supported. The plate initial temperature is assumed to be T_i. The temperature is uniformly raised to a final value T_f in which the plate buckles. The temperature change is $\Delta T = T_f - T_i$. The edge conditions are defined as

$$u = v = w = M_x = 0 \quad \text{on} \quad x = 0, a$$

$$u = v = w = M_y = 0 \quad \text{on} \quad y = 0, b \quad (14)$$

The prebuckling resultant forces are defined as

$$N^0_x = N^T$$

$$N^0_y = RN^T \quad (15)$$

$$N^0_{xy} = 0$$

The imperfections of the plate, considering the boundary conditions, are assumed as [1]

$$w^* = \mu h \sin \alpha_m x \sin \beta_n y \quad (16)$$

where

$$\alpha_m = \frac{m\pi}{a} \quad \beta_n = \frac{n\pi y}{b} \quad m, n = 1, 2, \dots \quad (17)$$

and μ is the imperfection coefficient which varies between 0 and 1. Also, m and n are the number of half waves in x and y-directions, respectively. In order to find the prebuckling deflection in z direction, w^0, which is an important parameter in imperfection sensitive buckling analysis, equilibrium Eqs. (6) are used.

Considering Eqs. (2), (5), and (15), the last three of Eqs. (6) are rewritten as

$$A_{44}(u^0_{1,x} + w^0_{,xx}) + A_{55}(v^0_{1,y} + w^0_{,yy})$$

$$+N^T(w^0_{,xx} + Rw^0_{,yy}) = 0$$

$$D_{11}u^0_{1,xx} + D_{12}v^0_{1,xy} + D_{66}(u^0_{1,yy} + v^0_{1,xy})$$

$$-A_{44}(u^0_1 + w^0_{,x}) = 0 \tag{18}$$

$$D_{66}(u^0_{1,xy} + v^0_{1,xx}) + D_{12}u^0_{1,xy} + D_{22}v^0_{1,yy}$$

$$-A_{55}(v^0_1 + w^0_{,y}) = 0$$

where superscript 0 is used to indicate prebuckling conditions. The following approximate solutions are assumed which satisfy both the Eqs. (18) and the kinematical boundary conditions

$$u^0_1 = u^0_{mn} \cos \alpha_m x \sin \beta_n y$$

$$v^0_1 = v^0_{mn} \sin \alpha_m x \cos \beta_n y \tag{19}$$

$$w^0 = w^0_{mn} \sin \alpha_m x \sin \beta_n y \quad m, n = 1, 2, \ldots$$

Substituting the above solutions in Eqs. (18), a set of non-homogeneous linear equations with respect to the constant coefficient u^0_{mn}, v^0_{mn}, and w^0_{mn} is established. Solving the set of equations, the pre-buckling deflection, w^0, may be obtained as following

$$w^0 = -\frac{(\alpha^2_m + R\beta^2_n)\Theta\mu h N^T}{\det[K] + (\alpha^2_m + R\beta^2_n)\Theta N^T} \times \tag{20}$$

$$\sin \alpha_m x \sin \beta_n y$$

where

$$\Theta = K_{21}K_{32} - K_{22}K_{31} \tag{21}$$

and $[K]$ is the coefficient matrix of the set of equations and its components are defined as

$$K_{11} = A_{44}\alpha_m$$

$$K_{12} = A_{55}\beta_n$$

$$K_{13} = A_{44}\alpha^2_m + A_{55}\beta^2_n$$

$$K_{21} = D_{11}\alpha^2_m + D_{66}\beta^2_n + A_{44}$$

$$K_{22} = (D_{12} + D_{66})\alpha_m\beta_n$$

$$K_{23} = A_{44}\alpha_m$$

$$K_{31} = (D_{12} + D_{66})\alpha_m\beta_n$$

$$K_{32} = D_{66}\alpha^2_m + D_{22}\beta^2_n + A_{55}$$

$$K_{33} = A_{55}\beta_n \tag{22}$$

4 THERMAL BUCKLING ANALYSIS

The last three of stability equations (9) and the stability equation (13), with the consideration of Eqs. (2), (5), (10), (15), (16), and (20), form a set of four equations with four incremental variables u^1_1, v^1_1, w^1, and Φ as follows

$$A_{44}(u^1_{1,x} + w^1_{,xx}) + A_{55}(v^1_{1,y} + w^1_{,yy})$$

$$+N^T(w^1_{,xx} + Rw^1_{,yy}) + \Phi_{,yy}(w^0_{,xx} + w^*_{,xx})$$

$$-2\Phi_{,xy}(w^0_{,xy} + w^*_{,xy}) + \Phi_{,xx}(w^0_{,yy} + w^*_{,yy}) = 0$$

$$D_{11}u^1_{1,xx} + D_{12}v^1_{1,xy} + D_{66}(u^1_{1,yy} + v^1_{1,xy})$$

$$-A_{44}(u^1_1 + w^1_{,x}) = 0 \tag{23}$$

$$D_{66}(u^1_{1,xy} + v^1_{1,xx}) + D_{12}u^1_{1,xy} + D_{22}v^1_{1,yy}$$

$$-A_{55}(v^1_1 + w^1_{,y}) = 0$$

$$(A_{22}\Phi_{,yyyy} - 2A_{12}\Phi_{,xxyy} + A_{11}\Phi_{,xxxx}) \times$$

$$\frac{1}{A_{11}A_{22} - A^2_{12}} + \frac{1}{A_{66}}\Phi_{,xxyy} - 2w^1_{,xy}(w^0_{,xy} + w^*_{,xy})$$

$$+w^1_{,xx}(w^0_{,yy} + w^*_{,yy}) + w^1_{,yy}(w^0_{,xx} + w^*_{,xx}) = 0$$

In order to solve this set of equations, the following approximate solutions for incremental variables are assumed

$$u^1_1 = u^1_{mn} \cos \alpha_m x \sin \beta_n y$$

$$v^1_1 = v^1_{mn} \sin \alpha_m x \cos \beta_n y \tag{24}$$

$$w^1 = w^1_{mn} \sin \alpha_m x \sin \beta_n y$$

$$\Phi = F_{mn} \sin \alpha_m x \sin \beta_n y \quad m, n = 1, 2, \ldots$$

where u^1_{mn}, v^1_{mn}, w^1_{mn}, and F_{mn} are constant coefficients that depend on m and n. Substituting the approximate solutions (24) in Eqs. (23) results in four non-zero residues R_1, R_2, R_3, and R_4. According to the Galerkin's method, the residues are made orthogonal with respect to the approximate solutions as [3]

$$\int_0^a \int_0^b R_1 \cos \alpha_m x \sin \beta_n y \, dx \, dy = 0$$

$$\int_0^a \int_0^b R_2 \sin \alpha_m x \cos \beta_n y \, dx \, dy = 0$$

$$\int_0^a \int_0^b R_3 \sin \alpha_m x \sin \beta_n y \, dx \, dy = 0$$

(25)

$$\int_0^a \int_0^b R_4 \sin \alpha_m x \sin \beta_n y \, dx \, dy = 0$$

The determinant of the system of Eqs. (25) for the coefficients $u_{mn}^1, v_{mn}^1, w_{mn}^1$, and F_{mn} is set to zero, which yields the thermal buckling load

$$N^T = -\frac{\det[K] + (I_{mn})^{1/3}}{(\alpha_m^2 + R\beta_n^2)\Theta}$$

(26)

where

$$I_{mn} = \frac{1024\alpha_m^4\beta_n^4\Theta\,(\det[K])^2\,\mu^2 h^2}{9\pi^4 m^2 n^2 K_{44}}$$

(27)

Using Eqs. (4) and (26), the buckling temperature change, ΔT, is obtained. ΔT_{cr} is the smallest value of ΔT when the numbers of half-waves (m,n) are changed.

5 ILLUSTRATION

In order to compare the results with the known data in the literature, we reduce Eq. (26) to the one for isotropic plates. The results are compared with the results, reported by Shariat and Eslami [10]. The material properties of the metallic plate are considered as

$$E_m = 70 GPa \qquad K_m = 204 W/mK$$

$$\alpha_m = 23e^{-6} \quad \nu_0 = 0.3 \quad h = 0.005 m$$

(28)

The variation of the thermal buckling load versus aspect ratio a/b is illustrated in Fig. (1). It is observed that the buckling load decreases by the increase of aspect ratio. It is seen that the results of the present work are in good agreement with the results of Ref. [10].

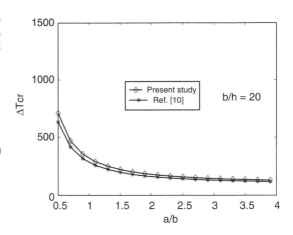

Figure 1. Critical buckling temperature change versus aspect ratio.

6 CONCLUSIONS

The thermal buckling load of imperfect laminated composite plates, based on the first order shear deformation theory, is obtained in closed-form solution for uniform temperature rise and is presented by Eq. (26). This equation indicates that the buckling load of an imperfect composite plate is increased in comparison with a perfect one. The increase in thermal buckling load is expressed by an imperfection term $(I_{mn})^{1/3}$, which directly depends on the imperfection size μ. Also, investigation of Eq. (27) shows that the imperfection term is affected by the material and geometrical properties of composite plate. The fact that the thermal buckling load of a plate is increased by existence of geometrical imperfections is fully explained by Morphy et al. [7]. The perfectly flat plate undergoes a symmetric pitchfork bifurcation at the buckling temperature. In contrast, the imperfect plate develops an asymmetric secondary state by means of a saddle-node bifurcation at the higher temperature [7]. The present study confirms this behaviour for laminated composite plates.

REFERENCES

Brush, D.O. and Almroth, B.O., Buckling of Bars, Plates, and Shells, McGraw-Hill, New York, 1975.

Thornton, E.A., Thermal Buckling of Plates and Shells, Applied Mechanics Review, Vol. 46, No. 10, pp. 485–506, 1993.

Turvey, G. and Marshall, I., Buckling and Postbuckling of Composite Plates, Chapman-Hall, New York, 1995.

Eslami, M.R. and Shariyat, M., Elastic, Plastic, and Creep Buckling of Imperfect Cylinders Under Mechanical and Thermal Loading, J. Pressure Vessel Technology, Transactions of the ASME, Vol. 119, No. 1, pp. 27–36, 1997.

Mossavarali, A., Peydaye Saheli, Gh. and Eslami, M.R., Thermoelastic Buckling of Isotropi and Orthotropic Plates with Imperfections, J. Thermal Stresses , Vol. 23, pp. 853–872, 2000.

Mossavarali, A. and Eslami, M.R., Thermoelastic Buckling of Plates with Imperfections Based On Higher Order Displacement Field, J. Thermal Stresses, Vol. 25, No. 8, pp. 745–771, 2002.

Murphy, K.D. and Ferreira, D., Thermal Buckling of Rectangular Plates, International Journal of Solids and Structures, Vol. 38, No. 22–23, pp. 3979–3994, 2001.

Eslami, M.R. and Shahsiah, R., Thermal Buckling of Imperfect Cylindrical Shells., J. Thermal Stresses, Vol. 24, No. 1, pp. 71–89, 2001.

Shariat, B.A.S. and Eslami, M.R., Buckling of Imperfect Functionally Graded Plates under In-plane Compressive Loading, Thin-Walled Structures, Vol. 43, No. 7, pp. 1020–1036, 2005.

Shariat, B.A.S. and Eslami, M.R., Thermal Buckling of Imperfect Functionally Graded Plates, International Journal of Solids and Structures, Vol. 43, No. 14–15, pp. 4082–4096, 2006.

Reddy, J.N. and Khdeir, A.A., Buckling and Vibration of Laminated Composite Plates Using Various Plate Theories, AIAA Journal, Vol. 27, No. 12, pp. 1808–1817, 1988.

Space, shell and hybrid structures

Steel and Composite Structures – Wang & Choi (eds)
© 2007 Taylor & Francis Group, London, ISBN 978-0-415-45141-3

Experimental and numerical investigations of beam-string structures during prestressing construction

X.Z. Zhao & Y.Y. Chen
Tongji University, Shanghai, China

J.X. Chen
East China Architectural & Design Institute, Shanghai, China

ABSTRACT: Beam-string structure (BSS), composed of beam or arch, cable and struts, obtains its overall rigidity by prestressing the cable. The geometric configuration and structural performance of BSS are highly pertinent to the process of prestressing. This paper presents the test results of 11 BSS specimens designed in light of the theory of orthogonal test. In these specimens, six parameters including the rigidity ratio of beam to string, the ratio of rise to span, κ, the number of struts, pre-tension in cable, construction sequence and the number of constructional supports are varied to investigate what parameters and how they affect the structural performance of BSS during prestressing. An analytical model is established and validated. Both the experimental and the numerical results indicate that (1) BSS achieves its overall rigidity at the moment when the critical pre-tension force in cable reach approximate $0.125/\kappa$ of the self-weight of the structure, and (2) the ratio of rise to span and the construction sequence have significant effects on the critical pre-tension force and structural rigidity, whereas, the rigidity ratio of beam to string and the number of constructional supports have less effect.

1 INTRODUCTION

Beam-string structure, composed of beam or arch, cable and struts, obtains its overall rigidity by pre-stressing the cable. By adding prestressed cable, beam-string structure gains additional stiffness and can span across large space with relatively less self-weight. Owing to the excellent properties, beam-string structures have been widely used in China in the past decade, such as the Shanghai Pudong International Airport Terminals (Figure 1), Guangzhou/Zhengzhou/ Harbin International Convention & Exhibition Center etc. Compared with the considerable research carried out on BSS structural performance in the service stage, little substantial information on the overall structural performance of the structure in construction stage is available. For the behavior of BSS during construction, six parameters including the rigidity ratio of beam to string, the ratio of rise to span, the number of struts, the pre-tension in cable, the construction sequence and the number of constructional supports should be considered. Attempts have been made to study numerically the influence of certain parameters on the behaviour of BSS (Bai et al. 2001, Chen et al. 1999, Masao et al. 1987), however, few of these numerical analysis have

Figure 1. Shanghai Pudong International Airport Terminals.

been validated by experiments or real structures. In addition, the sensitivity of the structural performance of beam-string structure on these parameters has not been clearly addressed.

In this paper, in total eleven BSS specimens were designed based on the theory of orthogonal test and prestressing tests were then carried out to investigate comprehensively the structural performance of BSS in construction stage. An analytical model was then established and validated. The effect of the above mentioned six parameters on the structural performance, the overall rigidity and the critical pre-tension was also studied.

Table 1. Details of test specimens.

Specimen	Number of struts	α	κ	Number of supports	Section type of arch	Sag of cable (mm)	Balance weight (N)
BSS-A	4	6.8×10^{-4}	0.1	1	B1	300	441.0
BSS-B	4	6.8×10^{-4}	0.15	3	B1	600	441.0
BSS-C	4	3.3×10^{-4}	0.1	3	B2	300	367.5
BSS-D	4	3.3×10^{-4}	0.15	1	B2	600	367.5
BSS-E	2	6.8×10^{-4}	0.1	1	B1	300	441.0
BSS-F	2	6.8×10^{-4}	0.15	3	B1	600	441.0
BSS-G	2	3.3×10^{-4}	0.1	3	B2	300	367.5
BSS-H	2	3.3×10^{-4}	0.15	1	B2	600	367.5
BSS-I	4	3.3×10^{-4}	0.1	1	B2	300	367.5
BSS-J	4	3.3×10^{-4}	0.1	1	B2	300	367.5
BSS-K	4	3.3×10^{-4}	0.1	1	B2	300	367.5

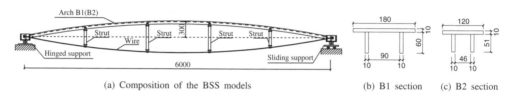

(a) Composition of the BSS models   (b) B1 section   (c) B2 section

Figure 2. Specimen of the BSS model test.

2 TESTS OF BSS DURING PRESTRESSING

2.1 Test scheme

In general, the main parameters that may affect the performance of the BSS during prestressing period are the rigidity ratio of beam to string, α, the ratio of rise to span, κ, the number of struts, the construction sequence and the number of constructional supports. α and κ are defined as

$$\alpha = E_b I_z / (E_c A_c L^2) \tag{1}$$

$$\kappa = f/L \tag{2}$$

where, E_b and E_c are the young's modulus of the rigid beam and the cable, respectively; I_z the moment of inertia around the principal axes perpendicular to the structural plane of the rigid beams, A_c the cross section area of the cable, f the rise of the structure, and L the span.

In order to reduce the number of models used for studying the various parameters, a test scheme was designed based on the theory of orthogonal tests and the details of the designed specimens are listed in Table 1.

The orthogonal test consists of eight specimens, test BSS-A to BSS-H, one beam-string structure in each test, to study the effect of the rigidity ratio of beam to string, the ratio of rise to span, the number of struts, the number of constructional supports on the structural performance in a orthogonal test of $L_8(2^7)$. Test BSS-I

and BSS-J are such designed to study the effect of the construction sequence. Both BSS-I and BSS-J consist of two BSS, but the braces between the two BSSs were assembled before prestressing being applied in BSS-I, while for BSS-J the prestressing was first performed to each BSS respectively prior to the assembling of the braces. The prestressing was applied until the displacement at the middle span of the structure reaches 1/60 of the span in test BSS-K, in order to study the behaviour of the structure under large deformation condition due to prestressing.

2.2 Specimens

Figure 2 illustrates a typical BSS test specimen. The span for all the specimens was 6 meters. Two types of cross section, B1 and B2, were used for the arch in different specimens, as listed in Table 1. Figure 2 shows the details of B1 and B2. One end of the specimen was fixed with a hinge support, which allows BSS in-plane rotation. Roller support was used at the other end where horizontal displacement and in-plane rotation are allowed. A high-strength steel bar with 7 mm diameter was used for the cable, which is connected to the arch through anchor cup at the two ends. Steel tube with a diameter of 48 mm and a thickness of 2.5 mm was used for the strut, which is connected to the rigid beam at the top end to satisfy the free rotation in plane and nearly rigidity out of plane, and connected to the cable at the other end through a special designed cable clamp to achieve ideal hinge. Spring was used as

Figure 3. Test specimen.

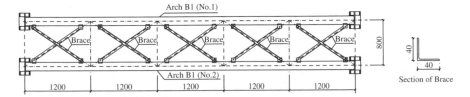

Figure 4. Specimen of BSS-I and BSS-J with diagonal braces.

Figure 5. Picture of BSS-I/BSS-J.

temporary supports to take into account the deformation of the constructional support due to self-weight in real structure. Figure 3 shows a picture of a typical specimen. In test BSS-I and BSS-J, the two BSSs were connected with a series of diagonal braces to form an integrated structure, as shown in Figure 4 and Figure 5.

2.3 Prestressing and loading

To simulate the fact that in real structure the rigid member is not able to carry self-weight before prestressing, balance weight of around two times the self-weight of the rigid member were applied on the rigid member by steel blocks, the value of which for each specimen is listed in Table 1. The prestressing was applied in two stages: before skew back, i.e., the rigid member still set on the constructional supports, and after skew back, i.e. after the rigid member leaves the constructional supports completely. At the first stage, prestressing force was increased by 600 N each time, which is then reduced to 200 N each time when the rigid member is

about to leave the constructional supports; at the second stage, prestressing force was increased in 300 N until the target displacement has been reached. Loading was applied by steel blocks at test BSS-I and BSS-J after the completion of the prestressing.

Measurements obtained from the tests include displacements (both vertical and horizontal), member force of the cable, the rigid member as well as the supports.

3 TEST RESULTS

3.1 Observations

Upon the start of the prestressing, the arch bends upwards and the sliding support (see Figure 2) moves slowly towards the middle of the span with increasing pre-tension. The arch departs from the springs when the pre-tension is adequately large. For the BSS with several constructional supports, the middle spring will depart from the arch prior to the other springs. After

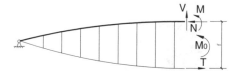

Figure 6. Mechanics analysis of BSS during prestressing.

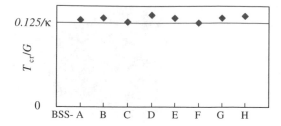

Figure 7. The critical pre-tension force obtained from tests.

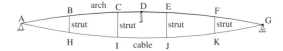

Figure 8. Analytical model of beam-string structure.

the arch has departed from the springs completely, the magnitude of the displacement increases rapidly at the same direction with continuing prestressing. A typical pre-tension force versus deflection response illustrates that the curve can be divided into two approximately linear parts at a critical point when the arch departs from the constructional supports completely and the BSS is supported on the two ends without any additional supports. The tension at the critical point is defined as critical pre-tension Tcr. At the moment when the prestressing reaches Tcr, the total deformation of BSS is nearly zero under the pre-tension applied and the self-weight of the structure. This actually means that the BSS gains its overall stiffness at this applied pre-tension load level.

3.2 The critical pre-tension T_{cr}

Figure 6 shows the mechanics analysis of BSS under the pre-tension load applied. The critical pre-tension T_{cr} can be calculated from the static equilibrium equation as:

$$T_{cr}/G \approx \frac{M_0}{Gf} = \frac{qL^2/8}{qLf} = \frac{1}{8f/L} = 0.125/\kappa \qquad (3)$$

Where, M_0 is the mid-span moment under self-weight, taking the structure as a simply supported beam; G and q is the self-weight and weight per unit length of the structure, respectively.

The measured values T_{cr}/G from tests BSS-A to BSS-H are given in Figure 7. The value predicted by Equation 3 agrees with the experimental results.

The critical pre-tension T_{cr} is one of the main parameters that can describe the overall structural properties of BSS during the prestressing process because it suggests the form of the structure.

4 NUMERICAL SIMULATION OF THE TESTS

4.1 Numerical model

An in-house program, called PABSS, was developed to simulate the orthogonal tests (BSS-A to BSS-H) using the numerical model shown in Figure 8. C.Oran beam-column element was chosen for the arch (rigid member) and the struts, catenary element for the cable and compression-only bar element for the constructional

supports. In the numerical model, the translation in three directions as well as torsion constraint were fixed at the left end, while at the other end vertical and lateral translation as well as torsion constraint were fixed. Vertical spring constraint was applied to the middle span of the model for the constructional supports.

4.2 Validation of the numerical model

The plots of the pre-tension force versus the vertical displacement at the middle of the span, ΔV (at point D in Figure 8), and the horizontal displacement, ΔH (at point G in Figure 8) obtained from both the test and the numerical simulation are given in Figure 9, where G is the self-weight of the structure including the balance weight, T the prestressing force. Curves labeled with T refer to the results from experiments; otherwise refer to that from numerical simulation. The upward vertical displacement is set to be positive, and horizontal displacement moving to right is positive. Good agreement is found between the experimental results and the numerical simulation.

The plots clearly show that the slopes before or after critical pre-tension T_{cr} are both approximately constant. To describe the overall structural property of BSS during prestressing process, the slope before and after the departure in the pre-tension versus vertical deflection curves, K_{by} and K_{ay}, are defined as the vertical rigidity of BSS. Similarly, K_{bx} and K_{ax}, represent the horizontal rigidity of BSS before and after the departure, respectively.

5 DISCUSSIONS

5.1 Orthogonal test analysis

To investigate how parameters, including number of struts, rigidity ratio of beam to string, ratio of rise to

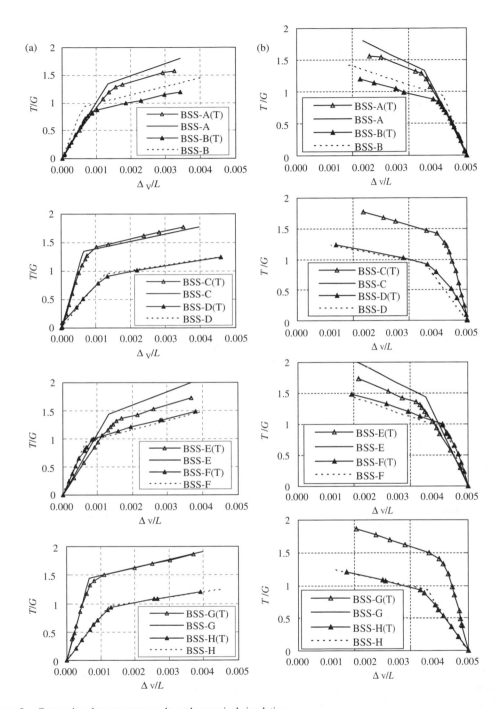

Figure 9. Comparison between test results and numerical simulation.

span and number of constructional supports, affect the structural performance represented by the critical pre-tension T_{cr} and pre-tension stiffness K_{by}, K_{ay}, K_{bx} and K_{ax}, eight specimens BSS-A ~ BSS-H were tested on

these four parameters (factors) with two levels each. The F-test method is used to study the significance of the parameters based on the orthogonal-test principles (Luan, 1987). A parameter, F, obtained from

Table 2. Test of significance of orthogonal test result.

Index		Parameter (factor)			
		$(1)^\#$	$(2)^\#$	$(3)^\#$	$(4)^\#$
T_{cr}	F	0.14	0.95	335.34	1.36
	Significance			**	
K_{by}	F	0.72	19.79	45.06	74.19
	Significance		*	**	**
K_{ay}	F	0.33	146.60	186.89	6.98
	Significance		**	**	(*)
K_{bx}	F	0.01	18.74	55.73	82.43
	Significance		*	**	**
K_{ax}	F	4.42	701.31	1222.5	71.96
	Significance	[*]	**	**	**

#(1) = number of struts,
(2) = rigidity ratio of beam to string,
(3) = ratio of rise to span,
(4) = number of constructional supports.

the variation arisen from different levels of parameters dividing variation arisen from test error, can be calculated. The results of the significance test analysis are listed in Table 2. If $F > F_{0.01}$, the influence of the parameter on the index is extremely significant, which is described as 'highly significant' and represented by symbols '**'; if $F_{0.01} \geq F > F_{0.05}$, the influence of the parameter on the index is significant, which is described as 'significant' and represented by '*'; if $F_{0.05} \geq F > F_{0.10}$, the influence of the parameter on the index is fairly significant, which is expressed as 'fairly significant' and represented by '(*)'; if $F_{0.10} \geq F > F_{0.25}$, the parameter may affect the index, thus 'influenceable but not significant' and '[*]' are used; if $F \leq F_{0.25}$, the parameter has no effect on the index, thus labeled as 'unnoticeable' and no symbols were used. $F_{0.01}, F_{0.05}, F_{0.10}, F_{0.25}$ equals to 34.1, 10.1, 5.54 and 2.02, respectively.

The results indicate that: the number of struts has not influence on the critical pre-tension force and structural rigidity, whereas, ratio of rise to span has high significant effect on them. The rigidity ratio of beam to string and the number of constructional supports have less effect on the critical pre-tension force but have significant effects on the structural rigidity.

5.2 Influence of construction sequence

The influence of construction sequence on the BSS structural performance was studied through test BSS-I and BSS-J. The pre-tension versus deflection of these two tests are shown in Figure 10. BSS-I1 and BSS-J1 are the results from No. 1 BSS, and BSS-I2 and BSS-J2 from No. 2. The shapes of each two curves (BSS-I1/BSS-J1, and BSS-I2/BSS-J2) are very similar, which indicates that construction sequence has

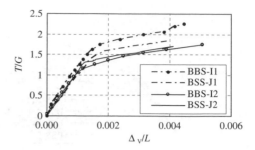

Figure 10. Comparison of vertical displacement at the arch mid-point for different construction sequence.

little effect on the rigidity of BSS. However, the values of T_{cr} for the same BSS in different test are different. In test BSS-I with diagonal braces installed prior to the pre-tension, No.2 BSS was assisted by No.1 BSS, which resulted in a lower pre-tension needed in No.2 BSS to make the arch depart from the springs and a larger pre-tension in No.1 BSS. This phenomenon is evidence of the existence of mutual interaction between two BSSs in prestressing after the braces are installed. In addition, the interaction causes additional forces in the braces, which should be taken into account in a real design.

6 CONCLUSIONS

Experimental and numerical investigation on beam-string structures during prestressing construction were carried out and presented in this paper. The study concentrates on how parameters, including the rigidity ratio of beam to string, the ratio of rise to span, the number of struts, pre-tension in cable, construction sequence and the number of constructional supports, affect the structural performance of BSS. Results show that (1) BSS achieves its overall rigidity at the moment when the critical pre-tension force in cable reach approximate $0.125/\kappa$ of the structural weight, and (2) the ratio of rise to span and the construction sequence have significant effects on the critical pre-tension force and structural rigidity, whereas, the number of struts has less effect on them; the rigidity ratio of beam to string and the number of constructional supports have less effect on the critical pre-tension but have significant effects on the structural rigidity.

ACKNOWLEDGEMENTS

The authors would like to thank National Natural Science Foundation of China under grant number 50108010 for providing fund for this work.

REFERENCES

Bai, Z.X., Liu, X.L. & Li, Y.S. 2001. Influence Analysis of Factors of Single Beam String Structure. *Steel Structure*, 16(3): 42–46.

Chen, J.X., Zhao, X.Z. & Chen, Y.Y. 2002. Structural Performance and Corresponding Research of Beam String Structure. *The 2nd National Seminar on Modern Civil Engineering, Maanshan, China.*

Chen, R.Y. & Dong, S.L. 2003. Prestressing Work of a Long-span Steel Truss String Structure. *Spatial Structures*, 9(2): 61–63.

Chen, Y.Y., Shen, Z.Y., Zhao, X.Z. et al. 1999. Experimental Study on a Full-scale Roof Truss of Shanghai Pudong International Airport Terminal. *Journal of Building Structures*, 20(2): 9–17.

Li, W.B., Shi, J. & Guo, Z.X. 2003. Research on Prestress stretching control of a large-span truss string structure. *Journal of Southeast University (Natural Science Edition)*, 33(5): 593–596.

Luan, J. 1987. *Skill and Method of Test Design*. Shanghai: Shanghai Jiaotong University Press.

Masao, S., et al. 1987. A Study on Structural Characteristic of Beam String Structure. *Summaries of Technical Papers of Annual Meeting Architectural Institute of Japan.* Tokyo: AIJ.

Steel and Composite Structures – Wang & Choi (eds)
© 2007 Taylor & Francis Group, London, ISBN 978-0-415-45141-3

Loading behavior of suspended domes considering glass panels

Chunli Xu & Yongfeng Luo
College of Civil Engineering, Tongji University, Shanghai, China

Gang Feng
Henan Provincial Architecture Design & Research Group, Zhengzhou, China

ABSTRACT: The glass roof supported with a suspend-dome system is found widely in practice. According to the current specifications, the co-working of glass panels and supporting structures isn't taken into account in structure design. The conventional simplified analysis doesn't accord with the actual loading behavior of the structure system. For further research, two finite element models of the suspend-dome supporting system with the glass panels and without glass panels are established in the paper. Static analysis and ultimate loading analysis of nonlinear stability are carried out. The numerical results are compared. The co-working behavior and the ultimate state of the supporting structure considering glass panels are summarized and presented. It is valuable for the structure design of similar steel suspend-dome structures.

1 INTRODUCTION

The daylighting glass roof supported with a suspend-dome system is found widely in engineering practice. According to the current design specification, only the loading capacity of the supporting structure is calculated in the structure design. The stiffness of glass panels isn't taken into account in structure design. There is no doubt that the conventional simplified analysis doesn't accord with the actual loading behavior of structures. In fact, the glass material has got quite great progress in strength, deformation performance, thermal stability and service safety nowadays. The thermal expansion coefficient of the glass is close to one of steel. The glass has also strong corrosion resistance. Therefore, it is feasible for glass panels to be used as structure members. Where a suspend-dome glass roof is constructed, a composite structure composed of glass panels set on a suspend-dome by claws and the supporting structure is formed if the stiffness of glass panels is considered. The glass panels have definite in-plane stiffness and contribute to the stiffness of the whole structure system. The panels not only enhance the integrity of the dome system, but also take part in the load bearing of the supporting system. For this reason, the contribution of the glass panels to the stiffness of supporting structures should be taken into account in the structure design of the suspend-dome glass roof. The self-weight of the dome will be reduced and the construction cost will be lowered if the glass panels are considered. At the same time, the configuration of supporting structures will be novel and flexible. However, few papers or research references about loading behavior of the structure considering glass panels have been found till now. The structure design without considering glass panels is not reasonable.

For further research, two finite element models of the suspend-dome, supporting system with glass panels and without glass panels, are established in the paper. The static analysis and the nonlinear stability analysis are carried out. The maximum displacements and critical loads of the two models under a certain load combination are obtained and compared. The co-working mechanism of glass panels and supporting structures, the loading behavior of different structure elements and the ultimate state characteristics of the composite structure are summarized and presented. It is a valuable method and a useful reference for the structure design considering the co-working of glass panels.

2 COMPUTATION MODELS

For suspend-domes, the cable pretension makes single-layer latticed shell members generate contrary internal forces with roof loads. The appropriate cable pretension can eliminate the tension of the circumferential members and reduce the horizontal push of supporting structures under loading. The cable pretension can adjust the internal force of outer members,

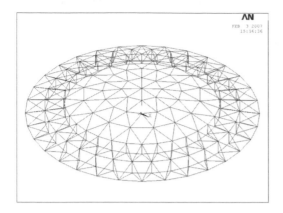

Figure 1. The model without glass panels.

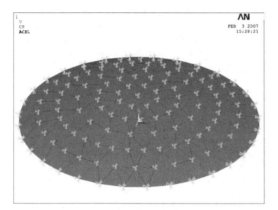

Figure 2. The model with glass panels.

while has less impact on the internal force of inner ring members. For this reason, the two radial and circumferential cables are arranged on the outer rings of suspend-domes considered in this paper to study the co-working of glass panels and the supporting structure. The suspend-dome with span of 35.4 m and rise of 4.6 m is pinned on the lower concrete ring beam. The single-layer latticed shell is K8-lamella reticulation. The members of the latticed shell are the steel pipes of $\Phi\ 133 \times 6$. The section of braces is $\Phi\ 87 \times 4$ and the section of cables is the steel strand of $6 \times 19\ \Phi\ 21.5$. The elastic modulus of steel pipes and cables is respectively 2.1E11N/m² and 1.8E11N/m². The thickness, design strength and the elastic modulus of tempering glass panels is respectively 6 mm, 58.5 Mpa and 7.2E10N/m².

Two finite element models of the suspend-dome supporting system with glass panels (Fig. 1) and without the glass panels (Fig. 2) are established in the paper. The numerical simulation is carried out by virtue of

the finite element software ANSYS and the mechanical properties of structures are analyzed. The beam, cable, bars and shell elements are used in two models. The couplings are applied on the joints between glass and latticed shell members. The rotational degrees of freedom are released. The uniform load of 2 kN/m² on glass panels is transformed into node concentrated loads acted on the nodes of the latticed shell.

3 CO-WORKING THEORECTICAL ANALYSIS OF SUSPEND-DOME WITH GLASS PANELS

In order to study the co-working characteristics of glass panels and the suspend-dome, the static analysis is firstly carried out under the condition of considering the influence of cable initial pretension and load grades on stiffness contribution of glass panels. Secondly, glass panels contribution to the overall stability of the suspend-dome supporting structures is studied by the linear, nonlinear (perfect and imperfect) and elasto-plastic stability analysis.

3.1 Relation between the cable initial tension and glass stiffness contribution

The influence of cable initial tension on stiffness of structures is obvious. Cable initial tension is usually easy to be changed. For this reason, the way of increasing cable initial tension is adopted to enhance the stiffness of structures. The initial tension of circumferential cables is assumed 20 kN, 40 kN, 60 kN, 80 kN and 100 kN in following numerical analysis. The load is divided into four grades to be applied on the structures. The stiffness contribution of glass panels is written as follows:

$$C_g = 1 - G_n / D_n \qquad (1)$$

Where G_n represents the maximal vertical displacement of the structure with glass panels at the different initial tension; D_n represents the maximal vertical displacement of the structure without glass panels at the different initial tension.

The maximal vertical displacements calculated at all load levels are listed in Table 1. The stiffness contribution of glass panels at different initial tension is listed in Table 2.

It is found from Table 1 and Figure 1 that the displacements increase and the stiffness contribution of the glass panels becomes stronger with the increment of load levels at a definite initial tension. For relatively small initial tension, the increasing rate of glass stiffness contribution tends to be stagnating and stable when loads exceed a definite value, such as the initial tension of 20 kN in Figure 3. From Table 2 and Figure 4, it is shown that the glass stiffness contribution

Table 1. Maximal displacements at all load levels (mm).

| Pretension | Load grades | | | |
	0.25	0.5	0.75	1
D20	4.028	4.621	5.413	6.406
G20	2.435	2.753	3.072	3.390
Cg (%)	39.5	40.4	43.2	47.1
D40	3.224	4.006	4.766	5.458
G40	2.090	2.418	2.747	3.075
Cg (%)	35.2	39.7	42.4	43.7
D60	2.659	3.491	4.274	4.798
G60	1.951	2.341	2.732	2.911
Cg (%)	26.6	32.9	36.1	39.3
D80	2.032	2.834	3.517	4.143
G80	1.659	2.110	2.459	2.730
Cg (%)	18.4	25.6	30.1	34.1
D100	1.315	2.091	2.917	3.579
G100	1.152	1.72	2.232	2.585
Cg (%)	12.4	17.7	23.5	27.8

Table 2. Maximal Mises stress and Stiffness contribution of glass panels at different Initial tension.

Pretension (kN)	20	40	60	80	100
Mises stress (Mpa)	3.22	3.25	3.33	3.40	3.52
Displacement (mm)	3.39	3.075	2.911	2.73	2.585
Contribution (%)	47.1	43.7	39.3	34.1	27.8

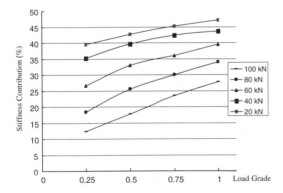

Figure 3. Curves of glass stiffness contribution with load grades.

is decreasing with the increasing of initial tension. The approximate linear relation can be found between the decreasing of glass stiffness contribution and the increasing of initial tension. The maximal Mises stress of glass panels is 3.52 Mpa which is less than the design strength of tempering glass, and the glass panels are unbroken.

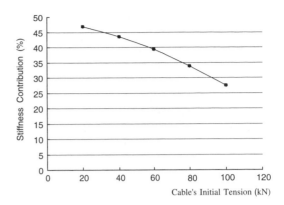

Figure 4. Curve of glass stiffness contribution with pretension.

In fact, glass stiffness contribution is closely related to the growing of displacement. The larger displacement, the larger stiffness contribution. Under the condition of the same initial tension, the glass stiffness contribution is increasing gradually with the rising of loads and the growing of displacement. Therefore, under the condition of the same loads, the larger cable initial tension, the larger structure stiffness, the smaller displacement, the smaller the glass stiffness contribution. The glass stiffness contribution is higher related to the displacement growing than to the cable initial tension. In one word, the structure displacements are dominant of glass stiffness contribution.

3.2 Influence of glass stiffness contribution on overall stability of structures

After the suspend-dome buckled, the load bearing capacity of structures descended rapidly. It is necessary for suspend-domes that the stability analysis is carried out. The large geometric nonlinearity exists in the structure due to the application of flexible cables and the finite stiffness of structures. So the geometric nonlinear (perfect and imperfect) and material nonlinear stability analysis are performed to study glass stiffness contribution to the overall stability of suspend-domes. The maximal initial geometric imperfection is taken as L/300 = 118 mm (L represents the span of the structure) and the first buckling mode is adopted as the imperfection shape. The ideal elasto-plastic model of stress-strain is adopted and the yield stress is equal to 235 N/mm². The cable initial tension is taken as 60 kN.

The critical loads, the maximal vertical node displacements and the maximal Mises stress of glass panels at the time of buckling are listed in Table 3. The geometric nonlinear buckling deformations of the imperfect structures are shown in Figure 5. The load-displacement curves of geometric nonlinear and

Table 3. Critical loads, maximal node displacements and maximal Mises stress at the time of buckling.

Critical loads	Linear	Perfect nonlinearity	Imperfect nonlinearity	plastic
Without glass	19.4	17.3	7.4	4.88
With glass	26.3	21.8	9.7	6.7
Contribution	36%	26%	31%	37%
Displacement	Linear	Perfect nonlinearity	Perfect nonlinearity	plastic
Without glass	–	93.3	187.7	160.5
With glass	–	69.7	171.6	105.8
Contribution	–	25%	9%	34%
Maximal stress	–	115	51.0	38.7

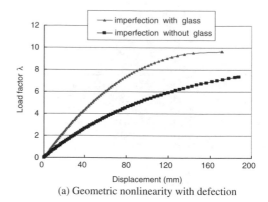

(a) Geometric nonlinearity with defection

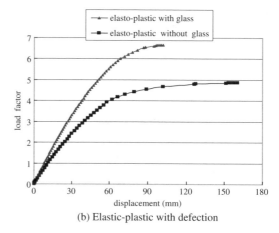

(b) Elastic-plastic with defection

Figure 6. Load-displacement Curves of the imperfect structures.

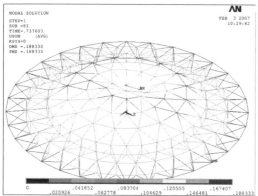

(a) Geometric nonlinear buckling deformation without glass

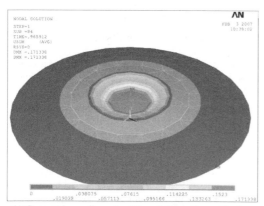

(b) Geometric nonlinear buckling deformation with glass

Figure 5. Buckling deformation of the suspend-domes.

elasto-plastic analysis of the imperfect structures are shown in Figure 6. The Mises stress distribution diagrams of geometric nonlinear analysis and elasto-plastic analysis of the imperfect structures are shown in Figure 7.

From the figures and tables above, it can be found that the suspend-dome without glass panels and with glass panels have the different buckling form, different critical loads and different node deformations. The area near the centre of the latticed shell is less curved and easy to buckle. The ultimate loads are larger in the structure with glass panels than in the structure without glass panels. The increasing of the ultimate loads is different under different analysis assumption. The largest increase occurs in the elasto-plastic analysis, increased by 37%. The stiffness contribution of glass panels also made the maximal vertical displacements decrease differently. The largest decrease of the displacements occurs in elasto-plastic analysis, decreased by 34%. It is obvious that the co-working of glass panels improves the stability behavior of the suspend-dome.

It is found from Table 3 and Figure 7 that the maximal Mises stress of glass panels is 115 Mpa at the geometric nonlinear analysis of the perfect structure which is greater than the design strength of tempering glass. The glass panels are unsafe. The maximal Mises stress of glass panels is respectively 51.0 Mpa

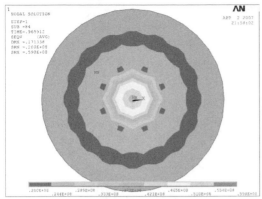

(a) Geometric nonlinearity with defection

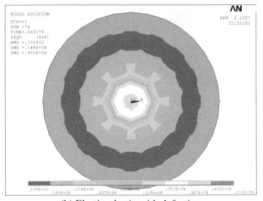

(b) Elastic-plastic with defection

Figure 7.   Mises stress diagrams of glass panels at buckling.

and 38.7 Mpa at the geometric nonlinear and elasto-plastic buckling analysis of the imperfect structures which are less than the design strength of tempering glass. The glass panels are unbroken.

4   CONCLUSIONS

(1) The stiffness contribution of the glass panels to the structure system depends on the displacement growing under loading. Under the condition of the same initial tension, the glass stiffness contribution is increasing step by step with the rising of loads and the growing of displacements. Under the condition of the same loads, the larger cable initial tension, the larger structure stiffness; the smaller displacements, the smaller the glass stiffness contribution.

(2) The linear, geometric nonlinear and elasto-plastic buckling analyses are carried out. The bearing capacity of the structure with the glass has different extents of increasing. The co-working of glass panels improves effectively and obviously the stability behavior of the suspend-dome.

(3) The maximal Mises stress of glass panels is 3.52 Mpa at the different initial tension. The glass panels are safe. For the buckling analysis, the maximal Mises stresses are both less than the design strength of tempering glass at geometric nonlinear and elasto-plastic buckling analysis of the imperfect structures. The glass panels are unbroken.

REFERENCES

Chen Zhihua. 2004. The Suspend-dome System and Characteristic Analysis. Building Structure, 34(5):38–41.
Feng Ruoqiang, et al. 2005. Static performance of single-layer cable net glass curtain in consideration of glass panels. Building Structure, 26(4):99–106.
Zhang Yigang, et al. 2005. Large Span Space Structure. Beijing: mechanical industry.
Technical Specification for Latticed Shell Structures (JGJ61–2003, J258–2003).
Chen Ji. 2003. Stability of steel structure theory and design. Beijing: science and technology press.

Steel and Composite Structures – Wang & Choi (eds)
© *2007 Taylor & Francis Group, London, ISBN 978-0-415-45141-3*

Computation method of buckling propagation of single layer reticulated shells

A.Y.F. Luo & X. Liu
College of Civil Engineering, Tongji University, Shanghai, China

B.W.M. Hong
Tongji Architecture Design & Research Institute, Shanghai, China

ABSTRACT: With the rapid development of the architecture, long span and large scale complex spatial steel structures are widely used in modern public buildings or structures. The buckling or instability and buckling propagation become more obvious and much important in the loading behavior of the spatial steel structures. However, the buckling behavior and its propagation have not been understood thoroughly till now. The current researches are mostly focused on the tracing strategy of the load-displacement path of the structures and the incremental-iterative computation techniques. Few papers or research references about the buckling propagation and the dynamic response of the structure induced by local buckling propagation are found. A practical computation method of the buckling propagation of the single layer reticulated shells (SLRSs) is proposed in this paper. The local buckling behavior and the buckling propagation of 2 single layer Kiewitt-6 (K6) reticulated shells are investigated by means of the method. The characteristics of the buckling propagation of the local buckling are concluded.

Keywords: computation method, reticulated shell, local buckling, buckling propagation, overall stability

1 INSTRUCTION

The single layer reticulated shells (SLRSs) are a kind of large span spatial lattice structures. They are widely used in various buildings because of their light self-weight, attractive architectural shapes, variable and novel system and better structural stiffness. The loading behavior of the single layer reticulated shells is similar to the continuous thin shells. Buckling is generally the controlling factor in the structural design. Although a lot of single layer reticulated shells have been built, the stability behavior of the structures has not been understood thoroughly till now.

The current researches on overall stability of spatial structures are mostly focused on the computation methods of the critical loads and buckling modes and the tracing strategy of the load-displacement path of the structures. Few papers or research references about the buckling propagation and the dynamic response of the structure induced by local buckling propagation are found. The researches about buckling propagation were studied early in 1970s and were mainly limited to the submarine pipelines[1]. The phenomenon of buckling propagation was firstly observed

in the model experiment performed in the Butle Columbus Lab of America in 1970. The researches of buckling propagation of reticulated shells were conducted late[2]. The buckling propagation of reticulated shells was initially considered by Lenza. Then, Gioncu and Balut conducted the study on structural vibration in the new equilibrium state induced by inertia forces which was generated by nodal snap-through buckling of SLRSs, and pointed out the vibration would possibly induce the overall buckling. Lenza analyzed the dynamic response of nodal snap-through of single layer hyperboloid shells with hinged joints. He presented that the maximum acceleration generated by dynamic nodal snap-through would reaches 16 times as gravity acceleration and may lead to the buckling propagation. Abedi and Parke explored the buckling propagation by transferring the dynamic energy into initial velocities in dynamic analysis[3]. Although there are some research papers on the buckling propagation recently, the researches are far from the need of practice engineering. Some difficult and deep researches should be conducted further from now, such as how to judge whether a kind of local buckling of a structure will lead to buckling propagation in the structure, whether the buckling propagation will induce the structural collapse further and which measures should be taken to prevent the possible buckling propagation of structures.

A practical computation method of the buckling propagation of the SLRSs is proposed in this paper. The local buckling behavior and the buckling propagation of 2 single layer Kiewitt-6 (K6) reticulated shells are investigated by means of the proposed method. The characteristics of the buckling propagation of the local buckling are concluded.

2 THE MECHANISM OF BUCKLING PROPAGATION

When local buckling occurs due to buckling deformation of one or several individual members or nodes of a structure, severer dynamic response may be induced by the unloading of buckling members or nodal snap-through[4]. The response may possibly result in the larger area buckling or overall buckling of the structure. The successive buckling phenomena of the larger area or the overall of the structure induced by dynamic effect are named 'buckling propagation' or progressive collapse. From the point of view of mechanics, when individual or local members buckle, the forces in the buckling members will probably lower rapidly, and the neighbor members must bear the additional load released by the buckling member or members. The inner forces in the structure will redistribute. If the neighbor members can bear the redistribution forces, the structure will be stable and the buckling will not propagate. Otherwise, the structure will be unstable and the buckling will propagate to other areas even all structure. In addition, when the snap-through buckling occurs in one or several nodes of a structure, the kinetic energy released during the buckling will transfer into the impulsion on the neighbor nodes. The dynamic effect is also possible to bring on the sudden change of the geometry of the structure. The buckling deformation similar to "Domino Effect" or even the overall buckling will be induced. The mechanism of buckling propagation in a structure is the dynamic transformation between critical equilibrium states or the dynamic transformation of strain energy of the structure. The final result of buckling propagation depends on the condition of the new equilibrium state after the change of the structure geometry.

Dynamic propagation of the local buckling is a dangerous dynamic buckling phenomenon for SLRS or the large span double layer thin reticulated shells, especially for shells with positive Gaussian curvature[5]. The local buckling is easier to induce the buckling propagation and brings on the overall buckling of structure finally.

3 THE CALCULATING AND JUDGING METHOD OF THE BUCKLING PROPAGATION

The forces of structure will often redistribute after the local buckling. The force redistribution will result in the transition or snap-through from one equilibrium state A to another equilibrium state B. The transition usually takes place momentarily. The dynamic effect induced by the transition should be taken into account for the structure stability. The instant dynamic analysis of the new equilibrium state with given initial displacements, velocity and acceleration should be conducted to ascertain whether buckling will propagate after the local buckling. If further buckling of new members or joints occurs in the new equilibrium state, the local buckling will result in the buckling propagation. Otherwise, the buckling will not propagate.

In general, with the assumption of given initial displacements, velocity and acceleration, the dynamic equilibrium equation of the structure with local buckling in state B is shown in Equation 1:

$$[M]\{\ddot{u}\} + [C]\{\dot{u}\} + [K]\{u\} = \{F(t)\} \qquad (1)$$

Where, [M] is mass matrix; [C] is damping matrix; [K] is stiffness matrix; {F} is impact load function caused by local buckling; {u} is nodal displacement vector; {$\dot{u}$} is nodal velocity vector; {$\ddot{u}$} is nodal acceleration vector.

In order to judge whether the buckling will propagate due to local buckling, the dynamic equilibrium equation (1) is not to be solved directly. The static nonlinear analysis of structure should be carried on firstly to ascertain the critical load, buckling mode, buckling characteristic nodes and the initial velocity generated by local buckling. Then the dynamic analysis under impact load is conducted to judge the behavior of buckling propagation. The steps in numerical analysis are shown in detail as followed:

1) Carry on the linear stability analysis and ascertain the buckling modes, buckling characteristic nodes and linear buckling factors (the upper limit of stability factors);
2) Trace the nonlinear load-displacement path and ascertain the first critical load of nonlinear local buckling and the specific location of buckling characteristic nodes before or after the buckling, and record the inner forces, displacements or the structural configuration under the critical equilibrium state;
3) Calculate the released energy in the local snap-through buckling according to the nonlinear load-displacement path of the structure and get the initial velocity for dynamic analysis further.
4) Use the nonlinear buckling critical state obtained in step 2 (include the forces and displacements of this

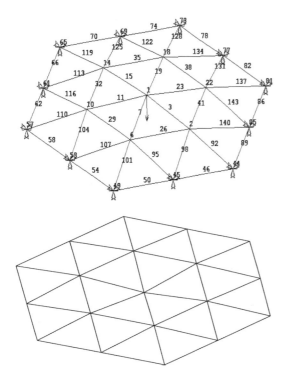

Figure 1. The FEM model and buckling mode of K6 SLRS under a single-joint load.

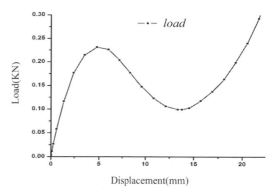

Figure 2. Nonlinear load-displacement curve of structure.

The yield strength is 180 N/mm2 and the density is 7850 kg/m3. The detail numbers of structural members and nodes are shown in Figure 1. The concentrated load is applied on the center node 1 of the shell.

The overall stability analysis of the reticulated shell is conducted. The critical load corresponding to the first buckling mode is 734.7 N. The first buckling mode is shown in Figure 1. The buckling deformation is downward of the middle parts of the shell. Node 1 is the buckling characteristic node. The nonlinear load-displacement path of the overall stability analysis is shown in Figure 2. The critical load of nonlinear overall stability is 232.4 N. Under the same critical load, the total displacement of center node 1 reaching to the new equilibrium state is 20.28 mm. The displacement during the unstable snap-through is 14.97 mm. The energy released in unstable snap-through is 1.1752 N·M. The corresponding initial velocity generated by the local buckling snap-through is 314.85 mm/s.

The dynamic analysis of snap-through local buckling of the structure in the equilibrium state corresponding to the first critical state is conducted. The time-increment displacement curve is shown in Figure 3. The Node 1 and Node 7 stand for the joint 1 and the joint 7 respectively. From the results of dynamic analyses, it can be found that the displacement of node 1 reaching to the new equilibrium state changes from 20.28 mm of static analysis to 23.11 mm because of the dynamic impact. The displacements of neighbor nodes are almost unchanged. The buckling of the structure is still found only in the adjacent of node 1. Node 7 does not buckle further under the dynamic impact induced by local buckling of node 1. The responses of other nodes are similar with node 7. The results show that the buckling doesn't propagate in the shell.

state) as the new computation model and carry on the dynamic characteristic analysis to get the first period T1;

5) Solve Equation 1 with the time increment $\Delta t = T1/100 \sim T1/20$ and get the dynamic responses of the buckling characteristic nodes and their neighbor nodes;

6) Judge whether the buckling will propagate due to the local buckling according to dynamic response, and ascertain the scope of buckling propagation and the effect on load-bearing capacity of the overall stability of the structure.

4 NUMERICAL EXAMPLES

Ansys is adopted to analyze two different K6 SLRSs under a single-joint load and under multi-joint loads in the article.

4.1 K6 SLRS under a single-joint load

A K6 single layer spherical shell with span of L = 1000 mm and rise of f = 40 mm is shown in Figure 1. Rectangle section of 4 mm × 4 mm for all members is used. The young's modulus and the Poisson ratio of material are 211.2 N/mm2 and 0.296 respectively.

4.2 K6 SLRS under multi-joint loads

Figure 4 shows a K6 single layer spherical shell. The section and the material of members are the same with the above example. The span of shell is still 1 m and

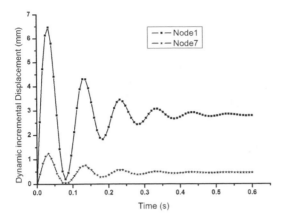

Figure 3. Time-increment displacement curve.

the rise is 40 mm. The numbers of lattices are increased for multi-joint loads. The circumferential grid numbers are changed from 2 to 4. The load mode is multi-joint loads (node 1 to node 7) which is shown in Figure 4.

The overall stability analysis of the reticulated shell is conducted. The critical load corresponding to the first buckling mode is 395.8 N. The first buckling mode is shown in Figure 4. The buckling deformation is downward of the middle parts of shell. Node 1 is the buckling characteristic node. The nonlinear load-displacement path of the overall stability analysis is shown in Figure 5. The critical load of nonlinear overall stability is 125.3 N. Under the same critical load, the total displacement of center node 1 reaching to the new equilibrium state is 26.10 mm. The displacements during the unstable snap-through are 19.30 mm. During the unstable snap-through, the total displacements of node 2 to node 7 in the new equilibrium state are all equal to 17.85 mm. The displacements during the unstable snap-through are 14.860 mm. The energy released in the unstable snap-through is 0.3747 N·M. The initial velocity of node 1 generated by the snap-through local buckling is 314.85 mm/s and the initial velocities of node 2 to node 7 are all equal to 186.40 mm/s.

The dynamic analysis of snap-through local buckling of the structure in the equilibrium state corresponding to the first critical load is conducted. The time-increment displacement curve is shown in Figure 6. The node 1, node 7 and node 19 stand for the joint 1, the joint 7 and the joint 19 respectively. From the results of dynamic analyses, it can be found that the dynamic responses of node1 and the nodes (2–7) in first circle are similar. At the meantime, when the structure reaches to the new stable equilibrium state, the buckling region extends to the second circle further because of the local dynamic snap-through buckling of center node and the nodes in first circle. Figure 7 and Figure 8 show the deformation of the structure in

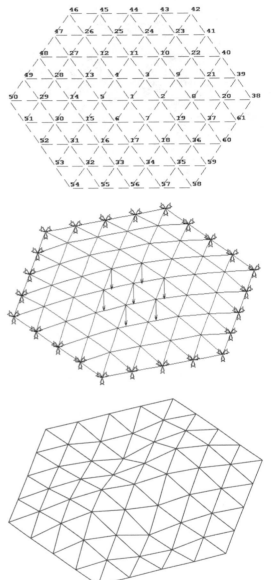

Figure 4. The FEM model and buckling mode of K6 SLRS under multi-joint loads.

the new stable equilibrium state in static analysis and the final structural configuration in dynamic analysis respectively. The results show that buckling propagates in the shell and the overall buckling of the structure is possible.

Two examples show that the dynamic effect generated by multi-node local buckling is larger than that of single-node local buckling. Therefore, the buckling propagation is more sensitive to the multi-node local buckling.

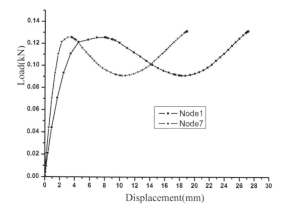

Figure 5. Nonlinear load-displacement curve of structure.

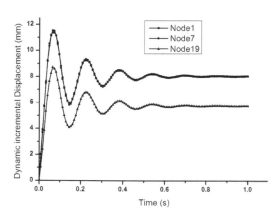

Figure 6. Time-increment displacement curve.

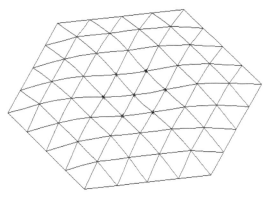

Figure 7. The static critical state of the structure.

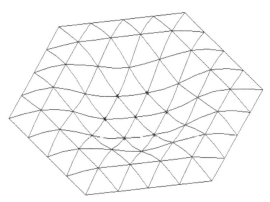

Figure 8. The critical state after buckling propagating.

5 CONCLUSION

The numerical results show that:

1) The method proposed in this paper is effective and reliable for the buckling propagation analysis of the reticulated shells;
2) Buckling propagation doesn't occur in any reticulated shell with local buckling;
3) The buckling propagation is more sensitive to the multi-node local buckling than the single-node local buckling when the shells reach the critical state.

REFERENCES

[1] Xia Kaiquan, Yao Weixing, Dong shilin. 2000 Dynamic Behavior of snap-through buckling in reticulated domes. The 9th National Conference on Spatial Structure.Zhejiang,.
[2] Victor Gioncu. Buckling of Reticulated Shells: State-of-the-Art. International Journal of Space Structures . 1995, Vol.10 No.1, pp1–46.
[3] K. Abedi, G. A. R. Parke. Progressive collapse of single-layer braced domes. International Journal of Space Structures, 1996, Vol.11 No.3, pp291–306.
[4] Qian Ruojun, Wang jian, Zeng Yinzhi. An investigation on modeling for instability analysis of reticulated shells. Journal of Building Structures. Jue.2003.Vol.24 No.3, pp10–15.
[5] Yang Lianping, Li Zhijian, Qian Ruojun. Concept design of reticulated shells. The 10th National Conference on Spatial Structure. Beijing, 2002.

Steel and Composite Structures – Wang & Choi (eds)
© *2007 Taylor & Francis Group, London, ISBN 978-0-415-45141-3*

Dynamic analysis of partial double-layer spherical reticulated shell structures

L.J. Li, Z. Yang & F. Liu
Faculty of Construction Engineering, Guangdong University of Technology, Guangzhou, China

ABSTRACT: A partial double layer reticulated shell structure is studied and its dynamic properties are analyzed in this paper. The dynamic responses of the partial double layer reticulated shell structure subjected to uni-dimensional, bi-dimensional and tri-dimensional earthquake motions are analyzed respectively, three cases of responses including the linear elastic, geometric nonlinear and both geometric and material nonlinear are considered. The nodal displacements and member stresses of the structure are simulated and analyzed. The characteristics of elastoplastic seismic responses of the structure under the seismic motion excitation are revealed and the basic rules are obtained. A vibration reduction method is proposed by adding viscous dampers in the reticulated shell structure. The rules of vibration reduction effects on the reticulated shell structure with different viscous damper system are acquired.

1 INTRODUCTION

Space structures are now widely used as factory buildings, sports gymnasiums, exhibition halls, etc, as their inherent rigidity of three-dimensional space structures and their ability to cover large spans with small weight are greatly appreciated by engineers. However the difficulties existed in analysis of such complicated systems limit their use. The purposes of this paper are to discuss the dynamic responses of partial double layer spherical reticulated shell structures under the earthquake motion excitation and to present an aseismic technique for these structures (Li et al. 2005).

Nonlinearity also arise when the stress-strain relationship of the material is nonlinear, this is called material nonlinearity. In this paper a bilinear material property is accepted which means that the structure behaves linearly until the first hinge has developed. With the increase of the load the structure continues to behave linearly, generally with a reduced stiffness, until a second hinge is formed. The same behavior continues with the increase of the load until sufficient hinges have developed to form the failure mechanism (Makowski 1993, Chang 2000 & Salazar et al. 2000).

2 CALCULATION MODEL

Reticulated shell structures are one type of space structures widely used in recent years. This paper chooses the optimum partial double layer reticulated shell structure as simulation model (Li et al. 2005). It

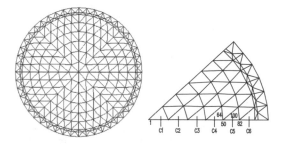

Figure 1. The plane layout of a partial double layer spherical reticulated shell.

is a 1/6 double-layer spherical reticulated shell structure, with a 0.23 height-to-span ratio, and has the most powerful load carrying capacity (Xie 2004). In order to investigate the seismic performance of spherical reticulated shell structure subjected to strong earthquake motion, the nodal displacement and stress of members are calculated by ANSYS program.

The geometry properties of the analyzed structure are: span L = 40 m, height H = 9.2 m, and radius R = 26 m. The partial double-layer spherical reticulated shell structure consists of simplified space beam elements for the double-layer part and space bar elements for the single layer part, the former is of ϕ 127 × 4 dimension, the latter is ϕ 104 × 3.5. The members used for the structure are steel tubes. There are 265 nodes and 792 members in the structure totally. The roof loads are taken as 2.5 kN/m². The structural model is shown in Figure 1.

3 BASIC ASSUMPTIONS

In order to analyze the structure simply and accurately, the basic assumptions are accepted as follows (Zhu 1994, Kani 1993, Liu et al. 1993):

(1) The material is an idealized elastic-plastic model with hardening phrase. The initial yield stress is 210 MPa, the failure stress 400 MPa, the Young's modulus 210 GPa and the tangential modulus is 19 GPa. The material model is a bilinear relationship isotropic hardening model, the von-Mises yield criteria is applied in calculation (Li 2003).
(2) The loads are assumed to be vertically and uniformly distributed over the dome surface. Self-weights of the members are treated as lumped mass concentrated at the nodes.
(3) Idealized beam and bar elements with rigid and pin joints are used respectively.

4 ANALYSIS PROGRAM AND MAIN RESEARCH CONTENTS

The numerical analysis is conducted by commercial finite element program ANSYS and FEAP to obtain linear, elastic geometric nonlinear and elastic-plastic solutions.

The paper focuses on three problems related to partial double-layer reticulated shell structure, they are:

(a) The property of the free vibration.
(b) Analysis of the reticulated shell structure subjected to the strong earthquake motion excitations.
(c) Vibration reduction analysis of the reticulated shell structure subjected to the strong earthquake motion excitation.

5 STRUCTURAL DYNAMIC CHARACTERISTICS

In this paper the important purpose is to calculate the dynamic responses of the reticulated shell structures subjected to the strong earthquake motion. The dynamic responses of structures are related to their properties, so the eigenvalues are calculated at first.

The modal analysis is performed for the spherical reticulated shell structures, from which the first 20 natural frequencies and vibration modes are obtained. It can be seen from Figure 2 and Figure 3 that the frequencies are very close and the vibration modes are fairly similar.

Figure 2 shows that the structure has a narrow band of the low natural frequencies, which are greatly concentrated. Figure 3 shows the first 4 modes of the structure. There are some modes with equal

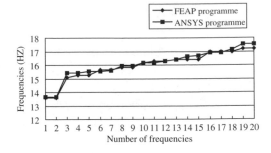

Figure 2. The comparison of the first 20 order natural frequencies.

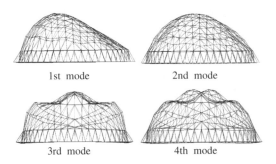

Figure 3. The first 4 vibration modes of the structure.

frequencies owing to the obvious geometric symmetry. For example, mode 1 and mode 2 have the same natural frequency, with dominantly horizontal displacements, while mode 3 has dominantly vertical displacements, and mode 4 behaves as coupled vibration modes in both vertical and horizontal direction. This kind of pattern is repeated for the higher modes. So the dynamic features of the structure are quite complex.

6 SEISMIC RESPONSE OF THE STRUCTURE

6.1 Mass matrix and damping matrix

The mass of the structure distributes averagely in the whole shell, and is often lumped to the nodes, which results in a diagonal mass matrix. It can be simplified by the static equivalent method and can fully meet the engineering requirement.

In a nonlinear analysis, the stiffness and damping of the structure will vary with the deformation of the structure. Generally, Rayleigh damping is employed in engineering as follows: $C = \alpha [M] + \beta [K]$, where M is the mass matrix of the structure, C is the damping matrix and K is the stiffness of the structure. The two parameters α and β can be obtained by the following equations.

$$\xi_1 = \frac{\alpha}{2\varpi_1} + \frac{\beta\varpi_1}{2} \tag{1}$$

$$\xi_2 = \frac{\alpha}{2\varpi_2} + \frac{\beta\varpi_2}{2} \tag{2}$$

where ω_1, ω_2 are the first two natural frequencies of the structure; and ς_1, ς_2 are the corresponding damping ratio of the structure.

It is difficult to get the damping ratio of the structure, and is generally regarded as 0.02 practically.

6.2 Choose of earthquake waves

Earthquake is a special kind of load which has some distinct characteristic such as low probability of occurrence and high intensity. Its action depends on such parameters as earthquake acceleration, frequency and duration, and has a significant role on calculation results. They are fully different among different earthquake motions. The characteristics of the earthquake motions affect greatly the structural dynamic response. Usually the real ground motion records are used as the kinematic excitation.

In the seismic criterion for an engineering structure, three or more ground motions are used to obtain the maximum response values of the structure. The motions usually include an artificial motion and two actual motions recorded. The accelerations of the earthquake motion should be adjusted accordance with the requirements given in the criterion or codes of design (GB50011-2001 2002). Three dimensional earthquake motions are considered in this paper, with two horizontal directions and one vertical direction. In this paper three directions are considered simultaneously and the ratio of the peak acceleration in X, Y and Z directions are adjusted to 400 gal, 340 gal and 260 gal respectively (Xue et al. 2003).

6.3 The Seismic response of the structure

The record of Tianjin earthquake motion is used to calculate the structural seismic response in this paper. The dynamic responses, including the structural linearity, the geometric nonlinearity and the geometric and material nonlinearity respectively, are calculated and considered in the paper. The representational nodes and members are shown in Figure 1.

6.3.1 Response of the structure subjected to undirectional earthquake motion

The representative simulation results are shown in Table 1 and Table 3. In these Tables, L, G and D represents linear, geometric nonlinear and both

Table 1. Maximum nodal displacements response.

Direction	Horizontal	Vertical
Linear	44.2	45.3
G nonlinear	46.9	49.7
D nonlinear	102.9	118

Table 2. Maximum nodal displacements response.

Direction	Horizontal	Vertical
Linear	45.4	46.9
G nonlinear	50.6	53.7
D nonlinear	155.7	198.4

Table 3. Maximum stresses response of members.

Direction		C1	C2	C3	C4	C5	C6
Rib	L	69	67	61	65	79	89
	G	94	92	84	81	96	99
	D	94	92	95	108	100	151
Circle	L	97	130	164	205	244	159
	G	123	147	181	228	254	176
	D	101	141	192	234	235	177
Oblique	L	77	102	82	88	82	66
	G	94	107	91	91	88	75
	D	95	105	83	92	111	67

Table 4. Maximum stresses response of members.

Direction		C1	C2	C3	C4	C5	C6
Rib	L	186	207	198	173	167	137
	G	190	211	198	173	170	140
	D	189	221	215	181	167	191
Circle	L	214	170	176	210	232	183
	G	226	198	209	237	249	197
	D	216	211	214	228	241	183
Oblique	L	192	225	98	99	93	116
	G	194	226	104	103	94	120
	D	198	210	99	118	142	117

geometric as well as material nonlinear (double nonlinear) respectively. The units of displacements and stresses are mm and MPa respectively.

6.3.2 Response of the structure subjected to three-dimensional earthquake motion

The representative computational results are shown in Table 2 and Table 4.

6.4 Comparison of response results

6.4.1 Displacement analysis

It is obvious that the results of linear, geometric non-linearity and the double nonlinearity are different. Generally, the results obtained by nonlinearity are larger than the linear results.

From the results of the maximum nodal displacements, the both geometric as well as material nonlinear results are by far the largest displacements of three cases, no matter whether the structure is subjected to the unidirectional or three-dimensional earthquake motion. The linear results and geometric nonlinear results are nearly similar. The increasing rate is about 15%. The main reason is that the earthquake motion selected for calculation is not strong enough and the influence of geometric nonlinearity is relatively small. But for the geometric and material nonlinearities, the situation has changed greatly. In linear situation the structure is regarded as linearity and elasticity, which is considered as small deformation and small strain problems. In geometric nonlinear situation, the deformation is larger than that of linearity, but the material is elastic, so it is considered as large deformation and small strain problems. In the both geometric and material nonlinearity, the material is regarded as idealized stress–strain relation, the plastic property is considered. A little change of its strain probably makes a large change of the structural rigid. So the structure has the largest responses with the mutual influence of the strain and deformation.

From the Table 1 and Table 3, it is clear that the dynamic responses of the structure subjected to three-dimensional earthquake are larger than that of the structure subjected to unidirectional earthquake. It is found from the calculation results that the structural responses caused by the different dimensional earthquake differ in about 5%, 8% and 55% respectively. From Table 2 and Table 4 we can draw the conclusion that the results of nonlinearity are larger than that of linearity. The stresses of the both geometric and material nonlinearity are more average than that of geometric nonlinearity.

For most rib members, the geometric nonlinear results increase about 35%, and the dual nonlinear results increase about 35%–65% compared with linear and elastic results. For most circle members the geometric nonlinear results increase about 15%–25%, and the dual nonlinear results increase about 5%–20%. For most oblique members the geometric nonlinear results increase about 5%–20%, and the dual nonlinear results increase about 5%–30%. The above conclusions are obtained according to the dynamic response of the structure subjected to unidirectional earthquake wave. In this situation only two members have yield.

In the linear analysis, the structural stiffness is regard as constant, but in the nonlinear analysis, the stiffness is regarded as a changing value that is decided

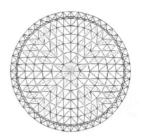

Figure 4. Oblique dampers Figure.

by the new structural deformation. Especially in both geometric and material nonlinearity the stiffness will change with the strain and the deformation when the material yields and the stress redistribution happens. When the material yields, its plastic deformation cannot return to the original situation when unloaded, the structural stiffness may decrease greatly. Sometimes, when the level of the materials yield is higher, the structural stiffness may decrease clearly. The period of the structure may become longer than ever, which make the period deviate the period of the earthquake wave, resulting in the response decrease. So in some situations the results of the geometric nonlinearity are larger than that of the both geometric and material nonlinearity if the members have not yield.

7 VIBRATION REDUCTION ANALYSIS

In this section the structure is analyzed by adding the viscous dampers to the reticulated shell. The structure is calculated and analyzed with different parameters as different position, different quantity and different damping coefficient.

The main virtue of the viscous dampers is that they can provide the system damping but don't add the system mass, so they don't change the structure period. Tests and studies indicate that the method is applicable to the reticulated shell structure under the dynamic loads (Lin 2003 & Fan 2003).

7.1 Different damper parameters

Two different positions of dampers in structure are considered, in first case the viscous damper is placed along oblique direction, in second case the viscous dampers are placed along circular direction. The configurations of the reticulated shell structure with the viscous dampers are shown in Figure 4 and Figure 5, and the dampers are expressed with red color.

7.2 Analysis of the result

In this part the influences of different dampers on the structure are discussed. Three damping coefficients,

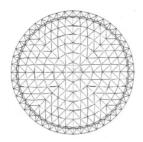

Figure 5.   Circle dampers.

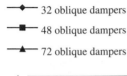

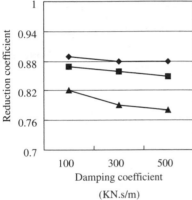

Figure 6.   Relationship of the horizontal displacement reduction coefficient and damping coefficient (node 82).

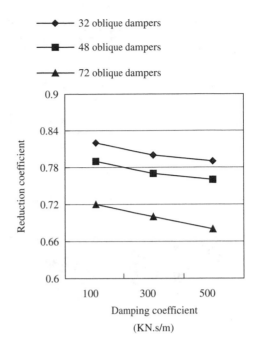

Figure 7.   Relationship of the vertical displacements reduction coefficient and damping coefficient (node 82).

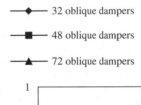

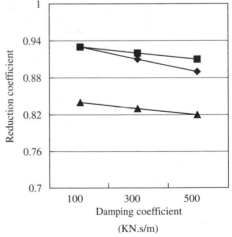

Figure 8.   Relationship of the stress reduction coefficient and damping coefficient (node 100).

which are 100 kNs/m, 300 kNs/m and 500 kNs/m, are calculated respectively. The nodes and members whose dynamic responses are largest in the whole structure are discussed, including the largest displacement node 82 and the largest stress member 100.

The relationship between the reduction coefficients and the different damping coefficients are shown in Figures 6–8. Owing to space limit, only results of oblique dampers are given here, the results of circle dampers can be referred to from Yang, 2005.

From the above figures we can draw the conclusion that the vibration reduction coefficients decrease with the increase of the damping coefficient, indicating that the vibration reduction effect is better when the damping coefficient increase.

Figure 9 and 10 is the comparison of damping coefficient for vertical displacements and stress respectively. Owing to space limit, the damping coefficient for horizontal displacements is not given here and can be referred to from Yang, 2005.

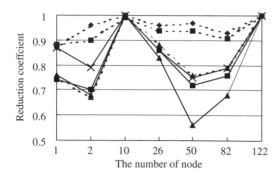

Figure 9. Comparison of vertical displacements damping coefficient on different position.

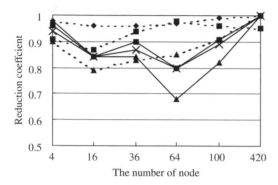

Figure 10. Comparison of stress damping coefficient on different position.

In the analysis for dampers in different positions, two cases are considered. One is the viscous dampers are along oblique direction. The other is the viscous dampers are along circle direction. Because of the different direction, it is impossible for the structure to keep the viscous dampers to have the same quantity in corresponding position, which give some difficulties in comparing the results of the two different cases. But we still can analyze the vibration reduction coefficient of the dampers to structure. The dampers coefficients used following are 500 kNs/m.

From Figures 9–10 it can be seen that different damper units and different damper positions will have different vibration reduction results. The oblique damper is more effective than the circle one. Generally, more dampers have good vibration reduction results than less one, Therefore the damper should be placed obliquely in practical operation.

7.3 Comparison of the dynamic response for different structures

Figure 11 and 12 is the comparison of the dynamic response for different structures. From the Figure 11

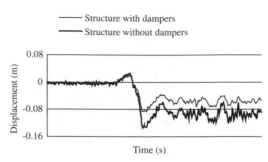

Figure 11. The horizontal displacement response at the 50th node.

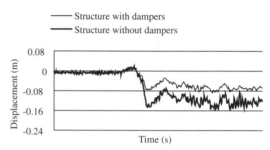

Figure 12. The vertical displacement response at the 50th node.

and Figure 12 we can know geometric and material nonlinearity responses of the structure with or without dampers have the similar curve, but the responses of the structure with viscous dampers are smaller by compare. That means the method by establishing the viscous dampers to structure to control the vibration of the structure is effective.

Method in this article only provides a method to reduce the response of the structure subjected to earthquake motion. It has advantages over traditional method, which is to strengthen the structure itself. The method also has the advantages of simple conception, explicit mechanical mechanism, and the safe credibility.

8 CONCLUSIONS

In this paper we investigate the dynamic properties of the partial double layer spherical reticulated shell structures under different dimensional earthquake motion, the paper focus on the effects of geometric and material nonlinearity on structural vibration responses based on numerical methods. Some conclusion can be summarized as follows.

Calculation results reveal that the natural frequencies of the spherical reticulated shell structure are concentrated greatly, which responds its dynamic features of such structures are complex.

When the structure are subjected to the strong earthquake motions, some nodal displacements may have a great increase owing to the stress redistribution. Some members in the structure respond in inelastic range, which will alter the dynamic characteristics of the structure. The effect of both geometric and material nonlinearities should be taken into account for seismic analysis, especially for the large span reticulated shell structures. The nonlinear responses are greater than the linear responses. At the same time, the multi-dimensional earthquake motion should be considered in the dynamic analysis.

Moreover, in the dual-nonlinear analysis the structure has fully different dynamic features. Stiffness decreasing, plasticity developing, stress redistributing and period changing probably make the structure represent complex behavior and performance.

The method adding the viscous dampers to structure is an effective measure to control the structural vibration under earthquake motion. By reducing vibration coefficients in the analysis we can find the approximate parameters suited for seismic design.

ACKNOWLEDGEMENTS

We would like to thank Guangdong Natural Science Foundation and Guangzhou Bureau of Science and Technology, Peoples' Republic of China, for partially supporting this project (Project number is 032489, 06104655, 03Z3-D0221)

REFERENCES

Chang, T. P. & Chang, H.C. 2000. Nonlinear vibration analysis of geometrically nonlinear shell structures. *Mechanics Research Communications*,27(2): 173–180.

Fan, F. & Shen, S.Z. 2003. Vibration reduction analysis of viscous damper on single-layer reticulated vaults. *World Earthquake Engineering* .6 19(2): 27–32.

GB 50011–2001, *Structural design code of anti-seismic*. Beijing: Chinese Construction Industry Press,2002 (in Chinese).

Hsiao K.M., Lin, J.Y, & Lin, W.Y. 1999. A consistent co-rotational finite element formulation for geometrically nonlinear dynamic analysis of 3-D beams. *Comput struct* (169): 1–18.

Kani, I.M. 1993. Combined non-linear analysis of shallow lattice domes using a beam super element. *Space structures the Fourth International Conference on Space Structures*. (1): 136–146.

Li, L.J., Xie, Z.H. & Guo, Y.C. F Liu. 2005. Antiseismic analysis of double-layer reticulated shell structures. *The 4th international Conference on Advances in Steel Structures,* Shanghai, China. *In: Advances in steel structures, (ICASS'05) ISBN: 0 00 8446 37 X*, Elsevier Ltd. Printed in Great Britain. 1199–1204.

Liu, X.L. & Yan, Z. 1993. Geometrically non-linear and experimental studies of single-layer aluminum braced domes. *Space structures, The Fourth International Conference on Space Structures* (1): 156–166.

Li, Q.S. & Chen, J.M. 2003. Nonlinear elastoplastic dynamic analysis of single-layer reticulated shells subjected to earthquake excitation. *Computers and Structures*, 177–188.

Lin,Y.Y. & Chang K.C. 2003. Study on damping reduction factor for buildings under the earthquake ground motions. *Journal of structure engineering*. 206–214.

Makowski, Z.S. 1993 Space structures-a review of the developments within the last decade. *Space structures. The Fourth International Conference on Space Structures*. (1). 1–8.

Salazar, A.R. & Haldar A. 2000. Structural response considering the vertical component of earthquakes. *Comput struct* (74). 131–145.

Xie, Z.H. 2004. Static and dynamic analysis of the rib-ring spherical shell. *Dissertation for Master of Engineering. Library in Guangdong University of Technology* (in Chinese).

Xue S.D., Zhao, J. & Gao, X.Y. 2003. *Structural seismic design* Beijing: Science Press. 342–348.

Yang, Z. 2005. Dynamic analysis of reticulated shell structures. *Dissertation for Master of Engineering. Library in Guangdong University of Technology*. (in Chinese).

Zhu, K.A., Bermani F.G.A. & Kitipornchai, S. 1994. Nonlinear dynamic analysis of lattice structures. *Comput struct*, 52(1): 9–15.

Steel and Composite Structures – Wang & Choi (eds)
© *2007 Taylor & Francis Group, London, ISBN 978-0-415-45141-3*

Numerical modeling of shells repaired using FRP

M. Batikha, J.F. Chen & J.M. Rotter

Institute for Infrastructure and Environment, Edinburgh University, Edinburgh, Scotland, UK

ABSTRACT: Extensive research has been conducted on the use of fibre reinforced polymer (FRP) composites to strengthen concrete, masonry and timber structures as well as metallic beams. The failure strength, rather than considerations of stability has been the main concern in these studies. This paper presents a numerical modelling study of isotropic thin metallic cylindrical shells that have been repaired using orthotropic FRP to increase the resistance to elephant's foot buckling. This form of buckling occurs under high internal pressure accompanied by axial forces in the shell structure, and is commonly found in earthquake damaged tanks and silos. The strengthening effect is shown to be sensitive to the amount of the FRP: both too little and too much FRP lead to a lower strength than the optimal amount.

1 INTRODUCTION

Thin metal cylindrical shells are widely used as containment structures, such as silos and tanks. These shells are sensitive to the magnitude of the imperfections which can cause elastic buckling, but under high internal pressure, this sensitivity is much reduced as shown by Rotter (1990, 1996, 2004). Teng & Rotter (1992) demonstrated that cylindrical shells fail near local imperfections by elastic buckling if the internal pressure is small. By increasing the internal pressure, yield of the wall leads to a local reduction of the flexural stiffness and the displacements increase. The circumferential membrane stress resultants are raised (Rotter, 1989) and elastic-plastic buckling occurs (Rotter 1990).

This elastic-plastic instability failure generally occurs at the base boundary condition and is known as elephant's foot buckling. Further, Rotter (1990) showed that, for thin cylinders, a clamped base provides a considerable increase in strength, whilst in thick cylinders, clamped and simple supports are similar. Chen et al. (2005, 2006) proposed to strengthen the shell against elephant's foot buckling by using a small ring stiffener. The optimal dimensions and location for this stiffener were derived. In this paper, fibre reinforced polymer (FRP) composites are used to strengthen the cylindrical shell against elephant's foot buckling.

Extensive research on the use of FRP for strengthening concrete structures has been undertaken since the 1990s (Teng et al. 2002). This FRP research has been extended to the strengthening of metallic beams, masonry and timber structures. In all these cases, strength, rather than stability, was the main concern. Teng & Hu (2004) applied FRP to circular steel tubes under axial loading alone to prevent elephant's foot buckling. They found that using FRP prevented outward elephant's foot buckling near the ends. In addition, increasing the thickness of the FRP increased both the ductility and ultimate load carrying capacity.

The aim of this paper is to investigate the application of FRP to increase the buckling strength of a thin metallic shell under internal pressure accompanied with axial load. Geometrically and Materially Non Linear Analysis, GMNA, has been used to obtain the optimal FRP sheet. Further more, fitted formulas of this optimal FRP sheet were produced.

2 FINITE ELEMENT MODELLING

Using ABAQUS (2006), a cylindrical shell was studied with a height h of 5000 mm, a radius R of 5000 mm and thickness t_s of 5 mm. This gives a radius to thickness ratio of 1000, and corresponds to a medium-length cylinder according to Eurocode3 Part 1.6 (2007).

The Young's modulus, E_s, and Poisson's ratio, ν_s, of the metal cylindrical shell are 200 GPa and 0.3 respectively. Figure 1 shows the stress-strain relationship of the metal.

The boundary condition at the bottom is simply supported (radial, axial and circumferential displacements restrained, but not the rotation) and for the top, the boundary is held circular (free in all directions except for rotation about the circumference).

A uniform internal pressure of p and a vertical load per unit circumference of N_z were applied to the shell.

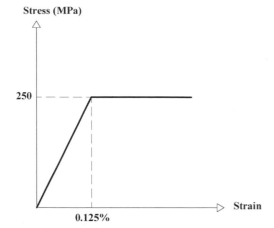

Stress (MPa)

250

0.125%

Strain

Figure 1. Stress-strain relationship for the metal shell.

The cylindrical shell was modelled using element SAX1. Element SAX1 is a 2-node axisymmetric general-purpose shell element with the effect of transverse shear deformation included. Each node has three degrees of freedom (radial, axial displacements and a rotation). Geometrically and Materially Non Linear Analysis, GMNA, was used and a mesh convergence study was conducted, leading to a mesh of 3.2 mm elements in a height of λ, the linear bending half-wavelength, above base given by:

$$\lambda = \frac{\pi}{(3(1-v_s^2))^{1/4}}\sqrt{Rt_s} \tag{1}$$

3 ANALYSIS OF BUCKLING STRENGTH

In shell, the stress resultants can be expressed in terms of displacements, w, as:

$$N_\theta = \frac{Et}{R}w + vN_z \tag{2a}$$

$$M_z = -D\frac{d^2w}{dz^2} \tag{2b}$$

$$M_\theta = vM_z \tag{2c}$$

$$Q_z = -D\frac{d^3w}{dz^3} \tag{2d}$$

in which D is the shell flexural rigidity given by:

$$D = \frac{Et^3}{12(1-v^2)} \tag{3}$$

The von Mises stress can be derived as:

$$\sigma_{vM} = \sqrt{(\sigma_{mz} + \sigma_{bx})^2 - (\sigma_{mz} + \sigma_{bx}) \times (\sigma_{m\theta} + v\sigma_{bx}) + (\sigma_{m\theta} + v\sigma_{bx})^2} \tag{4}$$

While the meridional membrane stress, σ_{mx}, is constant, it can be seen that the circumferential membrane stress, $\sigma_{m\theta}$, and the meridional bending stress, σ_{bx}, dominate the von Mises stress and produce the failure of elephant's foot buckling.

Rotter (1989) represented the axial stress at first yield at an elephant's foot buckle by an approximation to his exact solution as:

$$\sigma_{mx} = \sigma_{cl}\left[1 - \left(\frac{PR}{tf_y}\right)^2\right]\left[1 - \frac{1}{1.12 + \rho^{1.5}}\right]\left[\frac{\rho + \frac{f_y}{250}}{\rho + 1}\right] \tag{5}$$

$$\sigma_{vM_{max}} = \left(\frac{PR}{t} + 0.3\sigma_{mx}\right)\left(2.725 + \frac{1 + 0.55\sigma}{\sqrt{1-\sigma^2}}\right) - 2.65\frac{PR}{t} = f_y \tag{6}$$

in which:

$$\sigma = \frac{\sigma_{mx}}{\sigma_{cl}} \tag{7}$$

$$\sigma_{cl} = 0.605E\frac{t}{R} \tag{8}$$

$$\rho = \frac{1}{400}\frac{R}{t} \tag{9}$$

In addition, elastic buckling in other parts of the shell was described by Rotter (1997) and Eurocode3 Part1.6 (2007). Elastic buckling is related to the radius to thickness ratio and fabrication quality class as:

$$\sigma_{mx} = \sigma_{cl}.\alpha_x \tag{10}$$

where α_x is the meridional elastic imperfection factor, obtained from:

$$\alpha_x = \frac{0.62}{1 + 1.91(\Delta w_k / t)^{1.44}} \tag{11}$$

in which Δw_k is the characteristic imperfection amplitude, which Rotter (1997) treated as:

$$\frac{\Delta w_k}{t} = \frac{1}{Q}\sqrt{\frac{R}{t}} \tag{12}$$

where, Q is the meridional compression fabrication quality parameter as specified by Eurocode 3 Part 1.6 (2007) and Rotter (2004).

Table 1. The meridional compression fabrication quality parameter, Q.

Fabrication tolerance quality class	Description	Q
Class A	Excellent	40
Class B	High	25
Class C	Normal	16

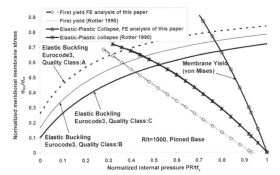

Figure 2. Collapse predictions for $R/t = 1000$ in a pinned-base cylindrical shell.

Figure 2 shows the collapse predictions of pinned cylindrical shell under internal pressure accompanied with vertical load according to the equations above.

The membrane von Mises stress, σ_{vM0}, can be found from Equation 4 when the bending moment is zero and the lateral displacement, w, is w_m, where w_m is the membrane theory normal deflection:

$$w_m = (p + \frac{v_s N_z}{r}) \frac{r^2}{E_s t_s} \qquad (13)$$

From Figure 2, it can be seen that elastic buckling occurs at low internal pressures, but elastic-plastic collapse (elephant's foot buckling), controls the failure under high internal pressure. Also, it can be seen that internal pressure increases the collapse strength of a cylinder under axial loads.

4 STRENGTHENING CYLINDRICAL SHELL USING FRP

For the cylindrical shell modelled in section 2, an FRP sheet with a height of h_f is bonded to the external surface of the cylinder starting at a distance x_f above the base. The FRP sheet was treated as orthotropic with Young's modulus in the circumferential direction, E_{fr}, of 230 GPa, and in the meridional direction, E_{fz}, of 3 GPa, with a Poisson's ratio in the circumferential direction, v_{fr}, of 0.35.

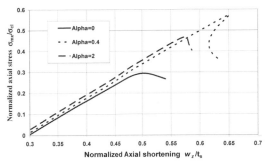

Figure 3. Effect of FRP thickness on load- axial shortening curves ($h_f/\lambda = 1$ and $x_f/\lambda = 0$).

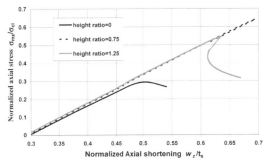

Figure 4. Effect of height ratio on load – axial shortening curves ($\alpha = 0.4$ and $x_f/\lambda = 0$).

An internal pressure of 0.2 N/mm^2 was applied. Different FRP thicknesses were explored using the height ratio $h_f/\lambda = 1$ and starting point ratio above the base, $x_f/\lambda = 0$. The quantity of FRP is here characterised by the stiffness parameter α, where $\alpha = E_{fr} t_f / E_s t_s$. Figure 3 shows the effect of changing this stiffness parameter α on load-axial shortening relationship for the shell, including the changing axial stress at bifurcation and collapse.

From Figure 3, it can be seen that the strength increases by adding FRP sheet, but the strengthening is shown to be sensitive to the thickness of the FRP: both too little and too much FRP lead to a lower strength.

Figure 4 shows the effect of changing the strip height ratio, h_f/λ, with $x_f/\lambda = 0$ and the stiffness parameter α at $\alpha = 0.4$, the optimal one in Figure 3.

Again, Figure 4 shows that the height ratio affects the strength and using a larger or smaller height ratio than the optimal reduces the strength.

5 OPTIMAL DIMENSIONS OF FRP SHEET FOR STRENGTHENING SILO AGAINST ELEPHANT'S FOOT COLLAPSE

Since the FRP sheet was added to function in circumferential tension at the area of maximum von Mises

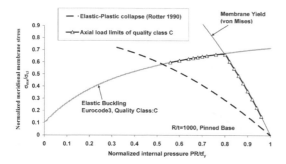

Figure 5. Membrane axial stress limits by using FRP.

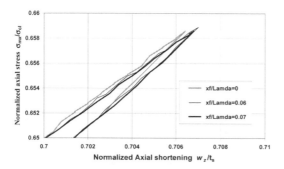

Figure 6. Effect of starting point ratio on load-axial shortening curves ($\alpha = 0.42$ and $(x_f + h_f)/\lambda = 0.71$).

stress, it is natural that the critical zone should be the area of elastic plastic collapse as indicated in Figure 2. Therefore, the principal aim of using FRP is to increase the axial load of elastic-plastic collapse to the value that is limited by elastic buckling. That is to say, the FRP cannot be usefully deployed to increase the membrane yield resistance of the whole shell, as that would require it all to be treated. Therefore, the limiting maximum axial load to which the FRP can usefully enhance the strength is shown in Figure 5 for fabrication quality Class C as an example.

It can be seen that changing the three parameters that determine the deployment of the FRP sheet t_f, h_f and x_f at the same time is very complicated. Therefore, the way was followed in this study was to fix two parameters and change the third one until an optimal value of each parameter was found.

For example, the same cylindrical shell above was studied. Figure 6 shows the changing buckling strength when a different starting point ratio x_f/λ is adopted, using a height ratio $(x_f + h_f)/\lambda = 0.71$ and stiffness ratio $\alpha = 0.42$. The results show that x_f/λ of 0.06 is the optimal starting point ratio for obtaining the height buckling strength.

By following the same procedure mentioned above, many examples were done and verification by using finite elements method was performed. On the basis of these calculations, the optimal values of the parameters, including FRP thickness, height, and starting point in relation to the internal pressure and axial force within the metal cylinder, can be found.

6 CONCLUSION

This paper has studied the use of FRP to strengthen a cylindrical shell under internal pressure accompanied by vertical load. Geometrically and Materially Non Linear Analysis, GMNA, has been used to obtain the optimal FRP sheet required to strengthen a cylindrical steel shell, using an example of radius to thickness ratio of 1000 and with fabrication quality Class C.

REFERENCES

ABAQUS. 2006. ABAQUS/Standard User's Manual, ABAQUS Inc.
Chen, J.F., Rotter, J.M. & Teng, J.G. 2005. Strengthening silos and tanks against elephant's foot buckling, Proc. of the Fourth Conference on Advances in steel structures, ICASS 05, Shanghai, China.
Chen, J.F., Rotter, J.M. & Teng, J.G. 2006. A simple remedy for elephant's foot buckling in cylindrical silos and tanks, Advances in Structural Engineering, Vol. 9, No. 3, pp 409–420.
EN 1993–1–6 (2007) Eurocode 3: Design of steel structures—Part 1–6: Strength and stability of shell structures. Brussels: CEN.
Rotter, J.M., 1989. Stress amplification in unstiffened cylindrical steel silos and tanks, Institution of Engineers, Australia, Civil Engrg Transactions,, Vol. CE31, No. 3, pp. 142–148.
Rotter, JM., 1990. Local collapse of axially compressed pressurized thin steel cylinders. Journal of Structural Engineering, Vol. 116, No. 7, pp. 1955–1970.
Rotter, J.M., 1996. Elastic plastic buckling and collapse in internally pressurised axially compressed silo cylinders with measured axisymmetric imperfections: interactions between imperfections, residual stresses and collapse, Proc. Int. Wkshp Imps Metal Silos, CA-Silo, Lyon, pp. 119–140.
Rotter, J.M 1997. Design standards and calculations for imperfect pressurised axially compressed cylinders. International conference on carrying capacity of steel shell structures. Brno. p. 354–360.
Rotter, J.M. 2004. Buckling of cylindrical shells under axial compression, in J.G. Teng & J.M. Rotter (eds), Buckling of Thin Metal Shells, Spon, London, pp. 42–87.
Teng, J.G., Chen, J.F., Smith, S.T. and Lam, L. 2002. FRP Strengthened RC Structures, John Wiley and Sons, Chichester, UK.
Teng, J.G. & Rotter, J.M. 1992. Buckling of pressurized axisymmetrically imperfect cylinders under axial loads, Journal of Engineering Mechanics, Vol. 118, No. 2, pp. 229–247.
Teng J.G & Hu Y.M 2004. Suppression of local buckling in steel tubes by FRP jacketing. 2nd International Conference on FRP Composites in Civil Engineering. Adelaide, Australia.

APPENDIX: NOTATION

N_z Axial load per circumferential (N/mm).

p internal pressure (N/mm2).

R radius of the silo (mm).

t_s thickness of the silo wall (mm).

h height of the silo (mm).

E_s Young's modulus of the metal silo (N/mm2).

v_s Poisson ratio of the metal silo.

t_f thickness of FRP sheet (mm).

E_{fr} Young's modulus of FRP in circumferential direction (N/mm2).

E_{fz} Young's modulus of FRP in meridional direction (N/mm2).

v_{fr} Poisson ratio of FRP in circumferential direction.

x_f height position of the FRP sheet above the base (mm).

h_f Height of FRP sheet (mm).

D Shell flexural rigidity, where $D = \frac{E_s t_s^3}{12(1-v_s^2)}$

w_m Normal deflection of membrane theory, where $w_m = (P + \frac{v_s N_z}{r})\frac{r^2}{E_s t_s}$

λ Meridional bending half wavelength, where $\lambda = \frac{\pi}{[3(1-v_s^2)]^{1/4}}\sqrt{rt_s}$

λ_b Meridional bending half wavelength of section B, where $\lambda_b = \frac{\pi}{[3(1-v_b^2)]^{1/4}}\sqrt{rt_b}$

σ_{vM} von Mises stress (MPa).

σ_{vM0} Membrane von Mises stress, when $w = w_m$ and the bending moment is zero (MPa).

σ_{mx} meridional membrane stress (MPa).

$\sigma_{m\theta}$ circumferential membrane stress (MPa).

σ_{bz} meridional bending stress (MPa).

σ_{cl} Classical buckling stress (MPa), where $\sigma_{cl} = 0.605 E\frac{t}{R}$.

f_y First yield stress (MPa).

P_n normalized internal pressure, where $P_n = \frac{pR}{t\sigma_{vM0}}$.

Steel and Composite Structures – Wang & Choi (eds)
© *2007 Taylor & Francis Group, London, ISBN 978-0-415-45141-3*

Masonry walls strengthening with innovative metal based techniques

D. Dubina, A. Dogariu & A. Stratan
Department of Steel Structures and Structural Mechanics, "Politehnica" University of Timisoara, Timisoara, Romania

V. Stoian, T. Nagy-Gyorgy, D. Dan & C. Daescu
Department of Civil Engineering, "Politehnica" University of Timisoara, Timisoara, Romania

ABSTRACT: Two innovative strengthening solutions for masonry walls are presented. First one consists in sheeting some steel or aluminium plates either on both sides or on one side of the masonry wall. Metallic plates are fixed either with prestressed steel ties, or using chemical anchors. The second one is derived from the FRP technique, but applies a steel wire mesh bonded with epoxy resin to the masonry wall. Both these techniques are described together with the experimental program carried out at the "Politehnica" University of Timisoara on the aim to validate them.

1 INTRODUCTION

1.1 *Importance of masonry buildings*

Masonry buildings are widely spread in Europe. Most of these structures represent historical constructions with symbol value for many towns or countries. Their functionality is diverse, including residential houses, hospitals, schools and other essential facilities. Therefore, these types of structures are important from many points of view: life safety, economical aspects and cultural heritage preservation.

Erected in a period when design methods where poor or missing, and the knowledge regarding seismic action was almost inexistent, these buildings need a structural upgrade in order to respect safety criteria of modern codes.

1.2 *Masonry behaviour*

Poor behaviour of masonry structures under seismic action is due to the lack of resistance, tensile stress mainly, small deformation capacity and low ductility. Moreover, under seismic action the masonry, because it is stiff and heavy, attracts significant inertial forces.

Common damage patterns for masonry buildings recorded during earthquakes can be classified in the following four categories:

– Out-of-plane damage or collapse of walls;
– In-plane shear or flexural cracking of walls;
– Loss of anchorage of walls to floor or roof diaphragms,
– Damage or collapse of corners.

Out-of-plane failure modes, e.g. falling down, can be a result of: load capacity exceeded due to inertial seismic forces, excessive deflection imposed on walls from diaphragm action, lack of anchorage, poor possibility of transferring deflection and inertial forces to horizontal elements.

In-plane damage can be a result of: diagonal cracking through masonry units due to excessive principal stress (tensile stress), shear sliding along bed joints, excessive toe compressive stress causing crushing (sliding shear), or tensile cracking normal to bed joints resulting in rocking (bending).

The interaction of in-plane and out-of-plane forces has as consequence failure of corners.

This paper will focus on strengthening techniques aiming to improve the in-plane behaviour of masonry panel. However, they obviously enhance the out-of-plane resistance, too.

1.3 *Objective of modern consolidation philosophy*

The objective of traditional consolidation techniques was mainly the local repair of damaged elements without a general strategy related to the global behaviour of the structure.

At present, not only the impact of local strengthening on the global response of the structure has to be considered, but also the reversibility of the used techniques and compatibility between materials, the added and existing ones (e.g. the "mixed" action) have to be analysed and evaluated.

The reversibility is very important because it offers the possibility to remove a solution when more advanced technology will be available.

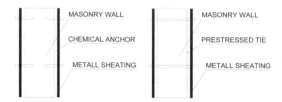

Figure 1. Proposed solution.

Figure 2. Weak area on masonry façade and location of SP or WM.

Figure 3. Wire mesh geometry and texture and chemical anchor.

The use of "mixed material based technology" enables to optimise the performance of retrofitted structure.

For this reasons, combining metal sheeting, which is resistant and ductile, with masonry, providing a proper connecting system, seems to be a suitable solution. The use of "dry" connection enables easy removal of metallic elements. Additionally, this solution offers the advantage of high mechanical properties, e.g. strength and ductility, without changing too much the initial rigidity.

This technique can provide a stable post-cracking behaviour to the masonry wall. Moreover, a performance based strengthening methodology could be developed.

1.4 Metallic based solutions

Two strengthening solutions were proposed and investigated within the research program. The solutions use steel (SSP) or aluminium (ASP) sheeting plates (see Figure 1), and steel wire mesh (SWM), respectively (see Figure 3).

Connection of the metal sheets plates to the masonry wall is realised in two ways: chemical anchors (CA) and prestressed ties (PT), placed at 200–250 mm. The wire mesh is glued using epoxy resin. Both systems can be applied on one side or both sides of the panel. It is expected that the system with metallic elements on both sides to perform better, but it isn't always possible due to architectural reasons.

Such a type of solution can be successfully applied in case of masonry walls, but it is not appropriate in case of masonry vaults and arches.

Observing the behaviour of a masonry wall with openings it is easy to identify the weak regions that need strengthening with metal plates (SP) or wire mesh (WM) (see Figure 2).

The application technology is rather simple. In the case of metallic plates they must be previously drilled. Afterwards the plate is placed on the wall, anchor holes are drilled in the masonry wall through the plate holes. The dust is blown away from the holes, followed by injection of epoxy resin and fixing of chemical anchors (see Figure 3). Prestressed ties are applied similarly,

but no resin is used, and the ties are tightened using a torque control wrench.

The mesh is produced either as galvanised steel or stainless steel bidirectional fabric. Spacing of the mesh is between 0.05 and 16 mm, while wire diameter is between 0.03 and 3.0 mm. Tensile strength reaches 650–700 N/mm^2, while elongation is about 45–55% in the case of stainless steel wires For galvanised steel wire, tensile strength is usually in the range of 400–515 N/mm^2.

Application of wire mesh (see Figure 3) requires a previous preparation of the walls to obtain a smooth surface. The preparation of resin is similar to the one used for Fiber Reinforced Polymers (FRP). The resin is applied in two steps: a fluid layer is applied first, and after it is dried, a second thick fluid layer is applied to embed the mesh. For large surfaces the mesh should be fixed to the wall with nails in order to keep plain its surface. It is important to mention that, by heating the resin layer, the wire mesh can be removed.

In order to validate the two solutions, an experimental program was carried out. It included:

– Material tests;
– Preliminary tests on 500 × 500 mm specimens;
– Full scale tests on 1500 × 1500 mm specimens, both under monotonic and cyclic loading.

Table 1. Summary of material tests.

Masonry component	Elastic modulus of masonry
	Compression test on brick
	Compression test on mortar
	Tension test on mortar
Steel wire mesh	Tensile test on wire
	Tensile test on mesh
Connectors	Tensile test on ties
Tensile test on steel plates	
Tensile test on aluminum plates	

2 CALIBRATION OF THE EXPERIMENTAL MODELS

2.1 Analytical calibration

Some simple numerical calculations have been performed to determine the thickness of steel shear plate in order to obtain a rational behaviour. On this purpose, three preliminary design criteria expressed in terms of stiffness, stability and strength have been used.

First material tests were performed in order to establish strength and stiffness parameters. They are summarized in Table 1.

First criterion is used to obtain comparable stiffness of the metallic sheeting plates with masonry panel, in order to provide a uniform distribution of stresses between wall and sheeting. To evaluate the rigidity of the wall and sheeting plate the following formulas have been used:

$$k_m = \frac{1}{\frac{h_{eff}^3}{E_m I_g} + \frac{h_{eff}}{A_v G_m}} \qquad (1)$$

where k_m = stiffness of masonry panel; h_{eff} = effective wall height; E_m = longitudinal elastic modulus of masonry; I_g = moment of inertia; A_v = shear area; and G_m = transversal elastic modulus of masonry;

$$k_{plate} = \frac{1}{\frac{h_{eff}}{A_v G_s}} \qquad (2)$$

where k_{plate} = stiffness of steel plate; h_{eff} = height of plate; A_v = shear area, and G_s = transversal elastic modulus of steel (Astaneh-Asl, 2001).

Considering known all material parameters and by equating the two relations, a 2.16 mm thickness demand for the steel sheeting was obtained.

Second condition follows to obtain a compact plate in order to prevent local buckling and assure dissipation of energy through plastic bearing work in connecting points only.

To establish the "non-compact" behaviour domain the following criterion was used:

$$1.10 \sqrt{\frac{K_v H}{F_{yw}}} \geq \frac{h}{t_w} \geq 1.37 \sqrt{\frac{K_v H}{F_{yw}}} \qquad (3)$$

where K_v = plate buckling coefficient; H = horizontal load of the panel; F_{yw} = yielding stress of steel; h = distance between connectors (imposed by masonry texture); and t_w = steel plate thickness.

From equation (3), the compactness criterion results as $t_w \geq 2.27$ (mm).

A more complex methodology, to evaluate the resistance of each component of the system, proposed by the producer of chemical anchor can be used. Three components govern the behaviour of the chemical connection, e.g. the matrix (masonry with epoxy resin), steel anchor and steel plates. It is believed that the most desirable failure mode is the bearing of the steel hole (e.g. in the connecting points). In order to obtain this failure mode, the bearing resistance should be less than the minimum between the shear resistance of connector and crushing resistance of matrix.

$$N_{bearing} \leq \min(N_{masonry}, N_{conector}) \qquad (4)$$

For chemical anchors, the design methodology suggested by producer (Hilti-Catalogue, 2005) has been adapted for masonry matrix e.g.

$$V_{Rd,c} = V_{Rd,c}^0 \cdot f_{BV} \cdot f_{\beta V} \cdot f_{AR,V} \qquad (5)$$

where $V_{Rd,c}$ = matrix edge resistance; $V_{Rd,c}^0$ = basic matrix edge resistance; f_{BV} = matrix strength influence; $f_{\beta V}$ = load direction influence; and $f_{AR,V}$ = spacing and edge coefficient.

Two cases were considered: ø8 and ø10 connector diameter. Corresponding plate thickness amounted to 2.20 and 2.48 mm.

It was decided to use a 3 mm thickness steel plate of S235 grade when applied on one side and 2 mm thickness plate of S235 grade when applied on both sides. Alternatively, 5 mm aluminium plates were used (99.5% Al 1050 H14 - $R_{p0.2\%}$ = 105 N/mm^2).

2.2 Experimental calibration

Because of the inherent approximations in design assumptions and the poor accuracy of analytical approach based on available formulas, it was decided to perform a series of test on small specimens in order to validate and calibrate the proposed techniques.

The tests on small specimens are summarized in Table 2.

Some preliminary tests were carried out on unreinforced masonry panels (brick unit strength of 10 N/mm^2 and mortar strength of 13 N/mm^2) to obtain reference values.

Table 2. Tests on small specimens.

Preliminary	Masonry panel		
Connection	Chemical anchor (CA)	ø8	
		ø10	
	Prestressed ties (PT)	ø10 – 0%	
		ø10 – 100%	
Diagonal tension test	Steel wire mesh (SWM)		
	Steel shear panel (SSP)	Chemical anchor	
		Prestressed ties	

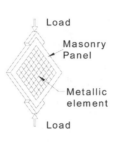

Figure 5. Experimental set-up for split test.

a)   b)

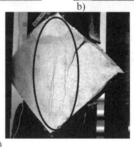

c)

Figure 6. Failure mode for SMW on both sides (a) ZC 0.4 × 1.0 (b) SS 0.4 × 0.5 and (c) SS 0.4 × 1.0.

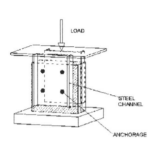

Figure 4. Experimental set-up and testing machine for connectors.

2.2.1 Connection tests

Connection tests were performed in order to establish the connector diameter and to assess the influence of prestress level of steel ties. The experimental set-up is presented in Figure 4.

2.2.1.1 Chemical anchors

Chemical anchors ø8 and ø10 diameters gr.5.8 have been tested. The failure mode for ø8 was the shear of connector and for ø10 the shear of connector and crushing of masonry. For the large specimen tests, a ø10 connector was chosen, due to the more efficient behaviour and resistance.

2.2.1.2 Prestressed ties

Two prestressing levels have been applied for the ø10 ties gr.5.8 (i.e. snug tightened ties (0% prestress) and full prestress (100%)). The failure mode was shear of ties, masonry specimens remaining almost intact. It was noted that the prestress level increases the resistance of connection due to confinement of masonry. In comparison with chemical anchors, prestressed ties lead more resistant and more rigid specimens.

2.2.2 System tests

Tests on systems was carried out in order to validate the analytical assumption in case of shear plates and to choose a proper steel wire mesh. The experimental set-up on small specimens and a sample test on unreinforced masonry panel are presented in the Figure 5.

Steel shear plates S235 grade of 2 mm thickness on both sides and 3 mm thickness on one side, connected with chemical anchors and prestressed ties were tested.

2.2.2.1 Steel wire mesh (SWM)

There are no analytical procedures to design the steel wire mesh reinforced masonry, therefore calibration was based on experimental test. The purpose of tests was to select the appropriate resin and wire mesh to be applied on large specimens. In the first step six types of wire mesh were tested.

Compared to FRP technique a thicker fluid resin was selected. In order not to change too many parameters and based on the experimental results, the following wire meshes were chosen: zinc coated (ZC) 0.4 × 1.0 ($D \times W$), stainless steel (SS) 0.4 × 0.5 and 0.4 × 1.0.

The failure modes are shown in Figure 6.

Table 3. Tests on large specimens.

Monotonic	Reference masonry wall test		*REF*
	Steel shear panel	Chemical anchor	*SSP-CA*
		Prestressed ties	*SSP-PT*
	Aluminum shear panel	Chemical anchor	*ASP-CA*
		Prestressed ties	*ASP-PT*
	Steel wire mesh		*SWM*
Cyclic	Reference masonry wall test		*REF-c*
	Steel shear panel	Chemical anchor	*SSP-CA-c*
		Prestressed ties	*SSP-PT-c*
	Aluminum shear panel	Chemical anchor	*ASP-CA-c*
		Prestressed ties	*ASP-PT-c*
	Steel wire mesh		*SWM-c*

- WM3 – sudden wire mesh rupture simultaneous with masonry crack – strength improvement (weak WM)
- WM5 – debounding of wire mesh, rupture in resin – strength improvement, energy dissipation due to the successive debounding (strong WM)
- WM6 – wire mesh yield – improvement of resistance and ductility (optimal).

Based on these observations, the stainless steel wire mesh 0.4×1.0 was chosen to be applied on large specimens.

3 TESTS ON LARGE MASONRY SPECIMENS

The experimental program on large specimens is summarised in Table 3.

The tests were carried out in two different experimental frames, one for monotonic loading and one for cyclic loading. The tests set-up is presented in Figure 7.

Loading was applied using displacement control, with lateral drift of the panel being used as control parameter. In case of cyclic loading the following loading protocol was used: one cycle at ± 0.5 mm, ± 1.0 mm, ± 1.5 mm, ± 2.0 mm, ± 3.0 mm, ± 5.0 mm, ± 7.0 mm, ± 9.00 mm, ± 11.00 mm, etc. The "yield" displacement, e_y, was considered when significant stiffness degradation was observed. After "yielding", three cycles at e_y, $1.5\ e_y$, $2\ e_y$, etc. were applied, until the failure of specimen occurred.

4 PRELIMINARY RESULTS

The interpretation of the experimental results is still in progress. Therefore, in present paper qualitative results are reported only (see Table 4).

Diagonal failure mode was observed for all specimens, both under monotonic and cyclic loading.

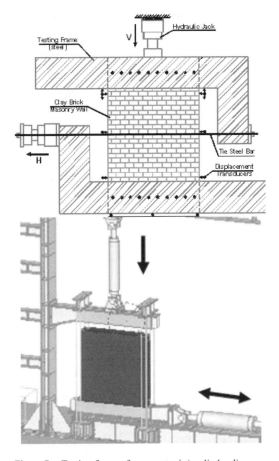

Figure 7. Testing frames for monotonic/cyclic loading.

Table 4. Large specimens' qualitative results.

	Monotonic		Cyclic	
	Resistance	Ductility	Resistance	Ductility
ASP–CA-1	→	↗	↗	↗
ASP–CA-2	↑	↑	↗	↗
ASP-PT-1	↗	↗	↗	↗
ASP-PT-2	↑	↑	↑	↑
SSP CA-1	→	↗	→	↗
SSP–CA-2	→	→	→	↗
SSP-PT-1	↗	↗	→	↗
SSP-PT-2	↗	↑	↗	↗
SWM-1	→	→	↗	→
SWM-2	↑	↗	↑	↗

Legend → slightly ↗ moderate ↑ high increase 1- one side; 2- both sides

Due to flexibility of testing frame used for cyclic loading, a more substantial damage at the corners of panel was observed in comparison with monotonic tests. However, for cyclic loaded specimens the

Figure 8. Failure mode for ASP-PT-2 m.

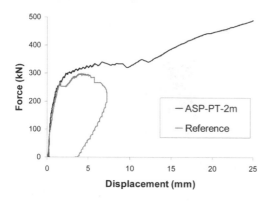

Figure 9. Monotonic test on aluminium shear panel.

Figure 10. Failure mode for ASP-PT-2C.

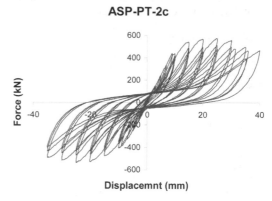

Figure 11. Cyclic test on aluminium shear panel connected with prestressed ties.

characteristic failure was also the diagonal shear, but with a small influence due to eccentric compression. A significant improvement in terms of ultimate displacemen t (that shows significant improvement in ductility), and also the increase in strength, with a slight increase in stiffness were recorded for all specimens. An overview of qualitative performance in terms of strength and ductility of tested specimens, related to unreinforced masonry, is presented in Table 4.

For the one side sheeting under cyclic loading a significant out of plane deformation was observed.

Even if the interpretation of tests results is still in progress, some numerical comparisons are available. For instance, the Force – Displacement relationships are presented for ASP-PT-2, monotonic and cyclic specimens (see Figure 9 and Figure 11), as well as their failure modes (see Figure 8 and Figure 10).

Due to large in-plane stiffness of masonry walls, the strengthening solution does not avoid completely damage to masonry. A limited amount of damage to masonry has to be allowed in order to take benefit from

ductility of the metal used for sheeting. Aluminium is believed to be particularly suitable in this case, due to a more favourable strength-to-stiffness ratio than steel.

It can be observed that, despite strengthening, the masonry panel cracks at almost the same force and displacement as reference panel. The mixed masonry-metallic plate system is activated only after masonry cracking. This can be observed also by the fact that the initial stiffness of both strengthened and reference panels does not change. This is an advantage for global behaviour of retrofitted building.

The monotonic curves (see Figure 9) show an important increase in terms of resistance, but the main advantage of this system seems to be the very large ultimate displacement that assures a very stable post-cracking behaviour and a large ductility.

Envelop for SMW-2 and ASP-PT-2

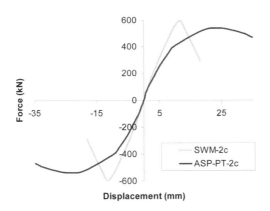

Figure 12.   Envelop curves for SMW-2c and ASP-PT-2c.

Also, for cyclic loading this system has proved his validity by increasing the resistance and obtaining a good hysteretic behaviour despite of significant pinching (see Figure 11).

In Figure 12 is shown the comparison of cyclic envelop curves for ASP-PT-2C and SWM-2C.

5   CONCLUDING REMARKS

The proposed strengthening solutions are an alternative to FRP technology enabling to obtain a ductile increase of strength, but without increasing the stiffness of the wall. It can be concluded that SP increases mainly the ductility, while WM increases the resistance. Both techniques are more efficient when applied on both sides. The prestressed tie connections seem to be more appropriate and the specimens sheeted with aluminium plates have shown a better behaviour than ones sheeted with steel.

It is expected that these strengthening solutions can be also applied in case of weak reinforced concrete diaphragms.

ACKNOWLEDGEMENT

This work was carried out in the CEMSIG Laboratory of the Department of Steel Structure and Laboratory of Department of Civil Engineering from the "Politehnica" University of Timisoara.

The proposed techniques are developed in the frame of PROHITECH (FP6 INCO-CT-2004-509119/2004 Earthquake **Pro**tection of **Hi**storical Buildings by Reversible Mixed **Tech**nologies).

REFERENCES

Astaneh-Asl, A. 2001. Seismic Behaviour and Design of Steel Shear Walls, *Steel TIPS, USA (2001)*.
IAEE/NICEE (2004). Guidelines for Earthquake Resistant Non-Engineered Construction, *First printed by International Association for Earthquake Engineering, Tokyo, Japan. Reprinted by the National Information Center of Earthquake Engineering, IIT Kanpur, India.*
Hilti-Catalogue (2005), Design Manual, Anchor Technology,(2005).

Steel and Composite Structures – Wang & Choi (eds)
© 2007 Taylor & Francis Group, London, ISBN 978-0-415-45141-3

Author index